FLUVIAL MEANDERS AND THEIR SEDIMENTARY PRODUCTS IN THE ROCK RECORD

Other publications of the International Association of Sedimentologists

Fluvial meanders and their sedimentary products in the rock record

Edited by

Massimiliano Ghinassi
Università degli Studi di Padova

Luca Colombera
School of Earth and Environment,
University of Leeds, UK

Nigel P. Mountney
School of Earth and Environment,
University of Leeds, UK

Arnold Jan H. Reesink
Lancing College, UK

SERIES EDITOR

Mark Bateman
Professor in Palaeoenvironments,
Deputy Head of Department,
Department of Geography,
Winter St., University of Sheffield,
Sheffield, UK

WILEY Blackwell

This edition first published 2019

Registered Office(s)
John Wiley & Sons, Inc., 111 River Street, Hoboken, NJ 07030, USA
John Wiley & Sons Ltd, The Atrium, Southern Gate, Chichester, West Sussex, PO19 8SQ, UK

Editorial Office
9600 Garsington Road, Oxford, OX4 2DQ, UK
For details of our global editorial offices, customer services, and more information about Wiley products visit us at www.wiley.com.

Wiley also publishes its books in a variety of electronic formats and by print-on-demand. Some content that appears in standard print versions of this book may not be available in other formats.

Library of Congress Cataloging-in-Publication Data

Names: Ghinassi, Massimiliano, editor. | Colombera, Luca, editor. | Mountney, Nigel P., editor. |
   Reesink, Arnold Jan H., editor. | International Association of Sedimentologists.
Title: Fluvial meanders and their sedimentary products in the rock record /
   edited by Massimiliano Ghinassi (Università degli Studi di Padova),
   Luca Colombera (School of Earth and Environment, University of Leeds, UK),
   Nigel P. Mountney (School of Earth and Environment, University of Leeds, UK),
   Arnold Jan H. Reesink (Lancing College, UK).
Description: Hoboken, NJ : John Wiley & Sons, Inc., [2019] | Series: Special publication ; 48 |
   Copyrighted by International Association of Sedimentologists. |
   Includes bibliographical references and index. |
Identifiers: LCCN 2018043693 (print) | LCCN 2018045025 (ebook) | ISBN 9781119424451 (Adobe PDF) |
   ISBN 9781119424321 (ePub) | ISBN 9781119424468 (hardcover)
Subjects: LCSH: Sedimentary rocks. | Meandering rivers.
Classification: LCC QE471 (ebook) | LCC QE471.F5785 2018 (print) | DDC 551.3/55—dc23
LC record available at https://lccn.loc.gov/2018043693

Cover Image: LiDAR topography of the Quinault River (Washington, USA), courtesy of Daniel Coe and the Washington Geological Survey.
Cover Design: Wiley

Set in 10/12pt Melior by SPi Global, Pondicherry, India
Printed and bound in Singapore by Markono Print Media Pte Ltd

10 9 8 7 6 5 4 3 2 1

Contents

List of contributors

Rolf Aalto
College of Life and Environmental Sciences,
University of Exeter, Exeter, UK

Koen Blanckaert
Technische Universität Wien Fakultät für
Bauingenieurwesen, Institute of Hydraulic
Engineering and Water Resources Management,
Wien, Austria

Dan S. Chaney
Department of Paleobiology,
National Museum of Natural History,
Smithsonian Institution,
Washington, DC, USA

Luca Colombera
Fluvial & Eolian Research Group,
School of Earth and Environment,
University of Leeds, Leeds,
UK

Neil S. Davies
Department of Earth Sciences,
University of Cambridge, Cambridge,
UK

William A. DiMichele
Department of Paleobiology,
National Museum of Natural History,
Smithsonian Institution, Washington,
DC, USA

Robert M. Dorrell
Institute of Energy and Environment,
University of Hull, Hull, UK

Paul R. Durkin
Department of Geological Sciences,
University of Manitoba, Winnipeg,
Manitoba, Canada

Milovan Fustic
Department of Geoscience, University of Calgary,
Calgary, Alberta, Canada

Fernando Garcia-Garcia
Sedimentary Reservoirs Workgroup,
Department of Stratigraphy and Palaeontology,
University of Granada, Spain

Massimiliano Ghinassi
Department of Geoscience, University of Padova,
Padova, Italy

Martin R. Gibling
Department of Earth Sciences,
Dalhousie University, Halifax,
Nova Scotia, Canada

Adrian J. Hartley
Department of Geology & Petroleum Geology,
University of Aberdeen, Aberdeen,
UK

Saturnina Henares
Sedimentary Reservoirs Workgroup,
Department of Stratigraphy and Palaeontology,
University of Granada, Spain

David M. Hodgson
Fluvial & Eolian Research Group,
School of Earth and Environment,
University of Leeds, Leeds, UK

John Holbrook
Texas Christian University, Fort Worth,
Texas, USA

Stephen M. Hubbard
Department of Geoscience, University of Calgary,
Calgary, Alberta, Canada

John Howell
Department of Geology and Petroleum
Geology, University of Aberdeen, Aberdeen,
UK

Alessandro Ielpi
Harquail School of Earth Sciences,
Laurentian University, Sudbury,
Canada

Shelby Johnston
Texas Christian University, Fort Worth,
Texas, USA

Rebecca Koll
Florida Museum of Natural History,
University of Florida, Gainesville,
USA

Kory Konsoer
Department of Geography and Anthropology,
Louisiana State University College of Humanities
and Social Sciences, Baton Rouge, Louisiana,
USA

Stefano Lanzoni
Department of Civil, Environmental and
Architectural Engineering, University of Padova,
Padova, Italy

Dale A. Leckie
Department of Geoscience, University of Calgary,
Calgary, Alberta, Canada

Sergio Lopez Dubon
Department of Civil, Environmental and
Architectural Engineering, University of Padova,
Padova, Italy

William J. Mcmahon
Faculty of Geosciences,
Utrecht University, Utrecht,
The Netherlands

Nigel P. Mountney
Fluvial & Eolian Research Group,
School of Earth and Environment,
University of Leeds, Leeds, UK

Andrew Nicholas
College of Life and Environmental Sciences,
University of Exeter, Exeter, UK

Amanda Owen
School of Geographical and Earth Sciences,
University of Glasgow, Glasgow, UK

Daniel Parsons
School of Environmental Sciences,
University of Hull, Hull, UK

Robert H. Rainbird
Department of Natural Resources, Geological
Survey of Canada, Ottawa, Canada

Arnold Jan H. Reesink
Lancing College, Lancing, West Sussex,
UK

Derek Richards
Department of Geography and Anthropology,
Louisiana State University College of Humanities
and Social Sciences, Baton Rouge, Louisiana,
USA

Catherine E. Russell
School of Geography, Geology and the
Environment, University of Leicester,
Leicester, UK

Arved Schwendel
School of Environmental Sciences,
University of Hull, Hull, UK;
College of Life and Environmental Sciences,
University of Exeter, Exeter, UK;
School of Humanities, Religion and
Philosophy, York St John University, York, UK

Louis Scuderi
University of New Mexico, Department of Earth
and Planetary Sciences, Albuquerque,
New Mexico, USA

Richard P. Sech
Chevron Energy Technology Company, Houston,
TX, USA

Michelle N. Shiers
CASP, West Bldg, Madingley Rise, Madingley
Road, Cambridge, UK

Sharane S.T. Simon
Department of Earth Sciences,
Dalhousie University, Halifax,
Nova Scotia, Canada

Derald G. Smith[†]
Department of Geography, University of Calgary, Calgary, Alberta, Canada

Rudy Strobl
EnerFox Enterprise, Calgary, Alberta, Canada

Alistair Swan
Department of Geology and Petroleum Geology, University of Aberdeen, Aberdeen AB24 3UE, UK

Christopher Turnipseed
Coastal Studies Institute, Louisiana State University, Baton Rouge, Louisiana, USA

Dario Ventra
Department of Earth Sciences, University of Geneva, Geneva, Switzerland;
Faculty of Geosciences, Utrecht University, Utrecht, The Netherlands

Daniele P. Viero
Department of Civil, Environmental and Architectural Engineering, University of Padova, Padova, Italy

César Viseras
Sedimentary Reservoirs Workgroup, Deptment of Stratigraphy and Palaeontology, University of Granada, Spain

Gary S. Weissmann
University of New Mexico, Department of Earth and Planetary Sciences, Albuquerque, New Mexico, USA

Brian J. Willis
Clastic Stratigraphy Team;
Earth Science Department, Chevron Energy Technology Company, Houston, Texas, USA

Clinton Willson
Department of Civil & Environmental Engineering, Louisiana State University, Baton Rouge, Louisiana, USA

Na Yan
Fluvial & Eolian Research Group, School of Earth and Environment, University of Leeds, Leeds, UK

Luis Miguel Yeste
Sedimentary Reservoirs Workgroup, Department of Stratigraphy and Palaeontology, University of Granada, Spain

Shuyu Zhang
Department of Geoscience, University of Calgary, Calgary, Alberta, Canada;
Petroleum University of China (East China), Qingdao, Shandong, China

---

[†]Deceased.

Acknowledgements

The editors gratefully acknowledge all the reviewers for their constructive comments, which significantly contributed to the quality of this volume.

Int. Assoc. Sedimentol. Spec. Publ (2018) **48**, 1–14.

Sedimentology of meandering river deposits: advances and challenges

MASSIMILIANO GHINASSI*, LUCA COLOMBERA[†], NIGEL P. MOUNTNEY[†] and ARNOLD JAN H. REESINK[‡]

* *Department of Geosciences, University of Padova, Padova, Italy*
[†] *Fluvial & Eolian Research Group, School of Earth and Environment, University of Leeds, Leeds, UK*
[‡] *Lancing College, Lancing, West Sussex, UK*

INTRODUCTION

Understanding of the form and origin of river meander bends received relevant contributions between the end of the 19th century and the mid-20th century (Thompson, 1876; Tower, 1904; Sellards *et al.*, 1923; Fisk, 1944; Sundborg, 1956; Wright, 1959). In particular, two fundamental contributions provided insights on the morphodynamics of river bends: i) Thompson (1876) provided the first description of the helical secondary flow structure in river bends; and ii) Fisk (1944) highlighted the temporal evolution of the lower Mississippi River, mapping its bends as mutable elements in unprecedented detail. Subsequent research efforts sought to link the helical flow pattern with the lateral mobility of meanders and culminated in the facies models by Allen (1963) and Bernard & Major (1963). These models associated the lateral shift of river bends with the development of clinostratified, fining-upward point-bar deposits. These theories were promptly supported by studies of modern rivers (e.g. Bluck, 1971; Jackson, 1975, 1976a, b; Nanson, 1980, 1981), validated with observations from the rock record (e.g. Allen, 1965; Puigdefàbregas & Vliet, 1978) and were also strengthened by the first direct measurements of flow velocities in river bends (Bathurst *et al.*, 1977). In parallel, new morphometric studies linked metrics of meandering channels (e.g. width, depth) and hydraulic parameters (Schumm, 1972; Ethridge & Schumm, 1978), providing the basis for development of palaeohydraulic reconstructions (Hajek & Heller, 2012; Hampson *et al.*, 2013). The first ICFS (International Congress of Fluvial Sedimentology) meeting – held in 1977 in Calgary, Canada – contributed to disseminating, applying and refining these models.

During the 1980s, facies models were further improved, notably with the recognition and classification of 'Inclined Heterolithic Stratification' (Thomas *et al.*, 1987) and reinforced through increasingly detailed comparisons between modern and ancient systems (e.g. Nanson, 1980; Dietrich & Smith, 1983; Smith, 1988; Willis, 1989). Since the early 1990s, the implementation of new technologies, including ground penetrating radar, acoustic Doppler current profilers, 3D outcrop imaging (e.g. using LiDAR and photogrammetry), 3D reflection seismic, together with enhanced approaches to numerical and laboratory experimental modelling, promoted development of new approaches for understanding meandering rivers. These developments now make it increasingly possible to consider the subject from complementary points of view (Kleinhans *et al.*, 2010) and to investigate more complex dynamic flow-form interactions over larger spatial and temporal scales. Thus, recent developments help improve our understanding and enable us to challenge long-held beliefs about meandering rivers.

This IAS Special Publication has arisen in part from contributions to a Special Session titled 'Fluvial meanders and their sedimentary products in the fossil record', which was held during the 32nd IAS International Meeting (May 2016, Marrakech, Morocco). This introductory paper outlines the key advances made in the study of meandering rivers and their deposits, and frames the scientific contributions of this volume within specific research themes. The resulting holistic view on meandering rivers provides insight to outstanding issues, which we hope will become the focus of follow-on studies that will seek to advance the state-of-the-science yet further.

The articles that form this volume demonstrate the breadth of scope in the research that is currently being undertaken in fluvial sedimentology. The organisation of the volume seeks to reflect

Fluvial Meanders and Their Sedimentary Products in the Rock Record, First Edition.
Edited by Massimiliano Ghinassi, Luca Colombera, Nigel P. Mountney and Arnold Jan H. Reesink.

how the research contributions variably focus on geological controls, processes and products (Fig. 1). Collectively these articles demonstrate how several connected strands of research contribute to a more integrated understanding of the sedimentology of meandering rivers, which is leading to the advancement of both fundamental and applied science. Within this field of research, four themes have been identified as being particularly topical; these are discussed below.

ESTABLISHED MODELS AND FORTHCOMING WORKS

Four fundamental research themes that capture the breadth of contributions to this volume (Fig. 2) have fascinated fluvial sedimentologists and geomorphologists working on meandering rivers since the early 1970s: i) channel-bend growth and related point-bar facies distribution; ii) mechanisms of meander-bend cutoff; iii) meandering river channels and vegetation cover; and iv) geometries of meander-belt sedimentary bodies. For each of these themes, the main research advances and contributions in this volume are summarised herein.

Channel-bend growth and related point-bar facies distribution

Previous studies

The advent of GPR investigations of the shallow subsurface sedimentary record marked a revolutionary step in linking the planform evolution of braided rivers with stratal patterns in their deposits (e.g. Bristow & Best, 1993; Lunt & Bridge, 2004; Lunt *et al.*, 2004). Ground Penetrating Radar was also used to investigate meandering river deposits (Bridge *et al.*, 1995). However, the loss of the electromagnetic signal in bar-top mud deposits limits its application principally to relatively sand-prone point bars (Kostic & Aigner, 2007), and methods such as parametric echo-sounders may need to be deployed as an alternative (e.g. Sambrook Smith *et al.*, 2016). The majority of modern point-bar sedimentary facies – including mud-prone deposits – were investigated through vibracoring, following the pioneering work of Smith (1988), who identified commonalities between sedimentary features of some modern tidally influenced rivers and those of the Cretaceous

McMurray Formation, which forms the Athabasca oil sands (Alberta, Canada). Development of the Athabasca oil sands – host of the largest heavy crude oil deposit in the world – strongly encouraged improved understanding of fluvial point-bar deposits, especially with regards to how sedimentary facies and architecture result from specific channel planform transformations.

Burge & Smith (1999) provided the first significant change to classical facies models by highlighting the common occurrence of translating meander bends (Daniel, 1971; Jackson, 1976a) and associated eddy-accretion deposits. This model was further refined by Smith *et al.* (2009, 2011), who investigated counter-point-bar and eddy-accretion deposits, and linked their development with specific conditions of outer bank erodibility. Despite their common occurrence in modern settings, translating (or 'downstream migrating', *sensu* Ghinassi *et al.*, 2016) point bars and related counter-point-bar and eddy-accretion deposits remain relatively poorly documented in the rock record (Ghinassi & Ielpi, 2015), except in rare cases for which high-resolution seismic data or planform exposures are available (Hubbard *et al.*, 2011; Ielpi & Ghinassi, 2014; Alqahtani *et al.*, 2015; Wu *et al.*, 2015). The noteworthy control of meander-bend planform transformations on spatial distribution of point-bar sedimentary facies has been recently highlighted through numerical simulations by Yan *et al.* (2017).

In the frame of understanding different styles of point-bar facies distribution, a special focus is often placed on deposits accumulated at the fluvial–marine transition zone, especially with the aim of unravelling the interaction between fluvial and tidal processes. Pioneering studies of Jones *et al.* (1993) and van den Berg *et al.* (2007) linked variations in fluvial discharge with sand-mud alternations in Inclined Heterolithic Stratification. Recent studies on the fluvial-tidal transition zone (see Ashworth *et al.*, 2015 for a review) have focussed their attention on the role of tidal currents in modulating fluvial point-bar sedimentation (Dalrymple & Choi, 2007; Martinius & Gowland, 2011; Shiers *et al.*, 2014; Carling *et al.*, 2015; Gugliotta *et al.*, 2016a, b) and have highlighted the different aspects of tidal signature on point-bar sedimentation. Choi *et al.* (2004) highlighted the spatial distribution of rhythmic tidal signatures in modern inclined heterolithic deposits. Jablonski & Dalrymple (2016) detected seasonality and climatic cyclicity

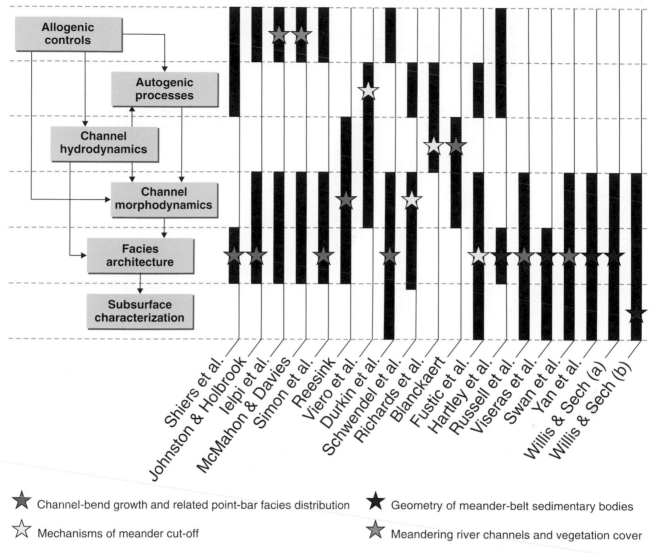

Fig. 1. Diagram that summarises the topics covered by each article in the volume. The flow chart on the left-hand side illustrates linkages between higher-order controls, processes, products and use of all related insight in applied contexts. For each article, as labelled, vertical black bars indicate which of these areas are covered and the position of stars indicates the primary focus of each. The four particular research themes discussed in more detail in the text are denoted by the coloured stars, as explained in legend. Allogenic controls (tectonics, eustatic sea-level changes, climate) are known to exert influence on fluvial systems over a range of timescales (e.g. through changes in sediment supply rate and calibre and in gradient). These factors are argued to affect fluvial systems through influences on both the behaviour of river systems and their long-term preservation in the stratigraphic record. Autogenic processes and river morphodynamics are distinguished, for convenience, because although certain processes will reflect the morphodynamic self-organisation of the river reach under study (e.g. neck cutoff), other processes might act independently (e.g. distant avulsions, shoreline progradation). Whereas studies of ancient successions allow consideration of what is ultimately preserved, studies of modern rivers permit observation of hydrodynamic processes and direct linkages of these to geomorphological and facies characteristics. It is therefore evident that integration is needed to obtain a more comprehensive understanding of the geological record of meandering rivers and to achieve improved predictions of subsurface fluvial successions. The order of articles in this volume largely follows the flow-chart in the figure and is indicated by the order in which the bars are laid out, from left to right.

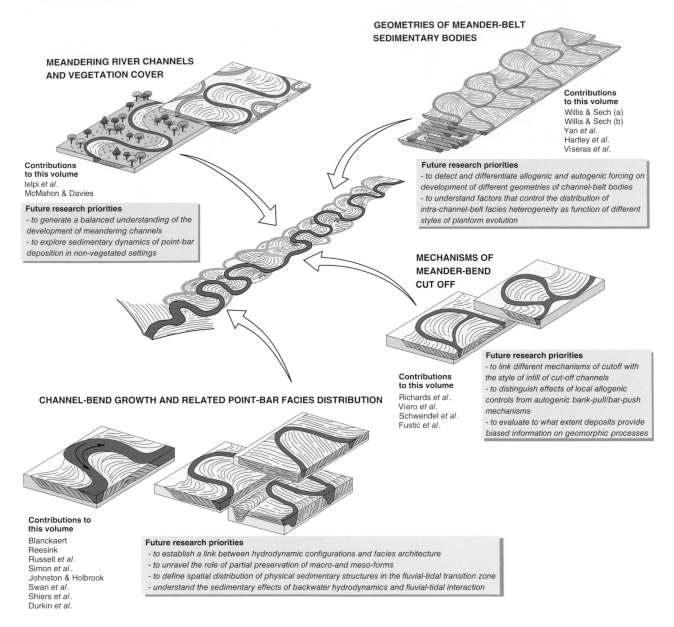

Fig. 2. Summary of the four main research themes and related future research priorities capturing the breadth of contributions to the present volume.

in tidally influenced point-bar deposits of the Cretaceous McMurray Formation. A spectrum of tidal influences spanning from the daily modulation of fluvial currents to the effects of tidal bores has been described by Martinius & Gowland (2011). The sedimentology of fluid muds in tide-dominated systems was investigated by Mackay & Dalrymple (2011). Olariu *et al.* (2015) highlighted the role of mutually evasive currents in modulating sedimentation in tidally influenced fluvial point bars.

Contributions to this volume

In this volume, contributions relating to point-bar sedimentary dynamics are provided by Blanckaert, Reesink, Russell *et al.*, Simon *et al.*, Johnston & Holbrook and Swan *et al.* Additionally, Shiers *et al.* and Durkin *et al.* focus on point-bar deposits accumulated in the distal part of alluvial plains, under the effects of processes that are characteristic of the fluvial-tidal transition zone.

Blanckaert reviews recent research on hydro-sedimentological processes and on their interaction in modern meandering rivers, highlighting and discussing the dominant controls on flow distribution. Reesink focuses on the bed-form scale and undertakes analyses of the preserved architecture of unit-bar cross strata in outcrops, highlighting how systematic measurements of these deposits may reveal valuable details of the formative fluvial palaeoenvironment. In seeking to address the discrepancy between the wide range of meander-bend planforms and the limited facies variability incorporated in facies models, Russell *et al.* present a new method to predict variable distribution of heterogeneities in point-bars, based on integration of meander-shape and meander scroll-bar pattern; the method is tested on Cretaceous deposits of the McMurray Formation (Alberta, Canada) and on Eocene deposits of the Montanyana Group (Spain). Mud-prone point-bar deposits are the focus of Simon *et al.* and Johnston & Holbrook. Simon *et al.* describe an exhumed point-bar element of the Permian upper Clear Fork Formation (Texas), highlighting the role of oblique accretion processes associated with suspended sediments plastering onto the steep inner channel bank. Johnston & Holbrook show mud-prone and sand-prone accretionary sets in a point-bar element of the Cretaceous Dinosaur Park Formation (Alberta, Canada) and link their formation with different styles of meander-bend transformations. Swan *et al.* illustrate the use of planform exposures of the Upper Jurassic Morrison Formation to reconstruct morphodynamic evolution of sandy fluvial point bars and to determine their internal facies distribution, highlighting how some sand-rich fluvial systems may previously have been interpreted incorrectly as deposits of braided rivers due to reliance on existing facies models of limited predictive capability. Focusing on the Campanian Neslen Formation (Utah, USA), Shiers *et al.* describe point-bar facies assemblages that only partially conform to those depicted in classical facies models and document and interpret substantial variability in point-bar architecture and internal facies distribution. Resultant models demonstrate how a range of interactions between allogenic (e.g. accommodation generation, fluvial discharge variations) and autogenic (e.g. backwater processes, presence of peat mires) processes can give rise to point-bar and related architectural elements with a variety of forms. Durkin *et al.* investigate transitions from point bars to counter-point bars along six modern river bends with varying channel scale, discharge and tidal influ-

ence. Results demonstrate downstream changes in net-to-gross ratio and provide criteria to detect counter-point bar deposits where a concave scroll pattern is not necessarily evident in planform.

Further developments

Although significant progress has been made in recent years in linking river-bend morphodynamics with related sedimentary products, there remains a need to improve our knowledge about how different hydrodynamic configurations are recorded in the facies architecture of point-bar deposits. Of particular importance is the role of partial preservation: most meander-belt deposits are lost to erosion and only a small portion of these deposits is ultimately preserved (Paola & Borgman, 1991; van de Lageweg *et al.*, 2013; Reesink *et al.*, 2015; Durkin *et al.*, 2017). The potential for deposition is linked to the sediment flux, which increases exponentially with discharge. Due to this exponential relationship, it is commonly assumed that sediment deposited by floods constitutes the majority of river-channel deposits. However, erosion and deposition are fundamentally controlled by the conservation of mass: along-stream changes in the transfer of sediment and not the absolute quantity of sediment transport, dictates the pattern of deposition. Consequently, erosion, deposition and sedimentary preservation must be affected greatly by increases in local gradients in water-surface slope and sediment transport, such as created by chute and neck cutoffs.

Furthermore, it is now known that the water-flow structure varies within the same meander bend at different flood stages (Kasvi *et al.* 2013; 2017), that overbank flood flows significantly modify the flow structure within bends (Loveless *et al.* 2000; Wormleaton *et al.* 2004) and that overbank deposition is a key control on the development of meanders (Van Dijk *et al.*, 2013a, b). Although the importance of these processes is widely acknowledged, the ways in which they are recorded in point-bar deposits and their ultimate preservation potential still needs to be further elucidated.

Although past research has offered notable insights into the detection of tidal influence in fluvial point-bar sedimentation at the bedform scale, less attention has hitherto been paid to understanding the role of tidal currents in shaping bar stratal architecture or controlling vertical and streamwise variations in sediment grain-size.

Assessment of these influences will require additional studies of modern meander bends, complemented with observations from tidal channels. Additionally, recognition of the location of channel deposits along the fluvial–marine transition zone (e.g. in terms of distance from a contemporaneous shoreline) is commonly attempted based on analysis of trace-fossil assemblages (e.g. Gingras *et al.*, 2012). A predictive tool that integrates knowledge of the spatial distribution of physical sedimentary structures is still lacking (Dalrymple *et al.*, 2015). Facies models developed for tidally influenced fluvial point bars should also be compared with those pertaining to muddy point bars formed far inland from tidal influence (Taylor & Woodyer, 1978; Jackson, 1981; Brooks, 2003). This comparison will contribute to the identification of distinctive features with which to detect tidally influenced fluvial point-bar deposits in the fossil record. In this context, the interference between backwater hydrodynamics and tidal–fluvial interaction was investigated in modern settings (Blum *et al.*, 2013) and efforts to detect their effects in terms of down-dip changes of architecture of distributary channel bodies were carried out by Colombera *et al.* (2016) and Fernandes *et al.* (2016). These studies provide a good starting point for further research into the diverse dynamics that characterise deposition in the zone of fluvial–marine transition.

Mechanisms of meander-bend cutoff

Previous studies

The process of abandonment of a channel reach, with the concomitant activation of a new river course (e.g. avulsion processes), has been widely documented in fluvial systems (Smith *et al.*, 1989; Slingerland & Smith., 1998, 2004; Morozova & Smith, 2000; Aslan *et al.*, 2005). In meandering rivers this process can occur at the bend scale and is known as meander-bend cutoff (Brice, 1974; Lewis & Lewin, 1983; Gagliano & Howard, 1984; Erskine *et al.*, 1992; Constantine & Dunne, 2008; Toonen *et al.*, 2012). The best-known of such processes is neck cutoff, whereby growing meanders intersect each other to cut off a meander loop (Lewis & Lewin, 1983). Less attention has been paid to chute cutoff, which occurs when a channel (i.e. chute channel) incises the inner side of the point bar (McGowen & Garner, 1970; Hooke, 2013). Chute channels can break through the

upstream edge of a meander neck during major floods (Johnson & Paynter, 1967) but they can also form on the downstream side of the bar and step progressively upstream (Gay *et al.*, 1998). The latter mechanism recently received attention through laboratory experiments (van Dijk *et al.*, 2012) and outcrop studies (Ghinassi, 2011). Constantine *et al.* (2010) showed that these two processes can occur together. Zinger *et al.* (2011) highlighted the importance of cutoff processes in river dynamics, demonstrating that extreme sediment pulses are released to the main channel after occurrence of a cutoff. Such local dynamics of enhanced erosion and deposition have great potential to be preserved in the sedimentary record. Indeed, channel-fill deposits generated by cutoff events are important elements within channel-belt bodies; they give rise to significant lithological heterogeneities, which can control both the lateral channel mobility of active channels (Smith *et al.*, 2010; Güneralp *et al.*, 2011; Bogoni *et al.*, 2017) and fluid flow through channel-belt deposits (Colombera *et al.*, 2017, and references therein). In this context, different types of oxbow-lake infills (Toonen *et al.*, 2012) have been demonstrated to exert a notable control on connectivity between point-bar bodies (Donselaar & Overeem, 2008). This has important implications for reservoir development and groundwater management.

Contributions to this volume

In this volume, articles that discuss mechanisms of meander-bend cutoff are provided by Richards *et al.*, Viero *et al.*, Schwendel *et al.* and Fustic *et al.* Richards *et al.* present a dataset of measurements of the three-dimensional flow through neck cutoffs with complex configurations that includes valuable observations on helical flows, recirculation and zones with stagnated flow. Viero *et al.* present a numerical modelling approach applied to two case studies (Sacramento River, California; Cecina River, Italy) and highlight the role of channelised flow inertia and of topographic and sedimentary floodplain heterogeneities in promoting chute cutoff processes. Schwendel *et al.* investigate the infill of abandoned chute channels and of channel segments that were abandoned after neck cutoff, from meanders of the Rio Beni (Bolivian Amazon basin). Results demonstrate how patterns of infill vary in relation to hydrological connectivity and distance to the main active channel. Fustic *et al.* describe channelised deposits encased within a large-scale point-bar element exposed in the McMurray Formation type section (Athabasca River, Alberta, Canada). These deposits are interpreted as relics of the infill of larger channel incisions that represent unsuccessful channel cutoffs or avulsions.

Further developments

Although significant advances have been made in understanding cutoff processes, a detailed model that attempts to link different mechanisms of cutoff with the style of infill of cutoff channels is yet to be developed. This is of particular importance because chute cutoff processes enable the transition from meandering to braiding (Kleinhans & Van den Berg, 2011). The lack of more sophisticated interpretative models is one of the reasons why interpretations of the rock record commonly take on a binary meandering-versus-braiding view, rather than allowing for transitional systems with individual flow-form characteristics.

Furthermore, the increased water-surface gradients created by cutoff process promote periods of accelerated planform change, increases in local sediment transport gradients and generate bed-scale pulses of sediment with effects that propagate both downstream and upstream, then eventually dissipate (Zinger *et al.*, 2011). Similarly, the consequence of shifting patterns of bed shear and sediment transport at confluences during large changes in the relative discharge of the upstream branches ought to lead to significant pulses of sediment redistribution within rivers. It is reasonable to assume that such local dynamics are recorded and preserved in the rock record; yet no diagnostic criteria exist for the distinction of such local allogenic controls from the migration of meanders through autogenic bank-pull and bar-push mechanisms (Parker *et al.*, 2011; van de Lagweg *et al.*, 2014). Consequently, it also remains unclear as to whether there is preferential preservation of specific morphological elements, or events, and therefore to what extent the deposits of a river provide biased information on the formative geomorphology.

Meandering river channels and vegetation cover

Previous studies

The relationship between the presence of vegetation cover and the development of meandering river channels has been the focus of considerable study by fluvial sedimentologists in recent years. Davies & Gibling (2010) noted a parallel between appearance of riparian vegetation and an increase of occurrence of deposits indicative of sinuous rivers in the rock record. Such a notion was in agreement with observations from a number of field-based studies (Ielpi *et al.*, 2015) and laboratory experiments (van Dijk *et al.*, 2013b), which indicated that the presence of vegetation favours the development of sinuous channels (Tal & Paola, 2007, 2010) by acting to stabilise river banks both through rooting and by encouraging retention of pedogenic cohesive mud. These notions supported the idea that pre-vegetation channels were dominantly shallow and braided in planform. This form, designated the 'sheet-braided' river style by Cotter (1978), has been considered representative of Precambrian fluvial styles.

However, other geological evidence supports the presence of plan forms indicative of meandering in some non-vegetated settings; such evidence includes the documentation of laterally accreting channels in pre-Devonian deposits (Long, 2011; Ielpi & Rainbird, 2016; Santos & Owen, 2016) and the presence of sinuous fluvial channels draining arid, non-vegetated areas (Matsubara *et al.*, 2015). Laboratory experiments by Smith (1998), Peakall *et al.* (2007) and van de Lageweg *et al.* (2014) also showed that sinuous channels were able to be produced and maintained on a non-vegetated substratum. The occurrence of meandering channels on extra-terrestrial surfaces (Lorenz *et al.*, 2008) further challenges the notion of a paucity of meandering channels in non-vegetated settings.

Contributions to this volume

In the present volume, integrating a review of pre-existing literature with field evidence, the papers by McMahon & Davies and Ielpi *et al.* summarise the two main views on interaction between vegetation growth and development of meandering river channels. McMahon & Davies, supporting their claims with field data from the 1 Ga Torridon Group (Scotland), argue that meandering planforms were less frequent on pre-vegetation Earth and that there is a tangible shift in the physical nature of global alluvium, coincident with the evolution of land plants. Ielpi *et al.* show laterally accreting deposits from five sedimentary rock units deposited on Laurentia between 1.6 to 0.7 Ga. Undertaking detailed sedimentary, architectural and palaeoflow analyses, they recognise the presence of lateral-accretion sets, a feature that was previously thought to be rare or absent in these deposits.

Further developments

The uncertainty in interpretations arising from complexity in the relationships between products and processes ensure that the relative roles of factors controlling the evolution of sinuous channels, including vegetation, remain of considerable research interest (Davies, 2017; Santos *et al.*, 2017a,b). Further architectural studies are needed to assess morphodynamic feedbacks and adequately explain the dynamics and preservation of point-bar deposits in pre-vegetation and extra-terrestrial river systems. A combination of deduction based on laboratory and numerical experiments, induction based on field-based studies of modern rivers in different environments and abduction based on analysis of preserved deposits present in the geological record (cf. Kleinhans, 2010) is needed in order to generate a balanced understanding of the development of meandering channels that is applicable to the full range of boundary conditions within which meanders are found.

Geometries of meander-belt sedimentary bodies

Previous studies

Channel-belt deposits generated by the lateral shift and avulsion of sinuous channels represent sedimentary bodies of primary interest as hydrocarbon reservoirs and aquifers (Hajek *et al.*, 2010). The width-to-thickness aspect ratios of these sedimentary bodies have been compared with those of braid belts by Gibling (2006) and Colombera *et al.* (2013), who provide criteria to distinguish between these sedimentary bodies. Recently, the internal architecture of channel-belt bodies has received significant attention and has been the focus of several studies mainly based on numerical simulations and laboratory experiments. Using numerical simulations, Willis & Tang (2011) showed that different styles of point-bar planform transformations exert a remarkable control in shaping the basal surface of channel-belt bodies and distributing facies heterogeneities. These studies also highlight how a combination of different styles of planform behaviour with a variable aggradation rate strongly controls intra-channel-belt connectivity. Laboratory experiments by van de Lageweg *et al.* (2013) established a relationship between preserved set thickness and morphology formed by a meandering channel. Numerical simulation by van de Lageweg *et al.* (2016) quantified the effects of bed

aggradation on the preservation of meandering channel morphologies and provided support to qualitative studies from the rock record (Ghinassi *et al.*, 2014).

Contributions to this volume

Geometries of meander-belt sedimentary bodies are analysed here by Willis & Sech (a, b), Yan *et al.*, Hartley *et al.* and Viseras *et al.* The two contributions by Willis & Sech are based on numerical simulations. Willis & Sech (a) predict the geometry and facies of channel belts by considering patterns of erosion and deposition during channel migration and underscore that facies models for channel belts need to better account for changes in the shape and position of channels, rather than present static views of river pattern. Willis & Sech (b) predict variations in fluid-flow patterns through subsurface hydrocarbon reservoirs and aquifers with improved consideration of 3D facies heterogeneity in channel-belt deposits. Yan *et al.* apply a 3D forward stratigraphic model, which is able to generate realistic architectural geometries and incorporate different types of facies heterogeneity, to a quantitative analysis of the static connectivity of point-bar sands based on data from geological analogues. Hartley *et al.* document amalgamated sandy meander belts from modern basins and the stratigraphic record, remarking that their recognition in the rock record is hindered by overlaps in facies characteristics between channel deposits of sandy meandering rivers and braided rivers. Viseras *et al.* present an outcrop/behind-outcrop multidisciplinary study of Triassic red beds from central Spain and make recommendations on how to identify and characterise poorly exposed ancient meander belts.

Further developments

Gaining improved understanding of intra-channel-belt facies heterogeneity has important applied implications, notably the characterisation of styles of compartmentalisation of sands by fine-grained deposits of different types (e.g. Colombera *et al.*, 2017; Yan *et al.*, 2017) and prediction of petrophysical heterogeneity (e.g. Burton & Wood, 2013; Nordahl *et al.*, 2014). At present, numerical modelling and laboratory experiments are the most powerful tools for understanding mechanisms controlling the internal architecture of channel-belt deposits formed by meandering channels, but improved remote sensing capabilities and the continuing efforts in capturing the variability in architectural styles from outcrop and modern analogues are also important sources of primary data. It is important that results from future research are translated to predictive tools that can be readily applied in subsurface studies.

A note on anthropogenic influences

Our future understanding of meandering rivers is contingent upon a multidisciplinary approach, which should be aimed at developing a new generation of quantitative fluvial facies models founded on datasets populated with information obtained from a broad range of investigations of modern and ancient rivers, laboratory experiments and numerical simulations. Although a comparison between these different datasets would be a fundamental step in advancing understanding in the discipline of fluvial sedimentology, it should be carried out considering the significance of anthropogenic effects on present-day fluvial systems. Nowadays, most rivers — whether they be considered to possess braided or meandering plan forms (or perhaps more usually combinations thereof) — are not hosted in pristine natural environments. The majority of present-day rivers are actively evolving under the influence of marked anthropogenic controls. Such controls have induced river behaviour and associated patterns of sediment erosion, transport and deposition that are difficult to predict. Therefore, understanding the continued evolution of meandering rivers in the Anthropocene represents an active and important field of research (e.g. Brooks *et al.*, 2003; Morais *et al.*, 2016; Munoz *et al.*, 2018). The effects of human-related activities (e.g. deforestation, loss of riparian vegetation, conversion of multi-channel systems to single-channel systems, channelisation [dredging] and bank revetments, flow regulation and damming, agricultural development, dispersion of pollutants, spreading of allochthonous aquatic faunas) need to be recognised in order to develop a new set of sedimentological models to assist with the management of rapidly evolving fluvial landscapes. Such models will enable valuable comparison of present-day fluvial deposits with the stratigraphic record and may, in turn, serve to predict the future effects of anthropogenic factors on river behaviour and patterns of erosion and sedimentation.

ACKNOWLEDGEMENTS

NPM and LC acknowledge FRG-ERG sponsors and partners AkerBP, Areva, BHPBilliton, Cairn India [Vedanta], ConocoPhillips, Murphy Oil, Nexen Energy, Petrotechnical Data Systems (PDS), Saudi Aramco, Shell, Tullow Oil, Woodside and YPF for financial support. LC has been supported by NERC (Catalyst Fund award NE/M007324/1; Follow-on Fund NE/N017218/1). MG has been supported by SID2016 project of Padova University (prot. BIRD168939). AR gratefully acknowledges the GSCO2 project for their support. We are grateful to Prof. Martin Gibling (Dalhousie University) for providing valuable comments that have helped improve this manuscript.

REFERENCES

Allen, J.R.L. (1963) The classification of cross-stratified units with notes on their origin. *Sedimentology*, **2**, 93–114.

Allen, J.R.L. (1965) The sedimentation and paleogeography of the Old Red Sandstone of Anglesy, North Wales. *Proceedings York. Geol. Soc.*, **35**, 139–185.

Alqahtani, F.A., Johnson, H.D., Jackson, C.A.L. and **Som, M.R.B.** (2015) Nature, origin and evolution of a Late Pleistocene incised valley-fill, Sunda Shelf, Southeast Asia. *Sedimentology*, **62**, 1198–1232.

Ashworth, P.J., Best, J.L. and **Parsons, D.R.** (2015) *Fluvial-tidal sedimentology*. Dev. Sedimentol., **68**, Elsevier, Amsterdam.

Aslan, A., Autin, W.J. and **Blum, M.D.** (2005) Causes of river avulsion: insights from the late Holocene avulsion history of the Mississippi River, USA. *J. Sed. Res.*, **75**, 650–664.

Bathurst, J.C., Thorne, C.R. and **Hey, R.D.** (1977) Direct measurements of secondary currents in river bends. *Nature*, **269**, 504–506.

Bernard, H.A. and **Major Jr, C.F.** (1963) Recent Meander Belt Deposits of the Brazos River: An Alluvial. *AAPG Bull.*, **47**, 350–350.

Bluck, B.J. (1971) Sedimentation in the meandering River Endrick. *Scot. J. Geol.*, **7**, 93–138.

Blum, M., Martin, J., Milliken, K. and **Garvin, M.** (2013) Paleovalley systems: insights from Quaternary analogs and experiments. *Earth-Sci. Rev.*, **116**, 128–169.

Bogoni, M., Putti, M. and **Lanzoni, S.** (2017) Modeling meander morphodynamics over self-formed heterogeneous floodplains. *Water Resour. Res.*, **53**, 5137–5157. https://doi.org/10.1002/2017WR020726

Brice, J.C. (1974) Evolution of Meander Loops. *Geol. Soc. Am. Bull.*, **85**, 581–586.

Bridge, J.S., Alexander, J.A.N., Collier, R.E., Gawthorpe, R.L. and **Jarvis, I.** (1995) Ground-penetrating radar and coring used to study the large-scale structure of point-bar deposits in three dimensions. *Sedimentology*, **42**, 839–852.

Bristow, C.S. and Best, J.L. (1993) Braided rivers: perspectives and problems. *Geol. Soc. London Spec. Publ.*, **75**, 1–11.

Brooks, A.P., Brierley, G.J. and Millar, R.G. (2003) The long-term control of vegetation and woody debris on channel and flood-plain evolution: insights from a paired catchment study in southeastern Australia. *Geomorphology*, **51**, 7–29.

Brooks, G.R. (2003) Alluvial deposits of a mud-dominated stream: the Red River, Manitoba, Canada. *Sedimentology*, **50**, 441–458.

Burge, L.M. and Smith, D.G. (1999) Confined meandering river eddy accretions: sedimentology, channel geometry and depositional processes. In: *Fluvial Sedimentology VI* (Eds N.D. Smith and J. Rogers), *Int. Assoc. Sedimentol. Spec. Publ.*, **28**, 113–130.

Burton, D. and Wood, L.J. (2013) Geologically-based permeability anisotropy estimates for tidally-influenced reservoirs using quantitative shale data. *Petrol. Geosci.*, **19**, 3–20.

Carling, P.A., Chateau, C.C., Leckie, D.A., Langdon, C.T., Scaife, R.G. and Parsons, D.R. (2015) Sedimentology of a tidal point-bar within the fluvial–tidal transition: River Severn Estuary, UK. In: *Fluvial-Tidal Sedimentology* (Eds P. Ashworth, J. Best and D. Parsons), *Dev. Sedimentol.*, **68**, 149–189, Elsevier.

Choi, K.S., Dalrymple, R.W., Chun, S.S. and Kim, S.P. (2004) Sedimentology of modern, inclined heterolithic stratification (IHS) in the macrotidal Han River delta, Korea. *J. Sed. Res.*, **74**, 677–689.

Colombera, L., Mountney, N.P. and McCaffrey, W.D. (2013) A quantitative approach to fluvial facies models: methods and example results. *Sedimentology*, **60**, 1526–1558.

Colombera, L., Mountney, N.P., Russell, C.E., Shiers, M.N. and McCaffrey, W.D. (2017) Geometry and compartmentalization of fluvial meander-belt reservoirs at the bar-form scale: Quantitative insight from outcrop, modern and subsurface analogues. *Mar. Petrol. Geol.*, **82**, 35–55.

Colombera, L., Shiers, M.N. and Mountney, N.P. (2016) Assessment of backwater controls on the architecture of distributary-channel fills in a tide-influenced coastal-plain succession: Campanian Neslen formation, USA. *J. Sed. Res.*, **86**, 476–497.

Constantine, J.A. and Dunne, T. (2008) Meander cutoff and the controls on the production of oxbow lakes. *Geology*, **36**, 23–26.

Constantine, J.A., McLean, S.R. and Dunne, T. (2010) A mechanism of chute cutoff along large meandering rivers with uniform floodplain topography. *Geol. Soc. Am. Bull.*, **122**, 855–869.

Cotter, E. (1978) The evolution of fluvial style, with special reference to the central Appalachians Paleozoic. In: *Fluvial Sedimentology* (Ed. A.D. Miall), *Mem. Can. Soc. Petrol. Geol.*, **5**, 361–384.

Dalrymple, R.W. and Choi, K. (2007) Morphologic and facies trends through the fluvial–marine transition in tide-dominated depositional systems: a schematic framework for environmental and sequence-stratigraphic interpretation. *Earth-Sci. Rev.*, **81**, 135–174.

Dalrymple, R.W., Kurcinka, C.E., Jablonski, B.V.J., Ichaso, A.A. and Mackay, D.A. (2015) Deciphering the relative importance of fluvial and tidal processes in the fluvial–marine transition. In: *Fluvial-Tidal Sedimentology* (Eds P. Ashworth, J. Best and D. Parsons), *Dev. Sedimentol.*, **68**, 3–45, Elsevier.

Daniel, J.F. (1971) *Channel movement of meandering Indiana streams. US Geol. Surv. Prof. Pap.*, **732-A**, 18 pp.

Davies, N.S. and Gibling, M.R. (2010) Cambrian to Devonian evolution of alluvial systems: the sedimentological impact of the earliest land plants. *Earth-Sci. Rev.*, **98**, 171–200.

Davies, N.S., Gibling, M.R., McMahon, W.J., Slater, B.J., Long, D.G.F., Bashforth, A.R., Berry, C.M., Falcon-Lang, H.J., Gupta, S., Rygel, M.R. and Wellman, C.H. (2017) Discussion on 'Tectonic and environmental controls on Palaeozoic fluvial environments: reassessing the impacts of early land plants on sedimentation'. *J. Geol. Soc. London*, **174**, 947–950. https://doi.org/10.1144/jgs2016-063. *J. Geol. Soc. London.*, doi: 10.1144/jgs2017-004.

Dietrich, W.E. and Smith, J.D. (1983) Influence of the point bar on flow through curved channels. *Water Resour. Res.*, **19**, 1173–1192.

Donselaar, M.E. and Overeem, I. (2008) Connectivity of fluvial point-bar deposits: An example from the Miocene Huesca fluvial fan, Ebro Basin, Spain. *AAPG Bull.*, **92**, 1109–1129.

Durkin, P.R., Hubbard, S.M., Holbrook, J. and Boyd, R. (2017) Evolution of fluvial meander-belt deposits and implications for the completeness of the stratigraphic record. *Geol. Soc. Am. Bull.*, **130**, 721–739.

Erskine, W., McFadden, C. and Bishop, P. (1992) Alluvial cutoffs as indicators of former channel conditions. *Earth Surf. Proc. Land.*, **17**, 23–37.

Ethridge, F.G. and Schumm, S.A. (1978) Reconstructing paleochannel morphologic and flow characteristics: methodology, limitations and assessment. In: *Fluvial Sedimentology* (Ed. A.D. Miall), *Can. Soc. Petrol. Geol. Mem.*, **5**, 703–722.

Fernandes, A.M., Törnqvist, T.E., Straub, K.M. and Mohrig, D. (2016) Connecting the backwater hydraulics of coastal rivers to fluvio-deltaic sedimentology and stratigraphy: *Geology*, **44**, 979–982.

Fisk, H.N. (1944) Geological Investigation of the Alluvial Valley of the Lower Mississippi River: U.S. Army Corps of Engineers Mississippi River Commission. 78 pp.

Gagliano, S.M. and Howard, P.C. (1984) The neck cutoff oxbow lake cycle along the Lower Mississippi River. In: *River Meandering* (Ed. C.M. Elliott). Proceedings of the Conference Rivers '83, American Society of Civil Engineers, 147–158.

Gay, G.R., Gay, H.H., Gay, W.H., Martinson, H.A., Meade, R.H. and Moody, J.A. (1998) Evolution of cutoffs across meander necks in Powder River, Montana, USA. *Earth Surf. Process. Land.*, **23**, 651–662.

Ghinassi M. (2011) Chute channels in the Holocene high sinuosity-river deposits of the Firenze plain, Tuscany, Italy. *Sedimentology*, **58**, 618–642.

Ghinassi, M. and Ielpi, A. (2015) Stratal Architecture and Morphodynamics of Downstream-Migrating Fluvial Point Bars (Jurassic Scalby Formation, UK). *J. Sed. Res.*, **85**, 1123–1137.

Ghinassi, M., Ielpi, A., Aldinucci, M. and Fustic, M. (2016) Downstream-migrating fluvial point bars in the rock record. *Sed. Geol.*, **334**, 66–96.

Ghinassi M., Nemec W., Aldinucci M., Nehyba S., Özaksoy V. and Fidolini, F. (2014) Plan-form evolution of ancient meandering rivers reconstructed from longitudinal outcrop sections. *Sedimentology*, **61**, 952–977.

Gibling, M.R. (2006) Width and thickness of fluvial channel bodies and valley fills in the geological record: a literature compilation and classification. *J. Sed. Res.*, **76**, 731–770.

Gingras, M.K., MacEachern, J.A. and Dashtgard, S.E. (2012) Chapter 16: Estuaries. In: *Trace Fossils as Indicators of Sedimentary Environments* (Eds D. Knaust and R.G. Bromley), *Dev. Sedimentol.*, **64**, 471–514.

Gugliotta, M., Flint, S.S., Hodgson, D.M. and Veiga, G.D. (2016a) Recognition criteria, characteristics and implications of the fluvial to marine transition zone in ancient deltaic deposits (Lajas Formation, Argentina). *Sedimentology*, **63**, 1971–2001.

Gugliotta, M., Kurcinka, C.E., Dalrymple, R.W., Flint, S.S. and Hodgson, D.M. (2016b) Decoupling seasonal fluctuations in fluvial discharge from the tidal signature in ancient deltaic deposits: an example from the Neuquén Basin, Argentina. *J. Geol. Soc.*, **173**, 94–107.

Güneralp, I. and Rhoads, B.L. (2011) Influence of floodplain erosional heterogeneity on planform complexity of meandering rivers. *Geophys. Res. Lett.*, **38**(14), https://doi.org/10.1029/2011GL048134

Hajek E.A. and Heller, P.L., 2012, Flow-depth scaling in alluvial architecture and nonmarine sequence stratigraphy: example from the Castlegate Sandstone, central Utah, USA. *J. Sed. Res.*, **82**, 121–130.

Hajek, E.A., Heller, P.L. and Sheets, B.A. (2010) Significance of channel-belt clustering in alluvial basins. *Geology*, **38**, 535–538. https://doi.org/10.1130/G30783.1

Hooke, J.M. (2013) Rivermeandering. In: *Treatise on Geomorphology* (Ed. in Chief J. Shroder and Ed. E. Wohl), Academic Press, San Diego, CA, *Fluvial Geomorphology*, **9**, 260–288.

Hubbard, S.M., Smith, D.G., Nielsen, H., Leckie, D.A., Fustic, M., Spencer, R.J. and Bloom, L. (2011) Seismic geomorphology and sedimentology of a tidally influenced river deposit, Lower Cretaceous Athabasca oil sands, Alberta, Canada. *AAPG Bull.*, **95**, 1123–1145.

Ielpi, A. and Ghinassi, M. (2014) Planform architecture, stratigraphic signature and morphodynamics of an exhumed Jurassic meander plain (Scalby Formation, Yorkshire, UK). *Sedimentology*, **61**, 1923–1960.

Ielpi, A., Gibling, M.R., Bashforth, A.R. and Dennar, C.I. (2015) Impact of Vegetation on Early Pennsylvanian Fluvial Channels: Insight From the Joggins Formation of Atlantic Canada. *J. Sed. Res.*, **85**, 999–1018.

Ielpi, A. and Rainbird, R.H. (2016) Highly variable Precambrian fluvial style recorded in the Nelson Head Formation of Brock Inlier (Northwest Territories, Canada). *J. Sed. Res.*, **86**, 199–216.

Jablonski, B.V. and Dalrymple, R.W. (2016) Recognition of strong seasonality and climatic cyclicity in an ancient, fluvially dominated, tidally influenced point bar: Middle McMurray Formation, Lower Steepbank River, northeastern Alberta, Canada. *Sedimentology*, **63**, 552–585.

Jackson, R.G. (1976) a. Depositional model of point bars in the lower Wabash River. *J. Sed. Res.*, **46**, 579–594.

Jackson, R.G. (1976) b. Largescale ripples of the lower Wabash River. *Sedimentology*, **23**, 593–623.

Jackson, R.G. (1981) Sedimentology of muddy fine-grained channel deposits in meandering streams of the American middle west. *J. Sed. Petrol.*, **51**, 1169–1192.

Jackson, R.G. (1975) Velocity–bed-form–texture patterns of meander bends in the lower Wabash River of Illinois and Indiana. *Geol. Soc. Am. Bull.*, **86**, 1511–1522.

Johnson, R.H. and Paynter, J. (1967) The development of a cutoff on the River Irk at Chadderton, Lancashire. *Geography*, **52**, 41–49.

Jones, B.G., Martin, G.R. and Senapti, N. (1993) Riverine–tidal interactions in the monsoonal Gilbert River fan delta, northern Australia. *Sed. Geol.*, **83**, 319–337.

Kasvi, E., Laamanen, L., Lotsari, E. and Alho, P. (2017) Flow patterns and morphological changes in a sandy meander bend during a flood—spatially and temporally intensive ADCP measurement approach: *Water*, **9**, 106.

Kasvi, E., Vaaja, M., Alho, P., Hyyppä, H., Hyyppä, J., Kaartinen, H. and Kukko, A. (2013) Morphological changes on meander point bars associated with flow structure at different discharges. *Earth Surf. Proc. Land.*, **38**, 577–590.

Kleinhans, M.G., Buskes, C. and de Regt, H. (2010) Philosophy of Earth Science, In: *Philosophies of the sciences* (Ed. F. Allhoff), Wiley-Blackwell, Chichester, UK, 213–236.

Kleinhans, M.G. and Van den Berg, J.H. (2011) River channel and bar patterns explained and predicted by an empirical and a physics-based method. *Earth Surf. Proc. Land.*, **36**, 721–738.

Kostic, B. and Aigner, T. (2007) Sedimentary architecture and 3D ground-penetrating radar analysis of gravelly meandering river deposits (Neckar Valley, SW Germany). *Sedimentology*, **54**, 789–808.

Lewis, G.W. and Lewin, J. (1983) Alluvial cutoffs in Wales and the Borderlands. In: *Modern and ancient fluvial systems* (Eds J.D. Collinson and J. Lewin), *Int. Assoc. Sedimentol. Spec. Publ.*, **6**, 145–154.

Long, D.G.F. (2011) Architecture and depositional style of fluvial systems before land plants: a comparison of Precambrian, early Paleozoic and modern river deposits. In: *From River to Rock Record: The Preservation of Fluvial Sediments and their Subsequent Interpretation* (Eds S. Davidson, S. Leleu and C.P. North), *SEPM Spec. Publ.*, **97**, 37–61.

Lorenz, R.D., Lopes, R.M., Paganelli, F., Lunine, J.I., Kirk, R.L., Mitchell, K.L., Soderblom, L.A., Stofan, E.R., Ori, G., Myers, M., Miyamoto, H., Radebaugh, J., Stiles, B., Wall, S.D., Wood, C.A. and The Cassini RADAR Team (2008) Fluvial channels on Titan: initial Cassini RADAR observations. *Planet. Space Sci.*, **56**, 1132–1144.

Loveless, J.H., Sellin, R.H.J., Bryant, T.B., Wormleaton, P.R., Catmur, S. and Hey, R. (2000) The effect of overbank flow in a meandering river on its conveyance and the transport of graded sediments. *Water and Environment Journal*, **14**, 447–455.

Lunt, I.A. and Bridge, J.S. (2004) Evolution and deposits of a gravely braid bar Sagavanirktok River, Alaska. *Sedimentology*, **51**, 1–18.

Lunt, I.A., Bridge, J.S. and Tye, R.S. (2004) A quantitative, three-dimensional depositional model of gravely braided rivers. *Sedimentology*, **51**, 377–414.

Mackay, D.A. and Dalrymple, R.W. (2011) Dynamic mud deposition in a tidal environment: the record of fluid-mud deposition in the Cretaceous Bluesky Formation, Alberta, *Canada. J. Sed. Res.*, **81**, 901–920.

Martinius, A.W. and **Gowland, S.** (2011) Tide-influenced fluvial bedforms and tidal bore deposits (late Jurassic Lourinhã Formation, Lusitanian Basin, Western Portugal). *Sedimentology*, **58**, 285–324.

Matsubara, Y., Howard, A.D., Burr, D.M., Williams, R.M.E., Dietrich, W.E. and **Moore, J.M.** (2015) River meandering on Earth and Mars: A comparative study of Aeolis Dorsa meanders, Mars and possible terrestrial analogs of the Usuktuk River, AK and the Quinn River, NV. *Geomorphology*, **240**, 102–120.

McGowen, J.H. and **Garner, L.E.** (1970) Physiographic features and stratification types of coarse-grained point bars: modern and ancient examples. *Sedimentology*, **14**, 77–111.

Morais, E.S., Rocha, P.C. and **Hooke, J.** (2016) Spatiotemporal variations in channel changes caused by cumulative factors in a meandering river: The lower Peixe River, Brazil. *Geomorphology*, **273**, 348–360.

Morozova, G.S. and **Smith, N.D.** (2000) Holocene avulsion styles and sedimentation patterns of the Saskatchewan River, Cumberland Marshes, Canada. *Sed. Geol.*, **130**, 81–105.

Munoz, S.E., Giosan, L., Therrell, M.D., Remo, J.W., Shen, Z., Sullivan, R.M., Wiman, C., O'Donnell, M. and **Donnelly, J.P.** (2018) Climatic control of Mississippi River flood hazard amplified by river engineering. *Nature*, **556**, 95–98.

Nanson, G.C. (1981) New evidence of scroll-bar formation on the Beatton River. *Sedimentology*, **28**, 889–891.

Nanson, G.C. (1980) Point bar and floodplain formation of the meandering Beatton River, northeastern British Columbia, Canada. *Sedimentology*, **27**, 3–29.

Nordahl, K., Messina, C., Berland, H., Rustad, A.B. and **Rimstad, E.** (2014) Impact of multiscale modelling on predicted porosity and permeability distributions in the fluvial deposits of the Upper Lunde Member (Snorre Field, Norwegian Continental Shelf). In: *Sediment-body Geometry and Heterogeneity: Analogue Studies for Modelling the Subsurface* (Eds A.W. Martinius, J.A. Howell and T.R. Good), *Geol. Soc. London Spec. Publ.*, **387**, 85–109.

Olariu, C., Steel, R.J., Olariu, M.I. and **Choi, K.** (2015) Facies and architecture of unusual fluvial–tidal channels with inclined heterolithic strata: Campanian Neslen Formation, Utah, USA. In: *Fluvial-Tidal Sedimentology* (Eds P. Ashworth, J. Best and D. Parsons), *Dev. Sedimentol.*, **68**, 353–394, Elsevier.

Paola, C. and **Borgman, L.** (1991) Reconstructing random topography from preserved stratification. *Sedimentology*, **38**, 553–565.

Parker, G., Shimizu, Y., Wilkerson, G. V., Eke, E. C., Abad, J. D., Lauer, J. W., Paola, C. and **Voller, V. R.** (2011) A new framework for modeling the migration of meandering rivers. *Earth Surf. Proc. Land.*, **36**, 70–86.

Peakall, J., Ashworth, P.J. and **Best, J.L.** (2007) Meander-bend evolution, alluvial architecture and the role of cohesion in sinuous river channels: a flume study. *J. Sed. Res.*, **77**, 197–212.

Puigdefàbregas, C. and **Van Vliet, A.** (1978) Meandering stream deposits from the Tertiary of the Southern Pyrenees. In: *Fluvial Sedimentology* (Ed. A.D. Miall), *Can. Soc. Petrol. Geol. Mem.*, **5**, 469–485.

Reesink A.J.H., Parsons D.R., Van den Berg J., Amsler M.L., Best J.L., Hardy R.J., Lane, S.N., Orfeo, O. and **Szupiany, R.** (2015) Extremes in dune preservation; controls on the completeness of fluvial deposits. *Earth-Sci. Rev.*, **150**, 652–665.

Sambrook Smith, G.H., Best, J.L., Leroy, J.Z. and **Orfeo, O.** (2016) The alluvial architecture of a suspended sediment dominated meandering river: the Rio Bermejo, Argentina. *Sedimentology*, **63**, 1187–1208.

Santos, M.G.M., Mountney, N.P. and **Peakall, J.** (2017a) Tectonic and environmental controls on Palaeozoic fluvial environments: reassessing the impacts of early land plants on sedimentation, *J. Geol. Soc.*, **174**, 393–404.

Santos, M.G.M., Mountney, N.P., Peakall, J., Thomas, R.E., Wignall, P.B. and **Hodgson, D.M.** (2017b) Reply to Discussion on 'Tectonic and environmental controls on Palaeozoic fluvial environments: reassessing the impacts of early land plants on sedimentation', *J. Geol. Soc. London*, **174**, 950–952. https://doi.org/10.1144/jgs2016-063.

Santos, M.G.M. and **Owen, G.** (2016) Heterolithic meandering-channel deposits from the Neoproterozoic of NW Scotland: Implications for palaeogeographic reconstructions of Precambrian sedimentary environments. *Precambrian Res.*, **272**, 226–243.

Schumm, S.A. (1972) Fluvial paleochannels. In: Recognition of Ancient Sedimentary Environments (Eds J.K. Rigby and W.K. Hamblin), *SEPM Spec. Publ.*, **16**, 98–107).

Sellards, E.H., Tharp, B.C. and **Hill, R.T.** (1923) Investigation on the Red River made in connection with the Oklahoma-Texas boundary suit. *Texas Bulletin*, **2327**, 174.

Shiers, M.N., Mountney, N.P., Hodgson, D.M. and **Cobain, S.L.** (2014) Depositional controls on tidally influenced fluvial successions, Neslen Formation, Utah, USA. *Sed. Geology*, **311**, 1–16.

Slingerland, R. and **Smith, N.D.** (1998) Necessary conditions for a meandering-river avulsion. *Geology*, **26**, 435–438.

Slingerland, R. and **Smith, N.D.** (2004) River avulsions and their deposits. *Ann. Rev. Earth Planet. Sci.*, **32**, 257–285.

Smith, C.E. (1998) Modeling high sinuosity meanders in a small flume. *Geomorphology*, **25**, 19–30.

Smith D.G. (1988) Modern point bar deposits analogous to the Athabasca oil sands, Alberta, Canada. In: *Tide-influenced sedimentary environments and facies* (Eds P.L. deBoer, A. VanGelder and S.D. Nio), D Reidel Publishing Company, 417–432 (530p).

Smith, D.G., Hubbard, S.M., Lavigne, J., Leckie, D.A. and **Fustic, M.** (2011) Stratigraphy of counter-point-bar and eddy-accretion deposits in low-energy meander belts of the peace-Athabasca Delta, Northeast Alberta, Canada. In: *From River to Rock Record: The Preservation of Fluvial Sediments and Their Subsequent Interpretation* (Eds S.K. Davidson, S. Leleu and C.P. North), *SEPM Spec. Publ.*, **97**, 143–152.

Smith, D.G., Hubbard, S.M., Leckie, D. and **Fustic, M.** (2009) Counter point bars in modern meandering rivers: recognition of morphology, lithofacies and reservoir significance, examples from Peace River, AB, Canada. *Sedimentology*, **56**, 1655–1669.

Smith, N.D., Cross, T.A., Dufficy, J.P. and **Clough, S.R.** (1989) Anatomy of an avulsion. *Sedimentology*, **36**, 1–23.

Sundborg, Å. (1956). The river Klarälven a study of fluvial processes. *Geogr. Ann.*, **38**, 125–316.

Tal, M. and Paola, C. (2007) Dynamic single-thread channels maintained by the interaction of flow and vegetation. *Geology*, **35**, 347–350.

Tal, M. and Paola, C. (2010) Effects of vegetation on channel morphodynamics: results and insights from laboratory experiments. *Earth Surf. Proc.*, **35**, 1014–1028.

Taylor, G. and Woodyer, K.D. (1978) Bank deposition in suspended-load streams. In: *Fluvial Sedimentology* (Ed. A.D. Miall). *Can. Soc. Petrol. Geol. Mem.*, **5**, 257–275.

Thomas, R.G., Smith, D.G., Wood, J.M., Visser, J., Calverley-Range, E.A. and Koster, E.H. (1987) Inclined heterolithic stratification and terminology, description, interpretation and significance. *Sed. Geol.*, **53**, 123–179.

Thomson, J. (1876) On the windings for rivers in alluvial plains, with remarks on the flow of water round bends in pipes. *Proc. Roy. Soc. London*, **25**, 5–8.

Toonen, W.H., Kleinhans, M.G. and Cohen, K.M. (2012) Sedimentary architecture of abandoned channel fills. *Earth Surf. Proc. Land.*, **37**, 459–472.

Tower, W.S. (1904) The Development of Cut-Off Meanders. *Bull. Am. Geogr. Soc.*, **36**, 589–599.

van de Lageweg, W.I., Van Dijk, W.M., Baar, A.W., Rutten, J. and Kleinhans, M.G. (2014) Bank pull or bar push: what drives scroll-bar formation in meandering rivers? *Geology*, **42**, 319–322.

van de Lageweg, W.I., Dijk, W.M. and Kleinhans, M.G. (2013) Channel belt architecture formed by a meandering river. *Sedimentology*, **60**, 840–859.

van de Lageweg, W.I., Schuurman, F., Cohen, K.M., Dijk, W.M., Shimizu, Y. and Kleinhans, M.G. (2016) Preservation of meandering river channels in uniformly aggrading channel belts. *Sedimentology*, **63**, 586–608.

Van den Berg, J.H., Boersma, J.R. and Van Gelder, A. (2007) Diagnostic sedimentary structures of the fluvial–tidal transition zone. Evidence from deposits of the Rhine Delta. *Neth. J. Geosci.*, **86**, 253–272.

van Dijk, W.M., Lageweg, W.I. and Kleinhans, M.G. (2012) Experimental meandering river with chute cutoffs. *J. Geophys. Res.: Earth Surf.*, **117**(F3).

van Dijk, W.M., Lageweg, W.I. and Kleinhans M.G. (2013b) Formation of a cohesive floodplain in a dynamic experimental meandering river. *Earth Surf. Proc. Land.*, **38**, 1550–1565.

van Dijk, W.M., Teske, R., van de Lageweg, W.I. and Kleinhans, M.G. (2013a) Effects of vegetation distribution on experimental river channel dynamics. *Water Resour. Res.*, **49**, 7558–7574.

Willis, B.J. (1989) Palaeochannel reconstructions from point bar deposits: a three-dimensional perspective. *Sedimentology*, **36**, 757–766.

Willis, B.J. and Tang, H. (2011) Three-dimensional connectivity of point-bar deposits. *J. Sed. Res.*, **80**, 440–454.

Wormleaton, P.R., Sellin, R.H.J., Bryant, T., Loveless, J.H., Hey, R.D. and Catmur, S.E. (2004) Flow structures in a two-stage channel with a mobile bed: *J. Hydraul. Res.*, **42**, 145–162.

Wright, M.D. (1959) The formation of cross-bedding by a meandering or braided stream. *J. Sed. Petrol.*, **29**, 610–615.

Wu, C., Bhattacharya, J.P. and Ullah, M.S. (2015) Paleohydrology and 3D facies architecture of ancient point bars, Ferron Sandstone, Notom Delta, south-central Utah, USA. *J. Sed. Res.*, **85**, 399–418.

Yan, N., Mountney, N.P., Colombera, L. and Dorrell, R.M. (2017) A 3D forward stratigraphic model of fluvial meander-bend evolution for prediction of point-bar lithofacies architecture. *Comput. Geosci.*, **105**, 65–80.

Zinger, J.A., Rhoads, B.L. and Best, J.L. (2011) Extreme sediment pulses generated by bend cutoffs along a large meandering river. *Nature Geoscience*, **4**, 675.

Int. Assoc. Sedimentol. Spec. Publ (2018) **48**, 15–46.

Controls on the depositional architecture of fluvial point-bar elements in a coastal-plain succession

MICHELLE N. SHIERS[†‡], NIGEL P. MOUNTNEY[†], DAVID M. HODGSON[†] and LUCA COLOMBERA[†]

[†] *School of Earth and Environment, University of Leeds, Leeds, UK*
[‡] *CASP, West Bldg, Madingley Rise, Madingley Road, Cambridge, UK*

ABSTRACT

The architecture and lithofacies organisation of fluvial point-bar elements record the spatio-temporal evolution of river channels. This study discusses the factors that control facies distributions and geometries of point-bar elements present in a fluvial succession that accumulated on a low-gradient coastal plain on the western margin of the Western Interior Seaway (Campanian Neslen Formation, eastern Utah, USA). Forty outcropping point-bar elements located within an established sequence stratigraphic framework have been examined through facies, architectural and palaeocurrent analyses. Point-bar elements increase in width-to-thickness aspect ratio vertically through the succession. Four point-bar element types are identified based upon their lithofacies assemblages and geometry. Two point-bar types conform to those depicted in traditional facies models; they are dominated by cross-bedded sandstone, with subordinate amounts of ripple-laminated and horizontally laminated sandstone. In contrast, the other two point-bar types exhibit unusually low proportions of cross-bedded sandstone and higher proportions of massive sandstone, horizontally laminated sandstone and ripple-laminated sandstone. The occurrence of these atypical point-bar assemblages is restricted to the marine-influenced lower and middle parts of the Neslen Formation. An up-succession increase in aspect ratio and degree of amalgamation of point-bar elements through the Neslen Formation may reflect a decrease in the rate of accommodation generation, an increase in the rate of sediment supply, or autogenic processes that operated on an overall prograding coastal plain. The accumulation of point-bar elements with lower proportions of cross-bedded sandstone in the lower Neslen Formation can be attributed to decreased stream power. Database-assisted analysis has been undertaken to compare the lithofacies and architecture of point-bar elements from the Neslen Formation to those in other humid-climate, coastal-plain successions. This comparison reveals that the geometry and facies observed in point-bar elements of the Neslen Formation might record an unusual set of combined allogenic (accommodation generation and fluvial discharge variations) and autogenic (backwater processes and presence of peat mires) process interactions.

Keywords: Fluvial, point-bar, Neslen Formation, marine influence, backwater

INTRODUCTION

Studies of point-bar elements in both fluvial and tidal environments (e.g. Visher, 1965; Allen, 1965; 1983; 1991; McGowen & Garner, 1970; Harms *et al.*, 1975; Barwis, 1977; Jackson II, 1976; 1978; Miall, 1977; 1985; 1988; Nanson, 1980; Nanson & Page, 1983; Smith, 1987; Cloyd *et al.*, 1990; Nio & Yang, 1991; Rasanen *et al.*, 1995; Galloway & Hobday, 1996; Fenies & Faugères, 1998; Leeder, 1999; Ghazi & Mountney, 2009, 2011; Brekke & Couch, 2011; Johnson & Dashtgard, 2014) have identified associations of commonly occurring bodies with predictable lithofacies and geometric arrangements. Such facies and architectural relationships are commonly summarised as facies

Fluvial Meanders and Their Sedimentary Products in the Rock Record, First Edition.
Edited by Massimiliano Ghinassi, Luca Colombera, Nigel P. Mountney and Arnold Jan H. Reesink.

models (Fig. 1; e.g. Cant & Walker, 1978; Nanson, 1980; Walker, 1984; Miall, 1985; Thomas *et al.*, 1987; Miall, 1988; Ghazi & Mountney, 2009; Smith *et al.*, 2009; Colombera *et al.*, 2013; Labrecque *et al.*, 2011; Fustic *et al.*, 2012; Musial *et al.*, 2012). There is a documented variability in the lithofacies assemblage of point-bar elements with a wide range of width-to-thickness aspect ratios (cf. Gibling 2006; Colombera *et al.*, 2017); this means that no single facies model can account for the known range of stratigraphic complexity in fluvial point-bar deposits.

The evolution of point-bar elements is controlled by the interplay of allogenic and autogenic parameters (Hampson, 2016). From an applied standpoint, understanding controls on point-bar evolution is important to enhance understanding of the distribution of facies (Russell *et al.*, this volume), including the occurrence of inclined heterolithic stratification (IHS; Fig. 1B) within point-bar elements (Weimer *et al.*, 1982; Demowbray, 1983; Thomas *et al.*, 1987; Shanley *et al.* 1992; Turner & Eriksson, 1999; Choi *et al.*, 2004; 2011a; 2011b; Dalrymple & Choi, 2007; Hovikoski *et al.*, 2008; Brekke & Couch, 2011; Sisulak & Dashtgard, 2012; Johnson & Dashtgard, 2014). Inclined heterolithic stratification has been observed in many fluvial successions (e.g. Cretaceous McMurray Formation, Alberta, Canada, Jablonski *et al.*, 2012; Fairlight Clay and Ashdown Beds Formation, Stewart, 1983; Cretaceous Wessex Formation, UK, Stewart, 1983) and is commonly associated with tidal processes (e.g. Weimer *et al.*, 1982; Thomas *et al.*, 1987; Shanley *et al.*, 1992; Choi *et al.*, 2004; Dalrymple & Choi, 2007; Hovikoski *et al.*, 2008; Sisulak & Dashtgard, 2012; Johnson & Dashtgard, 2014; Yan *et al.*, 2017), or with secondary or counter currents, notably during the development of counter point bars in exclusively fluvial settings (e.g. Smith *et al.*, 2009). The proportion and distribution of heterogeneities within fluvial successions exerts a major control on hydrocarbon reservoir behaviour and on water flow and contaminant transport in groundwater aquifers. Fine-grained deposits may act as baffles or even barriers to fluid flow (mesoscopic heterogeneity: Tyler & Finley, 1991; Miall, 2013; Colombera *et al.*, 2017). Recently, database approaches have proved to be a valuable tool to compare heterogeneity distribution, point-bar element geometries and internal facies arrangements from many studied successions. The Fluvial Architecture Knowledge Transfer System (FAKTS) is one such relational database that describes the anatomy of fluvial successions including lithofacies proportions and geometries of fluvial deposits from a wide variety of successions (Colombera *et al.*, 2012a, b; 2013).

The aim of this study is to discuss the controls that give rise to a range of facies distributions within point-bar elements present in a fluvial succession that accumulated on a low-gradient and low-relief coastal plain. To achieve this aim, a detailed outcrop-based study of point-bar elements in the Campanian Neslen Formation, Utah, USA, has been undertaken. This approach permits integration of detailed sedimentological data of multiple point-bar elements distributed laterally and vertically through a succession for which an established sequence stratigraphic framework is well constrained. Specific objectives of this study are as follows: (i) to describe the typical facies arrangements of point-bar elements within the Neslen Formation; (ii) to establish the stratigraphic and spatial distribution of point-bar elements through the formation; (iii) to use a quantitative database approach to compare and contrast the architecture and facies distributions of point-bar elements in the Neslen Formation to both previously proposed facies models and to other comparable successions; and (iv) to develop an understanding of the controls on the internal lithofacies within point-bar elements, the external geometry of the elements and their vertical stacking and connectivity.

GEOLOGICAL SETTING

The Campanian Neslen Formation accumulated in a low-gradient (Colombera *et al.*, 2016), low-relief (Cole & Cumella, 2003), lower coastal plain setting on the western margins of the Cretaceous Western Interior Seaway (WIS; Fig. 2A). The seaway formed in a foreland basin that was infilled with a wedge of siliciclastic strata shed from the Sevier Orogenic Belt to the west (Armstrong, 1968; Jordan, 1981). The seaway was characterised by relatively shallow water depths along its length, rarely exceeding 100 m, and by low gradient margins (Kauffman 1977). The climatic regime of Utah during the Cretaceous was humid and subtropical (Fillmore, 2010) with potentially monsoonal conditions (Fricke *et al.*, 2010; Foreman *et al.*, 2015).

The sequence stratigraphic framework of the Neslen Formation has been established previously

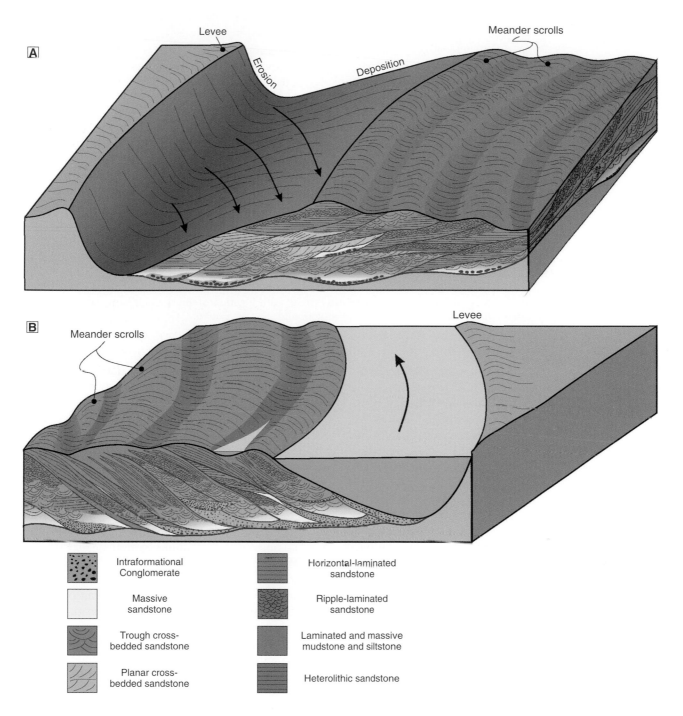

Fig. 1. Conceptual model of point-bar elements. (A) Three-dimensional depositional model of facies and architecture of a sandstone-dominated point-bar element (adapted after Allen, 1970; Cant & Walker, 1978; Nanson, 1980; Walker, 1984; Miall, 1985; Miall, 1988; Ghazi & Mountney, 2009; Colombera *et al.*, 2013). (B) Three-dimensional depositional model of facies and architecture of a heterolithic point-bar element (adapted in part from Thomas *et al.*, 1987; Smith *et al.*, 2009; Labrecque *et al.*, 2011; Fustic *et al.*, 2012; Musial *et al.*, 2012).

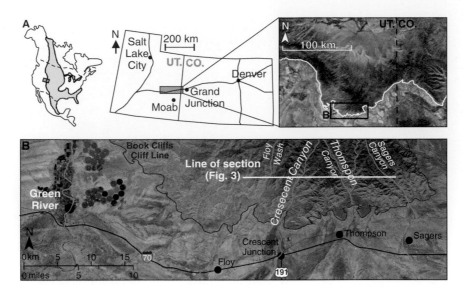

Fig. 2. Study location map. (A) Map of Western Interior Seaway (blue) with location of the study area in the Book Cliffs. (B) Location of the studied stratigraphy between Floy Canyon and Sagers Canyon. GoogleEarth ©. UT = Utah, CO = Colorado.

(Yoshida *et al.*, 1996; McLaurin & Steel, 2000; Hettinger & Kirschbaum, 2003; Shiers *et al.*, 2014; 2017; Fig. 3A). This framework has allowed each of the point-bar elements described here to be located within a specific systems tract, such that differences in point-bar internal facies distributions can be related to different accommodation settings (as described below).

The Neslen Formation comprises sandstones encased within argillaceous, commonly coal-bearing strata (Young, 1957; Fisher *et al.*, 1960; Willis, 2000; Cole, 2008; Colombera *et al.* 2016; Shiers *et al.*, 2014; 2017). The succession is interpreted as the accumulated sedimentary record of coastal delta-plain palaeoenvironments (Kirschbaum & Hettinger, 2004; Aschoff & Steel, 2011a; Shiers *et al.*, 2014; 2017). An overall westward transition from paralic to fluvial deposits across the coeval Neslen Formation and adjacent upper Castlegate Sandstone is associated with a lateral coarsening of lithologies toward the Sevier orogenic belt.

The lower Neslen Formation accumulated in brackish and fresh-water environments in a coastal-plain setting, which was characterised by various sub-environments including tidal flats, lagoons, bays, marshes and oyster reefs (Pitman *et al.* 1987; Chan & Pfaff, 1991; Shiers *et al.*, 2014; 2017) and is interpreted to represent a 4th order Transgressive Systems Tract (TST; Shiers *et al.*, 2017). Within the lower Neslen Formation, tidal and brackish-water influence is expressed by

bioturbation and ichnospecies common to brackish water, single and double drapes of fine-grained sediment on sandstone ripple foresets, laminations that show evidence of rhythmicity and sedimentary indicators of current reversal including successive sets in which ripple foresets show opposing dip directions (Shiers *et al.*, 2017). The upper Neslen Formation accumulated in an upper coastal-plain and lower alluvial-plain setting that was characterised by meandering rivers that traversed extensive floodplains (Pitman *et al.*, 1987) as part of a Highstand Systems Tract (HST).

Some tabular sandstone bodies within the Neslen Formation (Fig. 3) are interpreted be formed of wave reworked sandstone as part of a backstepping barrier system (Shiers *et al.*, 2017) and provide important correlation datums within the succession. One example, the Thompson Canyon Sandstone Bed (TCSB) separates the lower and upper Neslen Formation (Shiers *et al.*, 2014; 2017). The base of the TCSB can be mapped for at least 45 km in an east-west oriented depositional dip direction (Gualtieri, 1991) and it shows a marked facies dislocation that is interpreted to define a Maximum Flooding Surface (MFS). The Basal Ballard Sandstone Bed (BBSB) is a similar example, with a continuity of at least 18 km (Shiers *et al.*, 2017).

Laterally extensive coal beds are also present in the Neslen Formation (Cole, 2008; Shiers *et al.*, 2014; 2017). In the study region, the tabular

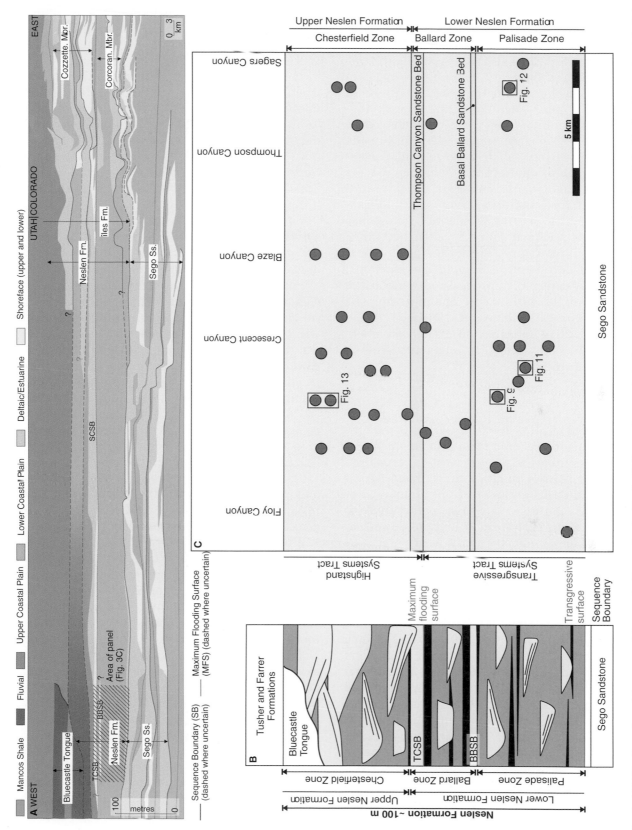

Fig. 3. (A) Stratigraphy of the upper Mesaverde Group in the Book Cliffs; study area is indicates (B) Simplified stratigraphy of the Neslen Formation (cf. Shiers *et al.*, 2014; 2017) and (C) the stratigraphic position of outcrops with studied point-bar elements. TCSB = Thompson Canyon Sandstone Bed. BBSB = Basal Ballard Sandstone Bed. SCSB = Sulphur Canyon Sandstone Bed. Part A based in part on Kirschbaum & Hettinger (2004).

sandstone bodies and the coal-bearing intervals collectively provide the basis for an informal subdivision of the Neslen Formation into 3 zones (Shiers *et al.*, 2014): from base to top, the Palisade Zone, the Ballard Zone and the Chesterfield Zone. The Palisade and Ballard Zones together make up the lower Neslen Formation, the Chesterfield Zone is the upper Neslen Formation (Fig. 3).

Three different types of channel-related element have been previously identified in the Neslen Formation: distributary-channel elements (or ribbon channel-fill elements) (Shiers *et al.*, 2014; 2017; Colombera *et al.*, 2016), sandstone-dominated point-bar elements, alternatively referred to as fluvial-channel sandstones (Willis, 2000; Kirschbaum & Hettinger, 2004; Cole, 2008; Shiers *et al.*, 2014; 2017; Colombera *et al.*, 2016), and heterolithic point-bar elements, alternatively referred to as inclined heterolithic strata or tidally influenced channel deposits (Willis, 2000; Kirschbaum & Hettinger, 2004; Shiers *et al.*, 2014; Olariu *et al.*, 2015; Colombera *et al.*, 2016). The point-bar elements are the focus of this study.

METHODS

Data have been collected from 40 ancient point-bar elements identified in the Neslen Formation, (Fig. 2). Point-bar elements were characterised at outcrop by their facies associations (Table 1), fining-upwards grain-size trends, occurrence of internal inclined surfaces and lenticular external geometry (Fig. 4), and by analysis of palaeocurrent indicators. These characteristics are used to interpret these sedimentary bodies as the product of meandering rivers, in accordance with previous interpretations for the Neslen Formation (Yoshida *et al.*, 1996; McLaurin & Steel, 2000; Hettinger & Kirschbaum, 2003; Willis, 2000; Kirschbaum & Hettinger, 2004; Cole, 2008; Shiers *et al.*, 2014; 2017; Burns *et al.*, 2017) and analogue deposits of modern meandering rivers. The lithological character of infill of fluvial channels by point-bar architectural elements is represented by 14 lithofacies (Fig. 5; Table 1). The facies are interpreted in terms of depositional or post-depositional processes.

Field study has focused on the identification and facies characterisation of sandbodies interpreted to have accumulated as point bars in response to the evolution of meandering fluvial channels. The stratigraphic position of studied point-bar elements was determined in relation to the top of the underlying Sego Sandstone, as well as their position relative to the marker beds (see also Colombera *et al.*, 2016).

The studied point-bar elements are distributed laterally and vertically throughout the studied interval of the Neslen Formation (Fig. 3): of the 16 located in the lower Neslen Formation, 13 are in the Palisade Zone and 3 are in the Ballard Zone (Fig. 3C). Of the 24 examples located in the upper Neslen Formation, 4 are in the lower Chesterfield Zone and 20 are in the upper Chesterfield Zone (Fig. 3C).

Measurements collected from the studied point-bar elements include the external geometry (length, width and thickness; cf. Colombera *et al.*, 2012b), internal lithofacies and bounding-surface orientations (Fig. 4). The relationship of point-bar elements to surrounding elements (e.g. the vertical and lateral relationship of point-bar elements with peat-mire and overbank elements; Table 1) was determined through the construction of measured architectural-element panels. Internal lithofacies distributions were described and logged in detail at one of more locations along the outcrops (the 40 studied elements were analysed using 65 measured logs in total with a cumulative measured section length of 400 m).

Trace fossils were ascribed to ichnological assemblages (Pemberton *et al.*, 1982; Bromley, 1996; Gingras *et al.*, 2012). Palaeocurrent readings (n = 1021) were determined from the dip direction of foresets of cross bedding and ripple lamination. The palaeocurrent database was augmented by the collection of 400 dip azimuths of bounding surfaces of lateral-accretion sets, from which bar-growth trajectories were determined. These data are displayed in rose diagrams which show both palaeocurrent and bar-growth trajectory indicators for individual point-bar elements. Note, however, that the rose diagrams do not express the spatial variability of palaeocurrents around the point bar, which is expected to arise due to the meandering nature of the formative river channels, for example in relation to potential secondary circulation. Channelised sandbodies in the upper Chesterfield Zone (Fig. 3) are highly amalgamated (Shiers *et al.*, 2014) and analysis of these bodies was restricted to single vertical logs and photographic stratigraphic panels because of the cliff-forming nature of the outcrop.

Table 1. Descriptions and interpretation of the facies observed in point-bar elements of the Neslen Formation.

Facies	Bed thickness (m)	Description	Interpretation
Massive mottled mudstone/siltstone (Fsm)	0.01–0.10	Massive silt/mud, grey to black in colour, often with roots and bioturbation, wood fragments and coal clasts.	Low-energy deposits by suspension settling, probably in the distal parts of crevasse splays (Collinson et al., 2006; Burns et al., 2017).
Laminated mudstone/siltstone (Fl)	0.01–0.50	Laminated and interbedded mudstone and siltstone, some ripples.	Overbank, abandoned-channel and/or waning-flood deposit
Massive-faintly laminated sandstone (Sm)	0.03–0.30	Fine-grained to medium-grained sandstone with lack of sedimentary structures, except occasional horizontal laminations. Bioturbation is commonly observed.	Deposits of sediment gravity flows or hyperconcentrated flows (Jones & Rust, 1983). Bioturbation may mask original sedimentary structures to produce a massive appearance.
Symmetrical ripple-laminated sandstone (Sw) Draped (Swd)	0.05–0.15	Observed on top surfaces and in cross section. Co-sets form thicknesses of decimetres to a few metres. Single and double drapes of mud are common, carbonaceous and detrital drapes are also observed.	Ripples formed as a product of either oscillatory flows or combined flows in a restricted marine environment, e.g. from waves (Collinson et al., 2006). Double drapes, often rhythmic, indicate a tidal influence.
Current ripple-laminated sandstone (Sr) Draped (Srd)	0.03–0.40	Observed on top surfaces and in cross section. Co-sets form thicknesses of decimetres to a few metres. Single and double drapes of mud are common, carbonaceous and detrital drapes are also observed.	Ripples formed in fluvial channels or bars with a unidirectional current. Drapes indicate periods of reduced energy in the system, possibly due to tidal fluctuations (Shanley et al., 1992).
Trough cross-stratified sandstone (St)	0.10–2.0	Very fine-grained to medium-grained sandstone, commonly associated with a lag and occasionally showing aligned intraformational clasts to cross-bedding sets. Commonly preserved stacked with reactivation surfaces.	Trough cross-bedding formed from migrating three-dimensional dunes under uni-directional currents in fluvial channels (Collinson et al., 2006).
Planar cross-stratified sandstone (Sp)	0.15–0.90	Very fine-grained to medium-grained sandstone with flat upper and lower bounding surfaces and approximately parallel cross-bedding	Formed from migrating two-dimensional dunes under uni-directional currents in fluvial channels (Collinson et al., 2006).
Horizontally bedded sandstone (Sh)	0.05–0.70	Interlaminated fine-grained to medium-grained sandstone.	Upper flow-regime plane bed conditions (Miall, 1992).
Low-angle cross-stratified sandstone (Sl)	0.20–0.60	Laminations within fine-grained to medium-grained sandstone dipping at <15°.	Formed from sediment under unidirectional flow which is transitional to upper flow-regime within the fluvial environment (Bridge, 1993).
Deformed sandstone (Sd)	0.10–1.0	Soft-sediment deformation in fine-grained to medium-grained sandstone beds. The most common expression is convolute lamination.	Loading and rapid deposition on water-logged sediment (Allen, 1977).
Interbedded sandstone and siltstone (Si)	0.10–0.50	Horizontally laminated beds exhibiting an alternation of mm-thick laminae of sandstone and siltstone (often carbonaceous). Alternations show occasionally rhythmic alternations in thickness.	Formed due to alternating current energies in the upper flow regime; siltstone deposited under low flow energy, possibly due to tidal forcing (Shanley et al., 1992).
Lenticular/wavy/flaser bedded sandstone (H)	0.05–0.50	Inter-laminated mud and sand arranged such that lenses of sand and mud can be preserved as lenticular, wavy, or flaser bedding.	Unidirectional currents producing ripples, which were draped in mud at slackwater due to fluctuations in flow energy, possibly of tidal origin (Reineck & Wunderlich, 1968; Shanley et al., 1992).
Intraformational conglomerate	0.05–0.30	Gravel-sized clasts generally a few centimetres in diameter, commonly occur in lags at the base of channels, lining scour surfaces or aligned to cross-bed surfaces. Clasts can occur in the form of mud-chip conglomerate or are composed of sideritised mudstone or sandstone.	Channel lags or scour fills of sediment which was reworked from nearby floodplain deposits (Bridge, 1993).

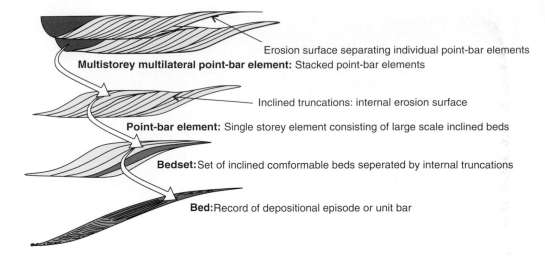

Erosion surface separating individual point-bar elements

Multistorey multilateral point-bar element: Stacked point-bar elements

Inclined truncations: internal erosion surface

Point-bar element: Single storey element consisting of large scale inclined beds

Bedset: Set of inclined comformable beds seperated by internal truncations

Bed: Record of depositional episode or unit bar

Fig. 4. Descriptive terminology for point-bar elements (cf. Bridge 1993); facies colours correspond to Fig. 6.

Quantitative analysis

The lithofacies and geometries for each studied point-bar element were quantified by coding collated data using the Fluvial Architecture Knowledge Transfer System (FAKTS; Colombera *et al.*, 2012a, b; 2013). Bespoke database queries allow analysis of facies characteristics and point-bar dimensions. Calculation of facies proportions within point-bar elements are recorded as a fraction of the logged sections within point-bar elements (Fig. 6), where the base and top of the element is well defined. Within sand-rich point-bar elements, lithological heterogeneity is defined as the thickness proportion of fine-grained lithofacies (silt and finer) relative to the total logged thickness. Point-bar elements with less than 10% fine material (mudstone and siltstone) are defined as being low heterogeneity; moderate-heterogeneity elements have 10 to 20% mudstone and siltstone; high-heterogeneity elements have > 20% mudstone and siltstone (Fig. 6A). Statistical techniques (ANOVA) are used to analyse the statistical significance of observed facies and geometry variations.

The aspect ratio of point-bar elements is calculated as a ratio between the minimum known width of the bar (measured at outcrop) and the maximum logged thickness (Fig. 6C). The true cross-stream width of the element was inferred where possible from 3D reconstructions through multiple exposures along spurs and re-entrants in the hillsides. An apparent width is defined where an outcrop lacked sufficient palaeocurrent indicators to establish the true width (cf. Geehan & Underwood, 1993).

Coding of studied examples within the FAKTS database facilitates quantitative comparison of similarities and differences between individual point-bar elements and has enabled the construction of quantitative facies models (cf. Colombera *et al.*, 2013). Moreover, the FAKTS database has additionally enabled point-bar elements from different systems to be compared to those studied in the Neslen Formation, as presented in the Discussion.

Dimensions of point-bar elements were used to extrapolate channel depth and estimates of fluvial discharge. Point-bar thickness measured at the outcrop is used as a proxy for the maximum bankfull channel depth. There is uncertainty using this method because any inference of maximum bankfull depth derived from bar thickness is a local estimate, due to changes in the geometry of a channel around a bend, from pool to riffle, and as the river sinuosity increases during channel migration (Willis & Tang, 2010).

RESULTS

Comparison of the average internal facies proportions and width-to-thickness aspect ratio of point-bar elements within the Neslen Formation (Figs 7 and 8) to other systems with similar climatic regimes and/or environmental settings analysed using FAKTS (Colombera *et al.*, 2012a; 2012b; 2013) exhibits overall similarity. However, our quantitative investigation allows us to separate the observed elements into four types based on their facies proportions and aspect ratio (Table 2).

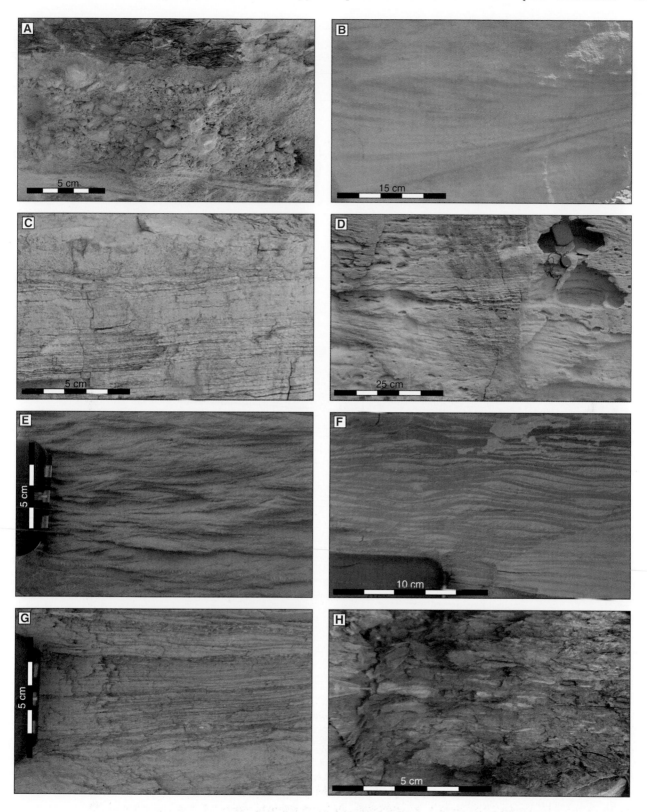

Fig. 5. Representative photographs of sedimentary facies observed within point-bar elements of the Neslen Formation. (A) Intraformational conglomerate (Gh) found at the base of elements and on erosional surfaces. (B) Trough cross-bedded sandstone (Sx). (C) Horizontally laminated sandstone (Sh). (D) Low-angle laminated sandstone (Sl). (E) Ripple cross-laminated sandstone (Sa). (F) Lenticular and wavy sandstone with intervening siltstone laminations (H). (G) Horizontally interbedded sandstone and siltstone (Si). H) Massive (Fsm) and laminated (Fl) mudstone and siltstone.

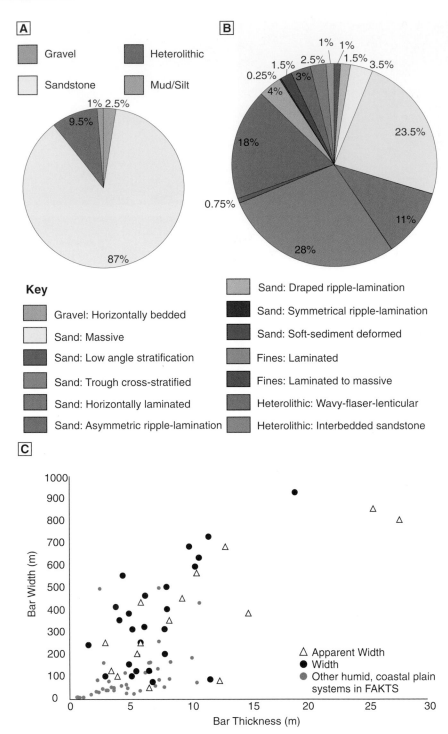

Fig. 6. Quantitative analysis of facies of point-bar elements of the Neslen Formation carried out using the FAKTS database. (A) Lithology of all studied point-bar elements in the Neslen Formation. (B) Facies proportions of all studied point-bar elements within the Neslen Formation. (C) Graph showing aspect ratio of all studied point-bar elements in the Neslen Formation in comparison to other bars analysed in the FAKTS database interpreted to have been deposited in successions accumulated in humid-climate and/or lower coastal-plain settings.

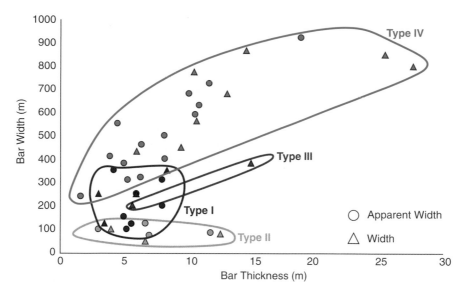

Fig. 7. Graph showing the aspect ratio of the different point-bar element types in the Neslen Formation.

Type I

Description

Type I point-bar elements (Table 2; Fig. 9) are characterised by high proportions of well-sorted ripple-laminated sandstone, mud, silt and carbonaceous draped ripple-laminated sandstone (17.6 to 68%; Fig. 9D). No cross-bedding is observed (Fig. 8). Thirteen elements of this type are identified. Elements are 3 to 12 m-thick and exhibit a wide range of width-to-thickness ratios (between 1:10 and 1:83; Table 2; Fig. 7). The proportion of horizontally laminated sandstone in these types of elements is 8 to 24%. The dip of point-bar accretion surfaces is 10 to 20°. Heterogeneity within this type of element is generally low (0 to 3.6%); two instances of higher heterogeneity also occur (16.7%, 17.6%).

Commonly, the lowermost 0.5 m of the fill exhibits wood fragments and intraformational conglomerate (Gh) within massive, fine-grained and medium-grained sandstone. *Teredolites* and *Thalassinoides* are observed in channel-lag deposits at the base of the elements. Ripples are dominantly asymmetrical (Fig. 5E) and are commonly draped by mud, silt or carbonaceous material; both single and double drapes are observed. Vertically, the changes in facies are subtle; ripple lamination gives way upwards to draped ripple lamination and heterolithic deposits of wavy and flaser bedding (H) and then interbedded siltstone and sandstone laminae (Si) with thin beds of massive and horizontally laminated sandstone (Fig. 9B). The facies proportions and their vertical and lateral transitions can be collated to produce a semi-quantitative depositional model for type I point-bar elements (Fig. 10A). There is a high angle between the palaeoflow direction indicated by ripple cross-laminae dip directions and the accretion direction of the bar-form elements demonstrated by the dip azimuth of inclined bar surfaces (Fig. 9F).

Interpretation

The dominantly lateral direction of accretion within these point-bar elements is demonstrated by the palaeocurrents (Fig. 9F). Rippled sandstone is the product of deposition by migrating asymmetrical ripples (Collinson *et al.*, 2006) under waning traction flows (Simons *et al.*, 1960; Visher, 1965). Thick stacks of rippled sandstone such as that observed in this type of point-bar element can also be indicative of sharp hydrographic variations and rapidly waning floods (cf. Ielpi *et al.*, 2014). Abundant drapes on ripple fore-sets indicate rapid fluctuations in current velocity, possibly influenced by tides (Table 1). Double drapes are specifically diagnostic of tidally influenced environments (Shanley *et al.*, 1992).

Point-bar elements that exhibit a large proportion of ripple-laminated sandstone include those within the Clear Fork Formation (Simon & Gibling 2017a; b). Horizontal laminated sandstone (Sh; Table 1; Fig. 5C) with primary current lineation was deposited from traction flows on the upper

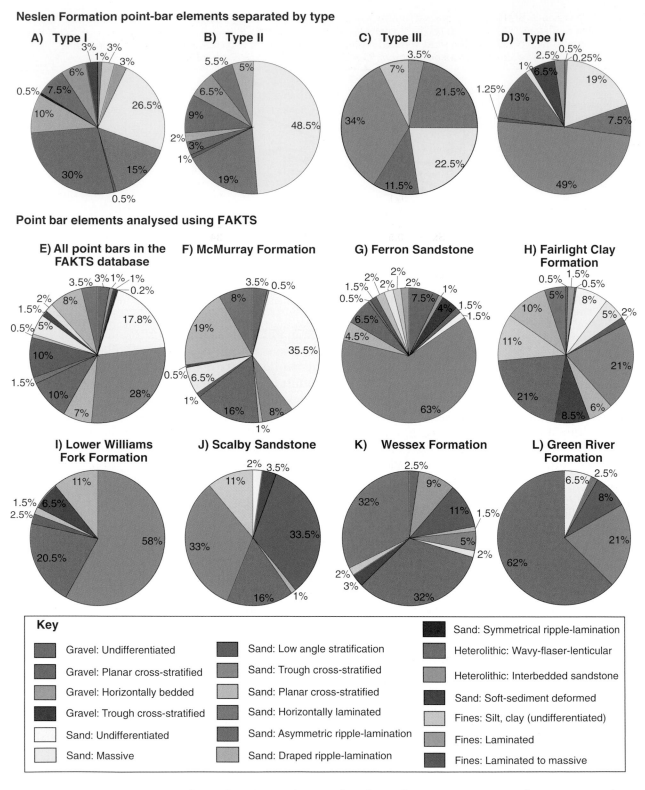

Fig. 8. Quantitative proportions of point-bar element facies within the Neslen Formation separated into interpreted types (A to D): (A) type I; (B) type II; (C) type III; (D) type IV. (E) Cumulative average constituent facies within all point-bar elements analysed in the FAKTS database B to H. Average facies proportions within point-bar elements for other modern and ancient successions (F to L) analysed using data present in the FAKTS database: (F) McMurray Formation facies proportions (Jablonski *et al.*, 2012); (G) Ferron Sandstone (Corbeanu *et al.*, 2004); (H) Fairlight Clay and Ashdown Beds Formation (Stewart, 1983); (I) Lower Williams Fork Formation (Pranter *et al.*, 2007); (J) Scalby Sandstone (Ielpi & Ghinassi, 2014), (K) Wessex Formation (Stewart, 1983); and (L) Green River Formation (Keighley *et al.*, 2003).

Table 2. Key variables for the interpretation of different point-bar element types within the Neslen Formation. Sm = massive sandstone, Sh = horizontally laminated sandstone, Sx = trough cross-bedded sandstone, Sa = asymmetrical laminated sandstone, H = heterolithic sandstone, Si = interbedded sandstone and siltstone, Gh = intraformational conglomerate. Te = *Teredolites*, Me = *Medousichnus*, Th = *Thalassinoides*, Ar = *Arenicolites*.

		Type I	Type II	Type III	Type IV
Number of elements		13	4	2	22
Thickness range (m) (average)		3–12.6 (6.4)	3.5–12.6 (5.8)	5.7–15 (10.35)	4–25 (10.6)
Aspect ratio range (average)		7–83 (36)	30–40 (20)	25–40 (30)	33–150 (71)
Lateral accretion angle (°)		10–20	5–10	10–15	<10
Heterogeneity range/% (average/%)		0–17.6 (7)	0.5–23.5 (10.5)	0.5–9 (4)	0–12.5 (0.75)
Average Facies proportions (%)	Sm	26.5	48.5	22.5	19
	Sh	15	19	21.5	7.5
	Sx	0.5	1	34	50.25
	Sa	40	5	11.5	13
	H	7.5	9	0	0
	Si	6	6.5	0	0
	Gh	3	0	3.5	2.5
Bioturbation		Ar, Te, Th throughout	Te and Th prevalent	Minor Te, Me at base	Minor Te at base of lowermost elements

flow-regime plane-beds. Finer-grained laminations (Si; Table 1) reflect minor energy fluctuations (Simons *et al.*, 1960; Visher, 1965; Fielding, 2006). Horizontally laminated sand deposits are characteristic of rivers subject to seasonal palaeoclimates and flashy discharge, (Fielding *et al.* 2006; Gulliford *et al.*, 2014; Plink-Björklund 2015). Horizontal lamination can also form in bartop areas during waning river stage. Heterolithic facies (H) indicates deposition under fluctuating flow energies (potentially modulated by tidal influence). There is a balance between deposition of mud from suspension and sand deposition either from suspension or as saltating bedload via migrating unidirectional ripples (Miranda *et al.*, 2009). Strong seasonal differences in river discharge and flashy river floods would be anticipated for the study area during the Campanian, in connection with the inferred dominance of a tropical, monsoonal climate (Fricke *et al.*, 2010).

The occurrence of heterolithic facies and the presence of *Teredolites* and *Thalassinoides* trace fossils can be used to infer marine influence in these deposits (Bromley, 1996).

Type II

Description

Type II point-bar elements (Table 2; Fig. 11) are classified based upon the dominance of massive sandstone (30 to 68%; Fig. 11G) and the small proportion of ripple-laminated sandstone (<12%). Four examples of this element are identified; they are 3.5 to 12.6 m-thick and have width-to-thickness ratios of 1:30 to 1:40 (Table 2). The dip of point-bar accretion surfaces is 5 to 10°. Examples of this type of point-bar element have a high proportion of mudstone and siltstone present compared to sandstone (up to 23.5%; Fig. 11G), arranged in heterolithic packages (Fig. 11B and E). These elements have the highest proportion of preserved fines (Figs 8B and 11G).

The proportion of horizontally laminated sandstones (Sh; Table 1) varies from 6 to 32%. Sandstone beds in these elements thin upward, from 0.2 to 0.8 m-thick at the base to 0.05 to 0.5 m-thick in the upper parts of the elements; however, sandstone beds do not exhibit strong fining-upward trends (Fig. 11B). The thickness of fine-grained beds increases upwards. Towards the base of an element, fine-grained beds are mm-thick to cm-thick beds. In the upper parts of elements individual fine-grained beds are up to 0.5 m-thick. Interbedded sandstone and siltstone beds (Si) and heterolithic sandstones (H) commonly exhibit subtle rhythmicity of lamina thicknesses. Massive sandstone occurs in beds, commonly with scoured bases. Trace fossils such as *Teredolites* and *Thalassinoides* are present in the basal-most deposits of some examples (e.g. Fig. 11C) as mono-specific assemblages. The semi-quantitative depositional model for these types of point-bar element (Fig. 10B) reflects the vertical successions and facies proportions observed.

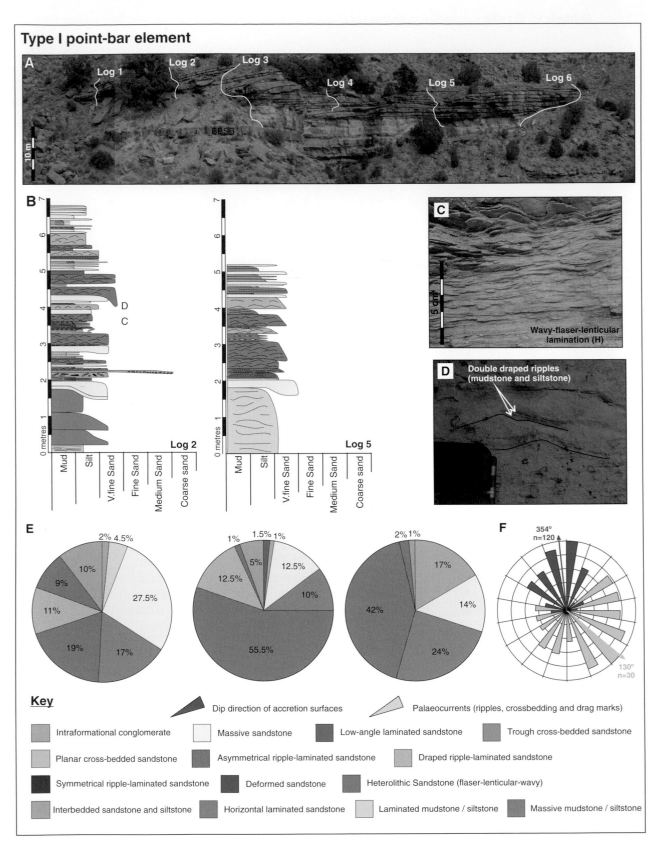

Fig. 9. Example of a type I point-bar element within the Neslen Formation. (A) Photograph panel of a representative type I point-bar element located at Crescent Canyon (Fig. 3); inclined surfaces dip at 10 to 15°. (B) Representative logged sections through the type I point-bar element in (A); location of logs are indicated. (C) Wavy-flaser-lenticular laminated sandstone (H), base shows horizontally interbedded sandstone and siltstone (Si). (D) Double draped ripple lamination. (E) Representative facies proportions within separate examples of type I point-bar elements. (F) Palaeocurrent data for the studied example shown in (A).

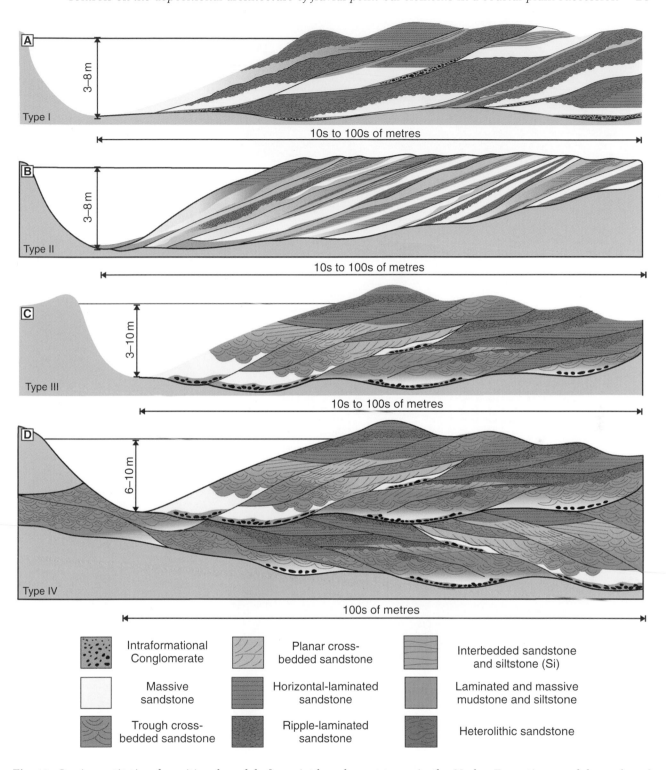

Fig. 10. Semi-quantitative depositional models for point-bar element types in the Neslen Formation; models are based upon accurate width-to-thickness ratios, facies proportions and observed vertical and spatial facies transitions: (A) type I; (B) type II; (C) type III; (D) type IV.

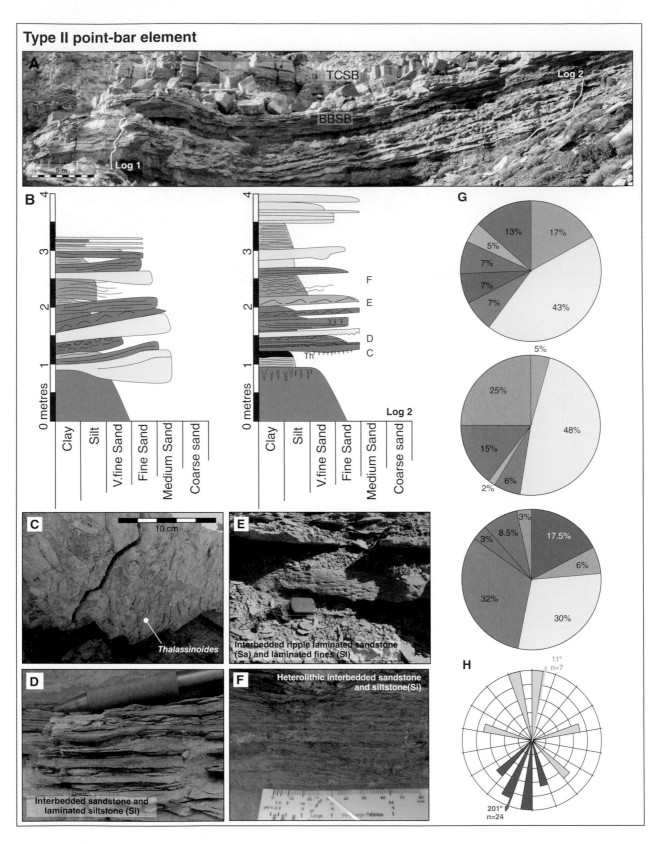

Fig. 11. Example of a type II point-bar element within the Neslen Formation. (A) Photograph panel of a type II point-bar element; inclined surfaces dip at 5 to 10° (Fig. 3) TCSB = Thompson Canyon Sandstone Bed, BBSB = Basal Ballard Sandstone Bed; location of logs are indicated. (B) Logged sections through type II point-bar element with alternating sandstone and siltstone beds. (C) Abundant Thalassinoides on the base of the point-bar element. (D) Horizontally laminated sandstone with interbedded siltstone (Sh and Si). (E) Ripple-laminated sandstone beds (Sa) with intervening laminated siltstones (Fl). (F) Pinstriped interbedded siltstone and sandstone (Si) within which sandstone laminae thicken upwards. (G) Constituent facies proportions for separate examples of type II point-bar elements. (H) Palaeocurrent data for the studied example shown in (A); there is a high angle between the azimuth of dip direction of accretion surfaces and that of ripple lamination indicating the dominance of lateral accretion. For key for facies colours refer to Fig. 9.

Interpretation

Similar to type I point-bar elements, these bodies are interpreted as laterally accreting point bars (Fig. 11H). The proportion of mudstone and siltstone, which occur alternating with sandstone beds in inclined packages within the elements, defines them as IHS. Although IHS is sometimes interpreted as the product of tidal influence (Shanley *et al.*, 1992; Dalrymple & Choi, 2007), such deposits can also occur in perennial and ephemeral fluvial systems (Thomas *et al.*, 1987; Lynds & Hajek, 2006; Archer, 1995; Kvale *et al.*, 1990; Kvale, 2011). Here, tidal influence is supported by the presence of mono-specific assemblages of trace fossils interpreted to reflect brackish or saline depositional environments (Bromley, 1996). Sedimentological evidence of tidal influence includes the presence of interbedded sandstone and siltstone (Si, H; Fig. 11D and E), repeating beds of which show subtle rhythmicity. The high proportion of massive sandstone (Sm; Figs 8B and 11) is interpreted to reflect rapid deposition of sediment with a narrow grain-size range, from concentrated flows, locally filling scours (Collinson *et al.*, 2006). Massive sandstone could also be the result of post-depositional modification through fluidisation, although no deformation structures commonly associated with fluidisation have been observed. Alternatively, intense burrowing could also result in a similar massive fabric. Low proportions of cross-bedding may be due to finer grain-size or a lower flow velocity than is required for sandstone to accumulate as dune-scale bedforms (Harms *et al.*, 1975).

The facies association of these bodies demonstrates that the sediment accumulated under fluctuating energy conditions, which may be due to marine influence.

Type III

Description

Type III point-bar elements (Table 2; Fig. 12) in the Neslen Formation are characterised by relatively high proportions of cross-bedded sandstone (over 30%) and low to medium heterogeneity (0.5 to 9% fines; Figs 8C and 12G). The two examples of this type have aspect ratios of 25 and 40, and element thicknesses of 5.7 m and 15 m. Beds have tangential geometries at the base, wedging out laterally over 3 to 5 m (Fig. 12A). The dip of point-bar accretion surfaces is 10 to 15°.

Beds thin and fine upwards from fine-grained and medium-grained sandstone at the base, to very fine-grained sandstone at the top (Fig. 10B). The basal deposits are gravel to pebble mud clasts (Figs 5A and 12B to C) and trace fossils (*Teredolites* and *Medousichnus*). Massive or cross-bedded sandstone beds (Sm/Sx) are common in the lowermost parts, with sets typically partitioned by erosion surfaces or multiple reactivation surfaces. Higher within the elements, cross-bedded sandstone dominates (Fig. 12B and D). The thickness of cross-bedded sandstone beds decreases upward (from ~ 0.5 m-thick beds towards the base to 0.2 m-thick in the upper parts of elements) and ripple laminated (Sr; Fig. 12B and E), massive (Sm) and horizontally laminated (Sh) sandstone facies become more common (Fig. 12G). These facies trends are shown in a semi-quantitative depositional model (Fig. 10C).

Interpretation

The high angle between the palaeoflow direction indicated by cross-laminae dip directions and the accretion direction in bar-form elements demonstrated by the azimuth of dipping bar surfaces supports a dominantly lateral direction of accretion (Fig. 12H; Bridge, 2006). Large proportions of trough cross-bedding (St; Fig. 5B) are interpreted as the record of deposition by migrating subaqueous dunes or unit bars (Table 1). Where cross-bedded sandstones exhibit multiple reactivation surfaces, they are interpreted as indicating variations in flow energy and/or direction, due to changes in river discharge or tidal processes (Shanley *et al.*, 1992). The presence of *Medousichnus* can indicate tidal influence (Howard & Frey, 1984; Gingras *et al.*, 2012) and *Teredolites* (bored wood) is typical of both marine and brackish environments. However, individual logs can be pushed (rafted) upstream within the fluvial-to-marine transition zone (Savrda, 1991). The proportion of trough and planar cross-bedded sandstone (26%) is similar to the amount preserved in all lateral-accretion elements recorded in the FAKTS database (35%; Fig.8E). The amount of ripple lamination in type III point-bar elements is 11% (Fig. 12D) and in all point-bar elements in the FAKTS database is 9% (Fig. 8E).

This element is interpreted as the preserved product of lateral accretion formed from fluvially dominated meandering channels that traversed the coastal plain (Kirschbaum & Hettinger, 2004; Aschoff & Steel, 2011b).

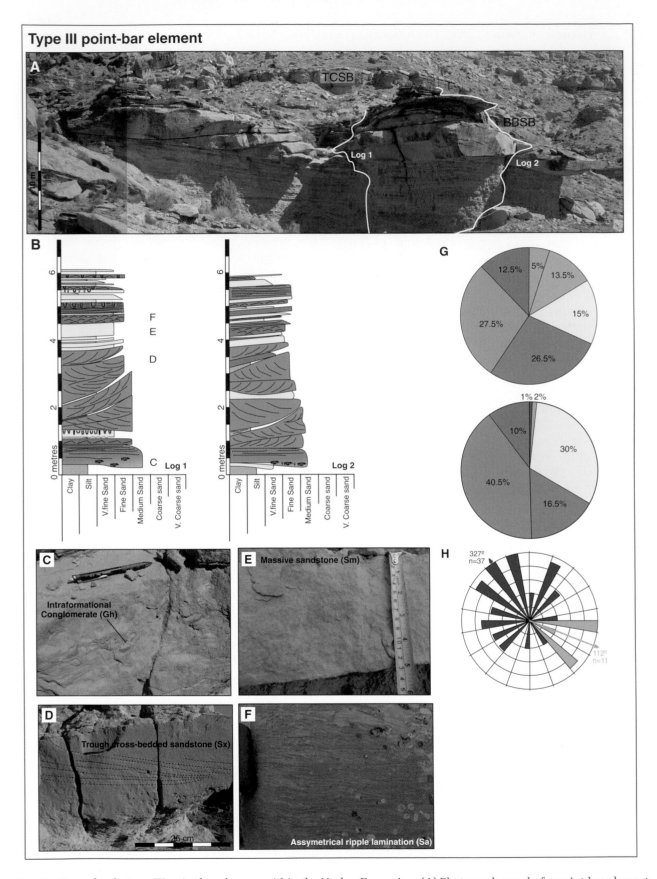

Fig. 12. Example of a type III point-bar element within the Neslen Formation. (A) Photograph panel of a point-bar element showing the sigmoidal shape of beds (Fig. 3), TCSB = Thompson Canyon Sandstone Bed; location of logs are indicated. (B) Logged sections through a type III point-bar element. (C) Photograph of intraformational conglomerate (Gh) found at the base of a type III point-bar element. (D) Trough cross-bedded sandstone (Sx). (E) Massive sandstone bed (Sm). (F) Asymmetrical ripple-laminated sandstone (Sa). (G) Constituent facies proportions for separate examples of type III point-bar elements. (H) Palaeocurrent data for the studied example shown in (A); there is a high angle between the azimuth of dip direction of accretion surfaces and that of ripple lamination indicating the dominance of lateral accretion. Key for facies colours refer to Fig. 9.

Type IV

Description

Type IV point bars (Table 2; Fig. 13) exhibit a high proportion of cross-bedded sandstone (49%). Twenty point-bars of this type are identified within the upper Chesterfield Zone. Elements are 4 to 25 m-thick (average: 10.2 m). The thickness of individual point-bar elements may be overestimated where the nature of the outcrop does not permit the identification of the thickest part of each individual point-bar element and possess high aspect ratios: 33 to 150 (average is 71; Fig. 7). Inclined accretion surfaces are less defined than in other point-bar types but where observed dip at moderate angles, up to 10° (Fig. 13A; Table 2). Where measurement has been possible, there is a high relative angle between lateral-accretion surfaces and the orientation of cross-bedding. Type IV elements generally have low heterogeneity; on average < 0.75%, with one example of 12.5% (Figs 8D and 13E; Table 2). Elements of this type are commonly vertically and laterally amalgamated, forming extensive sandstone belts (400 to 1000 m) in the upper Chesterfield Zone (Fig. 13; *sensu* Shiers *et al.*, 2014).

Erosion surfaces separating individual channel elements are commonly observed; with metre-scale relief and intraformational conglomerate preserved in the lowermost beds (Fig. 13B). Cross-bedded sandstone and massive sandstone dominates, passing upwards to ripple laminated and horizontally laminated sandstone (Fig. 13B); these relationships are shown in Fig. 10D. Bioturbation is not observed in these elements.

Interpretation

The facies assemblage within these bodies is similar to type III point-bar elements; however, they are distinguished by their thickness and aspect ratio (Table 2; Fig. 7), as well as the degree of amalgamation (Fig. 13A; Shiers *et al.*, 2014).

The high relative angle (80 to 150°) between the cross-bedding and lateral-accretion surfaces indicates a dominance of lateral accretion. The vertical succession of facies reflects lower velocity flows developing progressively through filling of the channel (Visher, 1965). The amalgamated nature of these point-bar elements is interpreted to reflect high energy channels eroding underlying floodplain and earlier channelised deposits (Leeder, 1977; Allen, 1978; Bridge & Leeder, 1979; Heller & Paola, 1996; Shiers *et al.*, 2014). The absence of marine indicators in these bodies might indicate deposition in a fully fluvial setting.

Distribution of point-bar element types

Type I point-bar elements are abundant throughout the Palisade, Ballard and lower Chesterfield zones. Type II elements are restricted to the Palisade Zone. Type III elements occur towards the middle of the formation (upper Palisade and lower Chesterfield zones). Type IV point bars occur exclusively within the upper Chesterfield Zone. The upward stratigraphic increase in width-to-thickness ratio of the point-bar elements (Fig. 14) means that the sandstone bodies are increasingly wide for a given thickness. Channel bodies become increasingly amalgamated upwards. The detailed analysis of the external geometry and internal facies character of each point-bar element type allows for the construction of semi-quantitative depositional models (Fig. 12).

The proportions of facies and the observed vertical transitions preserved in type III and IV point-bar elements (Figs 8C and D, 10, 12 and 13) are similar to many other point-bar elements that accumulated in humid subtropical settings, such as those of the Jurassic Scalby Formation (Nami & Leeder, 1977; Ielpi & Ghinassi, 2014; Fig. 8J) and the lower Williams Fork Formation (Pranter *et al.*, 2007; Fig. 8I). Type III and IV point-bar elements (Fig. 8C and D) have similar facies proportions to those elements analysed within the FAKTS database (Fig. 8E) and presented in published facies models (Fig. 1). The vertical transition of facies (Figs 12B and 13B) demonstrates that, as the point bar progressively developed, the preserved facies reflect vertically decreasing flow velocities on the inner bend of the migrating channel element.

Type I and II point-bar elements (Fig. 10A and B) are dissimilar to models presented in the literature (Fig. 1), as well as to the facies proportions of most successions analysed and stored in the FAKTS database (Fig. 8E). Examples exposed in the Wessex and Green River Formations (Fig. 8K, L; Keighley *et al.*, 2003; Stewart, 1983) also exhibit a dominance of ripple-laminated sandstone, although these elements also contain significant proportions of cross-bedded sandstone. Point-bar elements that are dominated by ripple strata are described by Miall (1985; his model 7) and are interpreted in that study as representative of

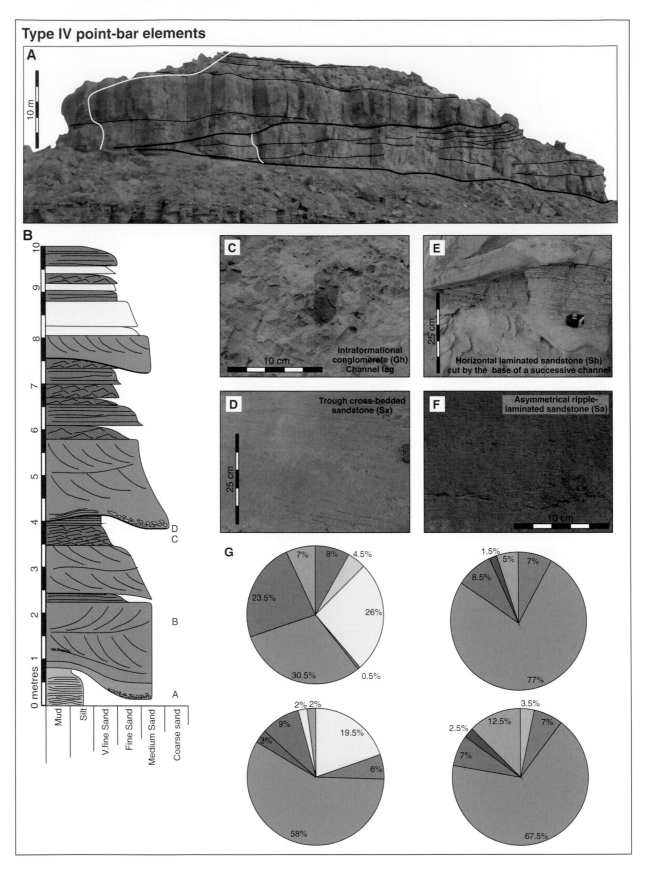

Fig. 13. Example of a type IV point-bar element within the Neslen Formation. (A) Photograph panel of representative type IV point-bar elements; thicker lines show the basal incision surface of individual point-bar elements (location shown in Fig. 3). (B) Representative logged section through typical type IV point-bar elements showing erosive surfaces at the base of each point-bar element and the vertical transition from cross-bedded to ripple laminated sandstone. (C) Intraformational conglomerate (Gh). (D) Trough cross-bedded sandstone (Sx). (E) Horizontally laminated sandstone (Sh) cut it into by an overlying point-bar element. (F) Asymmetrical ripple-laminated sandstone (Sa). (G) Representative facies proportions within separate examples of type IV point-bar elements. Key for facies colours refer to Fig. 9.

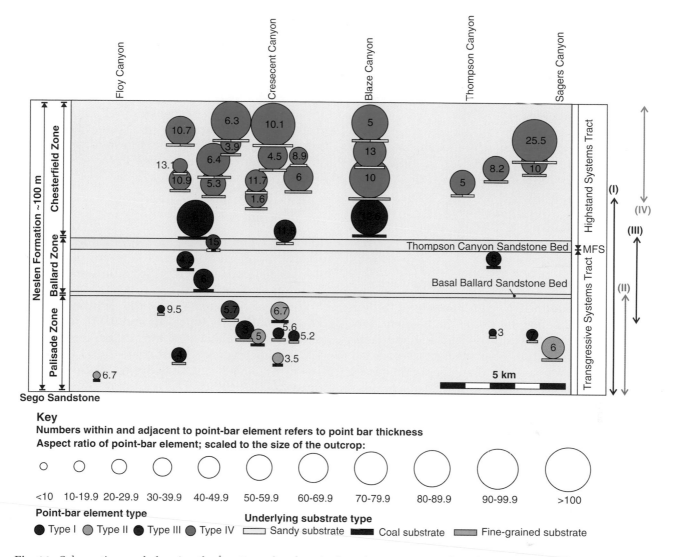

Fig. 14. Schematic panel showing the location of each point-bar element examined in this study in relation to the interpreted sequence stratigraphic framework. Each point-bar element is coloured to represent the point-bar element type (I, II, III or IV) and is shown in relation to underlying substrate, as well as scaled in proportion to the aspect ratio of the outcrop. Numbers within examples refer to the maximum logged thickness of the point-bar element in metres.

deposition from highly sinuous, suspended-load dominated rivers.

In the studied examples from the Neslen Formation, there is an increase in the amalgamation of sandstone bodies upwards. The aspect ratios of type I, II and III point-bar elements are similar to those of other point-bar elements in the FAKTS database (Fig. 7). The aspect ratio of type IV point-bar elements is higher than that of other systems. There is a statistically significant increase in the aspect ratio of point-bar elements from the lower to upper Neslen Formation with an abrupt

increase across the TCSB (tested using ANOVA: significance level = 0.05, p value = 0.001). This is probably due to the highly amalgamated nature of these sandbodies and their development on a substrate with limited cohesion. Therefore, the thickness of individual point-bar elements might be overestimated where the nature of the outcrop does not permit the identification of the base of each individual point-bar element. This uncertainty has been minimised as far as possible through the careful combined use of stratigraphic panels and sedimentary logs.

DISCUSSION

The presented results are discussed in two ways: (i) in terms of the vertical changes of point-bar character (geometry, facies and amalgamation); and (ii) in terms of the occurrence of atypical point-bar assemblages in the lower Neslen Formation.

Tectonism, climate change and eustasy influence point-bar lithofacies (Cecil *et al.*, 1993; Blum & Törnqvist, 2000; Hampson *et al.*, 2012; Shiers *et al.*, 2014), geometry and stacking patterns (Leeder, 1977; Bridge & Leeder, 1979; Bristow & Best, 1993; Mackey & Bridge, 1995; Heller & Paola, 1996).

Controlling Factors in the Neslen Formation

Vertically through the Neslen Formation, there is a systematic change in point-bar element type (Fig. 14). The controls on the stacking and facies assemblage of the point-bar elements are varied, encompassing allogenic and autogenic processes, as discussed below. The possibility that a range of these controls are responsible for the point-bar character observed in the Neslen Formation is considered below and further examined in relation to other successions that have been analysed using the FAKTS database.

Accommodation generation rate

In the lower Neslen Formation, interpreted as a TST (Shiers *et al.*, 2017), point-bar elements are predominantly type I and II elements. In the upper Neslen Formation, interpreted as the highstand systems tract (HST), there is a change from type I elements to type IV elements upwards. This change in element type is concurrent with an increase in amalgamation of the sandstone bodies. These changes reflect the interplay of eustasy, tectonics, sediment supply and compaction, which collectively control the stacking of accumulated fluvial sandbodies (cf. Leeder, 1977; Allen, 1978; Bridge & Leeder, 1979; Aitken & Flint, 1995; Heller & Paola, 1996; Currie, 1997; Sønderholm & Tirsgaard, 1998; Huerta *et al.*, 2011; Foix *et al.*, 2013). During periods of increased accommodation generation, a high proportion of overbank material is preserved, and reworking of fluvial deposits is limited (Wright & Marriott, 1993; Legarreta & Uliana, 1998). Periods of low accommodation generation promote extensive reworking of fine-grained overbank material due to lateral channel migration or avulsion (Posamentier & Vail, 1988; Holbrook, 1996), increasing stacking density of channel elements, and hence net:gross and connectivity. Although changes in the rate of accommodation generation can explain the change in aspect ratio and amalgamation of point-bar elements upwards through the Neslen Formation (Shiers *et al.*, 2014) it is difficult to reconcile this interpretation with the change in facies observed within point-bar elements.

Marine influence

Element types II, I, III, and IV exhibit progressively less marine influence and are inferred to have been deposited farther from the contemporaneous shoreline (Fig. 15).

Point-bar elements in the lower Neslen Formation (mostly types I, II and III) show moderate marine influence (Lawton, 1986; Pitman *et al.*, 1987; Kirschbaum & Hettinger, 2004; Gualtieri, 1991; Karaman, 2012; O'Brien 2015; Gates & Scheetz, 2015; Burton *et al.*, 2016; Shiers *et al.*, 2017). The overall regressive trend of the Neslen Formation through time is recorded by a shift in facies belts eastward (Shiers *et al.*, 2017). The upper Neslen Formation is interpreted to represent an environment up-dip of any discernible influence of marine or backwater processes (Shiers *et al.*, 2014; Fig. 15). Channel fills that might bear a record of the influence of backwater hydrodynamics (cf. ribbon channel fills of Colombera *et al.*, 2016) are recognised in stratigraphic proximity to the maximum flooding surface (the base of the TCSB; Fig. 3B). The stratigraphic location of ribbon channel fills is coincident with the change from isolated (types I, II and III) to amalgamated point-bar elements (type IV) and may correlate with the change in hydrodynamics of the channels through and out of the backwater zone, i.e. the part of the fluvial system downstream of the point where the streambed elevation drops below contemporary sea-level (Fig. 15). A reduction in the rate of lateral migration of fluvial channels is expected in a down-dip direction (through the backwater zone towards the shoreline) due to backwater control on sediment flux (Lamb *et al.*, 2012; Nittrouer *et al.*, 2012; Blum *et al.*, 2013). The decrease in point-bar heterogeneity and increase in bar width-to-thickness ratio through the overall progradational stratigraphy of the Neslen Formation can therefore be interpreted in

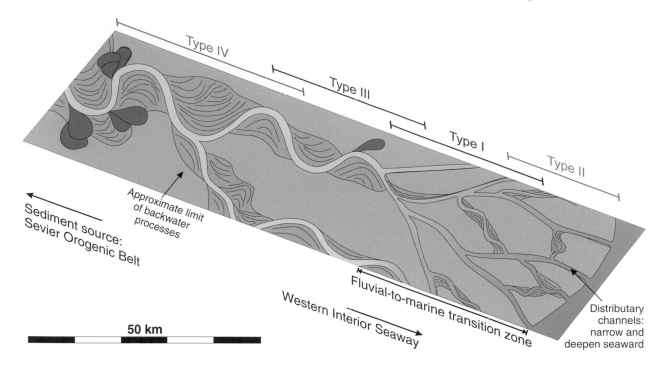

Fig. 15. Schematic diagram showing the possible spatial zones for deposition of the different type of point-bar elements with relation to the limit of marine influence (tidal and backwater processes).

terms of a progressive evolution of hydrodynamics and channel kinematics away from the shoreline, as recognised in some modern systems (cf. Blum *et al.*, 2013; Fernandes *et al.*, 2016).

The upstream end of the backwater zone is recognised as an area with increased avulsion frequency due to the change in hydrodynamics and its effect on streambed aggradation (cf. Chatanantavet *et al.*, 2012; Nittrouer *et al.*, 2012). Increased avulsion rate due to a shifting backwater zone is an autogenic explanation for the observed increase in point-bar amalgamation of type IV point-bar elements (Fig. 15). Alternatively, the change from dominantly type III to type IV point-bar elements in the upper Neslen Formation might be associated with a decrease in the rate of relative sea-level rise (or increase in sediment supply) from the TST to HST.

The influence of marine processes on facies assemblages within point-bar elements is well-documented (Weimer *et al.*, 1982; Thomas *et al.*, 1987; Shanley *et al.*, 1992; Choi *et al.*, 2004; Dalrymple & Choi, 2007; Hovikoski *et al.*, 2008; Sisulak & Dashtgard, 2012; Johnson & Dashtgard, 2014). Type I and II point-bar elements are interpreted to have been subject to marine influence. A greater marine influence is evident within type II point-bar elements. Type I and II elements also

exhibit abundant trace fossils indicative of deposition in a stressed, brackish water environment (Table 2; Bromley, 1996; Gingras *et al.*, 2012).The relative position of the deposition of the different types of point-bar elements on the floodplain can therefore be established (Fig. 15). Minor evidence of brackish water ichnofacies at the base of type III point-bar elements indicates that they might have accumulated at a site closer to the palaeo-coastline than type IV point-bar elements, which themselves apparently accumulated up-dip of the zone of marine and backwater influence (Fig. 15). The vertical changes in occurrence of point-bar element types could have originated through shifting of these facies belts through time according to Walther's Law.

Presence of coal beds

There is a decrease in the occurrence and quality of coal beds up-section in the Neslen Formation (Fig. 3). Changes in coal beds and point-bar element type may be related to the same allogenic changes. Alternatively, peat abundance may have been a causative factor influencing point-bar element type.

Laterally extensive, ombrotrophic mires in the lower Neslen Formation (Shiers *et al.*, 2017) was facilitated by an overall increase in the rate of accommodation generation such that it was balanced by the accumulation rate of peat (see discussions in Speiker, 1949; Young, 1955; Bloom, 1964; Rampino & Sanders, 1981; Tissot & Welte, 1984; Courel, 1989). The rapid compaction of peat can also generate localised topographic lows following accumulation (Bohacs & Suter, 1997), for channels located within these topographic lows, channel avulsion frequency would be limited.

Point-bar elements underlain by coal (type I or II elements; Fig. 14) have low aspect ratios and are associated with narrower channels than those with sandstone, siltstone or mudstone substrates (Fig. 14). Therefore, it is possible that channel evolution (through incision and accretion) was controlled by the presence of mires either adjacent to or underlying the channel. Incision and accretion of these narrower channels would have been limited by the higher relief of ombrotrophic mires (Shiers *et al.*, 2017) and their ability to withstand erosion (McCabe, 1985). The presence of a bounding mire would therefore potentially control the geometry of point-bar elements, as well as influence the morphodynamics of the river and hence the internal lithofacies character of the developing point bars.

Flow velocity

Stream power is directly proportional to discharge and gradient, and inversely proportional to the channel width (Flint, 1974). Higher stream power (i.e. flow velocity) may result in the preferential accumulation of sediment into dunes rather than ripple-scale features for a given grain size (Harms *et al.*, 1975). At finer grain-sizes it is not possible for sediment to accumulate into dune-sized features and ripples preferentially form. The palaeoflow velocity within the channels associated with ancient point-bar elements is difficult to determine. Numerous equations relate channel dimensions to sinuosity, water discharge and gradient (e.g. Leopold & Wolman, 1960; Schumm, 1963; 1972; Williams, 1986; Bridge & Mackey, 1993; Table 3). Although these relationships commonly ignore short-term discharge variations (e.g. single flood events), collectively they are a widely accepted approach to relate changes in hydraulic geometry to element thickness (e.g. Bridge & Tye 2000; Bridge *et al.*, 2000; Gouw & Autin 2008; Tewari *et al.*, 2012; Famubode & Bhattacharya 2016; Chakraborty *et al.*, 2017). Generally, the

calculated flow velocity (Table 3) for the channels associated with Neslen Formation point-bar elements increase upwards through the Neslen Formation. Moreover, the calculated annual discharge for type I and II point-bar elements (87.7 and 62.4 $m^3 s^{-1}$) is significantly lower than those calculated for type III and IV elements (470 and 510.8 $m^3 s^{-1}$).

The inferred low flow velocities of channels in the lower Neslen Formation, compared to the upper Chesterfield Zone and other units deposited on the margins of the WIS (Table 3), may be attributed to river deceleration in its backwater zone, or to allogenic controls, themselves linked to low floodplain gradients, or to times of very low river flow during a period of reduced water discharge (be it seasonal or longer term). There is no link between the proportion of preserved fines within the point bars and the flow velocity: the calculated flow velocities are time and depth averaged and fine-grained sediments are attributed to periods of low river flow.

An estimated floodplain gradient for the Neslen Formation of $\sim 2.5 \times 10^{-4}$ m/m (Colombera *et al.*, 2016) is inferred (subject to uncertainties in compaction and correlation) from the gradient of transgressive surfaces traced in the coastal-plain deposits (Aschoff & Steel, 2011a). This low basin gradient may have resulted from the interference between Sevier-style and Laramide-style tectonics (Armstrong 1968) and dynamic subsidence. No broad scale changes in tectonics (e.g. faulting or thinning) are recognised through the deposition of the Neslen Formation to account for any large changes in floodplain gradient.

A monsoonal climate is inferred to have operated during accumulation of the Neslen Formation (Fricke *et al.*, 2010); this would cause seasonal changes of flow within the channels of the Neslen Formation. The overall reduction in the presence, thickness, extent and continuity of coals from the lower to middle parts of the Neslen Formation to the upper part of the Neslen Formation may have arisen in response to a less humid climate (i.e. lower annual rainfall) through time. However, this is counter to what is suggested by the fluvial discharge variations determined throughout the Neslen Formation, as calculated from reconstructed channel dimensions (Table 3).

Type I and type II point-bar elements are attributed to channels within which the stream power was reduced (hence carrying a fine-grained sediment load) and hence sediment was not

Table 3. Quantitative analysis of channels in the Neslen Formation separated by element type, compared to other systems on the margins of the Western Interior Seaway (green) and humid-climate successions generally (orange). Maximum bankfull depth is interpreted from the average point-bar thickness assuming no compaction. Sources for equations: (1) Bridge & Mackey, 1993; (2) Williams, 1986; (3) Williams, 1986; (4) Leopold & Wolman, 1960; (5) Schumm, 1963; (6, 7, 8) Schumm, 1972.

	Unit	Equation		Neslen Fm.	Type I	Type II	Type III	Type IV	McMurray Fm.	Ferron Ss.	Lower Williams Fork Fm.	Scalby Ss.	Wessex Fm.	Green River Fm.
Maximum bankfull depth (d)	m	Measured		8.8	6.4	5.8	10.35	10.6	30–40	7	6.65	6.65	3.24	6.5
Mean bankfull depth (D)		$D = 0.57d$ Or measured	(1)	5.02	3.65	3.31	5.9	6.04						
Bankfull width (W)	m	$8.88 \times D^{1.82}$ Or measured	(2)	167.38	93.71	78.43	224.58	234.37	500–548	290	281	165	100	74
Channel belt width (Wm)	m	$148 \times D^{1.52}$	(3)	1719	1059	913	2198	2277	40,308	2850	2636	2636	883	2546
W:D (F)	/	W/D	(4)	33.3	25.7	23.7	38.1	38.8	13.9–18.3	41.4	42.3	24.8	30.86	11.4
Wavelength (λ)	m	$10.9 \times W^{1.01}$	(5)	1920	1069	893	2584	2698	5500–6400	3345	3241	1892	1141	842
Sinuosity (P)	/	$3.5F^{-0.27}$	(6)	1.36	1.46	1.49	1.31	1.3	1.6–2.4	1.3	1.2	1.5	1.4	1.8
Mean annual discharge (Qm)	m³ s⁻¹	$\dfrac{W^{2.43}}{18 \times F^{1.13}}$	(7)	267.9	87.7	62.4	470	510.8	10,978	796	719	361	83.5	123
Mean annual flood (Qma)	m³ s⁻¹	$16\left(\dfrac{W^{1.56}}{F^{0.66}}\right)$	(9)	4659	2236	1787	6743	7121	43,994	9511	8927	5534	2193	2645

capable of accumulating into dune-sized bed forms (Fig. 10). This situation could have arisen in channels located away from the main trunk channel belt, or within distributive channel splitting (Fig. 15). The increase in channel size and grain-size up-section can be explained by an increase in discharge. However, the equations used in determining the discharge of ancient river channels negate discharge variability; this means that any seasonality within the rivers is obscured. A change between seasonal and perennial discharges may explain the difference between type I and II point bar elements and type III and IV elements respectively. Type I and II point-bar elements might be related to a seasonal climate, due to the dominance of horizontally laminated sandstone (cf. Fielding *et al.*, 2009).

Comparison of the Neslen Formation to other similar depositional systems

Analysis shows that trends in point-bar element facies and their vertical stacking density through the Neslen Formation may be due to the increasing accommodation (influenced by the presence of coal and marine processes) and declining flow velocity towards the coast. Analysis has been undertaken using the FAKTS database to test if these trends can be observed in other ancient successions of humid-climate, coastal-plain settings.

Marine influenced successions

The change in the aspect ratio and amalgamation of point-bar elements within the Neslen Formation, as demonstrated above, can be attributed to their position on the floodplain in relation to the limit of backwater processes as part of the fluvial-to-marine transition zone. However, the link between point bar elements interpreted to have been modified by tidal processes and the lack of cross-bedded sandstone (but high proportion of ripple-laminated and horizontally laminated sandstone) is not observed within other formations (e.g. Weimer *et al.*, 1982; Thomas *et al.*, 1987; Shanley *et al.*, 1992; Bose & Chakraborty, 1994; Choi *et al.*, 2004; Dalrymple & Choi, 2007; van den Berg *et al.*, 2007; Hovikoski *et al.*, 2008; Matinius & Gowland, 2011; Sisulak & Dashtgard, 2012; Johnson & Dashtgard, 2014; Legler *et al.*, 2014). None of the successions interpreted as having been laid down in environments proximal to the marine realm show similar facies assemblages to the Neslen Formation. This indicates that, although marine

processes may have been responsible for the introduction of significant heterogeneities within point-bar elements, they cannot be shown to have been the dominant control on the occurrence of cross-bedding within these elements.

Successions containing appreciable amounts of coal

Other coal-bearing systems analysed using FAKTS do not exhibit similar facies assemblages to type I or II point-bar elements of the Neslen Formation. Other coal-bearing systems documented in the literature (e.g. Ferron Sandstone, Ryer, 1981; Raniganj coal measures, Casshyap & Kumar, 1987; Straight Cliffs Formation, Shanley *et al.*, 1992; Weisselster Basin, Halfar *et al.*, 1998; Lopingian coal measures, Wang *et al.*, 2011) do not display instances of facies assemblages similar to those in the type I or II point-bar elements of the Neslen Formation. This indicates that, although there is a strong relationship between the presence of coal substrates and the occurrence of type I and II facies assemblages (Fig. 14), this relationship has not hitherto been established in other successions.

Calculated discharge values within other successions

Systems associated with low mean annual discharge values ($<150\,m^3\,s^{-1}$; Table 3) have greater proportions of ripple-laminated sandstone within associated point-bar elements (e.g. Green River Formation: Keighley *et al.*, 2003; Wessex Formation: Stewart, 1983; Fig. 8). No other studied successions documented in the literature have a similarly low proportion of cross-bedded sandstone as the type I or II point-bar elements studied here. The high proportion of ripple-laminated and horizontally laminated sandstone are probably a product of sharp hydrographic variations and rapidly waning floods (Ielpi *et al.*, 2014), such as within tropical, monsoonal climates where rivers are subject to seasonal discharge variations (e.g. Fielding, 2006; Gugliotta *et al.*, 2015).

CONCLUSIONS

Quantitative analysis of 40 point-bar elements from the Cretaceous Neslen Formation has allowed four point-bar element types to be distinguished based on their internal facies types, proportions and geometry. Type I and II point-bar elements are characterised internally by a distinctive lack of cross-bedding and are instead dominated by ripple-laminated sandstone and massive and horizontally laminated sandstone, respectively. Type III and IV point-bar elements are similar to many examples from other successions, based on analysis undertaken using the FAKTS database; these types conform to traditional point-bar models.

Upwards through the Neslen Formation, passing from a transgressive systems tract to a highstand systems tract, there are a series of changes in the character of point-bar elements: (i) an increase in the width-to-thickness aspect ratio of point-bar elements; (ii) an increase in the thickness and amalgamation of point-bar elements; (iii) a decrease in internal heterogeneity (mud and silt content); (iv) a change from dominantly type I and II point-bar elements in the lower Neslen Formation to type III and IV point-bar elements in the upper part.

A vertical increase in sediment supply or a decrease in the rate of accommodation generation might have been the dominant controls on the vertical changes in channel-body stacking density. Similar changes might produce the upward decrease in the occurrence of coal beds. In the lower Neslen Formation, point-bar elements that exhibit an abundance of ripple and horizontally laminated and massive sandstone, and a corresponding absence of cross-bedded sandstone, are recognised. The deposition of these less common types of point-bar elements (i.e. types I and II) can be attributed to a combination of marine influence with inferred low stream power of autogenic (e.g. backwater-driven) or allogenic (e.g. climate-driven). Other important considerations in the deposition of unusual point-bar assemblages include the fine-grained nature of the sediment and the presence of mire deposits. This study emphasises the complicated interplay of depositional processes in channels within the fluvial-to-marine transition zone, the discernment of which requires a high-resolution sequence stratigraphic framework.

ACKNOWLEDGEMENTS

This research was funded by AkerBP, Areva, BHPBilliton, Cairn India (Vedanta), ConocoPhillips, Murphy Oil, Nexen, Petrotechnical Data Systems, Saudi Aramco, Shell, Tullow Oil, Woodside and YPF through their sponsorship and support of the Fluvial & Eolian Research Group at the University of Leeds. LC was supported by NERC (Catalyst Fund award NE/M007324/1; Follow-on Fund NE/N017218/1). Luke Beirne and Camille Dwyer are thanked for their help and assistance in the field. Catherine Russell and Howard Johnson are thanked for many helpful discussions. Reviews by Brian Willis and Alessandro Ielpi, and advice from Series Editor Mark Bateman have greatly improved the manuscript.

REFERENCES

Aitken, J.F. and Flint, S.S. (1995) The application of high-resolution sequence stratigraphy to fluvial systems: a case study from the Upper Carboniferous Breathitt Group, eastern Kentucky, USA. *Sedimentology*, **42**, 3–30.

Allen, G.P. (1991) Sedimentary processes and facies in the Gironde Estuary: a recent model for macrotidal estuarine systems. In: *Clastic Tidal Sedimentology* (Eds D.G. Smith, G.E. Reinson, B.A. Zaitlin and R.A. Rahmani), *Mem. Can. Soc. Petrol. Geol.*, **16**, 29–40.

Allen, J. (1970) Physical processes of sedimentation. Earth Science Series 1, Unwin University Books, George Allen and Unwin, 248pp.

Allen, J. (1978) Studies in fluviatile sedimentation: an exploratory quantitative model for the architecture of avulsion-controlled alluvial suites. *Sed. Geol.*, **21**, 129–147.

Allen, J. (1977) The possible mechanics of convolute lamination in graded sand beds. *J. Geol. Soc.*, **134**, 19–31.

Allen, J.R. (1965) A review of the origin and characteristics of recent alluvial sediments. *Sedimentology*, **5**, 89–191.

Allen, J.R. (1983) Studies in fluviatile sedimentation: bars, bar-complexes and sandstone sheets (low-sinuosity braided streams) in the Brownstones (L. Devonian), Welsh Borders. *Sed. Geol.*, **33**, 237–293.

Archer, A.W. and Greb, S.F. (2012) Hypertidal Facies from the Pennsylvanian Period: Eastern and Western Interior Coal Basins, USA. In: *Principles of Tidal Sedimentology*, 421–436. Springer.

Armstrong, R.L. (1968) Sevier orogenic belt in Nevada and Utah. *Geol. Soc. Am. Bull.*, **79**, 429–458.

Aschoff, J.L. and Steel, R.J. (2011a) Anatomy and development of a low-accommodation clastic wedge, upper Cretaceous, Cordilleran Foreland Basin, USA. *Sed. Geol.*, **236**, 1–24.

Aschoff, J. and Steel, R. (2011b) Anomalous clastic wedge development during the Sevier-Laramide transition, North American Cordilleran foreland basin, USA. *Geol. Soc. Am. Bull.*, **123**, 1822–1835.

Barwis, J.H. (1977) Sedimentology of some South Carolina tidal-creek point bars and a comparison with their fluvial counterparts. In: *Fluvial Sedimentology: Bedforms and Bars* (Ed. A.D. Miall). *Mem. Can. Soc. Petrol. Geol.*, **5**, 129–160.

Bloom, A.L. (1964) Peat accumulation and compaction in a Connecticut coastal marsh. *J. Sed. Res.*, **34**, 599–603.

Blum, M., Martin, J., Milliken, K. and Garvin, M. (2013) Paleovalley systems: insights from Quaternary analogs and experiments. *Earth-Sci. Rev.*, **116**, 128–169.

Blum, M.D. and **Törnqvist, T.E.** (2000) Fluvial responses to climate and sea-level change: a review and look forward. *Sedimentology*, **47**, 2–48.

Bohacs, K. and **Suter, J.** (1997) Sequence stratigraphic distribution of coaly rocks: fundamental controls and paralic examples. *AAPG Bull.*, **81**, 1612–1639.

Brekke, H. and **Couch, A.** (2011) Use of image logs in differentiating point bar and tidal bar deposits in the Leismer area: implications for SAGD reservoir definition in the Athabasca oilsands. In: *Extended Abstract, CSPG CSEG CWLS Convention.*

Bridge, J.S. and **Tye, R.S.** (2000) Interpreting the dimensions of ancient fluvial channel bars, channels, and channel belts from wireline-logs and cores. *AAPG Bull.*, **84**, 1205–1228.

Bridge, J.S., **Jalfin, G.A.** and **Georgieff, S.M.** (2000) Geometry, lithofacies, and spatial distribution of Cretaceous fluvial sandstone bodies, San Jorge Basin, Argentina: outcrop analog for the hydrocarbon-bearing Chubut Group. *J. Sed. Res.*, **70**, 341–359.

Bromley, R. (1996) *Trace Fossils: Biology, Taphonomy and Applications.* Chapman Hall. London, 361 pp.

Bridge, J.S. (1993) Description and interpretation of fluvial deposits: a critical perspective. *Sedimentology*, **40**, 801–810.

Bridge, J.S. and **Leeder, M.R.** (1979) A simulation model of alluvial stratigraphy. *Sedimentology*, **26**, 617–644.

Bridge, J. and **Mackey, S.** (1993) A theoretical study of fluvial sandstone body dimensions. In: *The geological modelling of hydrocarbon reservoirs and outcrop analogues* (Ed. I.D Bryant). *Int. Assoc. Sedimentol.*, Blackwell, 213–236.

Bridges, P.H. and **Leeder, M.R.** (1976) Sedimentary model for intertidal mudflat channels, with examples from the Solway Firth, Scotland. *Sedimentology*, **23**, 533–552.

Bristow, C. and **Best, J.** (1993) Braided rivers: perspectives and problems. *Geol. Soc. London Spec. Publ.*, **75**, 1–11.

Burns, C.E., **Mountney, N.P., Hodgson, D.M.** and **Colombera, L.** (2017) Anatomy and dimensions of fluvial crevasse-splay deposits: Examples from the Cretaceous Castlegate Sandstone and Neslen Formation, Utah, U.S.A. *Sed. Geol.*, **351**, 21–35.

Burton, D., **Flaig, P.P.** and **Prather, T.J.** (2016) Regional Controls On Depositional Trends In Tidally Modified Deltas: Insights From Sequence Stratigraphic Correlation and Mapping of the Loyd And Sego Sandstones, Uinta and Piceance Basins of Utah and Colorado, USA. *J. Sed. Res.*, **86**, 763–785.

Cant, D.J. and **Walker, R.G.** (1978) Fluvial processes and facies sequences in the sandy braided South Saskatchewan River, Canada. *Sedimentology*, **25**, 625–648.

Casshyap, S. and **Kumar, A.** (1987) Fluvial architecture of the Upper Permian Raniganj coal measure in the Damodar basin, Eastern India. *Sed. Geol.*, **51**, 181–213.

Cecil, C.B., **Dulong, F.T.** and **Cobb, J.C.** (1993) Allogenic and autogenic controls on sedimentation in the Central Sumatra Basin as an analogue for Pennsylvanian coal-bearing strata in the Appalachian Basin. *Geol. Soc. Am. Spec. Pap.*, **286**, 3–22.

Chakraborty, P. P., **Saha, S.** and **Das, K.** (2017) Record of continental to marine transition from the Mesoproterozoic Ampani basin, Central India: An exercise of process-based sedimentology in a structurally deformed basin. *J. Asian Earth Sci.*, **143**, 122–140.

Chatanantavet, P., **Lamb, M.P.** and **Nittrouer, J.A.** (2012) Backwater controls of avulsion location on deltas. *Geophys. Res. Lett.*, **39**, L01402.

Choi, K.S. (2011a) External controls on the architecture of inclined heterolithic stratification (IHS) of macrotidal Sukmo Channel: Wave versus rainfall. *Marine Geology*, **285**, 17–28.

Choi, K.S. (2011b) Tidal rhythmites in a mixed-energy, macrotidal estuarine channel, Gomso Bay, west coast of Korea. *Mar. Geol.*, **280**, 105–115.

Choi, K.S., **Dalrymple, R.W., Chun, S.S.** and **Kim, S.P.** (2004) Sedimentology of Modern, Inclined Heterolithic Stratification (IHS) in the Macrotidal Han River Delta, Korea. *J. Sed. Res.*, **74**, 677–689.

Cloyd, K.C., **Demicco, R.V.** and **Spencer, R.J.** (1990) Tidal channel, levee and crevasse-splay deposits from a Cambrian tidal channel system: a new mechanism to produce shallowing-upward sequences. *J. Sed. Res.*, **60**, 78–83.

Cole, R. (2008) Characterization of fluvial sand bodies in the Neslen and lower Farrer formations (Upper Cretaceous), Lower Sego Canyon, Utah. In: *Hydrocarbons Systems and Production in the Uinta Basin, Utah* (Eds M.W. Longman and C.D. Morgan). *RMAG-UGA Publication*, **37**, 81–100.

Cole, R. and **Cumella, S.** (2003) Stratigraphic architecture and reservoir characteristics of the Mesaverde Group, southern Piceance Basin, Colorado. In: *Piceance Basin 2003 guidebook* (Eds K.M. Petersen, T.M. Olsen and D.S. Anderson). Rocky Mountain Association of Geologists, 385–442.

Collinson, J.D., **Mountney, N.** and **Thompson, D.** (2006) *Sedimentary Structures.* Terra Publishing, Hertfordshire.

Colombera, L., **Felletti, F., Mountney, N.P.** and **McCaffrey, W.D.** (2012a) A database approach for constraining stochastic simulations of the sedimentary heterogeneity of fluvial reservoirs. *AAPG Bull.*, **96**, 2143–2166.

Colombera, L., **Mountney, N.P.** and **McCaffrey, W.D.** (2013) A quantitative approach to fluvial facies models: methods and example results. *Sedimentology*, **60**, 1526–1558.

Colombera, L., **Mountney, N.P.** and **McCaffrey, W.D.** (2015) A meta-study of relationships between fluvial channel-body stacking pattern and aggradation rate: Implications for sequence stratigraphy. *Geology*, **43**, 283–286.

Colombera, L., **Mountney, N.P.** and **McCaffrey, W.D.** (2012b) A relational database for the digitization of fluvial architecture concepts and example applications. *Petrol. Geosci.*, **18**, 129–140.

Colombera, L., **Mountney, N.P., Russell, C.E., Shiers, M.N.** and **McCaffrey, W.D.** (2017) Geometry and compartmentalization of fluvial meander-belt reservoirs at the bar-form scale: Quantitative insight from outcrop, modern and subsurface analogues. *Mar. Petrol. Geol.*, **82**, 35–55.

Colombera, L., **Shiers, M.** and **Mountney, N.** (2016) Assessment of backwater controls on the architecture of distributary channel fills in a tide-influenced coastal-plain succession: Campanian Neslen Formation, USA. *J. Sed. Res.*, **86**, 476–497.

Corbeanu, R.M., Wizevich, M.C., Bhattacharya, J.P., Zeng, X. and **McMechan, G.A.** (2004) Three-dimensional architecture of ancient lower delta-plain point bars using ground-penetrating radar, Cretaceous Ferron Sandstone, Utah. Regional to Wellbore Analog for Fluvial–Deltaic Reservoir Modeling, The Ferron of Utah. *AAPG Stud. Geol.,* **50**, 427–449.

Courel, L. (1987) Stages in the compaction of peat; examples from the Stephanian and Permian of the Massif Central, France. *J. Geol. Soc.,* **144**, 489–493.

Currie, B.S. (1997) Sequence stratigraphy of nonmarine Jurassic–Cretaceous rocks, central Cordilleran foreland-basin system. *Geol. Soc. Am. Bull.,* **109**, 1206–1222.

Dalrymple, R.W. and **Choi, K.** (2007) Morphologic and facies trends through the fluvial-marine transition in tide-dominated depositional systems: A schematic framework for environmental and sequence-stratigraphic interpretation. *Earth-Sci. Rev.,* **81**, 135–174.

Demowbray, T. (1983) The genesis of lateral accretion deposits in recent intertidal mudflat channels, Solway Firth, Scotland. *Sedimentology,* **30**, 425–435.

Famubode, O. and **Bhattacharya, J.** (2016) Hierarchical organization of strata within a high-frequency Milankovitch-scale fluvial sequence, cretaceous Ferron Notom delta, south central Utah, USA. In International Conference and Exhibition, Barcelona, Spain, 3-6 April 2016. *Society of Exploration Geophysicists and American Association of Petroleum Geologists.*

Fenies, H. and **Faugères, J.-C.** (1998) Facies and geometry of tidal channel-fill deposits (Arcachon Lagoon, SW France). *Mar. Geol.,* **150**, 131–148.

Fernandes, A.M., Törnqvist, T.E., Straub, K.M. and **Mohrig, D.** (2016) Connecting the backwater hydraulics of coastal rivers to fluvio-deltaic sedimentology and stratigraphy. *Geology,* **44**, 979–982.

Fielding, C.R. (2006) Upper flow regime sheets, lenses and scour fills: Extending the range of architectural elements for fluvial sediment bodies. *Sed. Geol.,* **190**, 227–240.

Fielding, C.R., Allen, J.P., Alexander, J. and **Gibling, M.R.** (2009) Facies model for fluvial systems in the seasonal tropics and subtropics. *Geology,* **37**, 623–626.

Fillmore, R. (2010) Geological Evolution of the Colorado Plateau of Eastern Utah and Western Colorado, Including the San Juan River, Natural Bridges, Canyonlands, Arches, and the Book Cliffs. University of Utah Press.

Fisher, D.J., Erdmann, C.E. and **Reeside, J.B.J.** (1960) Cretaceous and Tertiary formations of the Book Cliffs, Carbon, Emery and Grand Counties, Utah and Garfield and Mesa Counties, Colorado. *Geol. Surv. Prof. Pap.,* **332**, 1–79.

Flint, J.J. (1974) Stream gradient as a function of order, magnitude and discharge. *Water Resour. Res.,* **10**, 969–973.

Foix, N., Paredes, J.M. and **Giacosa, R.E.** (2013) Fluvial architecture variations linked to changes in accommodation space: Río Chico Formation (late Paleocene), Golfo San Jorge basin, Argentina. *Sed. Geol.,* **294**, 342–355.

Foreman, B. Z., Roberts, E. M., Tapanila, L., Ratigan, D. and **Sullivan, P.** (2015) Stable isotopic insights into paleoclimatic conditions and alluvial depositional processes in the Kaiparowits Formation (Campanian, south-central Utah, USA). *Cretaceous Res.,* **56**, 180–192.

Fricke, H.C., Foreman, B.Z. and **Sewall, J.O.** (2010) Integrated climate model-oxygen isotope evidence for a North American monsoon during the Late Cretaceous. *Earth Planet. Sci. Lett.,* **289**, 11–21.

Fustic, M., Hubbard, S.M., Spencer, R., Smith, D.G., Leckie, D.A., Bennett, B. and **Larter, S.** (2012) Recognition of down-valley translation in tidally influenced meandering fluvial deposits, Athabasca Oil Sands (Cretaceous), Alberta, Canada. *Mar. Petrol. Geol.,* **29**, 219–232.

Galloway, W.E. and **Hobday, D.K.** (1996) Fluvial Systems. In: *Terrigenous Clastic Depositional Systems*, pp. 60–90. Springer.

Gates, T.A. and **Scheetz, R.** (2015) A new saurolophine hadrosaurid (Dinosauria: Ornithopoda) from the Campanian of Utah, North America. *J. Systematic Palaeontology,* **13**, 711–725.

Geehan, G. and **Underwood, J.** (2009) The use of length distributions in geological modelling. In: *The Geological Modelling of Hydrocarbon Reservoirs and Outcrop Analogues* (Eds S.S. Flint and I.D. Bryant). *Int. Assoc. Sedimentol.,* Blackwell Publishing Ltd., Oxford, UK. doi: 10.1002/9781444303957.ch13.

Ghazi, S. and **Mountney, N.P.** (2009) Facies and architectural element analysis of a meandering fluvial succession: The Permian Warchha Sandstone, Salt Range, Pakistan. *Sed. Geol.,* **221**, 99–126.

Ghazi, S. and **Mountney, N.P.** (2011) Petrography and provenance of the Early Permian Fluvial Warchha Sandstone, Salt Range, Pakistan. *Sed. Geol.,* **233**, 88–110.

Gibling, M.R. (2006) Width and thickness of fluvial channel bodies and valley fills in the geological record: a literature compilation and classification. *J. Sed. Res.,* **76**, 731–770.

Gingras, M.K., MacEachern, J.A. and **Dashtgard, S.E.** (2012) The potential of trace fossils as tidal indicators in bays and estuaries. *Sed. Geol.,* **279**, 97–106.

Gouw, M. J. and **Autin, W. J.** (2008) Alluvial architecture of the Holocene Lower Mississippi Valley (USA) and a comparison with the Rhine–Meuse delta (The Netherlands). *Sed. Geol.,* **204**, 106–121.

Gualtieri, J.L. (1991) Map and cross sections of coal zones in the Upper Cretaceous Neslen Formation, North-Central part of the Westwater 30' * 60' quadrangle, Grand and Uintah counties, Utah, 133. *U.S. Geol. Surv.*

Gugliotta M.; Kurcinka C.E., Dalrymple R.W., Flint S.S. and **Hodgson D.M.** (2016) Decoupling seasonal fluctuations in fluvial discharge from the tidal signature in ancient deltaic deposits: An example from the Neuquén Basin, Argentina, *J. Geol. Soc.,* **173**, 94–107.

Gulliford A.R., Flint S.S. and **Hodgson D.M.** (2014) Testing applicability of models of distributive fluvial systems or trunk rivers in ephemeral systems: Reconstructing 3-D fluvial architecture in the Beaufort Group, South Africa, *J. Sed. Res.,* **84**, 1147–1169.

Halfar, J., Riegel, W. and **Walther, H.** (1998) Facies architecture and sedimentology of a meandering fluvial system: a Palaeogene example from the Weisselster Basin, Germany. *Sedimentology,* **45**, 1–17.

Hampson, G.J., **Royhan Gani, M.**, **Sahoo, H.**, **Rittersbacher, A.**, **Irfan, N.**, **Ranson, A.**, **Jewell, T.O.**, **Gani, N.D.**, **Howell, J.A.** and **Buckley, S.J.** (2012) Controls on large-scale patterns of fluvial sandbody distribution in alluvial to coastal plain strata: Upper Cretaceous Blackhawk Formation, Wasatch Plateau, Central Utah, USA. *Sedimentology*, **59**, 2226–2258.

Harms, J., **Southard, J.B.**, **Spearing, D.R.** and **Walker, R.G.** (1975) Stratification produced by migrating bed forms. In: *Depositional Environments as Interpreted from Primary Sedimentary Structures and Stratification Sequences* (Eds J. Harms, J. Southard, D. Spearing and R. Walker), **2**, pp. 45–61. SEPM.

Heller, P.L. and **Paola, C.** (1996) Downstream changes in alluvial architecture: an exploration of controls on channel-stacking patterns. *J. Sed. Res.*, **66**, 297–306.

Hettinger, R.D. and **Kirshbaum, M.A.** (2003) Stratigraphy of the Upper Cretaceous Mancos Shale (Upper Part) and Mesaverde Group in the Southern Part of the Uinta and Piceance Basins, Utah and Colorado. In: *Petroleum Systems and Geologic Assessment of Oil and Gas in the Uinta-Piceance Province, Utah and Colorado* (Ed. U.U.-P.A. Team), *U.S Geol. Surv.*, DDS-69-B, pp. 1–16.

Holbrook, J.M. (1996) Complex fluvial response to low gradients at maximum regression: A genetic link between smooth sequence-boundary morphology and architecture of overlying sheet sandstone. *J. Sed. Res.*, **66**, 713–722.

Hovikoski, J., **RÄSÄNen, M.**, **Gingras, M.**, **Ranzi, A.** and **Melo, J.** (2008) Tidal and seasonal controls in the formation of Late Miocene inclined heterolithic stratification deposits, western Amazonian foreland basin. *Sedimentology*, **55**, 499–530.

Howard, J.D. and **Frey, R.W.** (1984) Characteristic trace fossils in nearshore to offshore sequences, Upper Cretaceous of east-central Utah. *Can. J. Earth Sci.*, **21**, 200–219.

Huerta, P., **Armenteros, I.** and **Silva, P.G.** (2011) Large-scale architecture in non-marine basins: the response to the interplay between accommodation space and sediment supply. *Sedimentology*, **58**, 1716–1736.

Ielpi, A. and **Ghinassi, M.** (2014) Planform architecture, stratigraphic signature and morphodynamics of an exhumed Jurassic meander plain (Scalby Formation, Yorkshire, UK). *Sedimentology*, **61**, 1923–1960.

Ielpi, A., **Gibling, M.R.**, **Bashforth, A.R.**, **Lally, C.**, **Rygel, M.C.** and **Al-Silwadi, S.** (2014) Role of vegetation in shaping Early Pennsylvanian braided rivers: architecture of the Boss Point Formation, Atlantic Canada. *Sedimentology*, **61**, 1659–1700.

Jablonski, B.V. and **Dalrymple, R.W.** (2016) Recognition of strong seasonality and climatic cyclicity in an ancient, fluvially dominated, tidally influenced point bar: Middle McMurray Formation, Lower Steepbank River, north-eastern Alberta, Canada. *Sedimentology*, **63**, 552–585.

Jackson II, R.G. (1976) Depositional model of point bars in the lower Wabash River. *J. Sed. Res.*, **46**, 579–594.

Jackson II, R.G. (1978) Preliminary evaluation of lithofacies models for meandering alluvial streams. In: *Fluvial Sedimentology: Fluvial Facies Models* (Ed. A.D. Miall). *Mem. Can. Soc. Petrol. Geol.*, **5**, 543–576.

Johnson, S.M. and **Dashtgard, S.E.** (2014) Inclined heterolithic stratification in a mixed tidal-fluvial channel: Differentiating tidal versus fluvial controls on sedimentation. *Sed. Geol.*, **301**, 41–53.

Jones, B.G. and **Rust, B.R.** (1983) Massive sandstone facies in the Hawkesbury Sandstone, a Triassic fluvial deposit near Sydney, Australia. *J. Sed. Res.*, **53**, 1249–1259.

Jordan, T.E. (1981) Thrust loads and foreland basin evolution, Cretaceous, western United States. *AAPG Bull.*, **65**, 2506–2520.

Kauffman, E.G. (1977) Geological and biological overview: Western Interior Cretaceous basin. *The Mountain Geologist*, **14**, 75–99.

Karaman, O. (2012) Shoreline Architecture and Sequence Stratigraphy of Campanian Iles Clastic Wedge, Piceance Basin, CO: Influence of Laramide Movements in Western Interior Seaway. Masters Thesis, University of Texas at Austin, Texas, 137 pp.

Keighley, D., **Flint, S.**, **Howell, J.** and **Moscariello, A.** (2003) Sequence stratigraphy in lacustrine basins: a model for part of the Green River Formation (Eocene), southwest Uinta Basin, Utah, USA. *J. Sed. Res.*, **73**, 987–1006.

Kirschbaum, M.A. and **Hettinger, R.D.** (2004) Facies Analysis and Sequence Stratigraphic Framework of Upper Campanian Strata (Neslen and Mount Garfield Formations, Bluecastle Tongue of the Castlegate Sandstone and Mancos Shale), Eastern Book Cliffs, Colorado and Utah. *U.S. Geol. Surv. Digital Data Series.*

Kvale, E.P. (2011) Tidal Constituents of Modern and Ancient Tidal Rhythmites. In: *Principles of Tidal Sedimentology* (Eds A.R. Davis Jr and R.W. Dalrymple), Springer, Netherlands, 1–17.

Kvale, E.P. and **Archer, A.W.** (1990) Tidal deposits associated with low-sulfur coals, Brazil Fm.(Lower Pennsylvanian), Indiana. *J. Sed. Petrol.*, **60**, 563–574.

Labrecque, P.A., **Hubbard, S.M.**, **Jensen, J.L.** and **Nielsen, H.** (2011) Sedimentology and stratigraphic architecture of a point bar deposit, Lower Cretaceous McMurray Formation, Alberta, Canada. *Bull. Can. Soc. Petrol. Geol.*, **59**, 147–171.

Lamb, M.P., **Nittrouer, J.A.**, **Mohrig, D.** and **Shaw, J.** (2012) Backwater and river plume controls on scour upstream of river mouths: Implications for fluvio-deltaic morphodynamics. *J. Geophys. Res.; Earth Surface*, **117**, F01002.

Lawton, T. (1986) Fluvial systems of the Upper Cretaceous Mesaverde Group and Paleocene North Horn Formation, Central Utah: A record of transition from thin-skinned to thick-skinned deformation in the foreland region: Part III. Middle Rocky Mountains. In: *Paleotectonics and Sedimentation in the Rocky Mountain Region, United States* (Ed. J.A. Peterson). *AAPG Mem.*, **41**, 423–442.

Leeder, M. (1977) A quantitative stratigraphic model for alluvium, with special reference to channel deposit density and interconnectedness. In: *Fluvial Sedimentology* (Ed. A.D. Miall), *Can. Soc. Petrol. Geol.*, **5**, pp. 587–596.

Leeder, M. (1999) Sedimentology and Sedimentary Basins. From Turbulence to Tectonics. Blackwell Science Ltd., 594 pp.

Legarreta, L. and **Uliana, M.A.** (1998) Anatomy of hinterlanddepositional sequences: Upper Cretaceous fluvial strata, Neuquen basin, west-central Argentina. In: *Relative Role of Eustasy, Climate, Tectonism in*

Continental Rocks (Eds K.W.Shanley and P.J. McCabe), *SEPM Spec. Publ.*, **59**, 83–92. Society of Economic Paleontologists and Mineralogists,Tulsa, OK.

Legler, B., Hampson, G.J., Jackson, C.A., Johnson, H.D., Massart, B.Y., Sarginson, M. and **Ravnås, R.** (2014) Facies Relationships and Stratigraphic Architecture of Distal, Mixed Tide-and Wave-Influenced Deltaic Deposits: Lower Sego Sandstone, Western Colorado, USA. *J. Sed. Res.*, **84**, 605–625.

Leopold, L.B. and **Wolman, M.G.** (1960) River meanders. *Geol. Soc. Am. Bull.*, **71**, 769–793.

Lynds, R. and **Hajek, E.** (2006) Conceptual model for predicting mudstone dimensions in sandy braided-river reservoirs. *AAPG Bull.*, **90**, 1273–1288.

Mackey, S.D. and **Bridge, J.S.** (1995) Three-dimensional model of alluvial stratigraphy: theory and application. *J. Sed. Res.*, **65**, 7–31.

McGowen, J. and **Garner, L.** (1970) Physiographic features and stratification types of coarse-grained pointbars: Modern and ancient examples. *Sedimentology*, **14**, 77–111.

McLaurin, B.T. and **Steel, R.J.** (2000) Fourth-order nonmarine to marine sequences, middle Castlegate Formation, Book Cliffs, Utah. *Geology*, **28**, 359–362.

Miall, A. (2014) *Fluvial Depositional Systems*. Springer, 316 pp.

Miall, A.D. (1985) Architectural-element analysis: A new method of facies analysis applied to fluvial deposits. *Earth-Sci. Rev.*, **22**, 261–308.

Miall, A.D. (1988) Facies architecture in clastic sedimentary basins. In: *New perspectives in basin analysis* (Eds K.L. Kleinspehn and C. Paola). Springer, 67–81.

Miall, A.D. (1977) Lithofacies types and vertical profile models in braided river deposits: a summary. In: *Fluvial Sedimentology* (Ed. A.D. Miall), **5**, 597–604.

Miranda, M.C., De Fátima Rossetti, D. and **Carlos Ruiz Pessenda, L.** (2009) Quaternary paleoenvironments and relative sea-level changes in Marajó Island (Northern Brazil): Facies, δ13C, δ15N and C/N. *Palaeogeogr. Palaeoclimatol. Palaeoecol.*, **282**, 19–31.

Musial, G., Reynaud, J.-Y., Gingras, M.K., Féniès, H., Labourdette, R. and **Parize, O.** (2012) Subsurface and outcrop characterization of large tidally influenced point bars of the Cretaceous McMurray Formation (Alberta, Canada). *Sed. Geol.*, **279**, 156–172.

Nami, M. and **Leeder, M.** (1977) Changing channel morphology and magnitude in the Scalby Formation (M. Jurassic) of Yorkshire, England. *Mem. Can. J. Petrol. Geol.*, **5**, 431–440.

Nanson, G.C. (1980) Point bar and floodplain formation of the meandering Beatton River, northeastern British Columbia, Canada. *Sedimentology*, **27**, 3–29.

Nanson, G.C. and **Page, K.** (2009) Lateral accretion of fine-grained concave benches on meandering rivers. Modern and Ancient Fluvial Systems. *Int. Assoc. Sedimentol. Spec. Publ.*, **6**, 133–143.

Nio, S.D. and **Yang, C.S.** (1991) Sea-level fluctuations and the geometric variability of tide-dominated sandbodies. *Sed. Geol.*, **70**, 161–193.

Nittrouer, J.A., Shaw, J., Lamb, M.P. and **Mohrig, D.** (2012) Spatial and temporal trends for water-flow velocity and bed-material sediment transport in the lower Mississippi River. *Geol. Soc. Am. Bull.*, **124**, 400–414.

O'Brien, K.C. (2015) *Stratigraphic architecture of a shallow-water delta deposited in a coastal-plain setting: Neslen Formation, Floy Canyon, Utah, Colorado School of Mines*. Arthur Lakes Library, 77 pp.

Olariu, C., Steel, R.J., Olariu, M.I. and **Choi, K.S.** (2015) Facies and architecture of unusual fluvial-tidal channels with inclined heterolothic strata: Campanian Neslen Formation, Utah, USA. In: *Fluvial-Tidal Sedimentology* (Eds P.J. Ashworth, J.L. Best and D.R. Parsons). Elsevier, 1 edn, 68, pp. 634.

Pemberton, S.G., Flach, P.D. and **Mossop, G.D.** (1982) Trace fossils from the Athabasca oil sands, Alberta, Canada. *Science*, **217**, 825–827.

Pitman, J.K., Franczyk, K.J. and **Anders, D.E.** (1986) Marine and Nonmarine gas-bearing rocks in Upper Cretaceous Neslen and Blackhawk foramtions, Eastern Uinta Basin, Utah-Sedimentology, Diagenesis and Source rock potential. *AAPG Bull.*, **70**, 1052–1052.

Plink-Björklund, P. (2015) Morphodynamics of rivers strongly affected by monsoon precipitation: Review of depositional style and forcing factors. *Sed. Geol.*, **323**, 110–147.

Posamentier, H. and **Vail, P.** (1988) Eustatic controls on clastic deposition II—sequence and systems tract models. In: Sea-Level Changes – An Integrated Approach (Eds C. Wilgus, B. Hastings, H. Posamentier, J. Van Wagoner, C. Ross and C. Kendall). *SEPM Spec. Publ.*, **42**, 47–70.

Pranter, M.J., Ellison, A.I., Cole, R.D. and **Patterson, P.E.** (2007) Analysis and modeling of intermediate-scale reservoir heterogeneity based on a fluvial point-bar outcrop analog, Williams Fork Formation, Piceance Basin, Colorado. *AAPG Bull.*, **91**, 1025–1051.

Rampino, M.R. and **Sanders, J.E.** (1981) Episodic growth of Holocene tidal marshes in the northeastern United States: A possible indicator of eustatic sea-level fluctuations. *Geology*, **9**, 63–67.

Rasanen, M.E., Linna, A.M., Santos, J.C.R. and **Negri, F.R.** (1995) Late Miocene Tidal Deposits in the Amazonian Foreland Basin. *Science*, **269**, 386–390.

Reineck, H.E. and **Wunderlich, F.** (1968) Classification and origin of flaser and lenticular bedding. *Sedimentology*, **11**, 99–104.

Russell, C.E., Mountney, N.P., Hodgson, D.M. and **Colombera, L.** [this volume] Improving prediction of lithologic heterogeneity in fluvial point-bar deposits from analysis of meander morphology and scroll-bar pattern. *Int. Assoc. Sedimentol. Spec. Publ.*, **48**, 385–417.

Ryer, T.A. (1981) Deltaic coals of Ferron Sandstone Member of Mancos Shale: predictive model for Cretaceous coal-bearing strata of Western Interior. *AAPG Bull.*, **65**, 2323–2340.

Savrda, C.E. (1991) Teredolites, wood substrates and sea-level dynamics. *Geology*, **19**, 905–908.

Schumm, S.A. (1963) Sinuosity of alluvial rivers on the Great Plains. *Geol. Soc. Am. Bull.*, **74**, 1089–1100.

Schumm, S. and **Khan, H.** (1972) Experimental study of channel patterns. *Geol. Soc. Am. Bull.*, **83**, 1755–1770.

Shanley, K.W., McCabe, P.J. and **Hettinger, R.D.** (1992) Tidal influence in Cretaceous fluvial strata from Utah, USA: a key to sequence stratigraphic interpretation. *Sedimentology*, **39**, 905–930.

Shiers, M., **Mountney, N.**, **Hodgson, D.** and **Cobain, S.** (2014) Depositional Controls on Tidally Influenced Fluvial Successions, Neslen Formation, Utah, USA. *Sed. Geol.*, **311**, 1–16.

Shiers, M.N., **Hodgson, D.M.** and **Mountney, N.P.** (2017) Response of A Coal-Bearing Coastal-Plain Succession To Marine Transgression: Campanian Neslen Formation, Utah, USA. *J. Sed. Res.*, **87**, 168–187.

Simon, S.S. and **Gibling, M.R.** (2017b) Finegrained meandering systems of the Lower Permian Clear Fork Formation of north-central Texas, USA: Lateral and oblique accretion on an arid plain. *Sedimentology*, **64**, 714–746.

Simon, S.S. and **Gibling, M.R.** (2017b) Pedogenic Mud Aggregates Preserved In A Fine-Grained Meandering Channel In the Lower Permian Clear Fork Formation, North-Central Texas, USA. *J. Sed. Res.*, **87**, 230–252.

Simons, D.B. (1960) Sedimentary structures generated by flow in alluvial channels. Primary Sedimentary Structures, *SEPM Spec. Publ.*, **12**, 34–52.

Sisulak, C.F. and **Dashtgard, S.F.** (2012) Seasonal controls on the development and character of inclinced heterolithic stratification in a tide-influenced, fluvially dominated channel: Fraser River, Canada. *J. Sed. Res.*, **82**, 244–257.

Smith, D.G. (1987) Meandering river point bar lithofacies models: modern and ancient examples compared. In: *Recent developments in Fluvial Sedimentology* (Eds F.G. Ethridge, R.M. Flores and M.D. Harvey) *SEPM*, **39**, 83–91.

Smith, D.G., **Hubbard, S.M.**, **Leckie, D.A.** and **Fustic, M.** (2009) Counter point bar deposits: lithofacies and reservoir significance in the meandering modern Peace River and ancient McMurray Formation, Alberta, Canada. *Sedimentology*, **56**, 1655–1669.

Sønderholm, M. and **Tirsgaard, H.** (1998) Proterozoic fluvial styles: response to changes in accommodation space (Rivieradal sandstones, eastern North Greenland). *Sed. Geol.*, **120**, 257–274.

Speiker, E.M. (1949) The transition between the Colorado Plateaus and the Great Basin in central Utah. *Utah Geol. Soc. Guidebook*, **106**.

Stewart, D.J. (2009) Possible Suspended-Load Channel Deposits from the Wealden Group (Lower Cretaceous) of Southern England. In: *Modern and ancient fluvial systems* (Eds J.D. Collinson and J. Lewin). Blackwell Publishing, 369–384.

Tewari, R.C., **Hota, R.N.**, and **Maejima, W.** (2012) Fluvial architecture of Early Permian Barakar rocks of Korba Gondwana basin, eastern-central India. *J. Asian Earth Sci.*, **52**, 43–52.

Thomas, R.G., **Smith, D.G.**, **Wood, J.M.**, **Visser, J.**, **Calverleyrange, E.A.** and **Koster, E.H.** (1987) Inclined heterolithic stratification-terminology, description, interpretation and significance. *Sed. Geol.*, **53**, 123–179.

Tissot, B.P. and **Welte, D.H.** (1984) From kerogen to petroleum. In: *Petroleum Formation and Occurrence.* Springer, 160–198.

Turner, B. and **Eriksson, K.** (1999) Meander bend reconstruction from an Upper Mississippian muddy point bar at Possum Hollow, West Virginia, USA. *Fluvial Sedimentology*, **VI**, 363–379.

Tyler, N. and **Finley, R.J.** (1991) Architectural controls on the recovery of hydrocarbons from sandstone reservoirs. In: *The three-dimensional facies architecture of terrigenous clastic sediments and its implications for hydrocarbon discovery and recovery* (Eds N. Tyler and A. Miall). SEPM, *Concepts in Sedimentology and Paleontology*, **3**, 1–5.

van den Berg, J.H., **Boersma, J.R.** and **van Gelder, A.** (2007) Diagnostic sedimentary structures of the fluvial-tidal transition zone - Evidence from deposits of the Rhine and Meuse. *Geol. Mijnbouw*, **86**, 287–306.

Visher, G.S. (1965) Use of vertical profile in environmental reconstruction. *AAPG Bull.*, **49**, 41–61.

Walker, R.G. and **Cant, D.J.** (1984) Sandy fluvial systems. In: *Facies models.* Geoscience Canada Reprint Series, **1**, 71–89.

Wang, H., **Shao, L.**, **Hao, L.**, **Zhang, P.**, **Glasspool, I.J.**, **Wheeley, J.R.**, **Wignall, P.B.**, **Yi, T.**, **Zhang, M.** and **Hilton, J.** (2011) Sedimentology and sequence stratigraphy of the Lopingian (Late Permian) coal measures in southwestern China. *Int. J. Coal Geol.*, **85**, 168–183.

Weimer, R.J., **Howard, J.D.** and **Lindsay, D.R.** (1982) Tidal flats and associated tidal channels. Sandstone Depositional Environments: *AAPG Mem.*, **31**, 191–245.

Williams, G.P. (1986) River meanders and channel size. *J. Hydrol.*, **88**, 147–164.

Willis, A. (2000) Tectonic control of nested sequence architecture in the Sego Sandstone, Neslen Formation and upper Castlegate Sandstone (Upper Cretaceous), Sevier foreland basin, Utah, USA. *Sed. Geol.*, **136**, 277–317.

Willis, B.J. and **Tang, H.** (2010) Three-dimensional connectivity of point-bar deposits. *J. Sed. Res.*, **80**, 440–454.

Wright, P.V. and **Marriott, S.B.** (1993) The sequence stratigraphy of fluvial depositional systems: the role of floodplain sediment storage. *Sed. Geol.*, **86**, 203–210.

Yan, N., **Mountney, N.P.**, **Colombera, L.** and **Dorrell, R.M.** (2017) A 3D forward stratigraphic model of fluvial meander-bend evolution for prediction of point-bar lithofacies architecture. *Comput. Geosci.*, **105**, 65–80.

Young, R.G. (1957) Late Cretaceous cyclic deposits, Book Cliffs, Eastern Utah. *AAPG Bull.*, **41**, 1790–1774.

Young, R.G. (1955) Sedimentary facies and intertonguing in the Upper Cretaceous of the Book Cliffs, Utah, Colorado. *Geol. Soc. Am. Bull.*, **66**, 177–202.

Yoshida, S., **Willis, A.** and **Miall, A.D.** (1996) Tectonic control of nested sequence architecture in the Castlegate Sandstone (Upper Cretaceous), Book Cliffs, Utah. *J. Sed. Res.*, **66**, 737–748.

Int. Assoc. Sedimentol. Spec. Publ (2018) **48**, 47–80.

Toggling between expansion and translation: The generation of a muddy-normal point bar with an earthquake imprint

SHELBY JOHNSTON and JOHN HOLBROOK

Texas Christian University, Fort Worth, Texas, USA

ABSTRACT

Mud-dominated point bars are widely accepted as recording deposition by tidal influence, bar tails or counter point bars. Less understood are muddy point bars that lack these depositional origins and otherwise have the geometry of more traditional sandy point bars. This study seeks the cause of these 'muddy-normal' point bars by field examination of a point bar with features and architecture consistent with sandy deposits but containing mud-dominated internal lithofacies. This study examines a point bar in Late Cretaceous fluvial strata of the Dinosaur Park Formation in the Steveville badlands of Dinosaur Provincial Park, Alberta. Strikes and dips, palaeocurrents, photo panoramas and stratigraphic columns are used to determine accretion trajectories and lithologic trends throughout the bar with a focus on sand *vs.* mud accretion and its relation to accretion trajectory. The point bar alternates between sand-dominated and mud-dominated accretion sets, with mud comprising over 50% of the point bar by volume. Accretion sets are defined by consistency in accretion dips. Both mud and sand beds within accretion sets have current ripples and cross sets indicative of deposition by transport. This suggests that the mud layers were deposited by active accretion events and are not simple drapes. Muddy accretion sets consistently have orientations reflecting bar translation (parallel with palaeodip) and sandy accretion sets consistently have orientations consistent with bar expansion (normal to palaeodip). These data suggest that the muddy *vs.* sandy accretion sets record toggling between sand-favouring bar expansion and mud-favouring translational growth vectors. This bar toggle explains how to generate a mud-dominate point bar with general lobate geometry without imposing tidal drivers, bar abandonment, or asserting a fully counter-point-bar interpretation. This toggling between expansion and translation records periodic upstream to downstream shift in the point of attachment for flow momentum to the cutbank that is probably caused by unstable flood discharge trends over the life of the bar. Bar toggle should be considered a growth vector option for point bars along with more established translation and expansion. The point bar also has a strong palaeoseismic imprint in the form of liquefaction features and a large lateral spread near the tip that is contemporary with late bar growth. The source of this earthquake is uncertain but could record a currently unrecognised intraplate earthquake source or, intriguingly, could be associated with the nearby Bow City impact structure which occurred in the same time frame.

Keywords: Point Bar, Fluvial, Muddy, IHS, mud-dominated point bar, Expansion, Translation, Reservoir

INTRODUCTION

Point bars are typically considered sand-dominated deposits (Dixon, 1921; Allen, 1963; Allen, 1970; McGowen & Garner, 1970; Brice, 1974; Miall, 1978; Walker & Cant, 1984; Smith, 1987) but several authors (Thomas *et al.*, 1987; Smith, 1988; Smith *et al.*, 2009) cite examples of bars containing higher percentages of mud than sand. Mud-rich accretion sets usually have some sand and are collectively called inclined heterolithic strata (HIS; after Thomas *et al.*, 1987). IHS is a descriptive term and defines alternating layers of sandy, muddy strata arranged along subparallel *en*

Fluvial Meanders and Their Sedimentary Products in the Rock Record, First Edition.
Edited by Massimiliano Ghinassi, Luca Colombera, Nigel P. Mountney and Arnold Jan H. Reesink.
© 2019 International Association of Sedimentologists. Published 2019 by John Wiley & Sons Ltd.

échelon accretion surfaces at low (~10 degrees) angles to horizontal bedding that are typically attributed to muddy point bar development.

IHS is common and usually interpreted to record tidal point bars, bar tails, or counter point bars (Smith, 1987, 1988; Rahmani, 1988; Gingras *et al.*, 1999; Choi *et al.*, 2004; Smith *et al.*, 2009; Choi, 2010; Dashtgard & Johnson, 2014). IHS formation is attributed to seasonal or random changes in flow rates, or in a tidal environment it may be caused by seasonal and biweekly migration of turbidity maxima (Smith, 1985; De Boer *et al.*, 1988; Smith, 1988; Fustic *et al.*, 2012; and Blum, 2015). Though IHS is commonly attributed to tidal conditions (Thomas *et al.*, 1987; Smith, 1988; Schoengut, 2011; Fenies *et al.*, 2012), Thomas *et al.* (1987) made no indication that tidal influence is a required condition for IHS deposition. Indeed, non-tidal occurrences are common (Thomas *et al.*, 1987; Dalrymple *et al.*, 1992). In non-tidal meandering rivers, IHS forms in two ways. Multiple meander cut off events with consecutive lateral migrations over previous cut offs result in the preservation of amalgamated fragments of heterolithic channel fill and periodic growth of a single point may make HIS if mud deposition is generally high (Reineck & Wunderlich, 1968; Thomas *et al.*, 1987; and De Boer *et al.*, 1988). The cut-off examples typically are confined to the outer growth bands of the bar but the IHS reflective of periodic growth may permeate the bar. Flow variance in counter point bars tends to generate IHS throughout, but these bars also tend also to have a down-stream-elongated geometry and convex accretion surfaces (Smith, *et al.*, 2009).

Mud-dominant IHS point bars deposited owing to either tidal influence (Thomas *et al.*, 1987; Smith, 1988; Schoengut, 2011; Fenies *et al.*, 2012), waning bar tails (Willis & Tang, 2010; Smith *et al.*, 2011; and Ielpi & Ghinassi, 2014), or counter point bars (Brice, 1974; Hooke, 1984; Smith *et al.*, 2009; Fustic *et al.*, 2010) are common place and are modestly well understood. Less understood are point bars that lack characteristics of these depositional origins, yet are mud dominant. These 'muddy-normal' point bars bear no obvious tidal influence, have the lobate geometry of normal expansion-dominate sandy point bars but are mud-rich point bars dominated by IHS throughout.

The goal of the current study is to evaluate the process(s) of these muddy-normal point bars. This proceeds by examining the internal architecture of a heterolithic point bar deposit within the Cretaceous Belly River Group of Dinosaur Provincial Park, Canada that has the geometry and sedimentary features of a muddy-normal point-bar. This point bar also has the overprint of slumping and liquefaction that we here interpret to record a large lateral spread and liquefaction associated with a large earthquake that occurred late in point-bar development.

Point-bar processes and types

Point bars form as side-attached bars on the inside bend of single thread meandering rivers; and form mostly by accumulation of sand and gravel as the bar laterally accretes and the opposing cutbank retreats (Wolman & Leopold, 1957; Allen, 1965; Mertes *et al.*, 1996; Constantine & Dunne, 2008; Jo & Ha, 2013; Nardin, 2012; and van de Legeweg *et al.*, 2014). Bank pull (from rapid cutbank erosion compared to bar accretion) and bar push (from rapid bar accretion compared to cutbank erosion) cause channel migration towards the outer bank and an increase in sinuosity of the channel with continued bar growth (Constantine & Dunne, 2008; Willis & Tang, 2010; and Eke, 2013). Migration normal to a channel bank is expansion, whereas migration obliquely downstream relative to a channel bank is translation (Daniel, 1971; Jackson, 1976; Nanson, 1980; and Bridge & Jarvis, 1982).

Traditional models show point bars to fine upward owing to an upward decrease in bed shear stress with decreasing water depth. Multiple stacked upward fining trends of accretion packages may comprise a larger fining trend resulting from bar erosion and reactivation following major floods or changes in growth vector with channel rotation (Smith, 1987; Thomas *et al.*, 1987; Bridge, 2003; Constantine & Dunne, 2008; Willis & Tang, 2010; and Durkin *et al.*, 2015a & b.). These mechanisms produce scours, lateral-accretion bed sets and current structures exhibiting migration along lateral accretion faces which can be observed in most point bar deposits (Smith, 1987; Thomas *et al.*, 1987; Bridge, 2003; Constantine & Dunne, 2008; Willis & Tang, 2010; and Hubbard, 2011). Common point bar models include the sandy-normal point bar, counter point bar and tidal point bar.

The sandy-normal point bar records the typical 'text book' meandering point bar model. This model is simplified (Fig. 1) with subparallel lateral accretion surfaces, usually spanning from the top of the bar to the bottom of the bar (Sundborg, 1956; Allen, 1963, 1970; McGowen & Garner,

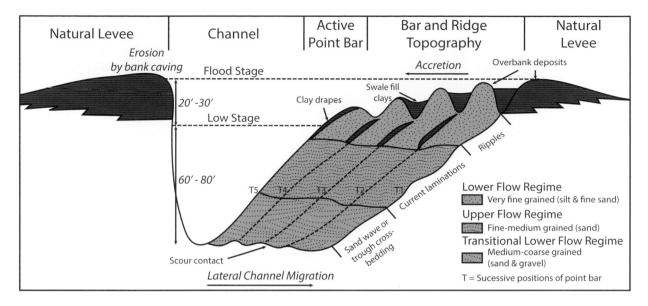

Fig. 1. (Above): From Saucier (1994), showing the internal sedimentary features of a sandy-normal point bar.

1970; Bluck, 1971; Bridge, 1975). These point bar deposits are typically sand-dominated with lower regime to some lower-upper regime sedimentary structures and minimal mud, most commonly in the form of relatively thin mud drapes between thicker sandy bed sets (Frazier & Osanik, 1961; Allen, 1970; and Walker & Cant, 1984). These bars are roughly the thickness of bank-full channel depth and form from a dominance of bar expansion compared to translation (Fig. 2) (Jackson, 1976; Nanson, 1980; and Bridge & Jarvis, 1982; Smith, 2006).

Counter point bars may be heterolithic with alternating accretion sets of sand and mud that form IHS deposits. Transition from normal point bar to counter point bar occurs across an inflection point separating the convex 'normal' point bar from the concave counter point bar during bar translation (Brice, 1974; Hooke, 1984; Smith *et al.*, 2009; and Fustic *et al.*, 2010). Translation pulls the channel from the concave bank downstream of the convex point bar and results in muddier bar-tail preservation (Smith *et al.*, 2009, 2011; Willis & Tang, 2010). Over time these bars accumulate elongate bodies with accretion sets that are concave to the channel in plane view (Fustic *et al.*, 2010; Brice, 1974; Hooke, 1984; and Smith *et al.*, 2009) (Fig. 3). Textural trends in counter point bars downriver from the crossover point include the thickening of silt-dominated facies, diminishing and fining of sand interbeds and general fining of grain-size within accretion sets

(Fustic *et al.*, 2010; Brice, 1974; Hooke, 1984; and Smith *et al.*, 2009). Counter point bars thicken at the expense of normal point bars as they build away from the inflection point and eventually reach a maximum thickness equal to the full point bar (Fustic *et al.*, 2010; Brice, 1974; Hooke, 1984; and Smith *et al.*, 2009).

Tidal point bars are also characterised by IHS. Tidal environments have rhythmic fluctuations in water levels and current velocities due to tidal cycles; as such, IHS deposition is common (Smith, 1985; Thomas *et al.*, 1987; and De Boer *et al.*, 1988). Tidally induced IHS typically are heterolithic or clean sands inter-fingering with mud clast breccias. Typically, these mud clast breccias lay under heterolithic accretion-stratified sands, with clay layers more abundant in the uppermost portion (Smith, 1985; De Boer *et al.*, 1988; and Fenies *et al.*, 2012). Other indictors of tidally influenced point bars include cross-bedding within lenticular and wavy bedding and reactivation surfaces on top of the cross-beds (Reineck & Wunderlich, 1968; de Mowbray, 1983; Smith, 1985; and De Boer *et al.*, 1988) and marine trace fossils along accretion sets (Smith, 1985; Pattison *et al.*, 2005; Desjardins *et al.*, 2012). Tidal IHS point bars otherwise form by the same mechanisms of expansive or translating lateral accretion as other point bars.

A muddy-normal point bar has similar lobate geometries and convex accretion sets to a sandy normal point bar, but it has subequal to more

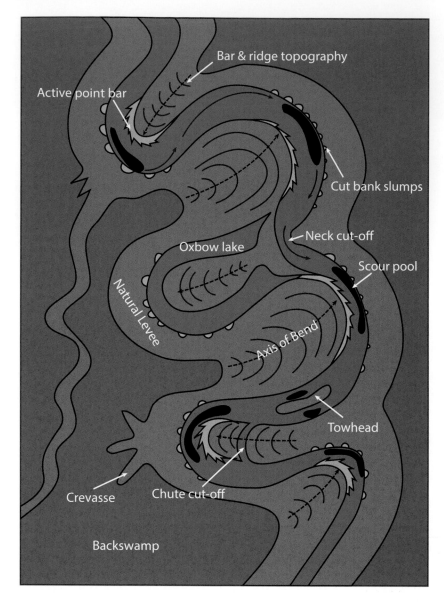

Fig. 2. (Left): From Saucier (1994), showing map view of a sandy-normal point bar. Shows expansion of point bars within a meandering system.

mud than sand in these accretion sets. Major accretion surfaces are subparallel and generally extend from the top of the bar to the bottom and, other than the higher mud content, these bars are similar to the sandy normal point bar. Likewise, these bars lack tidal features and heterogeneity cannot be attributed to tidal processes. The processes that form these deposits are largely uncertain, this paper therefore documents in detail the internal characteristics of a muddy-normal point bar to gain insights into the processes that form them.

Location, palaeogeography and stratigraphy

The muddy-normal point bar is located in the Steveville area of Dinosaur Provincial Park, AB, Canada (Fig. 3) and is found within the Dinosaur Park Formation of the Belly River Group (McLean, 1971; Hamblin, 1997b; and Therrien *et al.*, 2014). The channel belt is mostly confined above and below by floodplain strata and is locally incised into a similar channel belt below and incised by a similar channel belt above. The average thickness of the bar and its associated channel fill is 8 to 10 m.

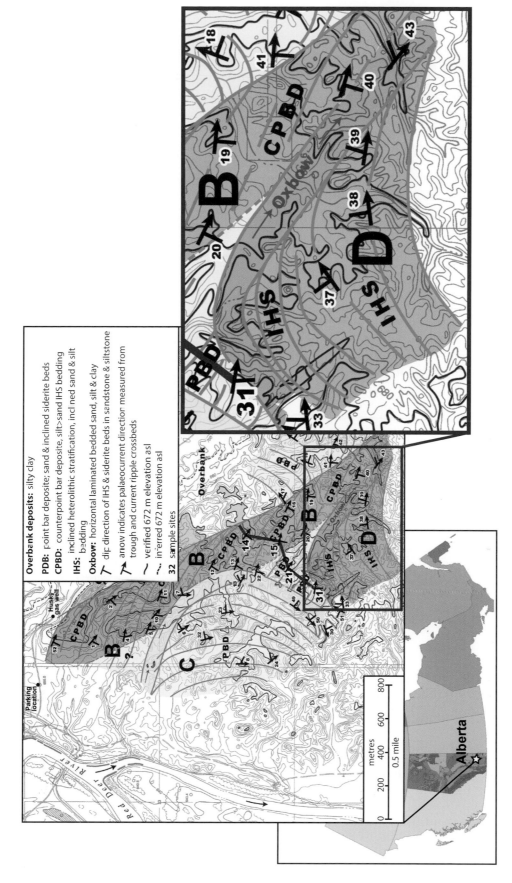

Fig. 3. Location of study area. Star signifies location of the Steveville area of Dinosaur Provincial Park. Zoomed portion shows a geologic map from Durkin *et al.* (2015b). Acronyms used by Durkin *et al.* (2015b) are as follows: point bar deposit (PBD), counterpoint bar deposit (CPBD) and inclined heterolithic stratification (IHS). Point bar for this study is outlined in red. Note: Letters (B, C, D) on figure were used in Durkin *et al.* (2015b) study and do not denote anything in this study.

The Steveville area provides excellent 3D exposures of the Dinosaur Park Formation across the entirety of the deposit owing to highly dissected badland topography. The Dinosaur Park Formation is approximately 128 m-thick and located at the top of the Belly River Group with the uppermost 60 to 100 m exposed at Dinosaur Provincial Park (McLean, 1971; Wood *et al.*, 1988; Ryan *et al.* 2001). The Dinosaur Park Formation is composed of numerous stacked channel-belt lenses composed of fine to medium grained sandstones and siltstones and local inclined heterolithic stratification within a muddy floodplain matrix with abundant petrified wood and dinosaur fossils throughout (Wood *et al.*, 1988; Brinkman, 1990; and Eberth & Hamblin, 1993).

Dinosaur Park Formation strata located in Dinosaur Provincial Park were deposited on the distal portion of a fluviodeltaic coastal plain during the middle Campanian between the Western Interior Seaway in the east and the contemporary Canadian Rocky highlands in the west (Williams & Stelck, 1975; Koster, 1983; Braman & Sweet, 1990; Eberth & Hamblin, 1993). The Dinosaur Park Formation is part of a dominantly eastward thinning fluvial clastic wedge located between the Lea Park and Bearpaw marine tongues (Fig. 4), implying deposition during a period of maximum regression (McLean, 1971; Jerzykiewicz, 1985; and Braman & Sweet, 1990). Marine assemblages are restricted to the uppermost horizons of the formation, stratigraphically higher than the target point bar (Braman & Sweet, 1990; Eberth & Hamblin, 1993; Eberth, 1996; Mallon, 2012). The palaeoclimate was probably temperate to subtropical and humid with perennial rivers that generally flowed in a south-east direction (Dodson, 1971; Jarzen, 1982; Koster, 1983; and Mallon, 2012).

Depositional environments from the succession found regionally in the Dinosaur Park Formation grade from fluvial to estuarine to facies of the Lethbridge Coal Zone (Koster & Currie, 1987; Eberth & Hamblin, 1993; Eberth, 1996; Mallon, 2012). The point bar of this study is located within the fluvial portion of this set. The precise distance upstream from the contemporary shoreline is uncertain but is at least on the order of several tens to well over a hundred kilometres (Fig. 5; Eberth, 2009). While not confirming the distance is too great for marine reach, this is consistent with the lack of tidal imprint in the form of traces or bi-directional indicators and the lack of tidal features or IHS in contemporary adjacent point bars. Laminations present do not show definitively rhythmic tidal patterns and similar inter-laminated sand and mud are common in fluvial systems (Wood, 1989). Likewise, the deposits lack reactivation surfaces, herringbone cross strata, or other definitive tidal sedimentary indicators. This muddy-normal point bar was also previously mapped by Durkin *et al.* (2015b) as a muddy heterolithic point bar with a lobate geometry based on the work of the late Derald Smith (Fig. 3), before the muddy-normal point bar was defined. All the point bars in the Smith map (Fig. 3) are within the same channel belt as the bar we studied; and none of the bars in the area have trace fossils or sedimentary features supporting tidal influence within this coeval set of channel loops. Tidal origin of the IHS is thus discounted.

METHODS

The point bar analysed in this study is exposed over 2.6 km². Seven detailed sedimentary logs were composed across the bar to document and assess internal features. Lithology descriptions, sedimentary structures, grain-size, sorting, roundness, organic material and palaeocurrent data were collected from this point bar. Ten less detailed sections were collected that catalogued sand *vs.* mud ratios, thickness; and strike and dip of accretion sets within locations where these data were not already collected in the more detailed sections. A Laser Range Finder attached to a Trimble Geo 7X was used to measure unit thicknesses and to obtain GPS readings for the location of each section. This data was stored using quick projects in Esri's ArcPad. All sections depict strikes and dips of the major accretion surfaces and use a generic template and colour scheme whereby sand and mud ratios were quantified as sand dominated (sand/mud >50%), mud dominated (sand/mud <50%) and mixed (sand/mud ~50%). These data were plotted on stereonets and in graphs according to set lithotype and accretion-set orientation. All accretion sets within the point bar were measured and measurements accounted for 100% of accretion sets within the bar.

Accretion surface data were collected via strike and dip measurements. Strikes and dips of major accretion surfaces, as well as accretion set boundaries, were collected using a Brunton Compass and recorded for every location at which

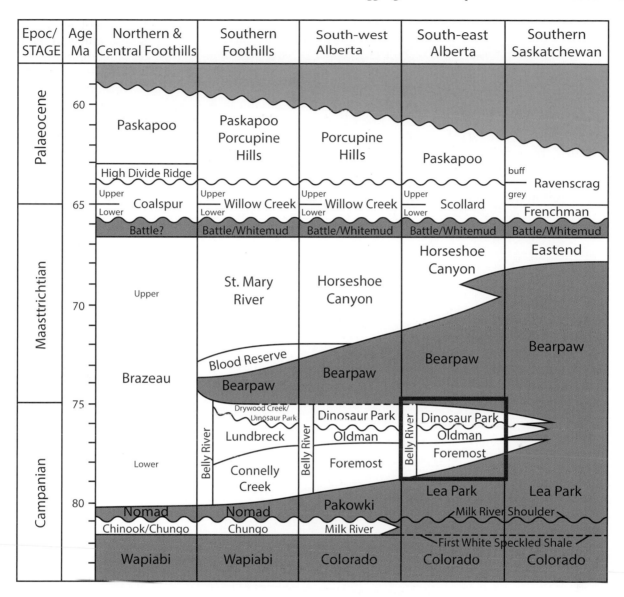

Fig. 4. Depiction of stratigraphic relationship and regional extent of members within the Belly River Group (modified from Hamblin, 1997a). Red box indicates the portion of the Dinosaur Park Formation studied.

a stratigraphic section was measured. Strikes and dips at accretion set boundaries were only recorded if their measurements differed from that of the major accretion surface above and below. Affiliation in orientation between set and boundary was noted at all sites.

Along with strike and dip data, pictures for photo panoramas were collected from 15 individual sites to aid in interpretation across the bar. Panoramas from the pictures were generated using Adobe's Photoshop version CC and exported to Adobe's Illustrator version CC for drafting surfaces measured in the field to produce graphics.

Panoramas were used to complete a simplified architectural-element analysis using the procedure of Holbrook (2001), as derived from Miaill (1985, 1986).

RESULTS

Architectural elements binding the muddy-normal lateral-accretion element

The meander loop including the muddy-normal point bar (the lateral-accretion element) comprises a thalweg element, channel-fill element and a

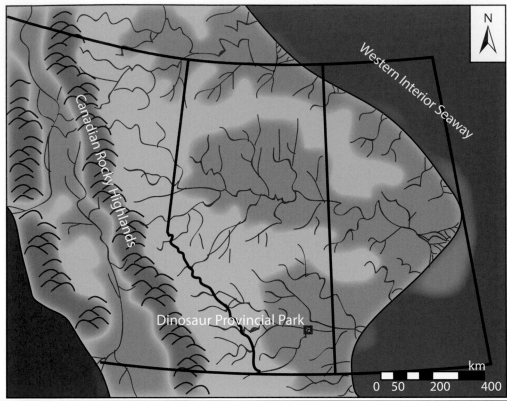

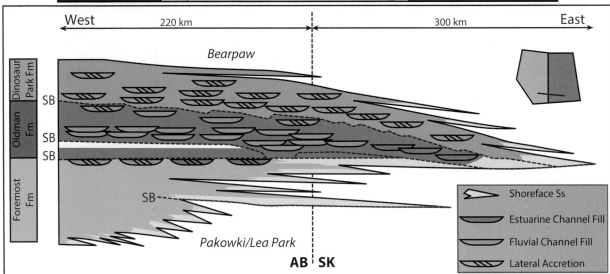

Fig. 5. Location of Dinosaur Provincial Park (red box) during the Middle Campanian to be approximately 200 km from the shoreline (modified from Eberth, 2009). Red box indicates where the park would have been during this time. Image below, also from Eberth (2009), shows cross section of area.

lateral-accretion element, bound by mudflat elements and locally older meander loops (Fig. 6). The thalweg element is semi-continuous across the bottom of the lateral-accretion element. The targeted muddy lateral-accretion element is continuous until it reaches the channel-fill

element to the N, NW and NE. The channel-fill element cuts northward into a sandy lateral-accretion deposit of an older adjacent lateral-accretion element (Fig. 6). In a small strip along the southern portion, part of a sandier lateral-accretion element incises slightly into the targeted

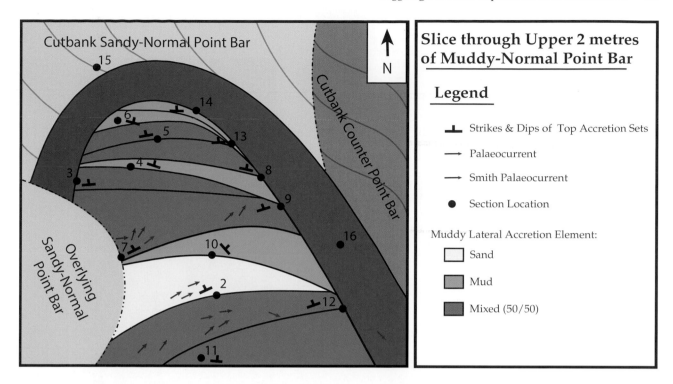

Fig. 6. Map projecting top-most accretion set of the muddy-normal point bar with accretion packages coloured based on content – mud (blue), sand (yellow), or mixed (red). Palaeocurrent data also displayed on map (purple arrows this study and red for palaeocurrent measurements by Durkin *et al.*, 2015b).

muddy lateral-accretion set from above (Fig. 6) and on the eastern side the thalweg element incises downward into an older sandy lateral accretion set. These elements are discussed below.

Channel-fill element

Recorded in this element is the abandoned channel fill associated with the muddy point bar. Abandoned channels are a result of meander cut-off or avulsion (Jordan & Pryor, 1992; Toonen *et al.*, 2012). At abandonment, infilling processes record palaeodischarge variations, palaeoflooding and fluvial style change (e.g. Vandenberghe, 1995; Page & Nanson, 1996; Macklin & Lewin, 2003; Erkens, 2009; Toonen *et al.*, 2012). Discharge processes can halt and reverse during this stage resulting in new semi-stable channel bifurcation and thus a possible alternation of sand and mud beds (Jordan & Pryor, 1992; Toonen *et al.*, 2012), as observed in the heterolithic strata of this channel fill.

The channel-fill element in this bar is bound above and below by floodplain mud-flat elements and cuts into older meander loops laterally. Palaeocurrent measurements indicate an overall average flow direction down channel from west to

east (Fig. 6). The channel fill is elongate and lenticular in cross section (Fig. 7). Beds and bedding planes within the channel fill are generally concave up at the scale of the channel fill. The bottom of the channel fill is marked by a scour surface topped by a mud drape containing a high amount of coalified plant debris in the toe of the drape.

Lithofacies for the channel fill in general alternate between beds of mud and sand (Figs 8 and 9). Grain-size fines up overall, as well as in the individual mud and sand beds throughout the point bar. Local iron staining is present randomly throughout the entirety of the channel fill. The mud throughout the channel fill for all sections is medium-bedded (~0.2 m to 0.45 m) to thick-bedded (>0.45 m) with upper fine grained silt. Small twig and leaf fragments, typically lying between bedding and coffee grounds (fine grained organic debris) are present in the mud beds. Sedimentary structures include low angle ripples and parallel laminations both of which ranged between thin (~0.05 m to 0.2 m) and very thin sets (<0.05 m). Sandstone beds are thin (~0.05 m to 0.2 m) to thick (>0.45 m) and are comprised of well sorted, moderately-cemented lower fine sand. Sedimentary structures include thin (<0.05 m) cross laminations

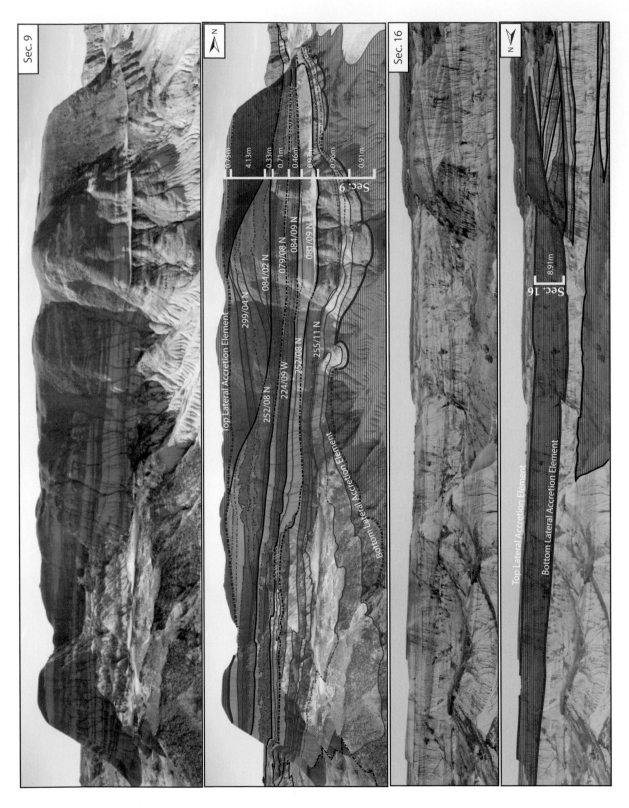

Fig. 7. Face interpretations from bar. Yellow lines indicate measured sections with unit thicknesses to the right in white. White numbers on accretion sets (black or grey lines) indicate orientations. Bold red lines indicate top and bottom of lateral accretion element, black lines indicate major accretion surfaces and grey lines indicate lateral-accretion subunits. Overlays of blue, yellow and red indicate mud, sand and mixed accretion packages accordingly. Orange, indigo and green indicate liquefaction, channel-fill and floodplain mud-flat.

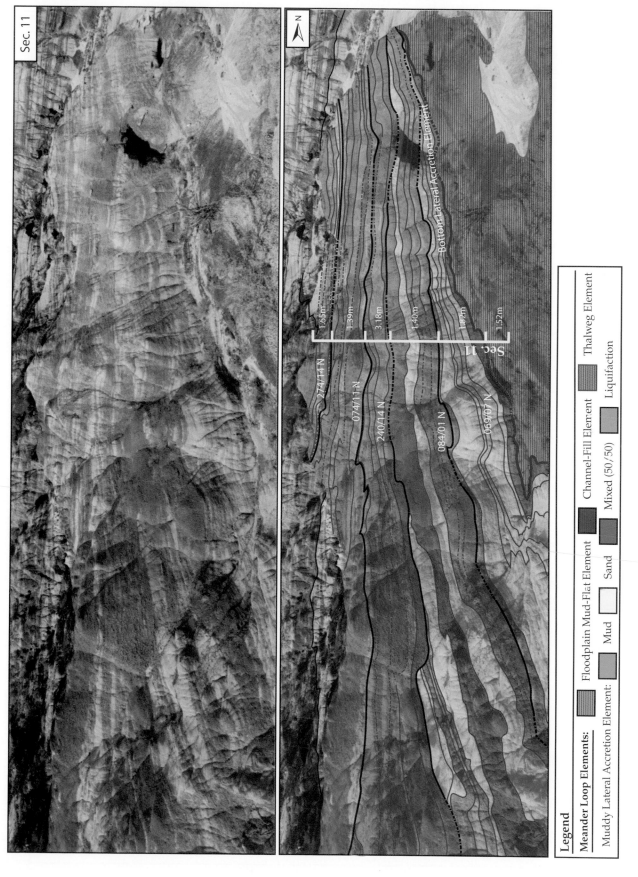

Sec. 11

Legend

Meander Loop Elements:

Floodplain Mud-Flat Element   Channel-Fill Element   Thalweg Element

Muddy Lateral Accretion Element:

Mud   Sand   Mixed (50/50)   Liquifaction

Fig. 7. (Continued)

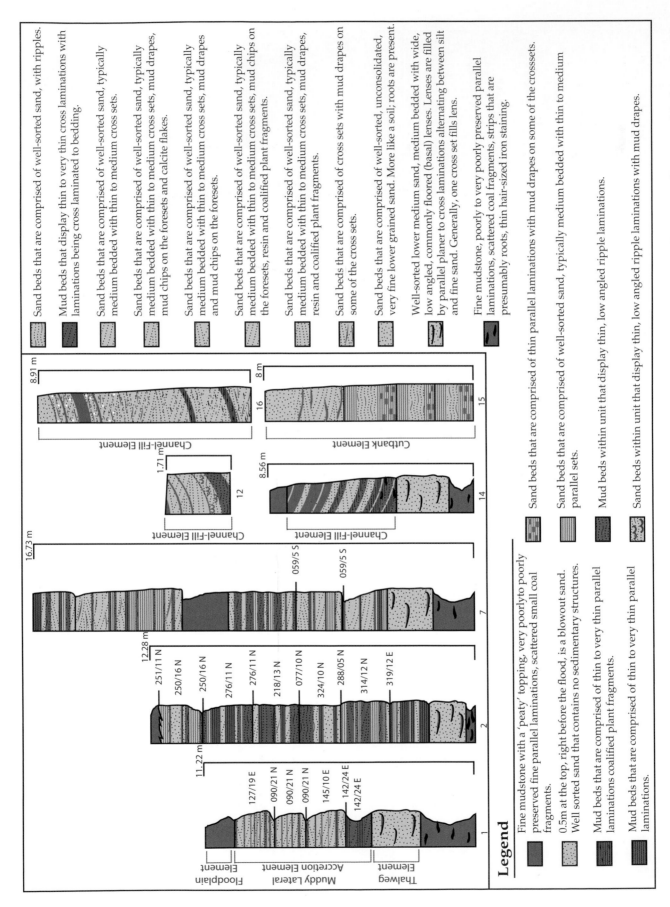

Fig. 8. Lithofacies of architectural-elements binding the muddy lateral-accretion element. Red dashes to the right of the columns represent major accretion surfaces and the numbers next to them are the associated strike and dip measurement.

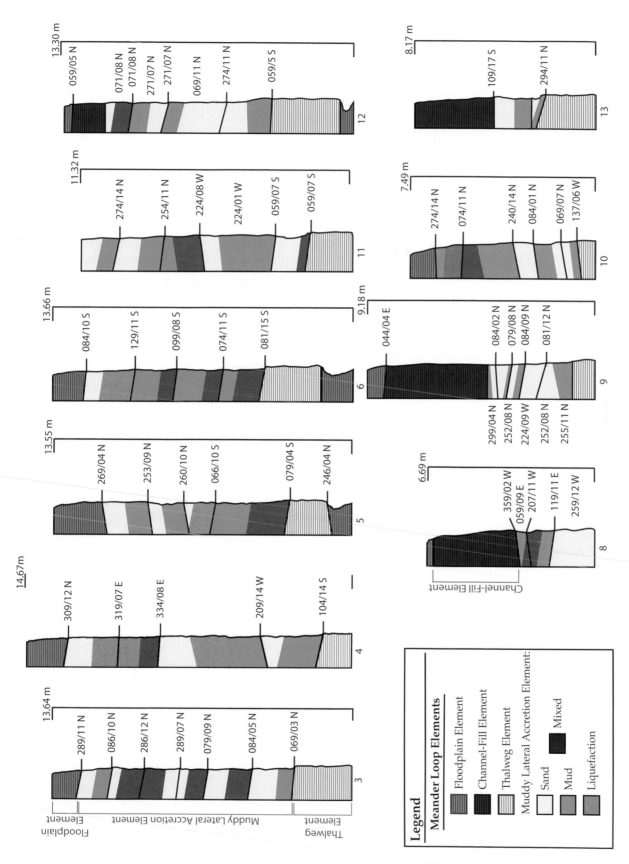

Fig. 9. Ten less detailed sections showing sand *vs.* mud ratios, thicknesses and strike and dip of accretion sets within locations where these data were not already collected in the more detailed sections (Fig. 8). Black lines within the sections mark major accretion surfaces and colours (blue, yellow and red) signify mud, sand and mixed content of accretion packages.

and low angle ripples. Locally, very thin mud drapes are present on cross-sets and ripples.

The downstream portion of the channel to the east is slightly coarser compared to the rest of the channel fill (Fig. 8). The authors presume that this means the downstream end of the loop remained open to channel flow longer and received a higher proportion of coarser sediment during abandonment. Mud beds are generally not present in the lower portion of the channel fill but do appear in the upper portion in the form of drapes. Where present, mud beds are thin-to-medium bedded (~0.05 m to 0.45 m) and have upper fine-grained silt fining to mudstone. Parallel laminations are present in the mud beds located at the bottom of the channel fill. Mud chips on foresets are also common, occurring heavily in upper sections and then decreasing in the lower areas of sections. Flakes of calcite occur locally but only in the upper portion. Organics, such as plant debris, coffee grains and resin, are in the lower and middle portion of this section. Finally, liquefaction features are present in the top 3 m of this section with a blow sand deposit in the top 0.5 m.

Floodplain mud-flat elements

The point bar, where it is not locally incised into an underlying meander loop or incised by a loop, is positioned between composite soils. This composite soil has poorly to very poorly preserved parallel laminations, scattered coal fragments, 1 to 2 cm root fragments and locally thin (3 to 5 cm) iron stained layers (Fig. 8). Generation of this soil occurred within aggrading subaerial floodplain mudflat strata.

The deposits are generally pedogenically stirred and lack primary depositional structures as well as distinct soil master horizons, which are indicative of long term depositional hiatus. The soil records slow and consistent subaerial aggradation on the floodplain. The composite soil includes a more organically rich soil, with lots of plant fragments, coffee grains and small coal fragments concentrated in a thin (3 cm) horizon approximately 15 cm above the top of the muddy lateral-accretion element.

Thalweg element

The thalweg element underlies the muddy lateral-accretion element and forms the base of the meander loop over about half of the bar area (Figs 10 and 11). Thalweg deposits form where the greatest velocities carve into the base of the channel. Thalweg lithofacies are medium bedded (~0.2 m to 0.45 m), lower medium grained, well-sorted and moderately-cemented quartz-arenite sandstone. Basal lenses within the thalweg element include fill with thick (>0.45 m) parallel and cross laminated alternating beds of silt and fine sand. These basal lenses, or scour fills, represent areas of local incision of scour holes at sites of the greatest amount of shear stress (Burge & Smith, 1999; Moore & Masch, 1962; Espey, 1963). The thalweg element is found in outcrop in about half of the point bar deposit (8 out of the 16 sections). There are no accretion sets within the thalweg element, making it easy to distinguish as a separate part of the point bar system.

The muddy lateral accretion element

Lithofacies

This element records accretion of a side-attached point bar inside a river meander bend (c.f. Miall, 1996; Bridge, 2009; Jordan & Pryor, 1992) and constitutes the defining element of the muddy-normal point bar. Dipping accretion surfaces record surfaces formed approximating the bar face between phases of growth. The element alternates between sandstone and mudstone dominance with changes in the orientation of major accretion surfaces being the separating factor between element subunits (Figs 7, 8 and 9).

Mudstone intervals throughout the element range from very thin-bedded (<0.05 m) to thick-bedded (>0.45 m). Sedimentary structures range between parallel laminations, cross laminations, wavy to planer laminations and low angle ripples, all of which range within thin (~0.05 m to 0.2 m) and very thin (<0.05 m) bed sets. Flat and angular mud rip ups are also found locally within bedding structures. Organics include small twig and leaf fragments, which typically lie between bedding plains, as well as coffee grounds. Iron concretions (1 to 10 cm) and some iron staining also are common in the mudstones.

Sandstone beds throughout the lateral-accretion element are thin-bedded (~0.05 m to 0.2 m) to thick-bedded (>0.45 m) and lower-medium-grained to lower-fine-grained, well sorted and moderately-cemented quartz arenite that fine up section. Sedimentary structures for the sandstone beds include thin-to-medium parallel laminations,

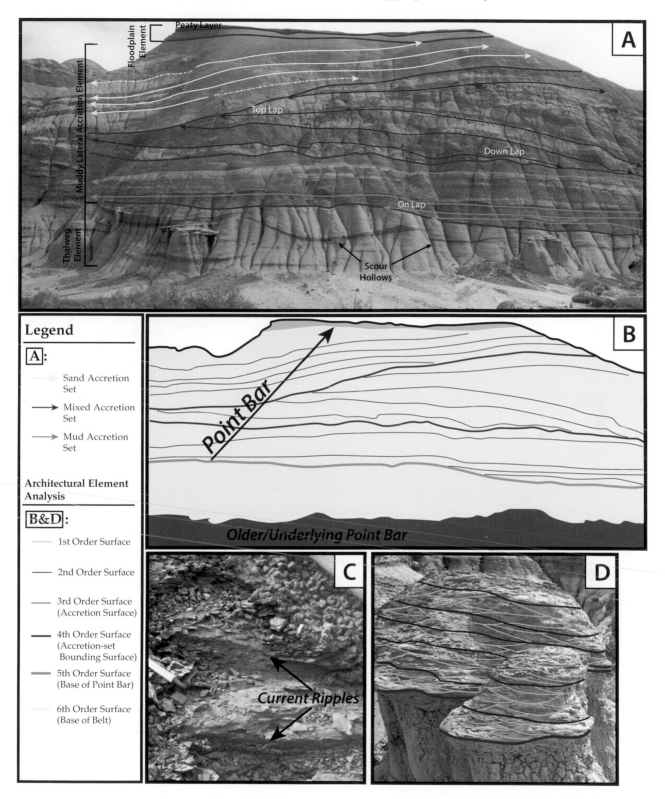

Fig. 10. (A, B, & D) The muddy lateral accretion element was separated using a simplified architectural-element analysis following the procedure of Holbrook (2001), as derived from Miaill (1985, 1986). A hierarchy of surfaces follows: 1st order, 2nd order, 3rd order (accretion surface), 4th order (accretion-set bounding surface), 5th order (base of point bar) and 6th order (base of belt). (C) Current ripples found in mud accretion package.

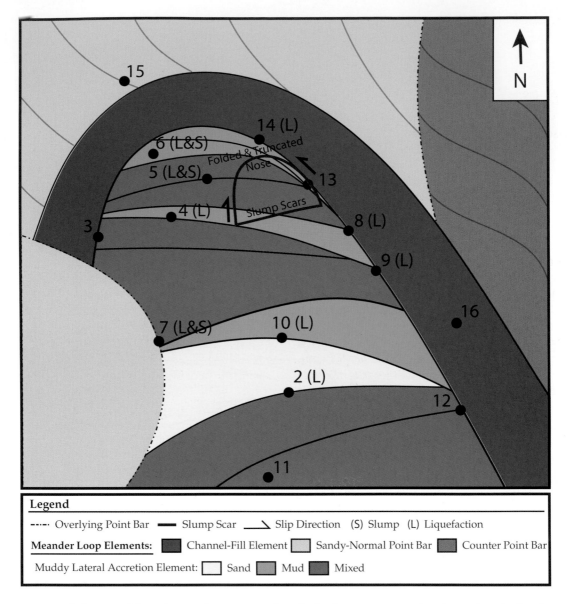

Fig. 11. Location of slump and liquefaction features are identified. Slump scar is outlined in red with slip direction (in black) approximately NW.

cross laminations and planer laminations. Organics, such as plant debris and coffee grounds, are present in the sandstone beds but are sparse and sporadic throughout and are not as prevalent as in the mudstone beds.

Mud content permeates the bar in its entirety and numerous vertical sections at different locations across the bar reveal no apparent trend in upward fining *vs.* upward coarsening in the bar at any location. This is because mud *vs.* sand accretion sets are stacked vertically, thus vertical sections encounter more than one accretion set vertically with no consistent pattern. Non-consistent

stacking of alternating muddy vs sandy accretion sets overwhelms any vertical upward fining signal typical point bars as generated by upward loss of bed shear stress with decreased depth.

Alternation of sand and mud packages suggest alternations in energy of deposition. Typically, mud deposition is associated with lags in flow and sand deposition is associated with more active flow conditions (Calverley, 1984; Smith, 1987; Thomas *et al.*, 1987; Makaske & Weerts, 2005). The mudstone deposition in this case does not appear to represent simple bar draping from suspension (c.f. Frazier & Osanik, 1961; Walker &

Cant, 1984). Flow-generated sedimentary struc- tures, such as current ripples, are prevalent within both mud and sand beds throughout the bar.

Accretion trends

Surfaces defining changes in orientation of major lateral-accretion surfaces were mapped to sepa- rate the element into lateral-accretion sets (Fig. 10). These accretion sets were between 3 to 4 m-thick and over 40 to 50 m-long horizontally and perpendicular to strike. Accretion sets have consistent orientation throughout but also tend to be distinctly mud or sand dominated. Sand and mud accretion subunits are both present through- out the areal distribution of the element and do not appear to reflect a consistent location within the element. Mud set distribution is not localised and thus is confined to late stages of growth in the overall bar formation. Accretion sets are also not consistently muddy or consistently oriented down palaeoslope, as would be expected if deposition was from counter point bar development.

Panoramas display how accretion sets are inter- nally arranged. Major accretion surfaces com- monly crosscut the predecessor accretion set, inferring erosion associated with reorientation of bar growth (Fig. 7). Accretion surfaces within an individual set typically approximate the orientation of the basal binding surface, but locally internal accretion surfaces may have a different orientation than their lower set boundary (Fig. 7). Lapping relations between set internal accretion surfaces and binding accretion surfaces thus were predominantly top lap against the upper surface with local basal onlap and downlap (Fig. 7).

Panoramas and sections show the relationship between orientation of accretion sets and set lithology. Accretion subunits with common orien- tation are categorised as mud-dominate (blue), sand-dominate (yellow) and mixed (40 to 60%; red) lithofacies assemblages (Fig. 7). Some sets were deformed by abundant liquefaction features (orange) and not assigned a separate lithology oth- erwise, though most are mud-dominant. Slump features and liquefaction features are prominent in the northern part of the lateral accretion ele- ment. The location of slump features tended to be near the toe of the element (Fig. 11) and liquefac- tion features occurred in approximately the upper 2 to 3 m throughout the meander loop (Fig. 7) with their occurrence increasing around slump features (Fig. 12). Calculations based on data collected

from panoramas confirm that the lateral accretion element is indeed mud dominate. Mud sets con- stitute approximately 38%, sand 24% and mixed 38% within this bar. Mud *vs.* sand across the full element by volume non-specific to accretion sets is 57% mud and 43% sand (Table 1).

Measurements of change in dip from one major accretion surface to the next up section quantifies bar growth trends and normalises changes in bar growth direction throughout the bar regardless of local bar orientation. If change in dip direction from the lower major accretion surface to the major accretion surface directly above is eastward (down current), the change is considered positive, and *vice versa* for westward shifts, and the magni- tude of the shift is measurable in each case. Data is plotted in a bar chart with colours assigned to each permutation of change in lithofacies assem- blage across the bounding surface. Orientation changes to the west tend to be sandy and orienta- tion changes east tend to be muddy (Fig. 13). This shows that changes to sandier from muddier set in this point bar is associated with a toggling of growth vector from down-dip to up-dip and *vice versa*. When compared to palaeoflow direction (Fig. 6), sand sets are grow in the direction of bar expansion and mud sets are associated with shifts toward bar translation.

Lithofacies assemblage *vs.* bar translation and expansion are additionally analysed by plotting strikes and dips of major accretion surfaces *vs.* lithologic dominance in aggregate (Fig. 14). Strikes and dips of each accretion set are plotted on a ster- onet and assigned a colour dependent on the lithofacies assemblage. Dip directions were then averaged for each lithofacies assemblage to obtain an average accretion direction for all sets with the same lithofacies assemblage. Local and composite averages are also plotted directly on a map-view of the bar (Fig. 15). Mud, sand and mixed surfaces are generally dispersed evenly across the point bar except at the bar tail where mud-dominate sets increase. Strikes and dips of major accretion surfaces have similar orientations within mud, sand and mixed sets. Average dip direction of sand orientations runs in a NW direction, mud sets are oriented in a slightly SE direction, mixed sets are in a slightly NE direction and then an overall direction of growth averaged among all sets is NE. Again, sand-set orientations are aligned closer to expansion, mud orientations align with bar translation and mixed orientations are between. The sets toggle between sandy expansion

Fig. 12. Slumps, scarps and nose all indicated above. Slumps and liquefaction are common features in the upper 3 m of point bar.

Table 1. This table shows the overall percentages of mud, sand and mixed (50/50) across the entire point bar. These values were determined using panoramas. 'Mud *vs.* sand volume' was calculated using data from the mud percentages shown per unit in the colourised portion.

Accretion Set		Pano/Sec. 2		Pano/Sec. 3		Pano/Sec. 4		Pano/Sec. 5		Pano/Sec. 6		Pano/Sec. 7		Pano/Sec. 8		Pano/Sec. 9		Pano/Sec. 10		Pano/Sec. 11		Pano/Sec. 12		Pano/Sec. 13	
Set	%	Type	%	Type	%	Type	%	Type	%	Type	%	Type	%	Type	%	Type	%	Type	%	Type	%	Type	%	Type	%
9	N/A	Floodplain																							Floodplain	N/A
8	0.33	Mud	N/A	Floodplain	N/A							Floodplain	N/A	Floodplain	N/A	Floodplain	N/A					Channel Fill	N/A	Channel Fill	N/A
7	0.75	Mud	0.50	Mud	0.50	Floodplain	N/A	Floodplain	N/A	Floodplain	N/A	Mud	0.50	Channel Fill	N/A	Channel Fill	N/A	Floodplain	N/A	Mud	0.50	Mud	0.8	Mud	0.50
6	0.67	Mud	0.75	Mud	0.75	Liq.	0.00	Mud	0.50	Liq.	0.00	Mud	0.75	Mud	0.50	Mud	0.25	Mud	0.75	Mud	0.50	Mud	0.67	Mud	0.50
5	0.25	Mud/Liq.	0.67	Mud	0.67	Mud	0.50	Mud	0.67	Mud	0.67	Mud	0.50	Mud	0.50	Mud	0.50	Mud	0.67	Mud	0.33	Mud	0.50	Mud	0.67
4	0.75	Mud	0.67	Mud	0.67	Mud	0.33	Liq.	0.00	Mud	0.67	Mud	0.25	Mud	0.00	Mud	0.00	Mud	0.67	Mud	0.75	Mud	0.25	Mud	0.67
3	1.00	Mud	0.50	Mud	0.50	Mud	0.67	Mud	0.50	Mud	0.67	Mud	0.25	Mud	0.50	Mud	0.67	Mud	0.50	Mud	0.50			
2	N/A	Thalweg	0.33	Thalweg	0.33	Thalweg	0.67	Thalweg	N/A	Thalweg	N/A	Thalweg	N/A			Thalweg	0.00	Thalweg	N/A	Thalweg	N/A	Other Point Bar	N/A	
1	N/A	Floodplain	N/A	Thalweg	N/A			Floodplain	N/A	Floodplain	N/A	Floodplain	N/A	Mud	0.00							Other Point Bar	N/A	

Dominate Accretion Sets

Sediment	Occurrence	%
Mud	19	0.38
Sand	12	0.24
50/50	19	0.38
Total	50	100

Mud *vs.* Sand Volume

Sediment	Occurrence	%
Mud	28.5	0.57
Sand	21.5	0.43
50/50	0	0
Total	50	100

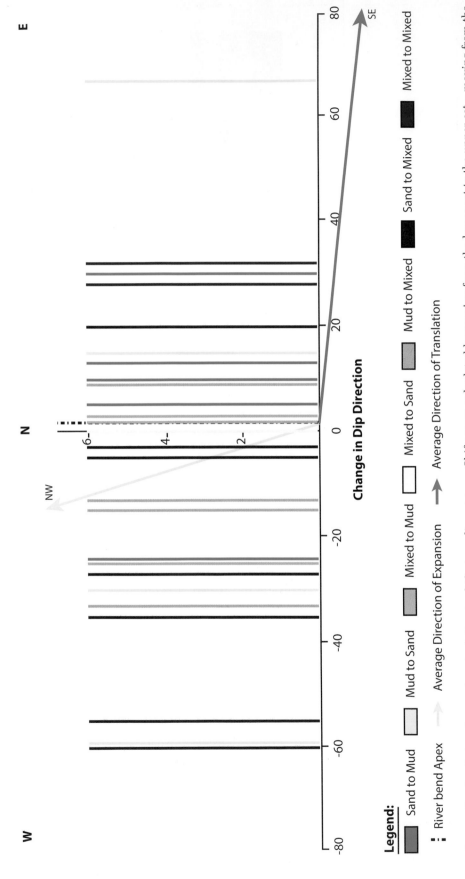

Fig. 13. Graph relating changes in orientation to shifts in sediment type. Shifts were calculated by moving from the lower set to the upper set – moving from the bottom of the bar up. Blue signifies a shift from sand to mud sets (becoming muddier), yellow a shift from mud to sand (becoming sandier), light blue from mud to mud (muddier), light yellow from mixed to sand (sandier), dark-blue from sand to mixed (muddier); and red from mixed to mixed (no change). Yellow arrow identifies the average direction of expansion of the bar and the blue arrow identifies the average direction of translation of the bar.

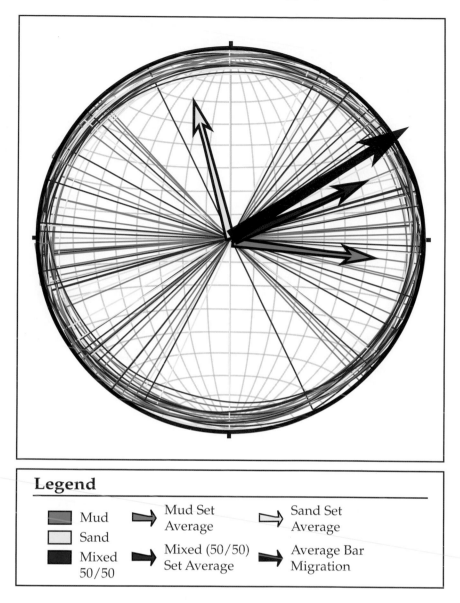

Legend

▢ Mud	⇨ Mud Set Average	⇨ Sand Set Average	
▢ Sand			
▢ Mixed 50/50	⇨ Mixed (50/50) Set Average	⇨ Average Bar Migration	

Fig. 14. Stereonet, plotted using Allmendinger's Stereonet 9.5.2 (Allmendinger *et al.*, 2013; Cardozo & Allmendinger, 2013), showing strikes and dips of major sets of accretion surfaces of this muddy-normal point bar. Stereonet also relates the strikes and dips of the major accretion surface sets to sediment type – mud (blue), sand (yellow); and mixed or 50/50 (red). Averaging dip directions reveal that sand sets are oriented NW, mud slightly SE, mixed slightly NE; and an overall meander direction of all sets is NE – supporting toggling between expansion and translation. The measurements include all parallel accretion sets within the bar.

orientations and muddy translation orientations across the expanse of the bar with no consistent repetitive pattern.

DISCUSSION

Depositional controls on mud and sand accretion sets in the muddy normal point bar

Bar accretion proceeds as an alternation between accretion sets that are muddy, sandy and/or mixed, resulting in an alternation of bar subunits forming the larger composite bar. Accretion alternates between mud-rich *vs.* sand-rich sets with no preference to location of mud or sand beds within or between accretion packages. The mud-rich sets contain alternating mud and sand beds parallel to accretion sets, preserving a bedding association commonly labelled Inclined Heterolithic Strata (IHS; after Thomas *et al.*, 1987). Each accretion set records a separate accretion direction and the binding surfaces between are commonly erosive of the sets beneath, as evidenced by top lap. These binding surfaces record resurfacing of the point

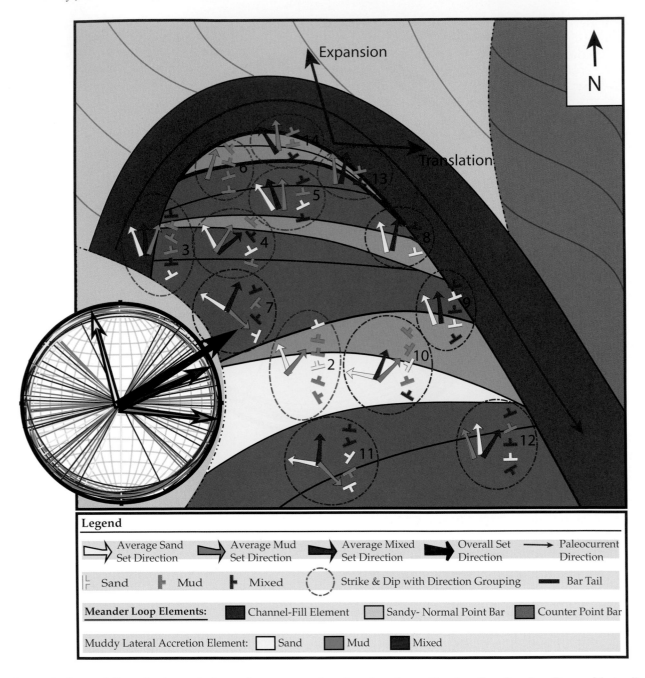

Fig. 15. Strikes and dips of major accretion surfaces were each assigned a colour – blue (mud), yellow (sand), or red (mixed). Average directions were determined for each grouping of strikes and dips. These were then compared with average directions calculated from the stereonet and palaeocurrent directions. This revealed that a NW orientation favours sand, slight SE direction favours mud; and slight NE direction favours mixed (50/50). In view of the palaeocurrent direction (thin black arrow) it can be determined that as sand is deposited, the bar expanded and as mud was deposited, the bar translated.

bar during episodes of change in bar migration direction. A similar higher-order bounding surface has been found in point bar deposits of the Horseshoe Canyon Formation by Durkin *et al.* (2015a) and is interpreted similarly in these rocks. The large-scale alternation of mud *vs.* sand sets

generates a lithologic heterogeneity in accretion sets that can also be considered IHS. However, this IHS exists at a higher order than the IHS within accretion sets and records larger-scale bar migration trends instead of discrete accretion events. This self-similarity in IHS between these two scales does record some level of fractal geometry in IHS deposition for this bar.

Since there is no consistent pattern of repetition for the accretion sets and no fining upward trend within sets exceeding the grain-size difference between sets, the point bar does not generate the classic fining upward grain-size profile typical of point bars (e.g. Nanz, 1954; Sundborg, 1956; Bersier, 1958; Allen, 1963, 1970; Visher, 1965; Bernard, *et al.*, 1970; McGowen & Garner, 1970; Bluck, 1971; Bridge, 1975). Any given vertical section encounters more than one accretion set. On average, the base of the bar is thus neither predictably muddier nor sandier than the top of the bar in any given section.

Mud beds in the muddy and mixed accretion sets appear to record active transport rather than slack-water draping. Mud in these sections was deposited as thick silty mud units with active bedforms (i.e. ripples) parallel to accretion surfaces by some semblance of bedload transport during flow events for sufficient sustained duration to accumulate thin to medium beds in single accretion events. Flume experiments by Schieber & Yawar (2009) support this notion. In their experiments, Schieber & Yawar (2009) demonstrated that mud suspensions can deposit as ripples in swift moving currents, 15 to 30 cm s^{-1}. Parallel laminations and cross-stratification also occurred locally within the mud beds, thus implying similar conditions as would deposit these same features in sand beds.

Sandy accretion sets in isolation have characteristics typical of sandy normal point bars. Structures in sand beds also attest to transport as bedload. Cross-stratification in sand beds record dune migration across accretion surfaces in fashion typical of point bar development and flow conditions associated with meandering systems (Sundborg, 1956; Harms *et al.*, 1963; McGowen & Garner, 1970; and Jackson, 1976). Parallel laminations are horizontal throughout cross sections of the point bar indicating combinations of depth and velocity favourable to upper plain bed deposition parallel of accretion surfaces, a feature also well established in point bars (Sundborg, 1956; Harms *et al.*, 1963; McGowen & Garner, 1970; and Jackson, 1976). Grain-size also fines upward along accretion surfaces in sandy sets in response to loss of bed shear stress with water depth, again, typical of normal point bar processes systems (Sundborg, 1956; Harms *et al.*, 1963; McGowen & Garner, 1970; Jackson, 1976; Smith, 1987).

Muddy and mixed accretion sets constitute IHS as packages between sandy bar sets. IHS deposits contained within this point bar appear unrelated to commonly attributed bar processes (tidal, counter, bar tail). Tidally influenced bars commonly have gravel, mud-balls and mud at the base and may have other tidal indicators reflective of by-directional flow (Jackson, 1978; Calverley; 1984; and Smith, 1987). Grain-size also commonly fines up section and downstream (Jackson, 1978; Calverley; 1984; and Smith, 1987). The point bar of this study lacks tidal indicators and includes more typical sandy bar sets. Most importantly, the bar lacks trace or body fossils distinctive of marine influence. Fossils are restricted to dinosaur bone fragments and plant debris, such as petrified logs, impressions of small twigs and small, indistinguishable coalified debris.

The precise distance upstream from the contemporary shoreline is uncertain, though is approximated to be between 150 to 200 km (Fig. 5; Eberth, 2009). The backwater length of a channel is the distance up-dip from the shore required for the channel elevation to climb one channel depth. Backwater length restricts the up-dip length for marine impacts; and tidal effects are generally restricted to much less than one backwater length (see Blum *et al.*, 2013). For the site to be within one backwater length from the shore at a distance of 150 km and the 8 m channel depth seen here requires that the slope to the sea is only 5×10^{-5}. This is shallow compared to the $\geq 10^{-4}$ slopes typical of similar coastal-plain rivers (Blum *et al.*, 2013). The point bar at the site was probably deposited well beyond one backwater length from the contemporary coast. Blum *et al.* (2013) also point out that large point bars are uncommon within one backwater length and generally found further up-dip. Point bars are well developed within the belt in which the studied bar is located (Fig. 3). Abundance of well-developed point bars with lack of tidal imprint in this or contiguous bars of this belt is consistent with the assertion that this bar was deposited over one backwater length up-dip of the contemporary shore. Tidal origin of the IHS is thus not probable for this study's point bar.

IHS deposition as either counter point bars or bar tails is unlikely because of observed inconsistencies between location within the bar and the relative extent of the IHS deposit. Counter point bar deposits transition from a sandier deposit to a muddier deposit across an inflection point (Fustic *et al.*, 2010; Brice, 1974; Hooke, 1984; Smith *et al.*, 2009). There is no analogous inflection point within the point bar of this study where a discrete zone of sand-dominate accretion sets transitions to muddier IHS accretion sets. Counter point bars will also be distinguished from normal point bars because of their comparatively elongate geometry, concave accretion sets and generally muddy/heterolithic lithology (Smith, 2006). Accretion set dip direction can also be used to infer slope perpendicular (expansion) to slope parallel (translation) bar migration (Ielpi & Ghinassi, 2014; Willis, 1993; Fustic *et al.*, 2012). The bar lacks the consistently translational growth vector and associated prismatic geometry produced by counter-point-bar development. However, the general translational process that produces the counter point bar is a component in the explanation for the generation of the muddy normal point bar as discussed below. Bar tail IHS results in slower discharge rates, triggered by onset of channel abandonment and occurs along the meander-bend axes (Willis & Tang, 2010; Smith *et al.*, 2011; and Ielpi & Ghinassi, 2014). While this does occur, and explains a small portion of the IHS deposits mapped (Fig. 15), IHS deposits in this point bar are found throughout the entirety of the deposit and not restricted to any particular portion.

Toggling between expansion and translation – the origin of the muddy-normal point bar

Expansion and translation are both common processes in the generation of point bars (Daniel, 1971; Jackson, 1976; Nanson, 1980; Bridge & Jarvis, 1982; and Holbrook, 2013). Brice (1974) defined expansion as the increase in the sinuosity of a meander bend by growth perpendicular to the belt axis and translation as the downstream movement of a meander bend. Sand tends to preferentially deposit on the apex of the point bar and finer-grained strata tends to accumulate downstream of the bar apex (Bridge & Jarvis, 1982; Smith *et al.*, 2009; Willis & Tang, 2010). Flow expansion around the apex of an expansional point bar causes loss of shear stress in the inner bend, deposition of bedload and preservation of point bar sand (Bridge & Jarvis, 1982; Smith *et al.*, 2009; Willis & Tang, 2010). In a translational bar, detachment of the flow from the downstream bar face commonly results in slow to backward accretion eddies, making pockets of weak flow and

supporting fine-grained heterolithic deposition on the downstream bar face (Leeder & Bridges, 1975). It follows that a purely expansional meander will produce a sandier/coarser point bar with a normal/lobate geometry and convex accretion sets (Fig. 16; Allen, 1963, 1970; Visher, 1965; Bernard, *et al.*, 1970; McGowen & Garner, 1970; Bridge, 1975; Willis & Tang, 2010). It also follows that a purely translational meander will tend to be a finer heterolithic counter point bar with elongate geometry and concave accretion sets, commonly forming downstream across an inflection from the convex sandy normal bar (Fig. 16; Daniel, 1971; Jackson, 1976; Nanson, 1980; Bridge & Jarvis, 1982; Smith *et al.*, 2009; Willis & Tang, 2010). Discrete meanders can grow through both expansional and translational phases (Calverley, 1984; Wood; 1985; and Smith, 1987).

The alternation of sandy *vs.* muddy accretion sets in the targeted muddy-normal point bar records high-frequency toggling between expansion and translation during bar migration and deposition of alternating sandy and muddy accretion sets. Sandy set orientations are typically NW, muddy deposit orientations were generally SE and mixed sets are oriented in between these two (Figs 14 and 15). Sandier deposits, with an average orientation NW, are more perpendicular to depositional palaeodip and show channel geometry indicative of growth by expansion. Pure expansion would produce a sandy normal point bar deposit with a lobate geometry. Pure translation would produce a muddy counter point bar deposit dominated by IHS with a more downstream-elongated geometry. Rivers producing muddy normal point bars, such as the one that produced the study loop, toggle in orientation between expansion and translation at high frequency, instead of migrating consistently perpendicular or parallel to slope. A mix of expansional and translational accretion sets preserve within the same bar element. The point bar produced has a roughly lobate geometry, similar to a sandy normal point bar, but can preserve abundant muddy counter point bar; and thus bears the traits of both bar types.

High frequency toggling is a newly reported growth vector for point bars that is first reported here and is yet to be attributed to a physical cause. Whilst the toggle vector is confirmable in the study bar from accretion directions, the flow characteristics that caused this are not. However, the drivers for this bar toggle can be speculated from available field information and modern analogues.

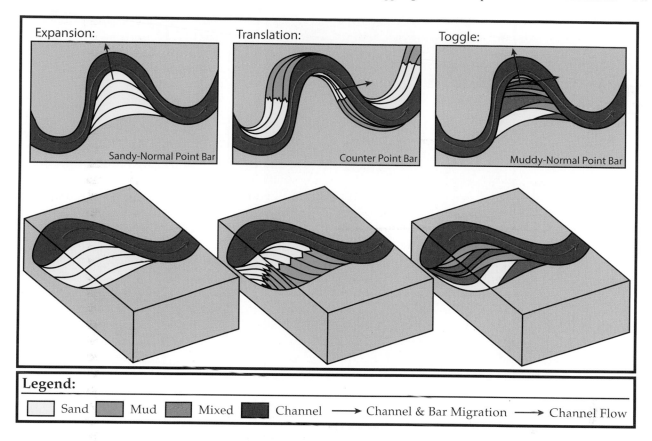

Fig. 16. Three different point bar models are displayed in both map and cross-sectional view. In a purely expansional meander system a normal (sandier/coarser) point bar will deposit. In a purely translational meander system a counter point bar will deposit. A muddy-normal point bar is made up of both normal (expansion-dominate) point bar deposits and counter (translation-dominate) point bar deposits in alternating succession. The alternation of sandy *vs.* muddy accretion sets records high-frequency alternation between expansion and translation during bar migration.

Differential bank material can alter growth directions for bars (Heitmuller & Hudson, 2009 and Limaye & Lamb, 2014). The muddy normal bar is mostly eroding into a consistently sandy normal point bar with minimal heterogeneity (Fig. 3). This would not readily explain the changes in growth vector observed and certainly would not explain the high frequency shifts.

The most logical reason for the toggled growth of the point bar is that the position of bank flow attachment shifted commonly between apical to downstream over the span of bar growth. Point bars expand or translate depending on whether the point of flow attachment to the cut bank is near the meander apex (expansion) or more downstream (translation) (Bridge & Jarvis, 1982; Smith *et al.*, 2009; Willis & Tang, 2010). This point of attachment shifted repeatedly over point bar development. While the cause of this shift is uncertain, it probably reflects an unstable, but repetitive, flood regime for the formative channel.

The channel probably experienced periods of weaker floods that permitted the channel to contain high flows around the meander bend whereby the flow momentum impacted the cutbank near the apex (Fig. 17). At this time, the bar grew by sandy expansion. Temporarily increased flood regime probably caused flows that crossed the upstream part of the point bar and impacted flow momentum on a more downstream point of flow attachment (Fig. 17). This reshaped the bar and introduced a phase of more muddy translation. Later return to more containable floods shifted the bar back to apical growth. Muddy normal point bars thus could be an indicator of unstable climate conditions.

Accretion sets on the outer bands (toe area) of this point bar are also muddier than the previous bands of the point bar (Fig. 15). This is a normal phenomenon associated with channel abandonment. Reductions in discharge provide for the accumulation of finer-grained beds as the bar

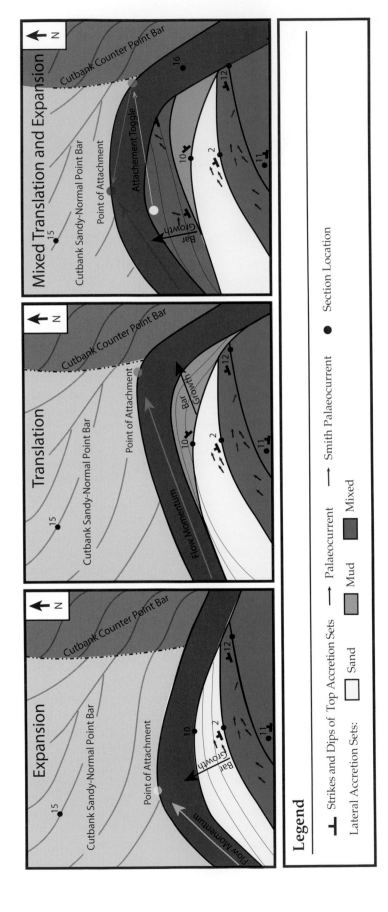

Fig. 17. Muddy-normal point bar growth model, showing how toggling between mud, sand and mixed occurred to deposit the point bar.

goes into abandonment phase (Wightman & Pemberton, 1997; Makaske & Weerts, 2005; and Willis & Tang, 2010). This process contributed to the general heterolithic characteristic of the bar but only explains the outward-most heterolithic deposits.

Evidence for a large syndepositional earthquake and possible implications

Liquefaction features are abundant throughout the upper 3 m of the studied point bar. Liquefaction features include ball and pillow structures, fluidised contorted bedding and sand blows in muds on the bar tops (Fig. 18). Liquefaction records the exceedance of the shear strength of sandy or heterolithic strata because of increased pore pressure induced by the passing of seismic surface waves (American Society of Civil Engineers, Geotechnical Engineering Division, committee on Soil Dynamics, 1978; Obermeier, 1989). Liquefaction is thus a common and strong indicator of palaeoseismicity (e.g. Gomberg, 1992; Johnston & Schweig, 1996; USGS, 2016). Liquefaction of sand is generally to a few metres in depth, is aided by high water tables in sandy substrates with capping hydraulically confining layers and records earthquakes on the order of 6.5 Mo or greater (Obermeier, 1989; Johnston & Schweig, 1996; Tuttle *et al.*, 2002). Thin massive sand sheets within the lower metre of the overlying composite soil element toward the southern and older part of the bar probably record sand blows locally preserved on the thin overbank mud that capped the point bar (Fig. 18). These sand blows record local breaching of the confining clay and extrusion of liquefied sand.

Slumps in the toe of the bar record a lateral spread associated with collapse of the point bar into the adjacent open channel. Lateral spreads are known from earthquakes like the 1965 Alaska earthquake (Waller, 1964; Senthamilkumar, 2009). They record the slip of a large generally flat mass of soft sediment along a glide plain laterally into an open space during a seismic event because of increased pore pressure along the base, decoupling of the overlying layers and sliding of these layers laterally into an open space where lateral support is absent (Bazair *et al.*, 1992; Palmer, 2006). The resultant lateral spread produces tensional backward rotating slump blocks in the back of the spread that usually evolves into multiple *en echelon* slumps throughout the mass as the spread fails. Folding in the forward toe of the spread record compression at the front of the advancing failing mass (Fig. 19). The slumps in the muddy point bar have these structural traits (Figs 15 and 20).

The liquefaction features and lateral spread that characterise this point bar resulted from a moderate to large (≥6.5 Mo) earthquake that deformed part of the bar just prior to its final abandonment. The liquefaction features, particularly the sand blows, are difficult to explain without seismic activity. While loading can cause liquefaction, it probably would not be confined to the upper bar and would not explain surface extrusion of sand in blows. An earthquake is supported by these structures. The lateral spread records collapse of the point bar into the lateral open space of the contemporary active channel. The lateral spread failed on a flat surface which negates slope failure as a cause and argues for seismic loading. Most of the overbank muds above the point bar are not deformed (Figs 12 and 20), meaning only a

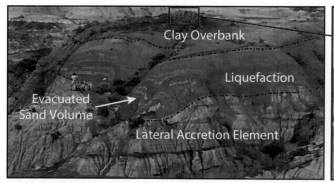 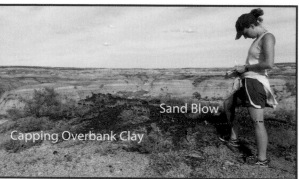

Fig. 18. The left image indicates evacuated sand volume and liquefaction features. The image on the right shows zoomed area of right image, showing the sand blow found on top of the clay overbank.

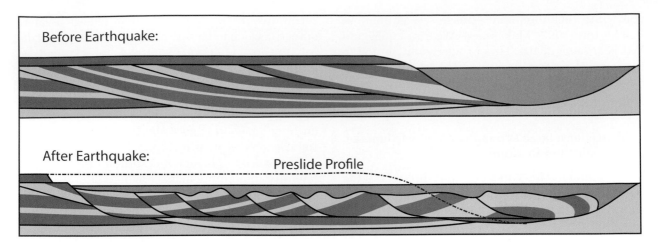

Fig. 19. Showing before (top) and after (bottom) earthquake of this muddy-normal point bar. The lateral spread produced a backward rotating slump in the back of the spread, which evolved into multiple *en echelon* slumps throughout (thin purple lines) and a compressively folded toe at the front as the spread failed.

small portion to none of these overbank deposits were there at the time. Likewise, the bar grew a short distance beyond the compressive fold that characterises the outer end of the spread (Fig. 20). The lack of accumulation of overbank mud at the time of deformation and the short growth length thereafter argue that the lateral spread occurred at the edge of the active channel shortly before the channel was finally abandoned. Co-occurrence of the lateral spread and liquefaction structures present a powerful case for a seismic event near the end of point bar deposition.

The cause for this earthquake is unknown as the Dinosaur Park region of this time is not near a plate boundary or otherwise in a known seismic region. Large intraplate earthquakes far from active plate margins are certainly common and known to disrupt contemporary fluvial systems (Schumm *et al.*, 2000; Holbrook, 2006). Such an intraplate earthquake is a possibility. One additional possibility is that the earthquake is associated with a bolide impact. Impacts of large meteorites and comets are known to generate great earthquakes (Alvarez *et al.*, 1998). A large crater of the approximate age of the targeted strata has recently been discovered near the study site (Glombick *et al.*, 2014; USGS, 2016). While the dates of the impact and the dates of seismic activity on the point bar are far too unconstrained to confirm any temporal association, the spatial co-occurrence and overlapping temporal error range does spark speculation and reasons for further consideration of their possible relationship.

Relation to petroleum and reservoir potential

Fluvial point bar deposits are common oil reservoirs but heterogeneities in bar architecture pose challenges to prediction of reservoir permeability trends. An example of this is offered from the Athabasca Oil Sands of north-eastern Alberta, Canada. Point bar deposits in these strata are one of the world's largest proven extra heavy crude oil reserves (Government of Alberta, 2014). Initial estimates of oil in place are 1.8 trillion bbl, most of which is within point bar deposits of the Lower Cretaceous McMurray Formation (Deutsch *et al.*, 2013). Despite the high success of proven reserves, production from the Athabasca Oil Sands is difficult owing to the high viscosity of the hydrocarbons and the naturally occurring heterogeneities found within point bar reserves.

Difficulties in interpretation and modelling of reservoir 3-dimensional stratigraphic architecture arise due to the stratigraphic and lithologic heterogeneities that occur at various scales and orientations in inclined heterolithic stratification (IHS) deposits (Dahl *et al.*, 2010). Advanced modelling, with grounding in geologic baseline studies, generates a clearer picture of reservoir connectivity and is used to interactively steer the drilling of production wells (Hein *et al.*, 2013). Further, baseline studies that integrate sedimentological and geochemical characterisation allow for predicting and mapping bitumen properties in the reservoir and assist in distinguishing between barriers and baffles (Fustic *et al.*, 2013).

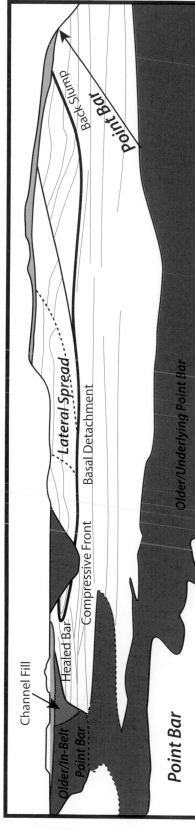

Fig. 20. Image of lateral spread structures seen in this study's point bar, with interpretation below.

Heterolithic properties within a point bar are predominately the result of IHS deposits. Tidally influenced IHS deposits and counter point bars are common in the McMurray Formation, but that does not explain all heterolithic occurrences (Demchuk *et al.*, 2007; Dolby *et al.*, 2013; Hubbard *et al.*, 2011; Fustic *et al.*, 2013; Hein *et al.*, 2013) the muddy-normal point bar is another bar type that may contribute an explanation for muddy point bars where other models fail. This is also true of other sites were IHS is not well explained by pre-existing tidal and counter-point-bar models. Additionally, better comprehensions of the internal structure of heterolithic point bars can always aid in further understanding point bar interconnectivity and how it affects oil and gas flow in reservoirs.

CONCLUSION

The muddy normal point bar forms due to toggling between meander bend expansion and translation orientations resulting in alternations between mud-dominated and sand-dominated accretion sets. The result is a point bar dominated by muddy IHS strata but with a lobate geometry similar to a normal sandy point bar that is not the result of the prior IHS explanations of counter point bar, bar tail, or tidally influenced deposition. The muddy normal model thus is a new explanation for deposition of point bars dominated by muddy IHS deposition. Toggling also records a new bar-growth vector that augments its previously reported and component vectors of expansion and translation. Toggling records repeated upstream *vs.* downstream shift in the point of attachment for flow momentum over deposition of the bar that probably derives from changing trends for discharge magnitude caused by unstable climate conditions.

Muddy, sandy and mixed accretion sets each record deposition by active flow and the resultant muddy-normal point bar formed from these alternating accretion lithologies does not show the upward fining trend typical of other point bars. Muddy, IHS, accretion sets form during translation and mud layers within are thick-bedded with current ripples indicative of deposition by transport rather than draping from suspension during slackwater episodes. Expansion sets are sandy, interspersed between translational IHS sets and have sand content and internal structure more consistent with typical sandy point bars. The sandy and muddy accretion sets alternate throughout the point bar and do not appear to reflect a locational pattern within the bar. The prevalence and high-frequency shift between muddy and sandy accretion sets means that vertical sections encounter set alternations and the fining upward trends typical of normal point bars do not occur here. Upward grain-size trends are thus somewhat random.

This particular muddy-normal point bar deposit shows signs of significant seismic activity based on pervasive liquefaction and slump features. The slump features record a large lateral spread associated with seismically induced collapse of the point bar into the adjacent open channel. Liquefaction features are contemporary with bar deposition and include plastically deformed beds and local sand blows. The cause of the earthquake is uncertain but could record intraplate seismicity from a currently unknown source. Intriguingly, the earthquake could also record a large earthquake associated with the Bow City Impact Crater impact which happed nearby and approximately at this time.

These muddy-normal point bars are common in hydrocarbon reservoirs. A better understanding of the muddy-normal point bar improves models of reservoir quality, as well as adding to the fundamental understanding of point bar processes.

ACKNOWLEDGEMENTS

First and foremost we dedicate this paper to Derald Smith. Derald laid the groundwork for this research and provided the insights and observations that lead to this product. We are saddened that his untimely death did not permit him the opportunity to see this work completed. His initial observation of the palaeoseismic features and his base map to the targeted channel belt made the work presented here possible. We would also like to thank deeply Laura and John Bucholz for their highly generous donation to the Gretchen Swanson Family Foundation Fund in Geology Research at Texas Christian University. Their timely support permitted it to move forward and constituted the project's principle support. We are grateful for insightful reviews by Dr. Amanda Owen and Dr. Paul Durkin that greatly improved this manuscript. We would also like to thank Tamie Morgan and Dr. Michael Slattery for their input and review of an earlier version of this document. Finally, we thank Matthew Palmer, Alex Torres, Blake Warwick, Julia Peacock and Rachel Gray for their input and endless support both in the field and in the lab.

REFERENCES

Allen, J.R. (1965) A review of the origin and characteristics of recent alluvial sediments. *Sedimentology*, **5**, 89–191.

Allen, J.R. (1970) Studies in fluviatile sedimentation: a comparison of fining-upwards cyclothems, with special reference to coarse-member composition and interpretation. *J. Sed. Petrol.*, **40**, 298–323.

Allen, J.R.L. (1963) Henry Clifton Sorby and the sedimentary structures of sands and sandstones in relation to flow conditions. *Geol. Mijnbouw*, **42**, 223–228.

Allmendinger, R.W., Cardozo, N.C. and **Fisher, D.** (2013) Structural Geology Algorithms: Vectors & Tensors: Cambridge, England, Cambridge University Press, p. 289.

Alvarez, W., Staley, E., O'Connor, D. and **Chan, M.** (1998) Synsedimentary deformation in the Jurassic of southeastern Utah—A case of impact shaking? *Geology*, **26**, 579–582.

American Society of Civil Engineers, Geotechnical Engineering Division, Committee on Soil Dynamics (1978) Definition of terms related to liquefaction. *Am. Soc. Civ. Eng. Proc., Journal of the Geotechnical Engineering Division*, **104**, 1197–1200.

Bazair, Dobry and **Alemo** (1992) Evaluation of lateral ground deformation using sliding block model. Earthquake Engineering, Tenth World Conference, Balkema, Rottordam.

Blakey, R. (2014) Paleogeography and paleotoctonics of the Western Interior Seaway, Jurassic-Cretaceous of North America. *AAPG* Search and Discovery Article #30392 (Availiable on-line at AAPG.org).

Blum, M. (2015) Scaling Relationships Between Fluvial Channel Fills, Channel-Belt Sand Bodies and Drainage Basins, With Implications for the Mannville Group, Alberta Foreland Basin. In: *AAPG Annual Convention and Exhibition.*

Blum, M., Martin, J., Milliken, K. and **Garvin, M.** (2013) Paleovalley systems: Insights from Quaternary analogs and experiments: *Earth-Sci. Rev.*, **116**, 128–169.

Braman, D.R. and **Sweet, A.R.** (1990) Overview of Campanian to Paleocene stratigraphy, southern Alberta Foothills. In: *Fieldguide to Uppermost Cretaceous Strata in Southern Saskatchewan and Alberta* (Eds D.R. Braman and A.R. Sweet). *Can. Soc. Petrol. Geol.*, Annual Convention Fieldtrip Guide, 66–70.

Brice, J.C. (1974) Evolution of meander loops. *Geol. Soc. Am. Bull.*, **85**, 581–586.

Bridge, J. and **Jarvis, J.** (1982) The dynamics of a river bend: A study in flow and sedimentary processes: *Sedimentology*, **29**, 499–540.

Bridge, J.S. (2009) *Rivers and floodplains: forms, processes and sedimentary record.* John Wiley & Sons.

Brinkman, D. (1990) Paleoecology of the Judith River Formation (Campanian) of Dinosaur Provincial Park, Alberta, Canada: Evidence from vertebrate microfossil localities. *Palaeogeogr. Palaeoecol. Palaeoclimatol.*, **78**, 37–54.

Burge, L. and **Smith, D.** (1999) Confined meandering river eddy accretions: sedimentology, channel geometry and depositional processes. *Spec. Publ. in Fluvial Sedimentology*, **28**, 113–130.

Constantine, J.A. and **Dunne, T.** (2008) Meander cutoff and the controls on the production of oxbow lakes. *Geology*, **36**, 23–26.

Dahl, M., Suter, J., Hubbard, S. and **Leckie, D.** (2010) Impact of inclined heterolithic stratification on oil sands reservoir delineation and management: an outcrop analog from the Late Cretaceous Horseshoe Canyon Formation, Willow Creek, Alberta: *Willow Creek, Alberta (abs.): Canadian Society of Petroleum Geologists Annual Convention GeoCanada.*

Desjardins, P., Buatois, L., Pratt, B. and **Mangano, M.** (2012) Sedimentological-ichnological model for tide-dominated shelf sandbodies; Lower Cambrian Gog Group of Western Canada: *Sedimentology*, **59**, 1452–1477.

Dixon, E.E.L. (1921) Geology of the South Wales Coalfield, Part 13. The country around Pembroke and Tenby. *Geol. Surv. Great Britain Mem.*, **220**, 103–136.

Durkin, P., Hubbard, S., Boyd, R. and **Leckie, D.** (2015a) Stratigraphic expression of intra-point-bar erosion and rotation. *J. Sed. Res.*, **85**, 1238–1257.

Durkin, P., Hubbard, S., Weleschuk, Z., Smith, D., Palmer, M., Torres, A. and **Holbrook, J.** (2015b) Spatial and temporal evolution of an ancient fluvial meanderbelt (Upper Cretaceous Dinosaur Park Formation, Southeastern Alberta, Canada) with emphasis on characterization of counter point bar deposits. *AAPG Annual Convention and Exhibition.*

Calverley, E. (1984) Sedimentology and geomorphology of the modern epsilon cross-stratified point bar deposits in the Athabasca upper delta plain. Unpublished M.Sc. thesis, Univ. of Calgary, Calgary, Alta., 116 pp.

Cant, D. (1982) Fluvial facies models and their application. In: *Sandstone depositional environments* (Eds P.A. Scholle and D.R. Spearing), *AAPG Mem.*, **31**, 115–137.

Cardozo, N. and **Allmendinger, R.W.** (2013) Spherical projections with OSXStereonet. *Comput. Geosci.*, **51**, 193–205.

Choi, K., Dalrymple, R., Chun, S. and **Kim, S.** (2004) Sedimentology of modern, inclined heterolithic stratification (IHS) in the macrotidal Han River Delta, *Korea. J. Sed. Res.*, **74**, 677–689.

Choi, K.S. (2010) Rhythmic climbing-ripple cross-lamination in inclined heterolithic strat- ification (IHS) of a macrotidal estuarine channel, Gosmo Bay, west coast of Korea. *J. Sed. Res.*, **80**, 550–561.

Constantine, J.A. and **Dunne, T.** (2008) Meander cutoff and the controls on the production of oxbow lakes. *Geology*, **36**, 23–26.

Dalrymple, R.W., Zaitlin, B.A. and **Boyd, R.** (1992) Estuarine facies models: conceptual basis and stratigraphic implications. *J. Sed. Petrol.*, **62**, no. 6, 1130–1146.

Daniel, J.F. (1971) Channel movement of meandering Indiana streams. *U.S. Geol. Surv. Prof. Pap.*, A1–A18.

Dashtgard, S.E. and **Johnson, S.M.** (2014) Inclined heterolithic stratification in a mixed tidal-fluvial channel. Differentiating tidal versus fluvial controls on sedimentation. *Sed. Geol.*, **301**, 41–53.

De Boer, P.L., Van Gelder, A. and **Nio, S.** (1988) *Tide-influenced sedimentary environments and facies.* D Reidel Publishing Company, Dordrecht.

de Mowbray, T. (1983) The genesis of lateral accretion deposits in recent intertidal mudflat channels, Solway Firth, Scotland. *Sedimentology*, **30**, 425–435.

Demchuk, T., Dolby, G., McIntyre, D. and **Suter, J.** (2007) The utility of palynofloral assemblages for the interpretation of depositional paleoenvironments and sequencestratigraphic systems tracts in the McMurray Formation at Surmont, Alberta. In: *Heavy oil and bitumen in foreland basins: From processes to products* (Eds J. Suter, D. Leckie and S. Larter), AAPG Hedberg Research Conference, Banff, Alberta, Canada, September 30–October 3, 5 p.

Deutsch, C.V., Hassanpour, M.M. and Pyrcz, M.J. (2013) Improved geostatistical models of inclined heterolithic strata for McMurray Formation, Alberta, Canada. *AAPG Bull.*, **97**, 1209–1224.

Dodson, P. (1971) Sedimentology and taphonomy of the Oldman Formation (Campanian), Dinosaur Provincial Park, Alberta (Canada). *Palaeogeogr. Palaeoclimatol. Palaeoecol.*, **10**, 21–74.

Dolby, G., Demchuk, D. and Suter, J. (2013) The significance of palynofloral assemblages from the Lower Cretaceous McMurray Formation and associated strata, Surmont and surrounding areas in north-central Alberta. In: *Heavyoil and oil-sand petroleum systems in Alberta and beyond* (Eds F.J. Hein, D. Leckie, S. Larter and J.R. Suter), *AAPG Studies in Geology*, **64**, 251–272.

Donselaar, M.E. and Overeem, I. (2008) Connectivity of fluvial point-bar deposits: An example from the Miocene Huesca fluvial fan, Ebro Basin, *Spain. AAPG Bull.*, **92**, 1109–1129.

Eberth, D.A. (1996) Origin and significance of mud-filled incised valleys (Upper Cretaceous) in southern Alberta, Canada. *Sedimentology*, **43**, 459–477.

Eberth, D.A. and Hamblin, A.P. (1993) Tectonic, stratigraphic and sedimentologic significance of a regional discontinuity in the upper Judith River Group (Belly River wedge) of southern Alberta, Saskatchewan and northern Montana. *Can. J. Earth Sci.*, Revue Canadienne Des Sciences De La Terre, **30**, 174–200.

Eke, E.C., Parker G. and Shimizu, Y. (2014) Numerical modeling of erosional and depositional bank processes in migrating river bends with self-formed width: Morphodynamics of bar push and bank pull. *J. Geophys. Res. Earth Surf.*, **119**, 1455–1483.

Erkens G. and Cohen K.M. (2009) Quantification of intra-Holocene sedimentation in the Rhine-Meuse delta: A record of variable sediment delivery. *Netherlands Geographical Studies*, **388**, 117–171.

Espey, W. Jr. (1963) A new test to measure the scour of cohesive sediment: Hydraulics Engineering Laboratory. Department of Civil Engineering, Technical Report HYD01-6301, The University of Texas, Austin, TX.

Féniès, H., Gingras, M.K., Labourdette, R., Musial, G., Parize, O. and Reynaud, J.Y. (2012) Subsurface and outcrop characterization of large tidally influenced point bars of the Cretaceous McMurray Formation (Alberta, Canada). *Sed. Geol.*, **279**, 156–172.

Frazier, D. and Osaniki (1961) *Point-bar deposits, Old River Locksite, Louisiana.* p. 121–137.

Fustic, M., Bennett, B., Hubbard, S., Huang, H., Oldenburg, T. and Larter, S. (2013) Impact of reservoir heterogeneity and geohistory on the variability of bitumen properties and on the distribution of gas- and water-saturated zones in the Athabasca oil sands, Canada. In: *Heavy-oil and oil-sand petroleum systems in Alberta and beyond* (Eds F. Hein, D. Leckie, S. Larter and J. Suter), *AAPG Stud. Geol.*, **64**, 163–205.

Fustic, M., Hubbard, S.M., Leckie, D.A., Smith, D.G. and Anonymous (2010) Predicting heterogeneity in meandering river deposits; the point bar to counter point bar transition. Abstracts: Annual Meeting, AAPG.

Gingras, M., Pemberton, S., Saunders, T. and Clifton, H. (1999) The ichnology of modern and Pleistocene brackish-water deposits at Willapa Bay, Washington: Variability in estuarine settings. *Palaios*, **14**, 352–374.

Glombick, P., Schmitt, D., Xie, W., Bown, T., Hathway, B. and Banks, C. (2014) The Bow City structure, southern Alberta, Canada: The deep roots of a complex impact structure? *Meteoritics & Planetary Science*, **49**, 872–895.

Government of Alberta (2015) Oil Sands: http://www.energy.alberta.ca/oilsands/791.asp (April 2015).

Hamblin, A.P. (1997b) Regional distribution and dispersal of the Dinosaur Park Formation, Belly River Group, surface and subsurface of southern Alberta. *Bull. Can. Petrol. Geol.*, **45**, 377–399.

Hamblin, A.P. (1997a) Stratigraphic architecture of the Oldman Formation, Belly River Group, surface and subsurface of southern Alberta. *Bull. Can. Petrol. Geol.*, **45**, 155–177.

Harms, J., Mackenzie, D. and McCubbin, D. (1963) Stratification in modern sands of the Red River, Louisiana. *J. Geol.*, **71**, 566–580.

Hein, F. (2013) Overview of heavy oil, seeps and oil (tar) sands, California. In: *Heavy-oil and oil-sand petroleum systems in Alberta and beyond* (Eds F. Hein, D. Leckie, S. Larter and J. Suter), *AAPG Studies in Geology*, **64**, 407–435.

Hein, F., Leckie, D., Larter, S. and Suter, J. (2013) Heavy oil and bitumen petroleum systems in Alberta and beyond: The future is nonconventional and the future is now. In: *Heavy-oil and oil-sand petroleum systems in Alberta and beyond* (Eds F. J. Hein, D. Leckie, S. Larter and J. Suter), *AAPG Studies in Geology*, **64**, p. 1–21.

Heitmuller, F. and Hudson, P. (2009) Downstream trends in sediment size and composition of channel-bed, bar and bank deposits related to hydrologic and lithologic controls in the Liano River watershed, central Texas, USA. *Geomorphology*, **112**, 246–260.

Holbrook, J. (2013) Funny things meanders do; a summary of the diversity of meander processes and morphology and implications for reservoir geometry and quality within channel belts. *Abstracts: Annual Meeting AAPG 2013*.

Holbrook, J. (2001) Origin, genetic interrelationships and stratigraphy over the continuum of fluvial channel-form bounding surfaces; an illustration from Middle Cretaceous strata, southeastern Colourado. *Sed. Geol.*, **144**, 179–222.

Hooke, J.M. (1984) Changes in river meanders: a review of techniques and results of analyses. *Prog. Phys. Geogr.*, **8**, 473–508.

Hubbard, S.M., Jensen, J.L., Labrecque, P.A. and Nielsen, H. (2011) Sedimentology and stratigraphic architecture of a point bar deposit, Lower Cretaceous McMurray Formation, Alberta, Canada. *Bull. Can. Petrol. Geol.*, **59**, 147–171.

Ielpi, A. and Ghinassi, M. (2014) Planform architecture, stratigraphic signature and morphodynamics of an exhumed Jurassic meander plain (Scalby Formation, Yorkshire, UK). *Sedimentology*, **61**, 1923–1960.

Jackson, R. (1978) Preliminary evaluation of lithofacies models for meandering alluvial streams. In: *Fluvial Sedimentology* (Ed. A.D. Miall), *Mem. Can. Soc. Pet. Geol.*, **5**, 543–576.

Jackson, R.G. (1976) Depositional model of point bars in the Lower Wabash River. *J. Sed. Petrol.*, **46**, 579–594.

Jarzen, D.M. (1982) Palynology of Dinosaur Provincial Park (Campanian), Alberta. *Syllogeus*, **38**, 69p.

Jerzykiewicz, T. (1985) Stratigraphy of the Saunders Group in the central Alberta Foothills - a progress report. *Geol. Surv. Can. Pap.*, **85-1B**, 247–258.

Jo, H.R. and **Ha, C.G.** (2013) Stratigraphic architecture of fluvial deposits of the Cretaceous McMurray Formation, Athabasca oil sands, Alberta, Canada. *Geosci. J.*, **17**, 417–427.

Johnston, A. and **Schweig, E.** (1996) The enigma of the New Madrid earthquakes of 1811–1812. *Annu. Rev. Earth Planet. Sci.*, **24**, 339–384.

Jordan, W. and **Pryor, W.** (1992) Hierarchical levels of heterogeneity in a Mississippi River meander belt and application to reservoir systems. *AAPG Bull.*, **76**, 1601–1624.

Koster, E.H. (1983) Sedimentology of the Upper Cretaceous Judith River (Belly River) Formation, Dinosaur Provincial Park, Alberta. Calgary. Canadian Society of Petroleum Geologists Conference, *The Mesozoic of Middle North America, Fieldtrip Guide Book* No. **1**, 121p.

Koster, E.H. and **Currie, P.J.** (1987) Upper Cretaceous coastal plain sediments at Dinosaur Provincial Park, southeast Alberta. In: *Rocky Mountain Sect. Geol. Soc. Am., Centennial Field Guide* (Ed. S.S. Beus), **2**, 9–14.

Leeder, M. and **Bridges, P.H.** (1975) Flow separation in meander bends, *Nature*, **253**, 338–339.

Limaye, A. and **Lamb, M.** (2014) Numerical simulations of bedrock valley evolution by meandering rivers with variable bank material. *J. Geophys. Res. Earth Surf.*, **119**, 927–950.

Macdonald, D.E., Ross, T.C., McCabe, E.J. and **Bosman, A.** (1987) An evaluation of the coal resources of the Belly River Group, to a depth of 400m in the Alberta plains. *Alberta Geological Survey, Open File Report 1987–8*, 76p.

Makaske, B. and **Weerts, H.** (2005) muddy lateral accretion and low stream power in a sub-recent confined channel belt, Rhine-Meuse delta, central Netherlands. *Sedimentology*, **52**, 651–668.

Mallon, J., Evans, D., Ryan, M. and **Anderson, J.** (2012) Megaherbivorous dinosaur turnover in the Dinosaur Park Formation (upper Campanian) of Alberta, Canada. *Palaeogeogr. Palaeoecol. Palaeoclimatol.*, **350–351**, 124–138.

McGowen, J. and **Garner L.** (1970) Physiographic features and stratification types of coarse-grained point bars: Modern and ancient examples. *Sedimentology*, **14**, 77–111.

McLean, J.R. (1971) *Stratigraphy of the Upper Cretaceous Judith River Formation in the Canadian Great Plains.* Saskatchewan Research Council, Geology Division, Report No. **11**, 96p.

Mertes, L.A., Dunne, T. and **Martinelli, L.A.** (1996) Channel-floodplain geomorphology along the Solimões-Amazon river, *Brazil. Geol. Soc. Am. Bull.*, **108**, 1089–1107.

Miall, A.D. (1978) Fluvial sedimentology: an historical review. In: *Fluvial Sedimentology* (Ed. A.D. Miall), *Can. Soc. Petrol. Geol. Mem.*, **5**, 1–47.

Miall, A.D. (1996) Ed., Fluvial Sedimentology. *Can. Soc. Petrol. Geol. Mem.*, **5**, 1–47.

Nanson, G.C. (1980) Point bar and floodplain formation of the meandering Beatton River, northeast-ern British Colombia, Canada. *Sedimentology*, **27**, 3–29.

Macklin M.G. and **Lewin J.** (2003) River sediments, great floods and centennial scale Holocene climate change. *J. Quatern. Sci.*, **18**, 101–105.

Moore, W. and **Masch, F.** (1962) Experiments on the scour resistance of cohesive sediments. *J. Geophys. Res.*, **67**, 1437–1449.

Nardin, T., Feldman, H. and **Carter, B.** (2012) Stratigraphic architecture of a large-scale point-bar complex in the McMurray Formation: Syncrude's Mildred Lake mine, Alberta, Canada. in: Heavy-oil and oil-sand petroleum systems in Alberta and beyond (Eds F.J. Hein, D. Leckie, S. Larter and J.R. Suter), *AAPG Studies in Geology*, **64**, 273–31.

Obermeier, S. (1989) The New Madrid earthquakes: An engineering-geologic interpretation of relict liquefaction features. *U.S. Geol. Surv. Prof. Pap.*, **1336-B**, 1–55.

Page, K.J. and **Nanson G.C.** (1996) Stratigraphic architecture resulting from Late Quaternary evolution of the Riverine Plain, south-eastern Australia. *Sedimentology*, **43**, 927–945.

Palmer, S. (2006) Assessment of the potential for earthquake induced lateral spreading. 2006 NZSEE Conference, p. 1–8.

Pattison, S. (2005) Significance of inner shelf turbidtic-rich channel-fill deposits, Gunnison butte to Tusher Canyon region, Book Cliffs, eastern Utah. Abstracts with Programs In: *Geologic Society of America*, **37**, p. 310.

Rahmani, R. (1988) Estuarine tidal channel and nearshore sedimentation of a Late Cretaceous epicontinental sea, Drumheller, Alberta, Canada. In: *Tide-Influenced Sedimentary Environments and Facies* (Eds P.L. deBoer, A. van Gelder and S.D. Nio), D. Reidel Publishing Co., The Netherlands, p. 433–471.

Reineck, H.E. and **Wunderlich, F.** (1968) Classification and origin of flaser and lenticular bedding. *Sedimentology*, **11**, 99–104.

Saucier, R.T. (1994) *Geomorphology and Quaternary Geologic History of the Lower Mississippi Valley.* U.S. Army Corps of Engineers, Mississippi River Commission, Vicksburg, Mississippi, p. 364.

Schieber, J. and **Southard, J.** (2015) Bedload transport of mud by floccule ripples – Direct observation of ripple migration processes and their implications. *Geology*, **37**, 483–486.

Schieber, J. and **Yawar, Z.** (2009) A new twist on mud deposition – mud ripples in experiment and rock record. *The Sedimentary Record*, **7**, 4–8.

Schoengut, J.A. (2011) *Sedimentological and ichnological characteristics of modern and ancient channel-fills, Willapa Bay, Washington.* University of Alberta, Alberta, Canada, p. 1–232.

Senthamilkumar, S. (2009) *Evaluation of liquefaction-induced lateral spread.* Indian Geotechnical Society Chennai Chapter, p. 55–60.

Smith, D. (1985) Modern analogues of the McMurray Formation channel deposits, sedimentology of meso-tidal-influenced meandering river point bars with inclined beds of alternating mud and sand: Alberta Oil Sands Technology and Research Authority, Final Rept. for Research Project.

Smith, D.G. (1987) Meandering river point bar lithofacies models: modern and ancient examples compared. In: *Recent Developments in Fluvial Sedimentology* (Eds F.G. Ethridge, R.M. Flores and M.D. Harvey), *SEPM Spec. Publ.*, **39**, 83–91.

Smith, D.G. (1988) *Modern point bar deposits analogous to the Athabasca Oil Sands, Alberta, Canada: Dordrecht, Netherlands* (NLD), D. Reidel Publ. Co., Dordrecht.

Smith, D.G., Hubbard, S.M., Leckie, D.A. and **Fustic, M.** (2009) Counter point bar deposits: lithofacies and reservoir significance in the meandering modern Peace River and ancient McMurray Formation, Alberta, Canada. *Sedimentology*, **56**, 1655–1669.

Sundborg, A. (1956) The river Klaralven: A study of fluvial processes. *Geogr. Ann.*, **38**, 217–316.

Therrien, F., Zelenitsky, D., Quinney, A. and **Tanaka, K.** (2014) Dinosaur trackways from the Upper Cretaceous Oldman and Dinosaur Park formations (Belly River Group) of southern Alberta, Canada, reveal novel ichnofossil preservation style. *Can. J. Earth Sci.*, **52**, 630–641.

Thomas, R.G., Smith, D.G., Wood, J.M., Visser, J., Calverley-Range, E.A. and **Koster, E.H.** (1987) Inclined heterolithic stratification; terminology, description, interpretation and significance. *Sed. Geol.*, **53**, 123–179.

Toonen, W., Kleinhans, M. and **Cohen, K.** (2012) Sedimentary architecture of abandoned channel fills. *Earth Surf. Proc. Land.*, **37**, 459–472.

Tuttle, M., Schweig, E., Sims, J., Lafferty, R., Wolf, L. and **Haynes, M.** (2002) The earthquake potential of the New Madrid seismic zone. *Bull. Seismol. Soc. Am.*, **92**, v. 6, 2080–2089.

USGS (2016) Historic Earthquakes: New Madrid 1811–1812 Earthquakes. Available at: https://earthquake.usgs.gov/earthquakes/states/events/1811-1812.php (November 2016).

Vandenberghe J. (1995) Timescales, climate and river development. *Quatern. Sci. Rev.*, **14**, 631–638.

van de Lageweg, W., Baar, A., Kleinhans, M., Rutten, J. and **van Dijk, W.** (2014) Bank pull or bar push: What drives scroll-bar formation on meandering rivers? *Geology*, **42**, 319–322.

Walker, R. and **Cant, D.** (1984) Sandy fluvial systems. *Facies models*, **1**, 71–89.

Wall, J., Sweet, A. and **Hills, L.** (1971) Paleoecology of the Bearpaw and Contiguous Upper Cretaceous Formations in the C.P.O.G. Strathmore Well, Southern. *Bull. Can. Petrol. Geol.*, **19**, 691–702.

Waller, R. (1964) Effects of the March 1964 Alaska earthquake on the hydrology of the Anchorage area. *Geol. Surv. Prof. Pap.*, **544-B**, 1–18.

Wightman, D. and **Pemberton, S.** (1997) The Lower Cretaceous (Aptian) McMurray Formation: an overview of the Fort McMurray area, northeastern, Alberta. *CSPG Special Publications*, **18**, 312–344.

Williams, G.D. and **Steick, C.R.** (1975) Speculations on the Cretaceous palaeogeography of North America. In: *Cretaceous System of the Western Interior of North America* (Ed. W.G.E. Caldwell), *Geol. Assoc. Can. Spec. Pap.*, **13**, 1–20.

Willis, B. and **Tang, H.** (2010) Three-dimensional connectivity of point-bar deposits. *J. Sed. Res.*, **80**, 440–454.

Wood, J., Thomas, R. and **Visser, J.** (1988) Fluvial processes and vertebrate taphonomy: The upper Cretaceous Judith River Formation, south-central Dinosaur Provincial Park, Alberta, Canada. *Palaeogeogr. Palaeoecol. Palaeoclimatol.*, **66**, 127–143.

Wood, J. (1989) Alluvial Architecture of the Upper Cretaceous Judith River Formation, Dinosaur Provincial Park, Alberta, *Canada. Bull. Can. Soc. Petrol. Geol.*, **37**, 169–181.

Wolman, M. and **Leopold, L.** (1957) River Flood Plains; some Observations on their Formation, Physiographic and hydraulic studies of rivers. *USGS Numbered Series*, Professional Paper **282**, U.S. Government Printing Office, Washington, D.C., 87–109.

Int. Assoc. Sedimentol. Spec. Publ (2018) **48**, 81–118.

Planform sinuosity of Proterozoic rivers: A craton to channel-reach perspective

ALESSANDRO IELPI*, MASSIMILIANO GHINASSI[†], ROBERT H. RAINBIRD[‡] and DARIO VENTRA[§a]

* *Harquail School of Earth Sciences, Laurentian University, Sudbury, Canada*
[†] *Department of Geosciences, University of Padova, Padova, Italy*
[‡] *Department of Natural Resources, Geological Survey of Canada, Ottawa, Canada*
[§] *Department of Earth Sciences, University of Geneva, Geneva, Switzerland*
[a] *Faculty of Geosciences, Utrecht University, Utrecht, The Netherlands*

ABSTRACT

Lacking evidence for fluvial lateral-accretion elements in early Palaeozoic systems has been ascribed to an absence of binding by rooted vegetation on subaerial landscapes. Transposing this thesis to earlier geological times, it has been proposed that, likewise, Precambrian landscapes could not have sustained highly sinuous fluvial networks. This paradigm has been hardly ever tested for the Proterozoic, a shortcoming addressed here through review of selected outcrop data and remote sensing of modern sinuous channel-flow configurations. Five sedimentary rock units deposited on Laurentia between 1.6 to 0.7 Ga record diverse palaeogeographic and tectonic settings and yield evidence of lateral accretion and planform sinuosity in fluvial channels over a full range of developmental stages. In the absence of vegetation, multiple processes interacted at craton to channel-reach scales, setting conditions favourable for self-sustained lateral accretion and thus sinuosity. Discharge modulation in perennial channels is interpreted to have had an overriding role, owing to craton-scale catchments capable of sustaining year-round flows or favourable climate settings. Steady sediment supply and local channel-bank strengthening limited braiding, allowing for narrow hydraulic profiles with flow configurations favourable to lateral accretion. Less than 15% of current literature on Proterozoic fluvial rocks reports reliable directional data on palaeo-flows and stratal accretion, a bias that undermines literature compilations aimed at gauging the relevance or insignificance of pre-vegetation lateral accretion. Fluvial deposits aggraded on unvegetated landmasses prior to the late Ordovician can only be assessed when comprehensive information on palaeoflow and bar accretion becomes available. The authors thus underline that a lack of evidence for early Palaeozoic lateral-accretion sets should not be used to support the inference that meandering fluvial planforms were a rare occurrence in earlier geological times.

Keywords: Precambrian, Meander, Laurentia, Remote sensing, Arctic Canada, Torridonian

INTRODUCTION

The global sedimentary record of meandering rivers is thought to be limited when compared to that of braided ones (Gibling, 2006), owing to their supply-limited nature and poor channel-floor aggradation (Ghinassi *et al.*, 2014; van de Lageweg *et al.*, 2015). The interpretation of meandering-channel planforms has been often based primarily on the recognition of abundant inclined hetero-lithic lateral-accretion packages and fine-grained channel fills (e.g. Allen, 1965, 1970; Embry & Klovan, 1976; Jackson, 1976; Cotter, 1978; Stewart, 1983; Turner & Eriksson, 1999; Davies *et al.*, 2011). While straightforward, this approach is to some extent simplistic because: (i) channel sinuosity develops irrespective of sediment grade (Lewin, 1976; Jackson, 1981; Ori, 1982; Arche, 1983;

Forbes, 1983; Brooks, 2003; Simon *et al.*, 2017), and its recognition in the rock record should not be based on the occurrence of inclined hetero-lithic packages; (ii) lateral-accretion packages occur over a range of fluvial styles (Ori, 1982; Bristow, 1987; Miall, 1996) and cannot be une-quivocally related to river-planform patterns unless a distinctive (palaeo)channel-flow configu-ration is demonstrated (Jackson, 1975, 1976; Bathurst *et al.*, 1977); and (iii) highly sinuous riv-ers display combinations of channel-bend expan-sion, translation and rotation (Daniel, 1971; Brice, 1974; Ikeda, 1989; Nicoll & Hickin, 2010), the sed-imentary records of which bear an architectural complexity higher than portrayed in classical facies models (Smith *et al.*, 2009, 2011; Willis, 1989, 1993; Fustic *et al.*, 2012; Yan *et al.*, 2017). Furthermore, highly sinuous fluvial channels are often treated in facies models as static planform configurations, yet they are characterised by con-tinuous change in lateral-accretion behaviour (Stølum, 1996; Timár, 2003; Thomas *et al.*, 2006) as meander bends continuously evolve through stages of development at different temporal scales (Keller, 1972; Seminara *et al.*, 2001; Hook & York, 2002; Pišút, 2002; Lanzoni & Seminara, 2006; Schook *et al.*, 2017). Hence, facies models for meandering rivers should consider their rock records as 'snapshots' of a dynamic state, rather than high-sinuosity end-members (Ghinassi & Ielpi, 2017). As a corollary, facies and architec-tural models based on heterolithic expansional point bars are increasingly deemed unreliable for the recognition of well-developed planform sinu-osity in the fluvial rock record. Bridge (1985) orig-inally called attention to the fact that fluvial palaeochannel patterns cannot be reconstructed based on facies analyses unaccompanied by a record of palaeocurrent readings and that observa-tion of extensive outcrops with fully preserved channel-belt deposits would be a necessary pre-condition. Brierley (1989) and Brierley & Hickin (1991), based on evidence from active systems, were among the first to confirm that channel plan-form cannot be inferred based on limited assem-blages of morphostratigraphic units. In his review, Ethridge (2011) warned against standard criteria for interpreting (palaeo)channel planform in the literature, as misconceptions still abound regard-ing the nature and dynamics of channel patterns even in active systems; and the overlap between sedimentological attributes of braided and mean-dering channel fills is greater than commonly

assumed. Hartley *et al.* (2015) applied facies and architectural analysis to large sandstone expo-sures deposited by a regional-scale distributive fluvial system, concluding that refined planform reconstructions are essential for the disambigua-tion of meandering from braided fluvial deposits. Likewise, Ghinassi & Ielpi (2015) described plan-form and cross-section exposures of downstream-migrating point bars characterised by complex internal interstratification, pointing to significant overlap with the depositional models for braided-channel deposits. These results suggest that fluvial sinuosity may be overlooked in the rock record unless rigorous architectural and palaeo-flow analysis is applied (Hartley *et al.*, 2015).

Nonetheless, semi-quantitative analyses of sin-uous fluvial systems demonstrated apparent secu-lar evolutionary trends of fluvial style in their relative abundance and their outcomes are widely debated (Davies & Gibling, 2010; Santos *et al.*, 2017a,b; Davies *et al.*, 2017). As such, establishing analytical approaches to elucidate relative degrees of sinuosity is particularly critical for rivers that flowed across continents devoid of evolved vascu-lar plants. In fact, a vast amount of experimental and field-based work indicates that vegetative rooting promotes sediment baffling, bank stabili-sation (Schumm, 1968; Smith, 1976; Cotter, 1978; Fielding *et al.*, 1997) and retention of pedogenic mud, thus providing bank cohesion (Corenblit & Steiger, 2009). These factors are commonly thought to demonstrate that the presence of vege-tation precludes channel braiding and favours the development of sinuosity (Tal & Paola, 2007). Transposing such information to ancient sedimen-tary environments, a widely held notion main-tains that pre-vegetation channels (i.e. older than late Ordovician; Gibling *et al.*, 2014) were uniquely shallow and wide, with a low-sinuosity planform (Schumm, 1968; Eriksson *et al.*, 1998, 2006; Bose *et al.*, 2012). This current paradigm corroborates the projection of early Palaeozoic outcrop evidence into deeper geological time, epitomised by the notion that Cambrian and Precambrian fluvial systems shared a sheet-braided style (Cotter, 1978). In this context, some authors theorised that pre-Devonian meanders would have been, at best, exceedingly rare (Davies & Gibling, 2010) or highly unstable and short-lived (cf. Santos & Owen, 2016). This assumption is potentially undermined by: (i) direct outcrop evidence of fluvial sinuosity in the Precambrian rock record (Long, 1978, 2011; Sarkar *et al.*, 2012;

Went, 2016; and case studies discussed in this paper), including the first documentation of Precambrian inclined heterolithic stratification within point bars (Santos & Owen, 2016); (ii) the occurrence of modern sinuous rivers flowing through landscapes with scarce to no vegetation (Matsubara *et al.*, 2015; Ielpi, 2016, 2017; Santos & Owen, 2016; Santos *et al.*, 2017a); (iii) flume experiments where sinuous channels are dynamically maintained in the absence of vegetation (Peakall *et al.*, 2007; van de Lageweg *et al.*, 2014), or where vegetation itself promotes braiding (Coulthard, 2005); and on a broader perspective, (iv) the occurrence of sinuous fluvial planforms on unvegetated extra-terrestrial surfaces (Howard *et al.*, 2005; Irwin *et al.*, 2005; Lorenz *et al.*, 2008).

These apparent contrasts in the analysis of pre-vegetation fluvial sinuosity can be reconciled with stricter comparisons between the flow configuration and accretionary patterns of modern and pre-vegetation channels. This research discusses the morphodynamics of modern and Precambrian sinuous-channel bends, offering parallels between their flow structure and accretionary patterns. Examples of Precambrian (Proterozoic) fluvial sandstones deposited within the Laurentian Craton between 1.6 and 0.7 Ga bear evidence of lateral accretion at different stages of meander-bend development (cf. Lanzoni & Seminara, 2006; Schook *et al.*, 2017), from incipient to mature. Through a combination of remote sensing, outcrop and 3D-photogrammetric observations, the authors aim to: (i) review some of the geometric relationships that allow for the quantification of fluvial sinuosity in the rock record; (ii) provide an integrated discussion on the occurrence and degrees of fluvial sinuosity in pre-vegetation settings from selected stratigraphic units over a range of palaeogeographic and basin settings; and (iii) reappraise the range of geological time during which well-developed fluvial sinuosity is expected to have occurred.

RATIONALE AND METHODS

The relationships between flow and bank accretion in a channel bend can be related to channel geometry (Miall, 1994, 1996; Long, 2006, 2011), irrespective of geological time, sediment grade and vegetation cover. Self-sustained lateral accretion is paramount for the development of planform sinuosity and is, by definition, referred to

the growth of a channel bank in a direction sub-orthogonal to flow (Thomson, 1876; Allen, 1965). This concept applies to bends occurring in both single-channel and multiple-channel systems. In the upstream part of a bend, water shoaling over a bank-attached bar and deepening over the pool lead to flow deflection towards the outer bank (Dietrich & Smith, 1983; Frothingham & Rhoads, 2003; Kasvi *et al.*, 2013). There, water super-elevation caused by centrifugal force induces a cross-channel pressure gradient at depth towards the inner bank (Bathurst *et al.*, 1977). A helicoidal flow develops just downstream of the bend-apex zone, where a circulation cell, secondary to the main flow direction, produces an asymmetric channel profile, with erosion along the outer bank and concomitant sediment transport towards the inner bank (Bluck, 1971; Fig. 1A). Up-bar climbing, cross-stratification verging in a direction oblique to the channel axis is a distinct signature of this helicoidal pattern of circulation in highly sinuous channel bends (Allen, 1965; Jackson, 1975, 1976). During formative-discharge events, stepwise outbuilding of sediment from the inner bend generates bar scrolls, the lateral stacking of which produces a distinctive bar-top ridge-and-swale topography with convex planform with respect to the inner bank (Nanson, 1980; Gibling & Rust, 1993; Ielpi & Ghinassi, 2014; Fig. 1A). Ridge-and-swale topography may nucleate from obstacles (including vegetation; Nanson, 1980), or from outsized bedforms (Ielpi & Ghinassi, 2014). Channel-flow impingement at high angles (typically higher than 90°) onto the outer bank downstream may result in hydraulic separation and reverse eddy-currents oriented upstream (Taylor *et al.*, 1971). The latter are associated with the accretion of point-bar-tail deposits with concave geometry with respect to the inner bank (counter-point bars; Smith *et al.*, 2009, 2011; Ielpi & Ghinassi, 2014; Ghinassi *et al.*, 2016). To analyse patterns of flow configuration, sediment distribution and bar accretion in sinuous channel reaches of both modern and ancient systems, the authors refer to the minimum angular value (<180°) between the bottom flow stressing the bar apron (bar flow) and the dip azimuth of bar strata as flow divergence (Δf; Fig. 1A). Flow divergence is an expression of the angular difference between the vector of fluid-flow stress upon the bar and the vector of accretion of resulting strata (i.e. dip-azimuth of the bar beds) (Fig. 1C and D); and can be employed to quantify the degree of channel-flow

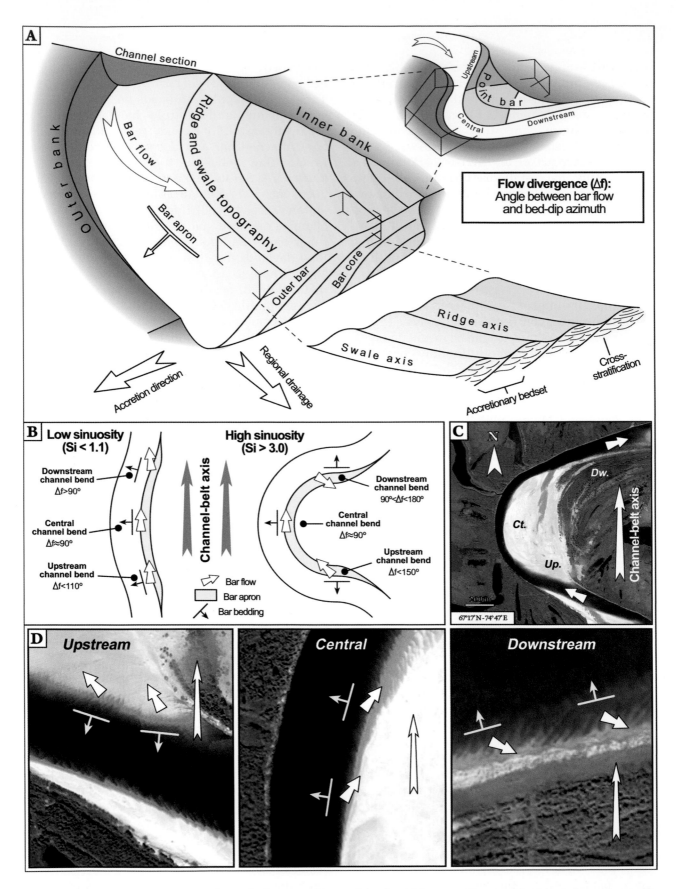

Fig. 1. Schematic representation and standard nomenclature of the fluvial features discussed in this study, with specific reference to flow divergence along sinuous channel bends. (A) Flow structure along a meander bend (cf. Jackson 1975, 1976). (B) Theoretical geometric relationships between the flow stressing a bar apron and the local direction of accretion in a low-sinuosity and high-sinuosity channel bend (see text for discussion). (C) Relationships between local flow and direction of stratal accretion in a nearly symmetrical expansional point bar in an unnamed tributary of the Pur River, Siberian Hills, Russia. (D) Insets of (C), showing higher flow divergence in the upstream and downstream segments of the channel bend.

separation in relation to planform sinuosity (Jablonski, 2012). Using remotely sensed imagery of modern rivers, the direction of migration of bedforms found in the middle-upper part of the bar can be approximated using the attitude and convexity of their crests as reference, while the vector of accretion of bar strata can be approximated to the strike of the bar edge itself. Remote-sensing approaches of this kind are to some extent imprecise given the nature of the approximations presented above, yet this limitation can be mitigated with collection of flow-divergence values from a large number of natural channel bends. On the other hand, flow divergence in the rock record can be estimated from any outcrop that exposes cross-bedded sets bounded by accretionary surfaces, although with different degrees of confidence depending on quality of exposures and quantity of measured palaeoflow and accretion indicators.

The present research is based on remote-sensing data of modern river channels and on a review and discussion of Proterozoic outcrop data. Remote sensing of modern rivers is based on open-access GeoEye-1™ satellite imagery with 60 cm of ground resolution, acquired in 2012–2016. The selected images were acquired from a near-nadir perspective with 0 to 15% of cloud coverage and are representative of flood stages when clear water allows examination of bedforms in critical areas such as near bar flanks (Fig. 1C and D). Depending on the amount of tannins and suspended load carried by the river, this approach can yield reliable results for channel tracts at least 5 m deep (Ielpi, 2016). Notably, this approach cannot be used to fully reconstruct channel-flow structure, although it allows to reliably depict bar-flow patterns along channel flanks. Proterozoic outcrop data was selected from five sedimentary units visited by the authors and is based on facies analysis and line-drawings of natural exposures complemented by directional indicators of macroform accretion and local palaeoflow (Jackson, 1976; Allen, 1982; Miall, 1994). Palaeoflow indicators were collected based on the azimuth of cross-strata recording accretion and migration of dunes. Directional indicators are shown in rose diagrams with average vector and 95%-confidence interval reported and are classified based on their spatial domains, including: bar cores, defined by the lower and bank-ward bar strata related to early stages of bar build-up and outgrowth; outer bars, defined by upper and channel-ward strata related to mature

stages of bar build-up and outgrowth; and channel fills (Fig. 1A). Full details for each stratigraphic unit, including basin setting and overall sedimentary architecture, are available elsewhere (Rainbird, 1992; Ielpi & Ghinassi, 2015, Ghinassi & Ielpi, 2017; Ielpi & Rainbird, 2015, 2016a; and Ielpi *et al.*, 2016). In some cases, outcrop data were complemented by photogrammetric models built using Agisoft™ PhotoScan and georeferenced in Midland Valley™ Move.

FLOW DIVERGENCE IN MODERN RIVERS

In an ideal low-sinuosity channel bend with limited curvature and wide radius, flow divergence is supposed to maintain values close to 90° throughout entire bends, indicating a quasi-perpendicular relationship between local flow structure along the channel bend and the direction of lateral stratal accretion (Fig. 1B). Conversely, in a markedly sinuous channel bend, flow divergence is predicted to attain higher values in the upstream and downstream sectors of the bend (cf. Long, 2006, 2011; Ielpi & Ghinassi, 2014) in response to bankward-directed flow and helicoidal circulation, respectively (Fig. 1B to D). In the central sector of the bend, the bar flow instead trails near-perpendicular to the bedding azimuth (cf. Miall, 1994), resulting in a flow divergence of ~90° (Fig. 1C and D). To corroborate the relationship between fluvial sinuosity and flow divergence irrespective of vegetation cover in natural, active channel bends, the authors report values of sinuosity index and maximum flow divergence from 104 bends of modern rivers characterised by different degrees of sinuosity and vegetation cover (Figs 2, 3 and 4; Table 1). Considered examples include the virtually unvegetated Rio Negro (Lençóis Maranhenses, northern Brazil; Fig. 4B), the sparsely grass-vegetated Fossálar River (southern Iceland; Fig. 4C) and both expansional and downstream-migrational meandering rivers with both grassland and arborescent vegetation established on their alluvial plains (including the Fort Nelson River, Alberta, Canada; an unnamed tributary of the Pur River, Siberian Hills, Russia; the Beaver River, Alberta Canada; and the Powder River, Montana, USA; Fig. 4D to G). The active state of meanders and associated accretionary banks along these rivers is demonstrated by the heterogeneity of surface texture and vegetation cover in proximity of the

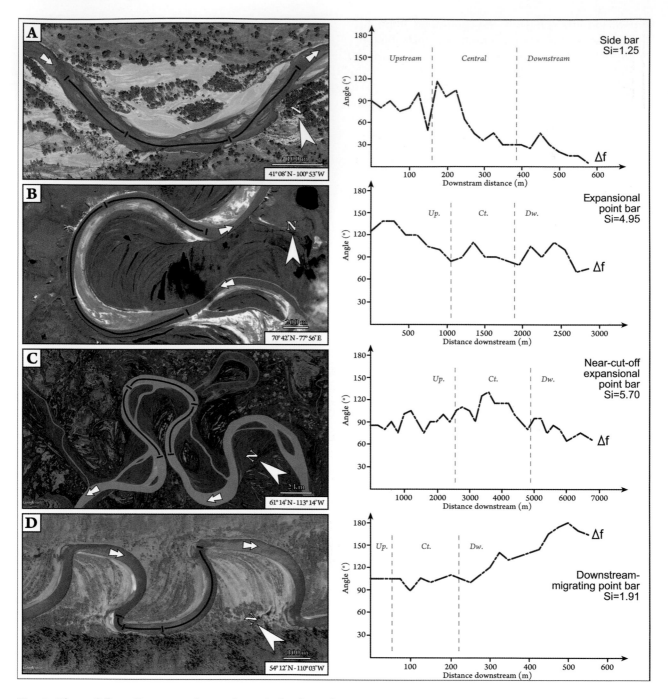

Fig. 2. Plots of flow divergence for modern single-channel systems. Upstream, central and downstream segments are defined based on the location of the channel pool facing the bend apex (cf. Smith *et al.*, 2011). Note higher flow divergence in the downstream segment of the channel bend in relation to stronger helicoidal circulation and flow separation. Examples include: (A) a side bar in a low-sinuosity bend of the Platte River, Nebraska, USA; (B) and expansional point bar with slight downstream rotation, tributary of the Pur River, Siberian hills, Russia; (C) a near-cut-off expansional point bar in the Slave River, Northwest Territories, Canada; and (D) a downstream-migrating point bar of the Beaver River, Alberta, Canada.

channels and by the variable width of channel bends, a trait usually associated with stages of active evolution of sinuous channel tracts (Luchi *et al.*, 2010).

Sinuosity index and maximum flow divergence show positive correlation ($r = 0.49$; $p < 0.01$; Fig. 4A), although with a significant scatter from individual rivers and the dataset as a whole. Such

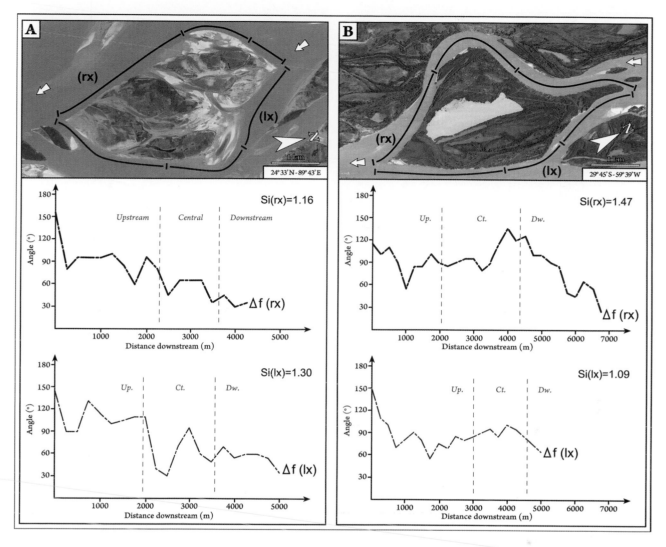

Fig. 3. Plots of flow divergence for modern multiple-channel systems. Upstream, central and downstream segments are defined based on the location of the channel pool facing the bend apex. Note comparatively lower flow divergence compared to the examples in Fig. 3. Examples include: (A) a mid-channel bar with low-curvature flanks in the Brahmaputra River main channel belt (Bangladesh); and (B) a mid-channel bar sided by a high-curvature (the river's right-hand) channel bend, Paraná River, Argentina.

scattering of values testifies that, although flow divergence may be represented in idealised geometrical models (Fig. 1B), the flow structure and accretion patterns of natural bends are expected to differ from such models (Fig. 1C and D), sometimes significantly. Basic reasons for this are that relationships between water flow and sediment movement as well as overall point-bar morphology vary with channel cross-sectional shape (Dietrich, 1987), sediment texture (Kawai & Julien, 1996) and hydrograph stage (Bennett *et al.*, 1998; Hooke, 2008). There is thus a greater variability of relations between local flow trends and patterns of stratal accretion than commonly assumed, both in

time and along a single river channel. Nonetheless, additional examples from modern single-channel and multiple-channel systems (Figs 2 and 3) demonstrate that, taken individually, channel patterns exhibit predictable flow divergence dependent on sinuosity. An example from the Platte River (Nebraska, US; Fig. 2A) depicts a bank-attached bar in a low-intermediate sinuosity bend. Flow divergence attains the highest values along the mid-upstream channel bend (up to 120°) and quickly diminishes to less than 50° along the central and downstream bend. Higher flow divergence in the upstream sector of the bar suggests that sediment is distributed over the bar top. Low flow

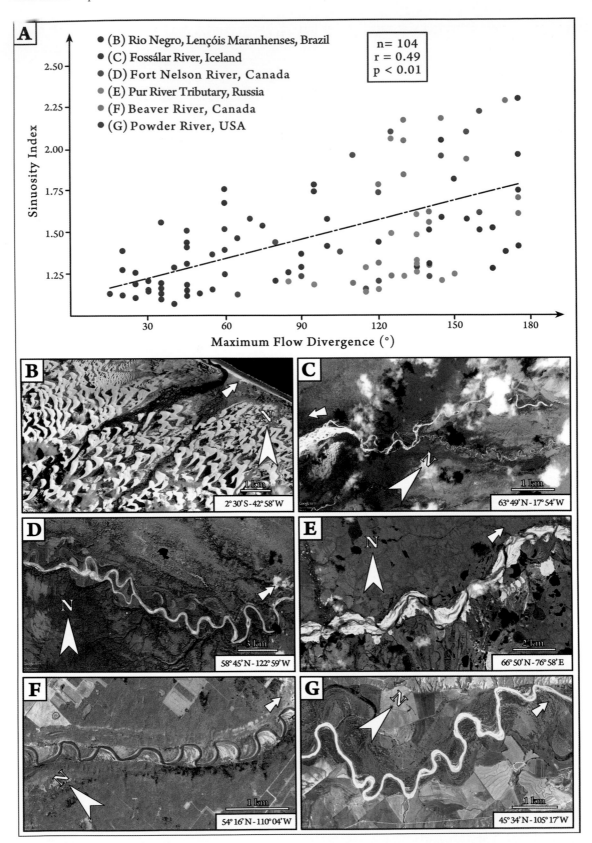

Fig. 4. Relationships between flow divergence and sinuosity. (A) Plot of maximum flow divergence against sinuosity for 104 channel bends contained in modern rivers with different degrees of vegetation cover on their alluvial plains, including: (B) Río Negro, Lençóis Maranhenses, Brazil; (C) Fossálar River, Iceland; (D) Fort Nelson River, Alberta, Canada; (E) tributary of the Pur River, Siberian Hills, Russia; (F) Beaver River, Alberta, Canada; and (G) Powder River, Montana, USA.

Table 1. Summary of the relationships between maximum flow divergence and sinuosity from 104 channel bends selected over a range of vegetation covers. Data is also graphically reported in Fig. 4.

River	Coordinates	Type and approximate cover of vegetation	Range of max. flow variance for individual bends	Range of sinuosity index
Rio Negro, Brazil	From 2°31'S−43°1'W to 2°29'S−42°58'W	None	15°−90°	1.10−1.75
Fossálar River, Iceland	From 63°51'N−51°1'W to 63°48'N−17°55'W	Grassland only, 60−80%	20°−160°	1.06−1.77
Powder River, U.S.A.	From 45°33'N−105°18'W to 45°34'N−105°17'W	Both arborescent and grassland, >80%	110°−175°	1.19−2.28
Beaver River, Canada	From 54°16'N−110°06'W to 54°15'N−110°02'W	Both arborescent and grassland, >80%	135°−175°	1.27−2.29
Fort Nelson River, Canada	From 58°47'N−123°10'W to 58°46'N−122°44'W	Arborescent only, >80%	65°−160°	1.12−3.73
Pur River tributary, Russia	From 66°49'N−76°49'E to 66°55'N−77°15'E	Both arborescent and grassland, 40−80%	85°−150°	1.13−2.05

divergence in the downstream bar is probably related to down-flow accretion in absence of helicoidal circulation. Examples of expansional meanders are reported from a left-hand tributary of the Pur River (Siberian hills, Russia; Fig. 2B) and from the Slave River (Northwest Territories, Canada; Fig. 2C). In both cases, flow divergence ranges from 80° to 140°, with higher values observed downstream of the impingement point of channel flow against the outer bank (Taylor *et al.*, 1971). The meandering Beaver River (Alberta, Canada) exhibits downstream point-bar migration associated with robust flow separation (Fig. 2D). Notably, flow divergence increases downstream as the helicoidal circulation strengthens, and attains limit values of 180° along the concave point-bar tail, where current deflection operated by channel-flow separation becomes dominant. Examples from mid-channel bars in multiple-channel systems with low to intermediate sinuosity (sinuosity index < 1.5; Brahmaputra River, Bangladesh, Fig. 3A; and Paraná River, Argentina, Fig. 3B) likewise show overall decreasing values of flow divergence in channel bends with weak helicoidal circulation and no significant flow separation. Only in channel bends with sinuosity index approximating 1.5 (e.g. the river's right-hand side of mid-channel bar in Fig. 3B) is helicoidal circulation strong enough to produce a spike in flow divergence along the tract with highest curvature. All of the illustrated examples refer to hydrographic stages where clear-water allows for the remote observation of bedforms and unit bars. However, the modulation of channel hydrographic levels on flow divergence has also to be considered.

Fig. 5 illustrates the variability of flow divergence in response to flood-cycle modulation, with five ideal stages identified, being respectively: 1) rising, low-level water; 2) rising, mid-level water; 3) near-peak flood, high-level water; 4) falling, mid-level water; and 5) falling, low-level water. High hydrographic levels are invariably associated with a decrease in channel-flow sinuosity, as floodwater overpasses bars instead of following curvilinear trajectories around them (Fig. 5B; Stage 3 of Fig. 5D). Straighter flow trajectories are in turn accompanied by a downstream shift in flow impingement against outer banks, with consequent down-channel migration of the point of onset of helicoidal circulation (Kasvi *et al.*, 2013; Fig. 5A, B and C). Therefore, flow divergence is supposed to reach a maximum in the upstream sectors of channel bends at peak stage (Stage 3 of Fig. 5D), to later diminish during waning flood stages (Stage 4 of Fig. 5D). Conversely, flow divergence in the downstream portion of channel bends reaches a minimum at the peak-flood stage and then increases as channel hydrographs decline during the following waning flood stages. Lateral accretion requires by definition high flow divergence but also sufficient flow strength to allow for bedform and unit-bar accretion (Stage 2 and 4 of Fig. 5D). While lateral accretion may take place during rising flood-water stages, its preservation potential will be nonetheless limited since it precedes the erosion-dominated and bypass-dominated peak flood stages. It follows that the ideal conditions for development of helicoidal circulation (and thus self-sustained lateral accretion and planform sinuosity) will take place during mid-waning flood stages, when bars are not entirely

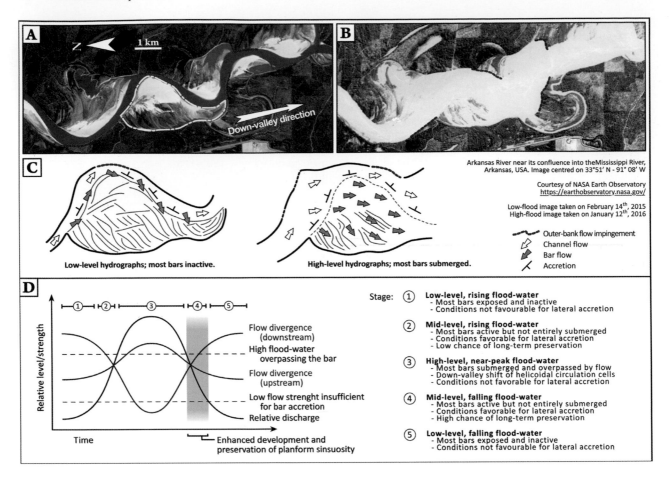

Fig. 5. Flood-cycle modulation of flow divergence along sinuous channel bends. (A and B) Satellite imagery displaying the Arkansas River (Arkansas, US) at low and high flood stages, (A) and (B), respectively. Note down-valley shifts of flow-impingement points onto outer banks at high-flood stages. (C) Interpretive sketch of a selected point bar – highlighted in yellow in (A) – showing different flow structure dependent on channel hydrographic levels. (D) Conceptual model (no temporal or spatial scales implied) showing the variation of flow divergence in the upstream and downstream segments of a sinuous channel bend during a hypothetical flood cycle with sinusoidal hydrograph. Stages 1 to 5 are defined based on flow strengths and emergence of bars. Note preferential development and preservation of lateral accretion and planform sinuosity in the mid-flood to post-flood stages.

overpassed by floodwaters, but at times of still sustained flow strength (Stage 4 of Fig. 5D).

THE PROTEROZOIC FLUVIAL RECORD OF LAURENTIA

Laurentia emerged as a coherent crustal block in the late Palaeoproterozoic (Hoffman, 1988; Zhao *et al.*, 2002) and is presently represented by the Canadian Shield and cratonic slivers recognised in Greenland and Scotland (Bleeker, 2003; Fig. 6A). This cratonic assemblage is covered by a number of extensive and multi-kilometre-thick Proterozoic sandstones, preserved within foreland, intracratonic sag, and rift basins (Rainbird

et al., 1996; Stewart, 2002; Davidson, 2008; Allen *et al.*, 2015). From 2.1 to 0.5 Ga, Laurentia witnessed two full Wilson cycles related to the assembly, outgrowth and breakup of supercontinents Nuna and, later, Rodinia (Li *et al.*, 2008; Pehrsson *et al.*, 2016). Prominent records of supercontinent assembly are respectively the 2.1 to 1.8 Ga Trans-Hudson Orogeny and 1.2 to 1.0 Ga Grenville Orogeny. The erosional unroofing of these orogens had profound influence on the fluvial systems draining Laurentia during the following phases of basin development related to crustal loading, later thermal insulation, and eventual rifting (Davidson, 2008; Rainbird & Young, 2009; Armitage & Allen, 2010). Young (1978) was among the first to theorise that the ancestral North American landmass

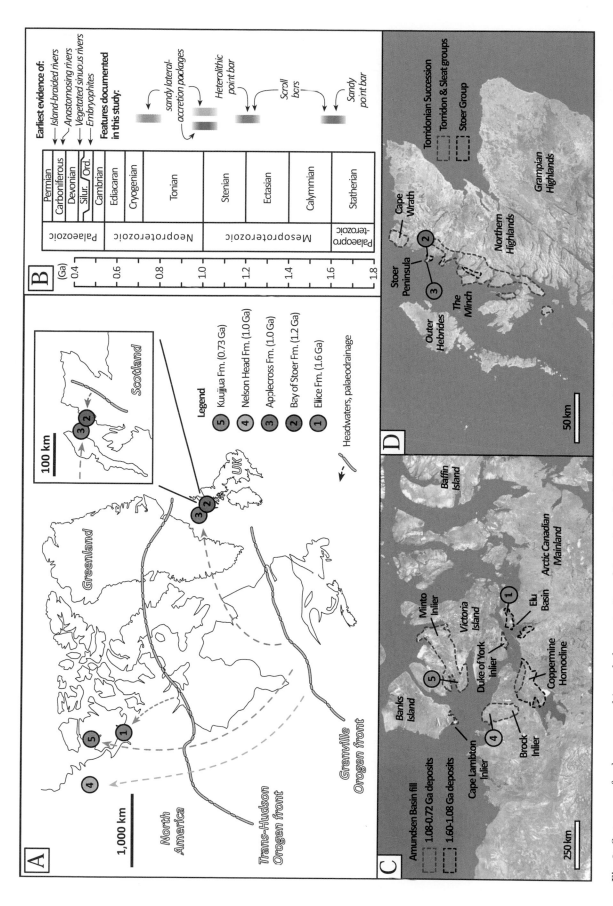

Fig. 6. Summary of palaeogeographic and chronostratigraphic settings for the Proterozoic rock units considered in this study. (A) Modern coastlines restored to the approximate configuration of the Laurentian Craton prior to the opening of the North Atlantic Ocean. The position of the selected rock units and the approximate path defined by their source drainage are reported. See full details in Rainbird (1992), Ielpi & Ghinassi (2015, 2017), Ielpi & Rainbird (2015, 2016a) and Ielpi *et al.* (2016). (B) Chronostratigraphic chart showing the temporal range covered by the selected rock units and distinctive features recorded therein (Figs 6 to 11). Key events for the co-evolution of land plants and fluvial style throughout the Palaeozoic are also reported for reference (cf. Gibling *et al.*, 2014). (C) Satellite inset of the region comprising Amundsen Basin, northern Canada, with location of the main Meso inliers, Neoproterozoic inliers and basins reported (cf. Rainbird *et al.*, 1996). (D) Satellite inset of northern Scotland (UK), with location of the main exposures of the Torridonian succession (cf. Stewart, 2002).

Table 2. Aspects of the Proterozoic rock units considered in this study. Sedimentological details are reported in Figs 7 to 12.

Rock Unit, Group	Approx. age	Basin type	Drainage pattern and distance from headwaters	Features documented
Ellice Formation, Elu Basin Group	1.6 Ga	Intracratonic sag	Away from orogenic belt, ~1000 km	Sandy point bars, scroll bars
Bay of Stoer Formation, Stoer Group	1.2 Ga	Rift	Basin-axial with lateral tributaries, ~60 km	Heterolithic point bar with scroll topography
Applecross Formation, Torridon Group	1.0 Ga	Foreland	Parallel to orogenic belt with lateral tributaries, ~100–2500 km	Upstream-lateral accretion packages
Nelson Head Formation, Rae Group	1.0 Ga	Intracratonic sag	Away from orogenic belt, ~3000 km	Downstream-lateral accretion packages
Kuujjua Formation, Shaler Supergroup	0.7 Ga	Intracratonic sag	Away from orogenic belt with local tributaries, <3000 km (?)	Lateral-accretion packages

had been at times transected by NW-draining rivers originated from large orogenic belts. Refined geochronology of detrital zircons from sandstones (Rainbird *et al.*, 1992, 2012, 2017; Rayner & Rainbird, 2013) and intercalated mudrocks (Rainbird *et al.*, 1997) offered independent tests and corroborated the hypothesis of craton-scale drainages flowing towards the NW and established in the aftermath of the Trans-Hudson and Grenville orogenies (Fig. 6). This model has also been applied to fluvial units preserved atop cratonic slivers rifted from the Laurentian core during the Phanerozoic – notably the Torridon Group of Scotland, for which an increasing amount of evidence points to a mature foreland drainage axial to the Grenville front (Nicholson, 1993; Rainbird *et al.*, 2001; Krabbendam *et al.*, 2008, 2017).

To represent the range of depositional settings recorded in the Proterozoic fluvial rocks of Laurentia, five stratigraphic units varying in age from 1.6 to 0.7 Ga are discussed here (Fig. 6B; Table 2): (i) the Ellice Formation of Elu Basin (Nunavut, Canada); (ii) the Bay of Stoer Formation and (iii) Applecross Formation, respectively part of the Stoer and Torridon groups ('Torridonian' succession) of Scotland, UK; and the (iv) Nelson Head Formation and (v) Kuujjua Formation of the Amundsen Basin (Northwest Territories, Canada). Relevant information on basin setting and alluvial architecture is provided for each unit, although full details on their sedimentology, basin setting and depositional evolution are available elsewhere (Rainbird, 1992; Ielpi & Ghinassi, 2015; Ielpi & Rainbird, 2015, 2016a; Ielpi *et al.*, 2016; Ghinassi & Ielpi, 2017). The authors describe selected fluvial forms related to a range of channel sinuosity and stages of meander-bend development, from incipient (e.g. lateral-accretion

packages and scroll bars in multiple-channel systems) to mature (e.g. point bars in single-channel systems) and compare their inferred patterns of palaeoflow and accretion to modern fluvial morphodynamics, with a particular focus on flow divergence.

Ellice Formation, Elu Basin, Nunavut (Canada)

The Ellice Formation includes nearshore-marine sandstones and dolomitic mudstones dated at ca. 1.6 Ga (Bowring & Grotzinger, 1992). Its type area is the Elu Basin, Nunavut, Canada (Fig. 6A and C), representing the erosional remnant of an intracratonic basin formed in response to thermal insulation of Nuna (Armitage & Allen, 2010) and sourced by the unroofing and erosion of the Trans-Hudson Orogen to the south-east (Rainbird & Young, 2009; Rainbird *et al.*, 2014). The Ellice Formation sits unconformably on older continental deposits (Ielpi & Rainbird, 2016b), or directly on Archean basement, and is overlain by shallow-water carbonate rocks of the Parry Bay Formation (Campbell, 1979).

Typical exposures of the Ellice Formation comprise low-lying steps (<5 m-tall) connected by horizontal platforms where the planform geometries of channel bodies can be appreciated (Fig. 7A and B). The lower portion of the formation is composed of a thin (<50 m) gravelly sandstone, interpreted to represent mixed-load fluvial deposits, overlain by ca. 400 m of quartzarenite interpreted to represent fluvial and subordinate aeolian deposits (Ielpi & Rainbird, 2015). Nearshore-marine quartzarenite, greywacke and dolomitic mudstone are predominant in the upper Ellice Formation. Analysis of depositional architecture in the fluvial tracts reveals the occurrence of amalgamated remnants of fluvial bars sided in places by sandstone-channel fills. These deposits probably composed a wedge of unconfined, low-gradient sandy alluvium sloping towards a marine shoreline (Campbell, 1979; Ielpi & Rainbird, 2015).

Sandy point-bar deposits

Sandy channel-belt deposits are exposed at 68°30'N to 105°42'W, between underlying Archean rocks and a cross-cutting Franklin (720 Ma) diabase dyke (Fig. 7A and B). The basal nonconformity is characterised by up to 50 m of erosional topography over 10 km along strike. Channel belts are floored by a flat erosional surface overlain by a gravelly lag (Fig. 7H) and include two architectural elements: wedge-shaped bars, <3 m-thick and <30 m-wide along strike; and lens-shaped channel fills, with erosional, concave-up lower surfaces and flat upper surfaces, <3 m-thick and <25 m-wide in a direction normal to their depositional axes (Fig. 7C). Both wedge-shaped bars and lens-shaped channel fills consist of compound cross-strata (Fig. 7C and F; cf. Haszeldine, 1983; Miall, 1988) with minor volumes of plane-bedded and ripple-laminated deposits (Fig. 7G). In the lowermost and better exposed channel-belt deposits, a bar displays growth in the SW-ward direction (Fig. 7D), as indicated by the attitude of accretionary surfaces. The direction of bar growth is normal to the NW-trending channel axis, based on the strike of channel banks exposed in plan-view and on palaeoflow indicators within a gravelly lag deposit (Fig. 7D and H). Palaeoflow data from channel fills overall indicate WNW-ward transport, while dispersed and polymodal indicators collected in the bar core and outer portion indicate on average NE-ward and ENE-ward transport, respectively.

Interpretation: Bar accretion normal to the channel axis indicates lateral accretion and specifically bend expansion (Fig. 7E). Relationships between the channel axis (consistent with the channel-lag palaeoflow indicators) and accretion vectors are consistent with a central portion of the bar. The southward palaeoflow rotation observed between the indicators collected in the channel lag and in the channel fill also is consistent with a slight bend rotation in an upstream direction during expansion. Bar palaeoflows are highly dispersed and may indicate interference of flood-waters re-entering the channel or over-passing the bar in response to over-spilling (Loveless *et al.*, 2000; Wormleaton *et al.*, 2004;

Shiono & Muto, 1998). Palaeoflow-vector averages and 95%-confidence arcs indicate that flow divergence ranged from 100° to 180° during bar-core accretion (Fig. 7E), consistent with the development of strong helicoidal circulation and flow separation due to impingement at high angle onto outer banks (Taylor *et al.*, 1971). Strong helicoidal re-circulation and flow separation are enhanced in well-entrenched or topographically confined channels (Fig. 3B and D), as expected for channel belts contained within prominent basement topography (cf. Long, 2017). Relationships between outer-bar palaeoflow and accretion suggest that flow divergence diminished to values of 50° to 145° in the later stages of bar growth, suggesting a weakening in helicoidal circulation. This was possibly related to reduced spatial confinement or locally reduced bend curvature after expansion.

Sandy scroll-bar deposits

Deposits of an amalgamated sandy channel belt are exposed in the central Elu Basin at 68°26'N to 106°12'W (Fig. 6C) and capped by Franklin diabase sills (Fig. 8A). These deposits lie roughly 350 m above the nonconformity with the Archean basement. The channel-belt deposits consist of amalgamated, wedge-shaped, bedsets that together compose a sheet-like, cross-stratified sandstone body (Fig. 8B and E). Although the channel-belt floor is not exposed, wedge-shaped bedsets attain a minimum thickness of ca. 2.5 m and are part of a compound bar exposed along a 120 × 70 m wave-cut platform. Both the core and lower-outer portions of the exposed bar yielded focused palaeoflow indicators towards the WSW (Fig. 8D). Within such a compound bar, cross-beds diminish in thickness upward and eventually transition into an overall flat surface with finely preserved depositional topography defined by ridges and swales (Fig. 8C). Ridges are defined by cross-sets prograding towards the swales (Fig. 8C), while swales are defined by symmetric scoop-shaped surfaces floored by thin cross-strata and ripple forms (Fig. 8F). The ridge-and-swale topography becomes less defined – and eventually disappears – towards the palaeo-river's left (the latter defined with respect to the lower-bar palaeoflow indicators; Fig. 8C). Swale axes are strongly oblique to sub-normal to palaeoflow indicators collected in the lower bar (Fig. 8D). Cross-sets coring the ridges display mildly dispersed

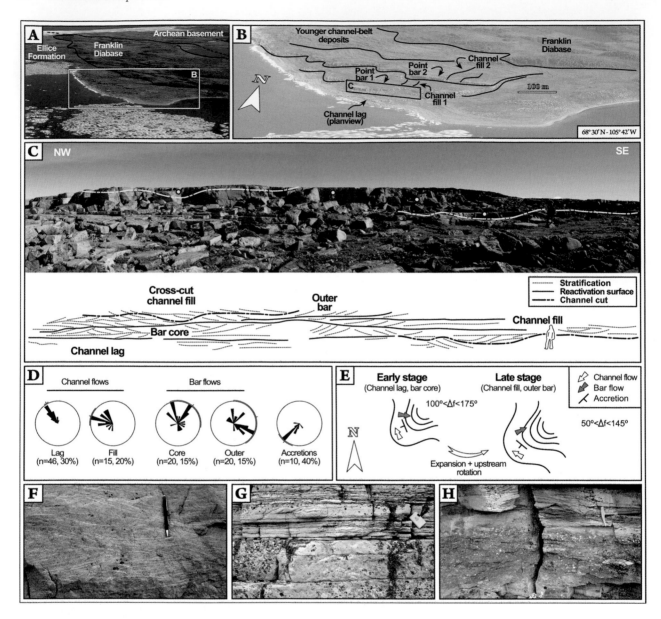

Fig. 7. Field aspects of sandy meandering-channel deposits in the Ellice Formation of Elu Basin, Nunavut (Canada). For scale reference, person in (C) is ca. 1.8 m-tall, pencil in (F) is ca. 10 cm-long, field book in (G) is ca. 20 cm-long and hammer in (H) is ca. 30 cm-long. (A) Aerial view of the study site, located to the west of Itibiak Lake, Elu Inlet. The foreground field of view is ca. 2 km-across. (B) Inset showing an aerial view of the studied channel-belt deposit and its relations to overlying rocks. (C) Field panorama and interpretative line-drawing of a point bar with attached channel fill; location of palaeocurrent and accretion-vector readings are reported with yellow dots. (D) Palaeoflow data for the studied outcrop; average vectors and 95%-confidence arcs are reported. (E) Morphodynamic model for the studied outcrop, with special reference to inferred flow divergence (no scale is intended); see text for discussion. (F to H) Key sedimentological attributes of the Ellice Formation at the site, including: cross-bedded quartzarenite (F); plane-parallel laminated quartzarenite with ripple forms preserved in places (G); and quartz-pebble-rich gravelly lags overlain by cross-bedded quartzarenite (H).

palaeoflow directions with overall westward transport; cross-bed sets flooring the swales show instead a focused SW-ward palaeoflow (Fig. 8D).

Interpretation: The occurrence of amalgamated bedsets composing a compound-bar sheet is suggestive of a mid-channel bar established within a multi-channel systems (Ashworth *et al.*, 2000);

however, no bounding channel fills were observed (Fig. 8G). Cross-sets diminishing in relief and thickness upward are consistent with water shoaling (Cant & Walker, 1978), indicating that the uppermost portion of the exposure preserves a bar-top flat with scroll topography defined by ridges and swales (Bridge *et al.*, 1998; Brookes,

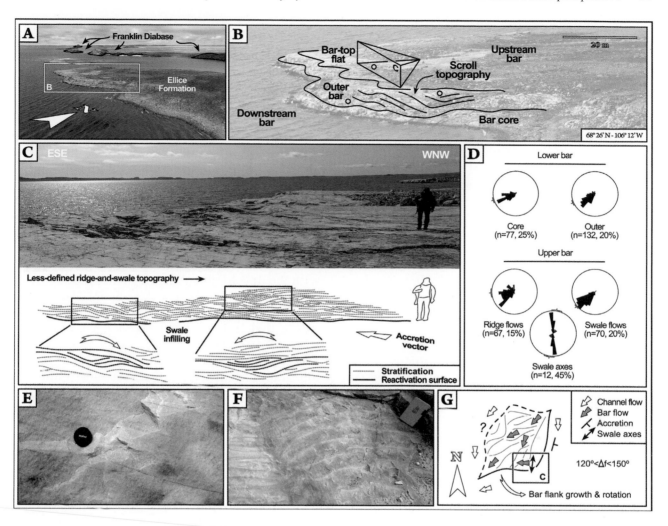

Fig. 8. Field aspects of sandy scroll bars in the Ellice Formation of Elu Basin, Nunavut (Canada). For scale reference, geologist in (C) is ca. 1.9 m-tall, lens cap in (E) is 5 cm-across and field book in (F) is ca. 20 cm-long. (A) Aerial view of the study site, located in central Elu Inlet. The foreground field of view is ca. 300 m-across. (B) Inset showing an aerial view of the studied channel-belt deposit; location of palaeocurrent and accretion-vector readings are reported with yellow dots. (C) Field panorama and interpretative line-drawing of a bar-top flat sided by scroll bars with ridge-and-swale topography. (D) Palaeoflow data for the studied outcrop; average vectors and 95%-confidence arcs are reported. (E and F) Key sedimentological attributes of the Ellice Formation at the site, including cross-bedded quartzarenite (E); and ripple forms preserved in relief within swales (F). (G) Morphodynamic model for the studied outcrop, with special reference to inferred divergence (no scale is intended; see text for discussion).

2003). Scroll bars trend in a direction normal to accretion; the latter, based on the depositional relief preserved by the ridge-and-swale topography, was probably towards the palaeo-river's right, i.e. to ESE (Fig. 8C and G). Bar-core palaeoflow indicators and swale axes define an angular range of 50° to 80°, corroborating the hypothesis of mixed lateral-downstream accretion of a bar flank. It follows that palaeoflow indicators recorded in both ridges and swales on the bar top point to bedform migration up-bar, with flow divergence comprised between 120° and 150° (Fig. 8G). Overall, this scenario implies the rotational expansion of a

bar flank (Fig. 8G), with helicoidal circulation responsible for bedform migration up-bar and bar scrolling generated along a bend apex (Richardson & Thorne, 1998), as in the example of Fig. 4B.

Bay of Stoer Formation, Stoer Group, Scotland (UK)

The Bay of Stoer Formation consists of continental sandstones with intervening mudstones dated at ca. 1.2 Ga (Turnbull *et al.*, 1996; Parnell *et al.*, 2011) and is best exposed along a coastal monocline at Stoer Peninsula (Scottish Highlands,

UK; Fig. 6D). The formation belongs to the Stoer Group, thought to represent the infill of a small (<100 km-wide) rift basin marginal to the Laurentian Craton, sourced by degradation of nearby Archean (Lewisian Gneiss Complex) uplands (Stewart, 2002). Together with the underlying Clachtoll Formation, the Bay of Stoer Formation is underlain by steep basement topography (~300 m over 2 km along strike, locally higher) and encompasses a range of terrestrial deposits including fluvial-aeolian strata (Ielpi et al., 2016), a prominent impact melt sheet (Stac Fada Member; Amor et al., 2007) and an uppermost lacustrine mudstone (Poll a'Mhuilt Member).

The bulk of the Bay of Stoer Formation consists of arkosic-lithic sandstones and wacke arranged in large clusters of fluvial channel bodies sided by gravelly fluvial fans, extra-channel floodbasin deposits and aeolian deposits (Ielpi et al., 2016). Clustered channel bodies point to deposition in confined and high-gradient, low-sinuosity streams. The degree of topographic confinement – and possibly gradients – diminished up-section, a trend inferred from the increasing volume of preserved floodbasin strata composed of tabular mudstones and heterolithic splay sandstones fed by isolated channel bodies.

Heterolithic point-bar deposits

The deposits of a channel body are preserved at 58°12'N to 5°20'W (Fig. 6D), immediately below the Stac Fada Member (Fig. 9). The channel body is < 2 m-thick and exposed along a N-S-oriented section; its lower erosional surface displays ca. 0.5 m of erosional relief over 20 m (Fig. 9A) and its upper surface is flat. The channel body rests on floodplain heterolithics (Fig. 9H) and contains a sandstone-dominated bar with minor mudstone partings (Fig. 9D). Mudstone partings are 1 to 5 cm-thick, up to 8 m continuous along strike and occur in inclined units with undulate geometry. The bar encloses compound cross-bed sets bounded by accretionary surfaces (Fig. 9B), with overall dip to the south-west and defining wedge-shaped accretionary macroforms (Fig. 9C). Accretionary surfaces are inclined yet planar in their lower portion and demonstrate an undulated topography upward. This irregular topography is defined by ridges and swales, the axes of which trend WNW-ward (Fig. 9C). Palaeoflow indicators from channel-lag and bar-core deposits point to SW-ward transport (Fig. 9C), whereas outer-bar deposits yield instead

N-ward palaeoflow direction (Fig. 9C). This geometric relationship is expressed in outcrop by cross-strata that climb atop the undulated accretionary surfaces (Fig. 9D, F and G).

Interpretation: The occurrence of a solitary channel body juxtaposed with floodplain strata is suggestive of a portion of the fluvial system relatively distal to its catchment, feeding the depressed portion of the alluvial plain (cf. Tunbridge, 1984). Palaeocurrent relationships between basal-channel lag and bar-core palaeo-flow indicators and accretion vectors indicate < 20° of flow divergence (Fig. 9C and E), consistent with downstream-growth mechanisms in the early stages of bar accretion (Fig. 3A). Bar outgrowth and build-up are more probably associated with the development of a ridge-and-swale topography, i.e. bar scrolls (Brookes, 2003). Bar scrolls are interpreted to have built in a direction opposite to that indicated by accretion vectors, with values of flow divergence reaching ~ 135° (Fig. 9C and E). Development of the ridge-and-swale topography with high flow divergence indicates enhanced helicoidal circulation during formative flow, in between episodes of lower flow strength when the deposition of intervening mud partings took place. The relationships among scroll axes, accretion vectors and channel-lag palaeoflow are consistent with expansion and downstream-rotation of a heterolithic point bar (Fig. 9E). An increase in bend sinuosity and rotational growth downstream may have been favoured by bank strength provided by muddy floodbasin strata (cf. Brooks, 2003; Labrecque et al., 2011), as documented elsewhere in the Torridonian succession (Santos & Owen, 2016).

Applecross Formation, Torridon Group, Scotland (UK)

The fluvial Applecross Formation is a gravelly sandstone unit dated at ca. 1.0 Ga (Turnbull et al., 1996; Rainbird et al., 2001; Parnell et al., 2011) and forms much of the iconic landscape of the North-western Scottish Highlands (Fig. 6D). The formation is over 3 km-thick and constitutes the bulk of the lower Torridon Group. Classically considered the infill of a rift basin (Williams, 1969; 2001; Stewart, 2002), the Torridon Group has been re-evaluated as the product of axial foreland deposition fed by unroofing of the Grenville Orogen (Nicholson, 1993; Rainbird et al., 2001; Kinnaird et al., 2007; Krabbendam et al., 2017). The

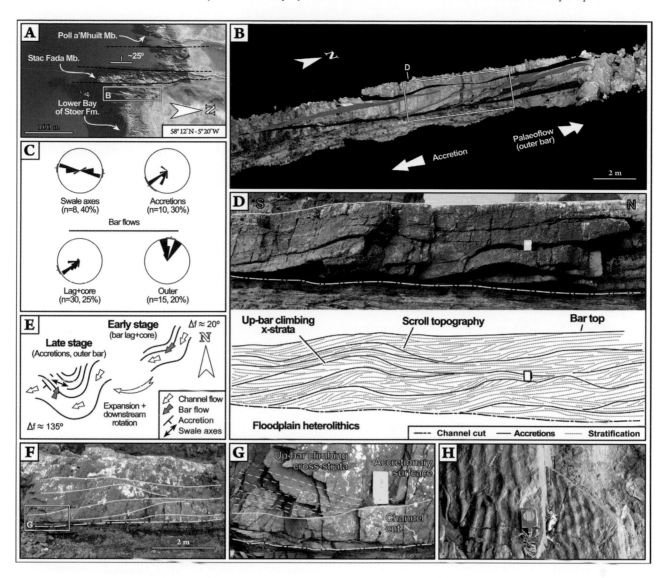

Fig. 9. Field and 3D-photogrammetric aspects of a heterolithic meandering-channel deposit in the Bay of Stoer Formation, Stoer Group, Scotland (UK). For scale reference, field book (D and G) is ca. 20 cm-long; compass (H) is ca. 10 cm-across. (A) Satellite view of the study site, along the southern shore of Stoer Peninsula, Scotland. (B) 3D photogrammetric model of the studied point bar, with highlighted accretionary surfaces; location of inset in (D) is also reported. (C) Palaeoflow data for the studied outcrop; average vectors and 95%-confidence arcs are reported; palaeocurrent and accretion-vector readings were collected in the area line-drawn in (B). (D) Field panorama and interpretative line-drawing of a scroll bars with preserved ridge-and-swale topography in cross-section. Field book for scale is ca. 20 cm-long. (E) Morphodynamic model for the studied outcrop, with special reference to inferred flow divergence (no scale is intended); see text for discussion. (F to H) Key sedimentological attributes of the Bay of Stoer Formation at the site, including: inclined strata pointing to lateral accretion (F); cross-strata climbing atop accretionary surfaces (H); and floodplain strata with preserved wave-ripple forms underneath the point-bar sandstone.

Applecross Formation rests conformably on the Diabaig Formation and unconformably onto older deposits of the Stoer Group or directly on Archean Basement (Stewart, 2002).

At Stoer Peninsula, both vertical and planform exposures are available along cliffs and toed by wave-cut platforms. There, the Applecross Formation is characterised by arkosic sandstone arranged in decametre-scale channel bodies. These channel bodies are bounded by sub-planar erosional surfaces, thought to represent the undercutting and later aggradation of individual channel belts contained within intermontane topography. Channel bodies are composed of three classes of architectural elements: large foreset bars generated in deep, open-water channels; compound-bar sheets, which recordbar braiding and reworking; and isolated channel fills, in places with preserved levees (Ielpi & Ghinassi, 2015).

Sandy upstream-lateral accretion deposits

The deposits of a channel body are exposed at 58°15'N to 5°24'W (storey 1.10 of Ielpi & Ghinassi, 2015; Fig. 6D) and include a lower sandstone overlain by a thin heterolithic sandstone-mudstone association (Fig. 10A). The channel body is < 7.2 m-thick and bounded by truncation surfaces floored by a thin intra-formational lag deposit composed of mud clasts (Fig. 10F). The lower truncation surface slopes towards the SW and displays 4 m of erosional relief over a distance of 45 m; the upper truncation surface is planar (Fig. 10B and C). Internally, the lower sandstone displays compound-cross stratification (Fig. 10G; cf. Haszeldine, 1983; Miall, 1988), defining gently inclined bedsets with attitude similar to the lower channel-body truncation surface (Fig. 10B). Accretionary surfaces bounding individual bedsets point to bar growth toward ENE (Fig. 10D). Palaeoflow indicators from channel-lag deposits and from overlying bedsets of the bar core indicate mildly dispersed SE-ward transport, toward the outer bedsets. Palaeoflow indicators become more uniform and progressively rotate towards the east (Fig. 10D). The uppermost heterolithic beds wedge out towards the NNE (Fig. 10H) and are composed of plane-parallel and rippled deposits (Fig. 10B and C). Palaeoflow indicators therein overall show ENE-ward transport (Fig. 10D).

Interpretation: Bar accretion in a direction strongly oblique to that of channel-body thinning indicates the expansion of a bank-attached bar (Fig. 10E). The thin heterolithic beds preserved at the top of the channel body are interpreted as a remnant of the parent-channel fill after progressive abandonment (Toonen et al., 2012), pointing to regional drainage in a direction opposite to accretion. Likewise, palaeoflow indicators collected in the lower sandstone are strongly oblique or opposite to accretion vectors (Fig. 10D and E), consistent with accretion of an upstream-bar portion (Bridge et al., 1995). The observed palaeoflow rotation from lag and bar-core deposits to outer-bar bedsets is consistent with an increase in flow divergence from ~ 135° to up to 180° (Fig. 10E). During early bend expansion, the channel flow followed a lower-sinuosity trajectory and was probably deflected along the bar, resulting in lower flow divergence and mixed upstream-lateral accretion (cf. Long, 2011). As bend expansion progressed, the channel flow was forced to trail along a higher-sinuosity trajectory (i.e. without significant flow deflection) and impacted a bar apron sloping in the opposite direction; this hypothesis is consistent with more focused palaeoflow readings in the outer macroforms. Occurrence of mud clasts and heterolithic channel fills suggests that muddy floodplain strata may have provided some degree of bank strengthening, as convincingly documented elsewhere in the Applecross Formation (Ielpi & Ghinassi, 2015; Santos & Owen, 2016).

Nelson Head Formation, Amundsen Basin, Northwest Territories (Canada)

The Nelson Head Formation is a fluvial sandstone unit with intervening shallow-marine sandstones and mudstones, dated at ca. 1.0 Ga (Rainbird et al., 2017). The Nelson Head is part of the Rae Group (lower Shaler Supergroup; Rainbird et al., 1996) and is exposed in several structural inliers that define the Amundsen Basin of north-western Canada (Young, 1981), including the Cape Lambton Inlier (type area of Miall, 1976), Minto and Duke of York inliers (Victoria Island) and Brock Inlier (Northwest Territories mainland; Fig. 6C). Correlative exposures comprise the Katherine Group in the Mackenzie Mountains (Northwest Territories; Young et al., 1979; Rainbird et al., 1996; Long & Turner, 2012). Between 1.08 and 0.72 Ga, the Amundsen Basin recorded intracratonic sagging of Rodinia, with detritus delivered from the erosion of the Grenville Orogen roughly 3000 km to the south-east (Cawood et al., 2007; Rainbird et al., 2012). The Nelson Head Formation is sandwiched, paraconformably in places, between carbonates of the Mikkelsen Island Formation (below) and the Aok Formation (above; Rainbird et al., 2015).

The alluvial architecture of the Nelson Head Formation is best appreciated along the Brock River Canyon, where the westward-dipping limb of an open syncline exposes the lower 120 m of the rock unit (Rainbird et al., 2015). There, the Nelson Head Formation is composed of 16 stacked channel bodies capped by a deltaic mudstone-sandstone. Each channel body is 7 to 12 m-thick and bounded by sharp erosional surfaces, recording the undercutting, aggradation and eventual abandonment of channel belts with

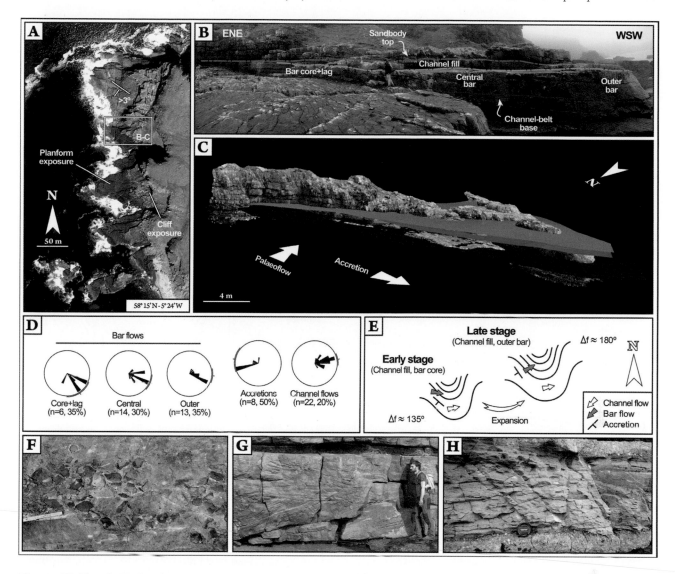

Fig. 10. Field and 3D-photogrammetric aspects of a bank-attached bar displaying evidence of upstream-lateral accretion in the Applecross Formation, Torridon Group, Scotland (UK). For scale reference, geologist in (B) is ca. 2 m-tall, stick in (F) is 10 cm-long, geologists in (G) is ca. 1.8 m-tall and lens cap in (H) is 5 cm-across. (A) Satellite view of the study site, along the western tip of Stoer Peninsula, Scotland. (B and C) Field view (B) and 3D-photogrammetric model of the studied bar and overlying channel-fill remnant, with critical surfaces highlighted. (D) Palaeoflow data for the studied outcrop, see (B) for locations of palaeocurrent and accretion-vector readings; average vectors and 95%-confidence arcs are reported. (E) Morphodynamic model for the studied outcrop, with special reference to inferred flow divergence (no scale is intended); see text for discussion. (F to H) Key sedimentological attributes of the Applecross Formation at the site, including: intra-formational pebble lag flooring the erosional boundaries of channel-belt deposits (F); compound cross-bedded sandstone organised in mutually truncating beds (G); and detail of heterolithic sandstone-mudstone related to channel filling and abandonment (H).

braided to wandering planform and variable degrees of valley confinement (Ielpi & Rainbird, 2016a). Aggradational phases are recorded by quartzarenite arranged in compound cross-bedded sheets, the remnants of amalgamated fluvial bars; abandonment phases are instead recorded by distinctively red mud-rich lenses, which typify channel fills.

Sandy downstream-lateral accretion deposits

A cliff at 69°21'N to 122°56'W exposes 11 stacked channel bodies, some of which preserve well-developed inclined accretion sets (Fig. 11A and B). A prominent example detailed here (channel body no. 7 from the section's base) includes a lower sandstone overlain by a heterolithic sandstone-mudstone association (Fig. 11C). The

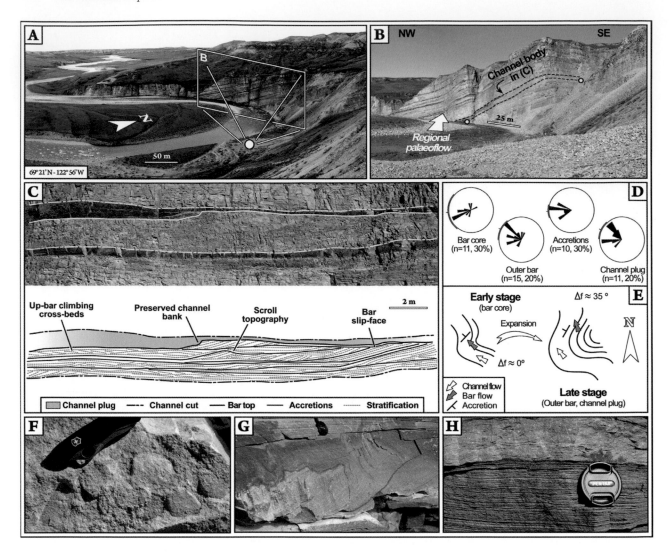

Fig. 11. Field aspects of downstream-lateral accretionary sets in the Nelson Head Formation, Amundsen Basin, Northwest Territories (Canada). For scale reference, pocket knife in (F) is ca. 10 cm-long and lens cap in (G, H) is 5 cm-across. (A) Panoramic view of the final Brock River gorge, with the Nelson Head Head Formation exposed along natural cliffs. (B) Inset of (A), showing the location of the studied channel-belt deposit; yellow dots indicate where palaeocurrent and accretion-vector readings were collected. (C) Field panorama and interpretative line-drawing of a bar composed of downstream-lateral accretion sets, in places with scroll bars preserved and an attached abandoned-channel fill. (D) Palaeoflow data for the studied outcrop; average vectors and 95%-confidence arcs are reported. (E) Morphodynamic model for the studied outcrop, with special reference to inferred flow divergence (no scale is intended); see text for discussion. (F to H) Key sedimentological attributes of the Nelson Head Formation at the site, including: pebble-lag deposit flooring erosional channel-belt bounding surfaces (F); cross-bedded sandstone, in this case with tangential bottom-set (G); and detail of heterolithic sandstone-mudstone related to channel filling and abandonment (H).

channel body is ~ 4 m-thick, bounded by a lower erosional surface sloping towards the SW and by an upper sub-planar erosional surface (Fig. 11B and C). Both bounding surfaces are overlain by a gravel lag (Fig. 11 F). Within the channel body, the lower sandstone consists of a bar form with compound cross-bed sets and preserved slip-faces (Fig. 11G) bounded by accretionary surfaces. Accretionary surfaces overall dip towards the

west (Fig. 11D) and define inclined bedsets with tangential terminations (Fig. 10C). Palaeoflow indicators from the bar core are highly dispersed and polymodal, defining an angular range of 235° with an average westerly palaeotransport (Fig. 11D). Palaeoflow indicators from the outer bar are slightly less dispersed, defining an angular range of 180° and overall indicating NW-ward transport (Fig. 11D). The dispersion in bar-palaeoflow indicators can be visually appreciated in Fig. 11C, where cross-beds with foresets dipping both down-bar and up-bar are preserved on either sides of sediment ridges defining locally a scroll topography. The upper heterolithic beds pinch out towards the east, in a direction opposite to the accretion vectors (Fig. 11C). Mildly dispersed palaeoflow indicators in the heterolithic beds display NW-ward transport (Fig. 11D).

Interpretation: Accretionary bedsets pinching out slightly obliquely to the downstream flow direction defined by channel fills (Fig. 11B and C) indicate that the preserved channel body is representative of a downstream and laterally accreting bar active before abandonment of the parent channel (Toonen *et al.*, 2012). The lower sloping surface bounding the channel body is suggestive of a bank-attached bar scenario. The average dip of accretionary surfaces defines flow divergence ranging from 0° to 35° with palaeoflow during early to mature phases of bar growth, respectively (Fig. 11D). Such relationships are in principle consistent with a mildly sinuous channel bend (Figs 2, 3A and 4). However, neither the high dispersion in bar-palaeoflow indicators nor the occurrence of up-bar-climbing bedforms (Fig. 11C) can be explained solely by downstream-directed sediment transport. Along the bar core, palaeoflow with an angular range of 235° can be explained by a combination of: (i) flow deflection from floodwater re-entering the channel after overspill (Gay *et al.*, 1998; Ghinassi *et al.*, 2013); (ii) flow deflection nearby a channel confluence (De Serres *et al.*, 1999); or (iii) eddy recirculation developed locally at the tail-end of a sharp channel bend (Wright & Kaplinski, 2011). Hypotheses (ii) and (iii) are consistent with the build-up of local scroll topography (Fig. 11C) on the bar apron, which can be reconciled with up-bar bedform climbing (Bridge *et al.*, 1998). Palaeoflow dispersion decreases along the outer bar, suggesting that either of the processes above diminished in intensity as the bar apron grew further away from the channel bank.

Kuujjua Formation, Amundsen Basin, Northwest Territories (Canada)

The Kuujjua (pronunciation: *ku-**gu**-uak*) Formation is a ca. 0.7 Ga fluvial sandstone with minor associated ephemeral-lacustrine dolomitic sandstones and mudstones (Rainbird, 1992), exposed in the south-western Minto Inlier, a regional structural window on western Victoria Island (Northwest Territories, Canada Fig. 6A and C; Young *et al.*, 1979). The Kuujjua is the uppermost formation of the Shaler Supergroup, the fill of the intracratonic Amundsen Basin (Young, 1981; Rainbird *et al.*, 1996). Based on detrital-zircon data, Rainbird *et al.* (1992) suggested that the Kuujjua Formation represents the fringe of a pan-continental drainage sourced from the Grenville Orogen, analogue to the Nelson Head Formation (cf. Rainbird *et al.*, 2017). However, a time gap of ~300 My between the Grenville Orogeny and the deposition of the Kuujjua Formation suggests that recycling of proximal intra-basinal detritus might have been important (cf. Rayner & Rainbird, 2013). The Kuujjua Formation conformably overlies nearshore-marine clastic, carbonate and evaporite rocks of the Kilian Formation and is capped by the ca. 0.72 Ga flood basalts of the Natkusiak Formation (Heaman *et al.*, 1992). Soft-sediment deformation and peperitic textures suggest that Kuujjua deposition was penecontemporaneous with volcanism (Rainbird, 1993).

The Kuujjua Formation is best exposed along a WSW-ENE-trending cuesta on the south side of the Kuujjua River (Fig. 12A) where it lies sub-horizontally in sections up to 120 m-thick (Jefferson, 1985; Rainbird, 1992). The formation is dominated by coarse-grained, cross-bedded sandstones with minor conglomerates organised into mainly tabular bedsets varying in thickness from < 1 to > 10 m, interpreted to record repeated undercutting and aggradation of fluvial-channel belts (Rainbird, 1992). Subordinate lens-shaped bodies, <3 km-wide and < 10 m-thick, composed of siltstones and muddy dolostones with evaporite pseudomorphs, also occur; these fine-grained deposits are related to overbank fines and evaporite-bearing deposits from ephemeral lakes.

Sandy lateral-accretion deposits

A 95 m-thick exposure at 71°10'N to 115°08'W (Fig. 12A) includes a 9 m-thick channel body (beds 2 to 4 in section 86 to 22 of Rainbird, 1992) bounded by sub-planar erosional surfaces. The channel body is floored by pebbly, massive to crudely cross-bedded sandstone (Fig. 12 F), overlain in turn by compound cross-bedded sandstone (Fig. 12G) organised in tabular inclined bedsets bounded by accretionary surfaces (Fig. 12B and E). Palaeoflow indicators collected from the basal pebbly sandstone point to mildly dispersed NW-ward transport (Fig. 12C). Likewise, palaeoflow

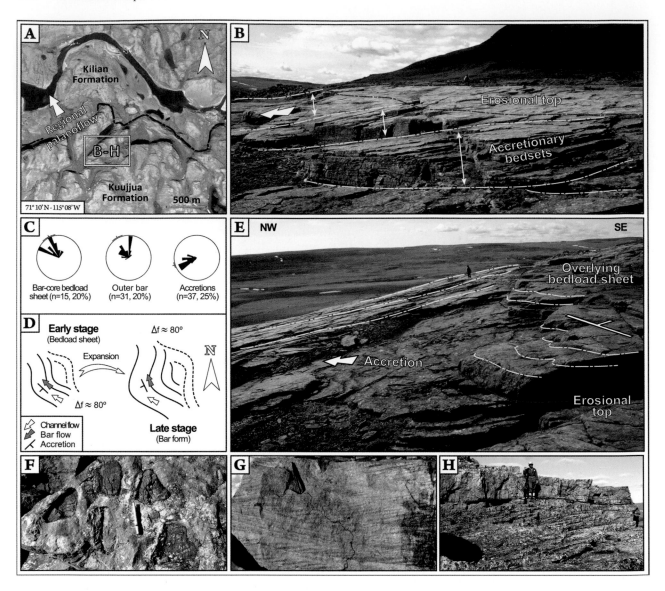

Fig. 12. Field aspects of laterally accreting bar deposits in the Kuujjua Formation, Amundsen Basin, Northwest Territories (Canada). For scale reference, geologist in (B, C, H) is ca. 1.8 m-tall and pencil in (F, G) is ca. 10 cm-long. (A) Satellite view of the study site, along the southern flank of the Kuujjua River valley, Victoria Island, Northwest Territories (Canada). (B) Particular of a channel-belt-top deposit, showing inclined bedsets composing accretionary units truncated at the top by a channel-belt bounding surface; palaeocurrent and accretion-vector readings were collected on the accretionary bedsets reported here. (C) Palaeoflow data for the studied outcrop; average vectors and 95%-confidence arcs are reported. (D) Morphodynamic model for the studied outcrop, with special reference to inferred flow divergence (no scale is intended); see text for discussion. (E) Alternative view of (B), similarly showing inclined accretionary bedsets containing compound cross-beds and truncated at the top. (F to H) Key sedimentological attributes of the Kuujjua Formation at the site, including: pebble-lag deposit flooring erosional channel-belt bounding surfaces (F); cross-bedded sandstone organised in compound sets bounded by erosional reactivation surfaces (G); and detail of sharp erosional boundaries truncating inclined accretionary bedsets (H).

indicators collected within the overlying compound cross-bedded inclined bedsets indicate NW-ward bedform migration, although with a stronger north-ward component (Fig. 12). Finally, surfaces bounding individual inclined beds display WSW-ward bar outgrowth (Fig. 12C).

Interpretation: The lower pebbly deposits are interpreted as preserved bedload sheets (Whiting *et al.*, 1988), overlain in turn by an outbuilding bar form. The lack of preserved abandoned-channel fills and the flat attitude of channel-belt bounding surfaces suggest that most bars in the

Kuujjua Formation developed within multiple-channel systems and were not attached to channel banks (cf. Ashworth *et al.*, 2000). The sub-normal angular relationships between bedload-sheet palaeoflow indicators and overlying accretionary surface point to an initial flow divergence approaching 80°, which is consistent with lateral expansion of a channel bend (Fig. 12D). Subsequently, relationships between accretion and bar palaeoflow remained essentially unchanged as the channel migrated laterally; flow divergence maintaining values of ~80° (Fig. 12D) indicates that no major flow deflection occurred during bar growth. Only the strong northward component of some palaeoflow indicators in bar strata suggests a locally higher flow divergence (<120°), yet no significant helicoidal recirculation is inferred to have developed, consistent with what is observed in central-channel bends with sinuosity index < 2 (Figs 3B to C and 4).

DISCUSSION

Current knowledge gaps and the ongoing debate on pre-vegetation fluvial style

The rock units described here predate the rise of land plants by 0.3 to 1.2 Gy and are representative of a wide range of palaeogeographic realms, from proximal and topographically confined alluvial plains to the distal fringes of pan-continental fluvial systems. The scale of the investigated channel-bar complexes is comparable to that of other channel bodies preserved in their respective formations, indicating that highly sinuous planforms were not the record of minor anabranches lateral to larger, low-sinuosity channel belts (cf. Rust, 1978). Notwithstanding, the question remains whether the available record of Proterozoic fluvial-planform sinuosity is representative of either spatially and temporally limited occurrences, or of morphodynamics far more common than previously realised. Our analysis is limited to six channel-belt records that may not be representative for their respective successions in their entirety. Similarly, documentation of fluvial-planform sinuosity in other pre-vegetation rock units (e.g. Long, 1978, 2011; Santos & Owen, 2016) is based on stratigraphically restricted packages, the global significance of which is challenged (Davies *et al.*, 2017). On a broader perspective, comprehensive literature compilations have been successfully employed to track the temporal occurrence of sinuous, vegetated rivers throughout the early

Palaeozoic, with a distinct emphasis on temporal parallels and process linkages with the evolution of deeply rooted and arborescent land plants (Davies & Gibling, 2010; Davies *et al.*, 2011). Recently, compilation approaches of this kind were extended into the Archean and Proterozoic records (McMahon & Davies, 2016; Davies *et al.*, 2017) to test the hypothesis that Precambrian and early Palaeozoic fluvial systems possibly shared key features such as limited mud content in their floodplains (a feature, in turn, typically related to limited resistance to erosion and thus negligible sinuosity; Davies *et al.*, 2017).

Such conjectures may, in principle, be consistent with the as yet scattered evidence of Proterozoic fluvial sinuosity but caution is warranted regarding literature compilations for the Precambrian fluvial rock record. Table 3 lists 322 selected peer-reviewed articles containing sedimentological data from Proterozoic fluvial units, sorted based on whether they contain reliable information for inferring planform style (Fig. 13). While the Archean and early Palaeozoic rock records are cautiously not considered here, the same analysis could as well be applied to fluvial successions of such ages. While the overall number of publications on Proterozoic fluvial rock units per year increased steadily in the last seven decades (Fig. 13B), only slightly more than half of the listed literature (53%, or 170 out of 322 publications) report palaeoflow data (Fig. 13A), in most cases limited to a few palaeocurrent measurements collected at the scale of entire stratigraphic units. Even more striking is that only 14% of the literature (46 out of 322 publications) report both palaeoflow and depositional-architecture data, including accretionary styles (Fig. 13A). This shortcoming is worsened by the fact that, in multiple instances, several contributions reporting convincing palaeoflow and depositional-architecture data focused selectively on a limited number of well-known, well-exposed and easily accessible successions such as the Torridon Group of Scotland (Stewart, 2002) or the Pretoria Group of South Africa (Bose *et al.*, 2012). Comparably well-exposed rock units, such as those from the Canadian Arctic described in this study, have received far less attention and thus highlight an obvious bias that limits inferences concerning global patterns of fluvial planform style for the Proterozoic. Over a range of methodological approaches to fluvial sedimentology, the critical evaluation of the relationships between depositional architecture, palaeoflow and accretion

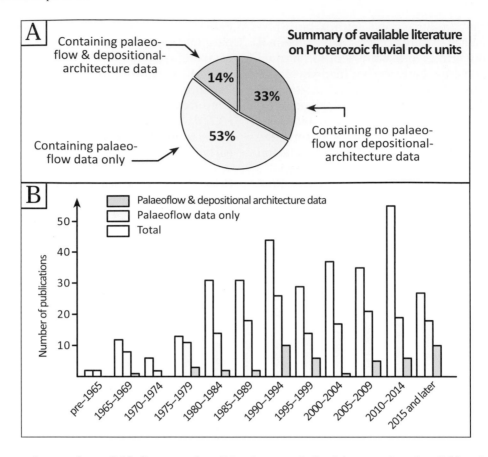

Fig. 13. Summary of currently available literature describing Proterozoic fluvial successions (see Table 3 for full details).

patterns remains a *conditio sine qua non* for the proper interpretation of fluvial planforms (e.g. Long, 1978, 2011; Allen, 1982; Miall, 1985, 1996; Bridge, 1993; Willis, 1993; Ethridge, 2011; Soltan & Mountney, 2016). Conversely, any compilation not based on such factual data remains potentially flawed when acknowledging or ruling out the relevance of fluvial sinuosity in pre-vegetation landscapes. The authors propose a conceptual model that explores the relative controls on the sinuosity of Proterozoic rivers.

Pre-vegetation planform sinuosity: the overriding role of discharge modulation

Previous work has proposed that, due to the recurrence of intense runoff and flood events not buffered by vegetation, Precambrian fluvial landscapes could have not sustained highly sinuous channel planforms even given favourable conditions of slope, substrate cohesion and accommodation space (Eriksson *et al.*, 1998; Sonderholm & Tirsgaard, 1998; Els, 1998; Almeida *et al.*, 2016).

However, evidence of Proterozoic depositional architecture and flow divergence presented here reveals a continuum of developmental stages affecting planform sinuosity, including incipient (Kuujjua Formation), intermediate (Applecross and Nelson Head formations) and mature (Ellice and Bay of Stoer formations). The recognition of these stages is consistent with the dynamic nature of fluvial planform growth, suggesting that more attention should be focused on recognising transitional and co-existent rather than singular end-member planform configurations (cf. Thomas *et al.*, 2006; Ielpi & Rainbird, 2016a). As a starting point, sedimentological characterisations for each rock unit (Rainbird, 1992; Ielpi & Ghinassi, 2015; Ielpi & Rainbird, 2015, 2016a; Ielpi *et al.*, 2016; Ghinassi & Ielpi, 2017) indicate prevalent accumulation in streams with low to intermediate sinuosity, probably with braided planform along distributary systems, although evidence of lateral accretion is prominent at specific stratigraphic intervals. In these examples, the dominance of cross-bedded deposits related to lower-flow-regime conditions and the overall scarcity of upper-flow-regime structures such as plane or antidunal beds can indicate relatively limited discharge fluctuations and/or prolonged waning flood stages, hydrographic features ascribed to perennial discharge regimes (Eriksson & Simpson, 1993; Leclair & Bridge, 2001). It can be postulated that limited discharge variability was an essential and overriding factor for the development of pre-vegetation lateral accretion and planform sinuosity. This notion is consistent with what is observed in both ancient post-vegetation and modern settings, where the recurrence of destructive floods favours planform instability, channel widening and bar braiding (Schumm & Lichty, 1963; Bluck, 1974; Gupta & Fox, 1974; Abdullatif, 1989; Huckleberry, 1994; Montgomery & Buffington, 1998; Takagi *et al.*, 2007), or prevents the full development of macroforms and complex in-channel architectures (Hassan, 2005; Singer & Michaelides, 2014), elements that together militate against lateral-bank accretion (cf. Tal & Paola, 2007).

Over a range of latitudinal climate belts and tectonic settings, low discharge variability is favoured in fluvial systems fed by wide and well-integrated catchments capable of supplying a baseline runoff throughout the year (Syvitski *et al.*, 2000; Syvitski & Milliman, 2007). This scenario fits well with fluvial successions accumulated in large and slowly subsiding intracratonic sags (Rainbird *et al.*, 2014), or with mature fluvial systems sourced from extensive orogenic belts capable of influencing regional climate belts (Zhisheng *et al.*, 2001; Rainbird *et al.*, 2012, 2017; Krabbendam *et al.*, 2017; Ielpi *et al.*, 2017), as per the Ellice, Applecross, Nelson Head formations and possibly the Kuujjua Formation. By comparison, it is improbable that a small rift basin sided by basement topography such as the one that hosted the Bay of Stoer Formation was part of a craton-scale catchment. In the latter case, notwithstanding the probable lack of a large catchment network, the abundance of poorly drained floodbasin strata and wet-aeolian systems (Stewart, 2002; Ielpi *et al.*, 2016; Lebeau & Ielpi, 2017) is suggestive of permanently high groundwater tables, a feature that can be related to positive water budgets (Mountney & Russell, 2009; Assine *et al.*, 2015) and thus overall humid conditions. As such, it appears that even small-scale catchments were capable, given a favourable climate (Döll & Schmied, 2012), of sustaining perennial discharge regimes with relatively limited hydrographic variability in pre-vegetation settings.

Despite multi-km-thick successions found atop several Precambrian cratons bear degrees of preservation and stratigraphic completeness comparable to Phanerozoic examples (Eriksson *et al.*, 1998; Allen *et al.*, 2015), the relative controls of Precambrian tectonic settings and climate on terrestrial clastic sedimentation are still not fully delineated (Grotzinger & McCormick, 1988; Catuneau *et al.*, 2005; Young, 2013a, b). Climate controls are particularly hard to disentangle, given large uncertainties in palaeogeography and plate-motion trajectories throughout the Proterozoic (Pehrsson *et al.*, 2016); and related inferences on latitudinal climate belts. That being said, speculations can be made regarding the controls on fluvial sinuosity at scales varying from craton to individual channel bends, as summarised in Fig. 14. Such speculations take into consideration that, by the end of the Palaeoproterozoic, cratonic sedimentary systems featured basin settings and patterns of bedrock weathering, sediment production, maturation, and routing overall comparable to Phanerozoic ones (Eriksson *et al.*, 1998; Cawood *et al.*, 2006; Pease *et al.*, 2008), thus allowing first-order parallels between their sedimentary styles.

Sudden supply of bedload is known to enhance braiding (Smith & Smith, 1984; Ashmore, 1991; Hoey & Sutherland, 1991), and it follows that increase in planform sinuosity requires a dynamic interplay of discharge and sediment supply to take place (Hooke, 2007). Previous work theorised that high-bedload yield would have been the norm in pre-vegetation environments, owing to the lack of stabilising roots and associated mechanisms of sediment production (Schumm, 1968; Fuller, 1985). However, as these authors acknowledged, sediment yield and bypass in response to floods would have quickly reached a threshold simply dependent on substrate erodibility, hence the hypothesis of recurring high bedload-yield events in aggrading conditions and through prolonged geological timespans is unsound. By comparison, steady sediment yield was again a condition more probably achieved in mature systems fed by well-integrated catchments subject to vigorous weathering (Carling, 1988; Went, 2005; Manners *et al.*, 2015) and was probably facilitated by the lack of mature soils buffering bedrock erosion (Retallack, 1985; Driese & Foreman, 1992; Algeo & Scheckler, 1998). This is a condition more probably established following episodes of supercontinent amalgamation, where extensive and poorly confined intracratonic basins received steady discharge and sediment yield from the dismantling of large

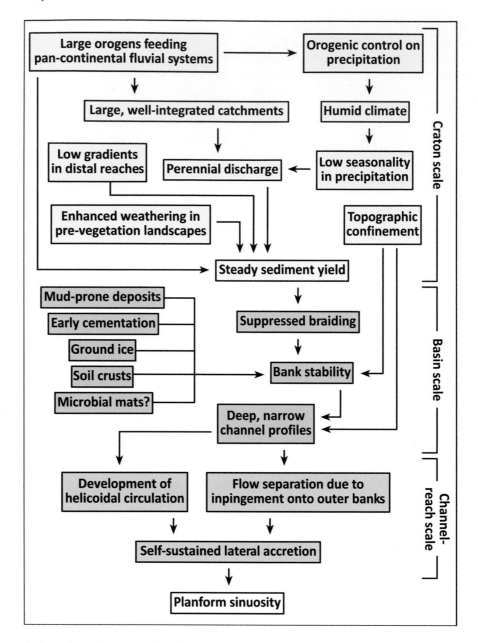

Fig. 14. Flow chart linking the various postulated controls on pre-vegetation fluvial-planform sinuosity at different scales, from craton to channel-reach. See text for discussion.

orogenic belts (Rainbird & Young, 2009; Rainbird *et al.*, 2012, 2017). Bed roughness required for the development of channel-bend curvature was relatively more significant in low-gradient reaches (Ferguson & Ashworth, 1991; Lazarus & Constantine, 2013). As such, it is not surprising that, among our case studies, lateral accretion is prominent in rock units recording either mature and low-gradient fluvial drainages located thousands of kilometres from their headwaters (e.g. scrolls bars in the Ellice Formation, or lateral accretion packages in the Nelson Head and

Kuujjua formations) or in terminal reaches of fluvial systems feeding low-gradient basinal depressions (e.g. point-bar element in the Bay of Stoer Formation). Santos & Owen (2016), in their documentation of point bars with inclined heterolithic stratification elsewhere in the Applecross Formation, similarly concluded that low-gradient settings facilitated the development of highly sinuous channel patterns. On the other hand, lateral accretion in proximal alluvial plains may have been aided by topographic confinement provided by nearby bedrock uplands (cf. Long,

2017), as testified by the point-bar elements preserved in the lowermost Ellice Formation and by upstream-lateral accretion packages in the Applecross Formation. More local factors at the scale of individual channels or single channel bends may have contributed, and their possible control on channel-flow structure is discussed below (Fig. 14).

Local factors: channel hydrographs and bank stability

Recurrent, yet short-lived lateral accretion has been described in large braided channel belts (Bristow, 1987; Best *et al.*, 2003), although braiding and reworking from adjacent channels hampers the development of highly sinuous channel-planform patterns. Transposing this notion to the Precambrian case studies treated here, and given favourable conditions of discharge, sediment supply, gradient and topographic confinement, the lateral growth of highly sinuous channel bends in pre-vegetation settings had still to depend on local (channel system to channel reach) factors to suppress destructive bar braiding. Channel geometry and bank stability are both considered primary controls on the flow configuration of sinuous channels (Osman & Thorne, 1988; Ikeda, 1989; Millar & Quick, 1993; Darby *et al.*, 2002). Recent modelling has suggested that well-established vegetation along channels is functional not only to the onset and maintenance of planform sinuosity but also to the spatial and temporal heterogeneity in bank resistance and accretion, contributing to the dynamic nature of channel-flow configuration during lateral accretion (e.g. Camporeale *et al.*, 2013; Van Oorschot *et al.*, 2016). However, the evidence from depositional architecture and flow divergence presented here demonstrates that a dynamic behaviour in lateral bank-accretion patterns can be recognised also for channel bodies pre-dating the Palaeozoic diffusion of riparian macrophytes. Irrespective of vegetation, the flow configuration of individual channel bends is governed by the discharge and geometry of the channel itself (Hooke, 1975; Dietrich & Smith, 1983; McLelland *et al.*, 1999; Ferguson *et al.*, 2003) and the strength and geometry of helicoidal circulation depend on the hydraulic section, since mature helicoidal cells do not develop in shallow, wide channels (McLelland *et al.*, 1999). Higher flow divergence was more probably attained in channels that, owing to stable banks (see below), incised their

alluvium instead of widening in response to floods (cf. Schumm & Lichtly, 1963; Ielpi *et al.*, 2017). Again, flood-cycle modulation exerts profound controls on the development of flow divergence along different segments of a channel bend (Fig. 5), with the most favourable conditions for development and preservation of planform sinuosity being achieved in the mid, post-flood waning stages, when helicoidal circulation strengthens and the flow is still sustained enough to allow for bar growth. Arguably, the modulation of hydrographic levels on the development of helicoidal circulation might have been even more important in pre-vegetation settings, where lack of rooting and sediment baffling may have resulted in lower-lying bar surfaces over channel thalwegs, as documented in modern rivers with sparse to no vegetation (Matsubara *et al.*, 2015; Ielpi, 2016, 2017). A corollary to these arguments is that the abrupt drops in hydrographic levels that typify ephemeral discharge regimes can only be related to episodic and short-lived lateral accretion (cf. Stear, 1983). By comparison, higher flow divergence and development of stronger helicoidal circulation in downstream-channel bends would have been preferentially achieved in channels with prolonged waning flood stages and comparatively narrow and deep hydraulic section, i.e. stable banks.

Bank stability is considered to have been weak in pre-vegetation alluvium, owing to the absence of plant rooting and the dominance of sand-grained, non-cohesive alluvium (Davies & Gibling, 2010; Davies *et al.*, 2011). Notwithstanding, several factors have been proposed to justify local strengthening of alluvial banks, including the presence of mud-rich strata (cf. Ikeda, 1989) in either poorly drained floodplains (e.g. Bay of Stoer Formation; Fig. 9) or abandoned-channel fills (e.g. Applecross and Nelson Head formations; Figs 10 and 11). In both scenarios, cohesion provided by mud, even if overall subordinate, may have restricted episodes of destructive channel-belt migration (cf. Peakall *et al.*, 2007), at times even favouring channel embankment (Ielpi & Ghinassi, 2015). Production and entrainment of vast volumes of mud is inferred to have occurred in relation to aggressive chemical weathering due to higher atmospheric pCO2 before the rise of land plants (Kaufman & Xiao, 2003; Godderis & Joachimski, 2004), the record of which includes mineralogically mature yet physically immature quartzarenites (Donaldson & de Kemp, 1998; Young, 2013) and strongly illitised and sericitised

Precambrian regoliths (Rainbird *et al.*, 1990; Gall, 1992, 1994; Fedo *et al.*, 1995; Young, 1999; Ielpi *et al.*, 2015). The retention and preservation of muddy alluvium was instead determined by favourable basin conditions that would have hampered ocean-ward transport of fines, either by river effluents or by wind deflation (cf. Dott *et al.*, 1986; Dott, 2003). A number of accessory factors contributing to bank strengthening, including ground ice, soil crusts and early cementation of alluvial beds, have been also proposed (Eriksson *et al.*, 2000; Fairén, 2010; Fairén *et al.*, 2014; Ielpi *et al.*, 2017). The role of microbial mats in providing substrate cohesion (cf. Prave, 2002; Brasier *et al.*, 2017) remains hard to quantify and recent work on late Precambrian-Cambrian strata has suggested negligible contributions to bank stability (McMahon *et al.*, 2017). However, despite a remarkable wealth of knowledge on Precambrian microbial ecosystems (Noffke, 2010), the specific study of their interactions with fluvial processes is still in its infancy. Thus, the potential of microscopic life to affect pre-vegetation fluvial morphodynamics should not be as of yet disregarded. In summary, the continuum of developmental stages in planform sinuosity documented here cannot be discounted as a stochastic oddity or as the product of an individual triggering mechanism. Rather, the authors propose that it represents the dynamic sedimentological expression of a set of favourable, concomitant factors operating at scales varying from that of entire cratonic landmasses to that of local channel reaches.

CONCLUSIONS

A combination of remote sensing and outcrop data have been employed on both modern and ancient fluvial systems to review herein the sedimentology and palaeoflow pattern of selected Proterozoic river-channel deposits preserved in sedimentary successions covering the Laurentian Craton (Fig. 6). Lateral-bank growth was accomplished by several mechanisms and aided the development of lateral-accretion packages and scrolls bars in both single-channel and multiple-channel systems, as well as the formation of mature point bars with morphodynamics and flow structure fully comparable to those of modern rivers. The authors conclude that lateral accretion was evidently part of the range of morphodynamic and depositional processes in Precambrian rivers, enhancing the

development of channel sinuosity in landscapes still devoid of macroscopic plant life. The range of Proterozoic sinuous-fluvial forms documented here and elsewhere in the literature cannot be simplistically explained as a stochastic oddity, warranting a shift in paradigm regarding the recognised planform geometry and dynamics of pre-vegetation rivers. While discharge modulation probably had an overriding control on pre-vegetation sinuosity, a multitude of other factors both extrinsic and intrinsic to alluvial-plain dynamics cooperated at different spatial scales (Fig. 13), from that of entire cratons to single channel tracts.

Large knowledge gaps remain regarding the palaeoflow directions and macroform accretion patterns of Proterozoic fluvial channels (Fig. 14) and the relevance of sinuous planforms cannot presently be quantified with confidence. Inferences on the morphodynamics of pre-vegetation, Archean and Proterozoic fluvial channels need to be informed by detailed palaeoflow and architectural analyses of corresponding deposits, in order to prevent the establishment of a sedimentological dogma on the presumed inevitability of Precambrian braided river planforms (Dickinson, 2009). Altogether, the fluvial systems considered here stretch a time span of almost one billion years and drained continental landmasses that witnessed fundamental planetary changes, including two entire Wilson cycles. Such vast temporal and spatial scales go well beyond what can be deduced from the Phanerozoic rock record. Hence, debates on the concurrent evolution of vegetation and fluvial style throughout the Palaeozoic should not be constructed as proof of absence for highly sinuous Precambrian river channels. The authors thus recommend that outcrop evidence from the early Palaeozoic rock record not be projected back in Precambrian time: lacking evidence of sinuous channel forms in the early Palaeozoic record is not evidence of lacking sinuous channel forms in the Precambrian record.

ACKNOWLEDGEMENTS

The authors feel indebted to reviewers Martin Gibling and Mauricio Santos, together with volume editor Luca Colombera, for constructive comments that greatly improved the manuscript. Fieldwork was conducted by: AI and RHR, Ellice and Nelson Head formations; AI and MG, Applecross Formation; AI, DV and MG, Bay of

Table 3 List of 322 selected peer-reviewed pieces of literature on Proterozoic fluvial rocks, updated to end-April 2017. PF: palaeoflow data. DA: Depositional-architecture data.

Authors	Year	PF	DA	Authors	Year	PF	DA
Almeida *et al.*	2016	Yes	Yes	Collinson *et al.*	2008	Yes	No
Anderson	1987	No	No	Cox et al.	2002	No	No
Andrews-Speed	1986	Yes	No	Daly & Unrug	1983	Yes	No
Andrews-Speed	1989	No	No	Damagala *et al.*	2000	No	No
Armstrong *et al.*	2005	No	No	Davis	2006	No	No
Aspler & Chiarenzelli	1996	Yes	No	Dawes	1997	No	No
Aspler & Chiarenzelli	1997	No	No	Day *et al.*	2004	No	No
Aspler & Donaldson	1984	No	No	de Borba *et al.*	2004	Yes	No
Aspler & Donaldson	1985	No	No	De *et al.*	2016	Yes	No
Aspler & Donaldson	1986	No	No	De Grey & Dehler	2005	Yes	No
Aspler *et al.*	1989	Yes	No	Dehler *et al.*	2010	Yes	No
Aspler *et al.*	1994	Yes	No	Dey	2015	No	No
Aspler *et al.*	2004	Yes	No	Dilliard *et al.*	1999	No	No
Austin	1971	No	No	Dott	1983	No	No
Banks	1973	Yes	No	Dott & Dalziel	1972	Yes	No
Banks	1973	No	No	Driese *et al.*	1995	No	No
Basu *et al.*	2014	No	No	Eliwa *et al.*	2010	No	No
Bauer *et al.*	2013	No	No	Elmor	1984	Yes	No
Becker *et al.*	2005	No	No	Eriksson	1988	Yes	No
Beraldi-Campesi *et al.*	2014	No	No	Eriksson & Catuneau	2004	Yes	No
Betts *et al.*	1999	No	No	Eriksson & Simpson	1993	Yes	No
Beyer *et al.*	2011	No	No	Eriksson & Vos	1979	Yes	Yes
Bhattacharyya & Morad	1993	Yes	Yes	Eriksson *et al.*	1989	Yes	No
Biswas	2005	No	No	Eriksson *et al.*	1993	Yes	No
Bonsor *et al.*	2010	Yes	No	Eriksson *et al.*	1993	No	No
Bose & Chakraborty	1994	Yes	Yes	Eriksson *et al.*	1995	Yes	Yes
Bose *et al.*	2008	Yes	Yes	Eriksson *et al.*	2006	Yes	No
Bostock	1983	No	No	Eriksson *et al.*	2008	Yes	No
Bradshaw *et al.*	2000	No	No	Eriksson *et al.*	2009	No	No
Brasier *et al.*	2016	No	No	Ethridge *et al.*	1984	Yes	No
Brett	1955	Yes	No	Fedo *et al.*	1997	No	No
Bryant & Reed	1970	No	No	Fowler & Osman	2013	No	No
Bull *et al.*	2011	No	No	Fralick	1999	Yes	No
Bumby *et al.*	2001	Yes	No	Fralick & Miall	1989	Yes	No
Bumby *et al.*	2001	Yes	No	Fralick & Zaniewski	2012	No	No
Bumby *et al.*	2002	Yes	No	Freitas *et al.*	2011	Yes	Yes
Callaghan *et al.*	1991	No	No	Gosh & Chatterjee	1994	No	No
Campbell	1978	Yes	No	Gracie & Stewart	1967	Yes	No
Campbell	1979	Yes	No	Greentree *et al.*	2006	Yes	No
Campbell	1981	Yes	No	Gresse *et al.*	1993	No	No
Campbell	1986	No	No	Gresse *et al.*	1996	No	No
Campbell & Cecile	1979	Yes	No	Grey *et al.*	2005	No	No
Campbell & Cecile	1981	Yes	No	Grotzinger & Gall	1986	No	No
Casshyap	1968	Yes	No	Grotzinger & McCormick	1988	Yes	No
Catuneau & Eriksson	2002	No	No	Grotzinger *et al.*	1989	Yes	No
Chakraborty	1994	Yes	No	Hadlari & Rainbird	2000	Yes	No
Chakraborty	1999	Yes	Yes	Hadlari & Rainbird	2006	Yes	No
Chakraborty	2006	No	No	Hadlari & Rainbird	2010	Yes	No
Chakraborty & Chaudhuri	1993	Yes	Yes	Hadlari *et al.*	2006	Yes	No
Chakraborty & Paul	2014	Yes	Yes	Heness *et al.*	2014	Yes	No
Chakraborty *et al.*	2009	Yes	No	Hiatt & Kyser	2005	No	No
Chakraborty *et al.*	2017	Yes	No	Hiatt *et al.*	2003	No	No
Chandler	1986	No	No	Hjellbakk	1993	Yes	Yes
Chandler	1988	No	No	Hjellbakk	1997	Yes	Yes
Chandler	1988	Yes	No	Hoffman	1969	Yes	No
Chaudhuri	2003	No	No	Hollings *et al.*	2007	Yes	No
Cheney *et al.*	1990	No	No	Höy	1993	Yes	No
Collinson	1983	No	No	Hu *et al.*	2014	No	No
Collinson *et al.*	1989	No	No	Ielpi & Ghinassi	2015	Yes	Yes

(Continued)

Table 3 (Continued)

Authors	Year	PF	DA	Authors	Year	PF	DA
Ielpi & Rainbird	2015	Yes	No	Mazumder *et al.*	2012	No	No
Ielpi & Rainbird	2015	Yes	Yes	Mazumder *et al.*	2015	Yes	No
Ielpi & Rainbird	2016	Yes	Yes	McCormick & Grotzinger	1988	Yes	No
Ielpi & Rainbird	2016	Yes	Yes	McCormick & Grotzinger	1992	Yes	No
Ielpi *et al.*	2015	Yes	No	McCormick & Grotzinger	1993	Yes	No
Ielpi *et al.*	2016	Yes	Yes	McManus & Bajabaa	1998	No	No
Ielpi *et al.*	2017	No	No	Medaris *et al.*	2003	No	No
Jackson & Rawlings	2000	No	No	Merk & Jirsa	1982	No	No
Jackson *et al.*	1984	No	No	Miall	1976	Yes	No
Jackson *et al.*	1990	Yes	Yes	Miller	1993	No	No
Johnson *et al.*	1978	Yes	No	Mitchell & Sheldon	2016	No	No
Johnson *et al.*	2011	No	No	Modie	1996	Yes	No
Johnson *et al.*	2013	No	No	Morey	1967	Yes	No
Kalliokoski	1982	No	No	Morey	1974	No	No
Karlstrom *et al.*	1983	Yes	No	Morey	1984	Yes	No
Khalaf	2010	No	No	Morey & Ojakangas	1982	No	No
Khalaf	2012	No	No	Mossman & Harron	1984	No	No
Khalaf	2013	No	No	Mukhopadhyay *et al.*	2014	Yes	No
Khalaf	2013	No	No	Mukhopadhyay *et al.*	2016	Yes	Yes
Kidder	1992	No	No	Murkute & Joshi	2014	No	No
Kingsbury-Stewart *et al.*	2013	Yes	No	Narbonne & Aitken	1995	No	No
Kingsley	1987	Yes	No	Neumann *et al.*	2009	No	No
Knight & Jackson	1994	Yes	No	Nicholson	1993	Yes	Yes
Köykkä	2011	Yes	No	Nystuen	1980	No	No
Köykkä	2011	Yes	No	Ojakangas	1965	Yes	No
Krabbendam *et al.*	2008	No	No	Ojakangas & Morey	1982	Yes	No
Krabbendam *et al.*	2017	No	No	Ojakangas & Weber	1984	Yes	No
Krassay *et al.*	2000	No	No	Ojakangas *et al.*	2001	No	No
Kröner & Correia	1980	No	No	Owen	1995	Yes	Yes
Kumpalainen	1980	Yes	No	Owen	1996	Yes	Yes
Laajoki & Corfu	2007	No	No	Owen & Santos	2014	No	No
Laajoki *et al.*	1989	Yes	No	Patranabis-Deb *et al.*	2012	No	No
Lamminen & Köykkä	2010	No	No	Pauley	1990	Yes	No
Lamminen *et al.*	2015	No	No	Pedreira & De Waele	2008	No	No
LeCheminant *et al.*	1996	Yes	No	Pereira *et al.*	2011	No	No
Levell	1980	Yes	Yes	Petrov	2011	No	No
Link & Christie-Blick	2011	No	No	Petrov	2014	Yes	No
Long	1978	Yes	Yes	Polito *et al.*	2005	No	No
Long	1978	Yes	No	Polito *et al.*	2006	No	No
Long	2002	Yes	Yes	Prave	2002	No	No
Long	2004	Yes	No	Preiss	2000	No	No
Long	2006	Yes	Yes	Pufahl *et al.*	2013	No	No
Long	2007	Yes	Yes	Pulvertaft	1985	Yes	No
Long	2011	Yes	Yes	Rainbird	1992	Yes	Yes
Long	2017	Yes	Yes	Rainbird & Davis	2007	Yes	No
Long & Turner	2012	Yes	Yes	Rainbird & Donaldson	1988	Yes	No
Long & Turner	2012	Yes	Yes	Rainbird & Hadlari	2000	Yes	No
Long *et al.*	2008	No	No	Rainbird & Ielpi	2015	No	No
Lund *et al.*	2003	No	No	Rainbird *et al.*	1992	Yes	No
Magalhães *et al.*	2014	Yes	No	Rainbird *et al.*	1992	Yes	No
Magalhães *et al.*	2015	Yes	No	Rainbird *et al.*	1994	No	No
Manier *et al.*	1993	No	No	Rainbird *et al.*	1996	No	No
Martins *et al.*	2000	Yes	No	Rainbird *et al.*	1997	Yes	No
Martins-Neto	1994	Yes	No	Rainbird *et al.*	2001	No	No
Martins-Neto	1996	Yes	No	Rainbird *et al.*	2003	Yes	No
Martins-Neto	2000	Yes	No	Rainbird *et al.*	2010	No	No
Mazumder	2005	Yes	No	Rainbird *et al.*	2012	No	No
Mazumder & Arima	2009	Yes	No	Rainbird *et al.*	2015	No	No
Mazumder & Sarkar	2004	Yes	No	Rainbird *et al.*	2008	No	No
Mazumder & Van Kranendonk	2013	Yes	No	Ramaekers	1980	No	No

Table 3 (Continued)

Authors	Year	PF	DA	Authors	Year	PF	DA
Ramaekers & Catuneanu	2004	Yes	No	Stewart	2002	Yes	No
Rice & Townsend	1996	No	No	Stewart	2005	Yes	No
Richards & Eriksson	1988	Yes	Yes	Stewart & Donnellan	1992	No	No
Rieu *et al.*	2006	No	No	Strand	2005	Yes	Yes
Røe	1987	Yes	Yes	Strand	2012	No	No
Røe & Hermansen	1993	Yes	Yes	Sutton & Watson	1964	Yes	No
Røe & Hermansen	2006	Yes	Yes	Sweet	1988	No	No
Rogala *et al.*	2007	Yes	No	Tankard *et al.*	1982	No	No
Ross	1983	No	No	Taylor & Middleton	1990	Yes	No
Ross & Villeneuve	2003	No	No	Timmons *et al.*	2005	Yes	No
Ross *et al.*	1992	No	No	Tirrul & Grotzinger	1990	Yes	No
Ryan & Buckley	1998	Yes	No	Tirsgaard & Øxnevad	1998	Yes	Yes
Samanta *et al.*	2011	No	No	Toghill & Chell	1984	No	No
Santos & Owen	2016	Yes	Yes	Tuke *et al.*	1966	No	No
Santos *et al.*	2013	Yes	No	Turner	2011	No	No
Santos *et al.*	2015	Yes	No	Unrug	1984	Yes	No
Sarkar *et al.*	2012	Yes	Yes	van der Neut & Eriksson	1999	No	No
Saylor *et al.*	1995	No	No	van der Neut *et al.*	1991	Yes	No
Scarpelli	1991	No	No	Van Kranendonk *et al.*	2015	Yes	No
Schreiber & Eriksson	1992	No	No	Verma & Shukla	2015	No	No
Schwab	1977	No	No	Verma *et al.*	2013	No	No
Selley	1965	Yes	No	von der Borch *et al.*	1988	No	No
Selley	1969	No	No	Vos & Eriksson	1977	Yes	Yes
Sherman *et al.*	2002	No	No	Walter *et al.*	1995	No	No
Simpson & Eriksson	1991	Yes	Yes	Wang & Zhou	2014	No	No
Simpson & Eriksson	1993	Yes	Yes	Ware & Hiscott	1985	Yes	No
Simpson *et al.*	2013	No	No	Webers *et al.*	1992	No	No
Smith & Minter	1980	Yes	Yes	Webers *et al.*	1992	No	No
Smith *et al.*	2004	No	No	Williams	1966	Yes	Yes
Soegaard & Callahan	1994	No	No	Williams	1969	Yes	No
Sønderholm & Tirsgaard	1998	Yes	No	Williams	1992	Yes	No
Southgate *et al.*	2000	No	No	Williams	2001	Yes	No
Southgate *et al.*	2006	No	No	Williams	2005	Yes	No
Southwick & Mossler	1984	No	No	Williams & Stevens	1969	No	No
Southwick *et al.*	1986	Yes	No	Wills & Stern	1988	No	No
Sovetov	2011	Yes	No	Winston	1978	No	No
Stear	1977	Yes	No	Winston	2016	Yes	Yes
Stewart	1966	No	No	Wygralak *et al.*	1988	No	No
Stewart	1982	Yes	No	Young	1999	No	No
Stewart	1988	Yes	No	Young	1983	No	No
Stewart	1990	No	No	Young & Long	1977	Yes	No
Stewart	1991	No	No	Young *et al.*	2001	No	No

Stoer Formation; RHR, Kuujjua Formation. The authors acknowledge economical support from Discovery Grants by the Natural Sciences and Engineering Research Council of Canada (AI and RHR), the second phase of the Geo-mapping for Energy and Minerals program (AI and RHR), the Strategic Investments in Northern Economic Development program (AI) and in-kind support from the Canada-Nunavut Geoscience Office (AI and RHR). Ed Weerts, the Polar Continental Shelf Project, TMAC Resources, Discover Mining Services, Great Slave Helicopters, Air Tindi, Summit Air, Parks Canada and the Paulatuk Visitor Center are thanked for logistical support.

REFERENCES

Abdullatif, O.M. (1989) Channel-fill and sheet-flood facies sequences in the ephemeral terminal River Gash, Kassala, Sudan. *Sed. Geol.*, **63**, 171–184.

Algeo, T.J. and **Scheckler, S.E.** (1998) Terrestrial–marine teleconnections in the Devonian: links between the evolution of land plants, weathering processes and marine anoxic events. *Phil. Trans. Roy. Soc. London B*, **353**, 113–130.

Allen, J.R.L. (1965) A review of the origin and character of recent alluvial sediments. *Sedimentology*, **5**, 89–191.

Allen, J.R.L. (1982) Sedimentary structures. *Dev. Sedimentol.*, **30**, 593 pp.

Allen, J.R.L. (1970) Studies in fluviatile sedimentation: a comparison of fining-upward cyclothems, with special reference to coarse-member composition and interpretation. *J. Sed. Petrol.*, **40**, 298–323.

Allen, P.A., Eriksson, P.G., Alkim, F.F., Betts, P.G., Catuneanu, O., Mazumder, R., Meng, Q. and **Young, G.M.** (2015) Classification of basins, with special reference to Proterozoic examples. In: *Precambrian Basins of India: Stratigraphic and Tectonic Context* (Eds R. Mazumder and P.G. Eriksson), *Geol. Soc. London Mem.*, **43**, 5–28.

Almeida, R.P., Marconato, A., Freitas, B.T. and **Turra, B.B.** (2016) The ancestors of meandering rivers. *Geology*, **44**, 203–206.

Amor, K., Hesselbo, S.P., Porcelli, D., Thrackrey, S. and **Parnell, J.** (2007) A Precambrian proximal ejecta blanket from Scotland. *Geology*, **36**, 303–306.

Arche, A. (1983) Coarse-grained meander lobe deposits in the Jarama River, Madrid, Spain. In: *Modern and Ancient Fluvial Systems* (Eds J.D. Collinson and J. Lewin), *Int. Assoc. Sedimentol. Spec. Publ.*, **6**, 313–321.

Armitage, J.J. and **Allen, P.** (2010) Cratonic basins and the long-term subsidence history of continental interiors. *J. Geol. Soc. London*, **167**, 61–70.

Ashmore, P. (1991) Channel morphology and bedload pulses in braided, gravel-bed streams. *Geogr. Ann.*, **73A**, 37–52.

Ashworth, P.J., Best, J.L., Roden, J.E., Bristow, C.S. and **Klaasen, G.J.** (2000) Morphological evolution and dynamics of a large, sand braid-bar, Jamuna River, Bangladesh. *Sedimentology*, **47**, 533–555.

Assine, M.L., Merino, E.R., Pupim, F.N., Maced, H.A. and **Santos, M.G.M.** (2015) The Quaternary alluvial systems tract of the Pantanal Basin, *Brazil. Brazil. J. Geol.*, **45**, 475–489.

Bathurst, J.C., Hey, R.D. and **Thorne, C.R.** (1979) Secondary flow and shear stress at river bends. *J. Hydraul. Div.*, **105**, 1277–1295.

Bathurst, J.C., Thorne, C.R. and **Hey, R.D.** (1977) Direct measurements of secondary currents in river bends. *Nature*, **269**, 504–506.

Bennett, S.J., Simon, A. and **Kuhnle, R.A.** (1998) Temporal variations in point bar morphology within two incised river meanders, Goodwin Creek, *Mississippi. Water Res. Eng.*, **1-2**, 1422–1427.

Best, J.L., Ashworth, P.J., Bristow, C.S. and **Roden, J.** (2003) Three-dimensional sedimentary architecture of a large, mid-channel sand braid bar, Jamuna River, Bangladesh. *J. Sed. Res.*, **73**, 516–530.

Bleeker, W. (2003) The late Archean record: a puzzle in ca. 35 pieces. *Lithos*, **71**, 99–134.

Bluck, B.J. (1971) Sedimentation in the meandering River Endrick. *Scot. J. Geol.*, **7**, 93–138.

Bluck, B.J. (1974) Structure and directional properties of some valley sandur deposits in southern Iceland. *Sedimentology*, **21**, 533–554.

Bose, P.K., Eriksson, P.G., Sarkar, S., Wright, D.T., Samanta, P., Mukhopadhyay, S., Mandal, S., Banerjee, S. and **Altermann, W.** (2012) Sedimentation patterns during the Precambrian: A unique record? *Mar. Petr. Geol.*, **33**, 34–68.

Bowring, S.A. and **Grotzinger, J.P.** (1992) Implications of new chronostratigraphy for tectonic evolution of Wopmay Orogen, northwest Canadian Shield. *Am. J. Sci.*, **292**, 1–20.

Brasier, A.T., Culwick, T., Battison, L., Callow, R.H.T. and **Brasier, M.D.** (2017) Evalutating evidence from the Torridonian Supergroup (Scotland, UK) for eukaryptic life on land in the Proterozoic. In: *Earth System Evolution and Early Life: a Celebration of the Work of Martin Brasier* (Eds A.T. Brasier, D. McIlroy and N. McLoghlin), *Geol. Soc. London Spec. Publ.*, **448**, doi: 10.1144/SP448.13.

Brice, J.C. (1974) Evolution of meander loops. *Geol. Soc. Am. Bull.*, **85**, 581–586.

Bridge, J.S. (1993) Description and interpretation of fluvial deposits: a critical perspective. *Sedimentology*, **40**, 801–810.

Bridge, J.S. (1985) Paleochannel patterns inferred from alluvial deposits: a critical evaluation. *J. Sed. Petrol.*, **55**, 579–589.

Bridge, J.S., Alexander, J., Collier, R.E., Gawthorpe, R.L. and **Jarvis, I.** (1995) Ground-penetrating radar and coring used to study the large-scale structure of point-bar deposits in three dimensions. *Sedimentology*, **42**, 839–852.

Bridge, J.S., Collier, R.E. and **Alexander, J.** (1998) Large-scale structure of Calamus river deposits revealed using ground-penetrating radar. *Sedimentology*, **45**, 977–985.

Brierley, G.J. (1989) River planform facies models: the sedimentology of braided, wandering and meandering reaches of the Squamish River, British Columbia. *Sed. Geol.*, **61**, 17–35.

Brierley, G.J. and **Hickin, E.J.** (1991) Channel planform as a non-controlling factor in fluvial sedimentology: the case of the Squamish River floodplain, British Columbia. *Sed. Geol.*, **75**, 67–83.

Bristow, C.S. (1987) Brahmaputra River: Channel migration and deposition. In: *Recent Developments in Fluvial Sedimentology* (Eds F.G. Ethridge, R.M. Flores and M.D. Harvey), *SEPM Spec. Publ.*, **39**, 63–74.

Brooks, G.R. (2003) Alluvial deposits of a mud-dominated stream: the Red River, Manitoba, Canada. *Sedimentology*, **50**, 441–458.

Brookes, I.A. (2003) Palaeofluvial estimates from exhumed meander scrolls, Taref Formation (Turonian), Dakhla Region, Western Desert, Egypt. *Cretaceous Res.*, **24**, 97–104.

Campbell, F.H.A. (1979) Stratigraphy and sedimentation in the Helikian Elu Basin and Hiukitak Platform, Bathurst Inlet-Melville Sound, Northwest Territories. *Geol. Surv. Can. Pap.*, **79–8**, 19 pp.

Camporeale, C., Perucca, E., Ridolfi, L. and **Gurnell, A.M.** (2013) Modeling the interaction between river morphodynamics and riparian vegetation. *Rev. Geophys.*, **51**, 379–414.

Cant, D.J. and **Walker, R.G.** (1978) Fluvial processes and facies sequences in the sandy braided south Saskatchewan River, Canada. *Sedimentology*, **25**, 625–648.

Carling, P. (1988) The concept of dominant discharge applied to two gravel-bed streams in relation to channel stability thresholds. *Earth Surf. Proc. Land.*, **13**, 355–367.

Catuneanu, O., Martins-Neto, M.A. and Eriksson, P.G. (2005) Precambrian sequence stratigraphy. *Sediment. Geol.*, **176**, 67–95.

Cawood, P.A., Kröner, A. and Pisarevsky, S. (2006) Precambrian plate tectonics: Criteria and evidence. *GSA Today*, **16**, 4–11.

Cawood, P.A., Nemchin, A.A., Strachan, R., Prave, T. and Krabbendam, M. (2007) Sedimentary basin and detrital zircon record along East Laurentia and Baltica during assembly and breakup of Rodinia. *J. Geol. Soc. London*, **164**, 257–275.

Corenblit, D. and Steiger, J. (2009) Vegetation as a major conductor of geomorphic changes on the Earth surface: toward evolutionary geomorphology. *Earth Surf. Proc. Land.*, **34**, 891–896.

Cotter, E. (1978) The evolution of fluvial style, with special reference to the central Appalachians Paleozoic. In: *Fluvial Sedimentology* (Ed. A.D. Miall), *Mem. Can. Soc. Petrol. Geol.*, **5**, 361–384.

Coulthard, T.J. (2005) Effects of vegetation on braided stream pattern and dynamics. *Water Resour. Res.*, **41**, W04003.

Daniel, J.F. (1971) Channel movement of meandering Indiana streams. *U.S. Geol. Surv. Prof. Pap.*, **732A**, 1–18.

Darby, S.E., Alabyan A. and Van de Wiel, M.J. (2002) Numerical simulation of bank erosion and channel migration in meandering rivers. *Water Resour. Res.*, **38**, 2-1–2-21.

Davidson, A. (2008) Late Paleoproterozoic to mid-Neoproterozoic history of northern Laurentia: An overview of central Rodinia. *Precambrian Res.*, **160**, 5–22.

Davies, N.S. and Gibling, M.R. (2010) Cambrian to Devonian evolution of alluvial systems: the sedimentological impact of the earliest land plants. *Earth-Sci. Rev.*, **98**, 171–200.

Davies, N.S., Gibling, M.R., McMahon, W.J., Slater, B.J., Long, D.G.F., Bashforth, A.R., Berry, C.M., Falcon-Lang, H.J., Gupta, S., Rygel, M.R. and Wellman, C.H. (2017) Discussion on 'Tectonic and environmental controls on Palaeozoic fluvial environments: reassessing the impacts of early land plants on sedimentation'. *J. Geol. Soc.*, 10.1144/jgs2016-063. *J. Geol. Soc.*, doi: 10.1144/jgs2017-004.

Davies, N.S., Gibling, M.R. and Rygel, M.C. (2011) Alluvial facies evolution during the Palaeozoic greening of the continents: case studies, conceptual models and modern analogues. *Sedimentology*, **58**, 220–258.

De Serres, B., Roy, A.G., Biron, P.M. and Best, J.L. (1999) Three-dimensional structure of flow at a confluence of river channels with discordant beds. *Geomorphology*, **26**, 313–335.

Dickinson, W.R. (2009) The place and power of myth in geoscience: an associate editor's perpective. *Am. J. Sci.*, **303**, 856–864.

Dietrich, W. (1987) Mechanics of flow and sediment transport in river bends. In: *River Channels: Environment and Processes* (Ed. K.S. Richards). Blackwell, Oxford, 179–227.

Dietrich, W.E. and Smith, J.D. (1983) Influence of the point bar on flow through curved channels. *Water Resour. Res.*, **19**, 1173–1192

Döll, P. and Schmied, H.M. (2012) How is the impact of climate change on river flow regimes related to the impact on mean annual runoff? A global-scale analysis. *Environ. Res. Lett.*, **7**, 014037.

Donaldson, J.A. and de Kemp, E.A. (1998) Archean quartz arenite of the Canadian Shield, examples from Superior and Churchill Province. *Sed. Geol.*, **120**, 153–176.

Dott, R.H., Jr. (2003) The importance of eolian abrasion in supermature quartz sandstones and the -paradox of weathering on vegetation-free landscapes. *J. Geol.*, **111**, 387–405.

Dott, R.H., Jr., Byers, C.W., Fielder, G.W., Stenzel, S.R. and Winfree, K.E. (1986) Aeolian to marine transition in Cambro-Ordovician cratonic sheet sandstones of the northern Mississippi valley, U.S.A. *Sedimentology*, **33**, 345–367.

Driese, S.G. and Foreman, J.L. (1992) Paleopedology and paleoclimatic implications of Late Ordovician vertic paleosols, Juniata Formation, southern Appalachians. *J. Sed. Petrol.*, **62**, 71–83.

Els, B.G. (1998) The auriferous late Archaean sedimentation systems of South Africa: unique palaeo-environmental conditions. *Sed. Geol.*, **120**, 205–224.

Embry, A.F. and Klovan, J.E. (1976) The Middle-Upper Devonian clastic wedge of the Franklinian Geosyncline. *Bull. Can. Soc. Petrol. Geol.*, **24**, 485–639.

Eriksson, P.G., Bumby, A.J., Brumer, J.J. and van der Neut, M. (2006) Precambrian fluvial deposits: enigmatic palaeohydrological data from the c. 2–1.9Ga Waterberg Group, South Africa. *Sed. Geol.*, **190**, 25–46.

Eriksson, P.G., Condie, K.C., Tirsgaard, H., Mueller, W.U., Altermann, W., Miall, A.D., Aspler, L.B., Catuneanu, O. and Chiarenzelli, J.R. (1998) Precambrian clastic sedimentation systems. *Sed. Geol.*, **120**, 5–53.

Eriksson, P.G., Simpson, E.L., Eriksson, K.A., Bumby, A.J., Steyn, G.L. and Sarkar, S. (2000) Muddy roll-up structures in siliciclastic interdune beds of the ca. 1.8Ga Waterberg Group, South Africa. *Palaios*, **15**, 177–183.

Ethridge, F.G. (2011) Interpretation of ancient fluvial channel deposits: review and recommendations. In: *From River to Rock Record: The Preservation of Fluvial Sediments and Their Subsequent Interpretation* (Eds S.K. Davidson, S. Leleu and C.P. North), *SEPM Spec. Publ.*, **97**, pp. 9–35.

Fairén, A.G. (2010) A cold and wet Mars. *Icarus*, **108**, 165–175.

Fairén, A.G., Stokes, C., Davies, N.S., Schulze-Makuch, D., Rodríguez, J.A.P., Davila, A.F., Uceda, E.R., Dohm, J.M., Baker, V.R., Clifford, S.M., McKay, C.P. and Squyres, S.W. (2014) A cold hydrological system in Gale Crater, *Mars. Planet. Space Sci.*, **93–94**, 101–118.

Fedo, C.M., Nesbitt, H.W. and Young, G.M. (1995) Unraveling the effects of potassium metasomatism in sedimentary rocks and Paleosols, with implications for paleoweathering conditions and provenance. *Geology*, **23**, 921–924.

Ferguson, R. and Ashworth, P. (1991) Slope-induced changes in channel character along a gravel-bed stream: the Allt Dubhaig, Scotland. *Earth Surf. Proc. Land.*, **16**, 65–82.

Ferguson, R.I., Parsons, D.R., Lane, S.N. and **Hardy, R.J.** (2003) Flow in meander bends with recirculation at the inner bank. *Water Resour. Res.*, **11**, 1322, doi: 10.1029/2003WR001965.

Fielding, C.R., Alexander, J. and **Newman-Sutherland, E.** (1997) Preservation of in situ, arborescent vegetation and fluvial bar construction in the Burdekin River of north Queensland, Australia. *Palaeogeogr. Palaeoclimatol. Palaeoecol.*, **135**, 123–144.

Forbes, D.L. (1983) Morphology and sedimentology of a sinuous gravel-bed channel system: Lower Babbage River, Yukon coastal plain, Canada. In: *Modern and Ancient Fluvial Systems* (Eds J.D. Collinson and J. Lewin), *Int. Assoc. Sedimentol. Spec. Publ.*, **6.**, pp. 195–206.

Frothingham, K.M. and **Rhoads, B.L.** (2003) Three-dimensional flow structure and channel change in an asymmetrical compound meander loop, Embarras River, Illinois. *Earth. Surf. Prof. Land.*, **28**, 625–644.

Fustic, M., Hubbard, S.M., Spencer, R., Smith, D.G., Leckie, D.A., Bennet, B. and **Larter, S.** (2012) Recognition of down-valley translation in tidally influenced meandering fluvial deposits, Athabasca Oil Sands (Cretaceous), Alberta, Canada. *Mar. Petrol. Geol.*, **29**, 219–232.

Gall, Q. (1992) The early Proterozoic Thelon paleosol as part of the Matonabee unconformity in the northwestern Canadian Shield. In: *Mineralogical and Geochemical Records of Paleoweathering* (Eds J.-M. Schmitt and Q. Gall), *ENSMP Mém. Sc. de la Terre*, **18**, 163–174.

Gall, Q. (1994) The Proterozoic Thelon paleosol, Northwest Territories, *Canada. Precambrian Res.*, **68**, 115–137.

Gay, G.R., Gay, H.H., Gay, W.H., Martinson, H.A., Meade, R.H. and **Moody, J.A.** (1998) Evolution of cutoffs across meander necks in Powder River, Montana, *USA. Earth Surf. Process. Land.*, **23**, 651–662.

Ghinassi, M., Billi, P., Libsekal, Y., Papini, M. and **Rook, L.** (2013) Inferring fluvial morphodynamics and overbank flow control from 3D outcrop sections of a Pleistocene point bar, Dandiero Basin, Eritrea. *J. Sed. Res.*, **83**, 1065–1083.

Ghinassi, M. and **Ielpi, A.** (2017) Precambrian snapshots: Morphodynamics of Torridonian fluvial braid-bars revealed by 3D-photogrammetry and outcrop sedimentology. *Sedimentology*, doi: 10.1111/sed.12389.

Ghinassi, M. and **Ielpi, A.** (2015) Stratal architecture and morphodynamics of downstream-migrating fluvial point bars (Jurassic Scalby Formation, U.K.). *J. Sed. Res.*, **85**, 1123–1137.

Ghinassi, M., Ielpi, A., Aldinucci, M. and **Fustic, M.** (2016) Downstream-migrating fluvial point bars in the rock record. *Sed. Geol.*, **334**, 66–96.

Ghinassi, M., Nemec, W., Aldinucci, M., Nehyba, S., Özaksoy, V. and **Fidolini, F.** (2014) Plan-form evolution of ancient meandering rivers reconstructed from longitudinal outcrop sections. *Sedimentology*, **61**, 952–977.

Gibling, M.R. (2006) Width and thickness of fluvial channel bodies and valley fills in the geological record: a literature compilation and classification. *J. Sed. Res.*, **76**, 731–770.

Gibling, M.R., Davies, N.S., Falcon-Lang, H.J., Bashforth, A.R., DiMichele, W.A., Rygel, M.C. and Ielpi, A. (2014) Palaeozoic co-evolution of rivers and vegetation: a synthesis of current knowledge. *Proc. Geol. Assoc.*, **125**, 524–533.

Gibling, M.R. and **Rust, B.R.** (1993) Alluvial ridge-and-swale topography: a case study from the Morien Group of Atlantic Canada. In: *Alluvial Sedimentology* (Eds M. Marzo and C. Puigdefábregas), *Int. Assoc. Sedimentol. Spec. Publ.*, **17**, 133–150.

Godderis, Y. and **Joachimski, M.M.** (2004) Global change in the Late Devonian: modelling the Frasnian–Famennian short-term carbon isotope excursions. *Palaeogeogr. Palaeoclimatol. Palaeoecol.*, **202**, 309–329.

Grotzinger, J.P. and **McCormick, D.S.** (1988) Flexure of Early Proterozoic lithosphere and the evolution of Kilohigok Basin (1.9 Ga), Northwest Canadian Shield. In: *New Perspective on Basin Analysis* (Eds K.L. Posamentier and C. Paola), *Front. Sed. Geol.*, **Part IV**, 405–430.

Gupta, A. and **Fox, H.** (1974) Effects of high-magnitude floods on channel form: A case study in Maryland Piedmont. *Water Resour. Res.*, **10**, 499–509.

Hartley, A.J., Owen, A., Swan, A., Weissmann, G.S., Holzweber, B.I., Howell, J., Nichols, G. and **Scuderi, L.** (2015) Recognition and importance of amalgamated sandy meander belts in the continental rock record. *Geology*, **43**, 679–682.

Hassan, M.A. (2005) Characteristics of gravel bars in ephemeral streams. *J. Sed. Res.*, **75**, 29–42.

Haszeldine, R.S. (1983) Descending tabular cross-bed sets and bounding surfaces from a fluvial channel in the Upper Carboniferous coalfield of Northeast England. In: *Modern and ancient fluvial systems* (Eds J.D. Collinson and J. Lewin), *Int. Assoc. Sedimentol. Spec. Publ.*, **6**, pp. 449–456.

Heaman, L.M., LeCheminant, A.N. and **Rainbird, R.H.** (1992) Nature and timing of Franklin igneous events, Canada: implications for a late Proterozoic mantle plume and the break-up of Laurentia. *Earth Planet. Sci. Lett.*, **109**, 117–131.

Hoey, T.B. and **Sutherland, A.J.** (1991) Channel morphology and bedload pulses in braided rivers: a laboratory study. *Earth Surf. Proc. Land.*, **16**, 447–462.

Hoffman, P.F. (1988) United plates of America, the birth of a craton: Early Proterozoic assembly and growth of Laurentia. *Annu. Rev. Earth Planet. Sci.*, **16**, 543–603.

Hooke, J.M. (2007) Complexity, self-organization and variation in behaviour in meandering rivers. *Geomorphology*, **91**, 236–258.

Hooke, J.M. (2008) Temporal variations in fluvial processes on an active meandering river over a 20-year period. *Geomorphology*, **100**, 3–13.

Hooke, R. (1975) Distribution of sediment transport and shear stress in a meander bend. *J. Geol.*, **83**, 543–565.

Hooke, R. and **Yorke, L.** (2011) Channel bar dynamics on multi-decadal timescales in an active meandering river. *Earth Surf. Proc. Land.*, **36**, 1910–1928.

Howard, A.D., Moore, J.M. and **Irwin, R.P., III** (2005) An intense terminal epoch of widespread fluvial activity on Mars: 1. Valley network incision and associated deposits. *J. Geophys. Res.*, **110**, E12S14.

Huckleberry, G. (1994) Contrasting channel response to floods on the middle Gila River, Arizona. *Geology*, **22**, 1083–1086.

Ielpi, A. (2017) Controls on sinuosity in the sparsely vegetated Fossálar River, southern Iceland. *Geomorphology*, **286**, 93–109.

Ielpi, A. (2016) Lateral accretion of modern unvegetated rivers: fluvial-aeolian morphodynamics and perspectives on the Precambrian rock record. *Geol. Mag*, **154**, 609–624.

Ielpi, A. and Ghinassi, M. (2014) Planform architecture, stratigraphic signature and morphodynamics of an exhumed Jurassic meander plain (Scalby Formation, Yorkshire, UK). *Sedimentology*, **61**, 1923–1960.

Ielpi, A. and Ghinassi, M. (2015) Planview style and palaeodrainage of Torridonian channel belts: Applecross Formation, Stoer Peninsula, *Scotland. Sed. Geol.*, **325**, 1–16.

Ielpi, A. and Rainbird, R.H. (2015) Architecture and morphodynamics of a 1.6 Ga fluvial sandstone: Ellice Formation of Elu Basin, Arctic Canada. *Sedimentology*, **62**, 1950–1977.

Ielpi, A. and Rainbird, R.H. (2016a) Highly variable Precambrian fluvial style recorded in the Nelson Head Formation of Brock Inlier (Northwest Territories, Canada). *J. Sed. Res.*, **86**, 199–216.

Ielpi, A. and Rainbird, R.H. (2016b) Reappraisal of Precambrian sheet-braided rivers: Evidence for 1.9 Ga deep-channelled drainage. *Sedimentology*, **63**, 1550–1581.

Ielpi, A., Rainbird, R.H., Greenman, J.W. and Creason, C.G. (2015) The 1.9 Ga Kilohigok paleosol and Burnside River Formation, western Nunavut: stratigraphy and gamma-ray spectrometry. *Canada-Nunavut Geoscience Office, Summary of Activities*, **2015**, 1–10.

Ielpi, A., Rainbird, R.H., Ventra, D. and Ghinassi, M. (2017) Morphometric convergence between Proterozoic and post-vegetation rivers. *Nature Comm.*, **8**, 15250.

Ielpi, A., Ventra, D. and Ghinassi, M. (2016) Deeply channelled Precambrian rivers: Remote sensing and outcrop evidence from the 1.2 Ga Stoer Group of NW Scotland. *Precambrian Res.*, **281**, 291–311.

Ikeda, S. (1989) Sedimentary controls on channel migration and origin of point bars in sand–bedded meandering rivers. In: *River Meandering* (Eds S. Ikeda and G. Parker), *Water Resour. Monogr.*, **12**, 51–68.

Irwin, R.P., III, Howard, A.D., Craddock, R.A. and Moore, J.M. (2005) An intense terminal epoch of widespread fluvial activity on early Mars: 2. Increased runoff and paleolake development. *J. Geophys. Res.*, **110**, E12S15.

Jablonski, B.V.J. (2012) Process sedimentology and three-dimensional facies architecture of a fluvially dominated, tidally influenced point bar: Middle McMurray Formation, Lower Steepbank River area, northeastern Alberta, Canada. Unpub. M. Sc. Thesis, Queen's University, 371 p.

Jackson, R.G., II (1976) Depositional model of point bars in the Lower Wabash River. *J. Sed. Petrol.*, **46**, 579–594.

Jackson, R.G., II (1981) Sedimentology of muddy fine-grained channel deposits in meandering streams of the American Middle West. *J. Sed. Res.*, **51**, 1169–1192.

Jackson, R.G., II (1975) Velocity–bed-form–texture patterns of meander bends in the lower Wabash River of Illinois and Indiana. *Geol. Soc. Am. Bull.*, **86**, 1511–1522.

Kasvi, E., Vaaja, M., Alho, P., Hyyppä, H., Hyyppä, J., Kaartinen, H. and Kukko, A. (2003) Morphological changes on meander point bars associated with flow structure at different discharges. *Earth. Surf. Proc. Land.*, **38**, 577–590.

Kaufman, A.J. and Xiao, S. (2003) High CO2 levels in the Proterozoic atmosphere estimated from analyses of individual microfossils. *Nature*, **425**, 279–282.

Kawai, S. and Julien, P.Y. (1996) Point bar deposits in narrow sharp bends. *J. Hydraul. Res.*, **34**, 205–218.

Keller, E.A. (1972) Development of alluvial stream channels: A five-stage model. *Geol. Soc. Am. Bull.*, **83**, 1531–1536.

Krabbendam, M., Bonsor, H., Horstwood, M.S.A. and River, T. (2017) Tracking the evolution of the Grenvillian foreland basin: Constraints from sedimentology and detrital zircon and rutile in the Sleat and Torridon groups, Scotland. *Precambrian Res.*, **295**, 67–89.

Krabbendam, M., Prave, T. and Cheer, D. (2008) Orogen foreland basin implications for Torridon-Morar Group correlation and the Grenville A fluvial origin for the Neoproterozoic Morar Group, NW Scotland. *J. Geol. Soc.*, **165**, 379–394.

Labrecque, P.A., Hubbard, S.M., Jensen, J.L. and Nielsen, H. (2011) Sedimentology and stratigraphic architecture of a point bar deposit, Lower Cretaceous McMurray Formation, Alberta, Canada. *Bull. Can. Soc. Petrol. Geol.*, **59**, 147–171.

Lanzoni, S. and Seminara, G. (2006) On the nature of meander instability. *J. Geoph. Res. Earth Surf.*, **111**, F000416, doi: 10.1029/2005JF000416.

Lazarus, E.D. and Constantine, J.D. (2013) Generic theory for channel sinuosity. *P. Natl Acad. Sci. USA*, **110**, 8447–52.

Lebeau, L.E. and Ielpi, A. (2017) Fluvial channel-belts, floodbasins, and aeolian ergs in the Precambrian Meall Dearg Formation (Torridonian of Scotland): Inferring climate regimes from pre-vegetation clastic rock records. *Sed. Geol.*, **357**, 53–71.

Lewin, J. (1976) Initiation of bed forms and meanders in coarse-grained sediment. *Geol. Soc. Am. Bull.*, **87**, 281–285.

Li, Z.X., Bogdanova, S.V., Collins, A.S., Davidson, A., De Waele, B., Ernst, R.E., Fitzsimons, I.C.W., Fuck, R.A., Gladkochub, D.P., Jacobs, J., Karlstrom, K.E., Lu, S., Napatov, L.M., Pease, V., Pisarevski, S.A., Thrane, K. and Vernokovski, V. (2008) Assembly, configuration and break-up history of Rodinia: A synthesis. *Precambrian Res.*, **160**, 179–210.

Long, D.G.F. (2011) Architecture and depositional style of fluvial systems before land plants: a comparison of Precambrian, early Paleozoic and modern river deposits. In: *From River to Rock Record: The Preservation of Fluvial Sediments and their Subsequent Interpretation* (Eds S. Davidson, S. Leleu and C.P. North), *SEPM Spec. Publ.*, **97**, 37–61.

Long, D.G.F. (2006) Architecture of pre-vegetation sandybraided perennial and ephemeral river deposits in Paleoproterozoic Athabasca Group, northern Saskatchewan, Canada as indicators of Precambrian fluvial style. *Sed. Geol.*, **190**, 71–95.

Long, D.G.F. (2017) Evidence of flash floods in Precambrian gravel dominated ephemeral river deposits. *Sed. Geol.*, **347**, 53–66.

Long, D.G.F. (1978) Proterozoic stream deposits: some problems of recognition and interpretation of ancient sandy fluvial systems. In: *Fluvial Sedimentology* (Ed. A.D. Miall), *Mem. Can. Soc. Petr. Geol.*, **5**, 313–342

Long, D.G.F. and **Turner, E.C.** (2012) Formal definition of the Neoproterozoic Mackenzie Mountains Supergroup (Northwest Territories) and formal stratigraphic nomenclature for terrigenous clastic units of the Katherine Group. *Geol. Surv. Can. Open File*, **7113**, 40 p.

Lorenz, R.D., Lopes, R.M., Paganelli, F., Lunine, J.I., Kirk, R.L., Mitchell, K.L., Soderblom, L.A., Stofan, E.R., Ori, G., Myers, M., Miyamoto, H., Radebaugh, J., Stiles, B., Wall, S.D., Wood, C.A. and the **Cassini RADAR Team** (2008) Fluvial channels on Titan: initial Cassini RADAR observations. *Planet. Space Sci.*, **56**, 1132–1144.

Loveless, J.H., Sellin, R.H.J., Bryant, T.B., Wormleaton, P.R., Catmur, S. and **Hey, R.** (2000) The effect of overbank flow in a meandering river on its conveyance and the transport of graded sediments. *Water Environ. J.*, **14**, 447–455.

Luchi, R., Hooke, J.M., Zolezzi, G. and **Bertoldi, W.** (2010) Width variations and mid-channel bar inception in meanders: River Bolin (UK). *Geomorphology*, **119**, 1–8.

Manners, R.B., Wilcox, A.C., Kui, L., Lightbody, A.F., Stella, J.C. and **Sklar, L.S.** (2015) When do plants modify fluvial processes? *Plant-hydraulic interactions under variable flow and sediment supply rates. J. Geophys. Res. Earth Surf.*, **120**, 325–345.

Matsubara, Y., Howard, A.D., Burr, D.M., Williams, R.M.E., Dietrich, W.E. and **Moore, J.M.** (2015) River meandering on Earth and Mars: A comparative study of Aeolis Dorsa meanders, Mars and possible terrestrial analogs of the Usuktuk River, AK and the Quinn River, NV. *Geomorphology*, **240**, 102–120.

McLelland, S.J., Ashworth, P.J., Best, J.L., Roden, J. and **Klaassen, G.J.** (1999) Flow structure and transport of sand-grade suspended sediment around an evolving braid bar, Jamuna River, Bangladesh. In: *Fluvial Sedimentology VI* (Eds N.D. Smith and J. Rogers), *Int. Assoc. Sedimentol. Spec. Publ.*, **28**, 43–57.

McMahon, W.J. and **Davies, N.S.** (2016) The 'Forgotten' Torridonian: Alluvial (upper flow regime) and aeolian bedforms in the pre-vegetation Meall Dearg Formation, NW Scotland. Abstract of the 55th Annual General Meeting, British Sedimentological Research Group, Cambridge (UK), **45**.

McMahon, W.J., Davies, N.S. and **Went, D.J.** (2017) Negligible microbial matground influence on pre-vegetation river functioning: evidence from the Ediacaran-Lower Cambrian Series Rouge, France. *Precambrian Res.*, **292**, 13–34.

Miall, A.D. (1985) Architectural-element analysis: A new method of facies analysis applied to fluvial deposits. *Earth-Sci. Rev.*, **22**, 261–308.

Miall, A.D. (1988) Architectural elements and bounding surfaces in fluvial deposits: anatomy of the Kayenta Formation (Lower Jurassic), southwest Colorado. *Sediment. Geol.*, **55**, 233–262.

Miall, A.D. (1976) Proterozoic and Paleozoic geology of Banks Island, Arctic Canada. *Bull. Geol. Surv. Can.*, **258**, 74 p.

Miall, A.D. (1994) Reconstructing fluvial macroform architecture from two-dimensional outcrops: examples from the Castlegate Sandstone, Book Cliffs, Utah. *J. Sed. Res.*, **64**, 146–158.

Miall, A.D. (1996) The Geology of Fluvial Deposits. Springer Verlag, Berlin.

Millar, R.G. and **Quick, M.C.** (1993) Effect of bank stability on geometry of gravel rivers. *J. Hydraul. Eng.*, **119**, 1343–1363.

Montgomery, D.R. and **Buffington, J.M.** (1998) Channel processes, classification and response. In: *River Ecology and Management* (Eds R. Naiman and R. Bilby), Springer–Verlag, New York, 13–42.

Mountney, N.P. and **Russell, A.J.** (2009) Aeolian dune-field development in a water table-controlled system: Skeiðarársandur, southern Iceland. *Sedimentology*, **56**, 2107–2131.

Nanson, G.C. (1980). Point bar and floodplain formation of the meandering Beatton River, northeastern British Columbia, Canada. *Sedimentology*, **27**, 3–29.

Noffke, N. (2010) Microbial Mats in Sandy Deposits from the Archean Era to Today. Springer Verlag, Berlin (Germany), 193 p.

Ori, G.G. (1982) Braided to meandering channel patterns in humid-region alluvial fan deposits, River Reno, Po Plain (northern Italy). *Sed. Geol.*, **31**, 231–248.

Osman, A.M. and **Thorne, C.R.** (1988) Riverbank stability analysis. I: Theory. *J. Hydraul. Eng.*, **114**, 134–150.

Parnell, J., Mark, D., Fallick, A.E., Boyce, A. and **Thackrey, S.** (2011) The age of the Mesoproterozoic Stoer Group sedimentary and impact deposits, NW Scotland. *J. Geol. Soc. London*, **168**, 349–358.

Peakall, J., Ashworth, P.J. and **Best, J.L.** (2007) Meander-bend evolution, alluvial architecture and the role of cohesion in sinuous river channels: a flume study. *J. Sed. Res.*, **77**, 197–212.

Pease, V., Percival, J., Smithies, H., Stevens, G. and **Van Kranendonk, M.** (2008) When did plate tectonics begin? Evidence from the orogenic record. In: *When Did Plate Tectonics Begin on Planet Earth?* (Eds K.C. Condie and V. Pease), *Geol. Soc. Am. Special Papers*, **440**, 199–228.

Pehrsson, S.J., Eglington, B.M., Evans, D.A.D., Huston, D. and **Reddy, S.M.** (2016) Metallogeny and its link to orogenic style during the Nuna supercontinent cycle. In: *Supercontinent Cycles Through Earth History* (Eds Z.X. Li, D.A.D. Evans and J.B. Murphy), *Geol. Soc. London Spec. Publ.*, **424**, 83–94.

Pišút, P. (2002) Channel evolution of the pre-channelized Danube River in Bratislava, Slovakia (1712–1886). *Earth Surf. Proc. Land.*, **27**, 369–390.

Prave, A.R. (2002) Life on the land in the Proterozoic: evidence from the Torridonian rocks of northwest Scotland. *Geology*, **30**, 811–814.

Rainbird, R.H. (1992) Anatomy of a large-scale braid-plain quartzarenite from the Neoproterozoic Shaler Group, Victoria Island, Northwest Territories, Canada. *Can. J. Earth Sci.*, **29**, 2537–2550.

Rainbird, R.H. (1993) The sedimentary record of mantle plume uplift preceding eruption of the Neoproterozoic Natkusiak flood basalt. *J. Geol.*, **101**, 305–318.

Rainbird, R.H., Cawood, P. and **Gehrels, G.** (2012) The great Grenvillian sedimentation episode: record of supercontinent Rodinia's assembly. In: *Tectonics of Sedimentary Basins: Recent Advances* (Eds C. Busby and A. Azor), John Wiley & Sons, 583–601.

Rainbird, R.H., Heaman, L.H. and Young, G.M. (1992) Sampling Laurentia: Detrital zircon geochronology offers evidence for an extensive Neoproterozoic river system originating from the Grenvill orogen. *Geology*, **20**, 351–354.

Rainbird, R.H., Ielpi, A., Long, D.G.F. and Donaldson, J.A. (2014) Similarities and paleogeography of late Paleoproterozoic Terrestrial sandstone deposits on the Canadian Shield: product of Hudsonian orogenesis. *Geol. Soc. Am. Abstract with Programs*, **46**, 89.

Rainbird, R.H., Ielpi, A., Turner, E.C. and Jackson, V.A. (2015) Reconnaissance geological mapping and thematic studies of northern Brock Inlier, *Northwest Territories. Geol. Surv. Can. Open File*, **7695**, 10 p.

Rainbird, R.H., Jefferson, C.W. and Young, G.M. (1996) The early Neoproterozoic sedimentary Succession B of northwestern Laurentia: Correlations and paleogeographic significance. *Geol. Soc. Am. Bull.*, **108**, 454–470.

Rainbird, R.H., Nesbitt, H.W. and Donaldson, J.A. (1990) Formation and diagenesis of a sub-Huronian saprolith: Comparison with a modern weathering profile. *J. Geol.*, **98**, 801–822.

Rainbird, R.H., Rayner, N.M., Hadlari, T., Heaman, L.M., Ielpi, A., Turner, E.C. and MacNaughton, R.B. (2017) Zircon provenance data record lateral extent of pan-continental, early Neoproterozoic rivers and erosional unroofing history of the Grenville orogeny. *Geol. Soc. Am. Bull.*, doi: 10.1130/B31695.1.

Rainbird, R.H. and Young, G.M. (2009) Colossal rivers, massive mountains and supercontinents. *Earth*, **54**, 52–61.

Rayner, N.M. and Rainbird, R.H. (2013) U-Pb geochronology of the Shaler Supergroup, Victoria Island, Northwest Canada. *Geol. Surv. Can. Open File*, **7419**, 62 p.

Retallack, G.J. (1985) Fossil soils as grounds for interpreting the advent of large plants and animals on land. *Phil. Trans. Royal Soc. London B*, **309**, 105–142.

Richardson, W.R. and Thorne, C.R. (1998) Secondary currents around braid bars in Brahmaputra River, Bangladesh. *J. Hydraul. Eng.*, **124**, 325–328.

Rust, B.R. (1978) Depositional models for braided alluvium In: *Fluvial Sedimentology* (Ed. A.D. Miall), *Mem. Can. Soc. Petr. Geol.*, **5**, 605–625.

Santos, M.G.M., Mountney, N.P. and Peakall, J. (2017a) Tectonic and environmental controls on Palaeozoic fluvial environments: reassessing the impacts of early land plants on sedimentation. *J. Geol. Soc. London*, **174**, 393–404.

Santos, M.G.M., Mountney, N.P., Peakall, J., Thomas, R.E., Wignall, P.B. and Hodgson, D.M. (2017b) Reply to Discussion on 'Tectonic and environmental controls on Palaeozoic fluvial environments: reassessing the impacts of early land plants on sedimentation', *J. Geol. Soc. London*, https://doi.org/10.1144/jgs2016-063. *J. Geol. Soc.*, doi: 10.1144/jgs2017-031.

Santos, M.G.M. and Owen, G. (2016) Heterolithic meandering-channel deposits from the Neoproterozoic of NW Scotland: Implications for palaeogeographic reconstructions of Precambrian sedimentary environments. *Precambrian Res.*, **272**, 226–243.

Schook, D.M., Rathbun, S.L., Friedman, J.M. and Wolf, J.M. (2017) A 184-year record of river meander migration from tree rings, aerial imagery, and cross sections. *Geomorphology*, **293**, 227–239.

Schumm, S.A. (1968) Speculation concerning paleohydrologic controls of Terrestrial Sedimentation. *Geol. Soc. Am. Bull.*, **79**, 1573–1588.

Schumm, S.A. and Lichty, R.W. (1963) Channel widening and flood-plain construction along Cimarron River in Southwestern Kansas. *U. S. Geol. Surv. Prof. Pap.*, **352-D**, 71–88

Seminara, G., Zolezzi, G., Tubino, M. and Zardi, D. (2001) Downstream and upstream influence in river meandering. Part 2. Planimetric development. *J. Fluid. Mech.*, **438**, 213–230.

Shiono, K. and Muto, Y. (1998) Complex flow mechanisms in compound meandering channels with overbank flow. *J. Fluid. Mech.*, **376**, 221–261.

Simon, S.S.T. and Gibling, M.R. (2017) Fine-grained meandering systems of the Lower Permian Clear Fork Formation of north-central Texas, USA: Lateral and oblique accretion on an arid plain. *Sedimentology*, **64**, 714–746.

Singer, M.B. and Michaelides, K. (2014) How is topographic simplicity maintained in ephemeral dryland channels? *Geology*, **42**, 1091–1094.

Smith, D.G. (1976) Effect of vegetation on lateral migration of anastomosed channels of a glacier meltwater river. *Geol. Soc. Am. Bull.*, **87**, 857–860.

Smith, D.G., Hubbard, S.M., Lavigne, J.R., Leckie, D.A. and Fustic, M. (2011) Stratigraphy of counter-point-bar and eddy-accretion deposits in low-energy meander belts of the Peace-Athabasca Delta, northeast Alberta, Canada. In: *From River to Rock Record: The Preservation of Fluvial Sediments and Their Subsequent Interpretation* (Eds S. Davidson, S. Leleu and C.P. North), *SEPM Spec. Publ.*, **97**, 143–152.

Smith, D.G., Hubbard, S.M., Leckie, D.A. and Fustic, M. (2009) Counter point bar deposits: lithofacies and reservoir significance in the meandering modern Peace River and ancient McMurray Formation, Alberta, Canada. *Sedimentology*, **56**, 1655–1669.

Smith, N.D. (1974) Sedimentology and bar formation in the upper Kickinh Horse River, a braided outwash stream. *J. Geol.*, **82**, 205–223.

Smith, N.D. and Smith, D.G. (1984) William River: An outstanding example of channel widening and braiding caused by bed-load addition. *Geology*, **12**, 78–82.

Soltan, R. and Moutney, N.P. (2016) Interpreting complex fluvial channel and barform architecture: Carboniferous Central Pennine Province, northern England. *Sedimentology*, **63**, 207–252.

Sønderholm, M. and Tirsgaard, H. (1998) Proterozoic fluvial styles: response to changes in accommodation space (Rivieradal sandstones, eastern North Greenland). *Sed. Geol.*, **120**, 257–274.

Stear, W.M. (1983) Morphological characteristics of ephemeral stream channel and overbank splay sandstone bodies in the Permian Lower Beaufort Group, Karoo Basin, South Africa. In: *Modern and Ancient Fluvial Systems* (Eds J.D. Collinson and J. Lewin), *Int. Assoc. Sedimentol. Spec. Publ.*, **6**, 405–420.

Stewart, A.D. (2002) The later Proterozoic Torridonian rocks of Scotland: their sedimentology, geochemistry and origin. *Geol. Soc. London Mem.*, **24**, 136 p.

Stewart, D.J. (1983) Possible suspended-load channel deposits from the Wealden Group (Lower Cretaceous) of southern England. In: *Modern and Ancient Fluvial Systems* (Eds J.D. Collinson and J. Lewin), *Int. Ass. Sedimentol. Spec. Publ.*, **6**, 369–384.

Stølum, H.H. (1996) River meandering as a self-organization process. *Science*, **271**, 1710–1713.

Syvitski, J.P. and Milliman, J.D. (2007) Geology, geography and humans battle for dominance over the delivery of fluvial sediment to the coastal ocean. *J. Geol.*, **115**, 1–19.

Syvitski, J.P., Morehead, M.D., Barh, D.B. and Mulder, T. (2000) Estimating fluvial sediment transport: The rating parameters. *Water Resour. Res.*, **36**, 2747–2760.

Takagi, T., Oguchi, T., Matsumoto, J., Grossman, M.J., Sarker, M.H. and Matin, M.A. (2007) Channel braiding and stability of the Brahmaputra River, Bangladesh, since 1967: GIS and remote sensing analyses. *Geomorphology*, **85**, 294–305.

Tal, M. and Paola, C. (2007) Dynamic single-thread channels maintained by the interaction of flow and vegetation. *Geology*, **35**, 347–350.

Taylor, G., Crook, K.A.W. and Woodyer, K.D. (1971) Upstream dipping foreset cross-stratification for paleoslope analyses. *J. Sed. Petrol.*, **41**, 578–581.

Thomas, R.G., Williams, B.P.J., Morrissey, L.B., Barclay, W.J. and Allen, K.C. (2006) Enigma variations: the stratigraphy, provenance, palaeoseismicity and depositional history of the Lower Old Red Sandstone Cosheston Group, south Pembrokeshire, Wales. *Geol. J.*, **41**, 481–536.

Thomson, J. (1876) On the windings for rivers in alluvial plains, with remarks on the flow of water round bends in pipes. *Proc. Roy. Soc. London*, **25**, 5–8.

Timár, G. (2003) Controls on channel sinuosity changes: a case study of the Tisza River, the Great Hungarian Plain. *Quatern. Sci. Rev.*, **22**, 2199–2207.

Toonen, W.H.J., Kleinhans, M.G. and Cohen, K.M. (2012) Sedimentary architecture of abandoned channel fills. *Earth Surf. Proc. Land.*, **37**, 459–472.

Tunbridge, I.P. (1984) Facies model for a sandy ephemeral stream and clay playa complex; the Middle Devonian Trentishoe Formation of North Devon, U.K. *Sedimentology*, **31**, 697–715.

Turnbull, M.J.M., Whitehouse, M.J. and Moorbath, S. (1996) New isotopic age determinations for the Torridonian, NW Scotland. *J. Geol. Soc. London*, **153**, 955–964.

Turner, B.R. and Eriksson, K.A. (1999) Meander bend reconstruction from an Upper Mississippian muddy point bar at Possum Hollow, West Virginia, USA. In: Fluvial Sedimentology VI (Eds N.D. Smith and J. Rogers), *Int. Assoc. Sedimentol. Spec. Publ.*, **28**, 363–379.

van de Lageweg, W.I., Schuurman, F., Cohen, K.M., Dijk, W.M., Shimizu, Y. and Kleinhans, M.G. (2015) Preservation of meandering river channels in uniformly aggrading channel belts. *Sedimentology*, **63**, 586–608.

van de Lageweg, W.I., Van Dijk, W.M., Baar, A.W., Rutten, J. and Kleinhans, M.G. (2014) Bank pull or bar push: what drives scroll-bar formation in meandering rivers? *Geology*, **42**, 319–322.

van Oorschot, M., Kleinhans, M., Geerling, G. and Middelkoop, H. (2016) Distinct patterns of interaction between vegetation and morphodynamics. *Earth. Surf. Proc. Land.*, **41**, 791–808.

Went, D.J. (2016) Alluvial fan, braided river and shallow-marine turbidity current deposits in the Port Lazo and Roche Jagu formations, Northern Brittany: relationships to andesite emplacements and implications for age of the Plourivo-Plouézec Group. *Geol. Mag.*, doi: 10.1017/S0016756816000686.

Went, D.J. (2005) Pre-vegetation alluvial fan facies and processes: an example from the Cambro-Ordovician Rozel Conglomerate Formation, Jersey, Channel Islands. *Sedimentology*, **52**, 693–713.

Willis, B.J. (1993) Interpretation of bedding geometry within ancient point-bar deposits. In: Alluvial Sedimentation (Eds M. Marzo and C. Puigdefábregas), *Int. Assoc. Sedimentol. Spec. Publ.*, **17**, 101–114.

Willis, B.J. (1989) Paleochannel reconstructions from point bar deposits: a three-dimensional perspective. *Sedimentology*, **36**, 757–766.

Wormleaton, P.R., Sellin, R.H.J., Bryant, T., Loveless, J.H., Hey, R.D. and Catmur, S.E. (2004) Flow structures in a two-stage channel with a mobile bed. *J. Hydraul. Res.*, **42**, 145–162.

Wright, S.A. and Kaplinski, M. (2011). Flow structure and sandbar dynamics in a canyon river during a controlled flood, Colorado River, Arizona. *J. Geophys. Res. Earth Surf.*, **116**, F01019.

Yan, N., Mountney, N.P., Colombera, L. and Dorrel, R.M. (2017) A 3D forward stratigraphic model of fluvial meander-bend evolution for prediction of point-bar lithofacies architecture. *Comput. Geosci.*, **105**, 65–80.

Young, G.M. (1999) A geochemical investigation of palaeosols developed on Lewisian rocks beneath the Torridonian Applecross Formation, NW Scotland. *Scott. J. Geol.*, **35**, 107–118.

Young, G.M. (2013b) Evolution of Earth's climatic system: Evidence from ice ages, isotopes, and impacts. *GSA Today*, **23**, 4–10.

Young, G.M. (1978) Proterozoic (<1.7 b.y.) stratigraphy, paleocurrents and orogeny in North America. *Egypt. J. Geol.*, **22**, 45–64.

Young, G.M. (2013a) Secular changes at the Earth's surface; evidence from palaeosols, some sedimentary rocks and palaeoclimatic perturbations of the Proterozoic Eon. *Gondwana Res.*, **24**, 453–467.

Young, G.M. (1981) The Amundsen embayment, Northwest Territories; relevance to the upper Proterozoic evolution of North America. In: *Proterozoic Basins of Canada* (Ed. F.H.A. Campbell), *Geol. Surv. Can. Pap.*, **81-10**, 203–211.

Young, G.M., Jefferson, C.W., Delaney, G.D. and Yeo, G.M. (1979) Middle and late Proterozoic evolution of the northern Canadian Cordillera and Shield. *Geology*, **7**, 125–128.

Zhao, G., Cawood, P.A., Wilde, S.A. and Sun, M. (2002) Review of global 2.1–1.8 Ga orogens: implications for a pre-Rodinia supercontinent. *Earth-Sci. Rev.*, **59**, 125–162.

Zhisheng, A., Kutzbach, J.E., Prell, W.L. and Porter, C. (2001) Evolution of Asian monsoons and phased uplift of the Himalaya–Tibetan plateau since Late Miocene times. *Nature*, **411**, 62–66.

Int. Assoc. Sedimentol. Spec. Publ (2018) **48**, 119–148.

The shortage of geological evidence for pre-vegetation meandering rivers

WILLIAM J. MCMAHON[†‡] and NEIL S. DAVIES[†]

[†]*Department of Earth Sciences, University of Cambridge, Downing Street, Cambridge, UK*
[‡]*Present Address: Faculty of Geosciences, Utrecht University, Princetonlaan 8a, Utrecht, The Netherlands*

ABSTRACT

Meandering rivers have been hypothesised to have been rare, or even absent, prior to the evolution of land plants. This is supported by experimental models that demonstrate how bank stability is essential for the development of sustained channel meanders and observations of modern rivers where many causes of bank stability involve direct or indirect biological stabilisation by vegetation. Further support is found in the geological record because overbank fines, heterolithic architectural complexity and laterally-accreting inclined heterolithic stratification (LA-IHS) are rarely observed in Precambrian and earliest Palaeozoic alluvium. An alternative view is that the development of meandering planforms in unvegetated tidal creeks, incised rivers and on other telluric planets and moons such as Mars and Titan, shows that meandering is possible in the absence of vegetation and, by extension, that meandering rivers may not have been uncommon on pre-vegetation Earth. Further evidence for the existence of pre-vegetation meandering channels has come from a recent report of classic meandering fluvial facies in the Neoproterozoic Allt-na-Béiste Member of the Torridon Group, Scotland. This paper reviews how different types of evidence have been used to evaluate the commonness of pre-vegetation meandering rivers, in order to identify where they are in agreement, and presents new field data from the Allt-na-Béiste Member that quantifies the dimensions and frequency of occurrence of classic meandering fluvial facies within the unit. The universal conclusion from different strands of evidence is that meandering rivers, particularly those with small-sized to moderate-sized channel dimensions, were much less common on Earth prior to the evolution of land plants. The pre-vegetation example of classic meandering river facies in the Allt-na-Béiste Member is shown to be a minor component of regional stratigraphy that records the deposits of channels that were <0.5 m-deep and probably less than 3 m-wide. It is anomalous in comparison with contemporaneous fluvial facies worldwide and appears to have developed due to highly localised controls in a near-filled lake basin, where relatively high mud content and a low gradient mimicked the physical effects of plants.

Keywords: Precambrian, Lateral Accretion, Inclined Heterolithic Stratification, Meandering Facies, Palaeozoic, Torridon Group

INTRODUCTION

Geological studies of ancient alluvium aim to describe preserved sedimentary characteristics and give an explanation for their formation. Frequently, this explanation involves an attempt to interpret the geomorphic planform of ancient river channels because modern 'end-member' river types (e.g. braided and meandering planforms) are known to accrete sediment in different ways due to differences in flow conditions. It has long been noted that certain ancient alluvial facies (bearing sedimentary characteristics traditionally interpreted as reflecting meandering river deposition) increase in their abundance and frequency in strata that post-date the evolution of land plants

Fluvial Meanders and Their Sedimentary Products in the Rock Record, First Edition.
Edited by Massimiliano Ghinassi, Luca Colombera, Nigel P. Mountney and Arnold Jan H. Reesink.
© 2019 International Association of Sedimentologists. Published 2019 by John Wiley & Sons Ltd.

(Cotter, 1978; Davies & Gibling, 2010a,b). Yet while some researchers have used these observations to suggest that, without land plants, it was not possible for rivers to meander (Eriksson *et al.*, 1998; Els, 1998), others have proposed that vegetation is not crucial for meander development (e.g. Bridge, 2006; Santos *et al.* 2017). This paper aims to compare and contrast the different strands of evidence that have led to conflicting conclusions concerning the frequency of meandering rivers on pre-vegetation Earth. Evidence from: 1) modern geomorphological observations; 2) experimental modelling; and 3) sedimentary geology are considered, along with a case study from a rare example of pre-vegetation alluvium that conforms to the classic meandering fluvial model in the Neoproterozoic Allt-na-Béiste Member of Scotland.

Interpreting pre-vegetation channel planforms

It is not possible to see a pre-vegetation river; as geomorphic entities, they went extinct when the first embryophytes evolved at least c.473 Ma (Rubinstein *et al.*, 2010). This crucial fact has occasionally been overlooked in studies of pre-vegetation alluvium (see discussion in McMahon & Davies, 2018b) but underlines that, when aiming to understand pre-vegetation geomorphology, any hypotheses can only be approached indirectly. Different ways of understanding pre-vegetation rivers have incorporated evidence from: 1) geomorphological observations of the role that plants play in shaping modern fluvial landscapes; 2) modelling of how rivers may have behaved in the absence of any plant-related controlling parameters; and 3) geological observations of the pre-vegetation alluvial rock record. Each of these approaches primarily employs a different type of reasoning (induction, deduction or abduction) and this is partly responsible for the contradictory assertions regarding the commonness of pre-vegetation meandering rivers. The types of reasoning employed are summarised in Fig. 1 (after Kleinhans *et al.* (2009, 2010); see original papers for further detail). The process of combining evidence from any two corners of the illustrated triangle, in order to determine the third, involves reasoning that varies depending on which corners are already known: 1) Using 'causes' and 'laws' to determine 'effects' involves deduction; 2) Using 'causes' and 'effects' to determine 'laws' involves induction and; 3) Using 'effects' and 'laws' to determine 'causes' involves abduction. With specific reference to the question 'were there pre-vegetation meandering rivers?': 1) geomorphological studies primarily utilise inductive reasoning because they combine observable 'causes' (e.g. bank stability, flow resistance, gradient) and 'effects' (meandering) to determine the 'laws' governing river planform; 2) modelling studies primarily utilise deductive reasoning because they combine different 'causes' (i.e. model variables) and known 'laws', to determine 'effects' (e.g. experimentally reproducing a meandering channel); and 3) geological studies primarily utilise abductive reasoning because they combine 'effects' (i.e. sedimentary characteristics of preserved strata) and known 'laws' to infer 'causes' (e.g. depositional processes).

In the following sections we address the reasons why these different approaches have resulted in sometimes mutually-exclusive claims regarding the nature of pre-vegetation rivers (e.g. Eriksson *et al.*, 1998; Els, 1998; Retallack *et al.*, 2014, 2015; Ielpi, 2016; Santos & Owen, 2016; Santos *et al.*, 2017) and identify those conclusions which are universally agreed upon by all three types of reasoning.

Were there pre-vegetation meandering rivers? Inductive reasoning

An inductive approach can use modern geomorphological observations to infer the laws of nature that surround the role of vegetation in fluvial systems (Fig. 2A to C). However, modern rivers can never be perfect analogues for those operating on pre-vegetation Earth because: 1) there are no fully-unvegetated rivers at the present day (i.e. lacking even the smallest stature plants, such as mosses or liverworts); and 2) it is impossible to observe systems operating outside of the present Earth condition (i.e. a post-glacial planet with a specific tectonic configuration on which plants have existed and potentially accumulated and with palimpsest effects for at least 473 Ma).

Geomorphological studies demonstrate that, in modern small-sized to moderate-sized fluvial systems, meandering is promoted by bank stability and flow resistance, which together retard bank retreat and curtail the development of wider low-sinuosity channels (less efficient at transporting sediment and thus more probable to adopt a braided planform; e.g. Church, 2006; Lazarus & Constantine, 2013). Moreover, it has been shown that vegetation actively promotes these parameters,

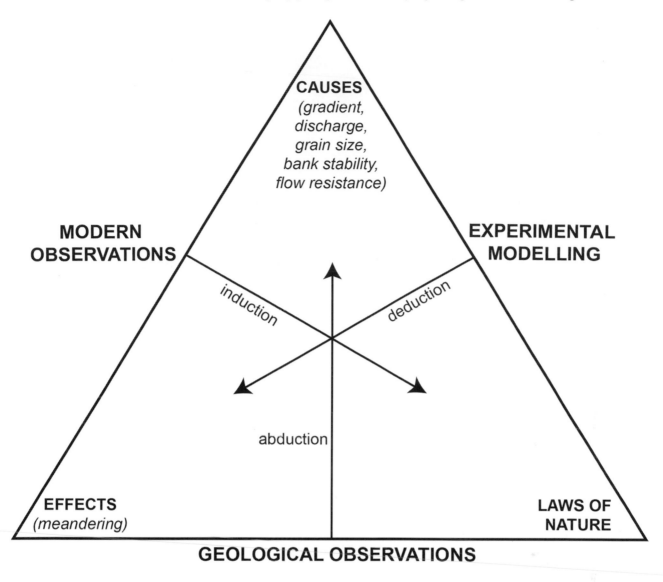

Fig. 1. Modified from Kleinhans *et al.* (2009). Diagram illustrating the three types of explanation based on causes, effects and laws, two of which are necessary to infer the third. Triangle framed with reference to the question 'Were there pre-vegetation meandering rivers?'

directly through rooting and indirectly through soil production and fines retention (e.g. Pollen-Bankhead & Simon, 2010; Bardgett *et al.*, 2014; Fig. 2D to F). In some instances, vegetation has been isolated as the direct trigger for meandering (e.g. Thorne, 1990; Simon & Collison, 2002; Tal *et al.*, 2004) and there are well-documented examples of rivers switching from a meandering to a braided channel pattern when riparian plants are removed or modified by human activity or flood (e.g. Brooks *et al.*, 2003; Erskine *et al.*, 2012; Horton *et al.*, 2017). By way of contrast, Bridge (2006) stated that 'there is no conclusive evidence

that vegetation (or early lithification) has a significant influence on channel pattern, as long as the flood flow is capable of eroding banks and transporting sediment'. Yet, in many natural settings, the presence of riparian vegetation is the dominant reason why bank strength and point bar cover has been elevated to a degree that flood flow cannot erode banks, carve chute cutoffs and transport sediment (Fig. 2D to E; Smith, 1976; Ferguson, 1987; Thorne, 1990; Simon & Collison, 2002; Brooks *et al.*, 2003; Tal *et al.*, 2004; Pollen-Bankhead & Simon, 2010; Kleinhans & van den Berg, 2011; Erskine *et al.*, 2012; Bardgett *et al.*,

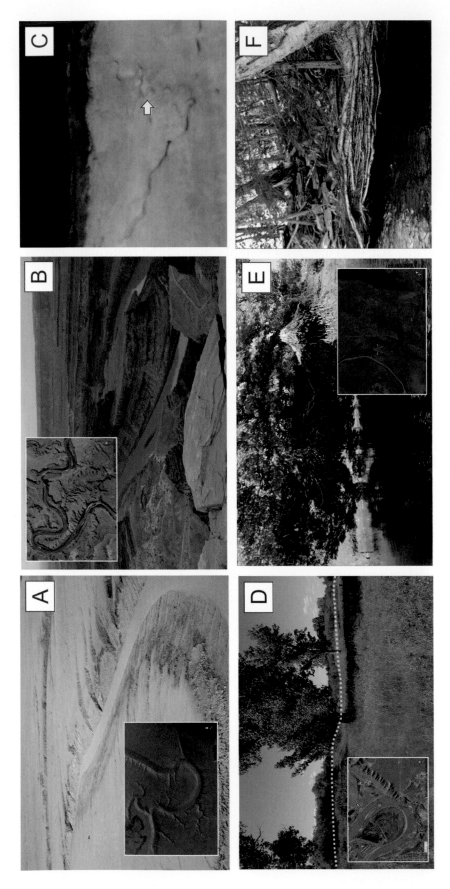

Fig. 2. Characteristics of modern channels which inform interpretations of pre-vegetation Earth. (A) Unvegetated meandering tidal channel in muddy intertidal deposits (channel submerged at high tide), Hopewell Rocks, New Brunswick, Canada. (B) Meandering planform (inset) to incised canyon of the Colorado River. Note that, while bedrock incision is the primary control on canyon sinuosity, within the alluvial deposits of the river, vegetation plays a role in the fluvial system by colonising banks and bars. Dead Horse Point, Utah, United States. (C) Meandering channels developed on unvegetated salt flats on the floor of a receding ephemeral lake. Photograph taken from an altitude of c. 10 to 30 m. Lake Eyre, South Australia. (D) Stabilisation of the top of a scroll bar by vegetation. Undulating surface of scroll bar morphology reflected in topography (dashed line), while stabilisation and flow dampening effects of vegetation discourage the development of chute cutoffs during flood events. South Saskatchewan River, Medicine Hat, Alberta, Canada. (E) Bank stability provided by tree roots at the outer bend of a meander loop (location illustrated with star on aerial photograph: flow from upper part of image). Despite being subject to high shear stresses, bank position is maintained by tree roots even after bank sediment has been removed. Musquodoboit River, Nova Scotia, Canada. (F) Example of part of an outer bend of a meandering river in which an entire portion of the river bank is formed by matted tree roots. Gaspereau River, Nova Scotia, Canada. Satellite images from Google Earth and Infoterra.

2014; Corenblit *et al.*, 2014; Horton *et al.*, 2017; Kleinhans *et al.*, 2018). On Earth, prior to the evolution of land plants, any vegetation-induced biostabilisation was wholly absent: accordingly, even without perfect modern analogue, it can be inductively reasoned that, with fewer potential causes, bank stability, bar cover and thus meandering was probably less frequent.

Yet inductive reasoning is also sometimes used to suggest that meandering rivers were unlikely to have been wholly absent on pre-vegetation Earth. Bank stability can also be afforded abiotically by: 1) naturally shear-resistant sediment (e.g. clays, salts or certain pedogenic chemical precipitates such as silcretes or calcretes; e.g. Ferguson, 1987; Magee *et al.*, 1995; Smith, 1998; Fig. 2C); or 2) ice (Walker & Arnborg, 1966; Costard *et al.*, 2003; Matsubara *et al.*, 2015). Neither of these causes occurs as frequently (across all modern fluvial systems) as does vegetation (and the former increased in abundance after the evolution of vegetation due to enhanced chemical weathering and mud retention [e.g. Morris *et al.*, 2015; McMahon & Davies, 2018a]) but they were present on pre-vegetation Earth. Additionally, while the role of vegetation in promoting bank stability holds true for small-sized to moderate-sized rivers, Earth's largest rivers do not share the same controlling variables (Ashworth & Lewin, 2012). The planform of 'big rivers' may more probably be controlled by regional climate, sea-level controls and tectonic entrenchment, rather than bank stability, discharge, slope, grain-size, or riparian vegetation (e.g. Miall, 1996, 2006; Bridge, 2006; Gibling, 2006). Finally, modern observations also show that sinuous patterns can develop within incised, erosional fluvial systems (e.g. bedrock canyons) independent of vegetation (Harden, 1990; Karlstrom *et al.*, 2013) (Fig. 2B).

Pre-vegetation rivers are sometimes considered in light of sinuous channel landforms that are less appropriate analogues because sinuosity is a convergent morphology that can arise from multiple unrelated processes within different settings (e.g. rivers, tidal channels, lava flows [e.g. Hulme, 1982; Komatsu & Baker, 1994], submarine channels [e.g. Damuth *et al.*, 1983; Imran *et al.*, 1999], possible methane-ethane conduits [Gilliam & Lerman, 2016]). Geomorphological inferences about pre-vegetation rivers should only be made with reference to modern rivers and it is recommended that the common use of two non-fluvial analogues in particular should be discouraged: 1) muddy

meandering tidal channels (e.g. Retallack *et al.*, 2014, 2015) (Fig. 2A); and 2) orbital imagery of sinuous channel patterns on the surface of other telluric planets and moons, primarily Mars and Titan (e.g. Santos *et al.*, 2017). While self-organised sinuous patterns in tidal systems bear many morphometric similarities to meandering rivers (Finotello *et al.*, 2018), they develop under the influence of many different parameters (i.e. periodic bidirectional flow punctuated by slack water intervals; diurnal fluctuations in water-level; variations in the balance of tidal, fluvial and wave processes affecting entrainment, transport and deposition; Hughes, 2011; Wang, 2011). Tidal channels also interact differently with vegetation: for example, their initial route of incision is determined as they avoid antecedent patches of vegetation (e.g. Temmerman *et al.*, 2007). Any analogy with Martian sinuous channels is currently premature because very little is presently known about the sedimentology and formative conditions of those large relict meandering channels that are observable only from orbital imagery (e.g. Matsubara *et al.*, 2015; Davis *et al.*, 2016). Even if such features record the deposits of 3 Ga-old water courses, the differences between fluvial processes operating on Mars and Earth would be significant, because the reduced acceleration due to gravity on Mars would minimise flow velocities (e.g. Kuhn, 2014) and the formation of sinuous planforms may have been less reliant on bank cohesion (Matsubara *et al.*, 2015). Channels seen on the surface of Titan have also been used as analogues for Earth's pre-vegetation rivers (Santos *et al.*, 2017) but these are also inappropriate (at present) because it is even less clear whether they were water conduits: they could as probably have been formed by the dissolution of ice by a concentrated solution of ammonium sulfate, or mechanical erosion by flow of liquid ammonia and ethane (Gillian & Lerman, 2016).

Inductive reasoning can also be applied to geological trends in the stratigraphic-sedimentary record in order to assess whether meandering rivers were common on pre-vegetation Earth. If it is accepted that recurring facies can be identified in the rock record then it is possible to ask whether the distribution of such facies has any relationship with stratigraphic age. Previous studies have used inductive reasoning to show that 'sedimentary facies interpreted to signify meandering river deposition' (an effect) are strongly linked to 'age relative to the evolution of land plants' (a cause) and have concluded that interpretations of meandering rivers are extremely rare in pre-vegetation strata: suggesting a linkage between cause and effect (a law of nature; Cotter, 1978; Davies & Gibling, 2010a). Davies & Gibling (2010a) acknowledged that the identification of ancient meandering rivers from facies could be ambiguous. However, they noted that, where previous authors had reached the conclusion that the deposits were 'meandering facies', this was a shorthand way of grouping a commonly recurring suite of sedimentary characteristics (a suggestion borne out by a more detailed analysis of some of these characteristics; Davies & Gibling, 2010a, their fig. 21). Thus the recognition of a vegetation-correlated stratigraphic shift in global facies characteristics is a tangible characteristic of Earth's sedimentary rock record (e.g. Davies *et al.*, 2017) but can only be considered to be circumstantial support for an absence of pre-vegetation meandering planforms until the validity of previous interpretations is considered (see Section 2.3).

Were there pre-vegetation meandering rivers? Deductive reasoning

Experimental models (using variable 'causes' and parameters based on known 'laws') can be used to deductively predict pre-vegetation channel planforms. However, such models are reliant on an accurate understanding of the pre-vegetation laws of nature, which are only inferable from the partial modern analogues discussed above.

Flume tank experiments suggest that bank stability, bar top stability and flow resistance are key factors in promoting meandering; without which experimental rivers undergo significant channel widening and ultimately adopt a braided planform (e.g. Parker, 1979; Ferguson, 1987; Schumm *et al.*, 1987; Ashmore, 1991; Paola, 2001; Xu, 2002; Kleinhans, 2010). In order to successfully model a self-sustaining meander, it is necessary to reduce or prevent the development of recurring chute cutoffs that dissect point bars on the inner bends of meanders, shortening and straightening a channel course and creating a multithreaded pattern that is, by definition, braided (Smith, 1998; Braudrick *et al.*, 2009; Van Dijk *et al.*, 2012, 2013). Chute cutoffs have, to date, been successfully curtailed in flume tank experiments by: 1) adding fine cohesive sediment to the model, which is sufficient to sustain meandering at least at the scale of these experiments (Smith, 1998; Peakall *et al.*,

2007; Van Dijk *et al.*, 2013); and 2) adding riparian vegetation which increases bank stability, reduces chute cutoffs and corrals channels into meandering patterns (Tal & Paola, 2007, 2010; Braudrick *et al.*, 2009). Flume tank experiments lacking mud or vegetation consistently have difficulties in reproducing sustained meandering channel patterns (e.g. Tiffany & Nelson, 1939; Friedkin, 1945; Schumm & Kahn, 1972; Smith, 1998; Tal & Paola, 2007; Van Dijk *et al.*, 2013).

There are inherent scaling problems with experiments that seek to construct a river in a flume tank. For example, grain-size cannot be scaled down by the same factor as channel features because cohesive sediment (<8 μm) behaves differently to granular media (Kleinhans, 2010). Likewise, experiments which have incorporated vegetation effects (e.g. Tal & Paola, 2007; Braudrick *et al.*, 2009) have been criticised because the root depths of the small plants used (usually alfalfa) often exceed the depth of modelled channels, so producing effects potentially dissimilar to natural environments (Santos *et al.*, 2017). Nonetheless, the results of these experiments suggest that with few or no plants or mud (both characteristics of the land on pre-vegetation Earth (McMahon & Davies, 2018a)), most meandering river planforms were probably transient landscape features due to their propensity for self-shortening and straightening. In the very rare instances where cohesive sediment was locally abundant in the pre-vegetation continental realm (McMahon & Davies, 2018a) the development of self-sustaining small meandering channels could have been feasible.

Were there pre-vegetation meandering rivers? Abductive reasoning

Using geological observations of sedimentary rock end-product ('effects'), combined with known laws of nature, to interpret no longer observable ancient processes ('causes') requires abductive reasoning. Of the three approaches outlined here, such geological investigations are the only ones to directly deal with a tangible physical product of pre-vegetation rivers. Yet the challenge involved with this approach is ascertaining what (if any) ancient geomorphic information is archived by the sedimentary rock record and whether any facies signatures are sufficiently diagnostic to confirm or refute the presence of ancient meandering rivers. In other words, abductive reasoning relies on inferring the 'best possible explanation'

(Kleinhans *et al.*, 2009) and so has an inherent risk of becoming overly qualitative when adjudging what 'best possible' means (particularly if the possibility of the true explanation is unknown or unknowable to a researcher).

The notion that ancient alluvial strata bear signatures that are diagnostic of different channel planforms can be traced back to precursor facies models such as those of Allen (1964, 1970), in which specific heterolithic vertical successions of the Anglo-Welsh Old Red Sandstone (ORS) were interpreted as having been deposited in point bars of sinuous streams (Fig. 3). Allen's (1964, 1970) interpretations were the result of detailed bed-by-bed lithological and palaeocurrent analysis but were originally only applied to particular ORS outcrops, with limited exposure (Fig. 3A). However, following the development of conceptual models for the deposits of 'typical' braided rivers (e.g. Cant & Walker, 1976; Miall, 1977), Allen's (1964, 1970) models became adopted as globally-applied archetypal counterpoints during the development of the 'facies model' paradigm (Walker, 1976; Fig. 3B and C) and have since been heavily reproduced in sedimentology textbooks. Such facies are hereafter referred to as the 'classic meandering facies model' (CMFM).

CMFM strata warrant further attention here because they comprise a suite of recurring facies characteristics that are a common sedimentary rock motif in the geological record, even though many legitimate questions have been raised about how well CMFM strata equate to meandering river deposits (e.g. Jackson, 1977; Bridge, 1985, 1993, 2006; Brierley & Hickin, 1991; Ethridge, 2011; Colombera *et al.*, 2013; Hartley *et al.*, 2015). CMFM strata are dominated by thick mudrock or heterolithic intervals (Fig. 3B) in association with fining-upwards channelised sandstones (Fig. 3C), laterally accreting inclined heterolithic stratification (LA-HIS; Fig. 3 to -E) and palaeocurrent variance indicating channel sinuosity (Bernard *et al.*, 1962; Allen, 1963; Ore, 1964; Thomas *et al.*, 1987). Of these characteristics, arguably the most commonly cited diagnostic criterion for determining CMFM strata are LA-IHS, which are composite architectural elements bearing hallmarks of both inclined heterolithic stratification (IHS) and lateral accretion (LA). IHS consist of packages of sigmoidal beds of alternating grain-size, inclined relative to the local tectonic dip (Fig. 4). If fully preserved, a gently dipping topset passes through an inflection point to a steeper dipping foreset and then through another inflection point to a gently dipping bottomset which terminates asymptotically against a basal erosion surface. Overlying strata often erosionally truncate topsets. Where the steepest part of the foresets of IHS can be determined to dip near orthogonal to neighbouring palaeocurrent directions, they are classified as heterolithic examples of the lateral accretion (LA) architectural elements (Miall, 1985; originally termed 'epsilon cross-stratification' by Allen, 1963).

Not all IHS are LA sets and not all LA sets are IHS (Fig. 5). Individually, both features can develop in low-sinuosity systems, but composite LA-IHS packages appear to require the existence of channel meanders in order to develop. This is because, in meandering channels, LA sets are deposited when flowing water is centrifugally deflected from the inner to outer bank of a curve, forcing helical overturn within the water body and inner bank accretionary sedimentation by the secondary current that moves obliquely up slope of the inner channel bend, creating a point bar (e.g. Leopold & Wolman, 1960). At the same time, IHS relates to varying discharge (Nami, 1976) and discontinuous point bar growth (Puigdefabregas & van Vliet, 1978; van de Lageweg *et al.*, 2014), receiving an additional coarser-to-fine couplet during every flood event (e.g. Bridge & Jarvis, 1976) or scroll bar passage (van de Lageweg *et al.*, 2014). In a one-dimensional vertical profile (e.g. in stratigraphic core or log), an erosional lower bounding surface (the original channel floor) is seen overlain by coarser-grained IHS bottomsets, succeeded by younger iterations of IHS foresets and ultimately finer-grained topsets that grade into overbank mudrocks.

The importance of LA-IHS to the CMFM is one reason why applying the CMFM in isolation cannot identify all meandering river deposits. Confident recognition of LA-IHS in the rock record requires appropriate exposure of strata: vertically, it must be equivalent to the depth (or erosionally truncated remnant) of the original channel and, laterally, it must extend for a recognisable fraction of the total point-bar length. As such the CMFM is most equipped to recognise meandering rivers whose channels were smaller than the outcrop in which they occur: therefore biasing the record of positively identified meandering rivers to the alluvium of small-sized to moderate-sized channels (Fig. 6). The CMFM cannot always identify large meandering rivers

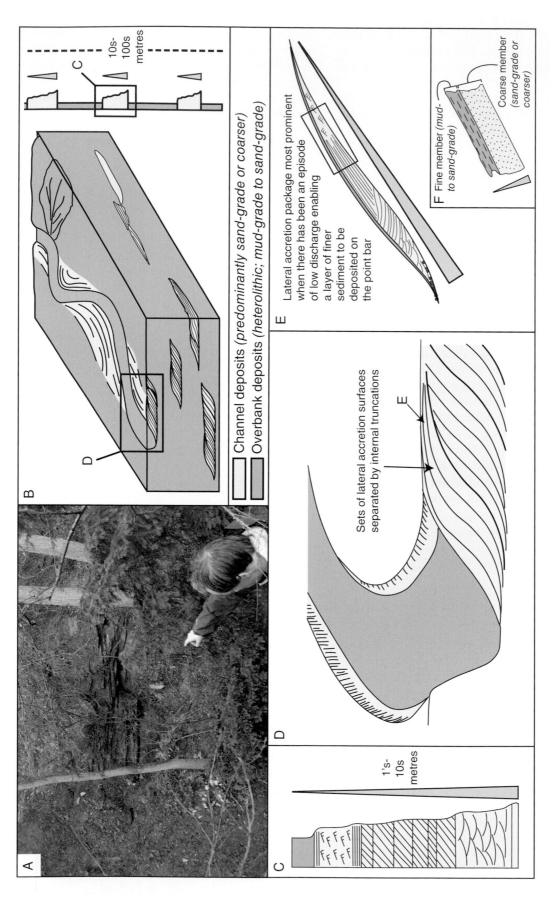

Fig. 3. (A) Typical exposure quality of the Anglo-Welsh Old Red Sandstone at Ludlow described by Allen (1964, 1970) (geologist is 187 cm-tall); (B) Hypothetical model illustrating the characteristics of the deposits of meandering streams as depicted by 'classic meandering facies models'; (C) Generalised vertical profile of a meandering river deposit as depicted by 'classic meandering facies models' (note profile based on model in Allen (1964, his fig. 5), originally intended to illustrate a typical cyclothem at Tugford, England); (D) Sedimentary model illustrating deposition of laterally-accreting inclined heterolithic stratification in a point bar of a sinuous stream; (E) Typical bed within a laterally-accreting package; (F) Individual inclined heterolithic stratification either consist of a normally graded bed or (more commonly) a distinct coarse-to-fine couplet (Thomas *et al.*, 1987).

Within panel B legend:

☐ Channel deposits (*predominantly sand-grade or coarser*)

▨ Overbank deposits (*heterolithic; mud-grade to sand-grade*)

Panel A label: **A**

Panel B labels: **B**, **D**, **C**, 10s–100s metres

Panel C labels: **C**, 1's–10s metres

Panel D labels: **D**, Sets of lateral accretion surfaces separated by internal truncations, **E**

Panel E labels: **E**, Lateral accretion package most prominent when there has been an episode of low discharge enabling a layer of finer sediment to be deposited on the point bar

Panel F labels: **F**, Fine member (*mud- to sand-grade*), Coarse member (*sand-grade or coarser*)

Fig. 4. Outcrop expressions of LA-IHS recording point bar deposits of various scales. In each image, dashed line shows bounding surfaces of the LA-IHS element, yellow arrow shows general direction of migration and pink bar represents one metre vertically: (A) Two superimposed LA-IHS packages in broadly opposed directions. Underlying mudrock contains vertisol structures; strata deposited in a fluvial floodplain environment. Late Silurian (Pridolian) Milford Haven Group, Llansteffan, Carmarthenshire, Wales; (B) Isolated LA-IHS within estuarine facies. Early Cretaceous Ashdown Formation, Fairlight, East Sussex, England; (C) Large scale LA-IHS with internal erosion surface, recording deposition within a tidally-influenced meandering point bar. Late Cretaceous Horseshoe Canyon Formation, Willow Creek, Alberta, Canada.

May be produced by:

- Point bar deposits
  (fluvial, tidal)
  or
- Delta Progradation

May be produced by:

- Point bar deposits
  (fluvial, tidal)
  or
- Lateral migration of side-channel and
  in-channel bars in low-sinuosity rivers

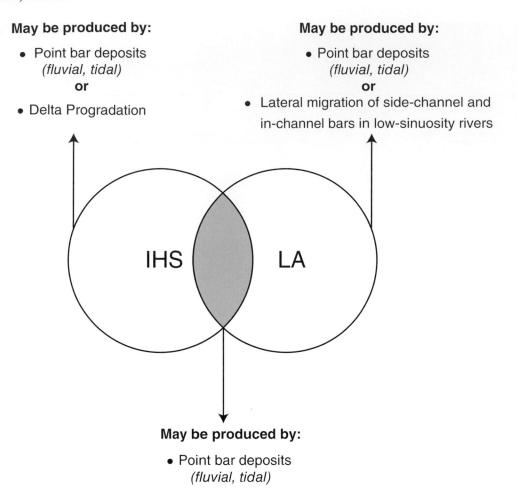

May be produced by:

- Point bar deposits
  (fluvial, tidal)

Fig. 5. Conceptual model illustrating the difference between inclined heterolithic stratification (IHS), lateral accretion (LA) and laterally accreting inclined heterolithic stratification (LA-IHS) and the depositional circumstances that each can potentially represent.

because the physical dimensions of rock outcrops are often inferior to the dimensions of many geomorphic components of contemporary large river systems (e.g. in the Mississippi River, meander-belts are up to 12 km-wide and associated with channel widths up to 1000 m and point-bars reaching up to 3750 m in length). Equally, not all small-sized to moderate-sized meandering streams necessarily leave LA-IHS as a signature: Hartley *et al.* (2015) documented examples of exhumed amalgamated meander belt deposits where point bar morphology was visible in planview, but LA sets comprised < 5% of the total outcrop area and showed no significant heterolithic component.

The use of geological evidence in isolation has unavoidable blind spots for understanding ancient Earth because there are presently few criteria for the recognition of big river meandering facies and

incised rivers (erosional environments) leave no sedimentary record of channel planform. Likewise, the CMFM is reliant on the identification of LA-IHS, which can only be identified when it occurs at sub-outcrop scale (i.e. the deposits of small-scale to moderate-scale channels) and it overlooks the deposits of meandering streams where mud and silt contribute insignificantly to the overall fraction of grains within a system. This is illustrative of the pitfalls of constraining abductive reasoning into facies models when interpreting the rock record. Yet while the recognition that there is a near absence of pre-vegetation CMFM strata (Section 2.1) cannot be used to imply that no meandering rivers at all existed before the evolution of land plants, it does reflect the fact that vegetation engineered profound geomorphic innovation within small- to moderate-sized, mixed load rivers.

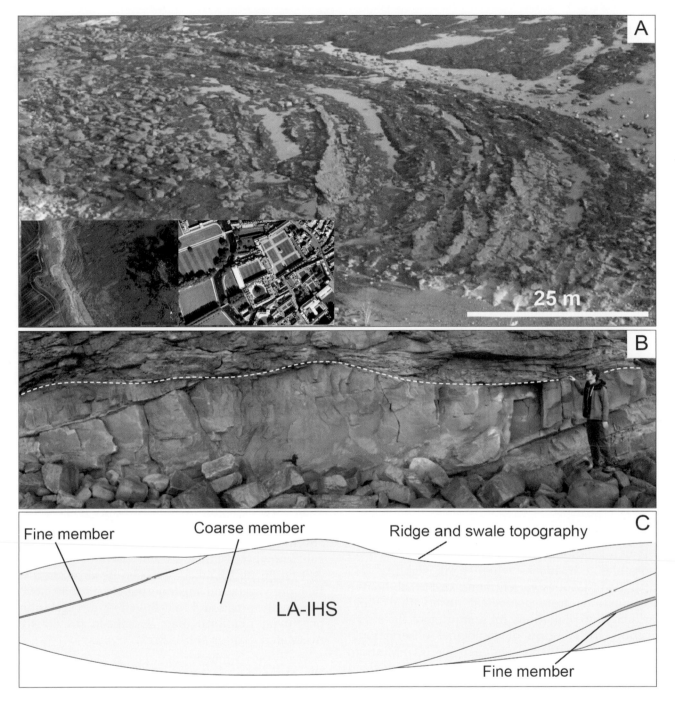

Fig. 6. Example of an exhumed meander plain in which LA-IHS can be related to ancient channel dimensions: (A) Exhumed meander plain. Jurassic Scalby Formation, Yorkshire, England (e.g. Nami, 1976; Leeder & Nuami, 1979; Alexander, 1992; Ielpi & Ghinassi, 2014; Ghinassi *et al.*, 2016). Inset satellite images (at same scale) show how the Scalby meander plain reflects a river with comparable dimensions to the moderate-sized lowland River Cam, eastern England, rather than a major river system. Images: Google Earth, Infoterra and Bluesky; (B) Vertical cross-section displaying inclined heterolithic stratification, with undulating scroll bar top. Jurassic Scalby Formation, Yorkshire, England; (C) Line tracing Fig. 6B. Scalby Formation IHS deposits have a maximum thickness of 4 metres which match estimated channel widths and bankfull depths of 21 and 4 metres, respectively (Nami, 1976). The fact that, even in such exceptionally exposed examples, clearly defined LA-IHS elements record only the deposits of relatively small rivers, emphasises that the sedimentary geological record and the CMFM are more suited for the positive recognition of deposits of small-sized to moderate-sized meandering rivers. Geologist is 180 cm-tall.

Were there pre-vegetation meandering rivers? A unified approach

The question 'were there pre-vegetation meandering rivers?' can only be answered indirectly and is better framed with respect to different types of meandering river. In terms of big rivers, modern geomorphic observations do not suggest a primary role for land plants, so inductive reasoning suggests that these may have existed prior to land plants. Modelling and geological approaches presently offer little evidence either way for big pre-vegetation meandering rivers: in the case of the former approach, this is because scaling issues prohibit accurate experimental modelling of such systems and in the case of the latter, existing facies models are largely considered unsatisfactory for the identification of their deposits. In terms of incised rivers, geomorphic observations suggest that such phenomena were possible before vegetation, but such systems, as erosional environments, will have left no geological evidence for original channel planform at outcrop. In terms of small-sized to moderate-sized rivers, the effects of land plant evolution appear to have been more fundamental: (1) inductive explanations based on modern geomorphic observations suggest that meandering rivers would have been less common before the evolution of land plants because plants play a major role in promoting bank and bar top stability (roles only rarely fulfilled by stability from factors such as ice or precipitates); (2) deductive explanations from modelling observations support these contentions, additionally suggesting that small meandering rivers might have been possible in instances where bank stability was provided by cohesive sediment; and (3) abductive explanations from geological observations (and inductive explanations from studies of the stratigraphic distribution of geological observations) find little conclusive evidence for CMFM strata (the deposits of small-sized to moderate-sized muddy meandering streams) in pre-vegetation alluvium.

The universal conclusion from these different strands of evidence is that meandering rivers, particularly those with small-sized to moderate-sized channel dimensions, were less common prior to the evolution of land plants. This biological influence on an essentially abiotic pattern is in accordance with hypotheses such as those set out by Dietrich & Perron (2006), who suggested that no landforms are diagnostic signatures of life (or its absence) but that their scale and frequency distribution can be heavily influenced by biology.

PRE-VEGETATION CMFM STRATA

If the rise of CMFM strata in the Silurian reflects a change in the frequency and dimensions of meandering rivers, rather than their 'invention' by land plants (Dietrich & Perron, 2006; Davies & Gibling, 2010b), then rare instances of CMFM strata should not be unexpected in pre-vegetation rocks. Yet there is presently a remarkable shortage of reports of such: in the 11 published geological interpretations of pre-vegetation meandering rivers to date, most are based upon limited or equivocal evidence (Table 1) and only one conforms to the CMFM: the Neoproterozoic Allt-na-Béiste Member, Scotland (Santos & Owen, 2016).

Santos & Owen (2016) showed that the Allt-na-Béiste Member contains the best facies evidence to date for a sinuous pre-vegetation fluvial system, making it a significant section for the study of pre-vegetation alluvium. However, Santos & Owen (2016) did not record how common CMFM strata were in comparison to adjacent strata of the member, or discuss their dimensions in detail: yet both of these factors are critical to the debate surrounding the impact the evolution of vegetation had on fluvial sedimentation (Davies *et al.* 2017; Santos *et al.*, 2017). The following case study revisits this key stratigraphic unit, quantifying the frequency and dimensions of CMFM strata. This is a significant knowledge gap at present, particularly as the Allt-na-Béiste Member has been used to suggest that pre-vegetation meandering rivers may have been widespread and that the effect of vegetation on fluvial planform may have been previously overstated (Santos *et al.*, 2017).

The Neoproterozoic Allt-na-Béiste Member, NW Scotland

The Allt-na-Béiste Member forms part of the Mesoproterozoic-Neoproterozoic Torridonian Supergroup (Fig. 7). It is a minor component of the overall stratigraphy: while it reaches a maximum thickness of 295 m at depth (out of the complete > 10 km-thick succession), its outcrop thickness is more limited and regionally variable (Fig. 8). It crops out (with certainty) at three locations: its type section at Diabaig (where it has a

Table 1. Previous interpretations of pre-vegetation meandering rivers, justification for the interpretations and problems with the criteria used.

Formation	Authors	Age (Ma)	Explanation for meandering fluvial interpretation	Potential issues with interpretation
Allt-na-Béiste Member	Santos & Owen (2016)	850–1000 Ma	LA-IHS deposits; presence of mudrock, interpreted as floodplain material; interpreted crevasse splay elements	Only 6 examples of IHS observed in c. 180 m of alluvium. Maximum thickness 41 cm. Sandy-bedforms dominant component of stratigraphy.
Hatches Creek Group	Sweet (1988)	1870–1846 Ma	1 fining up cycle; 'probable lateral accretion surfaces' capped with 3 m of mudstone	LA surfaces restricted to one 9 m-thick sandbody. 'Insufficient outcrop' exposure prohibits accurate understanding of the relationship between inclined foresets and their underlying surface. Surfaces may alternatively represent local lateral-accretion on an in channel bar within a sandy braided system (Long, 2011).
The Transvaal Sequence	Pretorius (1974)	2500–2100 Ma	None given	Conceptual model of system only
Nelson Head Formation (in the Brock Inlier).	Long (1978)	c. 1000 Ma	'Possible meandering stream deposit'. Palaeocurrents in sets of cross-stratified sandstone are at a high angle to underlying surface, and change systematically up section.	Minor overbank fines; no crevasse elements (Long, 2011)
Katherine Group	Long (1978)	c. 1000 Ma	LA surfaces associated with 3.17 m-thick CH element.	Minor overbank fines; no crevasse elements (Long, 2011)
Mount Currie Conglomerate	Long (2011)	Ediacaran-Lower Cambrian	iLA (McMahon *et al.*, 2017) can be traced across laterally extensive sheets.	No palaeoflow measurements could be obtained.
Orienta Formation	Morey & Ojakangas (1982)	c. 1000 Ma	Fining up cycles	Vertical sequence analysis only. No evidence for LA surfaces, levee facies, or systematic upsection deviation in palaeocurrent flow
Fond du Lac Formation	Morey & Ojakangas (1982)	c. 1000 Ma	171 fining up cycles, ranging from 0.3–18.6 m thick	Vertical sequence analysis only. No evidence for LA surfaces, levee facies, or systematic upsection deviation in palaeocurrent flow.
Solor Church Formation	Morey & Ojakangas (1982)	c. 1000 Ma	Fining up cycles	Vertical sequence analysis only. No evidence for LA surfaces, levee facies, or systematic upsection deviation in palaeocurrent flow. Palaeocurrent measurements are strongly unimodal (Morey, 1967).
Red Castle Formation	Wallace & Crittenden (1969)	Neoproterozoic	Fining up cycles	Vertical sequence analysis only. No evidence for LA surfaces, levee facies, or systematic upsection deviation in palaeocurrent flow.
Serpent Formation	Long (1978)	2200–2400 Ma	Santos *et al.* (2017) state that Long (1976) interpret the Serpent Formation as the product of a sandy meandering fluvial system.	Long (1978) states the Serpent Formation is 'interpreted as the product of deposition in a (braided) stream system with low to intermediate sinuosity'.

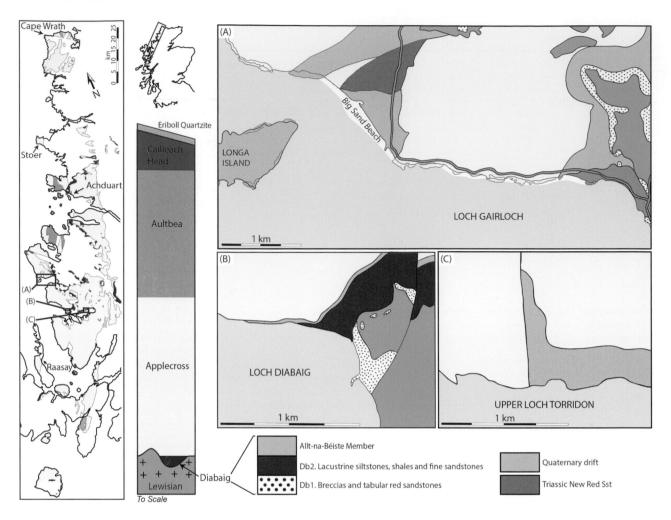

Fig. 7. Geological map and stratigraphic section of the Torridon Group, Scotland. Locations of study indicated in higher resolution inset maps: (A) Gairloch; (B) Diabaig; (C) Torridon.

thickness of 18 metres), Gairloch (c. 160 metres) and Torridon (95 metres) (Fig. 7). The type section is the only location where it can be studied in the sedimentary context of both the underlying Diabaig Formation lacustrine rocks and overlying Applecross Formation braided alluvium (Figs 8 and 9). At Gairloch, the base of the succession is faulted out whereas at Torridon, neither the section top nor base is exposed (Fig. 10).

Frequency of occurrence of LA-IHS

LA-IHS are only associated with rare, relatively mudrock-rich sections of the Allt-na-Béiste Member (Figs 9, 11 and 12). While mud clasts are locally common (Fig. 13C and D), *in situ* mudrock beds (siltstone, mudstone) (Fig.13A and

B) comprise an average of < 2% of the total thickness of the member. However, they vary in importance between outcrop locations, comprising a cumulative thickness of c. 4 metres of the 18 metres exposed at Diabaig (22%), c. 1 metre of the 160 m at Gairloch (<1%) and 3.5 metres of the 95 m at Torridon (4%) (Fig. 10). These finer lithologies occur as laterally discontinuous packages that are typically < 0.1 m-thick [maximum 2 m (Fig. 13A and B]). The packages have planar bases on top of underlying sandstones but their tops are usually erosional and irregular. In some horizons, cm-thick silty-sandstones alternate with cm-dm-thick medium-grained sandstones.

A total of six LA-IHS sets have been identified in association with mudrock packages (4 at Diabaig, 2 at Gairloch, 0 at Torridon). Individual

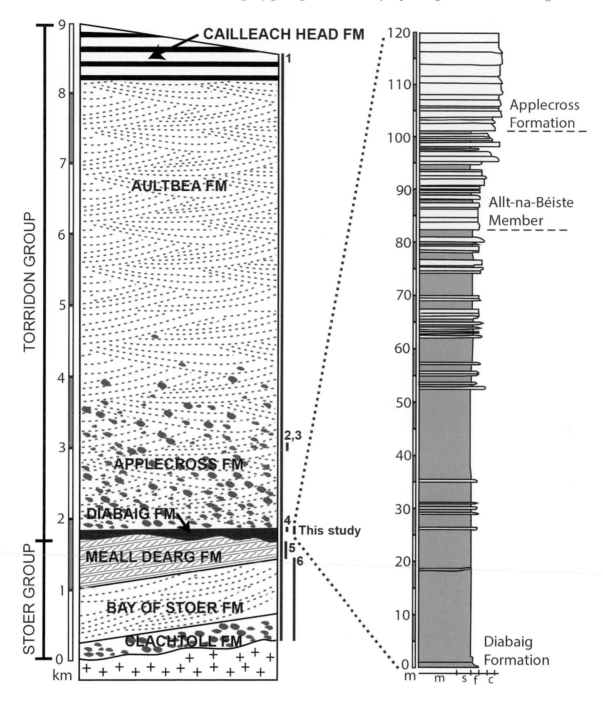

Fig. 8. Torridonian lithostratigraphy. Red lines mark the stratigraphic extent of recent sedimentological studies of Torridonian strata: 1) Stewart (2002); 2) Ielpi & Ghinassi (2015); 3) Ghinassi & Ielpi (2018); 4) Santos & Owen (2016); 5) McMahon & Davies (2018b); 6) Ielpi *et al.* (2016). Inset sedimentary log shows a measured section of the Diabaig Formation, Allt-na-Béiste Member and Applecross Formation at the Allt-na- Béiste types section at Diabaig.

Fig. 9. Sedimentary deposits at Diabaig: (A) Diabaig Formation lacustrine mudrocks; (B) The contact between Diabaig lacustrine facies and the Allt-na-Béiste Member; (C) Sandstone and mudstone dominated sedimentary facies preserved in the Allt-na-Béiste. Bag is 37 cm-high. Geologist is 187 cm-tall.

LA-IHS sets consist of a coarse-to-fine couplet (medium-grained sandstone and silty, very fine-grained sandstone; Figs 11 and 12B). Of the six LA-IHS sets, three are fully preserved (i.e. have distinct bottomset, foreset and topset components) (Fig. 11A). These examples display gradational contacts with overlying mudrocks that most probably represent overbank fines (Santos & Owen, 2016). In the other instances, LA-IHS deposits are erosionally truncated by overlying sandstones (Fig. 12B).

The widths of LA-IHS set range between 157 to 278 cm and heights range from 36 to 41 cm: in total they account for a combined vertical stratigraphic thickness of 2.1 m of the 295 m-thick member. The negligible contribution of LA-IHS and fine-grained

components to the overwhelmingly sandstone-dominated stratigraphy has been overlooked in previous studies of the member that have instead focused on the atypical characteristics (for example, the 4 LA-IHS at Diabaig are figured 6 times by Santos & Owen [2016]). Furthermore, earlier architectural interpretations are in part contradictory and require re-evaluation: a reinterpretation of the outcrop figured by Santos & Owen (2016; their figs 4 and 9) is shown in Fig. 11C.

Dimensions of LA-IHS

Complete set thicknesses (i.e. for those LA-IHS which are not top truncated) range between 36 to 41 cm and cosets of multiple LA-IHS are absent.

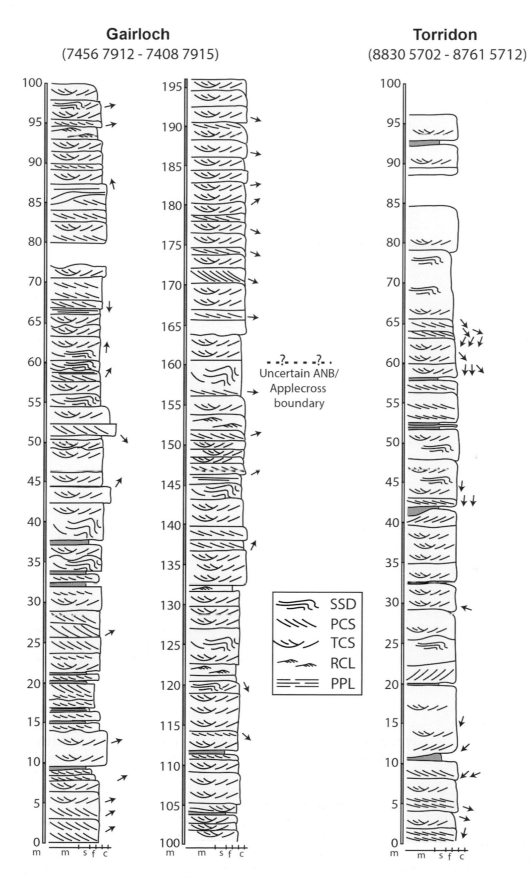

Gairloch
(7456 7912 - 7408 7915)

Torridon
(8830 5702 - 8761 5712)

Uncertain ANB/
Applecross
boundary

SSD
PCS
TCS
RCL
PPL

Fig. 10. Measured sedimentary logs of the Allt-na-Béiste Member at Gairloch and Torridon. SSD = Soft-sediment deformation; PCS = Planar cross-stratification; TCS = Trough cross-stratification; RCL = Ripple cross-lamination; PPL = Plane parallel lamination.

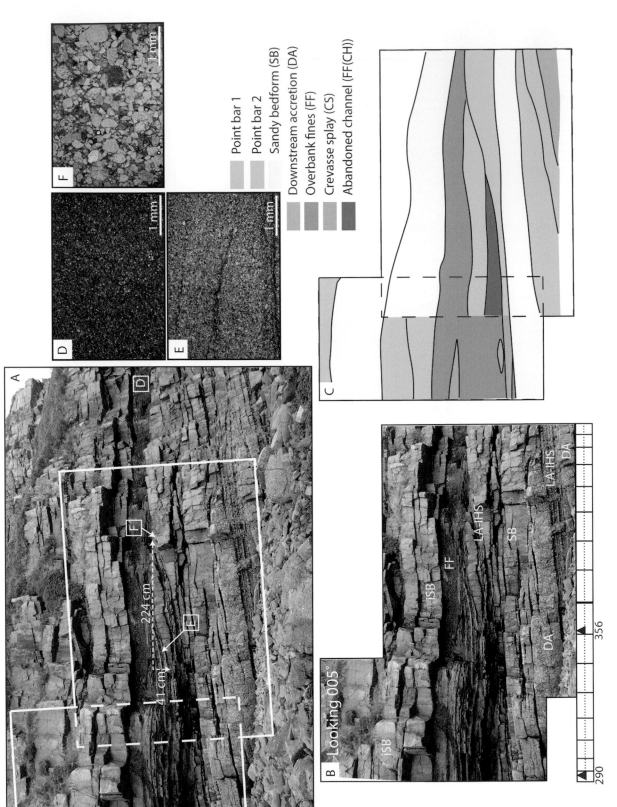

Fig. 11. Allt-na-Béiste Member laterally accreting inclined heterolithic stratification (LA-IHS) deposits: (A) 41 cm-thick LA-IHS deposit at Diabaig, overlain by muddy-siltstones interpreted as overbank material (Fig. 13B). Locations of thin sectioned samples in Fig. 11D to F highlighted. White boxes outline area corresponding to that illustrated in Figs 4 and 9 of Santos & Owen, 2016, and are detailed in Fig. 11B to C; (B) Interpretation of depositional architecture (see Table 2 for acronym descriptions). Graphic illustration of palaeoflow orientation and direction of accretion of inclined heterolithic stratification follows methodology presented by Davies *et al.* (2018). Red triangle indicates average palaeoflow orientation for unit SB (Fig. 11B). Blue triangle indicates average direction of accretion for LA-IHS deposits (Fig. 15B); (C) Previous interpretation of depositional architecture (redrawn from Santos & Owen, 2016 [their Figs 4 and 9]); (D) Thin section of mudstones-siltstones interpreted as overbank material; (E) Thin section of fine component of inclined heterolithic stratification (siltstone); (F) Thin section of coarse component of inclined heterolithic stratification (medium-coarse sandstone).

Fig. 12. Allt-na-Béiste Member succession at Gairloch. (A) Photograph indicating dominance of sand-grade or coarser sediment. See measured stratigraphic log (Fig. 10); (B) Laterally accreting inclined heterolithic stratification deposits at Gairloch. Inset demonstrates that fine-sand grade sediment constitutes the fine-grained component of the inclined heterolithic stratification couplet. Graphic illustration of palaeoflow orientation and direction of accretion of inclined heterolithic stratification follows methodology presented by Davies *et al.* (2018). Blue triangle indicates average direction of accretion for the LA-IHS deposit. Red triangle indicates average palaeoflow orientation for the underlying unit SB.

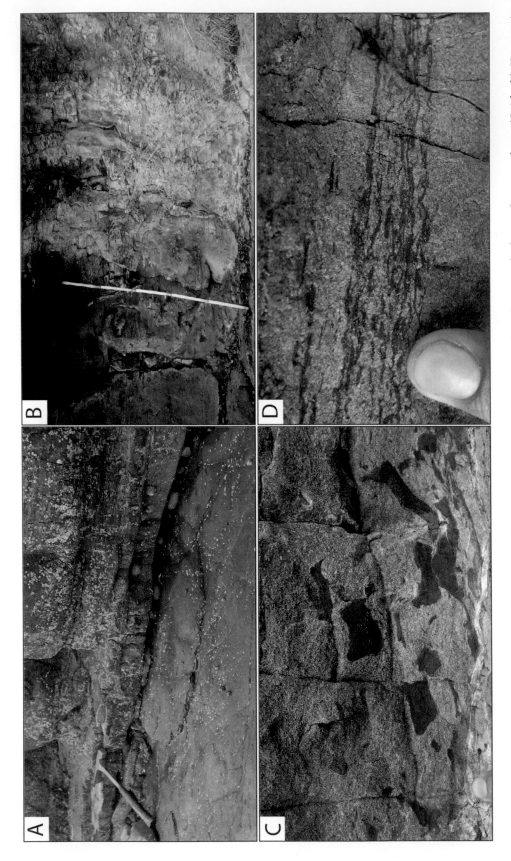

Fig. 13. Mudstone and siltstone sedimentary facies of the Allt-na-Béiste Member: (A) Intercalated sandstone and silty-mudstone packages (Gairloch). Hammer is 31 cm-long; (B) Silty-mudstone package (Diabaig). Metre rule for scale; (C, D) Intraformational mud clasts (both Gairloch).

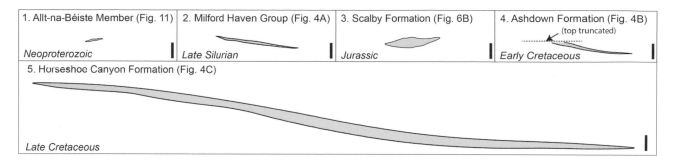

Fig. 14. Line tracing showing the comparative scale of examples of individual inclined heterolithic stratification sets from the: 1) Neoproterozoic Allt-na-Béiste Member; 2) Late Silurian Milford Haven Group; 3) Jurassic Scalby Formation; 4) Early Cretaceous Ashdown Formation; and 5) Late Cretaceous Horseshoe Canyon Formation. Red bar indicates 2 metres.

Whilst any palaeohydraulic reconstructions must be treated with caveats of uncertainty, thicknesses of 36 to 41 cm approximately equate to similar bankfull water-depths at the time of deposition (e.g. Allen, 1965; Bridge & Leeder, 1976; Miall, 2006). The channels that deposited the Allt-na-Béiste Member LA-IHS were therefore of negligible size and depth and unlikely to have been even minor fluvial trunk channels. This is apparent when their dimensions are directly compared with LA-IHS sets from other small-sized to moderate-sized meandering rivers preserved in Phanerozoic strata (Fig. 14; compare also Figs 4 and 6). The sigmoidal form of the majority of sets (4 of the 6) also indicates that the packages are not truncated and so may relate to gently curved meanders rather than high-sinuosity planforms (Bridge & Leeder, 1976). Additionally, the lack of lateral erosional amalgamation in 5 of the 6 LA-IHS examples suggests that the channels had limited lateral mobility (Thomas *et al.*, 1987).

Dominant Lithology of the Allt-na-Béiste Member

98% of the Allt-na-Béiste Member is composed of arkosic sandstone, presenting as concave-up trough cross-stratified sets (preserved thicknesses of 0.1 to 0.6 m), rarely associated with soft-sediment deformation (<5%) (Fig. 15). Planar cross-stratification is subordinate and occurs in either < 0.15 m sets (Fig. 15C) or larger 0.6 to 1.6 m sets which transition down-dip into planar stratification. Such planar cross-strata are usually top-truncated, either by low-angle to horizontal cross-stratification (Fig. 15C), further sets of planar-cross stratification, or tabular beds filled with trough-cross stratification. Low-angle

cross-stratified packages are typically several cm-thick and extend down-dip for up to 1.5 m. Locally, original dune topography is present (Fig. 15E), capped by cm-thick to dm-thick lenses of finer grained, ripple cross-laminated sandstone with siltstone intercalations (Fig. 15D). Architecturally, the sandstones are predominantly sandy-bedforms (SB of Miall, 1985) or interpreted sandy-bedforms (iSB of McMahon *et al.*, 2017; Table 2), presenting as tabular bodies (10 to 51 cm-thick). They are typically filled with trough cross-stratification, but packages displaying intercalations of small, planar cross-stratification and low-angle cross-stratification occur locally (Fig. 15C). Downstream-accretion elements (DA of Miall, 1985) are subordinate (Table 2) and comprise successive sets of large-scale planar cross-stratification which can be traced down-dip from foreset avalanche surfaces to planar-bedded stratification (Figs 16 and 17). Downstream-migration is indicated by the low palaeocurrent disparity between stratification sets and genetically related bounding surfaces (Miall, 1985) (Fig. 15). DA elements range in thickness between 0.3 to 2.8 m and can be traced up to 20 m in a down-flow direction.

The SB, iSB and DA sandstone elements of the Allt-na-Béiste Member provide no direct evidence for sinuous channel planforms. Widespread SB elements suggest deposition by migrating low amplitude three-dimensional dunes; their upwards transitions into ripple cross-laminae relating to long-term changes in aggradation rate and water depth shallowing. The tabular, stacked nature of the trough cross-stratified sets imply that sediment accreted largely by vertical aggradation. Preserved barforms also indicate modal downstream accretion and are ubiquitously

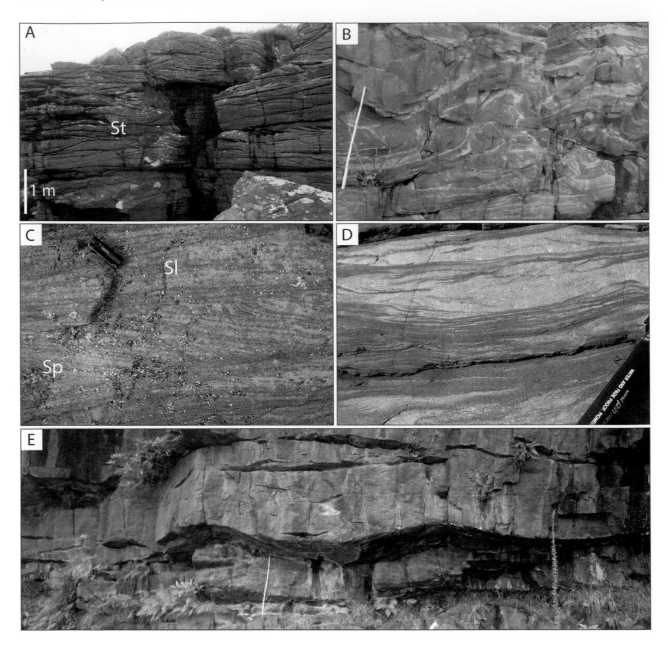

Fig. 15. Sandstone dominated sedimentary facies of the Allt-na-Béiste Member: (A) Trough cross-stratified sandstones (Gairloch); (B) Soft-sediment deformation (Gairloch). Metre rule for scale; (C) Small sets of planar cross-stratification and low-angle cross-stratification (Torridon). Pen lid is 4 cm-long; (D) Ripple cross-laminated sandstone with siltstone intercalations (Gairloch). Pen is 1 cm-wide; (E) Preserved dune topography (Torridon). Metre rule for scale. St = Trough cross-stratified sandstone; Sp = Planar cross-stratified sandstone; Sl = Low-angle cross-stratified sandstone.

top-truncated, such that the depth of the water body cannot be estimated.

The dominant facies of the Allt-na-Béiste Member are in accordance with classic facies models of low sinuosity streams, while those bearing signatures of the CMFM (crevasse splays, point bars, overbank deposits; Santos & Owen, 2016) are largely restricted to a singular outcrop at Diabaig and atypical of the Allt-na-Béiste Member as a whole.

Table 2. A comparison of the sedimentary characteristics of the Allt-na-Béiste Member and the Applecross Formation.

Element	Description	Allt-na-Béiste Member	Applecross Fm
Channels (CH)	Elements with erosional concave-up underlying geometry	Not recorded	Rare
Sandy-bedforms (SB)	Elements not genetically related to their underlying surface.	Common	Very common
'Interpreted' sandy-bedforms (iSB)	Packages where exposure prohibits an understanding of the relationship between inclined foresets and the underlying surface (McMahon *et al.*, 2017).	Common	Very common
Downstream-accretion element (DA)	Multiple cosets of downstream—orientated bedforms dynamically related to a common, underlying downstream-dipping surface (Miall, 1985).	Rare	Common
Homogeneous lateral-accretion deposit (LA)	Increments of gently dipping bedding planes with off-lapped upper terminations. Bedform migration 60–120° of underlying surface	Rare	Rare
Laterally-accreting inclined heterolithic stratification (LA-IHA)	Distinctive style of accretion characterized by alternating coarse-grained and fine-grained beds. Most commonly associated with lateral accretion. Tidal influence suspected in many reported cases.	6 recorded sets	Not recorded
Floodplain fines (FF)	Any fluvial element deposited outside the main fluvial channels	<2% total vertical thickness	<0.1% total vertical thickness

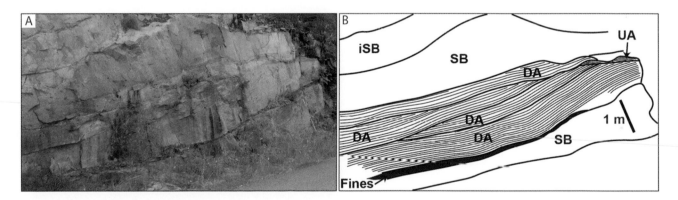

Fig. 16. Allt-na-Béiste Member downstream accretion elements at Torridon: (A) Field photograph. Metre rule for scale; (B) Interpreted architectural panel.

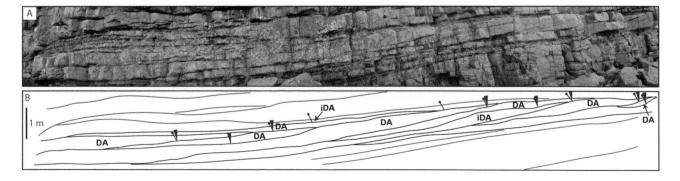

Fig. 17. Allt-na-Béiste Member downstream-accretion elements at Diabaig: (A) Field photograph; (B) Interpreted architectural panel. Blue rose diagram indicates palaeoflow measured from cross-bed foresets. Red pins indicate direction of dip of genetically underlying bounding surfaces. Bag for scale is 35 cm-high.

Interpretation of the Allt-na-Béiste Member LA-IHS

The rare LA-IHS in the Allt-na-Béiste Member represent the only unequivocal instance of pre-vegetation CMFM strata so far described from the global rock record (Santos & Owen, 2016). However, their small dimensions and rarity within the Allt-na-Béiste Member suggest that meandering channels were uncommon, localised and negligible-scale features. Although relatively insignificant conduits, the development of these pre-vegetation sinuous planforms requires additional explanation. While bank stability due to mud played a role in their formation, it is unlikely to be the sole trigger because other mudrock-rich intervals of pre-vegetation successions lack similar elements (even within specific stratigraphic intervals of the Torridonian Supergroup, such as in the Clachtoll Formation [Ielpi *et al.*, 2016], Poll a'Mhuilt Member [Stewart, 2002; Stueeken *et al.*, 2017]; and Cailleach Head Formation [Stewart, 2002]).

One probable explanation is that minor meandering channels formed as localised lacustrine basins became overfilled with sediment: a possibility that becomes apparent when the true stratigraphic placement of the Allt-na-Béiste Member is considered. Recently, the member has informally been taken to be the basal member of the overlying Applecross Formation (Stewart, 2002; Santos &

Owen, 2016), which records alluvium with a distal provenance that occurs across much of the regional outcrop area with little variation in lithology or palaeoflow (Nicholson, 1993; Stewart, 2002; Williams, 2001; Kinnaird *et al.*, 2007; Williams & Foden, 2011; Krabbendam *et al.*, 2017). However, formally, the Allt-na-Béiste is the uppermost member of the conformably underlying Diabaig Formation because it occurs below the notable stratigraphic hiatus (of unknown duration) that marks the base of the Applecross Formation (Peach *et al.*, 1907; British Geological Survey, 2017; Muirhead *et al.*, 2017). The sedimentary rocks of the Diabaig Formation are highly localised alluvial fan and lacustrine deposits, infilling palaeotopographic depressions in the Lewisian Gneiss basement and are characterised by a local Lewisian provenance and regional variation in lithology and thickness (Fig. 18). At the Allt-na-Béiste type locality, the Diabaig Formation consists of minor alluvial fan breccias and dominant lacustrine siltstones and very-fine-grained sandstones (Fig. 9A). Lacustrine facies are separated from the key outcrop of LA-IHS (Fig. 9C) by approximately 9 metres of thinly-bedded medium-grained arkosic sandstones (Fig. 9B).

When the Allt-na-Béiste Member is correctly considered as part of the Diabaig Formation rather than the Applecross Formation, two key characteristics of the member become readily explicable: 1) its variable palaeoflow indicators at each of its

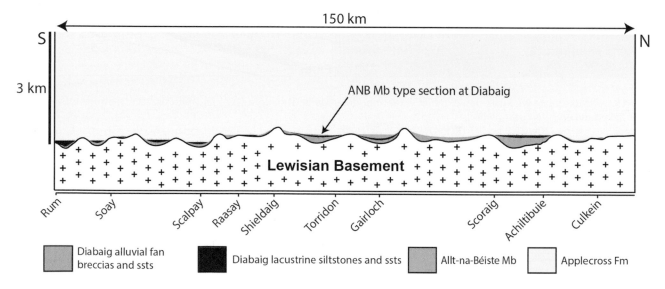

Fig. 18. Spatial distribution and representative thicknesses of Diabaig Formation, Allt-na-Béiste Member and Applecross Formation deposits across the Torridonian outcrop belt. Lewisian Basement Complex topography redrawn from Stewart (2002).

outcrop localities (Diabaig, $\theta = 331°$, n = 31; Gairloch, $\theta = 92°$, n = 56; Torridon, $\theta = 181°$, n = 42) compared with the uniform south-eastwards palaeoflow of the Applecross Formation (Nicholson, 1993); and 2) its marked thickness variation between outcrop areas, dependent on the extent to which palaeotopographic lows in the underlying Lewisian geology had previously been filled (Fig. 18). This also explains the absence of the member from those lows that are fully filled with older Diabaig Formation sediments (e.g. Rùm, Rasaay, Scoraig) and those areas lacking palaeotopography on top of the Lewisian (e.g. between Cape Wrath and the Stoer Peninsular [Williams, 1966]; Figs 18 and 19).

At the Allt-na-Béiste type section, Stewart (2002) estimated that the relief of the local palaeovalley was approximately 250 m. Diabaig Formation alluvial fan and lacustrine facies account for 216 m of the local stratigraphy (Fig. 8) meaning that such sediments would have previously filled most of the available accommodation space. Thus, by the time the Allt-na-Béiste LA-IHS were deposited, the local environment would have been typified by endorheic low-gradient alluvial plains, of limited geographic extent, where fine-grained cohesive sediment could sometimes accumulate. The 6 known < 41 cm-deep channels in the Allt-na-Béiste Member could therefore record small channels draining into the terminal Diabaig lakes, capable of developing short-lived, low mobility high sinuosity planforms due to the mud-rich, low gradient nature of the lake margins.

The rare instances of pre-vegetation LA IHS in the Allt-na-Béiste Member at Diabaig thus appear to be minor channels feeding a lake system in a near-filled localised palaeovalley and have less in common with the rest of the 295 m-thick member, other palaeovalleys, or with the 3000 metres of 'sheet-braided' alluvium (see McMahon & Davies, 2018b), which characterises the overlying Applecross Formation (Table 2). They attest to the possibility of meandering planforms being attained in specific instances on pre-vegetation Earth, but are atypical of pre-vegetation alluvium globally.

CONCLUSIONS

The possible existence, or not, of meandering rivers on Earth prior to the evolution of vegetation has been reviewed using evidence from geomorphology, experimental modelling and the geological record. Previous research may appear contradictory. However, when the differences between deductive, inductive and abductive reasoning are considered, it becomes clear that multiple methodologies all point towards similar conclusions. In small-sized to moderate-sized rivers, the effects of land plant evolution appear to have been fundamental: (1) inductive reasoning from modern geomorphic observations suggest that meandering rivers would have been less common before the evolution of land plants because plants play a major role in promoting bank and bar top stability (a role only rarely fulfilled by stability from factors such as ice or salt); (2) inductive reasoning from the stratigraphic record demonstrates an increase in the frequency and distribution of alluvial successions which conform to the

Fig. 19. Erosive contact between Diabaig Formation lacustrine facies and Applecross Formation braided alluvium at: (A) Achduart. Geologist is 196 cm-tall; (B) Raasay. Bag is 37 cm-high. Allt-na-Béiste Member absent.

'classic meandering river facies model', correlative with land plant evolution; (3) deductive reasoning from modelling observations support these contentions, additionally suggesting that small meandering rivers might have been possible in instances where bank stability was provided by cohesive sediment; and 4) abductive reasoning from geological observations find scant evidence for CMFM strata in pre-vegetation alluvium, which, while not being synonymous with all meandering rivers, do correlate strongly with the deposits of small-sized to moderate-sized muddy meandering streams. Unequivocal pre-vegetation CMFM strata are presently only known from the Neoproterozoic Allt-na-Béiste Member, Scotland. However, such deposits were probably deposited by localised, short-lived, small sinuous channels on a lake margin, rather than by meandering rivers on long-lived floodplains. While rare instances of larger-scale pre-vegetation CMFM strata (deposited by moderate-sized meandering rivers under localised favourable conditions) may not be unexpected, such facies signatures presently remain elusive in Earth's pre-Silurian sedimentary rock record.

ACKNOWLEDGEMENTS

Supported by Shell International Exploration and Production B.V under Research Framework agreement PT38181. We would like to thank Darrel Long, Mauricio Santos and Arjan Reesink for their constructive review of this manuscript. Maarten Kleinhans is also thanked for insightful discussion on the philosophical analysis of Earth science.

REFERENCES

Alexander, J. (1992) Nature and origin of a laterally extensive alluvial sandstone body in the Middle Jurassic Scalby Formation. *J. Geol. Soc.*, 149, 431–441.

Allen, J.R. (1963) The classification of cross-stratified units. With notes on their origin. *Sedimentology*, 2, 93–114.

Allen, J.R. (1965) A review of the origin and characteristics of recent alluvial sediments. *Sedimentology*, 5, 89–191.

Allen, J.R.L. (1970) Studies in fluviatile sedimentation: a comparison of fining-upwards cyclothems, with special reference to coarse-member composition and interpretation. *J. Sed. Res.*, 40, 298–323.

Allen, J.R.L. (1964) Studies in fluviatile sedimentation: six cyclothems from the Lower Old Red Sandstone, Anglowelsh Basin. *Sedimentology*, 3, 163–198.

Ashmore, P.E. (1991) How do gravel-bed rivers braid?. *Can. J. Earth Sci.*, 28(3), 326–341.

Ashworth, P.J. and Lewin, J. (2012) How do big rivers come to be different? *Earth-Sci. Rev.*, 114(1), 84–107.

Bardgett, R.D., Mommer, L. and De Vries, F.T. (2014) Going underground: root traits as drivers of ecosystem processes. *Trends in Ecology & Evolution*, 29, 692–699.

Bernard, H.A., LeBlanc, R.J. and Major, C.F. (1962) Recent and Pleistocene Geology of Southeast Texas: Field Excursion No. 3, November 10 and 11.

Braudrick, C.A., Dietrich, W.E., Leverich, G.T. and Sklar, L.S. (2009) Experimental evidence for the conditions necessary to sustain meandering in coarse-bedded rivers. *Proceedings of the National Academy of Sciences*, 106, 16936–16941.

Bridge, J.S. (2006) Fluvial facies models: Recent developments. In: *Facies Models Revisited* (Eds H.W. Posamentier and R.G. Walker), *SEPM Spec. Publ.*, 84, 85–170.

Bridge, J.S. (1985) Paleochannel Patterns Inferred From Alluvial Deposits: A Critical Evaluation Perspective: Perspective. *J. Sed. Res.*, 55(4).

Bridge, J.S. (1993) The interaction between channel geometry, water flow, sediment transport and deposition in braided rivers. *Geol. Soc. London Spec. Publ.*, 75, 13–71.

Bridge, J.S. and Jarvis, J. (1976) Flow and sedimentary processes in the meandering river South Esk, Glen Clova, Scotland. *Earth Surf. Proc. Land.*, 1, 303–336.

Bridges, P.H. and Leeder, M.R. (1976) Sedimentary model for intertidal mudflat channels, with examples from the Solway Firth, Scotland. *Sedimentology*, 23, 533–552.

Brierley, G.J. and Hickin, E.J. (1991) Channel planform as a non-controlling factor in fluvial sedimentology: the case of the Squamish River floodplain, British Columbia. *Sed. Geol.*, 75, 67–83.

Cant, D.J. and Walker, R.G. (1976) Development of a braided-fluvial facies model for the Devonian Battery Point Sandstone, Quebec. *Can. J. Earth Sci.*, 13(1), 102–119.

Brooks, A.P., Brierley, G.J. and Millar, R.G. (2003) The long-term control of vegetation and woody debris on channel and flood-plain evolution: insights from a paired catchment study in southeastern Australia. *Geomorphology*, 51, 7–29.

Church, M. (2006) Bed material transport and the morphology of alluvial river channels. *Annu. Rev. Earth Planet. Sci.*, 34, 325–354.

Colombera, L., Mountney, N.P. and McCaffrey, W.D. (2013) A quantitative approach to fluvial facies models: methods and example results. *Sedimentology*, 60(6), 1526–1558.

Corenblit, D., Steiger, J., González, E., Gurnell, A.M., Charrier, G., Darrozes, J., Dousseau, J., Julien, F., Lambs, L., Larrue, S. and Roussel, E. (2014) The biogeomorphological life cycle of poplars during the fluvial biogeomorphological succession: a special focus on Populus nigra L. *Earth Surf. Proc. Land.*, 39, 546–563.

Costard, F., Dupeyrat, L., Gautier, E. and Carey-Gailhardis, E. (2003) Fluvial thermal erosion investigations along a rapidly eroding river bank: application to the Lena River (central Siberia). *Earth Surf. Proc. Land.*, 28, 1349–1359.

Cotter, E. (1978) The evolution of fluvial style, with special reference to the central Appalachian Paleozoic. In: *Fluvial Sedimentology* (Ed. A.D. Miall)., *Can. Soc. Petrol. Geol. Mem.*, **5**, 361–383.

Damuth, J.E., Kolla, V., Flood, R.D., Kowsmann, R.O., Monteiro, M.C., Gorini, M.A., Palma, J.J. and Belderson, R.H. (1983) Distributary channel meandering and bifurcation patterns on the Amazon deep-sea fan as revealed by long-range side-scan sonar (GLORIA). *Geology*, **11**, 94–98.

Davies, N.S. and Gibling, M.R. (2010a) Cambrian to Devonian evolution of alluvial systems: the sedimentological impact of the earliest land plants. *Earth-Sci. Rev.*, **98**(3), 171–200.

Davies, N.S. and Gibling, M.R. (2010) Paleozoic vegetation and the Siluro-Devonian rise of fluvial lateral accretion sets. *Geology*, **38**, 51–54.

Davies, N.S., Gibling, M.R., McMahon, W.J., Slater, B.J., Long, D.G., Bashforth, A.R., Berry, C.M., Falcon-Lang, H.J., Gupta, S., Rygel, M.C. and Wellman, C.H. (2017) Discussion on 'Tectonic and environmental controls on Palaeozoic fluvial environments: reassessing the impacts of early land plants on sedimentation' . *J. Geol. Soc. London*, **174**(5), 947–950. https://doi.org/10.1144/jgs2016-063

Davies, N.S., Gibling, M.R. and Rygel, M.C. (2011) Alluvial facies evolution during the Palaeozoic greening of the continents: case studies, conceptual models and modern analogues. *Sedimentology*, **58**(1), 220–258.

Davies, N.S., McMahon, W.J. and Shillito, A.P. (2018) A Graphic Method For Depicting Horizontal Direction Data On Vertical Outcrop Photographs. *J. Sed. Res.*, **88**(4), 516–521.

Davis, J.M., Balme, M., Grindrod, P.M., Williams, R.M.E. and Gupta, S. (2016) Extensive Noachian fluvial systems in Arabia Terra: Implications for early Martian climate. *Geology*, **44**, 847–850.

Dietrich, W.E. and Perron, J.T. (2006) The search for a topographic signature of life. *Nature*, **439**, 411–418.

Els, B.G. (1998) The auriferous late Archaean sedimentation systems of South Africa: unique palaeo-environmental conditions?. *Sed. Geol.*, **120**, 205–224.

Eriksson, P.G., Condie, K.C., Tirsgaard, H., Mueller, W.U., Altermann, W., Miall, A.D., Aspler, L.B., Catuneanu, O. and Chiarenzelli, J.R. (1998) Precambrian clastic sedimentation systems. *Sed. Geol.*, **120**(1), 5–53.

Erskine, W.D., Saynor, M.J., Chalmers, A. and Riley, S.J. (2012) Water, wind, wood and trees: interactions, spatial variations, temporal dynamics and their potential role in river rehabilitation. *Geogr. Res.*, **50**, 60–74.

Ethridge, F.G. (2011) Interpretation of ancient fluvial channel deposits: review and recommendations. In: *From River to Rock Record: The Preservation of Fluvial Sediments and their Subsequent Interpretation* (Eds S. Davidson, S. Leleu and C.P. North), *SEPM*, **97**, 9–35.

Ferguson, R.I. (1987) Hydraulic and sedimentary controls of channel pattern. In: *River channels: Environments and processes*, Blackwell, 129–158.

Fielding, C.R. (2006) Upper flow regime sheets, lenses and scour fills: extending the range of architectural elements for fluvial sediment bodies. *Sed. Geol.*, **190**, 227–240.

Finotello, A., Lanzoni, S., Ghinassi, M., Marani, M., Rinaldo, A. and D'Alpaos, A. (2018) Field migration rates of tidal meanders recapitulate fluvial morphodynamics. *Proc. Nat. Acad. Sci.*, **115**, 1463–1468.

Friedkin, J.F. (1945) *Laboratory study of the meandering of alluvial rivers.* U.S. Waterways Engineering Experimental Station, Vicksburg.

Ghinassi, M. and Ielpi, A. (2018) Precambrian snapshots: Morphodynamics of Torridonian fluvial braid-bars revealed by three-dimensional photogrammetry and outcrop sedimentology. *Sedimentology*, **65**, 492–516.

Ghinassi, M., Ielpi, A., Aldinucci, M. and Fustic, M. (2016) Downstream-migrating fluvial point bars in the rock record. *Sed. Geol.*, **334**, 66–96.

Gibling, M.R. (2006) Width and thickness of fluvial channel bodies and valley fills in the geological record: a literature compilation and classification. *J. Sed. Res.*, **76**, 731–770.

Gilliam, A.E. and Lerman, A. (2016) Formation mechanisms of channels on Titan through dissolution by ammonium sulfate and erosion by liquid ammonia and ethane. *Planet. Space Sci.*, **132**, 13–22.

Harden, D.R. (1990) Controlling factors in the distribution and development of incised meanders in the central Colorado Plateau. *Geol. Soc. Am. Bull.*, **102**, 233–242.

Hartley, A.J., Owen, A., Swan, A., Weissmann, G.S., Holzweber, B.I., Howell, J., Nichols, G. and Scuderi, L. (2015) Recognition and importance of amalgamated sandy meander belts in the continental rock record. *Geology*, **43**, 679–682.

Horton, A.J., Constantine, J.A., Hales, T.C., Goossens, B., Bruford, M.W. and Lazarus, E.D. (2017) Modification of river meandering by tropical deforestation. *Geology*, **45**(6), 511–514.

Hughes, Z.J. (2011) Tidal channels on tidal flats and marshes. In: *Principles of Tidal Sedimentology* (Eds R.A. Davis Jr and R.W. Dalrymple), Springer Science and Business Media, 269–300.

Hulme, G. (1982) A review of lava flow processes related to the formation of lunar sinuous rilles. *Surveys in Geophysics*, **5**, 245–279.

Ielpi, A. (2016) Lateral accretion of modern unvegetated rivers: remotely sensed fluvial–aeolian morphodynamics and perspectives on the Precambrian rock record. *Geol. Mag.*, **154**, 609–624.

Ielpi, A. and Ghinassi, M. (2014) Planform architecture, stratigraphic signature and morphodynamics of an exhumed Jurassic meander plain (Scalby Formation, Yorkshire, UK). *Sedimentology*, **61**, 1923–1960.

Ielpi, A. and Ghinassi, M. (2015) Planview style and palaeodrainage of Torridonian channel belts: Applecross Formation, Stoer Peninsula, Scotland. *Sed. Geol.*, **325**, 1–16.

Ielpi, A. and Rainbird, R.H. (2016) Reappraisal of Precambrian sheet-braided rivers: Evidence for 1·9 Ga deep-channelled drainage. *Sedimentology*, **63**, 1550–1581.

Ielpi, A., Rainbird, R.H., Ventra, D. and Ghinassi, M. (2017) Morphometric convergence between Proterozoic and post-vegetation rivers. *Nature Comms*, **8**, 15250.

Ielpi, A., Ventra, D. and Ghinassi, M. (2016) Deeply channelled Precambrian rivers: Remote sensing and outcrop evidence from the 1.2 Ga Stoer Group of NW Scotland. *Precambrian Res.*, **281**, 291–311.

Imran, J., Parker, G. and Pirmez, C. (1999) A nonlinear model of flow in meandering submarine and subaerial channels. *J. Fluid Mech.*, **400**, 295–331.

Jackson II, R.G. (1977) Preliminary evaluation of lithofacies models for meandering alluvial streams. *AAPG Mem.*, **5**, 543–576.

Karlstrom, L., Gajjar, P. and Manga, M. (2013) Meander formation in supraglacial streams. *J. Geophys. Res.: Earth Surf.*, **118**, 1897–1907.

Kinnaird, T.C., Prave, A.R., Kirkland, C.L., Horstwood, M., Parrish, R. and Batchelor, R.A. (2007) The late Mesoproterozoic–early Neoproterozoic tectonostratigraphic evolution of NW Scotland: the Torridonian revisited. *J. Geol. Soc.*, **164**, 541–551.

Kleinhans, M.G. (2010) Sorting out river channel patterns. *Prog. Phys. Geogr.*, **34**, 287–326.

Kleinhans, M.G., Bierkens, M.F.P. and Van der Perk, M. (2010) HESS Opinions On the use of laboratory experimentation: Hydrologists, bring out shovels and garden hoses and hit the dirt. *Hydrology and Earth System Sciences*, **14**(2), 369–382.

Kleinhans, M.G., Buskes, C.J.J. and de Regt, H.W. (2009) Philosophy of Earth science. In: *Philosophies of the Sciences* (Ed. F. Allhoff), Wiley-Blackwell, N.Y, 213–235.

Kleinhans, M.G., de Vries, B., Braat, L. and van Oorschot in press. Muddy and vegetated floodplain effects on fluvial pattern in an incised river. *Earth Surf. Proc. Land.*, Article DOI: 10.1002/esp.4437, Internal Article ID: 15524064.

Kleinhans, M.G. and van den Berg, J.H. (2011) River channel and bar patterns explained and predicted by an empirical and a physics-based method. *Earth Surf. Proc. Land.*, **36**, 721–738.

Komatsu, G. and Baker, V.R. (1994) Meander properties of Venusian channels. *Geology*, **22**, 67–70.

Krabbendam, M., Bonsor, H., Horstwood, M.S. and Rivers, T. (2017) Tracking the evolution of the Grenvillian foreland basin: Constraints from sedimentology and detrital zircon and rutile in the Sleat and Torridon groups, Scotland. *Precambrian Res.*, **295**, 67–89.

Kuhn, N. (2014) *Experiments in Reduced Gravity: Sediment Settling on Mars*. Elsevier.

Lazarus, E.D. and Constantine, J.A. (2013) Generic theory for channel sinuosity. *Proc. Nat. Acad. Sci.*, **110**, 8447–8452.

Leeder, M.R. and Nami, M. (1979) Sedimentary models for the non-marine Scalby Formation (Middle Jurassic) and evidence for late Bajocian/Bathonian uplift of the Yorkshire Basin. *Proc. Yorkshire Geol. Polytech. Soc.*, **42**, 461–482.

Leopold, L.B. and Wolman, M.G. (1960) River meanders. *Geol. Soc. Am. Bull.*, **71**, 769–793.

Long, D.G. (1978) Proterozoic stream deposits: some problems of recognition and interpretation of ancient sandy fluvial systems. In: *Fluvial Sedimentology* (Ed. A.D. Miall), *Can. Soc. Petrol. Geol. Mem.*, **5**, 313–341.

Long, D.G.F. (2011) Architecture and depositional style of fluvial systems before land plants: a comparison of Precambrian, early Paleozoic and modern river deposits. In: *From River to Rock Record: The Preservation of Fluvial Sediments and their subsequent Interpretation* (Eds S. Davidson, S., Leleu and C.P. North), *SEPM*, 37–61.

Long, D.G.F. (1976) *The Stratigraphy and Sedimentology of the Huronian (Lower Aphebian) Mississagi and Serpent Formations*. PhD thesis. University of Western Ontario, London, ON.

Magee, J.W., Bowler, J.M., Miller, G.H. and Williams, D.L.G. (1995) Stratigraphy, sedimentology, chronology and palaeohydrology of Quaternary lacustrine deposits at Madigan Gulf, Lake Eyre, South Australia. *Palaeogeogr. Palaeoclimatol. Palaeoecol.*, **113**, 3–42.

Matsubara, Y., Howard, A.D., Burr, D.M., Williams, R.M., Dietrich, W.E. and Moore, J.M. (2015) River meandering on Earth and Mars: A comparative study of Aeolis Dorsa meanders, Mars and possible terrestrial analogs of the Usuktuk River, AK and the Quinn River, NV. *Geomorphology*, **240**, 102–120.

McMahon, W.J. and Davies, N.S. (2018a) Evolution of alluvial mudrock forced by early land plants. *Science*, **359**, 994–995.

McMahon, W.J. and Davies, N.S. (2018b) High-energy flood events recorded in the Mesoproterozoic Meall Dearg Formation, NW Scotland; their recognition and implications for the study of pre-vegetation alluvium. *J. Geol. Soc.*, **175**, 13–32.

McMahon, W.J., Davies, N.S. and Went, D.J. (2017) Negligible microbial matground influence on pre-vegetation river functioning: Evidence from the Ediacaran-Lower Cambrian Series Rouge, France. *Precambrian Res.*, **292**, 13–34.

Miall, A.D. (1985) Architectural-element analysis: a new method of facies analysis applied to fluvial deposits. *Earth-Sci. Rev.*, **22**, 261–308.

Miall, A.D. (1977) A review of the braided-river depositional environment. *Earth-Sci. Rev.*, **13**, 1–62.

Miall, A.D. (2006) How do we identify big rivers? And how big is big?. *Sed. Geol.*, **186**, 39–50.

Miall, A.D. (1996) *The geology of fluvial deposits: sedimentary facies, basin analysis and petroleum geology*. Springer.

Morey, G.B. (1967) RI-07 Stratigraphy and Petrology of the Type Fond du Lac Formation Duluth, Minnesota. *Minnesota Geol. Surv. Rept. Invest.*, **7**, 35.

Morey, G.B. and Ojakangas, R.W. (1982) 7D: Keweenawan sedimentary rocks of eastern Minnesota and northwestern Wisconsin. *Geol. Soc. Am. Mem.*, **156**, 135–146.

Morris, J.L., Leake, J.R., Stein, W.E., Berry, C.M., Marshall, J.E., Wellman, C.H., Milton, J.A., Hillier, S., Mannolini, F., Quirk, J. and Beerling, D.J. (2015) Investigating Devonian trees as geo-engineers of past climates: linking palaeosols to palaeobotany and experimental geobiology. *Palaeontology*, **58**, 787–801.

Muirhead, D.K., Parnell, J., Spinks, S. and Bowden, S.A. (2017) Characterization of organic matter in the Torridonian using Raman spectroscopy. *Geol. Soc. London Spec. Publ.*, **448**, 71–80.

Nami, M. (1976) An exhumed Jurassic meander belt from Yorkshire, England. *Geol. Mag.*, **113**, 47–52.

Nicholson, P.G. (1993) A basin reappraisal of the Proterozoic Torridon Group, northwest Scotland. *Tectonic Controls and Signatures in Sedimentary Successions, Int. Assoc. Sedimentol. Spec. Publ.*, **20**, 40, 183–202.

Ore, H.T. (1964) Some criteria for recognition of braided stream deposits. *Rocky Mountain Geology*, **3**, 1–14.

Paola, C., Mullin, J., Ellis, C., Mohrig, D.C., Swenson, J.B., Parker, G., Hickson, T., Heller, P.L., Pratson, L., Syvitski, J. and Sheets, B. (2001) Experimental stratigraphy. *GSA Today*, **11**(7), 4–9.

Parker, G. (1979) Hydraulic geometry of active gravel rivers. *Journal of the Hydraulics Division*, **105**, 1185–1201, cedb.asce.org

Peach, B.N., Horne, J., Gunn, W., Clough, C.T., Hinxman, L.W. and Teall, J.J.H. (1907) *The geological structure of the northwest highlands of Scotland*. Mem. Geol. Surv. Scotland.

Peakall, J., Ashworth, P.J. and Best, J.L. (2007) Meander-bend evolution, alluvial architecture and the role of cohesion in sinuous river channels: a flume study. *J. Sed. Res.*, **77**, 197–212.

Pretorius, D.A. (1974) Gold in the Proterozoic sediments of South Africa: Systems, paradigms and models. *University of the Witwatersrand, Economic Geology Research Unit, Information Circular*, **87**, 2.

Puigdefabregas, C. and Van Vliet, A. (1977) Meandering stream deposits from the Tertiary of the southern Pyrenees. In: *Fluvial Sedimentology* (Ed. A.D. Miall), *Mere. Can. Soc. Pet. Geol.*, **5**, 469–485.

Retallack, G.J., Gose, B.N. and Osterhout, J.T. (2015) Periglacial paleosols and Cryogenian paleoclimate near Adelaide, South Australia. *Precambrian Res.*, **263**, 1–18.

Retallack, G.J., Marconato, A., Osterhout, J.T., Watts, K.E. and Bindeman, I.N. (2014) Revised Wonoka isotopic anomaly in South Australia and Late Ediacaran mass extinction. *J. Geol. Soc.*, **171**, 709–722.

Rubinstein, C.V., Gerrienne, P., de la Puente, G., Astini, R.A. and Steemans, P. (2010) Early Middle Ordovician evidence for land plants in Argentina (eastern Gondwana). *New Phytologist*, **188**, 365–369.

Santos, M.G., Mountney, N.P. and Peakall, J. (2017) Tectonic and environmental controls on Palaeozoic fluvial environments: reassessing the impacts of early land plants on sedimentation. *J. Geol. Soc.*, **174**(3), 393–404.

Santos, M.G. and Owen, G. (2016) Heterolithic meandering-channel deposits from the Neoproterozoic of NW Scotland: Implications for palaeogeographic reconstructions of Precambrian sedimentary environments. *Precambrian Res.*, **272**, 226–243.

Schumm, S.A. and Khan, H.R. (1972) Experimental study of channel patterns. *Geol. Soc. Am. Bull.*, **83**, 1755–1770.

Schumm, S.A., Mosley, M.P. and Weaver, W. (1987) *Experimental fluvial geomorphology*. Chichester, NY: John Wiley & Sons.

Simon, A. and Collison, A.J.C. (2002) Quantifying the mechanical and hydrologic effects of riparian vegetation on streambank stability, *Earth Surf. Proc. Land.*, **27**, 527–546.

Smith, C.E. (1998) Modeling high sinuosity meanders in a small flume. *Geomorphology*, **25**, 19–30.

Smith, D.G. (1976) Effect of vegetation on lateral migration of anastomosed channels of a glacier meltwater river. *Geol. Soc. Am. Bull.*, **87**, 857–860.

Stewart, A.D. (2002) The later Proterozoic Torridonian rocks of Scotland: their sedimentology, geochemistry and origin. *Geol. Soc. London Mem.*, **24**.

Stueeken, E.E., Bellefroid, E., Prave, A.R., Asael, D., Planavsky, N. and Lyons, T. (2017) Not so non-marine? Revisiting the Stoer Group and the Mesoproterozoic biosphere. *Geochemical Perspectives Letters*, **3**, 221–229

Sweet, I.P. (1988) Early Proterozoic stream deposits: braided or meandering-evidence from central Australia. *Sed. Geol.*, **58**, 277–293.

Tal, M., Gran, K., Murray, A.B., Paola, C. and Hicks, D.M. (2004) Riparian vegetation as a primary control on channel characteristics in multi-thread rivers. *Riparian Vegetation and Fluvial Geomorphology*, 43–58.

Tal, M. and Paola, C. (2007) Dynamic single-thread channels maintained by the interaction of flow and vegetation. *Geology*, **35**, 347–350.

Tal, M. and Paola, C. (2010) Effects of vegetation on channel morphodynamics: results and insights from laboratory experiments. *Earth Surf. Proc. Land.*, **35**, 1014–1028.

Temmerman, S., Bouma, T.J., Van de Koppel, J., Van der Wal, D., De Vries, M.B. and Herman, P.M.J. (2007) Vegetation causes channel erosion in a tidal landscape. *Geology*, **35**, 631–634.

Thomas, R.G., Smith, D.G., Wood, J.M., Visser, J., Calverley-Range, E.A. and Koster, E.H. (1987) Inclined heterolithic stratification—terminology, description, interpretation and significance. *Sed. Geol.*, **53**(1–2), 123–179.

Tiffany, J.B. and Nelson, G.A. (1939) Studies of meandering of model-streams. *Eos, Am. Geophys. Union Trans.*, **20**, 644–649.

Thornes, J.B. (1990) *Vegetation and Erosion: Processes and Environments*. Wiley, Chichester.

Tyler, N. and Ethridge, F.G. (1983) Depositional setting of the Salt Wash Member of the Morrison Formation, southwest Colorado. *J. Sed. Res.*, **53**, 67–82.

van de Lageweg, W.I., van Dijk, W.M., Baar, A.W., Rutten, J. and Kleinhans, M.G. (2014) Bank pull or bar push: What drives scroll-bar formation in meandering rivers?. *Geology*, **42**, 319–322.

Van Dijk, W.M., Lageweg, W.I. and Kleinhans, M.G. (2012) Experimental meandering river with chute cutoffs. *J. Geophys. Res.: Earth Surf.*, **117**, F03023.

Van Dijk, W.M., Lageweg, W.I. and Kleinhans, M.G. (2013) Formation of a cohesive floodplain in a dynamic experimental meandering river. *Earth Surf. Proc. Land.*, **38**, 1550–1565.

Walker, R.G. (1976) Facies and facies models, general introduction. In: *Facies Models. Geoscience Canada Reprint Series* (Ed. R.G. Walker), 21–24, **3**(1).

Walker, H.J. and Arnborg, L. (1966) Permafrost and ice-wedge effect on riverbank erosion. In *Proceedings of permafrost international conference, Lafayette*, 164–171.

Walker, R.G. and Cant, D.J. (1979) Facies Models 3: Sandy Fluvial Systems. In: *Facies Models* (Ed. R.G. Walker), *Geol. Assoc. Can.*, 23–32.

Wallace, C.A. and Crittenden Jr, M.D. (1969) The stratigraphy, depositional environment and correlation of the Precambrian Uinta Mountain Group, western Uinta Mountains, Utah. *Geologic Guidebook of the Uinta Mountains: Utah's Maverick Range, Sixteenth Annual Field Conference*, 127–141.

Wang, P. (2012) Principles of sediment transport applicable in tidal environments. In: *Principles of Tidal Sedimentology* (Eds R.A. Davis Jr and R.W. Dalrymple), Springer Science and Buisness Media, 269–300.

Williams, G.E. (1966) *The Precambrian Torridonian sediments of the Cape Wrath district, north-west Scotland. PhD thesis*, University of Reading.

Williams, G.E. (2001) Neoproterozoic (Torridonian) alluvial fan succession, northwest Scotland and its tectonic setting and provenance. *Geol. Mag.*, **138**, 161–184.

Williams, G.E. and Foden, J. (2011) A unifying model for the Torridon Group (early Neoproterozoic), NW Scotland: Product of post-Grenvillian extensional collapse. *Earth-Sci. Rev.*, **108**, 34–49.

www.bgs.ac.uk/lexicon/lexicon.cfm?pub=TCD   (accessed 27th February, 2018).

Xu, J. (2002) River sedimentation and channel adjustment of the lower Yellow River as influenced by low discharges and seasonal channel dry-ups. *Geomorphology*, **43**, 151–164.

Int. Assoc. Sedimentol. Spec. Publ (2018) **48**, 149–172.

An exhumed fine-grained meandering channel in the lower Permian Clear Fork Formation, north-central Texas: Processes of mud accumulation and the role of vegetation in channel dynamics

SHARANE S.T. SIMON[†*], MARTIN R. GIBLING[†], WILLIAM A. DIMICHELE[‡], DAN S. CHANEY[‡] and REBECCA KOLL[§]

[†] *Department of Earth Sciences, Dalhousie University, Halifax, Nova Scotia, Canada*
[‡] *Department of Paleobiology, National Museum of Natural History, Smithsonian Institution, Washington, DC, USA*
[§] *Florida Museum of Natural History, University of Florida, Gainesville, FL, USA*
* *Corresponding author: Department of Geology, Allegheny College, Meadville, PA, USA*

ABSTRACT

Ancient fine-grained meandering channels are under-represented in the literature and their formative processes are rarely explored. The Montgomery Ranch 3 site of the Clear Fork Formation of Texas contains an exhumed fine-grained point bar that migrated for at least 50 m within a channel 2 m deep and 36 m wide. The point bar comprises thick inclined layers of unstratified mudstone intercalated with thin layers of fine-grained, ripple cross-laminated sandstone, with dips averaging nearly 16°. Rill casts and swept ripples on the sandstone surfaces indicate declining water levels. Petrographic analysis of the mudstone shows silt and clay laid down from suspension, but sand-sized mud aggregates transported as bedload (present at other sites in the formation) were not observed. The sandstone beds are attributed to lateral accretion on the point bar during periods of sustained flow, whereas the mudstone beds are attributed to oblique accretion as fine sediment draped the bar during waning and low-flow periods. Sandstone and mudstone units are composite units from numerous flow events and their alternation may reflect secular variation in flood frequency and intensity. In an associated abandoned-channel fill, weakly laminated mudstone with desiccation cracks contains leaves and seeds of *Evolsonia texana*, marattialean foliage and *Taeniopteris* sp., with root traces penetrating the leaves. Some taxa preferred high water tables and humid conditions, whereas others were dryland colonisers. This apparent discrepancy may reflect the persistence of wetter channel reaches within an otherwise dry setting. Despite the scarcity of preserved plant fossils, vegetation was probably sufficiently widespread to promote bank strength and local sediment accumulation.

Keywords: Exhumed point bar, fine-grained meandering channels, oblique accretion, lateral accretion, vegetation

INTRODUCTION

Fine-grained deposits laid down in meandering-fluvial channels are common in the geological record (Stewart, 1981; Edwards *et al.*, 1983; Thomas *et al.*, 1987; Mack *et al.*, 2003; Ghosh *et al.*, 2006; Rygel & Gibling, 2006; Simon & Gibling, 2017a; Dasgupta *et al.*, 2017). They have received less attention than sand-rich deposits, which are as economically important as hydrocarbon

Fluvial Meanders and Their Sedimentary Products in the Rock Record, First Edition.
Edited by Massimiliano Ghinassi, Luca Colombera, Nigel P. Mountney and Arnold Jan H. Reesink.

reservoirs and aquifers. Due to the erodible nature of the strata, cliff exposures are uncommon and exhumed channels and bars suitable for high-resolution architectural analysis represent an unusual opportunity to explore the processes responsible for their formation.

Among outstanding issues pertaining to fine-grained meandering systems, two are highlighted in this paper. Firstly, a thorough interpretation depends on discriminating between the processes that deposit fine sediment. Within active channels, these processes include lateral accretion of sand-sized mud aggregates transported as bedload (Rust & Nanson, 1989), oblique accretion of mud from suspension on bars and banks (Page *et al.*, 2003) and counter point bar accretion in the downstream parts of meander bends (Smith *et al.*, 2009). Discriminating mud deposited as bedload (lateral accretion) from mud deposited from suspension (oblique accretion) leads to an interpretation of active flow conditions or quiescence – a fundamental distinction in channel behaviour. Distinguishing these processes in the geological record is difficult because 1) they form end-members in a depositional spectrum, 2) desiccation and pedogenesis may destroy evidence of surface processes; and 3) diagenesis may preferentially alter the finer sediment (Sambrook Smith *et al.*, 2016; Simon & Gibling, 2017a,b). Criteria for discriminating between in-channel processes of mud deposition are addressed in the following section. Mud may also be deposited from suspension and as bedload aggregates on floodplains and from suspension in abandoned channels.

Secondly, many fine-grained fluvial deposits show features indicative of seasonal flow and periodic dryness, which allowed rooted vegetation to become established in the channel. Fielding *et al.* (2009) used in-channel vegetation as a criterion to identify tropical, seasonal fluvial systems in the ancient record. However, plants are typically poorly preserved in dryland fluvial deposits, leading to possibly erroneous conclusions that vegetation was sparse or absent. Where plant fossils are present within channel deposits, analysis of taxonomy and palaeoecology is crucial in establishing the capability of the vegetation to influence channel dynamics, as documented in many modern rivers (e.g. Millar, 2000; Eaton & Giles, 2009; Edmaier *et al.*, 2011). More analysis is required to supplement the modest number of existing studies for the ancient record (e.g. Demko

et al., 1998; Fielding & Alexander, 2001; Bashforth *et al.*, 2014; Ielpi *et al.*, 2015).

In the early Permian Clear Fork Formation of Texas, Simon & Gibling (2017a) documented fine-grained channel deposits exposed in low cliffs and exhumed on flat-lying ground. The inclined strata include layers rich in sand-sized mud aggregates and in silt and clay, allowing an assessment of accretion processes; and some localities have yielded exceptionally well-preserved plant fossils. Data from three localities at a similar stratigraphic level on Montgomery Ranch (MR1, MR2 and MR3) were included in these earlier analyses. The MR3 locality, documented in detail here, received additional study because 1) an exhumed channel body exposes inclined strata, the fill of an abandoned channel and the adjacent channel margin, allowing unusual certainty in reconstructing the channel geometry and accretion processes; and 2) an unusual and well-preserved assemblage of gigantopterids, marattialeans tree ferns and *Taenopteris*, as well as root traces, allows an assessment of habitat and the contribution of the vegetation to fluvial dynamics.

The integration of new information from MR3 with data from MR1 and MR2 contributes to addressing a central question: what processes govern sediment accumulation and morphodynamics in fine-grained meandering channels?

MUD ACCUMULATION IN ACTIVE FLUVIAL CHANNELS

As set out in Table 1, three main processes of mud accumulation have been recognised in active channels. Lateral accretion involves deposition of bedload on point bars during active flow (Jackson, 1981; Rust & Nanson, 1989; Sambrook Smith *et al.*, 2015). In contrast, oblique accretion involves deposition of suspended load on steep accretionary banks and bars during low-energy and waning flow (Page *et al.*, 2003). The former process is familiar from perennial rivers around the world and in seasonal rivers with a moderate duration of annual flow. The latter process has been documented from strongly seasonal channels where near-bankfull conditions are achieved with moderate to low flow strength. During waning flow, fine sediment drapes the progressively exposed banks and may extend to the channel floor if flow in the channel ceases entirely.

Table 1. Comparison of lateral accretion, oblique accretion and counter point bar accretion in modern, sinuous fine-grained channels, in terms of processes and deposits.

	Fine-grained Deposits of Modern Sinuous Channels		
	Lateral-Accretion Deposits	Oblique-Accretion Deposits	Counter Point Bar Deposits
Processes	Bedload deposits predominate, laid down on a migrating point bar during relatively prolonged flow in the channel. Point-bar advance commonly matches cutbank retreat. Relatively rapid rate of migration.	Suspended-load deposits predominate, draped on a prograding accretionary bank during low-energy and waning flow, commonly as banks are exposed. Accretion may narrow channel if cutbank erodes slowly. Relatively slow rate of migration.	Bed-load and suspended-load deposits laid down in the downstream parts of point bars that are migrating downvalley. Commonly subject to reverse eddy currents with upstream flow vectors.
Sediment	Coarser sand and sand-sized, pedogenic mud aggregates. Aggregates washed into channels from floodplains and banks.	Fine sand, silt and clay, commonly interbedded.	Silt predominant, with modest proportions of sand and minimal clay.
Structures and features	Trough and planar cross-beds, ripple cross-lamination, rare plane beds, erosion surfaces. Bedforms commonly directed up the inclined surfaces due to helicoidal flow. Upward fining to floodplain deposits.	Ripple cross-lamination (bedforms may be scarce), desiccation cracks, rill casts, bioturbation. May rest on point-bar deposits lower in the channel. Overlain by floodplain deposits.	Ripple cross-lamination, commonly climbing, and horizontal lamination.
Inclined surfaces	Dips modest, commonly < 10°, 4 to 15° in sandy and silty point bars (Allen, 1970; Sambrook Smith *et al.*, 2015) and 10 to 15° in muddy point bars (Jackson, 1981).	Dips steep, locally as high as 29° (Page *et al.*, 2003) and 30 to 40° (Brooks, 2003a). Sharp decrease in dip (oblique contact) of inclined surfaces on point bar or channel base.	Dips relatively steep, 16° and 22° in two examples (Smith *et al.*, 2009). Concave planforms on bar surfaces.
Vegetation	Plant material with bedload, rooted vegetation rare but present locally.	Plant debris common in fine drapes. Rooted vegetation commonly established low in channel during exposure.	Plant debris common.
Modern examples	Jackson, 1981; Sambrook Smith *et al.*, 2015 (for point-bars in general: Allen, 1970; Jackson, 1976; Parker *et al.*, 2011).	Bluck, 1971; Taylor & Woodyer, 1978; Nanson & Croke, 1992; Page *et al.*, 2003; Brooks, 2003a, b.	Woodyer, 1975; Hickin, 1979, Page & Nanson, 1982; Nanson & Page, 1983; Makaske & Weerts, 2005; Smith *et al.*, 2009.

For mud deposits in the geological record, criteria for discriminating between these processes include grain-size and type, the range of sedimentary features, the dip of inclined surfaces and the presence or absence of rooted vegetation low in the channel (Table 1). An important criterion is the presence or absence of sand-sized mud aggregates derived from reworking of palaeosols, a prominent bedload component in many distal dryland settings (Rust & Nanson, 1989; Maroulis &

Nanson, 1996). Although cryptic bedforms may be observed in mudstone outcrops (Èkes, 1993), aggregate identification usually requires petrographic analysis and imaging of these difficult materials (Müller *et al.*, 2004; Wolela & Gierlowski-Kordesch, 2007; Gastaldo *et al.*, 2013; Dasgupta *et al.*, 2017; Simon & Gibling, 2017b).

Mud may also accumulate in the downstream parts of meander bends through counter point bar accretion. Based on examples in the Peace River of Canada and elsewhere, diagnostic criteria (Table 1) include strata with a geometry that is concave to the adjacent channel, the predominance of ripple cross-laminated silt, modest sand, minimal clay and the relatively steep dips of inclined beds.

TECTONIC AND STRATIGRAPHIC SETTING

The Clear Fork Formation of Kungurian age was deposited on the Eastern Shelf of the Midland Basin, between 0° and 5°N of the equator on the western coastal zone of Pangea (Fig. 1A; Ziegler *et al.*, 1997; Scotese, 1999). The formation was deposited by rivers that flowed generally to the west, recycling Pennsylvanian deposits with contributions from the denuded Wichita, Arbuckle and Ouachita mountains (King, 1937; Oriel *et al.*, 1967; Houseknecht, 1983). During the early Permian, tectonic quiescence prevailed with intermittent strike-slip movement along the Matador and Red River uplifts (Regan & Murphy, 1986; Budnik, 1989; Brister *et al.*, 2002).

The Clear Fork Formation is conformably underlain by the Leuders Formation, a tidal deposit intercalated with marine carbonates (DiMichele *et al.*, 2006). The formation is conformably overlain by the San Angelo Formation of the Pease River Group, consisting of sandy fluvial deposits and muddy coastal-plain deposits (Nelson & Hook, 2005). Burial depth may not have exceeded 1100 m based on regional stratigraphic considerations and maturation data from Eastern Shelf strata (Bein & Land, 1983; Hackley *et al.*, 2009; Simon & Gibling, 2017a).

In the non-marine outcrop areas of the Clear Fork Formation in north-central Texas, the formation is 350 to 365 m-thick and is informally subdivided into lower, middle and upper units (Fig. 1B; Nelson *et al.*, 2013). The formation comprises fluvial and playa-lake deposits of red mudstone with subordinate sandstone, carbonate and evaporate; and the beds are nearly horizontally disposed with WNW dips of less than 1°(Olson, 1958; Wardlaw, 2005; Nelson *et al.*, 2013). Interestingly and unusually, many meandering-channel bodies are exhumed at the land surface (Edwards *et al.*, 1983; Simon & Gibling, 2017a). Fossils include seed plants, vertebrates and trackways of tetrapods and myriapods (Romer, 1928; Olson, 1958; Murry & Johnson, 1987; DiMichele *et al.*, 2006; Chaney & DiMichele, 2007; Anderson *et al.*, 2008; Lucas *et al.*, 2011; Milner & Schoch, 2013).

Based on regional studies across Euramerica (Tabor & Poulsen, 2008; Tabor, 2013), the western equatorial region of Pangea experienced semi-arid to arid conditions with seasonal precipitation. For the Clear Fork Formation (Tabor & Montañez, 2004; DiMichele *et al.,* 2006; Chaney & DiMichele, 2007; Simon, 2016), conditions became drier up-section, as shown by 1) an increased abundance of pedogenic carbonate and evaporite, with bedded dolomite and gypsum in the upper unit; 2) an upward change in cement and nodule composition from calcite, ankerite and dolomite in the lower and middle units to dolomite, gypsum and celestine in the upper unit; 3), a decrease in the proportion of channel bodies and increase of playa-lake deposits upwards; and 4) an upward disappearance of plant and vertebrate fossils, with the uppermost known occurrences just above the Burnet Dolomite.

Fig. 1. (A) Permo–Pennsylvanian palaeogeography of north-central Texas, showing the main uplifts and basin areas (modified from Tabor & Montañez, 2004). The study area is located in the Eastern Shelf and indicated by the pink box. (B) Composite log of the Clear Fork Formation along the Wichita River, showing the occurrence and relative thickness of the main channel bodies studied (modified from Nelson *et al.*, 2013). The Montgomery Ranch 3 (MR3) study site is located in the upper Clear Fork Formation. (C) Planform exposure of an exhumed point bar at MR3 with resistant, discontinuous sandstone scarps formed by inclined beds, abandoned channel, channel margin (cutbank) and adjacent floodplain deposits, visible on Google Earth (red arrow shows north). Yellow lines with triangles represent scarp strike and dip direction, respectively. The two scarps that dip north (green lines) are part of an underlying channel body. Rose diagram represents palaeoflow for all inclined surfaces based on ridge-and-furrow structures that represent cross-lamination. The blue lines and dots indicate the location of subsequent figures. Inset map shows the location of the three Montgomery Ranch sites. [MR1: 33°49′50.84″N, 99°37′35.55″W; MR2: 33°49′34.15″N, 99°37′43.42″W; MR3: 33°50′8.08″N, 99°37′5.28″W]. (D) Log of sandstone and mudstone units as measured in a strike exposure of inclined beds (element LA). (E) Log of mudstone unit underlying the exhumed point bar. On D and E, sand sizes are shown by VF, F and M (very fine, fine and medium, respectively).

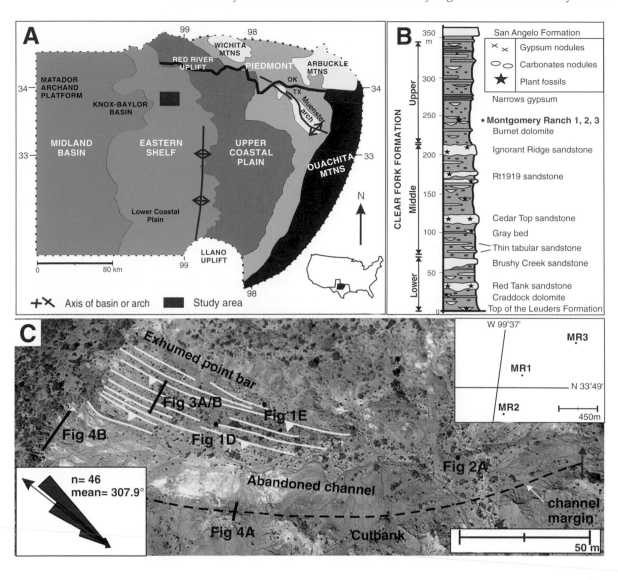

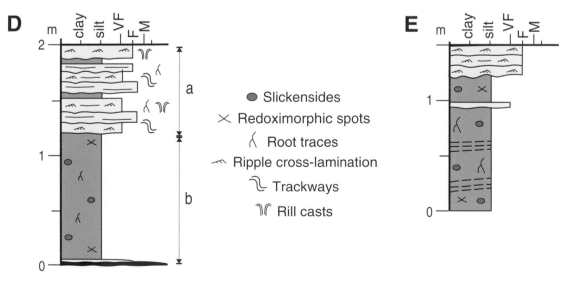

Fig. 1. (Continued)

Nelson *et al.* (2013) mapped the upper Clear Fork as a unit 165 m-thick, composed of mudstone and claystone with thin layers of siltstone and very fine-grained sandstone. Gypsum is abundant as nodules and beds, locally interlayered with dolomite and mudstone. Palaeosols are weakly developed. The study locality of Montgomery Ranch 3 (MR3) is one of three studied sites (other sites: MR1 and MR2) in the upper unit, approximately 1.5 km apart and within a stratigraphic range of less than 50 m (Fig. 1B and C). In view of the low dip, the precise stratigraphic relationship of the sites is difficult to ascertain. Exhumed accretion deposits are quartz-rich at MR2 and variably mud-rich and quartz-rich at MR1 and MR3 (Simon & Gibling, 2017a). Despite their proximity in time and space on the Permian landscape, these channel bodies show key differences that are discussed below.

METHODS

The MR3 locality was evaluated over two field seasons as part of a broader study of fluvial sites. Located in Foard County, the outcrops comprise partially exposed channel components with exhumed surfaces and cliff faces oriented parallel and perpendicular to local flow and they extend for 216 m across two hectares (Fig. 1C). Three elements were recognised: channel-base deposits (CD), inclined strata (LA) and massive mudstone (MM); formal element names are in italics where used in the later text (Table 2). Plant fossils were excavated from four sites at the locality.

To reconstruct the exhumed surfaces, eroded ridges up to 10 m-long formed by inclined sandstone bedsets were mapped using GPS readings accurate to ± 3 m. Depending on ridge length, at

Table 2. Elements for meandering-channel deposit at the MR3 site in the Clear Fork Formation. Facies codes from Miall (1996). Redoximorphic spots and Fe-oxide staining are prominent in all facies, which are red, brown and grey. Plants include woody fragments and leaves with *Diplichnites gouldi* trackways. VF and F = very fine and fine sand, respectively. Gm: massive or crudely bedded pebble conglomerate; Sr: ripple cross-laminated sandstone; Fl: finely laminated mudstone; Fm: massive to weakly ripple-cross laminated mudstone. A = abundant, C = common, R = rare.

Elements	Grain-size	Sedimentary features and fossils	Petrography	Interpretation
Channel-base deposits (CD)	Cemented siltstone, pebble conglomerate, VF-F sandstone.	Gm, Sr as lenses and mounds. Up to 0.3 m-thick.	Pebbles of finely crystalline dolomite, silt-sized quartz, and clay.	Basal cemented deposits laid down during or after incision; clasts reworked from local palaeosols.
Inclined strata (LA)	Bedsets of a) structureless to weakly ripple cross-laminated mudstone; and b) VF-F sandstone.	Sr, Fm. Climbing, swept, starved and symmetrical ripples; rill casts, trackways (R), root traces (R). Mudstone units < 1.5 m-thick and sandstone units < 0.4 m-thick.	Quartzose grains in clay and hematite matrix, with dolomite cement.	Lateral and oblique accretion deposits formed in a point bar. Bedload transport of sand-sized quartzose grains with suspension drapes on steep surfaces. Lower flow regime, with erosive flood events. Locally vegetated.
Massive mudstone (MM)	Weakly laminated to structureless mudstone, with some VF sandstone lenses.	Sr, Fl, Fm, red-brown to grey. Mud cracks, slickensides, charcoal, peds with clay coats, clastic dykes, mud chips. Plants (C), seeds (R), root traces (R to C), coprolites (R), bivalves (R). Up to 3 m-thick.	Quartzose grains in clay and hematite matrix, with dolomite cement.	Floodplain deposits and abandoned-channel fill with incipient to moderately developed soils. Low-energy flows and suspension settling. Modification due to exposure. Plants from vegetation growing in channel and riparian zone.

least three readings (upstream, medial and downstream sites) were recorded, encompassing grain-size and sedimentary features, strike/dip and palaeoflow. Using a tape or a rangefinder accurate to $\pm 1\,m$, ridge length and the distance between ridge crests were measured as a field check for the GPS. For cliff sections, key stratigraphic surfaces and sedimentological observations were recorded on photomosaics. One vertical profile (Fig. 1D) was measured through exhumed inclined bedsets and a second (Fig. 1E) was measured across a floodplain section adjacent to the channel body. Three fluvial elements were identified and the term 'mudstone' implies varied proportions of clay and silt (Table 1).

The exposure allows the direct measurement of channel-body dimensions. Channel-body thickness was estimated in two ways, as the maximum vertical extent of inclined surfaces and as the thickness of an abandoned-channel fill from the basal contact to the base of the overlying flood-plain deposit. Channel width was measured as the orthogonal distance between the last lateral-accretion surface to the north and the onlap of channel fill onto flat-lying floodplain deposits to the south, assuming that abandonment was concurrent with the termination of cutbank retreat. The width shortly before final filling of the channel is represented by the width of the concentric fill seen in a low cliff.

Five unpolished thin sections were examined under plane-polarised light and one polished thin section was studied using SEM/EDS (Scanning Electron Microscopy/Energy-Dispersive X-ray Spectroscopy; Table 3). X-ray diffraction (XRD) analysis was conducted on seven random powder mounts and three of these samples were selected for $< 2\,\mu m$ clay fraction analysis. Details of the sample preparation and analytical techniques are presented in Simon *et al.* (2016). The set includes four samples of massive mudstone, two from inclined strata and one from a channel-base deposit.

FLUVIAL ELEMENTS

Channel-base deposits (CD)

Description: Channel-base deposits consist of cemented siltstone as suites of discontinuous lenses $< 15\,cm$-thick with convex-up tops. The siltstone commonly drapes mounds with a core of red mudstone (MM; Fig. 2A). The mounds have a maximum height and length of 30 cm and 1 m, respectively, with crest-to-crest spacing of 1 to 3 m. The siltstone beds comprise cosets of red and grey cross-lamination with preserved ripple forms, some of which climb at a low angle (Fig. 2B and C). Adjacent to the siltstone is a 3 cm-thick layer of pebble conglomerate composed of carbonate clasts and very fine-grained to fine-grained ripple cross-laminated sandstone. XRD and petrographic analysis shows that the siltstone is

Table 3. Samples from the MR3 field area, with their elements, relative stratigraphic position and analyses conducted.

Sample numbers	Sediment type	Analyses			
		thin section	SEM/EDS	XRD (bulk)	XRD ($<2\,\mu m$)
MR3_1	Cemented siltstone bed (CD)	✓	–	✓	–
MR3_2	Structureless mudstone adjacent to the channel body (MM)	–	–	✓	✓
MR3_3	Structureless siltstone in the abandoned-channel fill (MM)	✓	✓	✓	–
MR3_4	Fossiliferous structureless siltstone in the abandoned-channel fill (MM)	✓	–	✓	✓
MR3_5	Cemented ripple-cross laminated sandstone in inclined bedsets of the exhumed point bar (LA)	✓	–	✓	
MR3_6	Structureless mudstone in inclined bedsets of the exhumed point bar (LA)	✓	–	✓	✓
MR3_7	Structureless mudstone underlying the channel body (MM)	–	–	✓	–

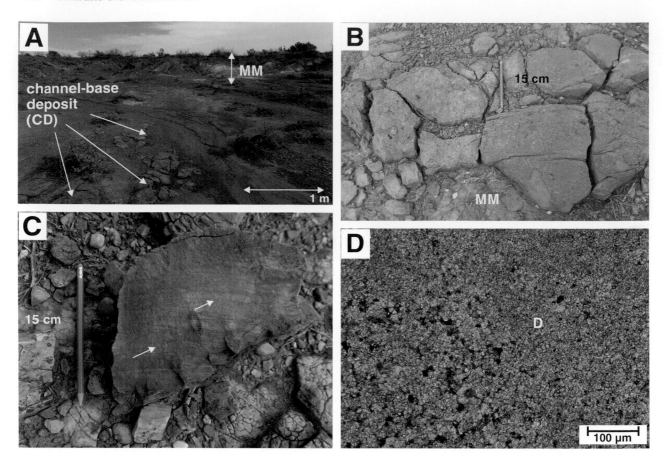

Fig. 2. Channel-base deposits (CD). (A) Channel-base deposits exposed across flatground, at the base of an abandoned channel filled with weakly laminated and structureless mudstone (MM). (B) Thick, structureless cemented siltstone with pebble conglomerate draped over mounds of structureless mudstone (MM), tapering down-current (to the left). (C) Cross-section through the siltstone showing the convex-upward surface and poorly-preserved ripple cross-laminated layers (yellow arrows). (D) Photomicrograph of cemented siltstone under cross-polarised light showing very finely crystalline dolomite 'D' with a few silt-sized quartz grains and a clay matrix.

composed of quartz grains, some clay matrix and very finely crystalline dolomite (Fig. 2D).

Interpretation: Channel-base deposits are interpreted as channel-lag deposits. The siltstone beds are interpreted as linear, flow-parallel mounds that formed by rapid deposition under lower-regime flow, probably downflow from obstacles that were not preserved. The texture of the carbonate clasts closely resembles that of carbonate-bearing palaeosols at other localities in the formation and the clasts are inferred to be pedogenic nodules that were eroded from soils and concentrated at the channel base. The dolomitic clast composition probably reflects early dolomitisation from shallow continental groundwater with an increased Mg/Ca ratio following calcite precipitation in an evaporative setting (Simon & Gibling, 2017a).

Inclined strata (LA)

Description: Inclined strata are heterolithic units (Thomas *et al.*, 1987) of mudstone and sandstone with an average dip of sandstone surfaces of $15.6 \pm 2.6°$ (n = 29) to the south (Figs 1D, 3A and B). Although stratification is poorly developed within the mudstone units, where visible it is parallel to that in the sandstones (Fig. 3C). The inclined units downlap onto coarse channel-base deposits or unstratified mudstone. Exposed in planview with large-scale curvature that is convex towards the cutbank, they form low, discontinuous scarps up to 2 m high, which can be traced across flat ground for up to 100 m along strike into nearby cliff exposures and for more than 50 m in the dip direction (Figs 1C, 3A and B). Exposures are mainly limited to the upper portions of the

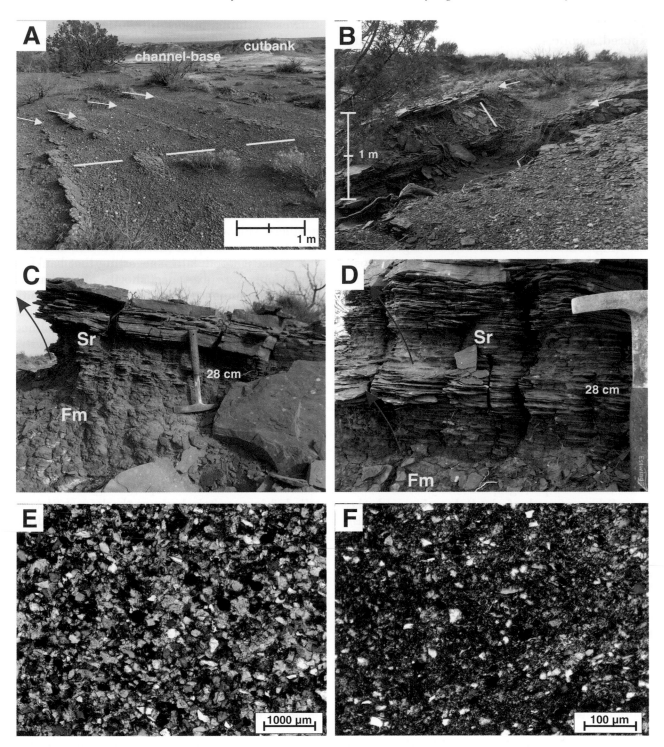

Fig. 3. Inclined strata (LA). (A, B) Planform exposure of resistant, discontinuous sandstone scarps (yellow arrows) interbedded with structureless mudstone (yellow bars). Scarps dip to the right in A and to the left in B, with palaeoflow angled up the inclined surfaces (see rose diagram in Fig. 1C). View in A is southward, across channel-base deposits (white area) to cutbank against older floodplain deposits in distance. (C) Vertical section through scarp. At the base of the sandstone bed, starved ripples are common and ripple sets progressively thicken upward (pink arrow) with less intervening mudstone into a well-developed sandstone layer. The ripple cross-laminated sandstone beds (Sr) dip to the right, overlying a structureless mudstone interval with redoximorphic spots and root traces (Fm). (D) Sandstone bed in the accretion deposits composed of two coarsening-upward beds (pink arrows). (E) Cemented sandstone composed of finely crystalline dolomite with detrital grains and clay matrix. (F) Mudstone composed of silt-sized quartz grains and ferruginous grains in a matrix of hematite and clay.

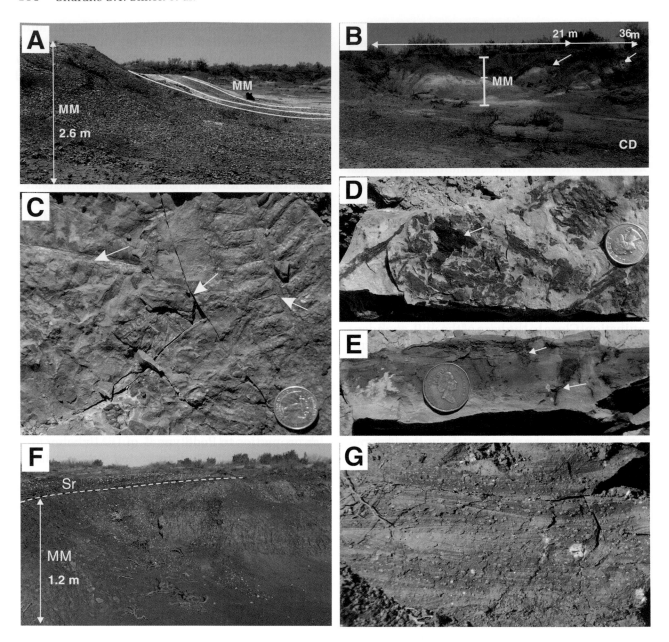

Fig. 4. Structureless mudstone (element MM). (A) Red and grey weakly laminated to structureless mudstone (MM at upper right) onlapping the steeply-dipping channel margin cut into red structureless mudstone (element MM at lower left), with rare root traces, slickensides, clastic dikes, wedge-shaped peds and abundant redoximorphic spots. (B) Abandoned channel containing concentric fill of element MM, which overlies channel-base deposits. The yellow arrows highlight the decreased width and depth of the abandoned channel as the fill advanced to the left. (C) Permineralised leaves preserved on the planes of weakly laminated mudstone with the yellow arrows highlighting the midrib of the leaves. Coin: 2.1 cm. (D) Charcoal fragments (yellow arrow) preserved alongside leaves. Coin: 2.1 cm. (E) Roots < 5 cm-long and 1 cm-wide preserved in weakly laminated mudstone, which commonly overlies the layers with plant fossils. Coin: 2.6 cm. (F/G) Structureless mudstone below the earliest recorded accretion deposit on the exhumed point bar (Sr, dipping to the north). The bed contains ripple cross-laminated sets up to 5 cm-thick with abundant redoximorphic spots.

The channel-base deposits are thin and relatively fine-grained and the carbonate clasts were derived from local palaeosols. These observations imply limited sediment availability on the alluvial plain and probably a long transport distance (Simon & Gibling, 2017a). The presence of mounds suggests that obstacles were present at the channel base, focusing sediment build-up.

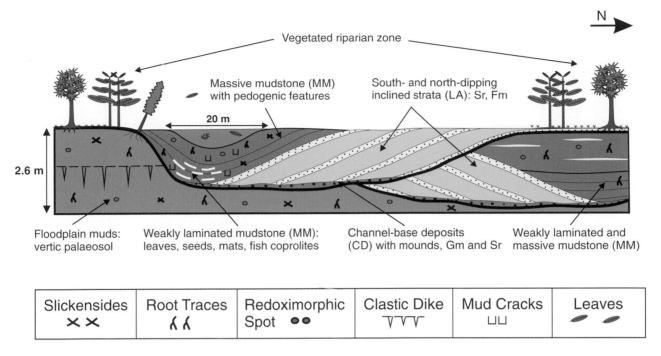

Fig. 5. Geometry, architectural elements and sedimentary features of the fine-grained meandering channel bodies at Montgomery Ranch 3 site. In the main channel body, the abandoned channel fill contains plant debris, roots and fish coprolites (not shown on the figure). Vertical scale is exaggerated.

With root traces present locally within the channel body, the obstacles may have been rooted vegetation, although no direct evidence supports this inference.

Inclined strata formed along the convex margin of a fine-grained point bar, which advanced over and in association with the basal deposits, with expansion in the dip direction. Sand was transported as bedload during the most active, near-bankfull conditions. During the final stages of these active periods, rill casts and swept ripples formed as the water-level fell. Mud was laid down during low-energy conditions with water levels approaching bankfull, as indicated by the extent of the mudstone layers from the channel base to the top of the ridges with inclined layers. Petrographic evidence supports deposition of mud largely from suspension, with no evidence of mud aggregates. Root traces within mudstone and sandstone units indicate that vegetation grew periodically within the channel.

The composite nature of many sandstone units and the probable composite nature of the thick mudstone units suggests that units of both types represent numerous flood events. The alternation of sediment types may reflect intrinsic events in the channel system, for example periodic access to a sand supply upstream. However, the correlation between lithology and flow strength suggests periodic variation in climatically controlled discharge conditions, including flood frequency and intensity, as documented on a decadal to century scale in the relatively arid south-west United States (Hereford, 2002; Harden *et al.*, 2010).

After migrating southward for at least 50 m, the channel avulsed to a new location. Water ponded within the abandoned channel where weakly laminated to structureless massive mudstone accumulated, with plant debris, bivalves and coprolites. The presence of plant beds penetrated by roots and overlain by disrupted, root-free mudstone implies that short-lived plants grew repeatedly within the channel, which was alternately water-filled then dry and exposed; some bed surfaces may represent a considerable period of time. The laminated mudstone at MR3 has a similar mineralogical composition to mudstone in other abandoned-channel fills in the formation but is generally coarser and lacks the regular, varicoloured laminae of other localities (Simon *et al.*, 2016). These differences may reflect more frequent exposure, precluding the formation and preservation of fine-scale lamination and influencing water chemistry.

An upward change to structureless mudstone, with evidence for desiccation, growth of vegetation and pedogenesis, probably represents a combination of sedimentation and drop in water level in the channel. The lithological change corresponds to a decrease in the channel width and depth to 21 m and 0.75 m, respectively. The width : depth ratio of 28, greater than the ratio of 18 at an early stage of abandonment, indicates that vertical accretion in the channel outpaced bank accretion. The preservation of a concentric fill of structureless mudstone (Fig. 4B) indicates that the channel was not reoccupied and the multi-storey body was fully abandoned in this area.

PLANT FOSSILS

Description: A flora characteristic of localities in the middle Clear Fork is preserved in the weakly laminated zones in the massive mudstone of the abandoned channel. Species richness is quite low. Three foliage taxa dominate the flora: *Evolsonia texana* (Mamay, 1989), marattialeans (possibly of more than one type) and *Taeniopteris* sp., in order of relative abundance. There are hints of other taxa represented only by scraps of foliage, possibly including pteridosperms (callipterids or medullosans) and *Sphenophyllum* sp. In addition, seeds are abundant on some bedding surfaces.

Evolsonia texana is the only species of this genus, which is attributed to the enigmatic plant family Gigantopteridaceae. The higher order affinities of the gigantopterids are probably with the seed plants, although relatedness to the Peltaspermales has been suggested (DiMichele *et al.*, 2005). The *Evolsonia* leaves are variable in size but locally large, >30 cm-long and >15 cm-wide. They are oval (Fig. 6A) with a complex, four-order venation (Fig. 6B and C); the 4th order forms an anastomosing network. The leaves were probably of stout construction with robust midvein and secondary veins (Fig. 6B and C) and vaulted between secondary veins (Fig. 6C and D).

The habit of *Evolsonia texana* is not known. Horizontally disposed axes with roots emanating from the lower side were found in a central part of the abandoned channel; the abundant *Evolsonia* leaves above this layer open the possibility that the plant may have had a ground creeping habit. The robust construction of the leaves suggests

long retention times on the plant. However, gigantopterids were among the early Permian tropical plants most heavily affected by insect herbivory, typically associated with lightly built leaves that have short life spans (Glasspool *et al.*, 2003; Beck & Labandeira, 1998; Schachat *et al.*, 2014). If the leaves were relatively short-lived, *Evolsonia* may have colonised disturbed habitats and had more rapid growth rates than if it retained its leaves for an extended period.

If the gigantopterids were seed plants, their seeds have not been confidently identified. However, asymmetrical, ovoid, flattened seeds (Fig. 6E) have been found consistently in association with gigantopterid remains in American deposits. Such seeds are common at MR3 and may have been produced by *Evolsonia*.

Marattialean fern foliage occurs in dense mats locally and is closely intermixed with other plant elements, commonly on the same bedding surfaces as *Evolsonia* (Fig. 7A and D). Both sterile (Fig. 7A) and fertile foliage (Fig. 7B and C) occur in the deposit. The fertile foliage suggests that more than one species of marattialean fern is present. Although iron encrustation has contributed to poor preservation, the synangia (fused groups of sporangia, the spore-producing organs) in the specimen illustrated in Fig. 7B appear to be elongated laterally with a second type seen in Fig. 7, which is circular. Both synangial structures are known in marattialeans in Pennsylvanian and Permian floras.

Late Palaeozoic marattialeans are generally thought to have been tree ferns with their trunks supported by a mantle of adventitious roots (Morgan, 1959), a habit reported among from early Permian species of this group (Rößler, 2000). They appear to have favoured habitats with high water tables, although not necessarily flooded; their roots contain abundant air spaces (aerenchyma; Ehret & Phillips, 1977), a feature frequently associated with plants subject to occasional to frequent flooding. Marattialeans produced copious spores and were excellent colonisers. They have been reported from Permian channel belts in far western Pangea and are among those 'wetland' plants, along with the calamitaleans, that have the broadest distribution in space and time, colonising landscapes that were otherwise seasonally dry and moisture-limited (Naugolnykh, 2005; DiMichele *et al.*, 2006; Tabor *et al.*, 2013).

Leaves of *Taeniopteris* sp. are also present. This is a genus of uncertain affinities and may include

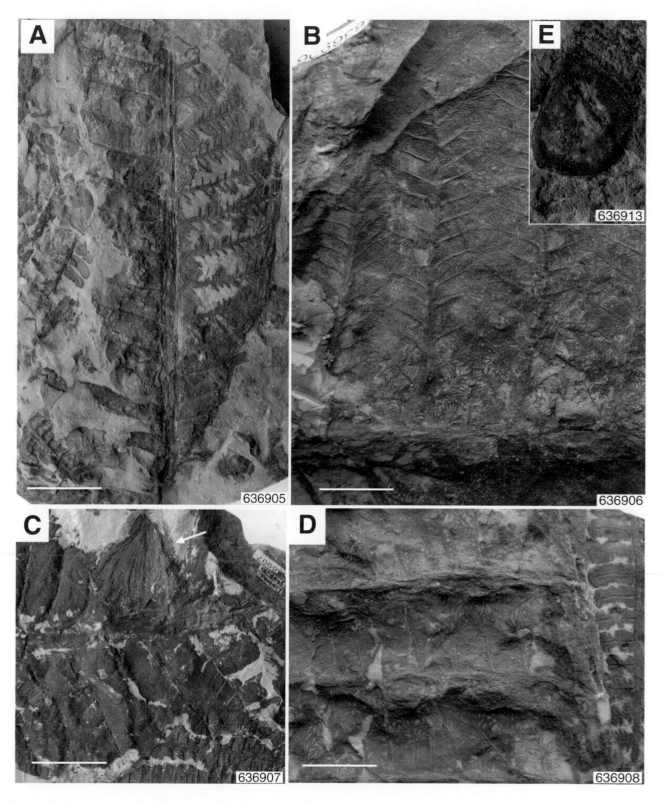

Fig. 6. *Evolsonia* from the MR3 channel body. (A) Leaf showing main architectural features of leaf shape and venation. Note co-occurring marattialean fern foliage. USNM specimen 636905. USNM locality 41382E. (B) Detail of leaf venation showing progressive size diminishment from midvein to secondary, tertiary and quaternary veins. The ultimate veins (quaternary) form a mesh. There is no suture vein, typical of some other gigantopterids. USNM specimen 636906. USNM locality 41382E. (C) Detail illustrating the three-dimensionality of the leaf surface. Note associated fertile marattialean foliage. Arrow points to an unidentified leaf, possibly *Sphenophyllum*. USNM specimen 686907. USNM locality 41382D. (D) Detail of venation; note particularly the ultimate reticulate mesh. Marattialean fern foliage in close association. USNM specimen 636908. USNM locality 41382E. (E) Asymmetrical seed of gigantopterid type. USNM specimen 636913. USNM locality 43866. Scale bars = 1 cm.

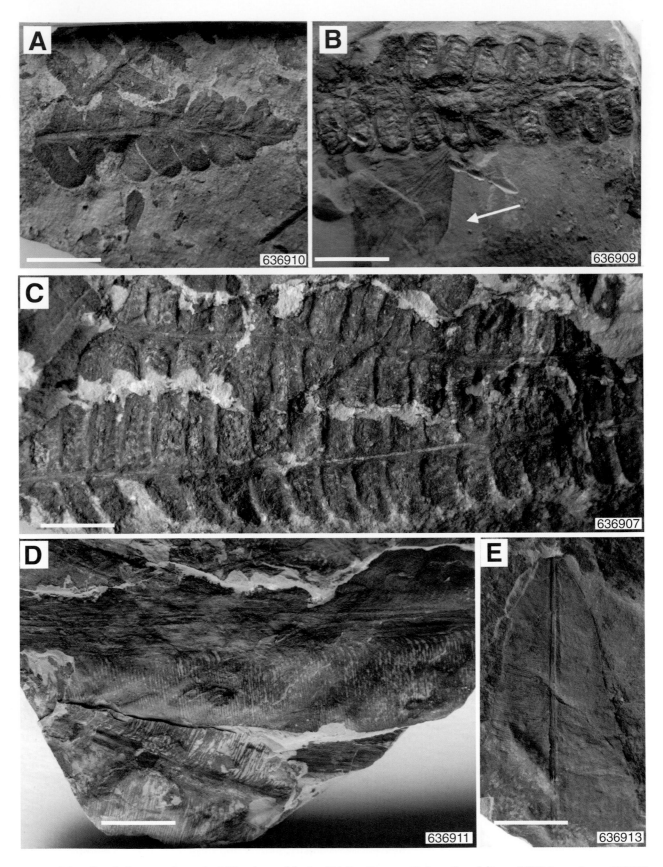

Fig. 7. Other floral elements from the MR3 channel body. (A) Sterile marattialean fern foliage. USNM specimen 636910. USNM locality 41382D. (B) Fertile marattialean fern foliage with elongate, oval synangia (Danaeites type). Arrow points to unidentified leaf, possibly *Sphenophyllum*. USNM specimen 636909. USNM locality 41382E. (C) Fertile marattialean fern foliage with round synangia (Asterotheca type). USNM specimen 636914. USNM locality 41382D. (D) *Taeniopteris* sp. Two specimens of different size illustrating basic architecture of the leaf. USNM specimen 636911. USNM locality 41382E. (E) *Taeniopteris* sp., leaf tip. USNM specimen 636912. USNM locality 41382E. Scale bars = 1 cm.

both ferns and seed plants; the taxonomy is complex and many species have been described (Remy & Remy, 1975). The larger leaves, such as those in the MR3 flora (Fig. 7D and E), are generally attributed to seed plants, possibly cycadophytes (Mamay, 1976). No reproductive organs have been found in association with these leaves.

The habit of *Taeniopteris* plants is uncertain. Whole plants with leaves attached have not been found in late Palaeozoic deposits. However, branch fragments with attached leaves are known from the western Pangean equatorial region (e.g. Koll & DiMichele, 2013) and the plant may have been a small tree. *Taeniopteris* is generally found in association with other floral elements that suggest seasonal drought, as reported from its earliest occurrence in central and western Pangea (Bashforth *et al.*, 2016).

There are indications of other kinds of plants in the MR3 assemblage, but they are represented by fragments or isolated leaves of uncertain affinity. Two such specimens are illustrated (arrows in Figs 6c and 7B). In each instance, the wedge-shaped lamina and multiple veins suggest that these may be leaves of the groundcover sphenopsid *Sphenophyllum*, perhaps of the S. *verticillatum* or S. *thonii* type.

Interpretation of life setting: Although the uncertain affinity and ecological tolerance of some taxa reduces their utility for palaeoenvironmental assessment, comparison with Pennsylvanian and early Permian assemblages across Euramerica allows a reasonable interpretation of the preferred habitat of the taxa found here. The assemblage is somewhat unusual in view of the ecological affinities broadly attributed to these plants, based on their patterns of occurrence in time and space. The assemblage includes plants typically associated with seasonal soil-moisture deficits, particularly *Evolsonia* and *Taeniopteris*, closely intermixed with plants generally associated with a persistently high water table and humid conditions, such as marattialean ferns. The ferns may have lived along the channel margin or even in parts of the channel where water levels remained high during drier seasons or periods. Marattialean foliage is not uniformly distributed in the channel deposits but, where observed, forms dense mats on bed surfaces. Overall, the flora suggests that the background conditions were seasonally dry, but that wet areas suitable for colonisation and growth of tree ferns persisted locally in or adjacent to the channel.

ADJACENT SITES MR1 AND MR2

Although in close proximity and within a stratigraphic interval of less than 50 m, the three sites at Montgomery Ranch (MR1, MR2 and MR3; Fig. 1C) display key differences, relevant to a broader landscape assessment. A summary of key features of MR1 and MR2 is given here (see Simon & Gibling, 2017a, for a fuller description).

Located 0.94 km south-west of MR3, the MR1 site has inclined beds of mudstone and sandstone with a vertical extent of 2.2 m and an average dip of $14.7 \pm 4.4°$ (n = 27). The mudstone is structureless to weakly ripple cross-laminated with rare preservation of trough cross-beds that angled up the inclined surfaces. Petrographic analysis showed that flattened sand-sized mud aggregates are present. The sandstones include preserved dune forms at the base of the channel with crests angled up the inclined surfaces, passing updip into ripples with crests angled down the surfaces. The updip orientation of dune crests in the mudstone and sandstone indicates helicoidal flow within the channel, whereas the down-dip oriented ripple crests above suggest re-entry of overbank water into the channel when bankfull conditions were exceeded; overbank flows are also supported by the presence of thin crevasse splays in adjacent floodplain deposits. Root traces are common in the inclined layers, with a few leaves preserved close to the base of the accretion deposits. Palaeoflow was to the south-south-west (203°).

Located 1.5 km south-west of MR3 and 400 m south-west of MR1, the MR2 site has inclined sandstone beds with 2 m of vertical extent and an average dip of $13 \pm 2.3°$ (n = 14). Trough and planar cross-sets attributed to dunes are angled up the inclined surfaces, indicating helicoidal flow. Mudstone layers are a minor component and leaves and root traces were not observed. Palaeoflow was to the west-north-west (285°).

DISCUSSION

Spectrum of channel accretion processes

All three sites (MR1, MR2 and MR3) exhibit exhumed point bars with inclined strata, with especially good planform exposure at MR3 where convex accretion surfaces face an abandoned channel and cutbank. The channels at all three sites were of modest size, a few metres deep and a few tens of metres wide. Palaeoflow vector means are 203°, 285° and 308°, respectively and the channel bodies are inferred to represent reaches of a westerly-directed meandering-fluvial system, in accord with other localities. The channels occupied a distal position on a dryland alluvial plain where coarser sediment was limited to fine quartzose sand and carbonate clasts and mud aggregates were reworked from palaeosols. The sand was sourced from outside the basin or derived from locally stored sediment that was excavated during flood events (Simon & Gibling, 2017a,b).

Based on the criteria of Table 1, lateral accretion was the predominant process for sandstone and mudstone in the MR1 and MR2 channel bodies. Quartzose sand is relatively abundant and intercalated mudstone beds are scarce in MR2. Dune forms and large-scale cross-beds low in the channel bodies indicate bedload transport under conditions of moderate flow strength and helicoidal flow. In the mudstones at MR1, cryptic trough cross-beds with mud aggregates indicate transport of much of the mud as bedload.

In contrast, lateral accretion and oblique accretion alternated in the more mud-rich MR3 channel body (Fig. 8). In the sandstone units, dune forms and large-scale cross-beds were not observed, but ripple cross-lamination indicates bedload transport, with local erosional events. In the mudstone units, neither bedforms nor sand-sized mud aggregates were observed and mixed silt and clay suggest deposition predominantly from suspension. We infer that the sandstone units represent lateral accretion during periods of relatively strong flow, whereas the relatively thick mudstone units represent oblique accretion during longer periods of low-energy flow, when suspended sediment draped the inclined surfaces, probably during repeated flood events.

One potential criterion for distinguishing lateral from oblique accretion is the dip of the inclined surfaces (Table 1). The average dip of surfaces at MR3 is 15.6°, only slightly steeper than an average of 13° at MR1 and 14.7° at MR3. At other localities (element LA-2 of Simon & Gibling, 2017a), mud-rich inclined strata show moderate dips of 12 to 15° and, as with MR3, the mudstones were interpreted as oblique-accretion deposits on the basis of predominant silt and clay and an apparent lack of aggregates. All three Montgomery Ranch sites have relatively steep surfaces in comparison with many modern point bars (Table 1) and in the dryland setting of western Pangea, case-hardening of mud may have contributed to bank strength and steep surfaces, along with the presence of vegetation (see below).

Oblique-accretion deposits in some modern channels dip much more steeply than those at MR3, up to 29° in the Murrumbidgee River (Page

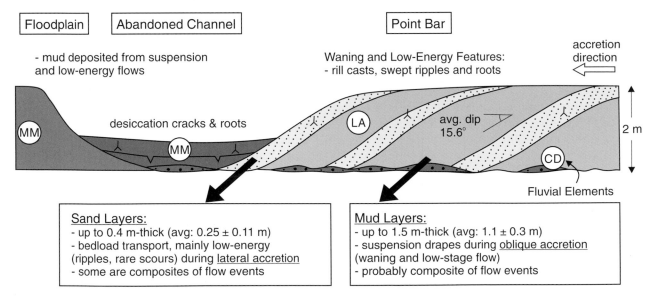

Fig. 8. Schematic diagram to show processes of mud and sand accumulation in fine-grained point bar at MR3. Within the inclined strata of the point bar, textural evidence suggests that, predominantly, the mud was laid down from suspension (oblique accretion) whereas the sand was transported as bedload (lateral accretion). The thickness of the mud layers and the observed composite nature of some sand layers suggests that they represent multiple flow events.

et al., 2003) and up to 40° in the Red River (Brooks, 2003a). The lower dips at MR3 may reflect the small dimensions and low bank height of the channel, in comparison to the Murrumbidgee (average 6 m-deep and 80 m-wide) and the Red River (up to 9.5 m-deep and 75 m-wide). Smaller modern systems with obliquely accreted mud and ripple cross-laminated sand may provide better analogues. Dips of 15° were recorded in the Channel Country of Australia (Gibling *et al.*, 1998) and dips of 8 to 12° were recorded in the upper Klip River of South Africa (Marren *et al.*, 2006), with channels in both instances 3 to 5 m-deep and 30 to 40 m-wide.

Counter point bar deposits were not identified at the three sites. As noted in Table 1, they would be expected to show concave planforms with respect to the adjacent channel and the Peace River examples described by Smith *et al.* (2009) consist mainly of rippled siltstone with upstream-directed bedforms and minimal clay.

In summary, the three Montgomery Ranch sites exhibit a spectrum of point-bar processes from predominantly lateral accretion to alternate episodes of lateral and oblique accretion. Based on the pronounced curvature of the accretion surfaces at MR3, the authors consider that the inclined strata were laid down on a point bar where mud drapes accumulated during periods of low-energy and waning flow, as on steeper accretionary banks in modern channels.

Overall, the river systems of Central Australia provide a reasonable analogue for the Montgomery Ranch sites. A supply-limited system, the Channel Country channels transport fine sediment from distant sources across a wide, low-gradient plain (Nanson & Croke, 1992; Maroulis & Nanson, 1996; Gibling *et al.*, 1998). Anabranching and avulsive channels commonly migrate slowly and fill through oblique and vertical accretion with local lateral accretion and more rapid migration. Their deposits contain quartz sand (locally recycled from the excavation of subsurface sands), sand-sized mud aggregates reworked from vertic soils, fine sediment deposited from suspension and organic material.

Water availability and contribution of vegetation to channel dynamics

The MR3 plant assemblage is dominated by gigantopterids (*Evolsonia*, a probable seed plant), marattialeans (tree ferns) and *Taeniopteris* (a possible cycadophyte, probably small trees), with other more cryptic taxa. *Evolsonia* and *Taeniopteris* are interpreted to have grown under conditions of seasonal soil-moisture deficit. *Evolsonia* may have had a creeping mode of life in the MR3 channel, although this is not certain and was capable of colonising disturbed habitats. In contrast, the marattialeans with adventitious roots and aerenchyma, could tolerate floods and preferred a high water table and high humidity, but were capable colonisers in seasonal settings.

At first sight, this habitat assessment suggests conflicting preferences for taxa in the same assemblage. A probable resolution is that the landscape was seasonally dry under semi-arid to arid conditions (Tabor & Montañez, 2004; DiMichele *et al.*, 2006), but that water was retained in wet areas of the channel and was accessible as groundwater below the channel belt for riparian water uptake, as in the Channel Country (Cendón *et al.*, 2010). Thus, taxa with different habitat tolerances could coexist within a channel reach. Demko *et al.* (1998) suggested that dryland valley fills would be biased towards riparian wetland habitats and Simon *et al.* (2016) suggested that moisture persisted locally in an oxbow lake, promoting plant growth and preservation on the Clear Fork dryland plain.

An initial reconnaissance might suggest that vegetation contributed little to channel dynamics in this early Permian formation. Upright trees and large logs are absent from the Montgomery Ranch sites and very rare in the formation as a whole and plant fossils are sparse and commonly poorly preserved. However, the excavation of exceptional plant-fossil localities over many years, especially in abandoned-channel fills (DiMichele *et al.*, 2006; Looy, 2013; Simon *et al.*, 2016), shows that a range of taxa was present near the channels, including trees with large leaves (Fig. 5). Examination of sandstone and mudstone beds shows that roots are widespread within the channel bodies and along their margins. These observations indicate that vegetation must be taken into account in assessing channel morphodynamics, although analysis will be largely qualitative.

As in modern settings and in ancient examples (Rygel *et al.*, 2004; Ielpi *et al.*, 2015), vegetation probably contributed significantly to bank stabilisation, along with abundant cohesive mud that would have hardened in the relatively arid setting (Gibling *et al.*, 1998; Matsubara *et al.*, 2015). With evidence for rooted vegetation within the MR3 channel, upright vegetation (especially of disturbance-tolerant taxa such as *Evolsonia* and marattialeans) and transported plant material may have nucleated sediment accumulation. No hard evidence for this was found at MR3 but wood fragments are present in channel-base mounds at other sites (Simon & Gibling, 2017a). The MR3 channel may have provided an especially suitable habitat for plant growth in view of the prolonged low-energy periods during point-bar migration, standing water following floods (as indicated by symmetric ripples) and standing water in the abandoned channel. Finally, a well-developed riparian zone would have promoted animal life, including the tetrapods for which the formation is famous.

CONCLUSIONS

A crucial issue in studying fine-grained meandering channels is to establish the processes that laid down the mud. This requires analysis of outcrop features and microscopic textures, both of which are probably cryptic and modified by weathering and diagenesis. In the early Permian Clear Fork Formation at Montgomery Ranch site MR3, exhumed point-bar deposits of a small channel (2 m-deep and 36 m-wide) comprise alternate, thin (<0.4 m) units of quartzose sandstone and thick (<1.5 m) of mudstone, laid down under a relatively arid climate. The sand was transported as bedload during active flow and the mud was laid down as suspension drapes of silt and clay that covered the inclined point-bar surfaces (with an average dip of nearly 16°) during waning and low-stage flow. Sand-sized mud aggregates and cryptic bedforms are present in channel mudstones elsewhere in the formation but, despite a careful search, were not identified in the MR3 mudstones, strengthening the case for deposition of mud from suspension rather than bedload. Thus, alternate periods of lateral accretion and oblique accretion were involved in the construction of the point bar (Fig. 8). Both sandstone and mudstone units are composites of numerous flows, suggesting a link to climatically controlled secular variation in discharge and flood parameters.

Plant fossils are generally scarce in the formation and may easily be overlooked in assessing channel dynamics. However, an associated abandoned-channel fill at MR3 contains an assemblage of gigantopterids (*Evolsonia texana*), marattialean

foliage and *Taeniopteris* sp. Root traces penetrate the leaves and are also present in the point-bar deposits. These observations suggest that, under conditions of seasonal flow in the channel system, vegetation grew widely in the channels and on the banks, with some taxa exploiting wetter sites in an otherwise moisture-deficient setting. The vegetation may have been sufficiently abundant to influence channel processes and geometry, although direct evidence is lacking. Fine-grained channels under a seasonal flow regime but with periodic standing water and wetter reaches, may be especially well suited to the establishment of vegetation.

ACKNOWLEDGEMENTS

The authors thank John Holbrook and Colin North for thoughtful reviews that greatly helped in focusing the manuscript. We thank James Edwards of Montgomery Ranch for property access to the outcrops. Thin sections were prepared at Dalhousie University by Gordon Brown and by Vancouver Petrographics Limited. The authors thank Georgia Pe-Piper and Xiang Yang at Saint Mary's University for assistance with SEM analysis and David Piper, Owen Brown, Lori Campbell and Jenna Higgins at Bedford Institute of Oceanography for assistance with X-ray diffraction. The authors are especially grateful to Neil Tabor, Timothy Myers and Lu Zhu for logistical support. Research was funded by a Discovery Grant to Martin Gibling from the Natural Sciences and Engineering Research Council of Canada (NSERC) and by research grants to Sharane Simon from the Society for Sedimentary Geology (SEPM) and Geological Society of America (GSA).

REFERENCES

Allen, J.R.L. (1970) Studies in fluviatile sedimentation: A comparison of fining upwards cyclothems with special reference to coarse-member composition and interpretation. *J. Sed. Petrol.*, **40**, 298–323.

Anderson, J.S., Reisz, R.R., Scott, D., Frobisch, N.B. and Sumida, S.S. (2008) A stem batrachian from the Early Permian of Texas and the origin of frogs and salamanders. *Nature*, **453**, 515–518.

Bashforth, A.R., Cleal, C.J., Gibling, M.R., Falcon-Lang, H.J. and Miller, R.F. (2014) Paleoecology of Early Pennsylvanian vegetation on a seasonally dry tropical landscape (Tynemouth Creek Formation, New Brunswick, Canada. *Review of Palaeobotany and Palynology*, **200**, 229–263.

Bashforth, A.R., DiMichele, W.A., Eble, C.F. and **Nelson, W.J.** (2016) Dryland vegetation from the Middle Pennsylvanian of Indiana (Illinois Basin): the dryland biome in glacioeustatic, paleobiogeographic and paleoecologic context. *J. Paleontol.*, **90**, 785–814.

Beck, A.L. and **Labandeira, C.C.** (1998) Early Permian insect folivory on a gigantopterid-dominated riparian flora from north-central Texas. *Palaeogeogr., Palaeoclimatol., Palaeoecol.*, **142**, 139–173.

Bein, A. and **Land, L.S.** (1983) Carbonate sedimentation and diagenesis associated with Mg-Ca-chloride brines; the Permian San Andres Formation in the Texas Panhandle. *J. Sed. Res.*, **53**, 243–260.

Bluck, B. (1971) Sedimentation in the meandering River Endrick. *Scot. J. Geol.*, **7**, 93–138.

Brister, B.S., Stephens, W.C. and **Norman, G.A.** (2002) Structure, stratigraphy and hydrocarbon system of a Pennsylvanian pull-apart basin in north-central Texas. *AAPG Bull.*, **86**, 1–20.

Brooks, G.R. (2003a) Alluvial deposits of a mud-dominated stream: the Red River, Manitoba, Canada. *Sedimentology*, **50**, 441–458.

Brooks, G.R. (2003b) Holocene lateral channel migration and incision of the Red River, Manitoba, Canada. *Geomorphology*, **54**, 197–215.

Budnik, R.T. (1989) *Tectonic structures of the Palo Duro Basin, Texas Panhandle.* Bureau of Economic Geology, University of Texas, Austin, 43 p.

Cendón, D.I., Larsen, J.R., Jones, B.G., Nanson, G.C., Rickleman, D., Hankin, S.I., Pueyo, J.J. and **Maroulis, J.** (2010) Freshwater recharge into a shallow saline groundwater system, Cooper Creek floodplain, Queensland, Australia. *J. Hydrol*, **392**, 150–163.

Chaney, D.S. and **DiMichele, W.A.** (2007) Paleobotany of the classic redbeds (Clear Fork Group – Early Permian) of north-central Texas. In: *Proc. XVth Int. Cong. Carb. Perm. Strat.* (Ed. T.E. Wong), pp. 357–366. University of Utrecht, The Netherlands.

DiMichele, W.A., Kerp, H., Krings, M. and **Chaney, D.S.** (2005) The Permian peltasperm radiation: evidence from the southwestern United States. *New Mex. Mus. Nat. Hist. Sci. Bull.*, **30**, 67–79.

DiMichele, W.A., Tabor, N.J., Chaney, D.S. and **Nelson, W.J.** (2006) From wetlands to wet spots: Environmental tracking and the fate of carboniferous elements in Early Permian tropical floras. In: *Wetlands through Time.* (Eds S.F. Greb and W.A. DiMichele), *Geol. Soc. Am. Spec. Publ.*, **399**, 223–248.

Edmaier, K., Burlando, P. and **Perona, P.** (2011) Mechanisms of vegetation uprooting by flow in alluvial non-cohesive sediment. *Hydro. Earth Syst. Sci.*, **15**, 1615–1627.

Edwards, M.B., Eriksson, K.A. and **Kier, R.S.** (1983) Paleochannel geometry and flow patterns determined from exhumed Permian point bars in north-central Texas. *J. Sed. Res.*, **53**, 1261–1270.

Ehret, D.L. and **Phillips, T.L.** (1977) *Psaronius* root systems — morphology and development. *Palaeontographica Abteilung B*, **161**, 147–164.

Èkes, C. (1993) Bedload-transported pedogenic mud aggregates in the Lower Old Red Sandstone in southwest Wales. *Geol. Soc. Lon, J*, **150**, 469–471.

Fielding, C.R. and **Alexander, J.** (2001) Fossil trees in ancient fluvial channel deposits: evidence of seasonal and longer-term climatic variability. *Palaeogeogr. Palaeoclimatol. Palaeoecol.*, **170**, 59–80.

Fielding, C.R., Allen, J.P., Alexander, J. and **Gibling, M.R.** (2009) A facies model for fluvial systems in the seasonal tropics and subtropics. *Geology*, **37**, 623–626.

Gastaldo, R.A., Pludow, B.A. and **Neveling, J.** (2013) Mud aggregates from the Katberg Formation, South Africa: additional evidence for Early Triassic degradational landscapes. *J. Sed. Res.*, **83**, 531–540.

Ghinassi, M., Nemec, W., Aldinucci, M., Nehyba, S., Özaksoy, V. and **Fidolini, F.** (2014) Plan-form evolution of ancient meandering rivers reconstructed from longitudinal outcrop sections. *Sedimentology*, **61**, 952–977.

Ghosh, P., Sarkar, S. and **Maulik, P.** (2006) Sedimentology of a muddy alluvial deposit: Triassic Denwa Formation, India. *Sed. Geol.*, **191**, 3–36.

Gibling, M.R., Nanson, G.C. and **Maroulis, J.C.** (1998) Anastomosing river sedimentation in the Channel Country of central Australia. *Sedimentology*, **45**, 595–619.

Glasspool, I., Hilton, J., Collinson, M. and **Wang, S.J.** (2003) Foliar herbivory in late Palaeozoic Cathaysian gigantopterids. *Rev. Palaeobot. Palynol.*, **127**, 125–132.

Hackley, P.C., Guevara, E.H., Hentz, T.F. and **Hook, R.W.** (2009) Thermal maturity and organic composition of Pennsylvanian coals and carbonaceous shales, north-central Texas: Implications for coalbed gas potential. *Int. J. Coal Geol.*, **77**, 294–309.

Harden, T., Macklin, M.G. and **Baker, V.R.** (2010) Holocene flood histories in south-western USA. *Earth Surf. Proc. Landforms*, **35**, 707–716.

Hereford, R. (2002) Valley-fill alluviation during the Little Ice Age (ca. A.D. 1400–1880), Paria River basin and southern Colorado Plateau, United States. *Geol. Soc. Am. Bull.*, **114**, 1550–1563.

Hickin, E.J. (1979) Concave-bank benches on the Squamish River, British Columbia, *Canada. Can. J. Earth Sci.*, **16**, 200–203.

Houseknecht, D.W. (1983) Tectonic-sedimentary evolution of the Arkoma Basin and guidebook to deltaic facies, Hartshorne sandstone. *SEPM Midcontinent Section*, **1**, 3–52.

Ielpi, A. and **Ghinassi, M.** (2014) Planform architecture, stratigraphic signature and morphodynamics of an exhumed Jurassic meander plain (Scalby Formation, Yorkshire, UK). *Sedimentology*, **61**, 1923–1960.

Ielpi, A., Gibling, M.R., Bashforth, A.R. and **Dennar, C.I.** (2015) Impact of vegetation on Early Pennsylvanian fluvial channels: insight from the Joggins Formation of Atlantic Canada. *J. Sed. Res.*, **85**, 999–1018.

Jackson, R.G., II (1976) Depositional model of point bars in the lower Wabash River. *J. Sed. Res.*, **46**, 579–594.

Jackson, R.G., II (1981) Sedimentology of muddy fine-grained channel deposits in meandering streams of the American Middle West. *J. Sed. Petrol.*, **51**, 1169–1192.

King, P.B. 1937. Geology of the Marathon region, Texas. *U.S. Geol. Surv., Prof. Pap.*, **187**, 148 p.

Koll, R. and **DiMichele, W.A.** (2013) A broad-leaved plant from the Abo Formation of northern New Mexico:

Lazarus taxon, paleobiogeographic anomay or convergent evolution? *New Mex. Mus. Nat. Hist. Sci. Bull.*, **60**, 175–177.

Looy, C.V. (2013) Natural history of a plant trait: branch-system abscission in Paleozoic conifers and its environmental, autecological and ecosystem implications in a fire-prone world. *Paleobiology*, **39**, 235–252.

Lucas, S.G., Voigt, S., Lerner, A.J. and Nelson, W.J. (2011) Late Early Permian continental ichnofauna from Lake Kemp, north-central Texas, USA. *Palaeogeogr. Palaeoclimatol. Palaeoecol.*, **308**, 395–404.

Mack, G.H., Leeder, M.R., Perez-Arlucea, M. and Bailey, B.D.J. (2003) Early Permian silt-bed fluvial sedimentation in the Orogrande basin of the Ancestral Rocky Mountains, New Mexico, *USA. Sed. Geol.*, **160**, 159–178.

Makaske, B. and Weerts, H.J.T. (2005) Muddy lateral accretion and low stream power in a sub-recent confined channel belt, Rhine-Meuse delta, central Netherlands. *Sedimentology*, **52**, 651–668.

Mamay, S.H. (1989) *Evolsonia*, a new genus of Gigantopteridaceae from the Lower Permian Vale Formation, north-central Texas. *Am. J. Bot.*, **76**, 1299–1311.

Mamay, S.H. (1976) Paleozoic origin of the cycads. *U.S. Geol. Surv. Prof. Pap.*, **934**, 1–48.

Maroulis, J.C. and Nanson, G.C. (1996) Bedload transport of aggregated muddy alluvium from Cooper creek, central Australia: A flume study. *Sedimentology*, **43**, 771–790.

Marren, P.M., McCarthy, T.S., Tooth, S., Brandt, D., Stacey, G.G., Leong, A. and Spottiswoode, B. (2006) A comparison of mud- and sand-dominated meanders in a downstream coarsening reach of the mixed bedrock-alluvial Klip River, eastern Free State, South Africa. *Sed. Geol.*, **190**, 213–226.

Matsubara, Y., Howard, A.D., Burr, D.M., Williams, R.M.E., Dietrich, W.E. and Moore, J.M. (2015) River meandering on Earth and Mars: A comparative study of Aeolis Dorsa meanders, Mars and possible terrestrial analogs of the Usuktuk River, AK and the Quinn River, NV. *Geomorphology*, **240**, 102–120.

Miall, A.D. (1996) *The geology of fluvial deposits: sedimentary facies, basin analysis and petroleum geology*. Springer, Berlin.

Millar, R.G. (2000) Influence of bank vegetation on alluvial channel patterns. *Water Resour. Res.*, **36**, 1109–1118.

Milner, A.R. and Schoch, R.R. (2013) *Trimerorhachis* (Amphibia: Temnospondyli) from the Lower Permian of Texas and New Mexico: cranial osteology, taxonomy and biostratigraphy. *N. J. Geol. Paläontol. Ab.*, **270**, 91–128.

Morgan, E.J. (1959) The morphology and anatomy of American species of the genus *Psaronius*. *Illinois Biol. Monogr.*, **27**, 1–108.

Müller, R., Nysuten, J.P. and Wright, V.P. (2004) Pedogenic mud aggregates and paleosol development in ancient dryland river systems: criteria for interpreting alluvial mudrock origin and floodplain dynamics. *J. Sed. Res.*, **74**, 537–551.

Murry, P.A. and Johnson, G.D. (1987) Clear Fork vertebrates and environments from the Lower Permian of north-central Texas. *Tex. J. Sci.*, **39**, 253–266.

Nanson, G.C. and Croke, J.C. (1992) A genetic classification of floodplains. *Geomorphology*, **4**, 459–486.

Nanson, G.C. and Page, K.J. (1983) Lateral accretion of fine-grained concave benches on meandering rivers. In: *Modern and Ancient Fluvial Systems* (Eds J. Collinson and J. Lewin), *Int. Assoc. Sedimentol. Spec. Publ.*, **6**, 133–143.

Naugolnykh, S.V. (2005) Permian *Calamites gigas* Brongniart, 1828: The morphological concept, paleoecology and implications. *Paleontol. J.*, **39**, 321–332.

Nelson, W.J. and Hook, R.W. (2005) Pease River Group (Leonardian-Guadalupian) of Texas: an overview. *New Mex. Mus. Nat. Hist. Sci. Bull.*, **30**, 243–250.

Nelson, W.J., Hook, R.W. and Chaney, D.S. (2013) Lithostratigraphy of the Lower Permian (Leonardian) Clear Fork Formation of north-central Texas. In: *The Carboniferous-Permian Transition* (Eds S.G. Lucas, W.A. DiMichele, J.E. Barrick, J.W. Schneider and J.A. Spielmann), *New Mexico Mus. Nat. Hist. Sci. Bull.*, **60**, 286–311.

Olson, E.C. (1958) Fauna of the Vale and Choza: 14. Summary, review and integration of the geology and the faunas. *Fieldiana Geol.*, **10**, 397–448.

Oriel, S.S., McKee, E.D. and Crosby, E.J. (1967) Paleotectonic investigations of the Permian system in United States. *US Geol. Surv. Prof. Pap.*, Washington, U.S., 2330–7102.

Page, K.J. and Nanson, G.C. (1982) Concave-bank benches and associated floodplain formation. *Earth Surf. Proc. Land.*, **7**, 529–543

Page, K.J., Nanson, G.C. and Frazier, P.S. (2003) Floodplain formation and sediment stratigraphy resulting from oblique accretion on the Murrumbidgee River, Australia. *J. Sed. Res.*, **73**, 5–14.

Parker, G., Shimizu, Y., Wilkerson, G.B., Eke, E.C., Abad, J.D., Lauer, J.W., Paola, C., Dietrich, W.E. and Voller, V.R. (2011) A new framework for modeling the migration of meandering rivers. *Earth Surf. Proc. Land.*, **36**, 70–86.

Regan, T.R. and Murphy, P.J. (1986) *Faulting in the Matador uplift area, Texas: Topical Report*. Stone and Webster Engineering Corp., Boston, MA.

Remy, W. and Remy, R. (1975) Beiträge zur Kenntnis des Morpho-Genus *Taeniopteris* Brongniart. *Argumenta Palaeobotanica*, **4**, 31–37.

Romer, A.S. (1928) Vertebrate faunal horizons in the Texas Permo-Carboniferous red beds. *Univ. Texas Bull.*, **2801**, 67–108.

Rößler, R. (2000) The late Palaeozoic tree fern *Psaronius*—an ecosystem unto itself. *Rev. Palaeobot. Palynol.*, **108**, 55–74.

Rust, B.R. and Nanson, G.C. (1989) Bedload transport of mud as pedogenic aggregates in modern and ancient rivers. *Sedimentology*, **36**, 291–306.

Rygel, M.C. and Gibling, M.R. (2006) Natural geomorphic variability recorded in a high-accommodation setting: fluvial architecture of the Pennsylvanian Joggins Formation of Atlantic Canada. *J. Sed. Petrol.*, **76**, 1230–1251.

Rygel, M.C., Gibling, M.R. and Calder, J.H. (2004) Vegetation-induced sedimentary structures from fossil forests in the Pennsylvanian Joggins Formation, Nova Scotia. *Sedimentology*, **51**, 531–552.

Sambrook Smith, G.H., Best, J.L., Leroy, J.Z. and **Orfeo, O.** (2016) The alluvial architecture of a suspended sediment dominated meandering river: the Río Bermejo, Argentina. *Sedimentology*, **63**, 1187–1208.

Schachat, S.R., Labandeira, C.C., Gordon, J., Chaney, D., Levi, S., Halthore, M.N. and **Alvarez, J.** (2014) Plant-insect interactions from early Permian (Kungurian) Colwell Creek Pond, north-central Texas: the early spread of herbivory in riparian environments. *Int. J. Plant. Sci.*, **175**, 855–890.

Scotese, C.R. (1999) PALEOMAP Animations 'Paleogeography'. In: *PALEOMAP Project*. Department of Geology, University of Texas, Arlington, TX.

Simon S.S.T. (2016) Sedimentology of the Fluvial Systems of the Clear Fork Formation in North-Central Texas: Implications for Early Permian Paleoclimate and Plant Fossil Taphonomy. Published PhD thesis, Dalhousie University, Halifax, Canada, 310 p.

Simon, S.S.T. and **Gibling, M.R.** (2017a) Fine-grained meandering systems of the Lower Permian Clear Fork Formation of north-central Texas, USA: Lateral and oblique accretion on an arid plain. *Sedimentology*, **64**, 714–746.

Simon, S.S.T. and **Gibling, M.R.** (2017b) Pedogenic mud aggregates preserved in a fine-grained meandering channel in the Lower Permian Clear Fork Formation, north-central Texas, U.S.A. *J. Sed. Res.*, **87**, 230–252.

Simon, S.S.T, Gibling, M.R., DiMichele, W.A., Chaney, D.S., Looy, C.V. and **Tabor, N.J.** (2016) An abandoned-channel fill with exquisitely preserved plants in redbeds of the Clear Fork Formation, Texas, USA: an Early Permian water-dependent habitat on the arid plains of Pangea. *J. Sed. Res.*, **86**, 944–964.

Smith, D.G., Hubbard, S.M., Leckie, D.A. and **Fustic, M.** (2009) Counter point bar deposits: lithofacies and reservoir significance in the meandering modern Peace River and ancient McMurray Formation, Alberta, Canada. *Sedimentology*, **56**, 1655–1669.

Smoot, J.P. (1991) Sedimentary facies and depositional environments of early Mesozoic Newark Supergroup basins, eastern North America. *Palaeogeogr. Palaeoclimatol. Palaeoecol.*, **84**, 369–423.

Smoot, J.P. and **Olsen, P.E.** (1988) Massive mudstones in basin analysis and paleoclimatic interpretation of the Newark Supergroup. In: *Triassic-Jurassic rifting, continental breakup and the origin of the Atlantic Ocean and passive margins* (Ed. W. Manspeizer), Elsevier, New York, 249–274.

Stewart, D.J. (1981) A meander-belt sandstone of the Lower Cretaceous of southern England. *Sedimentology*, **28**, 1–20.

Tabor, N.J. (2013) Wastelands of tropical Pangea: high heat in the Permian. *Geology*, **41**, 623–624.

Tabor, N.J., DiMichele, W.A., Montañez, I.P. and **Chaney, D.S.** (2013) Late Paleozoic continental warming of a cold tropical basin and floristic change in western Pangea. *Int. J. Coal. Geol.*, **119**, 177–186.

Tabor, N.J. and **Montañez, I.P.** (2004) Morphology and distribution of fossil soils in the Permo-Pennsylvanian Wichita and Bowie Groups, north-central Texas, USA: implications for western equatorial Pangean palaeoclimate during icehouse-greenhouse transition. *Sedimentology*, **51**, 851–884.

Tabor, N.J. and **Poulsen, C.J.** (2008) Palaeoclimate across the Late Pennsylvanian-Early Permian tropical palaeolatitudes: a review of climate indicators, their distribution and relation to palaeophysiographic climate factors. *Palaeogeogr. Palaeoclimatol. Palaeoecol.*, **268**, 293–310.

Taylor, G. and **Woodyer, K.D.** (1978) Bank deposition in suspended-load streams. In: *Can. Soc. Petrol. Geol. Mem.* (Ed. A.D. Miall), **5**, 257–275.

Thomas, R.G., Smith, D.G., Wood, J.M., Visser, J., Calverley-Range, E.A. and **Koster, E.H.** (1987) Inclined heterolithic stratification-terminology, description, interpretation and significance. *Sed. Geol.*, **53**, 123–179.

Tunbridge, I.P. (1984) Facies model for a sandy ephemeral stream and clay playa complex the Middle Devonian Trentishoe Formation of North Devon, U.K. *Sedimentology*, **31**, 697–715.

Wardlaw, B. (2005) Age assignment of the Pennsylvanian-Early Permian succession of north central Texas. *Permophiles*, **46**, 21–22.

Willis, B.J. (1989) Palaeochannel reconstructions from point bar deposits: a three-dimensional perspective. *Sedimentology*, **36**, 757–766.

Wolela, A.M. and **Gierlowski-Kordesch, E.H.** (2007) Diagenetic history of fluvial and lacustrine sandstones of the Hartford Basin (Triassic–Jurassic), Newark Supergroup, USA: *Sed. Geol.*, **197**, 99–126.

Woodyer, K.D. (1975) Concave-bank benches on the Barwon River, N.S.W. *Aust. Geogr.*, **13**, 36–40.

Ziegler, A.M., Hulver, M.L. and **Rowley, D.B.** (1997) Permian world topography and climate. In: *Late glacial and postglacial environmental changes: Pleistocene, Carboniferous-Permian and Proterozoic* (Ed. I.P. Martini), Oxford University Press, Oxford, 111–146.

Int. Assoc. Sedimentol. Spec. Publ (2018) **48**, 173–200.

Interpretation of cross strata formed by unit bars

ARNOLD JAN H. REESINK

Lancing College, Lancing, West Sussex, UK

ABSTRACT

This study illustrates how simple, systematic measurements of cross strata formed by unit-bars can reveal valuable information about the formative fluvial environment. The grain-size sorting in cross strata is created by two processes: 'pre-sorting' of the sediment arriving at the edge of the bar and 're-sorting' during deposition on the lee slope. Unit-bar cross strata commonly contain distinct pre-sorting patterns that are identified easily in cores, outcrops and high-resolution borehole resistivity data. Field studies and flume experiments have shown that measurements of pre-sorting patterns can be used to quantify the sizes of formative superimposed bedforms and analyse depositional increments associated with floods. This understanding has not yet been applied to the rock record. In order to fill this gap in our understanding and highlight opportunities for further research, this study systematically investigates unit-bar cross strata in outcrops. A sensitivity analysis of factors that control the dimensions of pre-sorting patterns indicates the dominance of only two controlling factors: i) the threshold that describes the transition from angle-of-repose cross strata to low-angle co-sets; and ii) the preservation potential of the host unit-bar sets. Despite the uncertainty created by these factors, the results show that pre-sorting patterns in unit-bar sets in outcrops lend themselves to interpretation of the sizes of superimposed dunes. The measurements of the cross strata indicate trends in dune development during the passage of formative floods and provide a viable proxy for the analysis of formative water depths. Most unit bars persist over multiple flood events. Unit-bar tops commonly approach bank-full water depths and angle-of-repose unit-bar cross strata are more abundant in the upper parts of channel deposits. This makes unit bar cross strata particularly sensitive to the magnitude and duration of floods and allows them to record short histories of flood events. This palaeo-hydraulic information may be used to investigate the dynamic evolution of unit bars during floods and to reconstruct palaeo-hydrographs for ancient river systems.

Keywords: Cross bedding, river deposits, fluvial sedimentology, dunes, bars, cross strata

INTRODUCTION

Cross strata formed by dunes and unit bars are ubiquitous in fluvial channel deposits (Miall, 1996; Bridge, 2003, Fig. 1A) including meandering rivers (Fig. 1B and C). The thickness, geometry and sorting of cross strata is controlled by flow and sediment transport conditions and can reveal valuable details about the palaeo-environment, including the co-evolution of host and superimposed bedforms (Hooke, 1968; Smith, 1972; McCabe & Jones, 1977; Reesink & Bridge, 2007, 2009).

Cross strata: patterns of pre-sorting and re-sorting

The grain-size sorting patterns that allow cross strata to be identified are created by repetitive changes in the rate and composition of the sediment arriving at the brink point as well as during deposition on the lee slope (Fig. 2A). Sorting of the arriving sediment before deposition is herein referred to as 'pre-sorting'. Pre-sorting can be related to changes in flow, such as tidal fluctuations in discharge and water depth (Nio & Yang,

Fluvial Meanders and Their Sedimentary Products in the Rock Record, First Edition.
Edited by Massimiliano Ghinassi, Luca Colombera, Nigel P. Mountney and Arnold Jan H. Reesink.

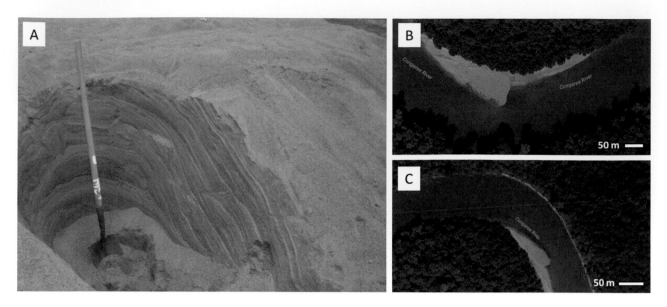

Fig. 1. (A) Trench in a unit-bar lee slope, showing angle-of-repose cross strata with pre-sorting patterns on the top of an exposed point bar in the meandering Congaree River, SC, USA. Shovel for scale. (B and C) Aerial photos of unit bars on point bars on the edge of Congaree National Park (Imagery ©2017 Google).

1991; Martinius & Van den Berg, 2011), or pulses of sediment created by the migration of superimposed bedforms (Hooke, 1968; Smith, 1972; McCabe & Jones, 1977). In river channel deposits, pre-sorting is pronounced when dunes are superimposed on bars (Reesink & Bridge, 2007, 2011).

Sorting during deposition on the lee slope is herein referred to as 're-sorting'. Resorting is commonly dominated by grainflows and settling of sediment from the flow (Kleinhans, 2004), although both mechanisms may be significantly affected by turbulence (Allen, 1982; Reesink & Bridge, 2009).

Detailed interpretations of cross stratification have been explored through experimental flume studies (Hooke, 1968; McCabe & Jones, 1977; Reesink & Bridge, 2007, 2009) and investigations of modern sand bars (Smith, 1972; Reesink & Bridge, 2011) but have not been explored systematically for the rock record. In order to test the applicability of the current understanding for analysis of the fluvial rock record, this paper presents a series of interpretations of easily recognised cross strata patterns: angle-of-repose cross strata with pre-sorting patterns that are formed by unit bars with superimposed dunes (Fig. 1).

Unit-bar deposits in meandering rivers

The unit-bar sets investigated in this study are mostly easily identified: they are thick sets with

distinct, angle-of-repose cross strata (Fig. 1A). These cross strata occur locally, are primarily associated with downstream accretion and are not distributed evenly across river channels. Studies of modern rivers show that unit-bar sets with angle-of-repose cross strata can make up as much as 20% of the total volume of the channel deposits and may locally approach the full thickness of the channel deposit (e.g. Reesink *et al.*, 2014). Unit-bars are more common and distinct in braided rivers and are therefore sometimes used as key evidence for braiding. However, observations from the rock record and modern rivers indicate that they are also prevalent in meandering rivers and that few sedimentary structures are exclusive to a specific planform (Fig. 1; Bridge, 2003). Moreover, lateral accretion surfaces, that are considered characteristic for meandering planforms, may vary in their abundance irrespective of planform (Hartley *et al.*, 2015; MacMahon & Davies, this volume). Similarly, it remains unclear how the presence of unit bars should be interpreted. What implications does the presence of unit bars have for interpretations of river planform and discharge regime? The scarcity of systematic investigations of unit-bar deposits hinders 1) the development of hypotheses on their formation and preservation; and 2) analyses of the relative importance of controls on river planform.

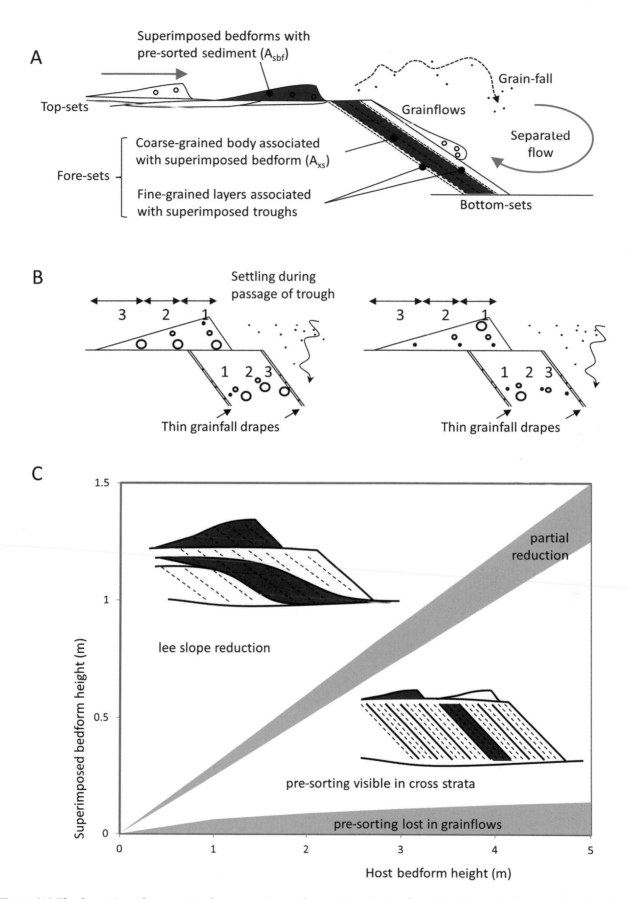

Fig. 2. (A) The formation of cross strata by pre-sorting and re-sorting during deposition by grain flows and settling from suspension. (B) Pre-sorted cross strata may contain evidence of the vertical sorting within the superimposed bedforms. (C) The existence of pre-sorted cross strata depends on the relative size of the host and superimposed bedforms.

The coexistence of dunes and unit-bars

Dunes and unit bars coexist, are both ubiquitous in alluvial rivers and both create cross-stratified sets. The vertical thickness of preserved cross-stratified sets reflects the size of the formative bedforms (Paola & Borgman, 1991; Leclair & Bridge, 2001; Bridge & Lunt, 2008; Reesink *et al.*, 2015). In theory, both dune and bar dimensions can be used to indicate formative flow depths, the thickness of the channel deposits and the size of the depositional system (Blum *et al.*, 2013; Holbrook & Wanas, 2014). However, the interpretation of set thickness distributions depends on the correct distinction of formative bedform types, which is not necessarily straightforward. Moreover, the quantitative relationships are known to vary spatially and cannot be used for an unknown proportion of river channel deposits (Reesink *et al.*, 2015). Pre-sorted cross strata indicate the volume of the formative superimposed bedforms and can therefore be used to interpret 1) partial preservation of the host bedform; and 2) the co-evolution of bedform types.

Reservoir characterisation and modelling

The study of pre-sorted cross strata is also of direct practical relevance. The grain-size sorting that allows most cross strata to be recognised affects the porosity and permeability of river channel deposits (Ritzi *et al.*, 2016; Gershenzon *et al.*, 2015). The geometry of cross strata affects flow paths, fluid dispersion and residual trapping of water, hydrocarbons and CO_2 in sandstone reservoirs (e.g. Brayshaw *et al.*, 1996; Krishnamurthy *et al.*, 2016). Furthermore, the large lateral extent and fine-grained bounding surfaces associated with unit bars may be critical to reservoir performance, especially in deposits with significant clay layers that form baffles or barriers to the flow (Smith *et al.*, 2009), or in deposits with gravel-rich layers that may form thief-zones (Lunt & Bridge, 2007; Guin *et al.*, 2010). However, at present, few quantitative data on unit-bar cross strata are available beyond those generated during a limited number of experimental and field studies (McKee, 1957; Hooke, 1968; Reesink & Bridge, 2007, 2009, 2011; Martinez De Alvaró, 2015). The systematic measurements of pre-sorting patterns in unit-bar sets in this study provide new data for the improvement of reservoir characterisations.

Objectives of this study

Improved understanding of cross strata has diverse applications. The primary objective of this paper is therefore to develop a comprehensive foundation for the analysis of unit-bar cross strata. Herein, the improved understanding of pre-sorting patterns is used to assess their value for: i) the distinction of dune sets from unit-bar sets, ii) quantitative interpretation of formative bedform size and water depths; and iii) interpretation of the formative hydrographs, environment and geomorphology. In order to begin to address some of these issues, this paper 1) summarises the relevant theory from published literature, 2) presents a new qualitative and quantitative dataset from four different outcrops, 3) illustrates different types of interpretations; and 4) highlights gaps in our understanding in order to help guide future research.

BACKGROUND

Cross strata

Sandy river-channel deposits are dominated by cross-stratified sets. Sets are centimetre-thick to metre-thick, roughly-horizontal layers of sediment. Sets are internally composed of millimetre to decimetre-thick diagonal layers, which are known as cross strata (Figs. 1A and 2A). The term 'cross strata' is used here in the strict sense to indicate angle-of-repose strata and excludes inclined co-sets (e.g. Haszeldine, 1983). Individual sets are formed by the migration of a single bedform and both dunes and unit bars create cross-stratified sets. Unit-bar 'deposits' are defined herein as composite units composed of bottomsets, foresets and topsets (cf. Gilbert, 1885). This paper is focussed on pre-sorted angle-of-repose cross strata (a subgroup of foresets) within unit-bar deposits.

The formation of cross strata by pre-sorting and re-sorting

The grain-size sorting in cross strata reflects sorting of the sediment that arrives at the brink point (pre-sorting) and the sorting that occurs during deposition (re-sorting; Fig. 2A). Angle-of-repose cross strata formed by unit-bars commonly contain distinct pre-sorting patterns that are created by the arrival of successive superimposed bedforms at the host lee slope (Fig. 2A; McKee, 1957; Hooke, 1968; Smith, 1972, 1974; Reesink & Bridge, 2007, 2009). Pre-sorted cross-strata are characteristically 0.01 to 0.2 m-thick (Reesink & Bridge, 2007, 2009) and can be formed by ripples on dunes or bars, dunes on large host dunes (e.g. Parsons *et al.*, 2005), or dunes on bars.

Unit bars with superimposed dunes are common and this combination creates distinct sorting patterns that are readily recognised in cores, outcrops and in high-resolution borehole geophysics such as Full-bore Formation MicroImager (FMI; e.g. Donselaar & Schmidt, 2005). The coarse-grained increments associated with the body of the superimposed dunes are internally stratified as a consequence of re-sorting by individual grain flows and are delineated by thin drapes of finer-grained sediment that settles out of suspension during the passage of the superimposed bedform trough (Fig. 2A). In some cases, sorting trends are observed within the coarse-grained increments that are associated with the internal sorting of the superimposed bedforms (Fig. 2B; Reesink & Bridge, 2009). Re-sorting by grain flows, modification of sorting processes by turbulence (Allen, 1982; Reesink & Bridge, 2009), stacking of small grain flows into larger compound grain flows (Hétu *et al.*, 1995; Koeppe *et al.*, 1997) and secondary failures of the lee slope (Hunter, 1985 a,b; Kleinhans 2005, 2006) can introduce additional complexities within the cross-stratification pattern. Re-sorting of pre-sorted sediment by grain flows during deposition on the lee slope also determines the resolution at which pre-sorting patterns can be quantified and therefore poses an important limitation to interpretations of pre-sorting patterns. Despite this additional complexity, angle-of-repose cross strata created by dunes on unit bars are among some of the easiest structures to identify in outcrops.

The cross-sectional areas of distinct pre-sorted cross strata (A_{xs}) matches the cross sectional area of their formative superimposed bedforms (A_{sbf}; Fig. 2A; Hunter & Rubin, 1982; Rubin & Hunter, 1983; Reesink & Bridge, 2007) following the relationship:

$$A_{sbf} = A_{xs} * C_1 \qquad \text{(Eq. 1)}$$

In which C1 is a constant that varies between 0 and 1 that represents the effects of: i) compaction, ii) partial erosion of the host set, iii) capture of suspended sediment on the host lee slope; and iv) loss of pre-sorting of smaller bedforms and incorporation of this sediment in another pattern during re-sorting on the lee slope. Furthermore, misinterpretation of the cross-sectional area may occur as a consequence of poor visibility of the pre-sorting pattern (e.g. due to pronounced weathering) and the visibility of the pre-sorting pattern can be affected by the three-dimensionality and interactions of the superimposed bedforms at the time of formation. Nonetheless, the relationship in Equation 1 provides the opportunity to quantitatively interpret preserved pre-sorting patterns. This paper investigates what the relative effects of the different controlling variables are on the preserved cross strata and therefore, to what extent preserved pre-sorting patterns can be used for quantitative palaeo-environmental interpretations.

Reactivation surfaces created by superimposed bedforms

Where the height of the superimposed bedforms (H_{sbf}) is large relative to the host bedform height (H_h), the host lee slope is temporarily reduced in angle during the passage of the superimposed bedforms, creating a low-angle inclined co-set composed of down-stream dipping sets (Fig. 2C; Rubin & Hunter, 1982, 1983; Haszeldine, 1983; Reesink & Bridge, 2007, 2009; Reesink *et al.*, 2015). Reactivation surfaces formed in this way cannot always be distinguished from reactivation surfaces that are caused by flows unsteadiness and both types are therefore described using the same term. The transition between angle-of-repose slopes versus reduction of the host lee slope by superimposed bedforms is given by:

$$H_{sbf} = C_2 * H_h \qquad \text{(Eq. 2)}$$

In which the constant C_2 was empirically found to be 0.25 to 0.3 by Reesink & Bridge (2007, 2009) and 0.11 by Warmink *et al.* (2014). The uncertainty about the magnitude of this threshold to reduction of the host lee slope is related to oversimplification of the problem: the analysis above concerns only downstream accretion, the values were established based on two-dimensional profiles from three-dimensional bedforms and the co-evolution of the shapes of the host and superimposed bedforms and the flow field are not taken into consideration. Nonetheless, measurements of sedimentary structures in the South Saskatchewan

River confirm that a value of 0.25 to 0.3 provides a reasonable estimate for the analysis of downstream accretion surfaces in exposed unit bars (Reesink & Bridge, 2011). It is clear that an increase in the size of superimposed bedforms relative to their hosts causes a transition from angle-of-repose cross strata to reactivation surfaces. This threshold of reduction (Eq. 2, Fig. 2C) makes it possible to predict a maximum bedform size for the development of pre-sorting patterns.

Dune and unit-bar morphology

Dunes and unit bars co-exist in alluvial rivers and both create cross-stratified sets. In order to interpret cross strata accurately, it is therefore necessary to define and distinguish dunes and unit bars. Furthermore, maximum dune and bar thicknesses depend on flow depth (Fig. 3A) and this dependency creates systematic trends in the sedimentary structures within river deposits (Fig. 3B).

Definition and geometric properties of dunes in rivers

Dunes are defined here as asymmetric, sinuous or three-dimensional, flow-transverse bedforms that scale to the flow depth (Allen, 1982; Ashley, 1990; Leeder, 1999; Bridge, 2003). The heights and lengths of dunes are linked by morpho-dynamic feedback mechanisms (Best, 2005). The dependency of equilibrium dune dimensions on flow depth has been captured quantitatively using a range of linear and non-linear relationships that may also include considerations of transport stage (e.g. Gill, 1971; Yalin, 1964, 1992, 2013; Allen 1968, 1970, 1982; Fredsoe 1980, 1982; Bradley & Venditti, 2016). Recent work illustrates that these quantitative relationships are sensitive to a range of common environmental variables such as sediment cohesion due to extracellular polymeric substances (EPS: Parsons *et al.*, 2016), grain-size (Tuijnder *et al.*, 2008); and the clay content of the flow (Baas *et al.*, 2002). Furthermore, dune-size varies spatially and temporally across rivers (Kleinhans *et al.*, 2007) and is subject to significant hysteresis (Julien & Klaassen, 1995; Kleinhans *et al.*, 2007; Martin & Jerolmack, 2013).

Dune height and length share a non-linear relation that is partially dependent on flow conditions (cf. Yalin 1972, 1992; Ashley, 1990; Naqshband *et al.*, 2014b). However, for the purpose of this study, the flow-parallel cross-sectional area of

superimposed dunes (A_{sd} [m]) can be captured by a simplified combination of their height, length and shape.

$$A_{sd} = H_{sd} * L_{sd} * C_3 / 2 \qquad \text{(Eq. 3)}$$

$$H_{sd} = C_4 * L_{sd} \qquad \text{(Eq. 4)}$$

In which H_{sd} (m) and L_{sd} (m) are the heights and lengths of superimposed dunes and C_3 is a constant, or 'shape factor' (Wilbers, 2004), that describes the deviation of the cross-sectional area of dunes from idealised triangles and which is assumed herein to be 1 for simplicity. The height-length ratio, C_4, is expected to be less than 0.06 for dunes (Bridge, 2003). Equations 3 and 4 provide a means to approximate the cross-sectional area of the formative superimposed bedforms and, hence, associate the dimensions of the cross strata to those of their formative bedforms.

Despite the complex relationship between water depth and dune-size and geometry, equilibrium dune heights are commonly related to water depth in open channels following a simple relation:

$$H_{d\text{-}e} = C_5 * d \qquad \text{(Eq. 5)}$$

In which $H_{d\text{-}e}$ is the equilibrium dune height in depth-limited flows, d is the water depth and C_5 is a constant that was empirically established to be 1/6th (Yalin, 1964), but which often approaches one third of the water depth in shallow flows such as in flumes. Despite the variability in dune dimensions (e.g. Leeder, 1999) and the uncertainty about the fundamental controls described above, Equation 5 performs a useful function in describing both the general trend (Fig. 3A) and providing an approximate magnitude of equilibrium dune height in relation to water depth.

Definition and geometric properties of unit bars in rivers

In contrast to dunes, unit bars are characteristic channel-scale lobate or oblique bedforms with heights that may approach bankfull flow depths (Fig. 3A). The term 'unit' is used herein to emphasise the coherent shape of unit bars, or the coherent sedimentology of unit-bar deposit (Walker, 1976; Bridge & Tye, 2000; Bridge, 2003). The coherent nature of unit bars distinguishes them from 'compound' bars such as bank-attached 'point bars' and 'mid-channel bars' that are formed by the amalgamation of multiple unit bars (cf. Bridge, 2003). This

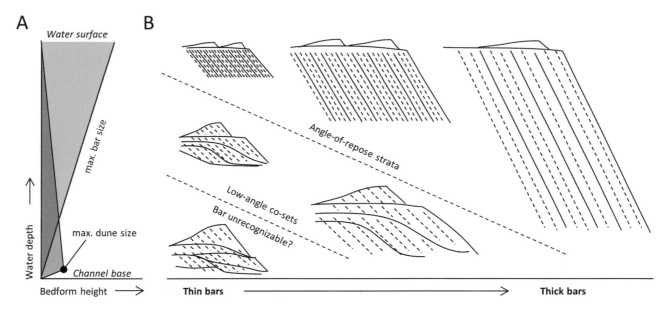

Fig. 3. (A) The maximum heights of dunes and bars vary systematically with flow depth. (B) Previous research (summary in Fig. 2C) indicates that these systematic changes in the dimensions of dunes and bars will create vertical trends in the types of sedimentary structures that are formed during downstream accretion of unit bars.

broad definition differs notably from morphology-based definitions such as transverse-bars, lobate-bars and scroll-bars (Sundborg 1956; Smith, 1971, 1972, 1974, 1977; Jackson 1976; Nanson, 1981; Ashmore, 1991) and process-based definitions such as free-bars and fixed-bars (cf. Tubino & Seminara, 1990; Tubino *et al.*, 1999; Rodriguez *et al.*, 2014). The term unit bar is preferred here because morphological and process-based definitions are hard to justify based on outcrops and cores. The choice of terminology herein evades morphological or genetic interpretations.

Definitions and description of unit bars (further below) are dominated by concerns about their morphology and relation to average conditions (e.g. Lanzoni & Tubino, 1999; Lanzoni, 2000 a,b; Nicholas, 2010). Their temporal evolution and relation to floods remains inadequately understood. Work by Bridge (1993, 2003) illustrates the fundamental importance of developing an understanding of bars that incorporates their spatial as well as their temporal variability and evolution. For example, it is probable that shifts in spatial gradients in bed shear stress that occur across channels as a consequence of temporally changing flow initiate the development of unit bars (cf. Carling, 1991; Bridge, 1993; Sear, 1996; Milan *et al.*, 2001). Unit-bars may persist over multiple floods once established (Collinson, 1980; Allen, 1982; Reesink & Bridge, 2011). Unit-bar cross-strata with pre-sorting patterns

record short histories of bar accretion, deactivation and reactivation and therefore provide a novel perspective on their temporal development.

Interpreting dune-bar and unit-bar deposits

Co-existence of dunes and unit bars: implications for sedimentary structures

The difference in the relative sizes of dunes and bars in any river constrains what structures are formed (Fig. 2C and 3). Large bars must have their tops in shallow water, which is where dunes are small (Fig. 3A). Dunes are largest in the deepest parts of the river (thalweg), where the heights of the unit bars are by definition low. In other words, bedforms exist between the maximum water depth and the deepest scour and this creates simple constraints that help interpret their co-existence and interactions (Fig. 3A).

The different trends in dune and bar size along the vertical create a distinct vertical trend in sedimentary structures that characterise river channel deposits (Fig. 3B). Thin bars with large superimposed dunes (deep in the river) either create low-angle reactivation surfaces (cf. Fig. 2C) or are unrecognisable in the rock record as unit-bar deposits. Furthermore, the settling of fine-grained sediment in the lee of small bars in a perennial thalweg may be too small to create recognisable bottomsets. Conversely, angle-of-repose unit bars are expected to be most common in the upper parts of the river channel where dunes are small. Angle-of-repose unit-bar cross strata are therefore probably characteristic for the upper parts of the channel deposits.

Interpreting and distinguishing dunes and unit bars

The geomorphological definition of a bedform does not precisely match the sedimentological definition of the same bedform: bedforms are defined as the shape of a sand bed at a specific point in time, whereas cross-stratified sets represent short histories of a deforming bed that are recorded in the sedimentary record. A mismatch between bedforms and their deposits is to be expected in attempts to relate them in a one-on-one basis, especially because bedforms develop though amalgamation and splitting (Raudkivi & Witte, 1990; Gabel, 1993).

The distributions of thicknesses of dune-sets and bar-sets must overlap. All sets have lateral terminations where thicknesses approaches zero. Set-thickness distributions that contain both dune sets and unit-bar sets contain a distinct kink in the alignment of the data in probability plots (Fig. 4). The bimodality in such set thickness distributions is not clear in histograms because the thinner

dune-sets are far more abundant than thicker unit bars. Moreover, preserved sets thicknesses are known to vary because of variations in river discharge, channel morphology, dynamic interactions during deposition and variable preservation potential within river channels (see Reesink *et al.*, 2015).

The fundamentally different hydro-dynamic controls on dunes and bars makes it possible to develop a set of guidelines for their distinction. Although no one criterion is truly unique to either dunes or bars, three diagnostic criteria have been suggested for their distinction (Reesink & Bridge, 2011; Reesink *et al.*, 2014): i) the character of bottomsets (see also Herbert *et al.*, 2015; Herbert & Alexander, 2018), ii) the character of pre-sorting patterns in the cross strata; and iii) the presence of characteristic reactivation surfaces (Collinson *et al.*, 1980; Allen, 1982; Reesink & Bridge, 2011).

Firstly, unit bars have deeper and longer troughs in comparison to dunes, which causes greater deceleration and creates a greater potential for sediment to settle in the trough. Thus, unit bars are expected to have significant finer-grained, ripple cross-laminated bottom-sets underneath the coarser-grained unit-bar cross strata (Reesink *et al.*, 2014; Herbert *et al.*, 2015; Herbert & Alexander, 2018). Secondly, bedload sediment transport over unit bars occurs in the form of migration of bedforms, which create pre-sorted

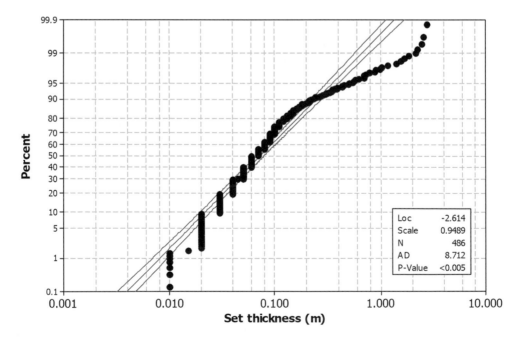

Fig. 4. Probability plots of angle-of-repose set thicknesses measured in cores (log-probability). Measurements include cross strata formed by dunes as well as bars. Data from the Rio Paraná, Argentina (anabranching planform).

cross strata as they arrive successively at the lee slope (Reesink & Bridge, 2007). Thirdly, the larger unit-bars respond slowly to changes in flow and are expected to be in disequilibrium with the flow for most of the time. Unit-bar sets commonly contain reactivation surfaces that indicate changes in sediment transport over the back of the unit bars and changes from slope-perpendicular to slope-parallel flow. Such distinct changes in flow are particularly characteristic for the upper half of the channel (Ashworth *et al.*, 1995) and are unlikely to occur over large dunes in the thalweg. Reactivation surfaces created by changes in flow are therefore expected to be common in unit-bar sets but not in large dunes that are found in the deeper parts of large rivers. The data collected in this paper are biased towards large sets with clear pre-sorting patterns and this bias prevents a critical test of the distinguishing criteria: further systematic research is needed to truly test the limitations of these guidelines.

Interpretation of pre-sorting patterns in unit-bar sets

Pre-sorting patterns created by dunes on unit bars provide several parallel lines of evidence. Firstly, the relationships in Equations 1, 3 and 4 (see also Fig. 1A) make it possible to quantify minimum formative superimposed bedform sizes. Secondly, Equation 2 (see also Fig. 2C) makes it possible to identify partial preservation of the host unit-bar. Finally, pre-sorting patterns do not occur in isolation but exist within sequences of 'bar-code' that indicate the development successive superimposed bedforms that arrive at the lee slope. These complementary lines of evidence are summarised in Table 1 and discussed below.

Constraining and quantifying host and superimposed bedform sizes

In cases where the topsets of the host unit bar can be traced into the foresets, we can assume a near-complete preservation of the host unit-bar. In all other cases, we must assume that the host unit bar is partially preserved and that the pre-sorting patterns indicate a minimum size of the superimposed bedforms (Table 1). We can quantify the partial preservation of the host bar in the case of thin unit-bar sets with pre-sorting patterns created by superimposed ripples because ripples have a narrow size distribution that is independent of depth and velocity (Kleinhans *et al.*, 2017). Due to the limited size of ripples, pre-sorted cross strata formed by ripples occur on unit bars that are 0.1 to 0.3 m-high. Thin unit bars and the prevalence of ripples are both characteristic for the shallow flows that are common on the top of compound bars. Consequently, pre-sorted cross strata formed by ripples on bars are a useful indicator of the upper parts of a channel. In most cases where bars have a significant size (>ca. 0.25 m), the pre-sorted cross strata are formed by dunes on bars. For the remaining cases, Equation 2 provides a means of testing the exceedance of the maximum

Table 1. Summary of different quantitative interpretations of host and superimposed bedforms.

		Superimposed bedforms	
		Size known	Formative size unknown
Host bedforms	Size known	Opportunity to quantify relationships between host and superimposed bedforms e.g. flume experiments (Reesink & Bridge, 2007; 2009, Warmink *et al.*, 2014)	Opportunity to quantify the development of superimposed bedforms, i.e. size distributions and record of hysteresis of formative superimposed dunes. e.g. field investigation of cross strata formed by dunes on bars (Reesink & Bridge, 2011)
Host bedforms	Formative size unknown	Opportunity to quantify preservation of the host unit-bar. e.g. field investigation of cross strata formed by ripples (with narrow size distribution; Reesink & Bridge, 2011)	Constrain bedform sizes: 1. Quantify minimum bedform sizes 　a. Superimposed bedform volume 　b. Host bedform height 2. Use (1a) and Equations 1, 3 and 4 to quantify superimposed bedform sizes 3. Use (2) and Equation 2 to check minimum host bedform size and assess partial preservation 4. Use (2) and Equation 5 to estimate minimum palaeo-water depths 5. Assess trends in superimposed bedform development

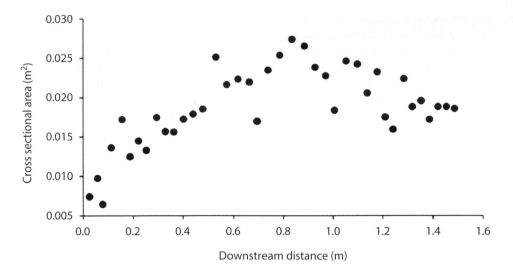

Fig. 5. Sequence of cross-sectional areas of pre-sorted cross strata formed by dunes on a unit bar measured in the braided part of the South Saskatchewan River, Canada (after Reesink & Bridge, 2011). Note that the strata indicate growth but not distinct decay in dune size.

height of the superimposed dune height relative to the host bedform height (see also Reesink & Bridge, 2009, 2011).

Interpreting sequences of pre-sorted cross strata

Systematic measurements made on pre-sorted cross strata in a single unit bar in the South Saskatchewan River, Canada, illustrated a pattern of growth of the successive dunes that migrate over the bar to the lee slope (Fig. 5; Reesink & Bridge, 2011). Little deposition was associated with the falling limb of the flood. This sequence provides a record of the delayed response of the dunes to changes in flow depth and velocity during the rise and peak of a flood: palaeo-dune hysteresis (Julien & Klaassen 1995; Ten Brinke *et al.*, 1999; Wilber & Ten Brinke, 2003; Kleinhans *et al.*, 2007; Martin & Jerolmack, 2013).

It is currently not known whether patterns of dune growth dominate or whether both growth and decays are observed within unit-bar strata. Patterns of growth and decay might both exist as a consequence of modifications of bar shape at different stages, such as widening and deformation during falling flow stages (Rodriguez *et al.*, 2014; Wintenberger *et al.*, 2016). Bar-tail and chute cut-off and the development of cross-bar channels are also common during falling flow stages (e.g. Ashmore, 1991; Zinger *et al.*, 2011, 2013). It is expected that such continued development of bars during falling flow stage is also recorded in

the rock record (Rice *et al.*, 2009). Pre-sorted cross strata may contain essential clues for the interpretation of such dynamics.

Finally, if unit-bar sets record a wide range of pre-sorting sequences from different flow stages, the information contained in these sequences may provide useful information on the recurrence, size and duration of formative floods. In turn, the nature of base-flow discharge that defines the perennial flow depth and the ephemeral part of the hydrograph that controls the bar-overtopping floods are linked to catchment size and climate (Lane *et al.*, 2008; Fielding *et al.*, 2009; Syvitski *et al.*, 2014; Plink-Björglund, 2015). The vertical composition of channel deposits (Fig. 3B) and sequences of pre-sorting patterns (Fig. 5) may provide a means of distinguishing long-duration floods in large and monsoonal rivers (Reesink *et al.*, 2014; Plink-Björglund, 2015) from short-duration floods in smaller and more ephemeral systems.

METHODS AND INVESTIGATED OUTCROPS

Measurements

Two geometric parameters of pre-sorted cross strata can be measured easily and objectively in outcrops and cores: the vertical thickness of the cross strata (th_v) and the vertical thickness of the unit-bar set (host set: H_h). The vertical direction is defined here as the formative vertical and the measurements must be corrected for any changes in dip due to tectonic tilting. Measurements of these variables also avoid the complexity of establishing the cross-sectional area from outcrops in which only an apparent dip is visible. The cross-sectional areas of the pre-sorting patterns are calculated here from the vertical thickness of the pre-sorting patterns (th_v) and the thickness of the preserved bar set (H_h) along the vertical. The calculation of the cross-sectional area (A_{xs}) from the vertical dimensions herein assumes an arbitrary value of 20% compaction, straight uniform thickness of the pre-sorted cross strata and a best-estimate of a formative lee slope (α) of 30 degrees:

$$th_p = th_v/C_6 * \cos(\alpha) \qquad \text{(Eq. 6)}$$

$$A_{xs} = H_h/C_6/\sin(\alpha)*th_p \qquad \text{(Eq. 7)}$$

In which th_p is the thickness of the cross strata measured perpendicular to their bedding plane (m), th_v is the vertical thickness of the cross strata (m), C_6 is the compaction factor (herein 0.8), α is the angle of the cross strata upon deposition and H_h is the height of the host unit bar (see also Fig. 2A). Different values for the lee slope and compaction are included in the sensitivity analysis. Calculation of the dimensions of the formative superimposed dunes are approximated using a shape factor (C3) of 1 and a height-length ratio (C4) of 1/10 for the superimposed dunes.

Field sites

Four field sites were investigated in this study: 1) The Cambrian Lamotte Sandstone, MO, USA (Fig. 6B), 2) the Carboniferous Chatsworth Grit at Gardom's Edge, Derbyshire, UK (Fig. 6A), 3) the Carboniferous Lower Brimham Grit at Brimham Rocks, Yorkshire, UK (Fig. 6C and E to H); and 4) the Carboniferous Fell Sandstone at Bowden Doors, Northumberland, UK (Fig. 6D).

The Lamotte Sandstone was investigated at Hawn State Park and Hickory Canyons Natural Area near Farmington, MO, USA. The chosen field sites are dominated by fluvial channel deposits, although marine and aeolian environments are also present (Ojakangas, 1960; Houseknecht & Ethridge 1978; Yesberger, 1982). Fourteen easily-accessible cross-stratified sets that were interpreted as unit bars were investigated along 3 different exposures: both sides of the road at Hickory Canyons and the cliffs next to the camping site at Hawn State Park.

The Lamotte Sandstone forms part of the Sauk Transgressive sequence on the Laurentian continent (Sloss, 1963; Collinson *et al.*, 1988) and is a contemporaneous deposit of the Mount Simon Sandstone, which is a target reservoir for CO_2 injection (e.g. Leetaru *et al.*, 2009 a,b; Frailey *et al.*, 2009; Freiburg *et al.*, 2014). The Lamotte Sandstone formed during the Cambrian (540 to 490 Ma), which implies that no land-based vegetation existed at the time. The absence of vegetation that lasted until the early Ordovician (470 Ma) would have greatly affected the formation and preservation of the channel deposits.

The Carboniferous Chatsworth Grit at Gardom's Edge is part of the Namurian (326 to 313 Ma), Marsden Formation. Five easily accessible unit-bar sets with distinct pre-sorted cross strata were investigated along Sheffield Road, near Baslow, UK.

The Carboniferous (Bashkirian; 323.2 to 315.2 Ma) Lower Brimham Grit at Brimham Rocks, UK, is the subject of the detailed investigation of Soltan & Mountney (2016). For the purpose of this study, the highly three-dimensional exposures at Brimham Rocks provided easy access to 19 well-exposed cross stratified sets that were interpreted as unit bars.

Seventeen unit-bar sets were investigates at Bowden Doors, which forms part of the Carboniferous Fell Sandstone Formation. The Bowden Doors has attracted attention because of interpreted mass movement deposits (e.g. Turner & Munroe, 1987; Martin & Turner, 1998) and also contains one of the more impressive exposures of a unit bar with distinct pre-sorted cross strata (Fig. 6D). Unfortunately, not all unit-bar sets are readily accessible: only a small section of the most prominent bar-form was investigated (Fig. 6D; located at about 6 m above the ground).

The four case studies allow comparison between the deposits of river systems of distinctly different sizes, climates and environments. The climate and environment of the unvegetated Cambrian world, the vegetated world and variable climates of the Carboniferous provide a clear contrast that should be reflected in the sedimentary structures within their deposits.

RESULTS AND INTERPRETATIONS

General observations of pre-sorted cross strata in the studied outcrops are presented first in order to develop a basic understanding of the pre-sorting patterns in these outcrops. A sensitivity analysis

Fig. 6. Pre-sorted cross strata in outcrops. (A) Pre-sorted cross strata grouped in flood deposits that are delineated by distinct finer-grained reactivation surfaces (weathered out: left of lower unit-bar set). Gardom's Edge, UK. (B) Distinct delineation of strata by fine-grained laminae. Lamotte Sandstone, MO, USA. (C) Unit-bar sets are commonly underlain by distinct bottomsets. Brimham Rocks, UK. (D) Clear visibility of internal lamination associated with individual grain flows. Bowden doors, UK. (F) Cross-stream exposures of pre-sorted cross strata illustrating lateral changes in the pre-sorting pattern from distinct to vague. (G) close-up of cross-stream exposure illustrating the laminated internal build-up of the patterns and evidence of occasional cross-stream flow. Brimham Rocks, UK. (H) Poor visibility of pre-sorting patterns due to dominance of a tafoni-type weathering pattern; Bowden Doors, UK.

of controlling factors is performed on combined measurements, after which the results are compared between the different field sites. This general analysis constrains the viability of pre-sorting patterns as indicators of formative bedforms and, therefore, provides the necessary justification for the subsequent quantitative interpretation of individual pre-sorted strata and sequences of pre-sorted cross strata.

Qualitative analysis of pre-sorting patterns in unit-bar sets

Observations of pre-sorted cross strata

The systematic investigation of unit-bar sets in the four selected outcrops indicates that the unit-bar sets contain clear pre-sorting patterns that can be interpreted as formed by superimposed dunes, although their visibility varied (Fig. 6). The grain-size of the cross strata varied from fine gravel to fine sand and no conglomerates that represent gravelly rivers were present. The characteristics of the pre-sorting patterns were in line with descriptions by Reesink & Bridge (2007, 2009; see also Fig. 2). The thickness and composition of bottomsets varied laterally (Fig. 6A to D). The preferentially-weathered finer-grained bottom sets were found under nearly all unit-bar foresets, with only a few exceptions. The internal lamination within these fine-grained bottomsets was not commonly visible in the outcrops.

The relative abundance of unit-bar sets, although hard to establish based on a limited number of field visits and exposures, varied considerably between the sandstone exposures. Unit-bar sets appeared most abundant at Brimham Rocks, UK (Soltan & Mountney, 2016). However, the occurrence of unit-bar sets varied spatially from 0% to near to 100% of the total height of the exposures, with smaller exposures regularly dominated by unit-bar sets.

A few thicker unit-bar sets (>0.8 m) did not contain repetitive pre-sorting patterns that can be interpreted as formed by series of successive superimposed bedforms. In such cases, the sorting patterns of the cross strata were irregular alternations of thick and thin sorting patterns. Larger bar cross strata also commonly contained distinct internal truncations that indicated changes in the orientation of the lee slope. Such truncations were often associated with local reworking of the slope by slope-parallel flows (Fig. 6G).

In contrast to the large unit-bar sets, reactivation surfaces that mark deactivation and periods of low flow within sequences of regular pre-sorting patterns sets of about 0.5 m thickness were not easily identified (cf. Reesink & Bridge, 2011), with some sequences only becoming obvious after the values of the cross-sectional areas were plotted. In between regular pre-sorting sequences, reactivation surfaces were primarily vaguely stratified sections that terminate in an enrichment of finer-grained sediment. The reactivation surfaces were occasionally marked by a subtle change in stratal slope that indicated a change in the orientation of the formative slope. Evidence of flow along the strike of the lee slope was only found locally and was more easily identified in cross-stream (Fig. 6F and G) and oblique exposures in comparison to stream-parallel exposures. No clear evidence of subaerial exposure was found in the bottomsets or foresets, although it is noted that such evidence might be difficult to identify in outcrops.

Cross-stream exposures (Fig. 6F and G) illustrated that the visibility of the pre-sorting patterns varied laterally and vertically and that junctions exist in the sorting pattern just as they exist in the crest lines of the formative superimposed bedforms. Furthermore, diagenetic patterns such as Liesegang bands and irregular weathering patterns such as honeycomb-style and tafoni-style weathering (Fig. 6H) obscured stratification patterns and hindered measurements in a few select locations.

Gradual transitions in which steepening low-angle co-sets are followed by angle-of-repose pre-sorted cross strata are commonly observed at the upstream end of unit-bar sets (cf. Haszeldine, 1983). A gradual reversed pattern was not observed. Repetitive transitions between low-angle-co-sets and angle-of-repose cross strata, as might be expected for cases of continued bar migration under fluctuating discharge, were not observed. In the investigated outcrops, unit-bar sets either thinned in the downstream direction until they were not identifiable, or they terminated in a truncation or reactivation surface that was followed by structures of a different type or with a different orientation.

Interpretation of observations of pre-sorted cross strata

Despite the local complexity within pre-sorting patterns and their variable visibility in outcrops, it is clear that pre-sorting patterns are present in the majority of angle-of-repose unit-bar sets and that they are easy to identify and measure. The qualitative observations of pre-sorting patterns match experimental studies of their formation (Reesink & Bridge 2007, 2009; Martinez de Alvaró, 2014). The uncertainty in the interpretation of formative superimposed bedforms from pre-sorting patterns decreases when more interpretative criteria are used: 1) regular and repetitive sequences of pre-sorting patterns, 2) clear and continuous fine-grained drapes, 3) internal sorting trends that are consistent with the fine-grained drapes; and 4) partial reactivation surfaces at the top of the set. However, not all pre-sorting patterns are equally clear and this commonly reflects the loss of pre-sorting patterns during resorting on the lee slope by grainflows. The clarity of sequences of pre-sorted cross strata formed by superimposed dunes decreases when unit-bar sets become thicker than 0.8 m in set thickness. The presence of irregular sorting patterns in the thicker bar sets indicates that pre-sorting is increasingly lost during re-sorting by grainflows when unit bars become taller. This approximate value coincides with the height of bars on which the cross-sectional area of grainflows (approx. 0.006 m-thick cf. Reesink & Bridge, 2009) matches the volume of small dunes (heights of 0.05 m and lengths of 0.5 m).

Breaks in sequences of pre-sorted cross strata linked to superimposed dunes were commonly poorly visible, which indicates that downstream accretion surfaces do not reliably record the falling limbs of the hydrographs. Precisely how unit-bars deform and where accretion takes place during different flow stages (Bridge, 1993; Rodriguez *et al.*, 2014; Wintenberger *et al.*, 2016) requires further systematic research.

Dimensional properties of pre-sorting patterns in unit-bar sets

Bulk dimensional properties of pre-sorted cross strata

For this study, the thicknesses of 1772 angle-of-repose cross strata were measured within a 56 unit-bar sets. The thicknesses of the pre-sorted cross strata had a narrow range from 0.01 to 0.2 m (Fig. 7A). The measurements in Fig. 7B show that the thicknesses of the cross strata do not increase with bar size (Fig. 7B).

The red lines in Figs 7B to E and F illustrate three thresholds for reduction of the host lee slope by superimposed bedforms ($C_2 = 0.11$, 0.25 and 0.3). Data points above these lines indicate that interpreted superimposed bedform dimensions are too large relative to the thickness of the preserved host set for the formation of angle-of-repose cross strata (cf. Fig. 2C) and hence, indicate that the unit-bar set is partially preserved. The different threshold factors, 0.11, 0.25 and 0.3, result in exceedance of the threshold of reactivation by 83, 22 and 10 percent of data. These values are the same for the strata thickness and cross-sectional area. The proportion of the data that exceeds the threshold also changes for different superimposed bedforms height-length ratios because the current empirical relationships are based on superimposed bedform height only. Height to length ratios of the superimposed bedforms of 1/10 to 1/15 and 1/20 change the percentages exceeding the threshold of reduction from 22 to 9 and 4 percent ($C_2 = 0.25$). Thus, uncertainty exists regarding the quantification of the threshold of reduction.

The cross-sectional areas of the pre-sorting patterns have a wide distribution that ranges from 0.002 m² to 0.97 m² that is heavily skewed (Fig. 7C) and significantly wider for larger bar sizes (Fig. 7D to E). Figs 7D and E illustrate that the cross-sectional area of the pre-sorting patterns increases with bar size. The cross-sectional areas of the pre-sorting patterns is partially dependent on bar size; bar height is used to calculate the cross-sectional area and bar height controls the minimum and maximum dimensions of the pre-sorting pattern that can be identified. Nonetheless, it is useful to plot these results against one another because they reflect the two most interesting formative factors: the sizes of the host and superimposed bedforms (Fig. 7D and E).

In addition to the analysis of the threshold of reduction, the thicknesses and cross-sectional areas of the pre-sorted cross strata are sensitive to wide variety of variables. The arrows in Figs 7D and E indicate the relative impact of realistic changes in these variables. Four arrows are placed at the 10th, 50th and 90th percentiles within the dataset (Fig. 7D and E). Firstly, partial preservation of the host set results in an underestimation of both the host bedform height and the cross sectional area of the superimposed bedforms (Fig. 7D and E: red arrow indicates 50% preservation of the original bar height). Secondly, the size of the superimposed bedforms may be overestimated in cases where significant volumes of suspended sediment and small bedforms are incorporated in

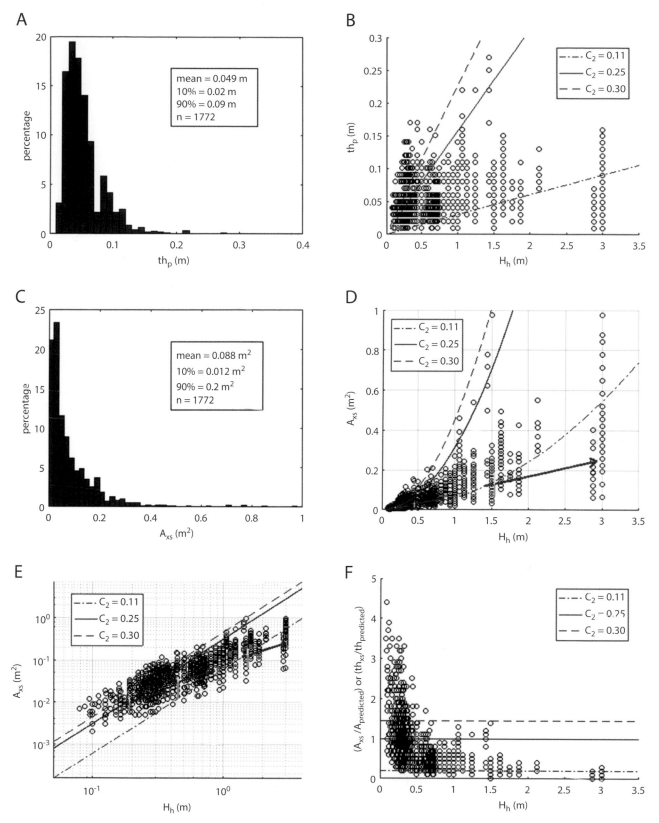

Fig. 7. (A) Histogram of the thickness of the pre-sorting patterns (th$_p$). (B) Thicknesses of the pre-sorting patterns for different set thicknesses illustrate no dependency on the host bar thickness. The red lines indicate different thresholds for lee-slope reduction (C$_2$) of the host bedform by superimposed dunes. (C) Histogram of the thickness of the pre-sorting patterns (A$_{xs}$). (D and E) linear and log plots of the preserved bar set thickness and the cross-sectional areas of the pre-sorted cross strata illustrate a clear dependency. This dependency may be partially related to partial preservation and the method of calculating the cross-sectional area. The arrows indicate the shift of the data points for a change in preservation (50%, red arrow), a lower compaction (10% as opposed to 20%: blue arrow), incorporation of additional bedload and suspended load (double volume, green arrow) and overestimation of the lee slope angle (20 degrees, yellow arrow). (F) The relative size of observed strata (compared against predicted maximum strata size) decreases sharply with an increase in bar size.

the pre-sorting pattern. The green arrows pointing down in Figs 7D and E indicate a doubling of the volume of sediment by inclusion of additional sediment (e.g. 50% bedload and 50% suspended load). Thirdly, the analysis assumes the formative angle of the host lee slope to be 30°. A reduction of the slope to 20° causes an increase in the cross-sectional area and hence the volume of the superimposed bedforms (Figs 7D and E, yellow arrow pointing up). Finally, a lower compaction of the sediment after deposition results in an overestimation of the both the host and superimposed bedform sizes (Figs 7D and E, blue arrow indicates a lower compaction of 10% as opposed to the 20% assumed as a default).

This sensitivity analysis indicates that partial preservation of the host set is the dominant factor that influences the position of the data-points (Fig. 7D to E, red arrows to right indicate 50% preservation). Thin bar sets can be the product of thin bars that have been almost entirely preserved or can be the result of partial erosion of larger bars. Partial preservation results in an underestimation of the dimensions of the host and superimposed bedforms and can account for most of data that exceed the threshold of lee-slope reduction, which occur in the thinner unit-bar sets (Fig. 7 D to F). Significant loss of the host set thickness due to later erosion matches expectations of partial preservation of unit-bar sets (Bridge & Lunt, 2008; Van de Lageweg *et al.*, 2013; Nicholas *et al.*, 2016).

The remaining variables have only a limited impact on the observed pattern. A lower compaction results in lower values of the cross-sectional area and host set thickness (Fig. 7D and E; blue arrows to the left indicate a lower (10%) compaction). Incorporation of by-passing sediment and the overestimation of the lee slope angle change the interpretation of the cross-sectional area of the superimposed bedforms only to a limited degree (Fig. 7D and E; downward green and upward yellow arrows, respectively).

Fig. 7F illustrates that the thickness of the pre-sorting pattern decreases sharply relative to their theoretical maximum when bar size increases (calculated here for $C_2 = 0.25$ and $H/L = 1/10$). Although absolute values change for different threshold models, this general trend remains the same: cross strata thicknesses and cross-sectional areas that exceed their theoretical maximum are common in thin unit-bar sets and rare in thick unit-bar sets.

Dimensional properties of pre-sorted cross strata in different formations

The unit-bar set thicknesses and the cross-sectional areas of the pre-sorting patterns differ systematically between the field sites (Fig. 8). The measurements were compared between the four formations: the Lamotte Sandstone, MO, USA (Fig. 8A); Gardom's Edge, in the Peak District, UK (Fig. 8B); Brimham Rocks, Yorkshire, UK (Fig. 8C); and Bowden Doors, Northumberland, UK (Fig. 8D). The cross sectional areas were: 0.024 ± 0.017 m^2, 0.087 ± 0.093 m^2, 0.11 ± 0.085 m^2; and 0.026 ± 0.035 m^2 respectively and the thicknesses of its unit-bar sets were 0.30 ± 0.14 m; 0.70 ± 0.43 m; 1.40 ± 0.68 m; and 0.34 ± 0.22 m respectively.

The abundance of unit-bar sets varies systematically between the studied outcrops. The outcrops at Brimham Rocks and Gardom's Edge contain a large proportion of unit-bar sets of considerable size. The unit-bar sets in the Lamotte Sandstone appeared spatially highly variable. The unit-bar sets at Bowden Doors only represented a relatively small proportion of the outcrop but contained exceptionally well-preserved laterally extensive unit-bar. Although it is possible that some thinner unit-bar sets at Gardom's Edge and Brimham Rocks (Fig. 8B and C) are under-sampled due to the abundance and greater visibility of the larger unit-bar sets, the general trend is one where the dimensions of the cross-strata and unit-bar set thicknesses change in unison. More of the data from the Lamotte Sandstone (Fig. 8A) plot above the threshold of reduction in comparison to the data from Bowden Doors (Fig. 8D), where a larger number of thick unit-bar sets were found.

Interpretation of dimensional properties of pre-sorting patterns

The observation that nearly all unit-bars in the selected outcrops have identifiable pre-sorting patterns indicates that superimposed bedforms are ubiquitous and that they dominate the sediment transport dynamics on top of unit bars. Pre-sorting patterns follow distinct distributions and trends in thickness and cross-sectional areas (Fig. 7). These relationships provide a rational basis for improvements in reservoir characterisation, as well as novel data for the investigation of the formation and preservation of unit-bar sets.

The formation of pre-sorting patterns is constrained by the relative sizes of the host and

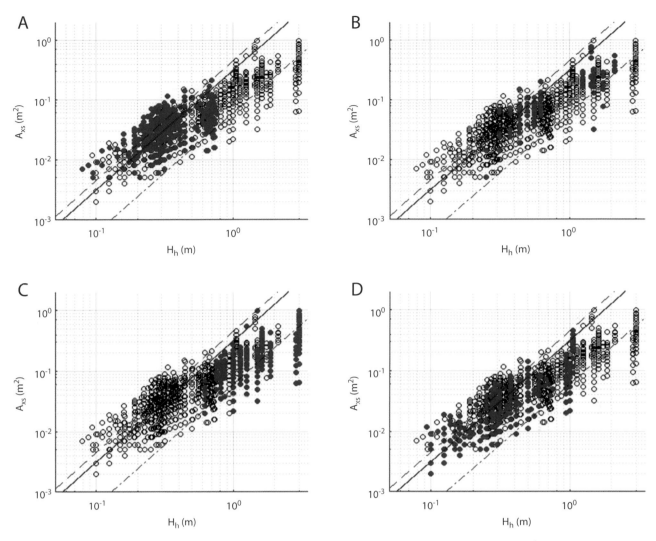

Fig. 8. Scatterplots of unit-bar heights and cross-sectional areas of the pre-sorting patterns that reflect superimposed bedform sizes for the full datasets (open black circles) and the individual outcrops (blue dots). (A) Lamotte Sandstone, MO, USA; (B) Gardom's Edge, Peak District, UK; (C) Brimham Rocks, Yorkshire, UK; (D) Bowden Doors, Northumberland, UK.

superimposed bedforms (Fig. 2C). Pre-sorting patterns can be seen as 'Goldilocks cross strata': they form when the conditions are just right. Small volumes of pre-sorted sediment are re-sorted during deposition and large superimposed bedforms create reactivation surfaces.

The cross-sectional area of the cross strata is calculated using the bar height and therefore partially dependent upon it (Fig. 7D and E). However, no clear increase in the thickness of pre-sorting patterns was seen with increasing bar size (Fig. 7B). This indicates that the dependency of the cross-sectional area on the thickness of the host set is not the primary control on the pre-sorting patterns and thus, that the pre-sorting patterns provide a

realistic proxy for superimposed bedform sizes. Furthermore, the sharp decrease in the cross-sectional areas relative to a theoretical maximum thickness (calculated based on host bar size) indicates that the size of the formative superimposed bedforms does not increase along with the host bar size (Fig. 7F). In larger bars, the pre-sorted strata volumes are far below their theoretical maximum – constrained by the threshold for lee slope reduction. The limited volumes of the cross strata indicate that the potential for recording larger pre-sorting patterns is not reached. The maximum size of superimposed bedforms is commonly limited by the elevated position of the unit bar top, where limited water depths tend to limit the superimposed bedform size (Fig. 3B). The overall trend of decreasing superimposed bedform size with increasing bar size matches the expectations laid out in Fig. 3.

The dimensional characteristics of pre-sorting patterns are controlled by multiple dependent variables and this complicates their quantitative interpretation. However, the analysis of the sensitivity of the data to the different controlling variables (Fig. 7; red lines and arrows) highlights that only two factors stand out as the key unknowns in the analysis: 1) the uncertainty about the threshold of reduction; and 2) partial preservation of the host set. A fuller understanding of the threshold of reduction (Fig. 7, red lines) is particularly important for forward modelling of sedimentary architecture (Figs 2 and 4). The partial preservation of the host set (Fig. 7D and E, red arrow) is particularly important for quantitative interpretation of formative bedform dimensions.

Fig. 8 illustrates that the unit-bar thicknesses and cross-sectional areas that reflect the size of the superimposed bedforms differ between river systems. The systematic clustering of data from different areas has multiple causes. Firstly, it is a consequence of the constraints on the formation of pre-sorting patterns. Larger bars can record larger superimposed bedforms, such that rivers with larger angle-of-repose unit bars will record larger superimposed dune volumes. In addition, the cross-sectional area is calculated based on preserved bar set thickness; the variables are partially dependent. However, the general trend – that larger rivers contain larger bars and larger dunes – matches expectations of system scale. The gap at the lower end of the measurements probably emphasises that, in rivers with large dunes, small bars are dominantly composed of low-angle reactivation surfaces as opposed to angle-of-repose strata with pre-sorting patterns. This creates a decrease in thin unit bar sets with pre-sorting patterns in larger systems (e.g. Brimham Rocks, Fig. 8C) relative to smaller systems (e.g. Lamotte Sandstone, Fig. 8A). Thus, results indicate that there is great potential for the development of (semi-) empirical relationships between the river system's size and behaviour and preserved unit-bar cross strata.

Quantitative and historical analysis of pre-sorted cross strata

The dimensional properties of the pre-sorting patterns indicate that they can be used to interpret the size of formative superimposed bedforms, albeit with a potentially significant loss to erosion. This enables two further analyses. Firstly, individual pre-sorted strata can therefore be used to approximate a minimum size of the superimposed dunes; and this enables the approximation of a minimum water depth over the unit bar. Secondly, sequences of pre-sorted cross strata can be used to infer trends in the development of successive superimposed bedforms that arrived at the lee slope.

Quantitative analysis of individual pre-sorted cross strata

In cases where formative superimposed bedforms can be interpreted reliably from pre-sorted cross strata, they can be used provide two different measures of water depth: 1) a measure of water depth above the unit bar (Fig. 9A); and 2) a minimum flow depth measured from the base of the bar (Fig. 9B). In this interpretation, the cross-sectional area of the cross strata is related to that of the superimposed bedforms and the superimposed bedforms are assumed to have predictable geometries that can be related to the water depth (Eqs 1, 3, 4 and 5).

The water depth above the unit bar is the flow depth associated with the formation of superimposed bedforms, most probably during floods. Fig. 9A and Table 2 show that these flood levels follow relatively narrow, coherent distributions. When the height of the host bar is added, the distributions widen considerably, which is attributed to the bases of the unit-bars commonly not extending into the deepest parts of the channel (cf. Fig. 3B). The distributions also clearly show that the different outcrops are characterised by different flood height distributions.

Sequences of pre-sorted cross strata

The dimensions of pre-sorting patterns reflect the dimensions of their formative superimposed bedforms and reflect both the natural variability and trends in bedform growth and decay in response to floods. Fig. 10 illustrates sequences of volumes of successive cross strata from all four field sites. The results highlight that unit-bars contain sequences of cross strata with clear trends, which can be interpreted as short histories of superimposed dune development. Whereas previous research primarily illustrated an upward trend in

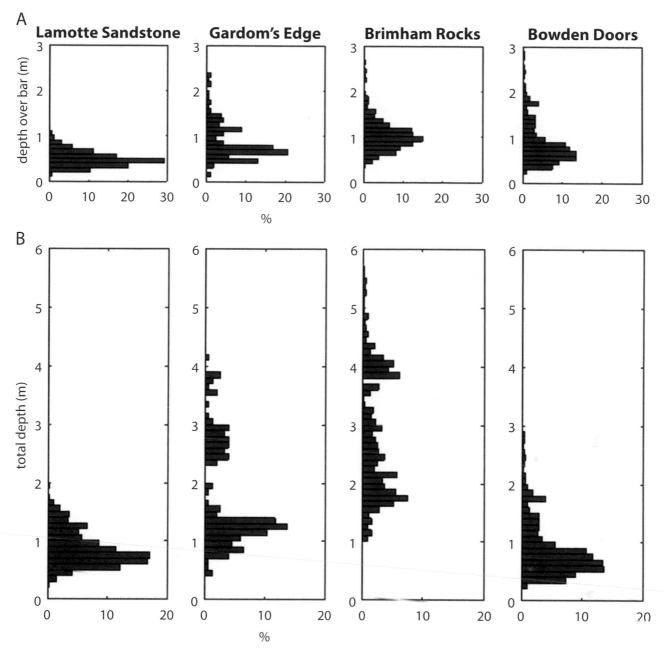

Fig. 9. Histogram of water depths calculated based on pre-sorting patterns. (A) Water depth over the bar top. (B) Water depths measured from the base of the host bar. See also Table 2.

Table 2. Overview of the unit-bar sets thicknesses, cross-sectional areas of pre-sorting patterns, and water depths calculated based on pre-sorted cross-strata for the four investigates formations.

	Lamotte	Gardom	Brimham	Bowden
Thickness of angle-of-repose unit-bar sets in m (± standard deviation)	0.3 (±0.14)	0.7 (±0.43)	1.4 (±0.68)	0.34 (±0.22)
Cross-sectional area of pre-sorting patterns in m² (± standard deviation)	0.024 (±0.017)	0.087 (±0.093)	0.11 (±0.085)	0.026 (±0.035)
Flow depth over the unit bar in m (± standard deviation)	0.49 (±0.17)	0.89 (±0.43)	1.04 (±0.36)	0.47 (±0.26)
Flow depth from the bar trough in m (± standard deviation)	0.87 (±0.3)	1.76 (±0.92)	2.78 (±1.06)	0.9 (±0.5)

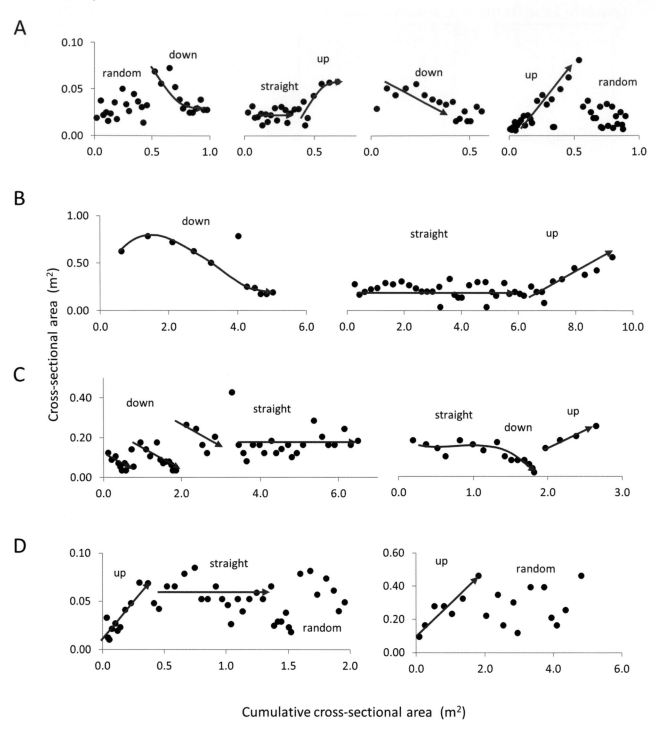

Fig. 10. Examples of sequences of pre-sorted cross strata. The dots indicate the volumes of individual pre-sorting patterns (cf Fig. 2A). Plotting up successive cross strata indicates changes in the size of superimposed dunes arriving at the lee slope. The results illustrate a variety of trends in the development of the formative superimposed bedforms (palaeo-hysteresis). This diversity is found in all outcrops: (A) Lamotte Sandstone, (B) Gardom's Edge, (C) Brimham Rocks, (D) Bowden Door. Note the differences in the cumulative cross-sectional areas for individual trends in different outcrops.

cross strata volume that indicates the growth of superimposed bedforms followed by the abandonment of the bar during low flow stage (Fig. 5; Reesink & Bridge, 2011), the analysis herein revealed a diverse range in trends (Fig. 10). The pre-sorting patterns contained: 1) upward trends of dune growth, 2) downward trends of dune decay, 3) straight trends of dune maintenance; and 4) highly irregular patterns than suggest the absence of a consistent superimposed bedform size. Upward and downward trends are identified when the change in cross-sectional area is gradual and when the slope of the trend is too large to be associated with a change in host bar height (e.g. double or half).

The proportion of the different trends varies among the outcrops (Table 2, Fig. 11). Level trends in which the average superimposed bedform heights remain stable dominate the pre-sorting patterns (Fig. 11; labelled straight). Upward and downward trends comprise comparable portions of the data (Fig. 11; labelled up and down), with a varied proportion of pre-sorting patterns in which no clear trends was observed (Fig. 11; labelled random).

In addition to the variable abundance of specific trends in superimposed bedform development, the outcrops vary in the typical cross-sectional area contained in individual trends (Fig. 10), which are attributed to individual flood events and interpreted as flood deposits. The Lamotte Sandstone is characterised by the smallest flood deposits at ~ 0.5 m². Gardom's Edge was characterised by the largest flood deposits of 4 to 6 m², Brimham Rocks contained similar sizes of unit-bars but had smaller flood deposits of 1 to 2 m²

and Bowden Doors had a wide range of flood deposits that ranged from 0.5 to 2 m.

Interpretation of individual and sequences of pre-sorted cross strata

Fig. 9 illustrates that cross strata yield realistic values for formative water depths. The approximations concern minimum values for water depth because of the probability of partial erosion of the host set. Future research will be needed to establish the limitations to this proxy for formative water depth, which can be achieved by comparing the values obtained from pre-sorted cross strata to those obtained by other means.

The relatively narrow distributions of the flow depths over the bars (Fig. 9A) indicate that the formation and migration of angle-of-repose unit bars may be restricted to a narrow range of local flood conditions. The water depth that is being estimated based on bar-overtopping flows is a flood-depth and not a mean water depth. The much wider distribution of water depths measured from the base of the bar sets (Fig. 9B) indicate that the troughs of the unit bars are located at different depths within the channel deposit.

All types of trends are observed in the pre-sorting patterns: upward, downward and straight trends, as well as apparently random distributions. The pre-sorting patterns indicate palaeo-dune development and hysteresis: the development of dune heights and lengths typically lags behind on changes in formative flow. Although the pre-sorting patterns may not indicate equilibrium dunes, it is clear that the magnitude of superimposed dunes is constrained by

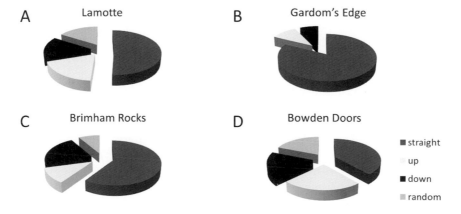

Fig. 11. The relative abundance of different trends in the pre-sorting patterns varies between the outcrops. All investigated outcrops had both upward and downward trends. The difference in relative abundance may be attributed to the singular properties of the individual formative river systems.

flow depth and velocity and therefore, that individual trends can be used for the interpretation of these dynamics.

The observation that all manners of trends exist is significant because different trends are related to restricted sets of formative conditions. Upwards trends as observed in Fig. 5 indicate the growth of superimposed dunes followed by the abandonment (deactivation) of the unit-bar. There is a general tendency of dunes to become stranded on the tops of unit bars during waning flow stages because: i) overall water-surface slopes and sediment transport rates are at lower during waning flow stages, ii) dune development lags behind discharge particularly in rapidly varying flows (Kleinhans *et al.*, 2007; Martin & Jerolmack, 2013); and iii) flow is increasingly diverted around the bars as the local water level decreases (Ashworth, 1996; Reesink *et al.*, 2015). Thus, upward trends and evidence of deactivation are expected in unit-bars that exist in the upper half of channel sequences.

In apparent conflict with the prevalence of conditions that promote and control upward trends, downward trends are also abundant (Fig. 11). These downward trends indicate a decrease in superimposed dune size and will occur only in areas where dunes keep pace with the decrease in flow. For example, downward trends may form on unit-bars lower in the channel, such as on the lower flanks of compound bars (Rodriguez *et al.*, 2014, Wintenberger *et al.*, 2015) or in settings where flow declines slowly such as in some modern monsoonal rivers. Alternatively, cross-bar channels and smaller anabranches are active during waning flow stages and commonly create splay-like deltas that are indistinguishable from other unit-bars (Ashmore, 1991; Egozi & Ashmore, 2008). Some sequences of pre-sorted cross strata contain multiple downward trends (e.g. Fig. 10). The absence of evidence of rising flow stages in between these successive sequences may indicate erosion, sediment bypassing and systematic shifts in the zones of deposition and erosion within the river channel during different flow stages (cf. Bridge, 1993). Due to the relation of downward trends to falling and lower flow stages, the relative occurrence of upward and downward trends may, in theory, be used to further develop hypotheses regarding the prevalence of primary and secondary unit bars, or 'cross-bar channel deltas' in a river channel system (Rice *et al.*, 2008). The dynamics of local, stage-dependent deposition in river channels merits further research.

The cross-sectional area of the flood deposits is sensitive to the preservation of the host set. However, the distinct differences in flood sequences between the systems with larger bars and superimposed bedforms (Gardom's Edge, Brimham Rocks) and the smaller system (Lamotte) indicate that there is significant scope for further research into parallel indicators of water depth and system size. In the case of the Lamotte sandstone, which was deposited before the development of land plants, the availability of an extra proxy of water depth is of particular importance because the width-depth ratios of pre-vegetated rivers may have been different from those formed during later eras.

DISCUSSION

Interpreting formative bedform sizes and water depths

The results illustrate how pre-sorted cross strata can be used to estimate the sizes of formative superimposed bedforms and hence water depths (Fig. 9). Although the visibility of pre-sorted cross strata may vary and further complexities arise when unit-bar sets exceed thicknesses of 0.8 m, the study highlights that the thicknesses of presorting patterns are very easy to measure in the field. The subsequent interpretations are based on (semi)-empirically established equations and therefore sensitive to the accumulation of assumptions, such as the height-length ratio and shape of the superimposed bedforms or the complex relationship between dune dimensions and flow depths. However, the results of this trial indicate that many of the complicating factors do not significantly change the outcome (Fig. 7D and E).

The systematic differences in the dimensions of unit-bar sets and pre-sorting patterns (Fig. 8) between river systems indicates that pre-sorting patterns and unit-bar sets provide a viable proxy for flow depth within a river system. Thus, simple measurements of pre-sorted cross strata can be used to improve characterisations of river systems.

When the superimposed dune dimensions are used to approximate the water depths over the unit bar, the expected water depth is probably underestimated because of partial preservation and because dune size tends to lag behind formative flow, especially in the upper parts of the flow where dunes are commonly abandoned on bars

(Reesink *et al.*, 2015). Nonetheless, the values obtained in such a quantitative interpretation provide a valuable parallel line of interpretation that can be compared against other proxies, such as dune set thickness distributions (Leclair & Bridge, 2001), the thickness of chronostratigraphic units of deposition (units of deposition over time, delineated by two erosional surfaces; Van de Lageweg *et al.*, 2013; Nicholas, 2016), the thickness of unit-bar deposits including low-angle co-sets (Bridge & Lunt, 2006), the overall thickness of the channel sandstones and the position of bar-top deposits and angle-of-repose unit bars within the channel sandstone.

Systematic composition of river channel deposits

River channel deposits contain distinct hierarchies, trends and groupings of structures. The angle-of-repose cross strata investigated in this study are a sub-group of the structures found in unit-bar deposits. Fig. 3 illustrates that the co-existence of dunes and bars leads to systematic trends in the deposits and that angle-of-repose unit bars are expected to dominate the upper parts of channel deposits. Due to their elevated position (Fig. 3B), angle-of-repose unit-bar sets in the upper parts of compound-bar deposits are sensitive to recurrence of erosion. Loss to erosion is often close to negligible in unit-bar fronts exposed in modern rivers (Fig. 1A) and such analyses may not accurately reflect ultimate preservation in the rock record. Moreover, pre-sorting patterns should be more abundant in emergent bars and cut-banks. It is probable that unit-bar sets are overrepresented in studies of exposed bars in modern rivers (e.g. Collinson, 1970; Reesink & Bridge, 2011), as it is far more labour intensive to sample the full depth of the deposits (e.g. Sambrook Smith *et al.*, 2006).

The transition between angle-of-repose cross strata and low-angle co-sets is controlled, in part, by the relative sizes of the host and superimposed bedforms (Reesink & Bridge, 2007, 2009; Warmink *et al.*, 2014). The lack of repetitions of angle-of-repose and low-angle co-sets may indicate that angle-of-repose lee slopes are relatively stable once established. Although detailed analysis of different types of bar foresets is outside the scope of the current investigation, future research may wish to quantify the relative abundance of different types of unit-bar foresets (low-angle, dominated by reduction and reactivation, angle-of-repose

without pre-sorting and angle-of-repose with pre-sorting), as well as the transitions between them.

The preservation potential of unit-bar sets

The quantitative interpretation of pre-sorted cross strata is far more sensitive to the preservation of the host set than to other variables (Fig. 7D to E, arrows). It is noted that the preservation of unit-bar sets may not be adequately described by the Kolmogorov (1951) model, which inherently assumes the chance of recurrence or erosion to be constant. Unit bars are large relative to their formative flow and actively steer the flow around them (Bridge, 1993, 2003; Ashworth, 1996) and, therefore, actively influence their accommodation space (Van de Lageweg, 2014). Thus, rather than forming part of a homogenous distribution of scour depths, successive unit-bar sets compete for vertical space in the channel deposit. With each unit bar deposit, the remaining accommodation space decreases. Due to this dynamic evolution, the distribution of bars is perhaps better approached using a variation on the 'broken stick model' (MacArthur, 1957; Macarthur & MacArthur, 1961), in which the channel only provides a finite accommodation space for a limited number of bar deposits. Finally, the process-based view of unit-bar deposits as lithostratigraphic units presented herein may contrast with the results of numerical and experimental chronostratigraphic studies (Van de Lageweg, 2013; Nicholas *et al.*, 2016). In view of these unresolved concerns, the precise nature of bar preservation will require further systematic study.

Relation of angle-of-repose unit bars to floods and meandering rivers

Unit-bars can form over time as a consequence of instabilities in sediment transport that are related to channel width independent of river planform and may develop without imposed changes in discharge (Yalin, 1972, 1992; Tubino *et al.*, 1999; Lanzoni & Tubino, 1999; Lanzoni, 2000 a,b; Nicholas, 2010, 2013). However, channel-scale patterns of shear stress and sediment transport respond to changes in discharge (Bridge, 1993, 2003); and discharge variability is known to alter the chronostratigraphic thickness distribution of bar-scale deposits (Nicholas *et al.*, 2016). Moreover, the ability to model unit bars that appear to have realistic dimensions in steady discharge does not imply that unsteady flow is irrelevant for the composition of those unit bars. Thus, the discharge variability, which can be revealed by investigating pre-sorted cross strata, can be used to tackle some of the more complex issues in fluvial sedimentology.

A significant gap remains in our understanding of precisely how unit-bar sets relate to river planform. Fig. 3B illustrates that unit-bar sets may hold a narrowly defined vertical position in river deposits. Facies models of meandering rivers emphasise the dominance of lateral accretion surfaces. The presence of downstream accretion surfaces on elevated unit-bar tops may vary in meandering rivers (e.g. Reesink & Bridge, 2011; example from the Cumberland Marshes), or may be more restricted in comparison to more flashy, braided systems. Systematic comparison of unit bar sets from different river systems will make it possible to identify to what degree the vertical trends in sedimentary structures are linked to river planform, or unique to each system.

The narrow distribution of water depths over the bars (Fig. 9A) seems to indicate that the dynamics of bar-overtopping floods may have comparable dynamics in different river systems. This observation matches the notion that flood depths only increase little with discharge due to the transfer of water from the channel to adjacent floodplains (e.g. Sambrook Smith *et al.*, 2010). The vertical trend in the abundance of unit-bar sets is almost certainly affected by the magnitudes of the perennial (base-flow) and ephemeral (floods) parts of the hydrograph and different between braided and meandering rivers. Sequences of pre-sorted cross strata provide information specific to the distribution of bar-overtopping flow depths (Fig. 9), which may be intimately linked to bankfull and geomorphologically effective discharges (Costa & O'Connor, 1995).

In addition to flood depth distributions, horizontal sequences of pre-sorting patterns reveal trends that indicate short histories of development of superimposed dunes that arrive at the lee slope and local flow diversion and deformation of the unit bar. Especially the latter may differ between meandering rivers, where the geometry of point-bars steers the flow at different stages and braided rivers in which bars are often shaped irregularly and channel migration is not necessarily dominated by a lateral migration of the thalweg (Ashmore, 1981; Gardner *et al.*, 2018). Thus, simple measurements of pre-sorted cross strata provide hard data on formative bedforms and flow. These data may help illuminate the dynamic dependencies between river flow, (unit) bar morphology and the ultimately preserved sediment deposits.

CONCLUSIONS

This study presents the first systematic analysis of pre-sorted cross strata in unit-bar sets, which are easily recognised in experiments, in outcrops and in high-resolution borehole geophysics. The analysis illustrates that simple measurements of pre-sorting patterns may reveal a great deal of useful information. The key points raised in this study are:

- The co-existence of dunes and bars probably causes angle-of-repose unit-bar strata to occur preferentially in the upper part of the channel.
- Quantitative analysis of pre-sorted cross strata has a resolution that is set by re-sorting of sediment by grainflows on the lee slope and is most sensitive to partial preservation of the host set.
- Significant gaps exist in our understanding of reduction of host bedforms by superimposed bedforms and the preservation of unit bars. These factors limit, at the moment, the accuracy of quantitative interpretation of unit-bar deposits.
- Pre-sorted cross strata reflect the superimposition of dunes on the host unit bar and may be used as an additional proxy in the interpretation of formative water depth over the bars and for the total water depth in the channel.
- Sequences of pre-sorted cross strata contain a variety of trends that indicate flood water depths and durations, and the dynamic deformation of the host bars.

ACKNOWLEDGEMENTS

This study and the data collection at the Lamotte Sandstone were supported by the GSCO2 project that is funded by the United States Department of Energy (DOE). Data collection at Gardom's Edge, Brimham Rocks and Bowden Doors was supported by the William George Fearnsides Fund of the Geological Society of London and greatly improved by the assistance of Colleen Wells. The data from the Río Paraná were collected under grant NE/E016022/1 from the UK Natural

Environment Research Council (NERC), awarded to Ashworth, Best, Lane, Parsons and Sambrook Smith. The manuscript was significantly improved based on comments from Jan Alexander, Wietse van de Lageweg and Luca Colombera. The author would like to thank, in particular, Jim Best for his support.

REFERENCES

Allen, J.R.L. (1970) A quantitative model of climbing ripples and their cross-laminated deposits. *Sedimentology*, **14**(1–2), 5–26.

Allen, J.R.L. (1973) Phase differences between bed configuration and flow in natural environments and their geological relevance. *Sedimentology*, **20**(2), 323–329.

Allen J.R.L. (1982) *Sedimentary structures, their character and physical basis* (Vol. **1**). Elsevier.

Allen, J.R.L. (1968) The nature and origin of bed form hierarchies. *Sedimentology*, **10**(3), 161–182.

Ashley, G.M. (1990) Classification of large-scale subaqueous bedforms: a new look at an old problem-SEPM bedforms and bedding structures. *J. Sed. Res.*, **60**(1).

Ashmore, P.E. (1991) How do gravel-bed rivers braid? *Can. J. Earth Sci.*, **28**(3), 326–341. 10.1139/e91-030

Ashworth, P.J.A. (1996) Mid-channel bar growth and its relationship to local flow strength and direction. *Earth Surf. Proc. Land.*, **21**, 103–123.

Baas, J.H. and Best, J.L. (2002) Turbulence modulation in clay-rich sediment-laden flows and some implications for sediment deposition. *J. Sed. Res.*, **72**(3), 336–340.

Barrell, J. (1917) Rhythms and the measurement of geologic time. *Geol. Soc. Am. Bull.*, **28**, 745–904.

Best, J. (2005) The fluid dynamics of river dunes: A review and some future research directions. *J. Geophys. Res. Earth Surf.*, **110**(F4).

Blum, M., Martin, J., Milliken, K. and Garvin, M. (2013) Paleovalley systems: Insights from Quaternary analogs and experiments. *Earth-Sci. Rev.*, **116**, 128–169.

Brayshaw, A.C., Davies, G.W. and Corbett, P.W.M. (1996) Depositional controls on primary permeability and porosity at the bedforms scale in fluvial reservoir sandstones. In: *Advances in Fluvial Dynamics and Stratigraphy* (Eds P.A. Carling and M.R. Dawson), John Wiley & Sons Ltd, 373–394.

Bridge, J.S. (2003) *Rivers and Floodplains; Forms, Processes and Sedimentary Record*. Blackwell Publishing, Oxford, U.K., 600 pp.

Bridge, J.S. (1993) The interaction between channel geometry, water flow, sediment transport and deposition in braided rivers. *Geol. Soc. London, Spec. Publ.*, **75**(1), 13–71.

Bridge, J.S. (1997) Thickness of sets of cross strata and planar strata as a function of formative bed-wave geometry and migration and aggradation rate. *Geology*, **25**(11), 971–974.

Bridge, J.S. and Lunt, I.A. (2006) Depositional models of braided rivers. Braided Rivers: Process, Deposits, Ecology and Management, *Int. Assoc. Sedimentol. Spec. Publ.*, **36**, 11–55.

Bridge, J.S. and Tye, R.S. (2000) Interpreting the dimensions of ancient fluvial channel bars, channels and channel belts from wireline-logs and cores. *AAPG Bull.*, **84**(8), 1205–1228.

Carling, P.A. (1991) An appraisal of the velocity-reversal hypothesis for stable pool-riffle sequences in the River Severn, England. *Earth Surf. Proc. Land.*, **16**(1), 19–31.

Claude N., Rodrigues S., Bustillo V., Bréheret J-G., Macaire J-J. and Jugé P. (2012) Estimating bedload transport in a large sand–gravel bed river from direct sampling, dune tracking and empirical formulas. *Geomorphology*, **179**, 40–57.

Collinson, C., Sargent, M.L. and Jennings, J.R. (1988) Illinois basin region. Sedimentary cover—North American craton, US: Boulder, Colorado. Geological Society of America, *Geology of North America*, **2**, 383–426.

Collinson, J.D. (1970) Bedforms of the Tana River, Norway. *Geogr. Ann.*, Series A: Physical Geography, **52**, p. 31–56.

Costa, J.E. and O'Connor, J.E. (1995) Geomorphically effective floods. *Natural and anthropogenic influences in fluvial geomorphology*, 45–56.

Donselaar, M.E. and Schmidt, J.M. (2005) Integration of outcrop and borehole image logs for high-resolution facies interpretation: example from a fluvial fan in the Ebro Basin, Spain. *Sedimentology*, **52**(5), 1021–1042.

Egozi, R. and Ashmore, P. (2008) Defining and measuring braiding intensity. *Earth Surf. Proc. Land.*, **33**, 2121–2138. doi:10.1002/esp.1658

Fernandez, R., Best, J. and Lopez, F. (2006) Mean flow, turbulence structure and bed form superimposition across the ripple–dune transition. *Water Resour. Res.*, **42**(5), p. 948–963.

Fielding, C.R., Allen, J. P., Alexander, J. and Gibling, M.R. (2009) Facies model for fluvial systems in the seasonal tropics and subtropics. *Geology*, **37**(7), 623–626.

Frailey, S.M., Damico, J. and Leetaru, H.E. (2011) Reservoir characterization of the Mt. Simon Sandstone, Illinois Basin, USA. *Energy Procedia*, **4**, 5487–5494.

Fredsøe, J. (1974) On the development of dunes in erodible channels. *J. Fluid Mech.*, **64**(01), 1–16.

Fredsoe, J. (1982) Shape and dimensions of stationary dunes in rivers. *Journal of the Hydraulics Division*, **108**(8), 932–947.

Freiburg, J.T., Morse, D.G., Leetaru, H.E., Hoss, R.P. and Yan, Q. (2014) *A Depositional and Diagenetic Characterization of the Mt. Simon Sandstone at the Illinois Basin-Decatur Project Carbon Capture and Storage Site, Decatur, Illinois, USA*. Illinois State Geological Survey, Prairie Research Institute, University of Illinois. Circular **583**.

Gabel, S.L. (1993) Geometry and kinematics of dunes during steady and unsteady flows in the Calamus River, Nebraska, USA. *Sedimentology*, **40**(2), 237–269. **DOI:** 10.1111/j.1365-3091.1993.tb01763.x

Gardner, T., Ashmore, P. and Leduc, P. (2018) Morpho-sedimentary characteristics of proximal gravel braided river deposits in a Froude-scaled physical model. *Sedimentology*, **65**(3), 877–896.

Gershenzon, N.I., Soltanian, M.R., Ritzi Jr., R.W., Dominic, D.F., Mehnert, E. and Okwen, R.T. (2015) Influence of small-scale fluvial architecture on CO2 trapping processes in deep brine reservoirs. *Water Resour. Res.* **51**, 8240–8256, http://dx.doi.org/10.1002/2015WR017638.

Gilbert, G.K. (1899) Ripple-marks and cross-bedding. *Geol. Soc. Am. Bull.*, **10**(1), 135–140.

Gilbert, G.K. (1885) The topographic features of lake shores. *U.S. Geol. Surv. Ann. Rep.* **5**, 75–123.

Gill, M.A. (1971) Height of sand dunes in open channel flows. *Journal of the Hydraulics Division*, **97**(12), 2067–2074.

Guin, A., Ramanathan, R., Ritzi, R.W. Jr., Dominic, D.F., Lunt, I.A., Scheibe, T.D. and Freedman, V.L. (2010) Simulating the heterogeneity in braided channel belt deposits: 2. Examples of results and comparison to natural deposits. *Water Resour. Res.*, **46**, W04516, doi:10.1029/2009WR008112

Hartley, A.J., Owen A., Swan A., Weissmann G.S., Holzweber B.I., Howell J., Nichols G. and Scuderi L. (2015) Recognition and importance of amalgamated sandy meander belts in the continental rock record. *Geology* **43**(8), 679–682.

Haszeldine, R.S. (1983) Descending tabular cross-bed sets and bounding surfaces from a fluvial channel in the Upper Carboniferous of North-East England. *J. Sed. Petrol.*, **53**, 1233–1247.

Herbert, C.M. and Alexander, J. (2018) Bottomset architecture in the troughs of dunes and unit bars. *J. Sed. Res.*, **88**(4), 522–553.

Herbert, C.M., Alexander, J., de Álvaro, M. and María, J. (2015) Back-flow ripples in troughs downstream of unit bars: Formation, preservation and value for interpreting flow conditions. *Sedimentology*. DOI: 10.1111/sed.12203

Hétu, B., Van Steijn, H. and Bertran, P. (1995) Le role des coulees de pierres seches dans la genese d'un certain type d'eboulis stratifies/The role of dry grainflow in the genesis of a type of stratified scree. *Permafrost Periglac. Process.*, **6**(2), 173–194.

Holbrook, J. and Wanas, H. (2014) A fulcrum approach to assessing source-to-sink mass balance using channel paleohydrologic paramaters derivable from common fluvial data sets with an example from the Cretaceous of Egypt. *J. Sed. Res.*, **84**(5), 349–372.

Hooke, R.L. (1968) Laboratory study of the influence of granules on flow over a sand bed. *Geol. Soc. Am. Bull.*, **79**, 495–500.

Houseknecht, D.W. and Ethridge, F.G. (1978) Depositional history of the Lamotte Sandstone of southeastern Missouri. *J. Sed. Res.*, **48**(2).

Hunter, R.E. (1985a) A kinematic model for the structure of lee side deposits. *Sedimentology*, **32**, 409–422.

Hunter, R.E. (1985b) Subaqueous sand flow cross strata. *J. Sed. Petrol.*, **55**, 886–894.

Hunter, R.E. and Rubin, D.M. (1983) Interpreting cyclic cross-bedding, with an example from the Navajo Sandstone, in Brookfield, M.E. and Ahlbrandt, T.S., eds., *Aeolian Sediments and Processes*, Amsterdam, Elsevier, 429–454.

Jackson, R.G. (1976) Largescale ripples of the lower Wabash River. *Sedimentology*, **23**(5), 593–623.

Jackson, R.G. (1975) Velocity–bed-form–texture patterns of meander bends in the lower Wabash River of Illinois and Indiana. *Geol. Soc. Am. Bull.*, **86**(11), 1511–1522.

Julien, P.Y. and Klaassen, G.J. (1995) Sand-dune geometry of large rivers during floods. *J. Hydraul. Eng.*, **121**(9), 657–663.

Kleinhans, M.G. (2005) Grain-size sorting in grainflows at the lee side of deltas. *Sedimentology*, **52**(2), 291–311.

Kleinhans, M.G. (2004) Sorting in grain flows at the lee side of dunes. *Earth-Sci. Rev.*, **65**(1), 75–102.

Kleinhans, M.G., Leuven, J.R., Braat, L. and Baar, A. (2017) Scour holes and ripples occur below the hydraulic smooth to rough transition of movable beds. *Sedimentology*. doi:10.1111/sed.12358

Kleinhans, M.G., Wilbers, A.W.E. and Ten Brinke, W.B.M. (2007) Opposite hysteresis of sand and gravel transport upstream and downstream of a bifurcation during a flood in the River Rhine, the Netherlands. *Geol. Mijnbouw*, **86**(3).

Koeppe, J.P., Enz, M. and Kakalios, J. (1997) Phase diagram for avalanche stratification of granular media. *Physical Review E* **58** (4), R4104–R4107.

Kolmogorov, A.N. (1951) Solution of a problem in probability theory connected with the problem of the mechanism of stratification. *American Mathematical Society*, 8 pp.

Krishnamurthy, P.G., Senthilnathan, S., Yoon, H., Thomassen, D., Meckel, T. and DiCarlo, D. (2016) Comparison of Darcy's law and invasion percolation simulations with buoyancy-driven CO_2-brine multiphase flow in a heterogeneous sandstone core. *J. Petrol. Sci. Eng.*, https://doi.org/10.1016/j.petrol.2016.10.022

Lane, S.N., Parsons, D.R., Best, J.L., Orfeo, O., Kostaschuk, R.A. and Hardy, R.J. (2008) Causes of rapid mixing at a junction of two large rivers: Río Paraná and Río Paraguay, *Argentina. J. Geophys. Res. Earth Surf.*, **113**, F02024, doi: 10.1029/2006JF000745.

Lanzoni, S. (2000a) Experiments on bar formation in a straight flume: 1. Uniform sediment, *Water Resour. Res.*, **36**(11), 3337–3349, doi:10.1029/2000WR900160.

Lanzoni, S. (2000b) Experiments on bar formation in a straight flume: 2. Graded sediment, *Water Resour. Res.*, **36**(11), 3351–3363, doi:10.1029/2000WR900161.

Lanzoni, S. and Tubino, M. (1999) Grain sorting and bar instability. *J. Fluid Mech.*, **393**, 149–174. DOI: https://doi.org/10.1017/S0022112099005583

Latrubesse, E.M. (2015) Large rivers, megafans and other Quaternary avulsive fluvial systems: A potential "who's who" in the geological record. *Earth-Sci. Rev.*, doi: 10.1016/j.earscirev.2015.03.004

Leclair, S.F. (2002) Preservation of cross-strata due to the migration of subaqueous dunes: an experimental investigation. *Sedimentology*, **49**(6), 1157–1180.

Leclair, S.F. and Bridge, J.S. (2001) Quantative interpretation of sedimentary structures formed by river dunes. *J. Sed. Res.*, **71**, 713–716.

Leeder, M.R. (1999). *Sedimentology and Sedimentary Basins*, 592 pp.

Leetaru, H.E., Frailey, S.M., Damico, J., Mehnert, E., Birkholzer, J., Zhou, Q. and Jordan, P.D. (2009) Understanding CO_2 plume behavior and basin-scale pressure changes during sequestration projects through the use of reservoir fluid modeling. *Energy Procedia*, **1**(1), 1799–1806.

Leetaru, H.E., Frailey, S., Morse, D., Finley, R.J., Rupp, J.A., Drahozval, J.A. and McBride, J.H. (2009) Carbon sequestration in the Mt. Simon Sandstone saline reservoir. In: *Carbon dioxide sequestration in geological media—State of the science* (Eds M. Grobe, J.C. Pashin and R.L. Dodge). *AAPG Stud. Geol.*, **59**, 261–277.

Lunt, I.A. and **Bridge, J.S.** (2004) Evolution and deposits of a gravelly braid bar, Sagavanirktok River, Alaska. *Sedimentology*, **51**(3), 415–432.

Lunt, I.A. and **Bridge, J.S.** (2007) Formation and preservation of open-framework gravel strata in unidirectional flows. *Sedimentology*, **54**(1), 71–87.

Lunt, I.A., **Bridge, J.S.** and **Tye, R.S.** (2004) A quantitative, three-dimensional depositional model of gravelly braided rivers. *Sedimentology*, **51**(3), 377–414.

MacArthur, R.H. (1957) On the relative abundance of bird species. *Proc. Natl. Acad. Sci.*, **43**, 293–295.

MacArthur, R.H. and **MacArthur, J.W.** (1961) On bird species diversity. *Ecology*, **42**, 594–598.

MacMahon W.J. and **Davies, N.S.** (2018) The shortage of evidence for pre-vegetation meandering rivers. *Int. Assoc. Sedimentol. Spec. Publ.*, **48**, this volume.

McCabe, P.J. and **Jones, C.M.** (1977) Formation of reactivation surfaces within superimposed deltas and bedforms. *J. Sed. Res.*, **47**(2), 707–715.

Martin, R.L. and **Jerolmack, D.J.** (2013) Origin of hysteresis in bed form response to unsteady flows. *Water Resour. Res.*, **49**(3), 1314–1333.

Martinez De Alvaro, M. (2015) Architecture and origin of fluvial cross-bedding based on flume experiments and geological examples field case studies: Rillo de Gallo, Spain and Northumberland, UK. Doctoral thesis, University of East Anglia, https://ueaeprints.uea.ac.uk/id/eprint/53370

McKee, E.D. (1957) Flume experiments on the production of stratification and cross-stratification. *J. Sed. Res.*, **27**(2).

McKee, E.D. and **Weir G.W.** (1953) Terminology for stratification and cross-stratification in sedimentary rocks. *Geol. Soc. Am. Bull.*, **64**(4), 381–390.

Miall, A.D. (1996) *The Geology of Fluvial Deposits: Sedimentary Facies, Basin Analysis and Petroleum Geology*. Springer-Verlag, Berlin, 589 pp.

Milan, D.J., **Heritage, G.L.**, **Large, A.R.G.** and **Charlton, M.E.** (2001) Stage dependent variability in tractive force distribution through a riffle–pool sequence. *Catena*, **44**(2), 85–109.

Nanson, G.C. (1981) New evidence of scroll-bar formation on the Beatton River. *Sedimentology*, **28**(6), 889–891.

Naqshband, S., **Ribberink, J.S.**, **Hurther, D.** and **Hulscher, S.J.M.H.** (2014) Bed load and suspended load contributions to migrating sand dunes in equilibrium. *J. Geophys. Res. Earth Surf.*, **119**(5), 1043–1063.

Nicholas, A.P. (2013) Modelling the continuum of river channel patterns. *Earth Surface Processes and Landforms*, **38**(10), 1187–1196.

Nicholas, A.P. (2010) Reduced-complexity modelling of free bar morphodynamics in alluvial channels. *J. Geophys. Res. Earth Surf.*, doi:10.1029/2010JF001774.

Nicholas, A.P., **Sambrook Smith, G.H.**, **Amsler, M.L.**, **Ashworth, P.J.**, **Best, J.L.**, **Hardy, R.J.**, **Lane, S.N.**, **Orfeo, O.**, **Parsons, D.R.**, **Reesink, A.J.H.**, **Sandbach, S.D.**, **Simpson, C.J.** and **Szupiany, R.N.** (2016) The role of discharge variability in determining alluvial stratigraphy. *Geology*, **44**(1), 3–6.

Nio, S-D. and **Yang C-S.** (1991) Diagnostic attributes of clastic tidal deposits: a review. *CSPG Spec. Publ.*, Clastic Tidal Sedimentology, Memoir **16**, 3–27.

Ojakangas, R.W. (1963) Petrology and sedimentation of the upper Cambrian Lamotte Sandstone in Missouri. *J. Sed. Res.*, **33**(4), 860–873.

Paola, C. and **Borgman, L.** (1991) Reconstructing random topography from preserved stratification. *Sedimentology*, **38**, 553–565.

Parsons, D.R., **Best, J.L.**, **Orfeo, O.**, **Hardy, R.J.**, **Kostaschuk, R.** and **Lane, S.N.** (2005) Morphology and flow fields of three-dimensional dunes, Río Paraná, Argentina: Results from simultaneous multibeam echo sounding and acoustic Doppler current profiling. *J. Geophys. Res. Earth Surf.*, **110**, F04S03, doi: 10.1029/2004JF000231.

Parsons, D.R., **Schindler, R.J.**, **Hope, J.A.**, **Malarkey, J.**, **Baas, J.H.**, **Peakall, J.**, **Manning, A.J.**, **Ye, L.**, **Simmons, S.**, **Paterson, D.M.**, **Aspden, R.J.**, **Bass, S.J.**, **Davies, A.G.**, **Lichtman, I.D.** and **Thorne, P.D.** (2016) The role of biophysical cohesion on subaqueous bed form size. *Geophys. Res. Lett.*, **43**, 1566–1573, doi:10.1002/2016GL067667.

Plink-Björklund, P. (2015) Morphodynamics of rivers strongly affected by monsoon precipitation: Review of depositional style and forcing factors. *Sed. Geol.*, **323**, 110–147.

Raudkivi, A.J., **Witte, H.H.**, (1990) Development of bed features. *Journal of Hydraulic Engineering*, **116**(9), 1063–1079. https://doi.org/10.1061/(ASCE)0733-9429(1990)116:9(1063)

Reesink A.J.H. and **Bridge J.S.** (2011) Evidence of bedform superimposition and flow unsteadiness in unit bar deposits, South Saskatchewan River, Canada. *J. Sed. Res.*, **81**(11), 814–840.

Reesink, A.J.H. and **Bridge, J.S.** (2009) Influence of bedform superimposition and flow unsteadiness on the formation of cross strata in dunes and unit bars—Part 2, further experiments. *Sed. Geol.*, **222**, 274–300.

Reesink, A.J.H. and **Bridge, J.S.** (2007) Influence of superimposed bedforms and flow unsteadiness on formation of cross strata in dunes and unit bars. *Sed. Geol.*, **202**, 281–296.

Reesink, A.J.H., **Ashworth P.J.**, **Sambrook Smith G.H.**, **Best J.L.**, **Parsons D.R.**, **Amsler M.L.**, **Hardy R.J.**, **Lane S.N.**, **Nicholas A.P.**, **Orfeo, O.**, **Sandbach S.D**, **Simpson C.J.** and **Szupiany R.N.** (2014) Scales and causes of heterogeneity in bars in a large multi-channel river: Río Paraná, Argentina. *Sedimentology*, **61**(4), 1055–1085.

Reesink A.J.H., **Parsons D.R.**, **Van den Berg J.**, **Amsler M.L.**, **Best J.L.**, **Hardy R.J.**, **Lane, S.N.**, **Orfeo, O.** and **Szupiany, R.** (2015) Extremes in dune preservation; controls on the completeness of fluvial deposits. *Earth-Sci. Rev.*, **150**, 652–665. https://doi.org/10.1016/j.earscirev.2015.09.008

Rice, S.P., **Church, M.**, **Wooldridge, C.L.** and **Hickin, E.J.** (2009) Morphology and evolution of bars in a wandering gravel-bed river; lower Fraser river, British Columbia, Canada. *Sedimentology*, **56**, 709–736. doi:10.1111/j.1365-3091.2008.00994.x

Ritzi R.W., **Freiburg J.T.** and **Webb N.D.** (2016) Understanding the (co)variance in petrophysical properties of CO2 reservoirs comprising fluvial sedimentary architecture. *Int. J. Greenhouse Gas Control*, **51**, 423–434. doi: 10.1016/j. ijggc.2016.05.001

Rodrigues, S., **Mosselman, E.**, **Claude, N.**, **Wintenberger C.L.** and **Juge, P.** (2014) Alternate bars in a sandy gravel bed river: generation, migration and interactions with superimposed dunes. *Earth Surf. Proc. Land.*, 1096–9837. Online DOI: 10.1002/esp.3657

Rubin, D.M. and **Hunter, R.E.** (1982) Bedform climbing in theory and nature. *Sedimentology*, **29**(1), 121–138.

Rubin, D.M. and Hunter, R.E. (1983) Reconstructing bedform assemblages from compound crossbedding. *Dev. Sedimentol.*, **38**, 407–427.

Sambrook Smith, G.H., Best, J.L., Ashworth, P.J., Lane, S.N., Parker, N.O., Lunt, I.A., Thomas, R.E. and Simpson, C.J. (2010) Can we distinguish flood frequency and magnitude in the sedimentological record of rivers? *Geology*, **38**(7), 579–582.

Sear, D.A. (1996) Sediment transport processes in pool–riffle sequences. *Earth Surf. Proc. Land.*, **21**(3), 241–262.

Sloss, L.L. (1963) Sequences in the cratonic interior of North America. *Geol. Soc. Am. Bull.*, **74**(2), 93–114.

Smith, D.G., Hubbard, S.M., Leckie, D.A. and Fustic, M. (2009) Counter point bar deposits: lithofacies and reservoir significance in the meandering modern Peace River and ancient McMurray Formation, Alberta, Canada. *Sedimentology*, **56**(6), 1655–1669.

Smith, N.D. (1974) Sedimentology and bar formation in the upper Kicking Horse River, a braided outwash stream. *J. Geol.*, **82**(2), 205–223.

Smith, N.D. (1977) Some comments on terminology for bars in shallow rivers. *Fluvial Sedimentology Memoir*, **5**, 85–88.

Smith, N.D. (1972) Some sedimentological aspects of planar cross-stratification in a sandy braided river. *J. Sed. Res.*, **42**(3).

Smith, N.D. (1971) Transverse bars and braiding in the lower Platte River, Nebraska. *Geol. Soc. Am. Bull.*, **82**(12), 3407–3420.

Soltan, R. and Mountney, N.P. (2016) Interpreting complex fluvial channel and barform architecture: Carboniferous Central Pennine Province, northern England. *Sedimentology*, **63**(1), 207–252.

Sundborg, Å. (1956) The River Klarälven: a study of fluvial processes. *Geogr. Ann.*, **38**(2), 125–237.

Syvitski, J.P., Cohen, S., Kettner, A.J. and Brakenridge, G.R. (2014) How important and different are tropical rivers?—An overview. *Geomorphology*, **227**, 5–17.

Tubino, M., Repetto, R. and Zolezzi, G. (1999) Free bars in rivers. *J. Hydraul. Res.*, **37**(6), 759–775.

Tubino, M. and Seminara, G. (1990) Free–forced interactions in developing meanders and suppression of free bars. *J. Fluid Mech.*, **214**, 131–159.

Tuijnder, A.P., Ribberink, J.S. and Hulscher, S.J. (2009) An experimental study into the geometry of supply-limited dunes. *Sedimentology*, **56**(6), 1713–1727.

Turner, B.R. and Monro, M. (1987) Channel formation and migration by mass-flow processes in the Lower Carboniferous fluviatile Fell Sandstone Group, northeast England. *Sedimentology*, **34**(6), 1107–1122.

Van den Berg, J.H. and Van Gelder, A. (2009) A new bedform stability diagram, with emphasis on the transition of ripples to plane bed in flows over fine sand and silt. In: *Alluvial Sedimentation* (Eds M. Marzo and C. Puigdefábregas), *Int Assoc. Sedimentol. Spec. Publ.*, **17**, 11, doi: 10.1002/9781444303995.ch2

Van de Lageweg, W.I., Van Dijk, W.M., Baar, A.W., Rutten, J. and Kleinhans, M.G. (2014) Bank pull or bar push: What drives scroll-bar formation in meandering rivers? *Geology*, **42**(4), 319–322.

Van de Lageweg, W.I., Van Dijk, W.M. and Kleinhans, M.G. (2013) Channel belt architecture formed by a meandering river. *Sedimentology*, **60**(3), 840–859.

Van Rijn, L.C. (1990) *Principles of fluid flow and surface waves in rivers, estuaries, seas and oceans.* Aqua Publications, Amsterdam. 900 pp.

Van Rijn, L.C. (1984) Sediment transport, part III: bed forms and alluvial roughness. *J. Hydraul. Eng.*, **110**(12), 1733–1754.

Venditti, J.G., Church, M. and Bennett, S.J. (2005a) Morphodynamics of small-scale superimposed sand waves over migrating dune bed forms. *Water Resour. Res.*, **41**(10).

Walker, R.G. (1976) Facies Model-3. Sandy Fluvial Systems. *Geosci. Can.*, **3** (2).

Warmink, J.J., Dohmen-Janssen, C.M., Lansink, J., Naqshband, S., Duin, O.J., Paarlberg, A.J., Termes, P. and Hulscher, S.J. (2014) Understanding river dune splitting through flume experiments and analysis of a dune evolution model. *Earth Surf. Proc. Land.*, **39**(9), 1208–1220.

Wilbers, A. (2004) *The development and hydraulic roughness of subaqueous dunes.* PhD thesis Utrecht, The Netherlands. NGS 323. 227 p.

Wilbers, A.W.E. and Ten Brinke, W.B.M. (2003) The response of subaqueous dunes to floods in sand and gravel bed reaches of the Dutch Rhine. *Sedimentology*, **50**(6), 1013–1034.

Wintenberger, C.L., Rodrigues, S., Claude, N., Jugé, P., Bréhéret, J.G. and Villar, M. (2015) Dynamics of nonmigrating mid-channel bar and superimposed dunes in a sandy-gravelly river (Loire River, France). *Geomorphology*, **248**, 185–204.

Yalin, M.S. (1964) Geometrical properties of sand waves. Journal of the Hydraulics Division, *Am. Soc. Civ. Eng.*, **90**(5), 105–119.

Yalin, M.S. (1972) *Mechanics of sediment transport.* Pergamon Press.

Yalin, M.S. (1992) *River Mechanics*, 219 pp. Pergamon Press.

Yesberger, W.L. (1982) Paleoenvironments and Depositional History of the Upper Cambrian Lamotte Sandstone in Southeast Missouri, MSc thesis, University of Missouri, Columbia, USA. University of Missouri, Master's Thesis, 287 p.

Zinger, J.A., Rhoads, B.L. and Best, J.L. (2011) Extreme sediment pulses generated by bend cutoffs along a large meandering river. *Nature Geoscience*, **4**(10), 675–678.

Zinger, J.A., Rhoads, B.L., Best, J.L. and Johnson, K.K. (2013) Flow structure and channel morphodynamics of meander bend chute cutoffs: A case study of the Wabash River, USA. *J. Geophys. Res. Earth Surf.*, **118**(4), 2468–2487.

Int. Assoc. Sedimentol. Spec. Publ (2018) **48**, 201–230.

Chute cutoffs in meandering rivers: formative mechanisms and hydrodynamic forcing

DANIELE P. VIERO[†], SERGIO LOPEZ DUBON[†] and STEFANO LANZONI[†]

[†]*Department of Civil, Environmental and Architectural Engineering, University of Padova, Padova, Italy*
The authors contributed equally to this work.

ABSTRACT

Chute cutoffs are autogenic mechanisms typical of many meandering rivers with wide cross sections, large curvature bends, high discharges and high overbank flow gradients. The shortening of the original meander loop through a bypassing channel produces a greater water-surface gradient and, hence, increases the overall transport capacity of the reach, enhancing the downstream sediment delivery. As a consequence, the mean channel width, as well as the planform shape of the meandering bends adjacent to that bypassed by the chute tend to progressively readjust. The occurrence of this type of cutoff is one of the most fascinating and less predictable events in the evolution of rivers, as a multiplicity of control factors are involved in chute cutoff formation. In the last decade, various researchers tried to shed light on the complex mechanisms that lead to chute incision and eventually determine the fate of the bypassed bend and the new chute channel. However, the subject is not yet settled and a systematic physics-based framework is still missing. In this contribution, formative mechanisms are reviewed and two different forcing factors leading to chute cutoffs are highlighted; the inertia and direction of the channelised flow and the topographic heterogeneity of the floodplain. The general features of the involved processes are investigated using a hydrodynamic finite-element model for the two-dimensional flow in the channel and over the floodplain. A linearised two-dimensional hydro-morphodynamic model is used to estimate the channel bed topography in the absence of field data. Two representative case studies are specifically considered, occurred in the Sacramento River (California) and in the Cecina River (Italy). The first concerns a chute cutoff driven by in-channel flow. The second deals with a chute cutoff due to overbank flow and the particular topography of the point bar sediment deposits placed inside the meander loop.

Keywords: Chute cutoffs, chute formation, floodplains, meandering rivers, scroll-bars

INTRODUCTION

The morphology of alluvial rivers is affected by several interrelated processes that contribute to shape the river bed, its planform configuration and the sedimentary structure of the surrounding floodplain (Howard, 1996; Frascati & Lanzoni, 2009; Toonen *et al.*, 2012; Słowik, 2016; van de Lageweg *et al.*, 2016; Bogoni *et al.*, 2017). The inherent correlation between the history of a river and of its floodplain is of fundamental importance

not only to predict reliably or reconstruct the evolution of the river but also to stratigraphic studies and to guide the research of sedimentologists (e.g. Erskine *et al.*, 1992; Ghinassi, 2011; Ghinassi *et al.*, 2016).

In general, the migration of an alluvial river is the result of the complex interplay between erosion at the outer bank of bends and point bar accretion at the inner bank (Eke *et al.*, 2014; Iwasaki *et al.*, 2016). Both processes are strongly related to the secondary helical flow circulations

Fluvial Meanders and Their Sedimentary Products in the Rock Record, First Edition.
Edited by Massimiliano Ghinassi, Luca Colombera, Nigel P. Mountney and Arnold Jan H. Reesink.

driven by the curvature of the channel axis and by the bed topography (Seminara, 2006), as well as by the variation of cross section width (Luchi *et al.*, 2011). The spatial structure of the flow field, in turn, controls the stresses transmitted to the channel banks and, consequently, the erosive processes that eventually lead to outer bank collapse (Rinaldi *et al.*, 2008) and sediment deposition responsible for the inner point bar accretion.

Several other processes add complexity to the system. Vegetation cover, if present, affects the erodibility of the outer bank and contributes to stabilise the inner point bar (Bertoldi *et al.*, 2014; Oorschot *et al.*, 2016; Zen *et al.*, 2016). Floodplain heterogeneities, both planimetric and stratigraphic (Ielpi & Ghinassi, 2014), exert a strong control on channel bank erodibility and, consequently, on meandering dynamics (Schwendel *et al.*, 2015; Bogoni *et al.*, 2017). Heterogeneities are usually generated by selective transport and deposition of sediments, as in the case of scroll-patterned surfaces (Nanson & Croke, 1992; van de Lageweg *et al.*, 2014). Sediment storage in abandoned channels and oxbow lakes (Slingerland & Smith, 2004; Gautier *et al.*, 2007), crevasse splay through levee breaches and overbank spilling of tie and tributary channels (Day *et al.*, 2008; Swanson *et al.*, 2008; David *et al.*, 2017) are other processes that concur to floodplain heterogeneity.

Abandoned channels and oxbow lakes (see, e.g., Fig. 1) not only favour the formation of preserved sedimentary features across the floodplain, but owing to the large accommodation space, they can also contribute to the overall sediment balance of the channel-floodplain system (Gay *et al.*, 1998; Constantine *et al.*, 2010; Grenfell *et al.*, 2012). Moreover, they result from morphodynamic processes as cutoffs, whereby a river limits its sinuosity and hence its planform complexity, ensuring the establishment of statistically stationary evolving planforms (Camporeale *et al.*, 2005; Frascati & Lanzoni, 2010). Two different types of cutoff are usually recognised in natural channels: neck and chute cutoffs. Neck cutoffs occur when the local sinuosity becomes so large that adjacent bends intersect each other, leading to the formation of an abandoned loop (oxbow lake) when sedimentation closes the loop ends (Hooke, 1995; Howard, 1996). Less clear is the genesis of chute cutoffs. They are relatively long flow diversions that occur when a meander loop is bypassed through a new channel that forms across the floodplain enclosed by the loop (Constantine *et al.*, 2010). They most frequently form in wide channels with large

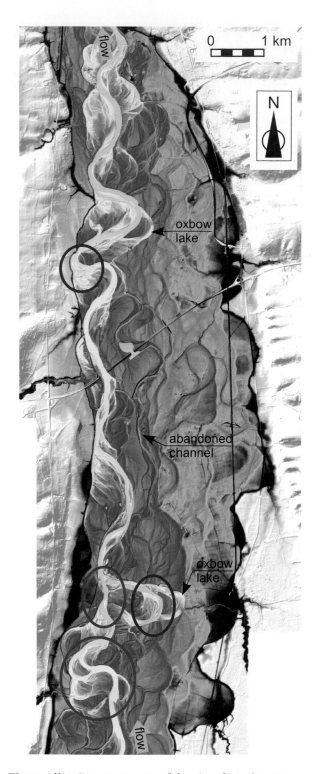

Fig. 1. Allier River upstream of the city of Moulins, France (river n. 112 in the dataset of Kleinhans & van den Berg, 2011; see also Van Dijk *et al.*, 2014 for a detailed study). The original Digital Elevation Model is here de-trended in order to highlight the topographical variability. The blue and magenta circles highlight chute cutoffs related to different formative mechanisms, as described later in the text (data courtesy of IGN-France and Ministère de l'Ecologie et du Développement Durable).

curvature bends, high discharges, poorly cohesive, weakly vegetated banks and high gradients (Howard & Knutson, 1984).

Commonly, chute cutoffs form in meandering channels when flood waters can no longer be contained within the main channel. A chute is thus incised in the floodplain and eventually conveys most of the river discharge. The new path is in fact shorter and with a steeper water surface than the original meander loop. Chute channels may develop over a long period (Gay *et al.*, 1998), with only flood-water flow at first, or during a single flood event (Iwasaki *et al.*, 2016). The original meander loop may either be filled with sediment to create a channel fill deposit or, less frequently, remain active along with the chute channel (e.g. as in the Strickland River, Fig. 2 and Grenfell *et al.*, 2012).

From a morphological point of view, chute cutoffs control the river sinuosity by counteracting the bend lengthening that is inherent in meander migration (Howard & Knutson, 1984; Tal & Paola, 2010; Ghinassi, 2011; Van Dijk *et al.*, 2012; Morais *et al.*, 2016). Moreover, they can be seen as a mechanism of transition from meandering channel to braided channel (Kleinhans & van den Berg, 2011; Zolezzi *et al.*, 2012; Iwasaki *et al.*, 2016). Their occurrence is related to floods as, unlike neck cutoffs, high water levels and high rates of bed load transport are required (Lewis & Lewin, 1983; Howard, 1996; Zinger *et al.*, 2011; Van Dijk *et al.*, 2014). The incision of chute channels causes the removal of relatively large amounts of floodplain materials, thus delivering intense sediment pulses (Zinger *et al.*, 2011) that stimulate downstream bar formation (Fuller *et al.*, 2003; Dieras *et al.*, 2013; Zinger *et al.*, 2013). This sediment delivery is also favoured by the shortening of the river path consequent to chute formation. The increase in water surface slope that occurs after the cutoff enhances the transport capacity of the river reach, that tends to get deeper and to adjust its width. More generally, individual cutoffs were shown to accelerate the river migration and to drive channel widening both upstream and downstream of the cutoff locations (Morais *et al.*, 2016; Schwenk & Foufoula-Georgiou, 2016).

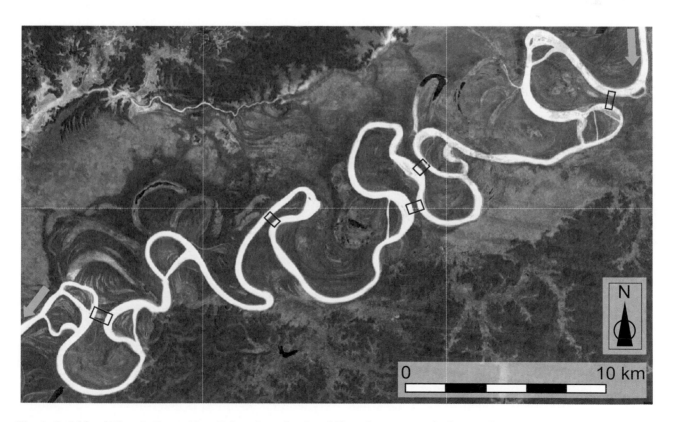

Fig. 2. Strickland River in Papua New Guinea (see also Grenfell *et al.*, 2012, 2014), about 50 km north-east of the confluence with the Fly River (aerial photo from Google Earth, 2017). Red rectangles denote locations probably characterised by large gradient advantage.

Given the important role of chute cutoffs in shaping both the river planform and the stratigraphy of the adjacent floodplain, their study has attracted increasing attention in the last years. Different mechanisms leading to the formation of chutes have been identified (e.g. Constantine *et al.*, 2010; Van Dijk *et al.*, 2014; Eekhout & Hoitink, 2015). However, due in part to the lack of extensive and systematic field observations (Micheli & Larsen, 2011) and especially to the multiplicity of mechanisms and controls involved in the process, the predictability of chute cutoff occurrence remains an open problem. Importantly, a systematic physics-based framework for chute cutoff occurrence is still missing.

In order to contribute to fill this gap, the present work investigates the hydrodynamic conditions that control the initiation of chute cutoffs. The aim is to identify the relevant features of the in-channel flow and of the overbank flow on the surrounding floodplain that enhance the triggering of chutes. Two different categories of mechanisms are considered. The first is related to the inertia and direction of the in-channel flow upstream of the chute channel. The second concerns the flow over the portion of floodplain contained within a meander loop and driven by the free-surface gradient. These mechanisms are investigated with reference to two field cases concerning the Sacramento River (US) and the Cecina River (Italy), respectively. We anticipate that in the first case the inception of a chute is controlled by the shear stress distribution at the outer bank downstream of a bend apex. This distribution is mainly associated with the inertia and direction of the in-channel flow driven by channel axis curvature and cross section width variations. In the second case, the free-surface gradient, the topographic irregularities and the sedimentological composition of the floodplain area within the meander loop play a major role in triggering the chute formation.

The paper is organised as follows. First, the processes and factors involved in the formation of chute cutoffs are reviewed based on the available literature. Then, a two-dimensional finite element model is used to investigate the hydrodynamic features of the flow field within the channel and over the floodplain, while a linearised two-dimensional hydro-morphodynamic model is adopted to reconstruct the channel bed topography when missing. The general features of the involved processes are analysed with reference to the considered case studies, paying specific attention to the hydrodynamic features of the flow field and the topographic signatures of the floodplain. Finally, a closure section summarises the main results of this study.

PROCESSES AND FACTORS CONTROLLING CHUTE CUTOFF DYNAMICS

Mechanisms of chute cutoff formation

Chute cutoffs occur when water flows from the main river channel to the adjacent floodplain. The flow can thus interact with the floodplain topography and the vegetation hosted on it. Eventually, a chute incises across the inner side of the point bar (i.e. a meander neck), or through exposed alluvial bars (McGowen & Garner, 1970; Erskine *et al.*, 1992; Gay *et al.*, 1998; Bridge, 2003; Ghinassi, 2011). These chutes, which typically form within the active channel belt (Hooke, 1995), are segments with length on the order of one meander wavelength (David *et al.*, 2017) linking together reaches of the same river.

Different mechanisms have been observed to cause the incision of a chute channel. With reference to the sketch of Fig. 3, (at least) three main types of mechanisms can be distinguished. The first consists of downstream extension, triggered by the concentration of in-channel flow velocity and shear stresses at the outer bank, downstream of a bend apex. The consequent localised erosion of bank material in some cases leads to the formation of a pre-chute erosional embayment (Constantine *et al.*, 2010). Subsequent floods extend the embayment downstream until it intersects the river course, forming a chute. During intermediate stages of the process, a bar can form at the downstream end of the chute incision (Kleinhans & van den Berg, 2011), which is completely eroded when the cutoff finally occurs. The second type of chute cutoff is driven by headward erosion of a channel (or gully) through the floodplain. The headcut migrates from the downstream end of the forming chute, progressively capturing an increasing fraction of the overbank flow and finally causing the cutoff of the original bend (Gay *et al.*, 1998; Zinger *et al.*, 2011, 2013). The third type of chute cutoff is determined by the gradual erosion of existing swales (or sloughs) within the scroll bar located inside of the bend. These swales

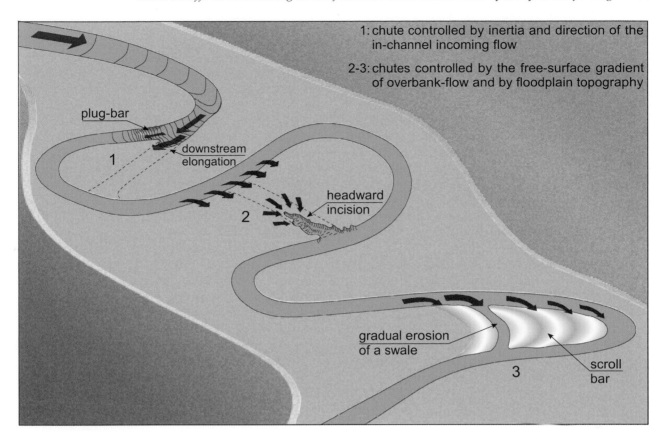

Fig. 3. Different mechanisms of chute cutoff formation in meandering rivers. 1) Downstream elongation of a chute, controlled by inertia and direction of the in-channel incoming flow, 2) headward incision; and 3) swale enlargement aligned to main scroll bar direction.

provide preferential flood routing paths, until most of the discharge is conveyed by the new channel, thus leading to a chute cutoff. This type of chute cutoff is actually the more widely recognised (Fisk, 1947; Bridge *et al.*, 1986) and is frequently found along meandering rivers that, during their migration, produce the ridge and swale topography typical of scroll-bars (Hickin & Nanson, 1975; Grenfell *et al.*, 2012). Possible preferential paths are also provided by abandoned channel reaches that are not yet completely filled with sediment deposited by overbank floods.

A chute may form by following one of these mechanisms or a combination of them (Kleinhans & van den Berg, 2011). Each mechanism requires specific conditions in order to occur. The formation of chutes through downstream extension, requires a sufficient stage difference across the floodplain (as recognised by Constantine *et al.*, 2010). In addition, the in-channel flow upstream of the outer bank where the chute starts must be suitably directed and possess a large enough momentum (Van Dijk *et al.*, 2012). On the other

hand, headward erosion of a chute is necessarily associated with overbank flow and is weakly affected by the upstream channel flow. Rather, the water level in the main channel at the downstream end of the forming chute must be sufficiently low, in order for the free-surface to drop at the headcut, thus enhancing headward erosion (Zinger *et al.*, 2011, 2013). Finally, in the case of a chute driven by an existing swale or a palaeo-channel that links an upstream and a downstream section of the river, the erosion power along the forming chute mainly depends on the stage difference and, in particular, on the so-called gradient advantage associated with the shorter path through the point-bar (Grenfell *et al.*, 2012). Van Dijk *et al.* (2014) observed that sloughs and hence chute channels, form where inner-bank attachment of scroll bars is interrupted.

The fate of the newly formed chute is another interesting issue that depends on many factors. Among the many, it is worth mentioning the intensity of helical flow due to the upstream channel curvature, which influences the bifurcation asymmetry and, hence, the division of bedload between the two branches (Kleinhans *et al.*, 2008). Due to the reduced water discharge, the sediment transport capacity in the original channel branch decreases. If the resulting transport capacity is still greater than the supply, the original channel can remain active along with the chute (Grenfell *et al.*, 2012). Alternatively, a sandy plug bar causes the closure of the original branch (Constantine *et al.*, 2010; Van Dijk *et al.*, 2012, 2014) which tends to be progressively filled by fine material supplied by overbank flow (Toonen *et al.*, 2012; Dieras *et al.*, 2013). The role of bars in determining the fate of bifurcations has been suggested by experiments conducted in straight channels (Bertoldi *et al.*, 2009) and by analytical findings (Redolfi *et al.*, 2016). Zinger *et al.* (2013) assessed the fate of an already formed chute, performing measurements of flow velocity and morphology. They also developed a conceptual model of chute-cutoff dynamics in which the upstream and downstream ends of a chute channel are treated as a bifurcation and confluence, respectively.

Controlling factors

The incision of chutes invariably needs overbank flow (Howard & Knutson, 1984; Bartholdy & Billi, 2002) and a significant stage difference between the upstream and the downstream ends of the forming chute ensuring a gradient advantage with respect to the original bend (Grenfell *et al.*, 2012). Various controlling factors can be identified with reference to specific features of the river and the floodplain.

In-channel flow features

Water levels normally rise during floods; and stage difference along the meander undergoing cutoff can be further increased by piling up of woody debris or ice (Keller & Swanson, 1979; Gay *et al.*, 1998) and by the formation of central-bars or plug-bars. Seminara (2006) observed that the formation of a central bar at the bend apex promotes a tendency of the stream to bifurcate into an outer and an inner branch, the latter being a potential precursor of chute cutoff (see also Jager, 2003). Recently, Eekhout & Hoitink (2015) described the formation of a chute cutoff triggered by the formation of a plug bar within the main channel, owing to a backwater effect. After the plug bar was deposited, an embayment (Type 1, Fig. 3) formed in the floodplain at a location where the presence of a former channel probably implied less consolidated sediment prone to erosion.

The presence of central-bars or point bars is often associated with longitudinal oscillations of the bankfull river width that shift in space and vary in time, strongly affecting the in-channel flow and, consequently, the rate of meander migration (Brice, 1975; Lagasse *et al.*, 2004; Luchi *et al.*, 2010; Zolezzi *et al.*, 2012). Meandering rivers exhibiting a larger bankfull width near the bend apexes are observed to migrate most rapidly, while rivers displaying an irregular distribution of width oscillation are more stable in terms of planform configuration. The nonlinear perturbation analysis carried out by Luchi *et al.* (2011) confirms that meander widening at the bend apex are more inclined to bend migration. This tendency is due to the nonlinear interactions between curvature and width oscillation that increase the velocity excess between the channel banks responsible for outer bank erosion.

Width oscillations can also occur as a consequence of the different rates at which, as a meander bend migrates, the outer bank is eroded and the inner bank progrades due to point bar accretion. When the outer bank retreat is faster than the inner bank progress (bank pull) the widest section tends to be located close to the bend apex. Conversely, smaller values of the ratio between erosion and accretion (bar push) imply a widening of the river towards the inflection points of consecutive bends.

The formation of central or plug bars as a driver of chute channel incision is an interesting feature, as it can lead to chute incision also under steady flow conditions and with water discharges that are unable to produce overbank flow. However, the occurrence of chute cutoffs under a constant water discharge has been, so far, observed only in laboratory or controlled experiments (Peakall *et al.*, 2007; Braudrick *et al.*, 2009; Van Dijk *et al.*, 2012; Visconti *et al.*, 2012). On the other hand, backwater effects are one of the factors promoting the formation of point-bars, central-bars and plug-bars within a river (Kasvi *et al.*, 2013). More generally, in-channel bars form due to a downstream reduction of bed shear stress, of which backwater effects are only one of the possible causes. Indeed, bed shear stress can reduce because of channel widening (Miori *et al.*, 2006), lengthening (Van Dijk *et al.*, 2014), curvature driven helical flows (Lanzoni *et al.*, 2006; Seminara, 2006; Kleinhans et al., 2008), presence of bars (Kleinhans *et al.*, 2011) and inlet steps (Bertoldi *et al.*, 2009).

Discharge variability can also influence the formation of chute cutoff (Schuurman *et al.*, 2016). Indeed, changes in the frequency of flood events were shown to be related to the frequency and the occurrence of chute cutoffs (Ghinassi, 2011; Micheli & Larsen, 2011; El Gammal, 2016; Li *et al.*, 2017).

Floodplain characteristics

Floodplain features such as local differences in topography, sediment composition and vegetation can alter locally the overbank flow, promoting or inhibiting erosion (Constantine *et al.*, 2010). Consequently, not only the type of chute incision but also its inception and subsequent development may vary considerably between different floodplain settings.

Actually, it is widely recognised that small channels crossing the point bar play a key role in the formation of chutes (Brierley, 1991), as they provide preferential pathways where overbank flow concentrates, increasing the shear stress and, hence, the erosion power. Besides scroll bar swales, also the presence of palaeo-meanders and abandoned channels, not completely filled by sediments, can promote the formation of chute cutoffs (Constantine *et al.*, 2010). An example is the chute channel that formed in 2011 along the Chixoy River, Guatemala (Fig. 4). It is worth noting that the presence of palaeo-meanders within the river belt is a widespread and common feature of meandering rivers, of which Fig. 2 and Fig. 4 are clear examples. David *et al.* (2017) performed a detailed reach-scale mapping of floodplains and found that up to 75% of channel reaches within a floodplain are probably palaeo-meander cutoffs.

Besides topography, floodplain erodibility is another crucial factor in determining the success or failure of a chute cutoff (Lewis & Lewin, 1983; Braudrick *et al.*, 2009; Constantine *et al.*, 2010; Tal & Paola, 2010; Micheli & Larsen, 2011; Dunne & Aalto, 2013; Grenfell *et al.*, 2014; Harrison *et al.*, 2015; Schuurman *et al.*, 2016; Słowik, 2016; El Gammal, 2016; Iwasaki *et al.*, 2016; David *et al.*, 2017).

River morphometry

Observed data suggest that chute cutoff occurrence is linked to specific morphometric features of the main river, first of all to relatively high slope reaches (Lewis & Lewin, 1983; El Gammal, 2016). In particular, the observations made by Lewis & Lewin (1983) indicate that chute cutoffs are more probable when the radius of bend curvature to channel width is in the range 1 to 2. The morphometric analysis carried out by Micheli & Larsen (2011) along the Sacramento River (CA, USA) confirms that the radius of bend curvature is one of the controlling factors for chute incision. A larger radius (smaller curvature) implies longer chute channel to form and, consequently, the need of a greater erosive power. Chute cutoffs were indeed found to occur most frequently in wide channels where bend curvature is strong (Howard & Knutson, 1984). Micheli & Larsen (2011) identified a specific range of sinuosity and angle of entrance for which chute cutoffs formed, but they also showed that chutes did not form in all the bends matching those characteristics. Then, the conditions for chute occurrence inferred by Micheli & Larsen (2011) were necessary but not sufficient for the incision of a chute.

The oscillations of channel width can also have possible implications for the occurrence of chute cutoffs, since river widening is known to promote

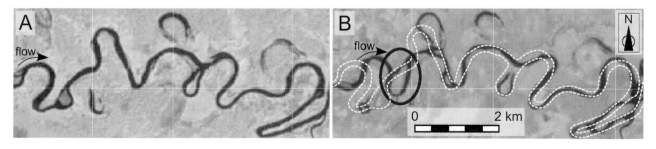

Fig. 4. Chute cutoff occurred in 2011 along the Chixoy River (Guatemala). The chute channel (blue circle in panel B) is clearly a palaeo-meander, still active in 1988 (panel A) and re-activated by overbank flow during a severe flood. The white dashed line in panel B represents the river configuration in 1988 (panel A).

the formation of central bars and, in turn, the tendency to originate channel bifurcations (Seminara, 2006; Camporeale *et al.*, 2008; Constantine *et al.*, 2010; Luchi *et al.*, 2010; Parker *et al.*, 2011; Zolezzi *et al.*, 2012; Frascati & Lanzoni, 2013; Grenfell *et al.*, 2014; Church, 2015; Eekhout & Hoitink, 2015). The analysis of three different rivers by means of binary logistic regression carried out by Grenfell *et al.* (2012) suggests that the only statistically significant predictor of chute initiation at a bend is the average rate of bend extension (i.e. the rate at which a bend elongates in a direction perpendicular to the overall channel axis trend). An increased probability of chute initiation seems to be related to a larger rate of bend extension, as typically observed in the presence of bend widening (Brice, 1975; Grenfell *et al.*, 2014; Słowik, 2016). Mathematical simulations confirmed that a faster migration favours the occurrence of chute cutoffs, which were instead not forming in the case of low migration rates (Schuurman *et al.*, 2016).

Sediments

Chutes are eroded in the floodplain provided that the energy of floods is sufficient to transport sediment delivered from the river to the floodplain, to remove sediments from the floodplain floor and to transport this whole sediment load downstream, back into the river (Constantine *et al.*, 2010). On one hand, the importance of sediment dynamics is obvious; on the other hand, the spatial and temporal patterns of erosion and deposition in topographically heterogeneous floodplains are extremely complex (Piégay *et al.*, 2008).

Interestingly, chute cutoffs are more frequently found in the presence of poorly cohesive banks (Howard & Knutson, 1984; Howard, 2009) and where infill rates are lower (Hooke, 1995; Dunne & Aalto, 2013). Note that the suspension of bed material is a key control on river morphology (Nicholas, 2013), as it limits the gravitational deflection of sediment in the direction of the local bed slope. Hence, in the long term, suspended sediment load leads to a more uniform filling of areas adjacent to the main course of the river, with repercussions on aggradation and resistance to erosion of a floodplain. Indeed, high rates of suspended sediment aid the connection of bars to floodplain (Braudrick *et al.*, 2009) and, by levelling out the point bar, they limit the presence of sloughs and swales that play a major role in chute formation by acting as preferential pathways to overflow.

Anthropogenic factors

Anthropogenic factors are also known to play a significant role in the evolution of rivers and, of course, in the formation of chute cutoffs (Parker & Andres, 1976; Simon & Robbins, 1987). An interesting example is the artificial cutoff produced in the Ucayali River, a tributary of the Amazon River that is extraordinarily active from a morphological point of view. In 1997, a 72 km-long, triple-lobed, meander bend was removed. The chute channel formed from a small shortcut channel that was carved some decades before by local people in order to reduce the canoe travel time along the river (Schwenk & Foufoula-Georgiou, 2016). Contrarily, anthropogenic modifications such as flood control structures and large impoundments, by diminishing the frequency and magnitude of floods and thus the river-floodplain connectivity, reduce the chance of chute cutoffs (Edwards *et al.*, 2016).

MATERIALS AND METHODS

A detailed examination of the hydraulic controls on chute incision is the first step towards a better understanding of how a chute cutoff occurs (Constantine *et al.*, 2010). In the following, the key hydraulic and topographic factors triggering the formation of chute channels are analysed on the basis of satellite and aerial photographs, digital terrain models and by means of two mathematical models. The first model is based on a linearised treatment of the two-dimensional equations governing mass and momentum conservation of the liquid phase and the sediment balance. In the absence of elevation data, it is used to estimate the steady bed topography and the corresponding two-dimensional (depth averaged) flow within the channel, except the bank region. The second model solves numerically, through finite elements, the two-dimensional shallow water equations (see Appendix A) and is here used to compute the in-channel and the overbank flow fields.

Linearised morphodynamic models, owing to their low computational costs, have been widely used to predict long-term migration of meandering rivers. Coupled with simplified laws to drive outer bank erosion and to account for neck cutoffs (e.g. Frascati & Lanzoni, 2010; Bogoni *et al.*, 2017), they typically model the water flow and the bed topography in the main channel but neglect the bank region where boundary layer effects cannot be described by a depth averaged treatment. In this type of models, effects of chute cutoffs have been accounted for only adopting a heuristic statistical framework, e.g. by considering a quasi-random occurrence of chute cutoffs (Howard, 1996).

On the other hand, more refined hydrodynamic models are needed to provide information about the water flow both in the main channel and over the adjacent floodplain and, therefore, to disclose the complex interactions associated with the transfer of momentum between the in-channel streamflow and the floodplain overflow (Shiono & Muto, 1998; Shiono *et al.*, 1999, 2009a; b; Patra & Kar, 2000; Wormleaton *et al.*, 2004; Patra *et al.*, 2004; Shiono & Shukla, 2008; Khatua & Patra, 2009; Liu *et al.*, 2014; Shan *et al.*, 2015). Although full three-dimensional (3D) models offer a more detailed description of the flow field, especially near to the banks, two-dimensional (2D) depth-averaged models have been found to reproduce accurately many important hydrodynamic features, with a sensible reduction of computational costs (Rameshwaran & Shiono, 2003; Shiono & Shukla, 2008; Riesterer *et al.*, 2016).

In the present contribution, the linearised hydro-morphodynamic model of Frascati & Lanzoni (2013) is used to estimate the channel bed topography antecedent to chute formation in the absence of reliable field data. Indeed, the model, although linearised, has been found to describe with a good accuracy the bed topography in mildly curved and long bends with weak width variations. The analytical character of the model allows to describe the two dimensional bed topography and the corresponding flow with the desired spatial resolution. More specifically, the model describes the steady, spatially varying flow and the sediment transport in channels with arbitrary distribution of channel axis curvature and channel width. The governing equations are linearised and solved by means of a two-parameter perturbation expansion technique, taking advantage of the fact that alluvial rivers often exhibit mild and long meander bends, as well as evident but relatively small width variations. The model, which accounts for the dynamic effects of secondary flows induced by both curvature and width variations, allows us to compute the steady two-dimensional in-channel flow field (i.e. water elevation, flow velocity and depth) and the equilibrium bed elevations corresponding to a given flow discharge.

The hydrodynamic interactions between the in-channel flow and the floodplain overbank flow are investigated by means of a finite-element model that solves the full two-dimensional shallow water equations on irregular grids, accounting for wetting and drying processes through a suitable parametrisation (Defina, 2000; Viero *et al.*, 2013, 2014; Viero & Valipour, 2017). The shallow water equations (see Appendix A) are solved using a semi-implicit staggered finite-element method, based on mixed Eulerian-Lagrangian approach (D'Alpaos & Defina, 2007; Viero & Defina, 2016). The depth integrated horizontal turbulent and dispersion stresses are evaluated using the Boussinesq approximation (Stansby, 2003) and the eddy viscosity according to Uittenbogaard & van Vossen (2004). In principle, the considered hydrodynamic model can be easily coupled with the two-dimensional Exner equations to describe channel bed dynamics (Defina, 2003). Nevertheless, the modelling of chute cutoffs, from chute inception to meander bypass and, possibly, abandoned channel infilling requires a more sophisticated approach, to include fundamental processes such as bank erosion (see, e.g. Iwasaki *et al.*, 2016) and graded sediment transport and deposition. In the following, attention is then restricted to the hydrodynamic and topographic factors that control the initiation of chute incisions.

CASE STUDIES

In this section, the modelling framework described above will be used to identify the hydraulic and topographic controls on chute cutoffs with reference to two case studies: the Sacramento River and the Cecina River. In these two examples, chute formation is driven by distinct mechanisms: in-channel flow in the former case and free-surface gradient in the second. This separation allows a clear characterisation of the different hydraulic and topographic controls.

Sacramento River

The Sacramento River is the largest river of California (USA). It collects precipitation and snowmelt runoff from the western slopes of the Sierra Nevada, the eastern slopes of the Coast Range and the southern Trinity and Klamath ranges; and finally discharges into the Pacific

Ocean via the San Francisco Bay (Micheli & Larsen, 2011). The river is about 480 km-long, flowing from north to south. The Sacramento River Valley is mainly composed of sedimentary rocks and recent alluvium, with a meander belt dominated by Pliocene–Pleistocene alluvium and fluvial deposits. Along the investigated reach, the Sacramento River is predominantly an unconstrained, single-thread, sinuous channel, with slope ranging from 0.0002 to 0.0007. The riverbed material is mainly sand and gravel with a median grain-size ranging from 5 to 35 mm. The average height of the top of the bank varies from 2 to 8 m with respect to the channel bed. The average channel width is about 250 m at bankfull conditions. The construction of the dam at Shasta Lake in the early 1940s, and of a number of flood control structures diverting excess flow into overflow catchment basins during peak floods, caused a sensible increase of return periods of bankfull and major peak floods (Micheli & Larsen, 2011).

The reach considered in this study is characterised by the presence of two consecutive bends (Fig. 5) and it is located about 15 km South of Red Bluff, between the river miles 233 and 236 (according to the 1991 Sacramento River Atlas, Army Corps of Engineers). The second of these bends experienced two distinct chute cutoffs, approximately in January 1978 (Fig. 5A) and in January–March 1995 (Fig. 5B).

The Digital Elevation Models (DEMs) shown in Fig. 5 are characterised by different resolution. The first (Fig. 5A) has a resolution of ~10 m (1/3 arc-second) and was extracted from the U.S. Geological Survey (USGS) National Elevation Dataset (NED). It refers to the year 1976, before the occurrence of a chute cutoff probably caused by a flood in January 1978 (see Fig. 6A) and has been selected after comparing together the planform paths attained by the river in the last 100 years, available in the form of shapefiles at the Sacramento River forum website (www. sacramentoriver.org).

The high-resolution (~3 m) DEM depicted in Fig. 5B, provided by the California Department of Water Resources, was acquired from the CVFED LiDAR survey. Although being useless for the present modelling purposes (it refers to a post-cutoff configuration), it provides an interesting picture of the topographic heterogeneity that characterises the Sacramento floodplain, with the presence of many scroll bars, floodplain channels and palaeo-meander cutoffs (David *et al.*, 2017). The gross

features of the floodplain are very similar to those emerging from the lower resolution DEM of 1976.

Fig. 5A indicates that both the considered meandering bends are strongly confined by relatively high banks on the outer side, with flattened elevations at the apex of the upstream bend and just after the apex of the downstream bend. One of the most interesting aspects of the 1976 channel configuration is the presence of an encroachment located downstream of the bend apex and upstream of the inflection point between the two bends, which shows similarities with that investigated by Constantine *et al.* (2010). Nevertheless, differently from Constantine *et al.* (2010), who linked the presence of embayments to nearly uniform floodplains, in the present case the floodplain is markedly irregular and the embayment is placed in correspondence of the remainder of an abandoned (and partially filled) palaeo-meander channel with a sinuous shape (Fig. 5A).

Fig. 6 reports the discharge hydrograph recorded at the Bend Bridge gauge station, upstream of Red Bluff, and the LandSat images showing different stages of the temporal evolution of the chute cutoff during the period 1977–1980. Near infrared band 7 and multispectral images are shown in order to emphasise water bodies and vegetation patterns. The chute seems to have been triggered by a relatively high flood (3000 m³ s⁻¹), which occurred in January 1978, and subsequently incised by a sequence of other flood events, the larger one (of about 3000 m³ s⁻¹) in February 1980.

The 2D finite-element model is applied to analyse the hydrodynamic features associated with the 1976 configuration, antecedent to the chute formation. The numerical grid is made up of about 30,000 nodes and 55,000 triangular elements, to cover a reach of about 13 km close to the bridge downstream of Los Molinos. Given the absence of reliable in-channel elevation data, the bathymetry within the 1976 meandering channel configuration (Fig. 7A) has been estimated through the linearised morphodynamic model of Frascati & Lanzoni (2013). The robustness of the estimated bed topography was preliminary tested by applying the model to the 2001 channel planform geometry and comparing the resulting bed topography with the bathymetric survey carried out in 2001 by the California Department of Water Resources (DWR). Fig. 7B shows the comparison of the computed cross section profiles with the available data. The overall agreement is reasonably good (relative error, scaled by the local flow depth,

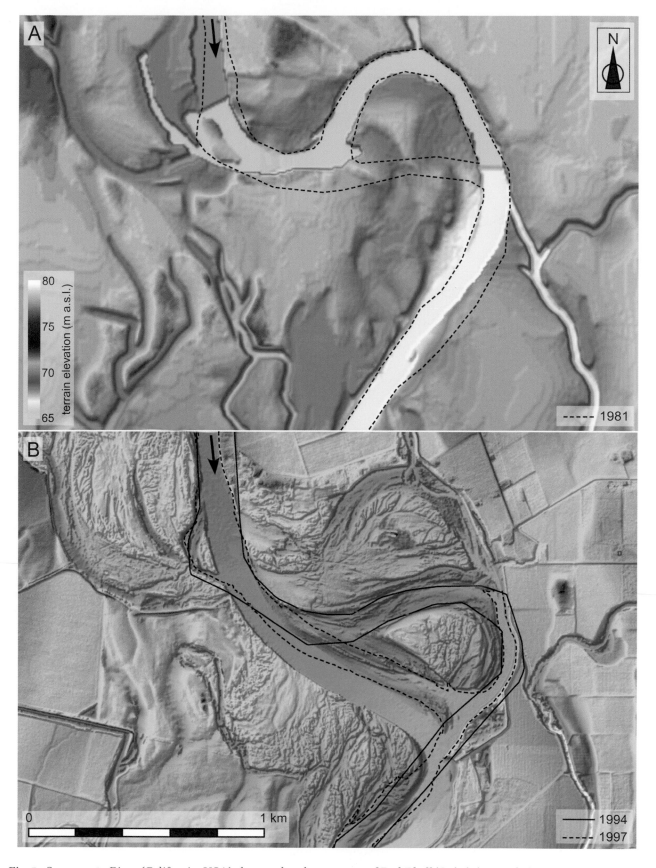

Fig. 5. Sacramento River (California, USA) about 15 km downstream of Red Bluff (CA). (A) Digital elevation model (DEM) from USGS National Elevation Dataset (NED), referring to 1976 (resolution ~ 10 m); (B) DEM from the CVFED LiDAR survey, provided by the California Department of Water Resources (DWR), referring to 2010 (resolution ~ 3 m). The black lines denote the channel geometry digitalised by DWR.

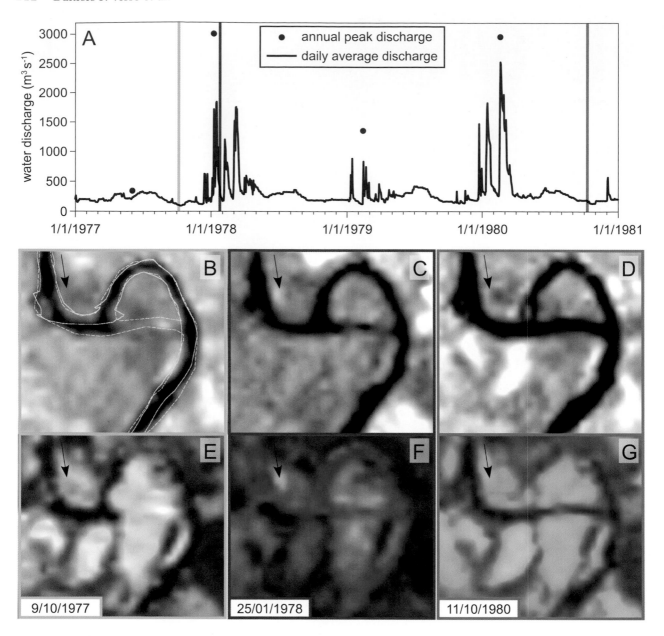

Fig. 6. (A) Discharge hydrograph of the Sacramento River recorded at USGS gauge station n. 11377100 (Bend Bridge upstream of Red Bluff) in the years 1977–1980. Vertical lines refer to the corresponding LandSat images in the panels below (band 7 in panels B to D, multispectral in panels E to G), which show significant stages of the chute incision.

~11%), given also the inability of the model to describe the boundary layer region close to the banks. In addition, the results of the one-dimensional hydraulic model of the Sacramento River implemented by DWR ('Sacramento and San Joaquin River Basins Comprehensive Study') were used to extract the rating curve to be applied as boundary condition at the downstream section of the computational domain and to calibrate the resistance parameters of the 2D finite-element

hydrodynamic model. A set of preliminary simulations, carried out by varying the model parameters within a reasonable range, suggests that the global hydrodynamic behaviour described in the following is not substantially affected by parameter uncertainties.

Fig. 8 shows the 2D flow field computed for a total discharge of 3000 m³ s⁻¹, which is the maximum discharge (Fig. 6) recorded in the period 1976–1981 at the Bend Bridge gauge station,

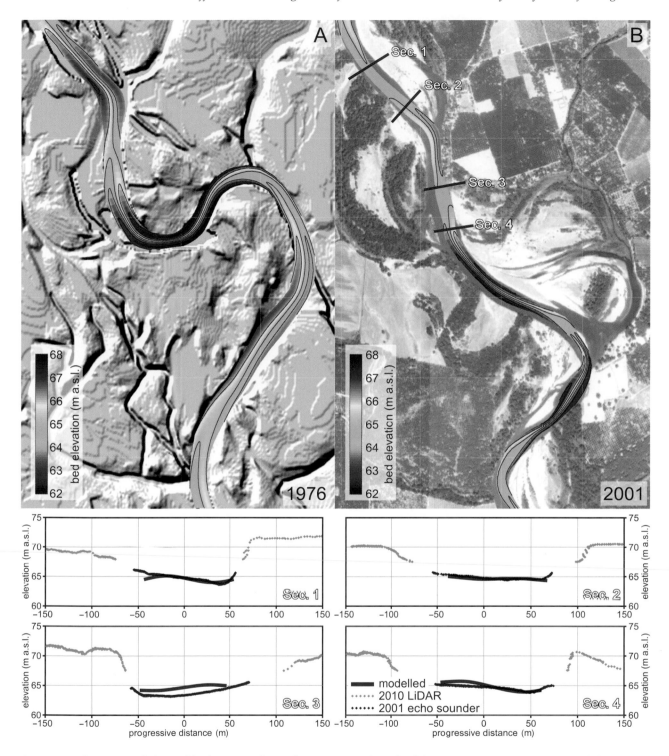

Fig. 7. Two-dimensional channel bed topography in the investigated reach of the Sacramento River, computed through the hydro-morphodynamic model of Frascati & Lanzoni (2013) for the planform configurations observed in 1976 (panel A) and 2001 (panel B). The bathymetry computed for the 2001 configuration has been compared with the echo-sounder data collected in the four cross sections marked in panel B. The agreement in the central region excluding the banks (where the model loses validity) is reasonably good (mean relative error, scaled by the local flow depth, ~11%). The flow discharge that ensures this fit is $270\,m^3\,s^{-1}$, about 80% of the mean annual discharge ($350\,m^3\,s^{-1}$), in accordance with the idea that the river topography provided by standard surveys is determined by flood events of moderate intensity, which recur much more frequently, rather than by extreme floods (Wolman & Miller, 1960; Lanzoni *et al.*, 2014).

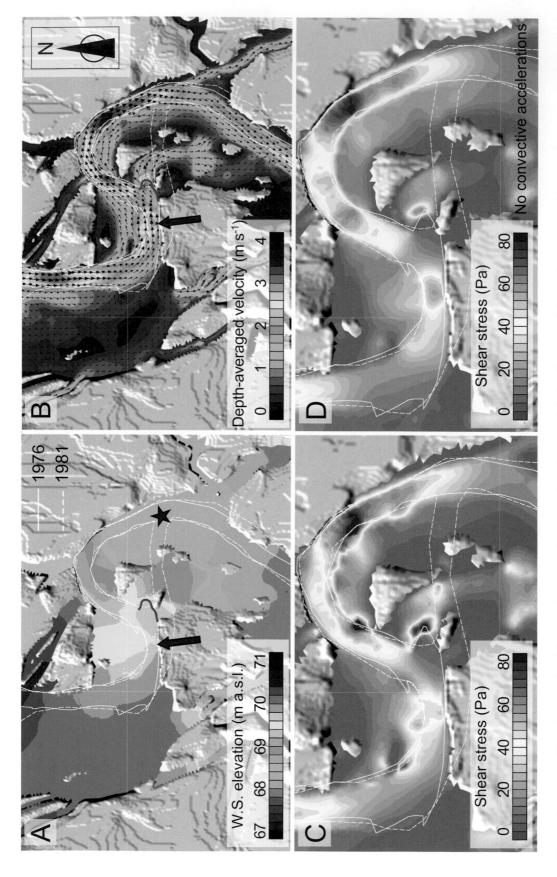

Fig. 8. Sacramento River in 1976. Model results in terms of water surface elevation (A), depth-averaged flow velocity (B), and bottom shear stress (C and D), for a total discharge of $3000 \, m^3 \, s^{-1}$. The red arrows denote the section narrowing at the apex of the first bend, the magenta lines locate the embayment. The red star in panel (A) denotes the end section of the chute that will be formed across the floodplain.

upstream of Red Bluff (USGS gauge station n. 11377100). The discharge confined within the upstream bend is 2760 m³ s⁻¹, while the remaining fraction flows over the floodplain. A discharge of 340 m³ s⁻¹ flows in correspondence of the embayment located after the bend apex of the upstream bend (magenta lines in Fig. 8), preluding the chute formation. It can be noted that the narrowing of the main channel nearby the bend apex (red arrows in Fig. 8) tends to concentrate the water flux and to increase the in-channel flow velocity (Figs 8B and 9).

As a consequence, a high momentum current is directed against the outer bank, that is impinged in the correspondence of the embayment. This favours a progressive erosion of the bank material and, hence, meander migration, as well as the progressive downstream extension of a chute (see the dashed lines in Figs 5A, 8 and 9). Remarkably, considering also the progressive migration of the outer channel bank, the direction of the chute tends to be aligned with that of the in-channel flow at the bend apex. This behaviour is promoted by the sharpness of the considered bend and the inertia of the incoming channelised flow, which is predominant with respect to the effects exerted by the floodplain characteristics (topography, vegetation, etc.). The palaeo-meander present on the floodplain, rather than driving the overbank flow, provides a more erodible portion of soil near to

the outer channel bank, where the flow shear stresses concentrate (Fig. 8C).

The importance of the in-channel flow inertia is supported by the additional simulation carried out by neglecting the convective acceleration terms in the 2D hydrodynamic model (i.e. the terms controlling the transport of momentum to a different position, see Appendix A). In the absence of these terms, the maximum flow velocity downstream of the encroachment is reduced by more than 20% (from 2.9 to 2.3 m s⁻¹), the bottom shear stress is reduced by 45% (from 90 to 50 Pa, Fig. 8C and D) and the discharge through the area where the chute is going to develop decreases by 30% (from 340 to 240 m³ s⁻¹).

To sum up, the initiation of the 1978 chute cutoff was due to the concomitant presence of two principal factors: i) a concentration of flow velocities (due to the river narrowing downstream the bend apex) directed against the outer bank; and ii) the presence of an embayment possibly associated with a palaeo-meander. The river cross-section enlargement caused by the embayment tends to reduce the flow velocity at the centre of the main channel (Figs 8B and 9A). This reduction, in turn, enhances sediment deposition and, consequently, favours the formation of a central-bar (Seminara, 2006) or a plug-bar (Eekhout & Hoitink, 2015) and, after all, the discharge diversion through the forming chute.

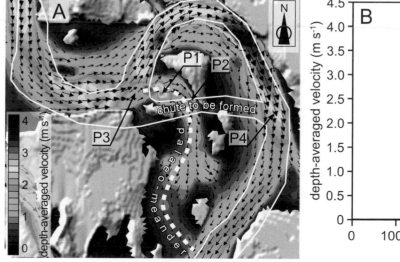
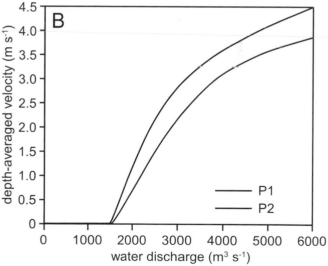

Fig. 9. Sacramento River in 1976. (A) Detailed view of the two-dimensional (i.e. depth-averaged) velocity field (shown in Fig. 8B). The white solid line denotes the 1981 channel geometry digitalised by DWR. (B) Depth-averaged velocity for different values of the total discharge at points P1 and P2 (see the black arrows in panel A), located at the edge of the embayment and at the apex of the palaeo-meander bend, respectively.

Note that these hydrodynamic features are associated with the presence of an embayment with size comparable with the local channel width (embayment mean width and length being ~60% and ~140% that of the local channel width), that is much larger than the bank embayments generated by the presence of slump blocks (Hackney *et al.*, 2015) or bank macro-roughness elements (Darby *et al.*, 2010).

Fig. 9B shows that the depth-averaged velocity in two relevant locations along the chute to be formed increases monotonically with the river discharge. Additional simulations showed that the overall hydrodynamic features of the flow remain remarkably similar for discharges in the range 2000 to $6000\,m^3\,s^{-1}$ (i.e. when overbank flow occurs), except for a gradual increase of water levels and flow velocity as the discharge grows.

The energy of the flood, as it interacts with the variable floodplain topography, is reflected in the variability of the free-surface slope (Figs 8A and 10). Importantly, the upstream and downstream ends of the future chute almost coincide with a local maximum and minimum of the overbank flow free-surface elevation, respectively. This confirms that local differences in the shape of the floodwater surface probably control the location of a chute and determine how a chute progressively advances (Constantine *et al.*, 2010).

In addition to the inertia and direction of the in-channel flow and the topographic controls of the overbank flow, the temporal sequence of flood events can have a fundamental role in determining chute inception. As depicted in Figs 5B (dashed line) and 11B to G, another chute cutoff occurred in the period January–March 1995. After the cutoff experienced in 1978, the river bend upstream of the chute progressively elongated and increased its curvature due to outer bank erosion. The discharges recorded at Bend Bridge gauge station (Fig. 11A), upstream of Red Bluff (USGS station n. 11377100) and a close inspection of satellite images (Fig. 11B to G) suggest that this second chute cutoff was triggered by a first flood (peak discharge of $2670\,m^3\,s^{-1}$) which occurred in January 1995 and developed completely during a second flood (peak discharge of $3030\,m^3\,s^{-1}$) which occurred in March 1995. This seems to confirm the fact that the occurrence of subsequent floods (i.e. discharge variability) plays a major role in chute formation (McGowen & Garner, 1970). As reported in Constantine *et al.* (2010), the formation of this second cutoff was preceded by the erosion of an embayment just downstream of the upstream bend apex (Fig. 11E). Similarly to 1978, the chute was initially curved (Fig. 11F), but eventually developed as a straight channel aligned with the direction of the incoming in-channel flow (Fig. 11G). The LandSat images reported in Figs 6 and 11 then suggest a remarkably similar behaviour of the 1978 and 1995 chute cutoffs.

Cecina River

The Cecina River is a single-tread gravel bed river in central Italy, with a catchment of about $900\,km^2$ and a total length of about 80 km. The flow regime is highly ephemeral, with flash flood flows induced mainly by intense cloudbursts (Bartholdy & Billi, 2002). Bed material exhibits a general, though irregular, downstream decrease in terms of median diameter, D_{50} (Billi & Paris, 1992). At the study reach (Fig. 12), which is located approximately 55 km from the source and 25 km from the outlet, D_{50} ranges between 15 and 30 mm, the bed elevation ranges between 34 and 38 m above sea-level and the mean slope is about $1.85 \cdot 10^{-3}$ (Fig. 13). In the considered reach, the Cecina River exhibited a remarkable increase of sinuosity starting from the 1990s (Bartholdy & Billi, 2002). During the 1950s and 1960s, a deficit in sediment supply due to change in land use in the upstream part of the basin was exacerbated by extensive mining of bed material, which was ultimately forbidden in 1978. The reduction in sediment supply and the consequent increase in mean channel slope led to a relatively straight, deep channel with alternate bars. After 1978, a consistent sediment load supply re-established due to the ceas-

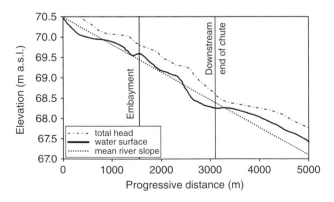

Fig. 10. Sacramento River (California, USA). Water surface (solid line) and total head (dash dotted line) profile extracted along the main channel centreline. The dotted lines denote the mean slope of the river.

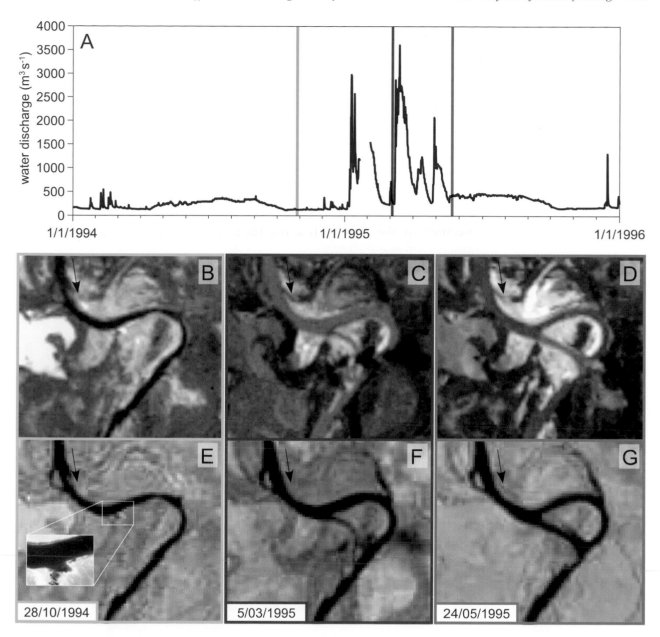

Fig. 11. (A) Discharge hydrograph of the Sacramento River recorded at USGS gauge station n. 11377100 (Bend Bridge upstream of Red Bluff) in the years 1994–1995. Vertical lines refer to the corresponding LandSat images in the panels below (RGB in panels B to D, multispectral in panels E to G), which show significant stages of the chute incision.

ing of mining and to bank erosion associated with channel widening (Rinaldi *et al.*, 2008; Luppi *et al.*, 2009; Nardi & Rinaldi, 2010; Rinaldi & Nardi, 2013).

Two consecutive meanders developed in the recent years (Fig. 12) due to pronounced bank erosion processes (Nardi *et al.*, 2013). The rapidity of the outer bank erosion, in particular at the first bend, led to the formation of a poorly developed (in term of surface elevation) point bar at the inner

bank (Luppi *et al.*, 2009). Moreover, vegetation had not enough time to completely cover and thus to protect, the newly formed inner point bar (Fig. 12, panel F). The analysis of aerial images (Fig. 12) and of the DEM derived from a LiDAR survey (Fig. 13), suggests that the flow was unable to remove from the channel bed the whole material eroded from the outer bank, thus leading to a relatively shallow and narrow river cross section. The inside of the upstream bend is characterised

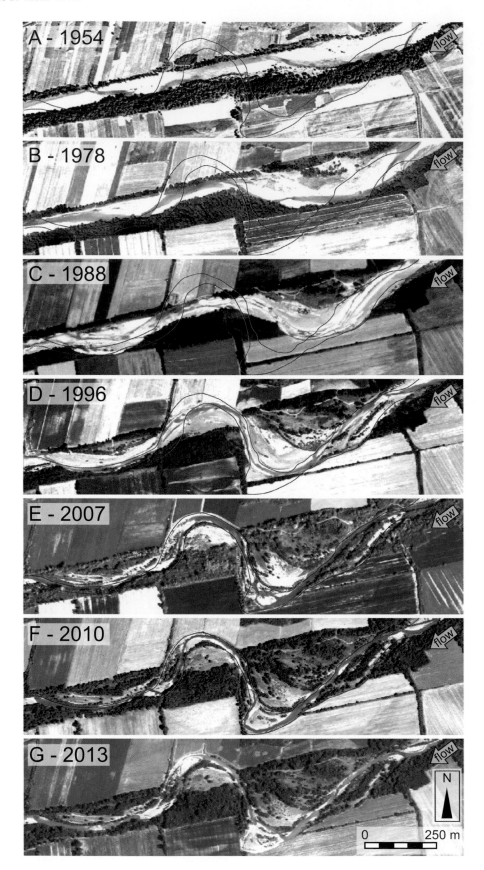

Fig. 12. Cecina River (Tuscany, Italy). Historical aerial images showing the recent evolution of the main channel. For comparison purposes, the red lines denote the channel banks of the 2013 configuration.

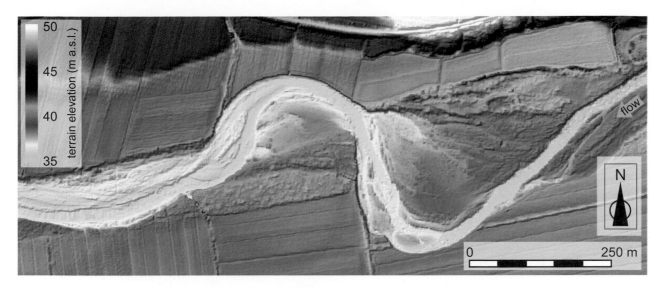

Fig. 13. Cecina River. Digital Terrain Model (resolution of 1×1 m) derived by a LiDAR survey (data source: 'Ministero dell'Ambiente e della tutela del Territorio e del Mare – Rilievi Lidar', courtesy of the Tuscany Region). The survey refers to year 2008.

by a bed topography degrading slowly toward the river and a sparse vegetation cover (Figs 13 and 14), which determines the presence of preferential pathways for overbank flow. During flood events, the incoming flow can then freely expand over the areas inside of the bend, reducing the mean velocity of the in-channel streamflow, its erosive power and sediment transport capacity as well.

It was in this context that, approximately between 2011 and 2013, the elongated meander bend shown in Fig. 12 F underwent a chute cutoff (Fig. 12, panel G). The mechanism that led to the formation of the chute can be ascribed to the flashy character of flood events, which caused the rapid erosion of the outer bank of the bend but an insufficient removal of the material fallen down in the channel bed.

The hydrodynamic conditions responsible for the chute cutoff have been assessed by applying the 2D finite-element model to the 2008 (Fig. 13) and 2010 (Fig. 14) configurations. The topography of this latter configuration, shown in Fig. 14, was reconstructed on the basis of a DEM obtained from a 2008 LiDAR survey and of an aerial image taken in 2010 (Fig. 12, panel F) in order to update the channelised path within the main channel. As a first approximation, bottom elevations of the portion of point-bar formed after 2008 were assigned assuming the same mean slope of the 2008 point-bar, while the channel bed bathymetry was assigned in analogy with that referring to 2008. Additional simulations, carried out by changing

the elevations of the channel bed and the point-bar within a reasonable extent (±0.5 m), suggest minor changes of the overall hydrodynamic behaviour described below and referring to the pre-chute 2010 configuration depicted in Fig. 12G.

The numerical grid used in the hydraulic computations is made up of about 26,000 nodes and 51,000 triangular elements and covers a reach long about 4 km (from 1 km upstream to 3 km downstream the chute cutoff). Since the topographical survey used to build the DEM was conducted during a low-flow period, most of the channel bed was actually covered by DEM data. The hydro-morphodynamic model of Frascati & Lanzoni (2013) was then used to estimate the bathymetry only in the few channelised portions of the river where no data were available.

Uniform flow condition were prescribed at the downstream section of the modelled river reach. Preliminary tests showed that the distance between the area interested by the cutoff and the downstream boundary section of the computational domain is large enough to rule out the uncertainties related to the adopted downstream boundary condition. The resistance parameters of the hydrodynamic finite-element model were chosen on the basis of granulometric and morphological characteristics of the considered reach (Nardi *et al.*, 2013). Although measured data are not available for a calibration of the model, a systematic set of simulations carried out to evaluate the sensitivity to changes in parameters, such as

Fig. 14. Cecina River. Bottom elevation of the numerical grid, reconstructed from the LiDAR survey (2008), the 2010 aerial image (shaded in background) and the model of Frascati & Lanzoni (2013).

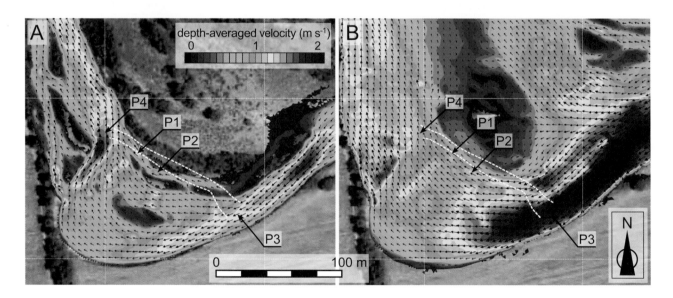

Fig. 15. Cecina River. Model results in terms of depth-averaged velocity for a total water discharge of 40 m³ s⁻¹ (A) and 200 m³ s⁻¹ (B), respectively. The white dashed lines denote the location of the chute occurred between 2010 and 2013.

friction, indicates that the overall hydraulic description provided by the model is quite robust.

The total discharge has been varied in the range between 5 and 600 m³ s⁻¹ (Nardi *et al.*, 2013). The simulated flow field turns out to change drastically with increasing discharge (Fig. 15), mostly in terms of velocity pattern in the downstream part of the considered bend and on the inner

point-bar. The most interesting and not obvious feature is that, contrarily to the Sacramento River case study (Fig. 9B), the flow velocity does not increase monotonically with the total discharge. This behaviour emerges clearly from Fig. 16, showing the modulus of the depth-averaged velocity in two points, P1 and P2, placed inside the area to be cut by the chute (see Fig. 15B). The flow

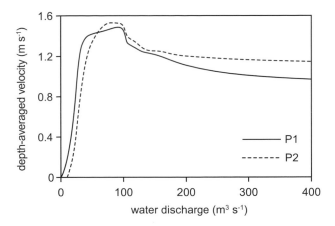

Fig. 16. Cecina River. Depth-averaged velocity for different values of the total discharge at points P1 and P2 (see the black arrows in Fig. 15), located where the chute is going to form.

velocity initially increases as the inside of the bend is flooded and suddenly drops as the discharge exceeds approximately $100 \, \mathrm{m^3 s^{-1}}$ (see Fig. 16). This rapid decrease is mainly due to backwater effects and to the spreading of streamflow throughout the point bar deposits inside of the bend (Fig. 15).

The mechanism that with great probability led to the chute incision in the Cecina River is not related to the intensity of the upstream discharge. Indeed, the in-channel flow is almost tangent to the point bar contained in the meander loop and the maximum flow velocities are not attained for the larger discharges (Fig. 16). Hence, the overbank flow is essentially controlled by the gradient advantage dictated by the point bar topography (see, the third type of formative mechanism depicted in Fig. 3). In particular, the gradient advantage is ensured by preferential paths on the point-bar deposits and the relatively shallow and narrow (and little conveying) channel section. The incision of a chute thus occurs for flow rates large enough to flood the point bar deposits, but relatively small if compared with the major floods of the Cecina River.

The analysis of the discharge time series, provided by the Hydrological Service of Tuscany Region (www.sir.toscana.it) for the Ponte di Monterufoli gauging section, located a few kilometres downstream of the study reach, shows that the two major floods in the period 2010–2013 were characterised by maximum discharges of about 220 and $170 \, \mathrm{m^3 s^{-1}}$, respectively. Such values are significantly less than the floods which occurred just before (6th January 2010, peak discharge $\sim 340 \, \mathrm{m^3 s^{-1}}$) and after (21st October 2013, peak discharge $\sim 540 \, \mathrm{m^3 s^{-1}}$) this 3-year period.

DISCUSSION

Hydrodynamic considerations and the numerical analyses highlight that the energy of the flow through the floodplain is in general due to two distinct, but strictly related, factors: the inertia and direction of the upstream in-channel flow; and the gradient of hydraulic head between the upstream and downstream ends of paths through the floodplain, denoted as gradient advantage. The first factor dominates when the flow, coming from an upstream relatively narrow bend, impinges against the bank of the meander undergoing a cutoff, first causing the incision of an embayment and then the progressive cut of the chute through the floodplain. When the second factor dominates, the chute forms either by progressive enlargement of existing swales (or sloughs) through the point/scroll bar deposits, or by headward incision of a gully. Headward erosion of a chute can also occur from its downstream end in the presence of an overland flow ensuring a large enough free-surface drop when the water re-enters into the channel.

Accordingly, with the aim of defining a physics-based framework for the inception of chute cutoffs, the different mechanisms of chute formation can be grouped in two macro-groups. The first includes chute incisions driven by overbank flow occurring downstream of the bend apex of a relatively narrow meander bend, as in the case study of the Sacramento River, where direction and inertia of the streamflow has been observed to drive the erosion process. The second macro-group concerns chute incisions driven by the gradient advantage of a preferential path through the area inside of the bend, owing to the presence of poorly developed point bar deposits and insufficient removal of previously eroded material from the channel bed (as in the case of the Cecina River). Existing preferential pathways can be also due to the ridge-and-swale sequence typical of scroll-patterned bar surfaces, as well as to topographic depressions related to older, not yet filled, meander paths (as in the case of the Chixoy River, Fig. 4).

In both the macro-groups, the occurrence of overbank flow is needed in order for a chute to form but the total discharge plays quite a different role in the two cases. The difference can be appreciated from Fig. 17, which reports the free-surface gradient (panel A) and the fraction of discharge (panel B) conveyed through the portion of floodplain to be cut by the chute, as a function of the total discharge. In the case of the Sacramento River, the free-surface gradient (Fig. 17A, solid lines) shows a general increase with the discharge up to a maximum. The in-channel flow is directed against the upstream section of the chute to be formed and convective acceleration enforces a rise in the free-surface elevation at the outer bank, downstream of the bend apex, where the formation

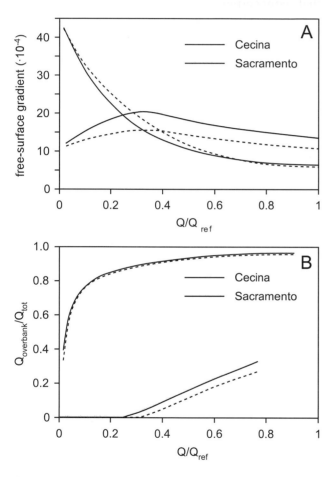

Fig. 17. (A) Free surface gradient (difference in free-surface elevation divided by the linear distance) between the upstream (P3) and downstream (P4) ends of the chute to be formed. The locations of the chute ends are those reported in Fig. 9 for the Sacramento River and in Fig. 15 for the Cecina River. (B) Overbank flow through the inside of the meander loop as a fraction of the total discharge. The total discharge, Q, is scaled with a reference value, Q_{ref}, equal to 6000 and 600 m³s⁻¹ for the Sacramento and the Cecina Rivers, respectively. The dashed lines refer to numerical simulations in which the convective accelerations have been neglected.

of an embayment precedes the chute incision. Contrarily, in the Cecina River, the increasing of discharge leads to a marked, monotonic decrease of the free-surface gradient through the floodplain. The incoming channel flow is not directed toward the inside of the bend and backwater effects reduce the free-surface gradient also for moderate discharges. The solid lines in Fig. 17B show that the ratio of overbank flow to the total discharge is sensibly larger in the Cecina River than in the Sacramento River. This is due to the shallowness of the channel and to the relatively small height of the point-bar in the Cecina River, such that overbank flow starts for small discharges (~10 m³s⁻¹).

Interestingly, convective accelerations are shown to play a different role in the two rivers. Neglecting convective accelerations (see dashed lines in Fig. 17) generally leads to a slight increase in the free-surface gradient across the chute to be formed in the Cecina River, while a significant reduction occurs in the Sacramento River (Fig. 17A). As a consequence, when convective accelerations are neglected, in the Sacramento River the overbank flow occurs for significantly larger values of the total discharge (~1,900 m³s⁻¹ instead of ~1,500 m³s⁻¹) and, in general, overbank flow discharge is lower. On the contrary, the discharge partitioning in the Cecina River is negligibly affected by convective accelerations.

The two case studies discussed above can be considered as end members and, hence, they show a clear prevalence of one of the two mechanisms that in general tend to coexist together. Possibly, the present analysis can be used to infer the possible tendency of river reach to develop a particular type of chute cutoff on the basis of hydrodynamic indicators such as the monotonic (Fig. 9B) or non-monotonic (Fig. 16B) dependence of the velocity on the point bar to be cut, as well as of the free-surface gradient (Fig. 17A).

It is worth noting that, in rivers where the regular and mild shape of meanders prevent the in-channel streamflow to impinge directly against the outer bank at some location downstream of the bend apex, the incision of chutes can be forced only by the gradient advantage associated with an overland flow (macro-group 2) and not by the inertia of the upstream in-channel flow. Accordingly, it can be conjectured that chutes assume pathways different from that of point/scroll bar swales, sloughs or abandoned palaeo-meanders only for quite sharp meander bends in which a localised narrowing of the cross section enhances a concentration of the shear stresses and, contemporaneously, directs markedly the flow towards the inside of the downstream meander loop.

According to the above considerations, chute cutoffs of both macro-groups can be detected in the reach of the Allier River depicted in Fig. 1. Magenta circles denote chute channels mainly forced by the inertia and direction of the in-channel flow, while blue circles locates the chute cutoffs possibly controlled by gradient advantage. On the other hand, in the case of the Strickland River reach depicted in Fig. 2, chutes are generally almost perpendicular to the in-channel flow direction (i.e. tangent to the point bar included in the meander loop) and tend to follow swale patterns across the scroll bars, independently of the gradient advantage associated with other short paths through the floodplain. Indeed, in this reach, many locations (red rectangles in Fig. 2) are probably characterised by values of the gradient advantage higher than those where chute cutoffs actually occurred, but the possible presence of relatively high natural levees (Aalto *et al.*, 2008) and the absence of any preferential floodplain path prevent the formation of chute cutoffs there. This behaviour is possibly linked also to the flow regime, which is far more uniform in the Strickland River than those of temperate rivers of similar size (Aalto *et al.*, 2008).

In conclusion, the floodplain topography plays a major role in controlling not only the patterns of meander migration (e.g. Bogoni *et al.*, 2017) but also the occurrence of chute cutoffs. In the case studies analysed here, it was observed that overbank flow can trigger chutes in the correspondence of topographic lows due to the typical ridge and swale sequence characterising scroll bars, possibly reinforced by the presence of a vegetation cover, or being the remainder of palaeo-meanders. Scroll patterned point bars are typical of medium-energy non-cohesive floodplains (Nanson & Croke, 1992) and consist of heterolithic stratification of interbedded sand and mud deposits with lateral accretion surfaces. On the other hand, palaeo-meander cutoffs can form up to the 75% of channel reaches within a floodplain (David *et al.*, 2017); they are lower in elevation and probably different in terms of stratigraphy, with respect to the surrounding floodplain. Hence, they are preferential pathways for overbank flow and prone to erosion given the presence of less consolidated sediment deposits with respect to the adjacent floodplain.

In addition, the hydrodynamic flow field within the main channel can play a crucial role in the occurrence of chute cutoffs. The application of the hydro-morphodynamic model developed by Frascati & Lanzoni (2013) to different rivers (Lopez Dubon *et al.*, 2016) indicates that a significant concentration of flow velocity and bottom shear stresses is always found in the main channel towards the bank where the upstream section of a chute is going to be incised. The analysis of the chute initiation in the Sacramento River confirms that convective acceleration of the in-channel flow crucially affects also the magnitude and the direction of velocities over the floodplain near to the location where the chute incision is going to occur. It can then be argued that the inertia and direction of the in-channel flow, possibly enhanced by a localised narrowing of the river and not only the water surface gradient of the overbank flow (Constantine *et al.*, 2010), are important factors controlling chute formation.

Clearly, a multiplicity of other factors concur to determine chute cutoff inception and progression and complicate the overall picture discussed above. The fluctuations in intensity and type of sediment transport within the river are responsible for adjustments of the bed topography and of the channel width which, in turn, strongly affect the in-channel flow. The height of the banks controls occurrence and intensity of overbank flow. Vegetation encroachments and sedimentary heterogeneities throughout the floodplain entail spatial variability in bank resistance to erosion (Konsoer *et al.*, 2016) and can drive preferentially the overbank flow, eventually determining where a chute forms. This complexity and multiplicity of controlling factors is exemplified by the Sacramento River itself, for which other reaches can be found where no chute incisions are observed despite the in-channel flow downstream of bend apexes shows values of the near bank velocity and of the shear stresses similar or even greater than those considered in the present study.

CONCLUSIONS

The numerical analyses carried out in the present study, which concentrated on the hydrodynamic forcing responsible for the chutes' initiation, allowed to distinguish between hydraulically driven and topographically driven chute cutoffs. In the first case, the incision on the floodplain is determined by the flow field within the main channel, upstream of the chute to be formed; in the second case, the incision is determined by the gradient advantage and the flow field over the floodplain. In hydraulically dominated cutoffs, the main stream is usually directed toward the outer bank due to pronounced channel curvature, thus significantly increasing the water-level against the outer bank and promoting the formation of embayments as precursors of chute incisions. This mechanism is fostered by hydrodynamic conditions (e.g. backwater effects) and by sediment dynamics (e.g. deposition and formation of plug-bars or central-bars within the main channel). In topographically driven chute cutoffs, a major role is played by the gradient advantage of flow paths through the floodplain with respect to the main channel and to the presence of ridge and swale topography or palaeo-channels.

In any case, the presence of preferential pathways on the floodplain has a great influence on the formation of chutes. This implies that, to accurately model chute cutoff inception, the actual configuration of the floodplain has to be accounted for and the flow field within the main channel and over the adjacent areas have to be solved jointly. Clearly, the presence of several controlling factors, interplaying and competing together, greatly increases the complexity of the problem. Heterogeneity of the floodplain associated with sediment sorting and preserved sedimentary landforms, as well as vegetation encroachment, may act either to enhance or to inhibit the occurrence of chute cutoffs.

Reliable predictions of chute cutoffs can be achieved by accurately modelling both the hydrodynamic flow field and the bed evolution within the main channel and over the floodplain. The flow field within the main channel, in fact, can significantly affect the overland flow on the floodplain. In addition, the possible formation of plug-bars and central-bars within the main channel can play a major role in bifurcation dynamics and, thus, in controlling the incision of chutes. Floodplain heterogeneity can also have an important impact on cutoff occurrence. The knowledge gathered from this modelling approach may be used not only to assess local behaviours but also to develop a general, physics-based statistical framework to be inserted in long-term meander evolution models.

ACKNOWLEDGMENTS

The authors are grateful to the volume editor, A.J.H. Reesink, to K. Konsoer and the other two anonymous reviewers for valuable and insightful suggestions that allowed them to significantly improve the paper.

APPENDIX A   BASIC EQUATIONS

The two-dimensional equations that govern the conservation of water momentum in free surface flows are the shallow water equations, also known as de Saint Venant equations. These equations, in order to be solved, must be coupled to the mass conservation of water (continuity equation) and, in the presence of a movable sediment bed, to the sediment balance equation (Exner equation).

The shallow water equations describe the depth-averaged flow field in free-surface water bodies where the longitudinal u_x and transverse u_y components of the mean (i.e., turbulence-averaged) velocity components are one order of magnitude higher than the vertical component u_z. This occurs when the spatial horizontal scales characterising the longitudinal and transverse flow variations are much larger than the flow depth. These assumptions are typical of long waves on shallow waters occurring, for example, during flood propagation in rivers and tidal propagation in estuaries, deltas, lagoons and near shore waters.

Under the shallow water assumption, the longitudinal and transverse momentum equations read (Vreugdenhil, 1994):

$$\frac{\partial(DU_x)}{\partial t} + \frac{\partial(DU_x^2)}{\partial x} + \frac{\partial(DU_xU_y)}{\partial y} + gD\frac{\partial H}{\partial x} + \frac{\tau_{bx}}{\rho}$$
$$= \frac{1}{\rho}\left[\frac{\partial}{\partial x}D\left(\bar{T}_{xx}^t + T_{xx}^d\right) + \frac{\partial}{\partial y}D\left(\bar{T}_{yx}^t + T_{yx}^d\right)\right]$$

$$(A1)$$

$$\frac{\partial(DU_y)}{\partial t} + \frac{\partial(DU_xU_y)}{\partial x} + \frac{\partial(DU_y^2)}{\partial y} + gD\frac{\partial H}{\partial y} + \frac{\tau_{by}}{\rho}$$
$$= \frac{1}{\rho}\left[\frac{\partial}{\partial x}D\left(\bar{T}_{xy}^t + T_{xy}^d\right) + \frac{\partial}{\partial y}D\left(\bar{T}_{yy}^t + T_{yy}^d\right)\right]$$

$$(A2)$$

where D is the local flow depth, U_x and U_y are the components of the depth averaged mean velocity, H is the water surface elevation with respect to a horizontal plane, τ_{bx} and τ_{by} are the components of the bed shear stress, g denotes gravity, ρ is the water density and the terms on the right hand side represent the residual dispersive stresses arising from the averaging of nonlinear acceleration terms.

On the other hand, the two-dimensional continuity equation takes the form

$$\frac{\partial D}{\partial t} + \frac{\partial (U_x D)}{\partial x} + \frac{\partial (U_y D)}{\partial y} = 0 \qquad (A3)$$

It is the immediate to observe that the first three terms on the left hand side of equations (A1) and (A2), with the help of the continuity equation (A3), can be written as:

$$D\frac{dU_x}{dt} = D\left(\frac{\partial U_x}{\partial t} + U_x\frac{\partial U_x}{\partial x} + U_y\frac{\partial U_x}{\partial y}\right)$$
$$D\frac{dU_y}{dt} = D\left(\frac{\partial U_x}{\partial t} + U_x\frac{\partial U_y}{\partial x} + U_y\frac{\partial U_y}{\partial y}\right) \qquad (A4)$$

This alternative form points out that the acceleration components dU_x/dt and dU_y/dt are given by the sum of local (or temporal) accelerations (e.g, $\partial U_x/\partial t$) due to the variation in time on the velocity at a given point and of the so called convective accelerations (e.g., $U_x\,\partial U_x/\partial x$) due to the transport of momentum to a different position (Batchelor, 1967).

Finally, the two-dimensional version of the Exner sediment balance equation reads (Seminara, 1998):

$$(1-p)\frac{\partial \eta}{\partial t} + \frac{\partial (DC)}{\partial t} + \frac{\partial q_{sx}}{\partial x} + \frac{\partial q_{sy}}{\partial y} = 0 \qquad (A5)$$

where η is the bed elevation with respect to an horizontal reference plane, p is the sediment bed porosity, C is the depth-averaged sediment concentration and q_{sx}, q_{sy} are the components of the vector sediment flux per unit width, including bedload and suspended load.

REFERENCES

Aalto, R., Lauer, J.W. and Dietrich, E. (2008) Spatial and temporal dynamics of sediment accumulation and exchange along Strickland River floodplains (Papua New Guinea) over decadal-to-centennial timescales. *J.Geophys.Res.*,**113**,F01S04,doi:10.1029/2006JF000627.

Bartholdy, J. and Billi, P. (2002) Morphodynamics of a pseudomeandering gravel bar reach. *Geomorphology*, **42**, 293–310, doi:10.1016/S0169-555X(01)00092-7.

Batchelor, G.K. (1967) *An Introduction to Fluid Dynamics*, Cambridge University Press, Cambridge, UK.

Bertoldi, W., Siviglia, A., Tettamanti, S., Toffolon, M., Vetsch, D. and Francalanci, S. (2014) Modeling vegetation controls on fluvial morphological trajectories. *Geophys. Res. Lett.*, **41**, 7167–7175, doi:10.1002/2014GL061666.

Bertoldi, W., Zanoni, L., Miori, S., Repetto, R. and Tubino, M. (2009) Interaction between migrating bars and bifurcations in gravel bed rivers. *Water Resour. Res.*, **45**, 1–12, doi:10.1029/2008WR007086.

Billi, P. and Paris, E. (1992) Bed sediment characterisation in river engineering problems. In: *Erosion and Sediment Transport Monitoring Programmes in River Basins* (Eds J. Bogen, D.E. Walling and T.J. Day), IAHS Press, Wallingford, UK, p. 11–20.

Bogoni, M., Putti, M. and Lanzoni, S. (2017) Modelling meander morphodynamics over self-formed heterogeneous floodplains. *Water Resour. Res.*, **53**, 5137–5157, doi:10.1002/2017WR020726.

Braudrick, C.A., Dietrich, W.E., Leverich, G.T. and Sklar, L.S. (2009) Experimental evidence for the conditions necessary to sustain meandering in coarse-bedded rivers. *Proc. Natl Acad. Sci. USA*, **106**, 16936–16941, doi:10.1073/pnas.0909417106.

Brice, J.C. (1975) *Airphoto Interpretation of the Form and Behavior of Alluvial Rivers*. Final report 1 July 197 – 31 December 1974, Washington University, St Louis.

Bridge, J.S. (2003) *Rivers and Floodplains: Forms, Processes and Sedimentary Record*. Blackwell, Malden, MA.

Bridge, J.S., Smith, N.D., Trent, F., Gabel, S.L. and Bernstein, P. (1986) Sedimentology and morphology of a low-sinuosity river: Calamus River, Nebraska Sand Hills. *Sedimentology*, **33**, 851–870, doi:10.1111/j.1365-3091.1986.tb00987.x.

Brierley, G.J. (1991) Floodplain sedimentology of the Squamish River, B.C.: relevance of element analysis. *Sedimentology*, **38**, 735–750.

Camporeale, C., Perona, P., Porporato, A. and Ridolfi, L. (2005) On the long-term behavior of meandering rivers. *Water Resour. Res.*, **41**, W04109, doi:10.1029/2005WR004109.

Camporeale, C., Perucca, E. and Ridolfi, L. (2008) Significance of cutoff in meandering river dynamics. *J. Geophys. Res. Earth Surf.*, **113**, 1–11, doi:10.1029/2006JF000694.

Church, M. (2015) Channel Stability: Morphodynamics and the Morphology of Rivers. In: *Rivers-Physical, Fluvial and Environmental Processess* (Eds P. Rowinski and A. Radecki-Pawlik), Springer International Publishing, Switzerland, p. 255–277.

Constantine, J.A., McLean, S.R. and Dunne, T. (2010) A mechanism of chute cutoff along large meandering rivers with uniform floodplain topography. *Geol. Soc. Am. Bull.*, **122**, 855–869, doi:10.1130/B26560.1.

D'Alpaos, L. and Defina, A. (2007) Mathematical modeling of tidal hydrodynamics in shallow lagoons: A review of open issues and applications to the Venice lagoon. *Comput. Geosci.*, **33**, 476–496, doi:10.1016/j.cageo.2006.07.009.

Darby, S.E., Trieu, H.Q., Carling, P.A., Sarkkula, J., Koponen, J., Kummu, M., Conlan, I. and **Leyland, J.** (2010) A physically based model to predict hydraulic erosion of fine-grained riverbanks: The role of form roughness in limiting erosion. *J. Geophys. Res.*, **115**, F04003. doi:10.1029/2010JF001708.

David, S.R., Edmonds, D.A. and **Letsinger, S.L.** (2017) Controls on the occurrence and prevalence of flood-plain channels in meandering rivers. *Earth Surf. Proc. Land.*, **42**, 460–472, doi:10.1002/esp.4002.

Day, G., Dietrich, W.E., Rowland, J.C. and **Marshall, A.** (2008) The depositional web on the floodplain of the Fly River, Papua New Guinea. *J. Geophys. Res.*, **113**, F01S02, doi:10.1029/2006JF000622.

Defina, A. (2003) Numerical experiments on bar growth. *Water Resour. Res.*, **39**, 1–12, doi:10.1029/2002WR001455.

Defina, A. (2000) Two-dimensional shallow flow equations for partially dry areas. *Water Resour. Res.*, **36**, 3251, doi:10.1029/2000WR900167.

Dieras, P.L., Constantine, J.A., Hales, T.C., Piegay, H. and **Riquier, J.** (2013) The role of oxbow lakes in the off-channel storage of bed material along the Ain River, France. *Geomorphology*, **188**, 110–119, doi:10.1016/j.geomorph.2012.12.024.

Dunne, T. and **Aalto, R.E.** (2013) Large River Floodplains. In: *Treatise on Geomorphology* (Ed. J. Shroder), Academic Press, 645–678. First Ed.

Edwards, B.L., Keim, R.F., Johnson, E.L., Hupp, C.R., Marre, S. and **King, S.L.** (2016) Geomorphic adjustment to hydrologic modifications along a meandering river: Implications for surface flooding on a floodplain. *Geomorphology*, **269**, 149–159, doi:10.1016/j.geomorph.2016.06.037.

Eekhout, J.P.C. and **Hoitink, A.J.F.** (2015) Chute cutoff as a morphological response to stream reconstruction: The possible role of backwater. *Water Resour. Res.*, **51**, 3339–3352, doi:10.1002/2014WR016539.

Eke, E., Parker, G. and **Shimizu, Y.** (2014) Numerical modeling of erosional and depositional bank processes in migrating river bends with self-formed width: Morphodynamics of bar push and bank pull. *J. Geophys. Res. Earth Surf.*, **119**, 1455–1483, doi:10.1002/2013JF003020.

El Gammal, E.S.A. (2016) On The Chutes and Chute Cutoff along the River Nile Within Egypt. *Arab. J. Sci. Eng.*, **1–14**, doi:10.1007/s13369-016-2074-x.

Erskine, W., McFadden, C. and **Bishop, P.** (1992) Alluvial cutoffs as indicators of former channel conditions. *Earth Surf. Proc. Land.*, **17**, 22–37, doi:10.1002/esp.3290170103.

Fisk, H.N. (1947) *Fine-grained alluvial deposits and their effects on Mississippi River activity.* Vicksburg, Miss. Waterways Experiment Station, 1947.

Frascati, A. and **Lanzoni, S.** (2013) A mathematical model for meandering rivers with varying width. *J. Geophys. Res. Earth Surf.*, **118**, 1641–1657, doi:10.1002/jgrf.20084.

Frascati, A. and **Lanzoni, S.** (2010) Long-term river meandering as a part of chaotic dynamics? A contribution from mathematical modelling. *Earth Surf. Proc. Land.*, **35**, 791–802, doi:10.1002/esp.1974.

Frascati, A. and **Lanzoni, S.** (2009) Morphodynamic regime and long-term evolution of meandering rivers. *J. Geophys. Res. Earth Surf.*, **114**, 2156–2202, doi:10.1029/2008JF001101.

Fuller, I.C., Large, A.R.G. and **Milan, D.J.** (2003) Quantifying channel development and sediment transfer following chute cutoff in a wandering gravel-bed river. *Geomorphology*, **54**, 307–323, doi:10.1016/S0169-555X(02)00374-4.

Gautier, E., Brunstein, D., Vauchel, P., Roulet, M., Fuertes, O., Guyot, J.L., Darozzes, J. and **Bourrel, L.** (2007) Temporal relations between meander deformation, water discharge and sediment fluxes in the floodplain of the Rio Beni (Bolivian Amazonia). *Earth Surf. Proc. Land.*, **32**, 230–248, doi:10.1002/esp.1394.

Gay, G.R., Gay, H.H., Gay, W.H., Martinson, H.A., Meade, R.H. and **Moody, J.A.** (1998) Evolution of cutoffs across meander neck in Powder River, Montana, USA. *Earth Surf. Proc. Land.*, **23**, 651–662.

Ghinassi, M. (2011) Chute channels in the Holocene high-sinuosity river deposits of the Firenze plain, Tuscany, Italy. *Sedimentology*, **58**, 618–642, doi:10.1111/j.1365-3091.2010.01176.x.

Ghinassi, M., Ielpi, A., Aldinucci, M. and **Fustic, M.** (2016) Downstream-migrating fluvial point bars in the rock record. *Sediment. Geol.*, **334**, 66–96, doi:10.1016/j.sedgeo.2016.01.005.

Grenfell, M., Aalto, R. and **Nicholas, A.** (2012) Chute channel dynamics in large, sand-bed meandering rivers. *Earth Surf. Proc. Land.*, **37**, 315–331, doi:10.1002/esp.2257.

Grenfell, M., Nicholas, A.P.P. and **Aalto, R.** (2014) Mediative adjustment of river dynamics: The role of chute channels in tropical sand-bed meandering rivers. *Sediment. Geol.*, **301**, 93–106, doi:10.1016/j.sedgeo.2013.06.007.

Hackney, C., Best, J., Leyland, J., Darby, S.E., Parsons, D., Aalto, R. and **Nicholas, A.** (2015) Modulation of outer bank erosion by slump blocks: Disentangling the protective and destructive role of failed material on the three-dimensional flow structure. *Geophys. Res. Lett.*, **42**, 10663–10670, doi:10.1002/2015GL066481.

Harrison, L.R., Dunne, T. and **Fisher, G.B.** (2015) Hydraulic and geomorphic processes in an overbank flood along a meandering, gravel-bed river: Implications for chute formation. *Earth Surf. Proc. Land.*, **40**, 1239–1253, doi:10.1002/esp.3717.

Hickin, E.J. and **Nanson, G.C.** (1975) The character of channel migration on the Beatton River, northeast British Columbia, Canada. *Geol. Soc. Am. Bull.*, **86**, 487–494, doi:10.1130/0016-7606(1975)86<487:TCOCMO>2.0.CO;2.

Hooke, J.M. (1995) River channel adjustment to meander cutoffs on the River Bollin and River Dane, north-west England. *Geomorphology*, **14**, 235–253, doi:10.1016/0169-555X(95)00110-Q.

Howard, A. and **Knutson, T.R.** (1984) Sufficient conditions for river meandering: A simulation approach. *Water Resour. Res.*, **20**, 1659–1667.

Howard, A.D. (2009) How to make a meandering river. *Proc. Natl. Acad. Sci. USA*, **106**, 17245–17246, doi:10.1073/pnas.0910005106.

Howard, A.D. (1996) Modeling channel evolution and floodplain morphology. In: *Floodplain Processes* (Eds M.G. Anderson, D.E. Walling and P.D. Bates), John Wiley and Sons Ltd., Chichester, UK, p. 15–65.

Ielpi, A. and **Ghinassi, M.** (2014) Planform architecture, stratigraphic signature and morphodynamics of an exhumed Jurassic meander plain (Scalby Formation, Yorkshire, UK). *Sedimentology*, **61**, 1923–1960, doi:10.1111/sed.12122.

Iwasaki, T., Shimizu, Y. and Kimura, I. (2016) Numerical simulation of bar and bank erosion in a vegetated floodplain: A case study in the Otofuke River. *Adv. Water Resour.*, **93**, 118–134, doi:10.1016/j.advwatres.2015.02.001.

Jager, H.R.A. (2003) Modelling planform changes of braided rivers. PhD Thesis.

Kasvi, E., Vaaja, M., Alho, P., Hyyppä, H., Hyyppä, J., Kaartinen, H. and Kukko, A. (2013) Morphological changes on meander point bars associated with flow structure at different discharges. *Earth Surf. Proc. Land.*, **38**, 577–590, doi:10.1002/esp.3303.

Keller, E.A. and Swanson, F.J. (1979) Effects of large organic material on channel form and fluvial processes. *Earth Surf. Proc. Land.*, **4**, 361–380, doi:10.1002/esp.3290040406.

Khatua, K.K. and Patra, K.C. (2009) Flow distribution in meandering compound channel. *ISH J. Hydraul. Eng.*, **15**, 11–26, doi:10.1080/09715010.2009.10514956.

Kleinhans, M.G., Cohen, K.M., Hoekstra, J. and Ijmker, J.M. (2011) Evolution of a bifurcation in a meandering river with adjustable channel widths, Rhine delta apex, The Netherlands. *Earth Surf. Proc. Land.*, **36**, 2011–2027, doi:10.1002/esp.2222.

Kleinhans, M.G., Jagers, H.R.A., Mosselman, E. and Sloff, C.J. (2008) Bifurcation dynamics and avulsion duration in meandering rivers by one-dimensional and three-dimensional models. *Water Resour. Res.*, **44**, 1–31, doi:10.1029/2007WR005912.

Kleinhans, M.G. and van den Berg, J.H. (2011) River channel and bar patterns explained and predicted by an empirical and a physics-based method. *Earth Surf. Proc. Land.*, **36**, 721–738, doi:10.1002/esp.2090.

Konsoer, K.M., Rhoads, B.L., Langendoen, E.J., Best, J.L., Ursic, M.E., Abad, J.D. and Garcia, M.H. (2016) Spatial variability in bank resistance to erosion on a large meandering, mixed bedrock-alluvial river. *Geomorphology*, **252**, 80–97, doi:10.1016/j.geomorph.2015.08.002.

Lagasse, P.F., Spitz, W.J., Zevengergen, L.W. and Zachmann, D.W. (2004) *Handbook for Predicting Stream Meander Migration*, NCHRP REPORT 533, Washington D.C.

Lanzoni, S., Luchi, R. and Bolla Pittaluga, M. (2015) Modeling the morphodynamic equilibrium of an intermediate reach of the Po River (Italy), *Adv. Water Resour.*, **81**, 95–102, doi:10.1016/j.advwatres.2014.11.004.

Lanzoni, S., Siviglia, A., Frascati, A. and Seminara, G. (2006) Long waves in erodible channels and morphodynamic influence. *Water Resour. Res.*, **42**, W06D17, doi:10.1029/2006WR004916.

Lewis, G.W. and Lewin, J. (1983) Alluvial Cutoffs in Wales and the Borderlands. In: *Modern and Ancient Fluvial Systems* (Eds J.D. Collinson and J. Lewin), Blackwell Publishing Ltd., Oxford, UK., DOI: 10.1002/9781444303773.ch11.

Li, Z., Yu, G.A., Brierley, G.J., Wang, Z. and Jia, Y. (2017) Migration and cutoff of meanders in the hyperarid environment of the middle Tarim River, northwestern China. *Geomorphology*, **276**, 116–124, doi:10.1016/j.geomorph.2016.10.018.

Liu, C., Wright, N., Liu, X. and Yang, K. (2014) An analytical model for lateral depth-averaged velocity distributions along a meander in curved compound channels. *Adv. Water Resour.*, **74**, 26–43, doi:10.1016/j.advwatres.2014.08.003.

Lopez Dubon, S.A., Viero, D.P. and Lanzoni, S. (2016) Chute cutoffs initiation and the flow field inside the main channel bed. In: *Atti del XXXV Convegno Nazionale Di Idraulica E Costruzioni Idrauliche, Bologna*, 757–759.

Luchi, R., Hooke, J.M., Zolezzi, G. and Bertoldi, W. (2010) Width variations and mid-channel bar inception in meanders: River Bollin (UK). *Geomorphology*, **119**, 1–8, doi:10.1016/j.geomorph.2010.01.010.

Luchi, R., Zolezzi, G. and Tubino, M. (2011) Bend theory of river meanders with spatial width variations, *J. Fluid Mech.*, **681**, 311–339, doi:10.1017/jfm.2011.200.

Luppi, L., Rinaldi, M., Teruggi, L.B., Darby, S.E. and Nardi, L. (2009) Monitoring and numerical modelling of riverbank erosion processes: a case study along the Cecina River (central Italy). *Earth Surf. Proc. Land.*, **34**, 530–546, doi:10.1002/esp.1754.

McGowen, J.H. and Garner, L.E. (1970) Physiographic features and stratification types of coarse-grained point bars – modern and ancient examples. *Sedimentology*, **14**, 77–111, doi:10.1111/j.1365-3091.1970.tb00184.x.

Micheli, E.R. and Larsen, E.W. (2011) River channel cutoff dynamics, Sacramento River, California, USA. *River Res. Appl.*, **27**, 328–344, doi:10.1002/rra.1360.

Miori, S., Repetto, R. and Tubino, M. (2006) A one-dimensional model of bifurcations in gravel bed channels with erodible banks. *Water Resour. Res.*, **42**, W11413, doi:10.1029/2006WR004863.

Morais, E.S., Rocha, P.C. and Hooke, J. (2016) Spatiotemporal variations in channel changes caused by cumulative factors in a meandering river: The lower Peixe River, Brazil. *Geomorphology*, **273**, 348–360, doi:10.1016/j.geomorph.2016.07.026.

Nanson, G.C. and Croke, J.C. (1992) A genetic classification of floodplains. *Geomorphology*, **4**, 459–486, doi:10.1016/0169-555X(92)90039-Q.

Nardi, L., Campo, L. and Rinaldi, M. (2013) Quantification of riverbank erosion and application in risk analysis. *Nat. Hazards*, **69**, 869–887, doi:10.1007/s11069-013-0741-8.

Nardi, L. and Rinaldi, M. (2010) Modelling riverbank retreat by combining reach-scale hydraulic models with bank-scale erosion and stability analyses. *River Flow 2010*, 1286–1291.

Nicholas, A. (2013) Morphodynamic diversity of the world's largest rivers. *Geology*, **41**, 475–478, doi:10.1130/G34016.1.

Oorschot, M. van, Kleinhans, M., Geerling, G. and Middelkoop, H. (2016) Distinct patterns of interaction between vegetation and morphodynamics. *Earth Surf. Proc. Land.*, **41**, 791–808, doi:10.1002/esp.3864.

Parker, G. and Andres, D. (1976) Detrimental effects of river channelization. In: Proc. Conf. Rivers, 1248–1266.

Parker, G., Shimizu, Y., Wilkerson, G. V., Eke, E.C., Abad, J.D., Lauer, J.W., Paola, C., Dietrich, W.E. and Voller, V.R. (2011) A new framework for modeling the migration of meandering rivers. *Earth Surf. Proc. Land.*, **36**, 70–86, doi:10.1002/esp.2113.

Patra, K.C. and Kar, S.K. (2000) Flow interaction of meandering river with floodplains. *J. Hydraul. Eng.*, **126**, 593–604, doi:0733-9429/00/0008-0593–0604/$8.00.

Patra, K.C., Kar, S.K. and Bhattacharya, A.K. (2004) Flow and Velocity Distribution in Meandering Compound Channels. *J. Hydraul. Eng.*, **130**, 398–411, doi:10.1061/(ASCE)0733-9429(2004)130:5(398).

Peakall, J., Ashworth, P. and Best, J. (2007) Meander-Bend Evolution, Alluvial Architecture and the Role of Cohesion in Sinuous River Channels: A Flume Study. *J. Sed. Res.*, **77**, 197–212, doi:10.2110/jsr.2007.017.

Piégay, H., Hupp, C.R., Citterio, A., Dufour, S., Moulin, B. and Walling, D.E. (2008) Spatial and temporal variability in sedimentation rates associated with cutoff channel infill deposits: Ain River, France. *Water Resour. Res.*, **44**, W05420, doi:10.1029/2006WR005260.

Rameshwaran, P. and Shiono, K. (2003) Computer modelling of two-stage meandering channel flows. *Proc. Inst. Civ. Eng. Water Marit. Eng.*, **156**, 325–339, doi:10.1680/maen.156.4.325.37931.

Redolfi, M., Zolezzi, G. and Tubino, M. (2016) Free instability of channel bifurcations and morphodynamic influence. *J. Fluid Mech.*, **799**, 476–504, doi:10.1017/jfm.2016.389.

Riesterer, J., Wenka, T. and Brudy-Zippelius, T. (2016) Bed load transport modeling of a secondary flow influenced curved channel with 2D and 3D numerical models. *J. Appl. Water Eng. Res.*, **4**, 54–66, doi:10.1080/23249676.2016.1163649.

Rinaldi, M., Mengoni, B., Luppi, L., Darby, S.E. and Mosselman, E. (2008) Numerical simulation of hydrodynamics and bank erosion in a river bend. *Water Resour. Res.*, **44**, 1–17, doi:10.1029/2008WR007008.

Rinaldi, M. and Nardi, L. (2013) Modeling Interactions between Riverbank Hydrology and Mass Failures. *J. Hydraul. Eng.*, **18**, 1231–1240, doi:10.1061/(ASCE)HE.1943-5584.0000716.

Schuurman, F., Shimizu, Y., Iwasaki, T. and Kleinhans, M.G. (2016) Dynamic meandering in response to upstream perturbations and floodplain formation. *Geomorphology*, **253**, 94–109, doi:10.1016/j.geomorph.2015.05.039.

Schwendel, A.C., Nicholas, A.P., Aalto, R.E., Sambrook Smith, G. H. and Buckley, S. (2015) Interaction between meander dynamics and floodplain heterogeneity in a large tropical sand-bed river: The Rio Beni, Bolivian Amazon. *Earth Surf. Proc. Land.*, **40**, 2026–2040. doi:10.1002/esp.3777.

Schwenk, J. and Foufoula-Georgiou, E. (2016) Meander cutoffs nonlocally accelerate upstream and downstream migration and channel widening. *Geophys. Res. Lett.*, **43**, 12437–12445, doi:10.1002/2016GL071670.

Seminara, G. (2006) Meanders. *J. Fluid Mech.*, **554**, 271–297, doi:10.1017/S0022112006008925.

Seminara, G. (1998) Stability and morphodynamics, *Meccanica*, **33**, 59–99, doi:10.1023/A:1004225516566.

Shan, Y., Liu, C. and Luo, M. (2015) Simple analytical model for depth-averaged velocity in meandering compound channels. *Appl. Math. Mech.*, **36**, 707–718, doi:10.1007/s10483-015-1943-6.

Shiono, K., Chan, T.L., Spooner, J., Rameshwaran, P. and Chandler, J.H. (2009a) The effect of floodplain roughness on flow structures, bedforms and sediment transport rates in meandering channels with overbank flows: Part I. *J. Hydraul. Res.*, **47**, 5–19, doi:10.3826/jhr.2009.2944-I.

Shiono, K., Chan, T.L., Spooner, J., Rameshwaran, P. and Chandler, J.H. (2009b) The effect of floodplain roughness on flow structures, bedforms and sediment transport rates in meandering channels with overbank flows: Part II. *J. Hydraul. Res.*, **47**, 20–28, doi:10.3826/jhr.2009.2944-II.

Shiono, K. and Muto, Y. (1998) Complex flow mechanisms in compound meandering channels with overbank flow. *J. Fluid Mech.*, **376**, 221–261, doi:10.1017/S0022112098002869.

Shiono, K., Muto, Y., Knight, D.W. and Hyde, A.F.L. (1999) Energy losses due to secondary flow and turbulence in meandering channels with overbank flows. *J. Hydraul. Res.*, **37**, 641–664, doi:10.1080/00221689909498521.

Shiono, K. and Shukla, D.R. (2008) CFD modelling of meandering channel during floods. *Proc. ICE - Water Manag.*, **161**, 1–12, doi:10.1680/wama.2008.161.1.1.

Simon, A. and Robbins, C.H. (1987) Man-induced gradient adjustment of the South Fork Forked Deer River, west Tennessee. *Environ. Geol. Water Sci.*, **9**, 109–118, doi:10.1007/BF02449942.

Slingerland, R. and Smith, N.D. (2004) River avulsions and their deposits. *Annu. Rev. Earth Planet. Sci.*, **32**, 257–285, doi:10.1146/annurev.earth.32.101802.120201.

Słowik, M. (2016) The influence of meander bend evolution on the formation of multiple cutoffs: Findings inferred from floodplain architecture and bend geometry. *Earth Surf. Proc. Land.*, **41**, 626–641, doi:10.1002/esp.3851.

Stansby, P.K. (2003) A mixing-length model for shallow turbulent wakes. *J. Fluid Mech.*, **495**, 369–384, doi:10.1017/S0022112003006384.

Swanson, K.M., Watson, E., Aalto, R., Lauer, J.W., Bera, M.T., Marshall, A., Taylor, M.P., Apte, S.C. and Dietrich, W.E. (2008) Sediment load and floodplain deposition rates: Comparison of the Fly and Strickland rivers, Papua New Guinea. *J. Geophys. Res.*, **113**, F01S03, doi:10.1029/2006JF000623.

Tal, M. and Paola, C. (2010) Effects of vegetation on channel morphodynamics: Results and insights from laboratory experiments. *Earth Surf. Proc. Land.*, **35**, 1014–1028, doi:10.1002/esp.1908.

Toonen, W.H.J., Kleinhans, M.G. and Cohen, K.M. (2012) Sedimentary architecture of abandoned channel fills. *Earth Surf. Proc. Land.*, **37**, 459–472, doi:10.1002/esp.3189.

Uittenbogaard, R. and van Vossen, B. (2004) Subgrid-scale model for quasi-2D turbulence in shallow water. In: *Shallow Flows*, Taylor & Francis, p. 575–582.

van de Lageweg, W.I., van Dijk, W.M., Baar, A.W., Rutten, J. and Kleinhans, M.G. (2014) Bank pull or bar push: What drives scroll-bar formation in meandering rivers? *Geology*, **42**, 319–322, doi:10.1130/G35192.1.

van de Lageweg, W.I., Schuurman, F., Cohen, K.M., van Dijk, W.M., Shimizu, Y. and Kleinhans, M.G. (2016) Preservation of meandering river channels in uniformly aggrading channel belts. *Sedimentology*, **63**, 586–608, doi:10.1111/sed.12229.

Van Dijk, W.M., Schuurman, F., Van de Lageweg, W.I. and Kleinhans, M.G. (2014) Bifurcation instability and chute cutoff development in meandering gravel-bed rivers. *Geomorphology*, **213**, 277–291, doi:10.1016/j.geomorph.2014.01.018.

Van Dijk, W.M., Van De Lageweg, W.I. and Kleinhans, M.G. (2012) Experimental meandering river with chute cutoffs. *J. Geophys. Res. Earth Surf.*, **117**, 1–18, doi:10.1029/2011JF002314.

Viero, D.P., D'Alpaos, A., Carniello, L. and Defina, A. (2013) Mathematical modeling of flooding due to river bank failure. *Adv. Water Resour.*, **59**, 82–94, doi:10.1016/j.advwatres.2013.05.011.

Viero, D.P. and Defina, A. (2016) Water age, exposure time and local flushing time in semi-enclosed, tidal basins with negligible freshwater inflow. *J. Mar. Syst.*, **156**, 16–29, doi:10.1016/j.jmarsys.2015.11.006.

Viero, D.P., Peruzzo, P., Carniello, L. and Defina, A. (2014) Integrated mathematical modeling of hydrological and hydrodynamic response to rainfall events in rural lowland catchments. *Water Resour. Res.*, **50**, 5941–5957, doi:10.1002/2013WR014293.

Viero, D.P. and Valipour, M. (2017) Modeling anisotropy in free-surface overland and shallow inundation flows. *Adv. Water Resour.*, **104**, 1–14, doi:10.1016/j.advwatres.2017.03.007.

Visconti, F., Stefanon, L., Camporeale, C., Susin, F., Ridolfi, L. and Lanzoni, S. (2012) Bed evolution measurement with flowing water in morphodynamics experiments. *Earth Surf. Proc. Land.*, **37**, 818–827, doi:10.1002/esp.3200.

Vreugdenhil, C. (1994) *Numerical Methods for Shallow-Water Flow*. Springer, Netherlands.

Wolman, M.G. and Miller, J.P. (1960) Magnitude and frequency of forces in geomorphic processes, *J. Geol.*, **68**, 54–74.

Wormleaton, P.R., Sellin, R.H.J. and Bryant, T. (2004) Conveyance in a two-stage meandering channel with a mobile bed. *J. Hydraul. Res.*, **42**, 493–506, doi:10.1080/00221686.2004.9641219.

Zen, S., Zolezzi, G., Toffolon, M. and Gurnell, A.M. (2016) Biomorphodynamic modelling of inner bank advance in migrating meander bends. *Adv. Water Resour.*, **93**, 166–181, doi:10.1016/j.advwatres.2015.11.017.

Zinger, J.A., Rhoads, B.L. and Best, J.L. (2011) Extreme sediment pulses generated by bend cutoffs along a large meandering river. *Nat. Geosci.*, **4**, 675–678, doi:10.1038/ngeo1260.

Zinger, J.A., Rhoads, B.L., Best, J.L. and Johnson, K.K. (2013) Flow structure and channel morphodynamics of meander bend chute cutoffs: A case study of the Wabash River, USA. *J. Geophys. Res. Earth Surf.*, **118**, 2468–2487, doi:10.1002/jgrf.20155.

Zolezzi, G., Luchi, R. and Tubino, M. (2012) Modeling morphodynamic processes in meandering rivers with spatial width variations. *Rev. Geophys.*, **50**, 1–24, doi:10.1029/2012RG000392.

Int. Assoc. Sedimentol. Spec. Publ (2018) **48**, 231–250.

Predicting heterogeneity in meandering fluvial and tidal-fluvial deposits: The point bar to counter point bar transition

PAUL R. DURKIN[†], STEPHEN M. HUBBARD[‡], DERALD G. SMITH[§+] and DALE A. LECKIE[‡]

[†] *Department of Geological Sciences, University of Manitoba, Winnipeg, Manitoba, Canada*
[‡] *Department of Geoscience, University of Calgary, Calgary, Alberta, Canada*
[§] *Department of Geography, University of Calgary, Calgary, Alberta, Canada*
[+] *Deceased*

ABSTRACT

Geomorphologic features and sediment distribution from meander-belts informs our understanding of ancient deposits, with specific application to predicting heterogeneity in petroliferous strata. The point bar to counter-point bar transition has been of recent interest and its common occurrence in modern fluvial environments suggests that they are often over-looked and under-recognised in ancient datasets. In this study, six point bar to counter-point bar transitions are examined along meander bends from rivers with varying channel scale, discharge and tidal influence. The data indicate that observed trends of grain-size fining from point bar to counter-point bar are consistent regardless of channel scale, discharge and tidal influence. High net sand to gross thickness (>0.7) point bars and low net-to-gross counter-point bars (<0.3) are documented. The average decrease in net-to-gross across the transition is 57%; the transition length scales to channel size and is approximately three times channel width. Tidally-influenced counter-point bar deposits are recognised despite the absence of concave scroll patterns in tidal flat areas. The lack of scroll bar topography contributes to the challenges in identifying counter-point bar deposits in tidally-influenced settings. Recognising and predicting heterogeneity related to the point bar to counter-point bar transition in ancient fluvial and tidal-fluvial deposits is considered, with specific implications for steam chamber growth during development of the Athabasca Oil Sands, Alberta, Canada.

Keywords: Tidal-fluvial, fluvial, point bar, counter point bar, meander bend

INTRODUCTION

Fluvial meander-belt deposits consist of a complex amalgam of depositional components with varied sedimentary characteristics. Associated strata comprise significant hydrocarbon reservoirs worldwide (Mossop & Flach, 1983; Butcher, 1990; Miall, 1996; Carter, 2003; Slatt, 2006). Many scales of heterogeneity are known to impact reservoir performance (e.g. Jones *et al.*, 1995; Strobl *et al.*, 1997; Lopez *et al.*, 2009; Pyrcz *et al.*, 2009; Hassanpour *et al.*, 2013; Alpak & Barton, 2014; Su *et al.*, 2017; Yan *et al.*, 2017; Willis & Sech, this volume) and as a result, controls on intra-meander-belt heterogeneity have been widely investigated (Jordan & Pryor, 1992; Willis & Tang, 2010; Colombera *et al.*, 2017). Bed-scale variation in grain-size, often associated with fluctuating fluvial discharge, is commonly considered (e.g. inclined heterolithic stratification; Thomas *et al.*, 1987; Jablonski & Dalrymple, 2016); larger-scale sediment distribution patterns within a point bar due to lateral accretion, flow separation and meander-bend migration style have also been widely documented (e.g. Allen, 1965; Bryce, 1974; Leeder, 1975; Jackson, 1976; Nardin *et al.*, 2013; Ielpi & Ghinassi, 2014). Over the last decade, the sand to mud differentiation within counter-point bar deposits have been of renewed interest (Smith *et al.*, 2009; Willis & Tang, 2010; Hubbard *et al.*, 2011; Ghinassi *et al.*, 2016). Point bars, counter-point bars, abandoned channel fills and floodplain deposits are key depositional elements within meander-belts. Predicting associated sediment

Fluvial Meanders and Their Sedimentary Products in the Rock Record, First Edition.
Edited by Massimiliano Ghinassi, Luca Colombera, Nigel P. Mountney and Arnold Jan H. Reesink.
© 2019 International Association of Sedimentologists. Published 2019 by John Wiley & Sons Ltd.

distribution patterns is essential to predicting heterogeneity in subsurface reservoirs (Fig. 1).

Concave-bank benches were recognised along modern meandering rivers at the downstream tail of convex point bars (Woodyer, 1975; Hickin, 1978; 1979; Page & Nanson, 1982). Lewin (1983) first introduced the term 'counter-point' to describe the deposit that resulted from concave-bank bench translation down-valley. Counter-point bar deposits are documented in most meander-belts, associated with rivers of varied discharges and grain-sizes (e.g. Fig 1; Hicken, 1979; Page & Nanson, 1982; Nanson & Page 1983; Lewin, 1983; Smith *et al.*, 2009; 2011). More recently, counter-point bar deposits have been recognised in ancient meander-belt deposits (Smith *et al.*, 2009; Hubbard *et al.*, 2011; Ielpi & Ghinassi, 2014; Durkin *et al.*, 2015; Raigemborn *et al.*, 2015; Durkin *et al.*, 2017; Martinus *et al.*, 2017); however, their common occurrence in modern fluvial environments suggests that they are perhaps often over-looked and under-recognised in ancient datasets. This is partially due to the common lack of planform exposure in outcrop and poor seismic resolution in the sub-surface (Ghinassi *et al.*, 2016), which enables ready identification of their diagnostic concave accretion patterns (Smith *et al.*, 2009). To date, relatively few studies have provided recognition criteria for facies-based data sets (i.e. core, outcrop).

Smith *et al.* (2009; 2011) provided vibracore-based transects from the Peace River in northern Alberta to characterise the deposits of the point bar to counter-point bar transition. In this study a more comprehensive overview of this transition is provided, covering a broader range of modern depositional environments from fluvial to tidal-fluvial. A starting hypothesis is that the previously documented change in grain-size across the point bar to counter-point bar is prevalent, regardless of scale, discharge or tidal influence. In order to test

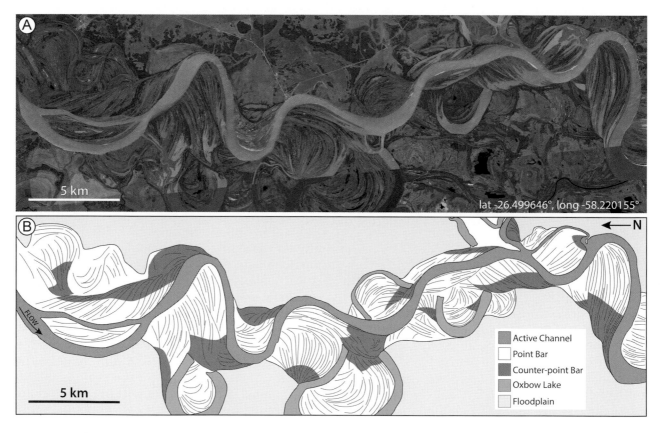

Fig. 1. Meander-belt deposit in Paraguay, South America (Lat/Long: -26.499646, -58.220155). (A) Satellite image of the meander belt from Google Earth (2017). (B) Line tracing of Part (A) highlighting the floodplain, including scroll patterns, oxbow lake fills and the active channel. Classification of point bar and counter-point bar deposits is based on convex (point bar) and concave (counter-point bar) scroll bar topography patterns. Counter-point bar deposits account for approximately 20 to 30% of meander-belt area.

this, grain-size changes, channel morphology, drainage area, discharge and degree of tidal influence in a series of modern and ancient data sets are documented. The results provide important constraints on this facies transition, with implications for subsurface prediction of lithofacies in hydrocarbon-bearing meander-belt deposits.

STUDY AREAS

Four study areas were selected to provide a range of point bar to counter point bar transition data. Three modern meandering channel study areas were selected with varying degrees of tidal influence (Peace, Chehalis and Willapa rivers) and the fourth area is an ancient meander-belt deposit from the Cretaceous Athabasca Oil Sands of Alberta, Canada (Fig. 2).

Two meander bends on the Chehalis River in south-western Washington State, U.S.A. were investigated (Fig. 2). The first site (CH1) is a ~1 km reach along the Chehalis River south of Montesano, WA. The meander bend is 26.5 river km upstream from the mouth of the estuary (Grays Harbour) at Aberdeen, WA. and the mean tidal range 4 to 2.5 m; average channel width is 130 m (range from 146–115 m) and average channel depth is 7.1 m. The second site (CH2) is a 1 km reach of the Chehalis River approximately 3 km upstream of Cosmopolis, WA. The meander bend is 13.6 river km upstream from the mouth of the estuary and at this location the mean tidal range is 3 m; average channel width is 213 m (range from 256–154 m) and average channel depth is 17.1 m. Both study sites were examined during the last week of July and first week of August 2009. The Chehalis River drainage basin includes 3351 km^2 and the main channel length is 185 km (USGS, 2017). The average discharge at Porter, WA (32.5 river kms upstream from CH1) from 1953 to 2016 is 116 m^3 s^{-1} (USGS, 2017) and the mean annual tidal range at Aberdeen, WA is 3.081 m (NOAA, 2017).

One meander bend on the Willapa River in south-western Washington State was investigated (Fig. 2). This meander bend was investigated by Smith (1987), although the bar tail was not considered in that study. The study site (WR1) is a 0.8 km reach approximately 1 km east of Raymond, WA and 19 river km from the mouth of the Willapa Bay estuary, between Tokeland

and South Bend, WA. The tidal range at this location is 3.4 m. Average channel width is 135.5 m (range from 175 to 95 m); average channel depth is 8.2 m. The Willapa River drainage basin includes 336 km^2 and the average discharge near Willapa, WA from 1949 to 2016 is 18.3 m^3 s^{-1}. Sampling took place during the second week of August 2009.

Two meander bends on the Peace River in north-eastern Alberta, Canada were investigated. The first site (PR1), featured in Smith *et al.* (2009), is a 3 km reach 14 km west of the boundary of Wood Buffalo National Park close to the village of Garden Creek. The meander bend is 253 river km upstream of the confluence with the Slave River and 686 river km from Great Slave Lake. The average channel width is 442.5 m (ranges from 480–410 m); the average bankfull depth is 15 m. PR1 sampling was conducted mid-July 2007. The second site (PR2), featured in Smith *et al.* (2011), is a 3 km reach just upstream of Carlson Landing. The meander bend is 36 river km from the confluence with the Slave River and 469 river km from Great Slake Lake. The average channel width is 443 m (range from 482 to 369 m); average bankfull depth is 14 m. PR2 sampling was conducted during the last two weeks of July 2008. The Peace River has a drainage area of 300,000 km^2 (Environment Canada, 1991) and an average discharge of 12,000 m^3 s^{-1} (Kellerhals *et al.*, 1972); this intracratonic river has no marine connection.

The fourth study area is an ancient meander-belt deposit from the subsurface McMurray Formation in north-eastern Alberta, Canada. The study area is located in Township 83 Range 07 West of the 4th Meridian and occurs at a depth of 350 to 500 m below the surface. The study area is covered by high-quality 3D seismic data and was recently investigated by Durkin *et al.* (2017; 2018). A point bar to counter-point bar transition is captured in 3D seismic data and investigated through four boreholes along a single accretion package. The formative channel associated with the meander-belt deposit, measured from the abandoned mudstone-filled channel, has an average width of 730 m (range from 475 to 1180 m). The average bankfull depth estimated from abandoned channel fill is 40 m. The estimated discharge of McMurray Formation channels in the study area is 15,000 m^3 s^{-1} and estimated distances to the shoreline are at least 150 to 200 km (Musial *et al.*, 2012; Durkin *et al.*, 2017).

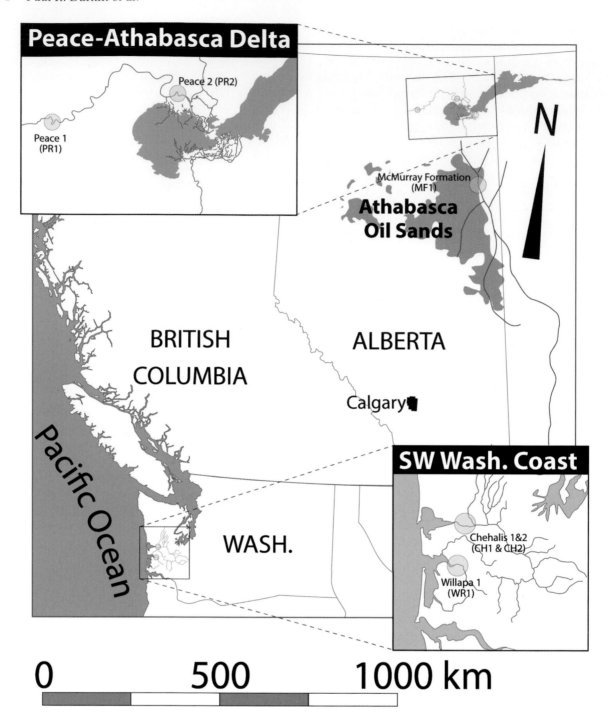

Fig. 2. Study area map of western Canada and north-western United States showing the location of each study area.

METHODS

Potential field sites were identified from previous publications (Smith *et al.*, 2009; 2011; Durkin *et al.*, 2017) and Google Earth. Meander bends with a convex to concave morphology on rivers with varying degrees of tidal influence were selected in order to characterise a variety of point bar to counter-point bar transitions.

Bathymetric profiles were collected as perpendicular cross sections to river flow to show channel width, depth, location of channel-bottom samples and geometry of channel. Depth to channel bottom was measured using a sonar depth finder and horizontal distance to riverbanks was measured using an Impulse laser range finder. Channel-bottom sediment samples were collected at the sediment-water interface using a Lane bottom drag sampler. Visual grain-size of sediment samples was investigated using a 20-power field microscope and stratigraphic grain-size chart.

Vibracoring was used to recover sediment cores from the subsurface along each meander bend. Sediment cores were vibrated out of tubes into acrylonitrile butadiene styrene (ABS) receiving troughs. Cores were cut in half and logged using standard stratigraphic logging techniques, observing characteristics including grain-size, bedding and sedimentary structures. Coring sites averaged 5 to 10 metres from the channel edge and water level was used as a datum. At tidally influenced locations, the low neap tide level was used as a datum. Composite sedimentary logs are constructed from vibracores and grain-sizes extrapolated from channel-bottom sediment samples at equivalent elevations to the deeper parts of sedimentary sequences not accessed by vibracoring.

In the McMurray Formation dataset, borehole gamma-radiation logs are used as a proxy for lithology, where low values are correlated to sandstone and high values are indicative of fine-grained units. Gamma radiation is well calibrated to core-based lithofacies in the area of investigation (Labrecque *et al.*, 2011a, b). The common occurrence of siltstone-clast breccia deposits results in a high GR reading; however, the lithology is sandstone with siltstone clasts. Meander-belt scroll-like patterns are interpreted from 3D seismic amplitude stratal slices that highlight the contrasting lithology between dipping lateral accretion layers (cf. Hubbard *et al.*, 2011; Durkin *et al.*, 2017).

Net sand to gross thickness calculations were made based on the total thickness of sand(stone) beds relative to the total thickness of composite sedimentary or borehole logs. In composite sedimentary logs, the use of channel-bottom grain-size samples provides only an estimation of true net-to-gross in the modern systems.

RESULTS

The Chehalis River 1 site (CH1) is characterised by a high sinuosity (2.29) meander bend with a vegetated floodplain, lack of evident scroll bar

topography and low-lying areas inundated by water at high tide (Fig. 3A). Human modification of the channel bank immediately downstream of the study area has impacted the planform morphology of the channel and the meander bend is approaching neck cut-off. A tributary to the Chehalis River supplies gravel to the river 8.5 km upstream of CH1. Based on successive satellite images, the meander-bend apex has migrated ~ 100 m from 2006 to 2016 (Fig. 3A: Google Earth, 2016). The bathymetric profile from the concave bank to the adjacent point bar (Fig. 4: A-A') is characterised by an asymmetrical, uniform channel with an aspect ratio (width:depth) of 15.5:1. Channel bottom sediment samples show coarse-grained sand, granules and pebbles in the thalweg of the channel, with fine sand along the concave bank and silt to coarse sand on the adjacent point bar.

Five vibracores were extracted along the meander bend, ranging from 1.1 m to 4.8 m of penetration (Fig. 5A). Composite sedimentary logs with net sand to gross thickness values are indicated (Fig. 5A). The upstream point bar is composed of granules and pebbles, with subordinate medium-grained to coarse-grained sand and very little fine-grained sediment (Fig. 5A: CH1-a, CH1-b). The overall coarse-grained nature of the upstream point bar sediments resulted in poor vibracore penetration. At the inflection point between upstream convex point bar to downstream concave counter-point bar, medium-grained to coarse-grained sand, granules and pebbles near the base are overlain by fine-grained to very-fine-grained sand and silt (Fig. 5A: CH1-c). The downstream, counter-point bar is composed of a lower 5 m-thick package with a similar grain-size distribution to the upstream point bar deposits, overlain by a 5 m-thick package of fine to very-fine sand interbedded with silt (Fig. 5A: CH1-d, CH1-e).

The Chehalis River 2 site (CH2) is characterised by a moderate sinuosity (1.52) meander bend surrounded by a vegetated floodplain, lack of scroll bar topography and sinuous tidal channels (Fig. 3B). The point bar top is exposed at low tide, a tidal drainage channel debouches into the main channel immediately downstream of the bend apex and a chute channel is present across the downstream point bar (Fig. 3B). The measured bathymetric profile from the concave bank to the adjacent point bar (Fig. 4: B-B') is characterised by a highly asymmetrical channelform with a steep concave bank and adjacent point bar clinoform dipping into the channel thalweg.

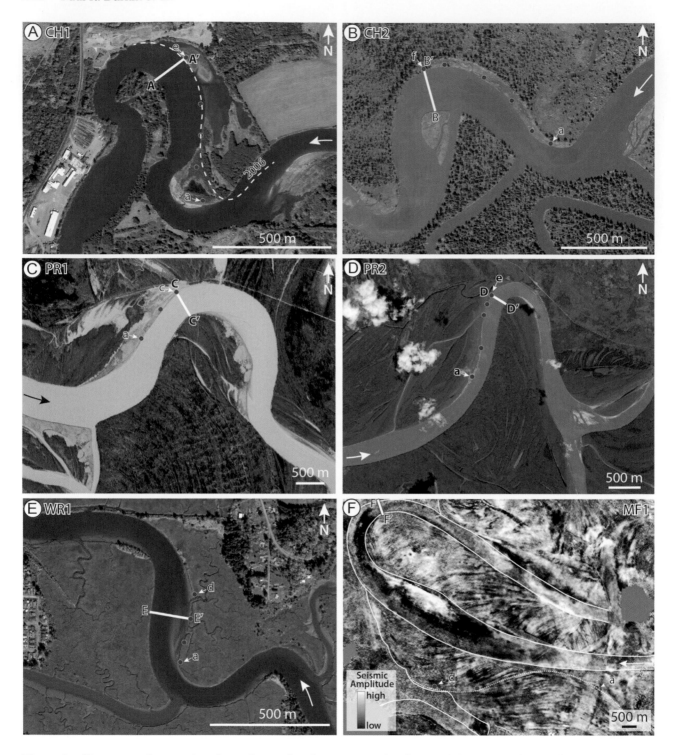

Fig. 3. Satellite images (A to E; Google Earth, 2016) and seismic amplitude stratal slice (F; Durkin *et al.*, 2017) from each of the study areas. (A) Chehalis River 1 (CH1) meander bend at Montesano, WA. (46.969408°, -123.593929°) (B) Chehalis River 2 (CH2) meander bend near Cosmopolis, WA. (46.952341°, -123.724037°) (C) Peace River 1 (PR1) meander bend near the village of Garden Creek, AB (58.718480°, -113.772874°). (D) Peace River 2 (PR2) meander bend near Carlson Landing, AB (58.965560°, -111.838111°). (E) Willapa River 1 (WR1) meander bend at the town of Raymond, WA (46.685852°, -123.712327°). (F) McMurray Formation 1 (MF1) meander bend deposit in the subsurface of north-eastern Alberta (Township 83, Range 07, West of the 4th Meridian).

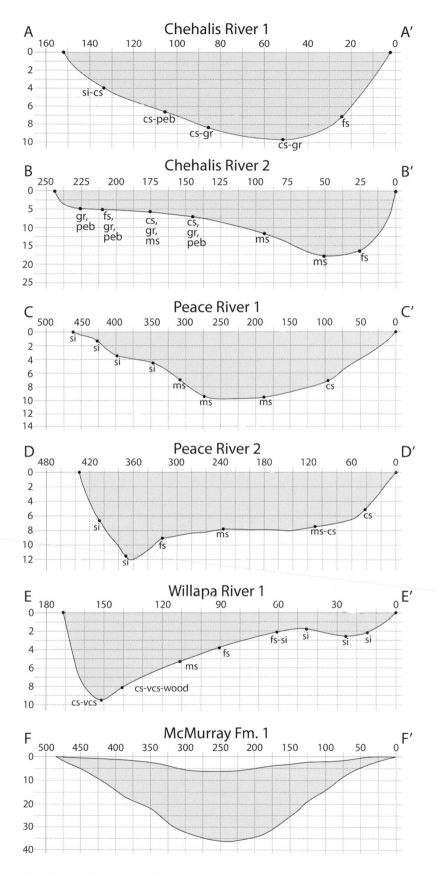

Fig. 4. Bathymetric profiles from each of the study areas. Locations of cross-sections are indicated on Fig. 3. Location and grain-size of channel-bottom sediment samples are also shown. F-F' profile is interpreted from 3D seismic data. Grain-size key: si = silt, fs = fine sand, ms = medium sand, cs = coarse sand, vcs = very coarse sand, gr = granules, peb = pebbles, wood = wood material and plant debris.

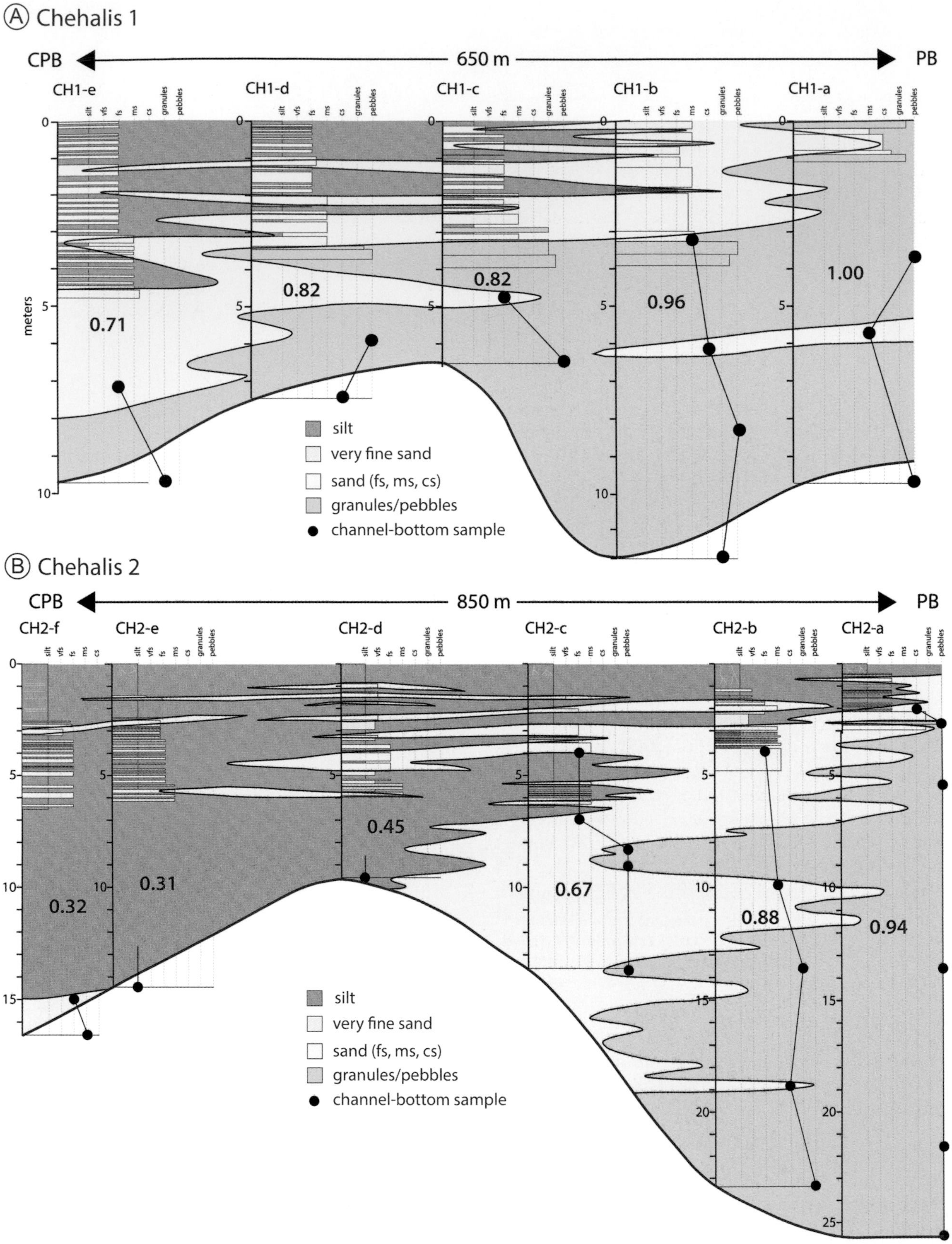

Fig. 5. Lithostratigraphic profiles from the Chehalis River, WA. (A) Five vibracores taken from the right bank at the CH1 study site. (B) Six vibracores taken from the right bank at the CH2 study area. Upstream (point bar) is on the right and downstream (counter-point bar) is on the left. Location of vibracores are shown on Fig. 3. Black circles indicate grain-size of channel-bottom sediment samples at the equivalent depth. Numbers ranging from 0 to 1 represent the ratio of net sand/granules/pebbles to gross thickness in the lithostratigrahic profiles. Red line represents the maximum depth of the channel at each profile location. Correlation of granules/pebbles (orange), sand (yellow) and silt (grey) characterise the distribution of sediment within the cross-section.

Six vibracores were extracted along the meander bend, ranging from 3.0 m to 6.5 m of penetration (Fig. 5B). The upstream cores reveal the point bar is composed of mainly coarse sand to granules with an upper 3 m of interbedded silt and sand (CH2-a, CH2-b). Channel bottom sediments (below 8 m) are dominated by pebbles. Across the meander-bend inflection point at CH2-c, channel-bottom sediments are silt-dominated, with minor fine-grained to medium-grained sand. The upper portions of the cores are composed of fine-grained to medium-grained sand with thick silt interbeds. Overall, counter-point bar cores are dominated by silt with some fine-grained to medium-grained sand interbeds.

The Peace River 1 site (PR1) is characterised by a moderate sinuosity (1.49) meander bend with a vegetated floodplain and evident scroll bar topography (Fig. 3C). The channel width varies significantly over the meander bend of interest; the counter-point bar scroll patterns are low-lying, inundated by water (ice). A bathymetric profile from the concave bank to the adjacent point bar shows a generally symmetrical channelform with a clear concave bank bench profile (e.g. Hickin, 1978).

Three vibracores were drilled along the meander bend, with an average of 8.5 m of penetration. In the upstream core (PR1-a), point bar deposits are characterised by thick beds of medium-grained and coarse-grained sand, overlain by 4 m of interbedded silt and fine-grained sand. In the middle and downstream cores (PR1-b, PR1-c) there is a significant decrease in sand-sized material with only minor fine-grained sand interbeds amongst thick beds of silt. Channel-bottom sediment samples indicate medium-grained to coarse-grained sand is present at the base of the counter-point bar. Refer to Smith *et al.* (2009) for a more detailed analysis of the Peace River 1 study area.

The Peace River 2 site (PR2) is a moderate sinuosity (1.61) meander bend, with a vegetated floodplain and well-defined scroll bar topography (Fig. 3D). A diminutive chute channel is present from the upstream point bar to the downstream counter-point bar. Evidence for meander bend translation is clear in the truncated scroll pattern along the upstream extent of the downstream-adjacent point bar. The bathymetric profile shows an asymmetrical channelform with a deep scour along the concave bank (Fig. 4D).

Five vibracores were extracted along the meander bend, ranging from 2.8 m to 6.4 m of penetration (Fig. 6B). The upstream-most cores (PR2-a, PR2-b) are dominated by medium-grained to coarse-grained sand with an increase in silt interbeds from PR2-a to PR2-b. At the meander-bend inflection point, PR2-c is characterised by 6 m of interbedded silt and medium-grained sand. The downstream cores (PR2-d, PR2-e) are dominated by silt beds, with minor fine-grained to medium-grained sand interbeds. Channel-bottom samples adjacent to PR2-d and PR2-e indicate silt at the base of the channel. Refer to Smith *et al.* (2011) for additional information about the Peace River 2 study area.

The Willapa River 1 site (WR1) is a low-sinuosity (1.31) meander bend, with abundant tidal drainage creeks and vegetated floodplain/tidal flats (Fig. 3E). No scroll bar topography is evident. The bathymetric profile, slightly downstream of the meander-bend inflection point, shows an asymmetrical channel with a deeper thalweg along the outer bend and a gently sloping inner bank with a subtle high 40 to 50 m from the bank.

Four vibracores penetrate the meander bend to a depth of 5.7 m to 6.8 m. The upstream core (WR1-a) is characterised by interbedded silt and fine-grained sand, with minor medium-grained sand near the base. The two cores at the inflection point are similar, characterised by interbedded silt and fine-grained sand, overlain by thick beds of silt in the upper 3 metres. There is a slight increase in very fine-grained sand and silt beds in WR1-c compared to WR1-b. The downstream core (WR1-d) is dominated by silt, with rare very fine-grained sand interbeds and organic material. Channel-bottom samples indicate fine-grained to medium-grained sand at the base of the channel.

The McMurray Formation (MF1) study area is characterised by ancient meander-belt deposits revealed in 3D seismic and well data. A convex to concave transition is identified and an associated abandoned channel is highlighted (Fig. 3F). The abandoned meander loop is highly sinuous (2.54) and the four borehole locations were chosen along a single scroll, capturing contemporaneous deposition along the meander bend. A bathymetric profile of the abandoned meander-loop fill was interpreted from 3D seismic data and shows a symmetrical channelform 480 m wide and 36 m deep (Fig. 4F).

Four boreholes with gamma-ray log profiles reveal a range in meander-belt thickness from 47 to 49 m. A 75 API cut-off was used to infer sandstone lithology, which was calibrated to core

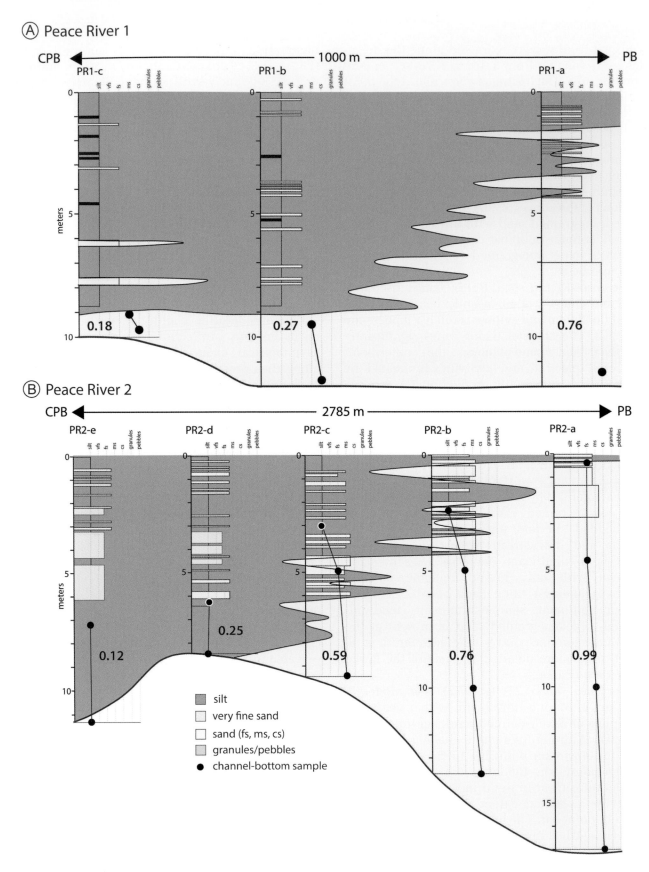

Fig. 6. Lithostratigraphic profiles from the Peace River, AB. (A) Three vibracores taken from the left bank at the PR1 study site. (B) Five vibracores taken from the left bank at the PR2 study area. Upstream (point bar) is on the right and downstream (counter-point bar) is on the left. Location of vibracores are shown on Fig. 3. Black circles indicate grain-size of channel-bottom sediment samples at the equivalent depth. Numbers ranging from 0 to 1 represent the ratio of net sand to gross thickness in the lithostratigrahic profiles. Red line represents the maximum depth of the channel at each profile location. Correlation of sand (yellow) and silt (grey) characterise the distribution of sediment within the cross-section.

data. The most upstream borehole (MF1-a) is characterised by a thick, blocky sandstone overlain by 10 metres of interbedded sandstone and fine-grained material (Fig. 7B). Note that this core is relatively further upstream than those featured in any of the other study areas. Downstream, just before the inflection point, MF1-b is characterised by a serrated gamma-ray profile interpreted to record interbedded sandstone and siltstone, with an overall upward fining profile. Siltstone-clast breccia is common in the McMurray Formation and results in a high gamma-ray value, representing a coarse-grained deposit. Siltstone clasts are more prevalent near the base of the meander belt (Durkin *et al.*, 2017). The borehole immediately downstream of the inflection point (MF1-c) is characterised by a thick, blocky sandstone at the base that is overlain by interbedded siltstone and sandstone with a thick fine-grained bed at the top. The furthest downstream borehole is dominated by siltstone with minor sandstone interbeds. The borehole does not reach the base of the channel; however, thick successions of fine-grained material are present at depths of up to 40 m. Refer to Durkin *et al.* (2017) for additional information about the counter-point bar deposits of the McMurray Formation 1 study area.

DATA INTEGRATION AND INTERPRETATION

To compare all of the study areas, ratios of net sand to gross thickness were calculated for all of the composite borehole logs along each cross section. Each log was defined as point bar or counter point bar based on location along the meander bend and plotted by study area (Fig. 8). Results show that point bar deposits have a high net-to-gross (>0.70), while most of the counter point bar deposits are characterised by significantly lower net-to-gross values (<0.30), with the exception of the Chehalis 1 locality. The shift in net-to-gross from point bar to counter point bar ranges from 0.30 to 0.82, with an average of 0.57. The change in net-to-gross from upstream to downstream is consistent across all study areas and coincident with the inflection point from convex to concave along the meander bend. The facies transition from point bar to counter point bar appears to be scaled to channel width and point bar area. As channel and point bar dimensions increase, the transition length from point bar to counter point bar also increases; transition length is approximately three times the width of the channel.

At Chehalis 1, the counter point bar is composed of medium sand, granules and pebbles at the base, overlain by interbedded sand and silt, resulting in a net-to-gross value of approximately 0.7 (Figs 5 and 8). The coarse-grained nature of the counter point bar is attributed to the nearby tributary channel, which supplies gravel and coarse sand only 8 km upstream from the bar (Fig. 2). While the CH1 counter point bar net-to-gross value is much higher than other study areas investigated, a significant grain-size shift across the point bar to counter point bar transition is documented. This is particularly well expressed at the bar top, which transitions from gravel-dominated to mud-dominated over 300 to 400 m.

The significant change in net-to-gross between the upstream, convex point bar to the downstream, concave counter point bar appears to be consistent regardless of river scale, discharge, sinuosity and tidal influence. Downstream fining along a meander bend has been widely observed (e.g. Jackson, 1976; Wood, 1989; Willis & Tang, 2010), attributed to flow separation from the inner bank and promotion of fine-grained sedimentation on the downstream end of sinuous meander bends (e.g. Leeder, 1975). A concave bank bench often develops due to translation of a sinuous meander bend, previously attributed to erosion-resistant cut-bank material, which inhibits lateral expansion and promotes down-dip migration of the meander bend (Smith *et al.*, 2009; Ghinassi *et al.*, 2016; Durkin *et al.*, 2017). The collected data indicates that the development of a fine-grained counter point bar can develop at a variety of scales under the influence of variable marine and non-marine processes. Notably, in each example investigated, bar morphology is largely controlled by fluvial processes. Net-to-gross of the counter-point bar is consistently between 20 to 30% sand whether in pure fluvial or marine-influenced settings (Fig. 8), cautioning interpretations of tidal influence based on bar lithology (e.g. presence of inclined heterolithic stratification (IHS) in the stratigraphic record.

DISCUSSION

The recognition of counter-point bar deposits in the rock record has proven difficult due to their fine-grained recessive nature, difficulty in discerning their diagnostic concave planform

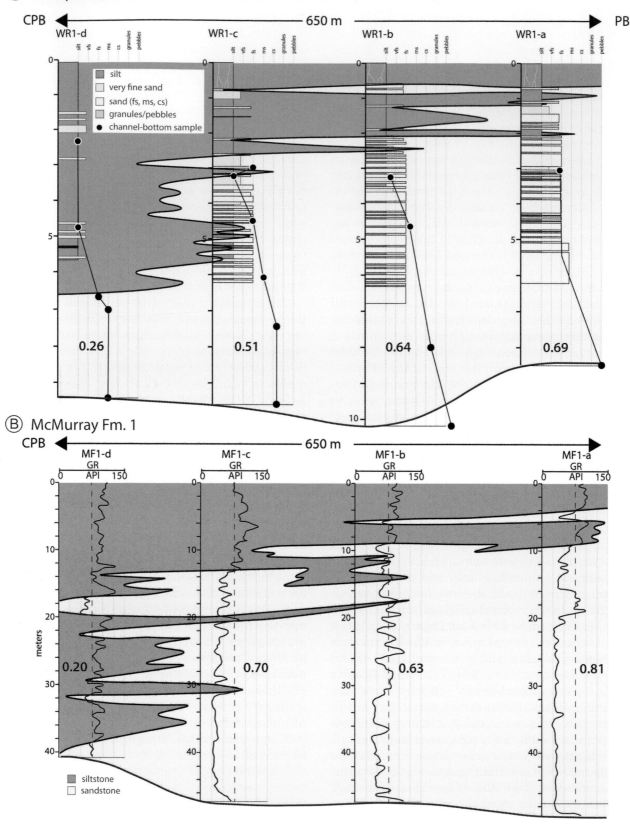

Fig. 7. Lithostratigraphic profiles from the Willapa River, WA and the subsurface McMurray Formation. (A) Four vibracores taken from the right bank at the WR1 study site. (B) Four borehole logs from the subsurface MF1 study area. Profiles are recorded by gamma-radiation logs scaled from 0 to 150 API, with lower API values calibrated to sandstone and higher API values calibrated to siltstone. A 75 API cutoff defines the boundary between sandstone and siltstone. Upstream (point bar) is on the right and downstream (counter-point bar) is on the left. Location of vibracores and boreholes are shown on Fig. 3. Black circles indicate grain-size of channel-bottom sediment samples at the equivalent depth. Numbers ranging from 0 to 1 represent the ratio of net sand to gross thickness in the lithostratigrahic profiles. Red line represents the maximum depth of the channel at each profile location. Correlation of sand (yellow) and silt (grey) characterise the distribution of sediment within the cross-section.

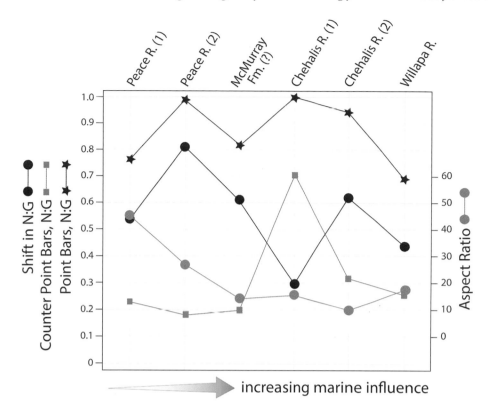

Fig. 8. A compilation graph of net sand to gross thickness data and channel aspect ratio, plotted by study area and ranked in order of increasing marine influence, from left to right. Scale for net-to-gross is on the left, from 0 to 1 and scale for aspect ratio is on the right, from 0 to 60. Aspect ratios are calculated from bathymetric profile data.

architecture and poorly defined facies-based recognition criteria (e.g. Smith *et al.*, 2009; Hubbard *et al.*, 2011; Ghinassi *et al.*, 2016). However, counter-point bars have proven to be common in a wide variety of fluvial meander belt deposits (Hickin, 1979; 1986; Page & Nanson, 1982; Makaske & Weerts, 2005; Smith *et al.*, 2009; Hooke & Yorke, 2011; Hubbard *et al.*, 2011; Willis & Tang, 2011; Durkin *et al.*, 2015; Ghinassi *et al.*, 2016). Recognition of counter-point bar deposits has important implications for palaeo-environmental interpretations, as well as reservoir heterogeneity prediction in subsurface reservoirs.

Tidally influenced counter point bar deposits

Tidal influence on fluvial meander belt processes has been widely documented in modern environments (e.g. Dorjes & Howard, 1975; Dalrymple & Choi, 2007; Van den Berg *et al.*, 2007; Dashtgard *et al.*, 2012; Carling *et al.*, 2015) and interpreted from the rock record (e.g. Smith *et al.*, 1987; Shanley *et al.*, 1992; Hubbard *et al.*, 2011; Reijenstein *et al.*, 2011; Martinius & Gowland, 2011; Feldman & Demko, 2015). However, specific

examples of tidally influenced counter-point bars are under-reported, with some exceptions (e.g. Hubbard *et al.*, 2011). In modern fluvial environments, the most diagnostic feature of the counter point bar is the concave scroll bar pattern on the floodplain, evident in satellite imagery (e.g. Fig. 1). In the tidally-influenced examples from this study, floodplains lack scroll bar topography and are characterised by low-lying, relatively flat, vegetated areas with sinuous tidal drainage creeks (e.g. Fig. 3B and E).

Despite a lack of concave scroll bar topography, the deposit architecture and facies observations from this study suggest counter point bars are an important component of tidally influenced meander belts. The lack of scroll bar topography makes accretion history difficult to discern and it is plausible that features like counter point bars have been overlooked as a result. A decrease in meander-bend migration rate in the direction of increasing marine influence is well documented along the tidal-fluvial transition (e.g. Hudson & Kessel, 2000; Dalrymple & Choi, 2007; Nittrouer *et al.*, 2012; Blum *et al.*, 2013). In the Chehalis (CH2) and Willapa River (WR1) examples (Fig. 3B and E), cuspate meanders exhibit a 'pinch and swale' morphology between bend apexes that suggests the presence of strong flood-tide currents (Li *et al.*, 2008; Hughes, 2012). During a strong flood tide, the concave bank may be subject to erosion, which would inhibit migration and thus limit fluvially driven downstream translation. Despite this, it is uncertain whether a similar accretionary product to a fluvial counter-point bar could be produced by a tidally influenced counter-point bar over a longer period of time. This has potential implications for reservoir delineation in major ancient fluvial systems such as the Cretaceous McMurray Formation in north-eastern Alberta, Canada and the Triassic Mungaroo Formation of the north-west Australian shelf, both of which have been interpreted to be deposited under tidal influence in some regions (Smith, 1988; Stoner, 2010; Hayes *et al.*, 2017). To date, recognition of counter-point bar deposits in the McMurray Formation has depended on concave scroll pattern recognition, inferred from heterolithic accretion sets imaged in seismic data (e.g. Hubbard *et al.*, 2011; Durkin *et al.*, 2017).

Although the data presented in this study confirms the presence of counter point bars in tidally influenced rivers, it is clear that additional penetrative studies (e.g. geophysical methods) are required from modern systems to fully characterise the internal architecture of counter point bars along the fluvial to tidal transition. With these additional data, a detailed facies model can be devised to enable more reliable interpretations of stratigraphic datasets.

Reservoir implications

Meander-belt deposit heterogeneity has been widely considered (e.g. Jordan & Pryor, 1992; Colombera *et al.*, 2017; Willis & Sech, this volume), yet few studies consider counter-point bar deposits, with some exceptions (e.g. Willis & Tang, 2010). With counter point bars composing 10 to 30% of many meander belts (Smith *et al.*, 2009), delineating their distribution has important implications for reservoir volume estimates, as well as well placement strategies (Columbera *et al.*, 2017). Due to a lack of stratigraphic models for counter point bar deposits, they undoubtedly have been widely (and understandably) overlooked. Reijenstein *et al.* (2011) studied Pleistocene meander-belt deposits from the Gulf of Thailand in 3D seismic reflection data (Fig. 9). Time slices reveal an up to 10 km-wide meander-belt deposit with well-defined scroll patterns and an abandoned channel (Fig. 9). Amplitude variations highlight

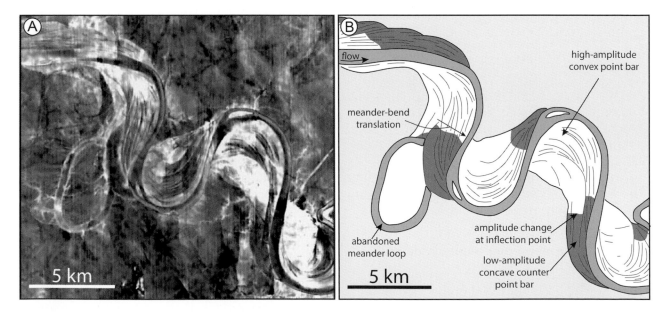

Fig. 9. 3D seismic time slice data of a Pleistocene meander-belt deposit from the Gulf of Thailand (modified from Reijenstein *et al.*, 2011). (A) Seismic amplitude time slice characterised by a relatively high-amplitude (white to light grey) meander-belt deposit incised into the surrounding low-amplitude (dark grey) floodplain. (B) Line-trace of part (A) highlighting meander-belt scroll patterns, abandoned channel, oxbow lake fills and associated floodplain deposits. Classification of point bar and counter-point bar deposits is based on convex (point bar) and concave (counter-point bar) scroll patterns. Link to part A to note the significant seismic amplitude change across the inflection point from point bar to counter-point bar. An interpreted change in lithology from sandstone to siltstone supports the results from this study.

the meander-belt deposit from the surrounding floodplain and illustrate the lithologic variation between bar deposits (sand-dominated) and abandoned channel deposits (mud-dominated) (Reijenstein *et al.*, 2011). Also evident is a clear change in seismic amplitude from bright, high values where the scroll pattern is convex, to dark, low values where the scroll pattern is concave, relative to palaeo-dip (Fig. 9). The change in amplitude is consistent with the inflection between convex point bar deposits and concave counter-point bar deposits. Overlooking counter point bars in this exquisite dataset would support a notion that associated deposits have been misidentified in countless subsurface and outcrop datasets.

The Athabasca Oil Sands in north-eastern Alberta, Canada are characterised by stacked Cretaceous meander-belt deposits, saturated with low-viscosity bitumen (Fustic *et al.*, 2013; Nardin *et al.*, 2013; Hein *et al.*, 2017). Extraction techniques involve production from two horizontal wells separated vertically by 4 to 6 m; the upper well injects steam into the reservoir and the lower well produces mobilised bitumen in a process called Steam Assisted Gravity Drainage (SAGD) (Strobl *et al.*, 1997; Gotwala & Gates, 2010). Production is strongly influenced by reservoir heterogeneity (e.g. Su *et al.*, 2013) and reservoir characterisation is essential to effective well placement strategies. Recently, Su *et al.* (2014) have demonstrated the sensitivity of horizontal well placement on oil recovery based on whether horizontal well bores are oriented parallel or perpendicular to accretion surfaces in immense point bar deposits of the McMurray Formation. Results of the study inform the impact of internal stratigraphic architecture and point bar heterogeneity on SAGD well-pad performance. Su *et al.* (2014) find that individual well pair performance is highly variable within a single point bar deposit and attribute the variation to internal heterogeneity, specifically the presence of thick, intervening siltstone beds. Although they did not consider counter-point bar deposits, from their analysis it is clear that the deposits would have a significant impact on well deliverability (Fig. 10). Previous studies of the Cretaceous McMurray Formation have identified counter-point bar deposits in 3D seismic data sets (Hubbard *et al.*, 2011) and in the McMurray Formation 1 area of this study (Durkin *et al.*, 2017). Rapid facies transitions from sandstone-dominated point bar to fine-grained counter-point bar deposits can occur over the scale of a single SAGD well-pad and identification of these transitions is essential for development of an effective development strategy (Fig 10).

CONCLUSIONS

Six study areas characterised by convex to concave meander bends from a range of fluvial to tidal-fluvial environments were analysed. The analyses indicate that the previously documented change in grain-size across the point bar to counter-point bar transition is consistent, regardless of river scale, discharge or tidal influence. Point bar deposits are characterised by high net sand to gross thickness (>0.7) and counter point bars are typically characterised by low net-to-gross value (<0.3); the average decrease in net-to-gross across the transition is 57%. The point bar to counter point bar transition length scales to channel size and is approximately three times channel width.

In tidally influenced meander-belts, counter-point bar deposits are present; however, the characteristic concave scroll pattern is not preserved. The lack of scroll bar topography makes identifying counter-point bar deposits challenging and this has contributed to their poor characterisation from tidally influenced settings to date. Additional penetrative methods are required to characterise counter-point bar deposits from tidally influenced settings, which will facilitate development of a comprehensive facies model applicable to stratigraphic interpretation.

Predicting heterogeneity in fluvial to tidal-fluvial meander-belt reservoirs is critical for reserve estimates and well placement strategies, particularly in heavy-oil reservoirs. The distribution of low net-to-gross counter point bar deposits has implications for fluid flow, with particular impact on the Steam Assisted Gravity Drainage process used to access the vast Athabasca Oil Sands of northeastern Alberta, Canada. The prevalence of counter-point bar deposits in the meander belts investigated suggests that they are a commonly overlooked element in tidal-fluvial reservoirs globally.

ACKNOWLEDGEMENTS

This work was supported by Nexen Inc. (grant to Smith) and the Natural Sciences and Engineering Research Council (NSERC) of Canada (Discovery Grant RG-PIN/341715-2013 to

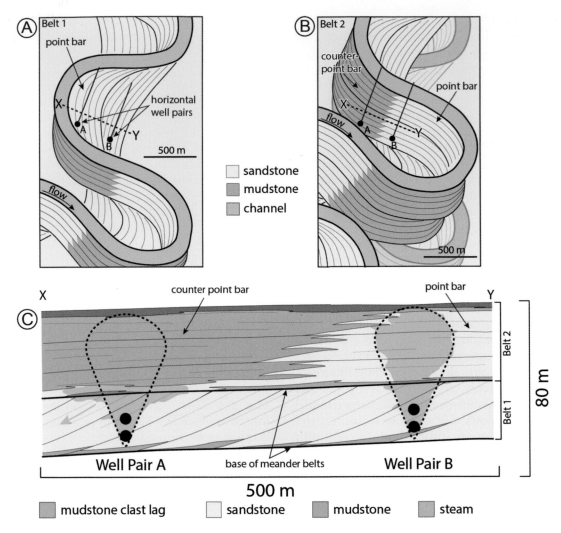

Fig. 10. Hypothetical planform maps and cross-section of two vertically stacked meander-belt deposits from the Athabasca Oil Sands in north-eastern, Alberta, Canada. (A) A planform map of a meander-belt deposit with two horizontal well pairs (A and B) in a sandstone-dominated point bar. (B) A planform map of a meander belt stacked on top of the deposit from Part A. Meander belt 1 (Part A) is shown in light grey; the general sedimentological trends from this study (upstream point bar is sand-dominated, downstream counter-point bar is mud-dominated) are conceptualised. (C) Cross-section of the two, stacked meander-belt deposits along X to Y from Parts A and B. Two Steam Assisted Gravity Drainage (SAGD) horizontal well pairs oriented into the page are shown. The difference in steam chamber development is compared across the point bar to counter-point bar transition. Where point bar deposits from meander belt 2 are stacked vertically above meander belt 1, the steam chamber is fully developed. Where the counter-point bar deposit from meander belt 2 is present, the steam chamber is restricted to the lower meander belt 1 deposit and is under-developed; drainage of heated bitumen may be directed away from the producing well. An array of a SAGD development wells commonly covers on the order of 1 km² and the transition between point bar and counter point bar occurs over similar or smaller areas.

Hubbard). Additional support was provided by ConocoPhillips (post-doctoral fellowship for Durkin) and the industrial sponsors of the McMurray Consortium (BP, Cenovus, Husky, Nexen and Woodside). The authors are grateful to reviewers Kyungsik Choi and anonymous for their suggestions to improve the manuscript.

REFERENCES

Allen, J.R. (1965) A review of the origin and characteristics of recent alluvial sediments. *Sedimentology*, **5**, 89–191.

Alpak, F.O. and **Barton, M.D.** (2014) Dynamic impact and flow-based upscaling of the estuarine point-bar stratigraphic architecture. *J. Petrol. Sci. Eng.*, **120**, 18–38.

Blum, M., Martin, J., Milliken, K. and Garvin, M. (2013) Paleovalley systems: Insights from Quaternary analogs and experiments. *Earth-Sci. Rev.*, **116**, 128–169.

Brice, J.C. (1974) Evolution of meander loops. *Geol. Soc. Am. Bull.*, **85**, 581–586.

Butcher, B.P. (1990) Northwest Shelf of Australia. In: *Divergent/Passive Margin Basins* (Eds J.D. Edwards and P.A. Santogrossi), *AAPG Mem.*, **48**, 81–115.

Carling, P., Chateau, C., Leckie, D., Langdon, C., Scaife, R. and Parsons, D. (2015) Sedimentology of a tidal point-bar within the fluvial–tidal transition: River Severn estuary, UK. In: *Fluvial-Tidal Sedimentology: Developments in Sedimentology* (Eds P.J. Ashworth, J.L. Best and D.R. Parsons), **68**. Elsevier, 149–192.

Carter, D.C. (2003) 3-D seismic geomorphology: Insights into fluvial reservoir deposition and performance, Widuri field, *Java Sea. AAPG Bull.*, **87**, 909–934.

Colombera, L., Mountney, N.P., Russell, C.E., Shiers, M.N. and McCaffrey, W.D. (2017) Geometry and compartmentalization of fluvial meander-belt reservoirs at the bar-form scale: Quantitative insight from outcrop, modern and subsurface analogues. *Mar. Petrol. Geol.*, **82**, 35–55.

Dalrymple, R.W. and Choi, K. (2007) Morphologic and facies trends through the fluvial–marine transition in tide-dominated depositional systems: a schematic framework for environmental and sequence-stratigraphic interpretation. *Earth-Sci. Rev.*, **81**, 135–174.

Daniel, J.F. (1971) Channel movement of meandering Indiana streams. *US Geol. Surv. Prof. Pap.*, **732-A**, 18 p.

Dashtgard, S.E., Venditti, J.G., Hill, P.R., Sisulak, C.F., Johnson, S.M. and La Croix, A.D. (2012) Sedimentation across the tidal–fluvial transition in the Lower Fraser River, Canada. *SEPM, The Sedimentary Record*, **10**, 4–9.

Dorjes, J. and Howard, J.R. (1975) Fluvial-marine transition indicators in an estuarine environment, Ogeechee River-Ossabaw Sound. *Senckenbergiana Maritima*, **7**, 137–179.

Durkin, P.R., Hubbard, S.M., Boyd, R.L. and Leckie, D.A. (2015) Stratigraphic expression of intra-point bar erosion and rotation. *J. Sed. Res.*, **85**, 1238–1257.

Durkin P.R., Hubbard, S.M., Boyd, R.L., Shultz, A.W. and Blum, M. (2017) Three-dimensional reconstruction of meander-belt evolution, Cretaceous McMurray Formation, Alberta Foreland Basin, Canada. *J. Sed. Res.*, **87**, 1075–1099.

Durkin, P.R., Hubbard, S.M., Holbrook, J. and Boyd, R. (2018) Evolution of fluvial meander-belt deposits with implications for the completeness of the stratigraphic record: GSA Bulletin, doi:/10.1130/B31699.1.

Environment Canada (1991) Historical Streamflow Summary 1990. *Water Survey of Canada*, Ottawa, Canada, 629 p.

Feldman, H. and Demko T. (2015) Recognition and prediction of petroleum reservoirs in the fluvial/tidal transition. International Association of Sedimentologists. In: *Fluvial-Tidal Sedimentology: Developments in Sedimentology* (Eds P.J. Ashworth, J.L. Best and D.R. Parsons), Elsevier, **68**, 483–528.

Flach, P.D. and Mossop, G.D. (1985) Depositional environments of Lower Cretaceous McMurray Formation, Athabasca oil sands, Alberta. *AAPG Bull.*, **69**, 1195–1207.

Fustic, M., Bennett, B., Hubbard, S.M., Huang, H., Oldenburg, T. and Larter, S. (2013) Impact of reservoir heterogeneity and geohistory on the variability of bitumen properties and on the distribution of gas- and water-saturated zones in the Athabasca Oil Sands. In: *Heavy-Oil and Oil-Sand Petroleum Systems in Alberta and Beyond* (Eds F.J. Hein, D. Leckie, S., Larter and J.R. Suter), *AAPG Studies in Geology*, **64**, 163–205.

Fustic, M., Hubbard, S.M., Spencer, R., Smith, D.G., Leckie, D.A., Bennett, B. and Larter, S. (2012) Recognition of down-valley translation in tidally influenced meandering fluvial deposits, Athabasca oil sands (Cretaceous), Alberta, Canada. *Mar. Petrol. Geol.*, **29**, 219–232.

Ghinassi, M., Ielpi, A., Aldinucci. and Fustic, M. (2016) Downstream-migrating fluvial point bars in the rock record. *Sed. Geol.*, **334**, 66–96.

Gotawala, D.R. and Gates, I.D. (2010) On the impact of permeability heterogeneity on SAGD steam chamber growth. *Nat. Resour. Res.*, **19**, 151–164.

Hassanpour, M.M., Pyrcz, M.J. and Deutsch, C.V. (2013) Improved geostatistical models of inclined heterolithic strata for McMurray Formation, Alberta, Canada. *AAPG Bull.*, **97**, 1209–1224.

Hayes, D.A., Timmer, E.R., Deutsch, J.L., Ranger, M.J. and Gingras, M.K. (2017) Analyzing dune foreset cyclicity in outcrop with photogrammetry. *J. Sed. Res.*, **87**, 66–74.

Hein, F.J. (2017) Geology of Bitumen and Heavy Oil: An Overview. *J. Petrol. Sci. Eng.*, in press.

Hein, F.J. (2016) The Cretaceous McMurray oil sands, Alberta, Canada: A world-class, tidally influenced fluvial estuarine system – An Alberta government perspective. International Association of Sedimentologists In: *Fluvial-Tidal Sedimentology: Developments in Sedimentology* (Eds P.J. Ashworth, J.L. Best and D.R. Parsons), Elsevier, **68**, 561–621.

Hein, F.J. and Cotteril, D.K. (2006) The Athabasca oil sands – A regional geological perspective, Fort McMurray area, Alberta, Canada. *Nat. Resour. Res.*, **15**, 85–102.

Hickin, E.J. (1986) Concave-bank benches in the floodplains of Muskwa and Fort Nelson rivers, British Columbia. *The Canadian Geographer/Le Géographe Canadien*, **30**, 111–122.

Hickin, E.J. (1979) Concave-bank benches on the Squamish River, British Columbia, Canada. *Can. J. Earth Sci.*, **16**, 200–203.

Hickin, E.J. (1978) Meandering channels. *Sedimentology*, 703–709.

Hickin, E.J. (1974) The development of meanders in natural river-channels. *Am. J. Sci.*, **274**, 414–442.

Hickin, E.J. and Nanson, G.C. (1975) The character of channel migration on the Beatton River, Northeast British Columbia, Canada. *Geol. Soc. Am. Bull.*, **86**, 487–494.

Hooke, J.M. (1995) River channel adjustment to meander cutoffs on the River Bollin and River Dane, northwest England. *Geomorphology*, **14**, 235–253.

Hooke, J.M. and Yorke, L. (2011) Channel bar dynamics on multi-decadal timescales in an active meandering river. Earth Surface Processes and Landforms, **36**, 1910–1928.

Hubbard, S.M., Smith, G.D., Nielsen, H., Leckie, A.D., Fustic, M., Spencer, J.R. and Bloom, L. (2011) Seismic geomorphology and sedimentology of a tidally influenced river deposit, lower Cretaceous Athabasca oil sands, Alberta, Canada. *AAPG Bull.*, **95**, 1123–1145.

Hudson, P.F. and Kesel, R.H. (2000) Channel migration and meander-bend curvature in the lower Mississippi River prior to major human modification. *Geology*, **28**, 531–534.

Hughes, Z.J. (2012) Tidal Channels on Tidal Flats and Marshes. In: *Principles of Tidal Sedimentology* (Eds R.A. Davis, Jr. and R.W. Dalrymple), 269–300, Springer, Netherlands.

Ielpi, A. and Ghinassi, M. (2014) Planform architecture, stratigraphic signature and morphodynamics of an exhumed Jurassic meander plain (Scalby Formation, Yorkshire, UK). *Sedimentology*, **61**, 1923–1960.

Jablonski, B.V.J. and Dalrymple, R.W. (2016) Recognition of strong seasonality and climatic cyclicity in an ancient, fluvially dominated, tidally influenced point bar: Middle McMurray Formation, lower Steepbank River, north-eastern Alberta, Canada. *Sedimentology*, **63**, 552–585.

Jackson, R.G. (1976) Depositional model of point bars in the lower Wabash River. *J. Sed. Res.*, **46**, 579–594.

Jones, A., Doyle, J., Jacobsen, T. and Kjønsvik, D. (1995) Which sub-seismic heterogeneities influence waterflood performance? A case study of a low net-to-gross fluvial reservoir. *Geol. Soc. London Spec. Publ.*, **84**, 5–18.

Jordan, D.W. and Pryor, W.A. (1992) Hierarchical levels of heterogeneity in a Mississippi River meander belt and application to reservoir systems: Geologic Note. *AAPG Bull.*, **76**, 1601–1624.

Labrecque, P.A., Hubbard, S.M., Jensen, J.L. and Nielsen, H. (2011a) Sedimentology and stratigraphic architecture of a point bar deposit, Lower Cretaceous McMurray Formation, Alberta, Canada. *Bull. Can. Soc. Petrol. Geol.*, **59**, 147–171.

Labrecque, P.A., Jensen, J.L. and Hubbard, S.M. (2011) Cyclicity in Lower Cretaceous point bar deposits with implications for reservoir characterization, Athabasca Oil Sands, Alberta, Canada. *Sed. Geol.*, **242**, 18–33.

La Croix, A.D. and Dashtgard, S.E. (2015) A synthesis of depositional trends in intertidal and upper subtidal sediments across the tidal–fluvial transition in the Fraser River, Canada. *J. Sed. Res.*, **85**, 683–698.

Leeder, M. and Bridges, P.H. (1975) Flow separation in meander bends. *Nature*, **253**, 338–339.

Lewin, J. (1983) Changes of channel patterns and floodplains. In: *Background to Palaeohydrology* (Ed. K.J. Gregory) Chichester, UK, John Wiley & Sons, 303–319.

Li, C., Chen, C., Guadagnoli, D. and Georgiou, I.Y. (2008) Geometry-induced residual eddies in estuaries with curved channels: Observations and modeling studies. *J. Geophys. Res.*, **113**, 2156–2202, DOI: 10.1029/2006JC004031

Lopez, S., Cojan, I., Rivoirard, J. and Galli, A. (2009) Process-based stochastic modelling: meandering channelized reservoirs. Analogue Number Model Sediment System: From Understanding to Prediction. In: *Int. Assoc. Sedimentol. Spec. Publ.*, **40**, Wiley-Blackwell, Oxford doi: 10.1002/9781444303131.ch5

Makaske, B. and Weerts, H.J. (2005) Muddy lateral accretion and low stream power in a sub-recent confined channel belt, Rhine-Meuse delta, central Netherlands. *Sedimentology*, **52**, 651–668.

Martinius, A.W. and Gowland, S. (2011) Tide-influenced fluvial bedforms and tidal bore deposits (Late Jurassic Lourinhã Formation, Lusitanian Basin, Western Portugal). *Sedimentology*, **58**, 285–324.

Martinius, A.W., Fustic, M., Garner, D.L., Jablonski, B.V.J., Strobl, R.S., MacEachern, J.A. and Dashtgard, S.E. (2017) Reservoir characterization and multiscale heterogeneity modeling of inclined heterolithic strata for bitumen-production forecasting, McMurray Formation, Corner, Alberta, Canada. *Mar. Petrol. Geol.*, **82**, 336–361.

Miall, A.D. (1996) The Geology of Fluvial Deposits: Sedimentary Facies, Basin Analysis. *Petrol. Geol.*, Springer-Verlag, New York, 582.

Mossop, G.D. and Flach, P.D. (1983) Deep channel sedimentation in the Lower Cretaceous McMurray Formation, Athabasca oil sands, Alberta. *Sedimentology*, **30**, 493–509.

Musial, G., Reynaud, J.Y., Gingras, M.K., Fenies, H., Labourdette, R. and Parize, O. (2012) Subsurface and outcrop characterization of large tidally influenced point bars of the Cretaceous McMurray Formation (Alberta, Canada): *Sed. Geol.*, **279**, 156–172.

Nanson, G.C. and Page, K.J. (1983) Lateral accretion of fine-grained concave benches on meandering rivers. *Int. Assoc. Sedimentol. Spec. Publ.*, **6**, 133–143.

Nardin, T.R., Feldman, H.R. and Carter, B.J. (2013) Stratigraphic architecture of a large-scale point-bar complex in the McMurray Formation: Syncrude's Mildred Lake Mine, Alberta, Canada. In: *Heavy-Oil and Oil-Sand Petroleum Systems in Alberta and Beyond* (Eds F.J. Hein, D. Leckie, S. Larter and J.R. Suter), *AAPG Studies in Geology*, **64**, 273–311.

Nittrouer, J.A., Shaw, J., Lamb, M.P. and Mohrig, D. (2012) Spatial and temporal trends for water-flow velocity and bed-material sediment transport in the lower Mississippi River. *Geol. Soc. Am. Bull.*, **124**, 400–414.

NOAA (2017) Tides & Currents: Datums for 9441187, Aberdeen WA. https://tidesandcurrents.noaa.gov/datums.html?units=1&epoch=0&id=9441187&name=Aberdeen&state=WA, Nov. 22, 2017.

Page, K. and Nanson, G. (1982) Concave-bank benches and associated floodplain formation. *Earth Surf. Proc. Land.*, **7**, 529–543.

Pyrcz, M.J., Boisvert, J.B. and Deutsch, C.V. (2009) ALLUVSIM: A program for event-based stochastic modeling of fluvial depositional systems. *Comput. Geosci.*, **35**, 1671–1685.

Raigemborn, M.S., Matheos, S.D., Krapovickas, V., Vizcaino, S.F., Bargo, M.S., Kay, R.F., Fernicola, J.C. and Zapata, L. (2015) Paleoenvironmental reconstruction of the coastal Monte Leon and Santa Cruz formations (Early Miocene) at Rincon del Buque, Southern Patagonia: A revisisted locality. *J. S. Am. Earth Sci.*, **60**, 31–55.

Reijenstein, H.M., Posamentier, H.W. and Bhattacharya, J.P. (2011) Seismic geomorphology and high-resolution seismic stratigraphy of inner-shelf fluvial, estuarine, deltaic and marine sequences, Gulf of Thailand. *AAPG Bull.*, **95**, 1959–1990.

Shanley, K.W., McCabe, P.J. and Hettinger, R.D. (1992) Tidal influence in Cretaceous fluvial strata from Utah, USA: a key to sequence stratigraphic interpretation. *Sedimentology*, **39**, 905–930.

Slatt, R.M. (2006) *Stratigraphic reservoir characterization for petroleum geologists, geophysicists and engineers.* Elsevier, **61**.

Smith, D.G. (1987) Meandering river point bar lithofacies models: modern and ancient examples compared. In: *Recent Developments in Fluvial Sedimentology, SEPM Special Publication*, **39**, 83–91.

Smith, D.G. (1988) Tidal bundles and mud couplets in the McMurray Formation, northeastern Alberta, Canada. *Bull. Can. Soc. Petrol. Geol.*, **36**, 216–219.

Smith, D.G., Hubbard, S., Leckie, D. and **Fustic, M.** (2009) Counter point bars in modern meandering rivers: Recognition of morphology, lithofacies and reservoir significance, examples from Peace River, AB, Canada: *Sedimentology*, **56**, 1655–1669, doi:10.1111/j.1365-3091. 2009.01050.x.

Smith, D.G., Hubbard, S.M., Lavigne, J.R., Leckie, D.A. and **Fustic, M.** (2011) Stratigraphy of counter-point bar and eddy-accretion deposits in low energy meander belts of the Peace-Athabasca delta, northeast Alberta, Canada. In: *From River to Rock Record: The Preservation of Fluvial Sediments and Their Subsequent Interpretation* (Eds S.K. Davidson, S. Leleu and C.P. North), *SEPM Spec. Publ.*, **97**, 143–152.

Stoner, S.B. (2010) Fluvial architecture and geometry of the Mungaroo Formation on the Rankin Trend of the Northwest Shelf of Australia (Doctoral dissertation, The University of Texas at Arlington).

Strobl, R.S., Wightman, D.M., Muwais, W.K., Cotteril, D.K. and **Yuan, L.P.** (1997) Geological modelling of McMurray Formation reservoirs based on outcrop and subsurface analogues. In: Petroleum Geology of the Cretaceous Mannville Group, Western Canada (Eds S.G. Pemberton and D.P. James), *Can. Soc. Petrol. Geol. Mem.*, **18**, 292–311.

Su, Y., Wang, J.J. and **Gates, I.D.** (2014) Orientation of a pad of SAGD well pairs in an Athabasca point bar deposit affects performance. *Mar. Petrol. Geol.*, **54**, 37–46.

Su, Y., Wang, J. and **Gates, I.D.** (2017) SAGD Pad performance in a point bar deposit with a thick sandy base. *J. Petrol. Sci. Eng.*, **154**, 442–456.

Thomas, R.G., Smith, D.G., Wood, J.M., Visser, J., Calverly-Range, E.A. and **Koster, E.H.** (1987) Inclined heterolithic stratification — Terminology, description, interpretation and significance. *Sed. Geol.*, **53**, 123–179.

USGS (2017) United States Geologic Survey: National Water Information System. 12013500 Willapa River near Willapa, WA. Accessed June 25th 2017.

USGS (2017) United States Geologic Survey: National Water Information System. 12031000 Chehalis River at Porter, WA. Accessed June 25th 2017.

Van den Berg, J.H., Boersma, J.R. and **Gelder, A.V.** (2007) Diagnostic sedimentary structures of the fluvial-tidal transition zone–Evidence from deposits of the Rhine and Meuse. Netherlands Journal of Geosciences/Geol. Mijnbouw, **86**(3).

Willis, B.J. and **Sech, R.** (2017) Quantifying impacts of fluvial intra-channel-belt heterogeneity on reservoir behaviour. *Int. Assoc. Sedimentol. Spec. Publ.*, **48**.

Willis, B.J. and **Tang, H.** (2010) Three-dimensional connectivity of point-bar deposits. *J. Sed. Res.*, **80**, 440–454.

Wood, J.M. (1989) Alluvial architecture of the Upper Cretaceous Judith River Formation, Dinosaur Provincial Park, Alberta, Canada. *Bull. Can. Soc. Petrol. Geol.*, **37**, 169–181.

Woodyer, K.D. (1975) Concave-bank benches on Barwon River, NSW. *Australian Geographer*, **13**(1), 36–40.

Yan, N., Mountney, N.P., Colombera, L. and **Dorrell, R.M.** (2017) A 3D forward stratigraphic model of fluvial meander-bend evolution for prediction of point-bar lithofacies architecture. *Comput. Geosci.*, **105**, 65–80.

Int. Assoc. Sedimentol. Spec. Publ (2018) **48**, 251–272.

Fill characteristics of abandoned channels and resulting stratigraphy of a mobile sand-bed river floodplain

ARVED SCHWENDEL[†‡§], ROLF AALTO[‡], ANDREW NICHOLAS[‡] and DANIEL PARSONS[†]

[†] *School of Environmental Sciences, University of Hull, Hull, UK*
[‡] *College of Life and Environmental Sciences, University of Exeter, Exeter, UK*
[§] *School of Humanities, Religion and Philosophy, York St John University, York, UK*

ABSTRACT

Floodplains are diverse sedimentary environments where infill processes of abandoned channels interact with overbank sedimentation and bank erosion. The result, particularly in river systems with high suspended load and rapid channel migration, is a complex three-dimensional mosaic of deposits with spatial variability in terms of grain-size, age, organic carbon content and resistance to erosion. Abandoned channels represent a significant deposition volume in fluvial systems that can accommodate large proportions of the equivalent material mobilised during their abandonment. However, time scales and fill processes vary between different kinds of abandoned channels and the sediment calibre involved and are not fully understood, particularly in respect to highly dynamic sand-bed rivers. This study investigates time scales and spatio-temporal patterns of infill of abandoned chute channels and abandoned channel segments left behind following neck cutoff of meander bends. The study focuses on the Rio Beni, a large, tropical, sand-bed river in the Bolivian Amazon basin. Electrical resistivity ground imaging is used to elucidate the stratigraphy of floodplains and satellite imagery is employed to investigate contemporary fill processes and rates. Given suitable bend migration patterns, chute channels may remain stable for several years but are eventually abandoned and rapidly filled with bed material during a single flood season. Smaller scroll sloughs can convey coarse bedload across point bars and, when filled, present stratigraphic bodies similar to chute fills. Abandoned meander bends tend to develop plug bars at both ends immediately after cutoff. Of these bars, the downstream plug aggrades at a faster rate due to the often larger diversion angles with the main channel and efficiently seals off the bend. The subsequent infill of the channel is a function of hydraulic connectivity and distance to the active channel as well as rate of lake deposition. Considerable overbank deposition can increase the spatial sedimentological heterogeneity of these floodplains, which needs to be taken into account in floodplain evolution models.

Keywords: Meander cutoff, overbank sedimentation, floodplain evolution, channel fill, electrical resistivity ground imaging

INTRODUCTION

Alluvial floodplains comprise an assemblage of deposits including fines from overbank sedimentation, splay sediments and fill of abandoned channels. The interplay between floodplain sedimentation and channel migration effectively dominates alluvial floodplain stratigraphy (Nanson & Hickin, 1986; Salo *et al.*, 1986; Lauer & Parker, 2008). This relationship is also a key control on channel conveyance capacity, the dynamics of floods, long-term floodplain morphodynamics, avulsion; and the ecological functioning of the floodplain environment (Ward *et al.*, 2002; Gueneralp & Marston, 2012; Hajek & Edmonds, 2014). Understanding this feedback and interplay is fundamental in building a set of process-product relationships that explain floodplain sedimentology and the stratigraphic record. It is also key to comprehending a range of

Fluvial Meanders and Their Sedimentary Products in the Rock Record, First Edition.
Edited by Massimiliano Ghinassi, Luca Colombera, Nigel P. Mountney and Arnold Jan H. Reesink.

phenomena, including unlocking longer-term climate signals preserved in the deposits (Aalto *et al.*, 2003) and their significance as hydrocarbon reservoirs (e.g. Smith *et al.*, 2009). The latter can develop in coarse-grained lithofacies while sedimentary bodies dominated by silt or clay will decrease percolation of fluids and therefore limit the extent of these reservoirs. Despite the potential significance of these feedbacks and controls on sedimentation, few studies have quantified process-product linkages between channel evolution and floodplain construction in dynamic settings with rapid channel migration histories and high overbank deposition rates of suspended sediments during floods (Schwendel *et al.*, 2015).

The heterogeneity in sedimentation results in spatial variability in bank erodibility and hydraulic connectivity of the channel to the floodplain, which in turn influences channel migration, overall planform development and meander belt width as observed in contemporary and ancient river systems (Allen, 1965; Hajek *et al.*, 2010; Hajek & Wolinsky, 2012).

Bank erosion and avulsion processes often result in bend cutoffs, which can occur by formation of chute channels or formation of channels across the neck of a bend, with the subsequent abandonment of chutes and bifurcate channels (Schwendel *et al.*, 2015) and their incorporation into the floodplain (Fig. 1). The style and rate of channel cutoff is known to have important implications for floodplain evolution. For example, Schwendel *et al.* (2015) showed that at the Rio Beni the migration rate of bends eroding into a cutbank made out of channel or oxbow lake deposits (58.8 m a^{-1} and 98.8 m a^{-1} respectively) is much higher than that of bends migrating into point bar deposits (43.3 m a^{-1}). Thus, point bars with a high number of deep chute channels will experience a more variable erodibility than predicted by a model assuming uniform erosion rates.

Cutoff processes are known to modify sediment supply, e.g. meander cutoffs and chute channel formation can mobilise substantial amounts of sediment (Zinger *et al.*, 2011), whilst abandoned channels represent an important storage volume (Lauer & Parker, 2008). Deposition in abandoned channels results in deposits that will be significantly different from the surrounding floodplain (Fisk, 1944; Allen, 1965). The fill of chute channels and scroll sloughs on the point bar can also form coarse bedload sediment bodies of varying depth placed within finer point bar and counter-point bar deposits which subsequently have varying resistance to meander migration (Fig. 1). The timescales of abandoned channel evolution, from inception to sedimentary fill, will influence the style of fill and the deposit produced. The influence of hydraulic and sedimentary connectivity and timescales is also a poorly constrained control on deposit characteristics (e.g. Rowland *et al.*, 2005). Rapid bedload plugging of the entry and/or channel exit points will rapidly reduce sediment exchange with the abandoned channel sections. Overbank deposition will interact with abandoned channels as they fill; however, the style of this interaction will control the fractionation of the suspended sediment and bedload components that comprise the fill (Rowland *et al.*, 2005; Citterio & Piégay, 2009). As such, the abandonment timescale and the evolution of hydraulic connectivity will form an important, but hereto largely unquantified, control on floodplain stratigraphy (Gautier *et al.*, 2007). This study aims to constrain fill processes of abandoned channels on the floodplain in regard to their timescale and spatial extent. This is particularly relevant at very mobile rivers with high overbank sedimentation rates where alluvial stratigraphy is complex and frequently modified.

Herein, we employ a combination of geographic information system (GIS) analysis of satellite imagery and direct field observations, including transects of sediment cores and geophysics along 3200 m of floodplain across a recently abandoned floodplain section of the Rio Beni in the Bolivian Amazon. This combination allows us to track the development of the abandoned channel and map its evolution and fill timescale. The geophysics surveys and core analysis subsequently allow us to correlate the deposit characteristics with the fill processes.

Study site and methods

The Rio Beni drains 283,350 km^2 of the High Andes, their eastern piedmont and lowland basin; and is a major tributary of the Rio Madeira (Roche & Fernandez, 1988; Fig. 2). It provides >70% of the sediment load of the Rio Madeira, which in turn is the major contributor of sediment to the Amazon River (Guyot *et al.*, 1999). The upper catchment comprises semi-arid basins of the altiplano with highly erodible Tertiary-Quaternary sedimentary series, humid valleys (rainfall up to 6000 mm a^{-1}) on the eastern slopes of the Andes

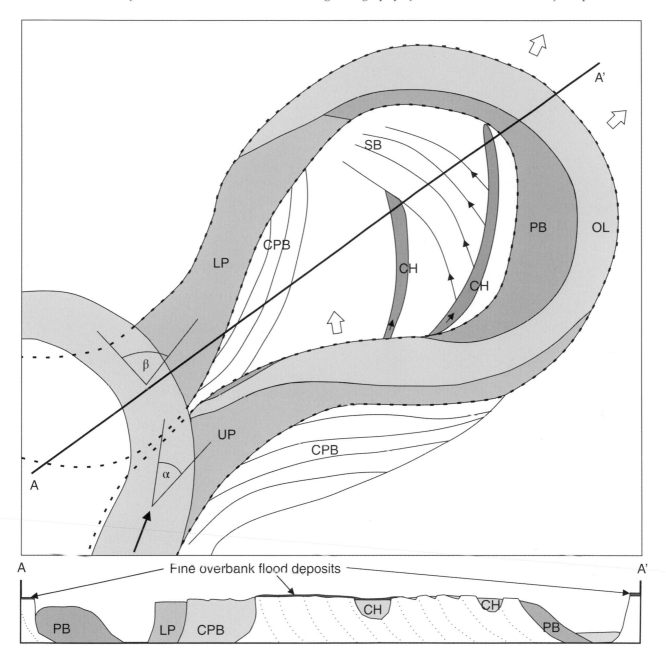

Fig. 1. Schematic depiction of floodplain sediment heterogeneity around a recently cutoff meander. Usually the lower diversion angle β is greater than the upper diversion angle α which results in a blocking of the lower limb of the abandoned channel with a large lower plug bar (LP). The resulting oxbow lake (OL) may be still partially connected to the new channel via a flood channel across the upper plug bar (UP). The floodplain is an assemblage of counter point bar deposits (CPB), point bar sediments (PB), scroll bars (SB) and former chute channels (CH). The latter two may be filled with coarse bed deposits as indicated by the arrows. The white arrows indicate deposition pathways of fines during floods.

and the piedmont with Palaeozoic rocks and Tertiary deposits. The lower catchment consists of vast plains of alluvial Late-Miocene and Quaternary sediments in the Andean foredeep basin (Dumont, 1996; Guyot *et al.*, 1999). The Rio Beni has an Austral tropical pluvial flood regime with a distinct flood season from December to March (Gautier *et al.*, 2007). The sediment load of the Rio Beni amounts to 219 Mt a^{-1} (Latrubesse & Restrepo, 2014) and its clay mineral composition indicates fresh Andean sources (Guyot *et al.*, 2007).

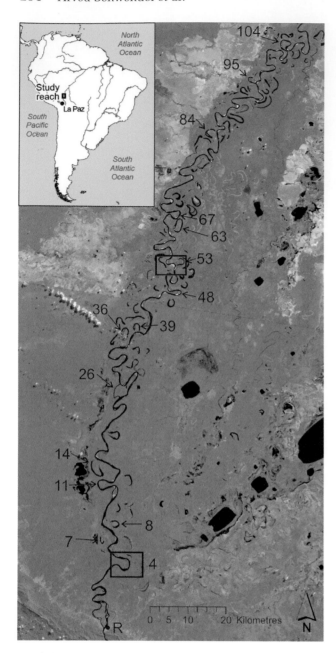

Fig. 2. Study reach of the Rio Beni and its location within South America (insert). The areas covered by Fig. 4 and Fig. 9 are outlined by rectangular boxes and the locations of meanders in the text are shown with their number. R indicates the town of Rurrenabaque on the foothills of the Andes. The background Landsat 5 image was taken during the dry season in 2011.

The 375 km study reach is situated in the fore-deep basin just downstream of the piedmont fan (Fig. 2) where the bed material transitions from cobble-gravel to sand. Within the study reach bed-load size is fine sand (D_{50}: 0.01 to 0.15 mm) while suspended load is medium silt (D_{50}: 0.011 mm)

(Dumont, 1996). The single-thread channel, with an average width of 430 m, is freely meandering in an extensive forested floodplain and essentially unaffected by human influence or Holocene sea-level changes. The course of the channel belt has shifted over the Holocene in a counter-clockwise direction from a north-east to a northerly orientation (Dumont, 1996; Dumont & Hannagarth, 1993; Plafker, 1964), which responds to tectonic forcing from south-west to north-east striking differential subsidence and uplift (Allenby, 1988; Plafker, 1964); and a north-striking fault line that marks the western margin of the recent channel belt. The gradient of the study reach decreases from 0.0002 m m^{-1} at the toe of the fan to 0.00007 m m^{-1} at the lower end (Schwendel *et al.*, 2015). The channel sinuosity can be related to floodplain heterogeneity and varies between 1.3 and 2.7 in different sub-reaches (Schwendel *et al.*, 2015, see also Dumont, 1996; Gautier *et al.*, 2007). The reach experiences high overbank deposition and meander migration rates (0.1 channel widths a^{-1}) (Aalto *et al.*, 2003; Schwendel *et al.*, 2015) with relatively short periods for meander evolution from bend inception to neck cutoff (down to 10 a). As a result, within the time documented by aerial imagery (e.g. since 1960), 33 cutoffs across the neck have been observed (Gautier *et al.*, 2007; Schwendel *et al.*, 2015).

The hydrological regime of the study reach can be characterised using data from a gauging station located 20 km upstream of Rurrenabaque (Fig. 2) which indicates a mean discharge of 2300 m^3 s^{-1} with flood peaks over 20,000 m^3 s^{-1} (Schwendel *et al.*, 2015). At discharges larger than 6000 m^3 s^{-1}, floodplain inundation begins (Aalto *et al.*, 2003; Schwendel *et al.*, 2015). Overbank sedimentation is a function of the timing of the flood peak, floodplain topography, distance from the channel and flood level (Aalto *et al.*, 2003; Gautier *et al.*, 2010).

The evolution of abandoned channels and potentially related factors such as annual bend migration rates and channel curvature were elucidated from geo-referenced multi-spectral Landsat imagery and earlier aerial photography for the time between 1960 and 2015 (Schwendel *et al.*, 2015). These were processed and analysed in ArcGIS (ESRI, Redlands, USA). Bends were given numbers which increased in down-valley direction (Fig. 2). The persistence of chute channels and the influence of bend migration styles on abandonment were investigated at 27 bends while another set of 17 bends were used to

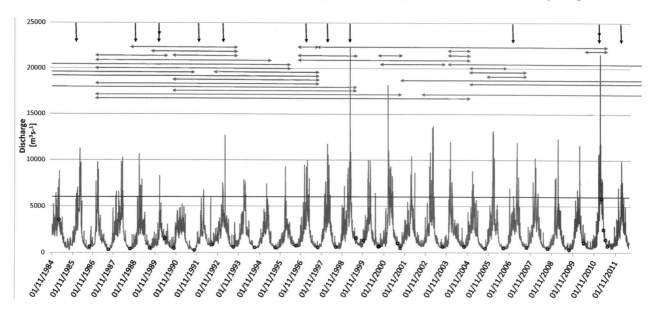

Fig. 3. Hydrograph at Rurrenabaque between 1984 and 2012 (courtesy of ORE HYBAM) with dates of the satellite imagery shown in circles and approximate time of neck cutoffs shown by vertical black arrows at top. The presence of stable chute channels is indicated by horizontal grey lines while the red line indicates bankfull discharge (6000 m³ s⁻¹). Note that stable chute channels might still be visible on the floodplain on the imagery following abandonment because vegetation might take a year to fully establish.

assess the infill characteristics of channels immediately after a neck cutoff. Herein only ratios of plug bar areas were compared over time in order to account for the varying water levels when the photos were taken (Fig. 3).

The sedimentary and geomorphological characteristics of abandoned chute channels and an infilled meander bend have been investigated in detail at bend 4 (Fig. 2). This bend has been formed prior to 1960 but did not undergo strong migration (20 m a⁻¹) while rotating down-valley before 1975 (Fig. 4). Between 1975 and 1999 it extended laterally (43 m a⁻¹) in SE direction and around 2000 the bend extension axis rotated to NE in several stages, probably as a result of contact with a resistant and clay-rich oxbow fill in the East. From 2004 to 2010 migration to the NE was dominant, which had an expanding effect on the entire bend. More recently migration to the South has commenced resulting in bend expansion in both directions.

The floodplain surface and sub-surface have been investigated at three linear transects at this bend using Electrical Resistivity Ground Imaging (ERGI, see details below). Data were obtained in 2011. The point bar transects T1 (1280 m) and T2 (1120 m) were oriented in WNW-ESE direction (~296°) and NE-SW direction (~52°), respectively, in order to align with the bend extension direction

between 1975 and 1987 and between 2004 and 2010 (Fig. 4F). A third transect, T3, with a length of 850 m, of which 800 m underwent electrical resistivity measurements, is located on the cutbank side of the bend. It is oriented in a SW-NE direction (~47°) which was chosen in order to traverse an oxbow channel perpendicularly.

The extent and stratigraphy of infilled channels on the floodplain was investigated using two-dimensional Electrical Resistivity Ground Imaging (Banton *et al.*, 1997). A Tigre 128 electrical resistivity system in combination with ImagerPro2006 software (Allied Associates Geophysical Ltd, Dunstable, UK) was employed with 64 electrodes in a roll-along survey mode (Loke, 2000). The electrodes were spaced 5 m apart and used in a Wenner α array configuration (Reynolds, 2011) which provides high sensitivity to vertical resistivity gradients (Loke, 2000). With the employed setup the maximal investigation depth was approximately 50 m at the centre of the array with a spatial resolution of 2.5 m (Ward, 1990). ERGI is regarded as an efficient means to elucidate subsurface information from surface surveys using electrical resistivity tomography (ERT; Baines *et al.*, 2002) and has been used widely in alluvial settings (Chambers *et al.*, 2012; Clifford & Binley, 2010; Crook *et al.*, 2008; Gourry *et al.*, 2003; Hausmann *et al.*, 2013;

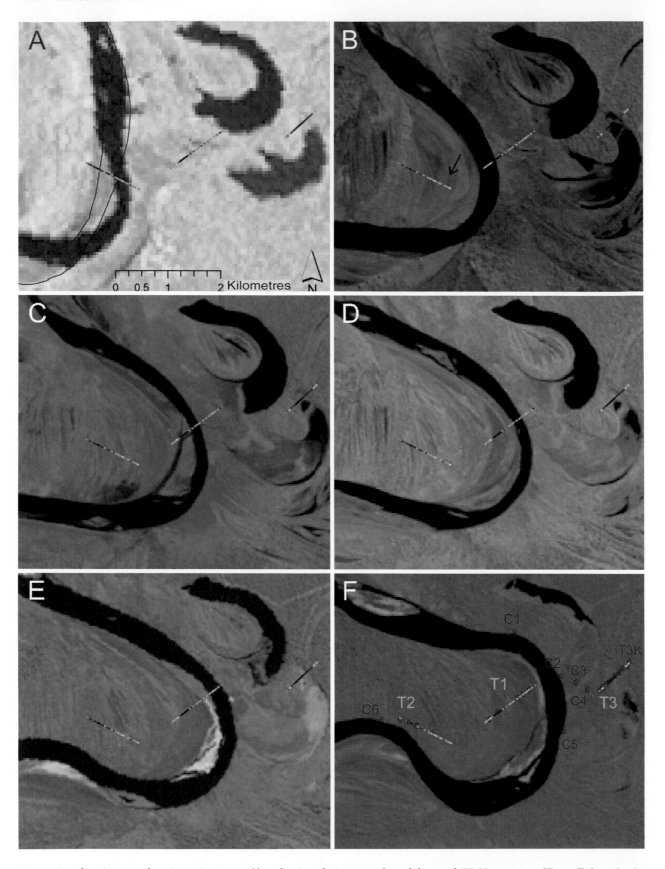

Fig. 4. Landsat images showing migration of bend 4 in relation to oxbow lakes and ERGI transects (T1 to T3) at six time steps: (A) 1975 with the 1960 channel outlined in red, (B) 1984, (C) 1991, (D) 1995, (E) 2001; and (F) 2011. In the latter image the location of cores (red dots) is shown with those cores not on transects labelled. The colour code of the transects represents surface resistivity and corresponds to the colour scale in Fig. 6A.

Leopold *et al.*, 2013). The raw apparent resistivity data were checked for obvious measurement errors such as isolated unrealistic high values which could be often traced to an individual malfunctioning electrode. These were removed and the resulting gaps were filled by linear interpolation. Inversion modelling was carried out in Res2DInv (Geotomo Software, Gelugor, Malaysia) using a robust model constrain technique (Loke *et al.*, 2003) to produce a 2D model of the subsurface. Such a technique is most suitable because sharp boundaries between different sedimentological bodies were expected (Loke, 2000). The iterative modelling process was stopped when the absolute error between calculated and apparent resistivity did not improve significantly (Geotomo, 2011). The chosen models at transects T1, T2 and T3 were reached after 3, 3 and 2 iterations with absolute errors of 4.3, 7.2 and 3.7 Ωm respectively. Auger cores were taken at points of interest along the transects in 2013 in order to provide ground-truthing (Baines *et al.*, 2002). The cores were sampled at 1m depth intervals and at any obvious stratigraphic changes down core.

Floodplain sedimentation rates and soil characteristics were assessed from push cores of up to 3m in depth which were collected along the cut-bank of 18 bends and straight channel reaches, at regular distances away from the cutbank and along transects (Fig. 4F). The cores were photographed and described in the field and selected 2cm depth intervals were extracted in the laboratory. These samples were analysed for grain-size distribution using a Sedigraph 5100 (Micromeretics Instrument Corp., Norcross, USA) and clay-normalised absorbed excess ^{210}Pb activity following the CIRCAUS method outlined in Aalto & Nittrouer (2012). This method allows the detection of individual episodic deposition events as well as steady accumulation on decade-to-century time scales. These data provided spatially distributed floodplain sedimentation rates across the study site. In addition, the grain-size distribution of bed samples along an active chute channel was assessed at bend 4 (Fig. 4F).

RESULTS

Stable chute channels

Chute channels across point bars are common along the Rio Beni and, while many are abandoned or shift location within a year, some remain stationary over many years (up to 25 years in some cases). The bed grain-size along an active chute channel at bend 4 was in the silt range with a mean D_{50} of 45µm and did not show any longitudinal trend. These channels usually develop on unvegetated point bars and are then incorporated into the vegetated floodplain when the bend migrates away. Cutoff of meanders via these stable chute channels has not been observed at the Rio Beni since 1975; however, on four occasions cutoffs across the neck of a bend via channels of tributaries to the Rio Beni, or relatively recently abandoned channels that connected across the neck, have been observed.

Thirty-two chute channels, which were stable for more than one year, were analysed at 27 bends. The abandonment of stable chute channels coincided with greater than average percentages of extending (increase in bend amplitude, 33%), expanding (increase in meander wavelength, 25%) and stationary bends (29%) (Fig. 5). Although some bends showed an increase in migration rate at the time of abandonment, no general relation between the inception and abandonment of these channels and bend migration rate could be established. The abandonment of stable chutes is clustered in certain years such as 1993, 1997, 1999, 2004 and 2007; however, these years show hydrologically different patterns (Fig. 3). While the flood characteristics in 1999 are extreme, the floods in 1993 and 1997 succeed periods of low cumulative discharge over the wet season, flood magnitude and flood duration (Gautier *et al.*, 2010) and therefore the floods in these years might be morphologically more relevant. In general, no clear relationship between flood characteristics and abandonment of stable chute channels has been found even when flood data from the previous three years have been used, as suggested by Schwendel *et al.* (2015).

Transect 2

Transect 2 (T2) encompasses pre-1960 deposits between approximately 825m and the north-western end which are located at the tail end of the then SE oriented meander (Figs 4A and 6A). As indicated by ERGI and the cores T2D and T2E (Fig. 6B), these deposits of up to 14m in depth are of low resistivity (e.g. up to 50Ωm) and high silt content which corresponds to relative fine counter point bar deposits. By 1975 the channel edge has only shifted about 100m to the SE but, with the

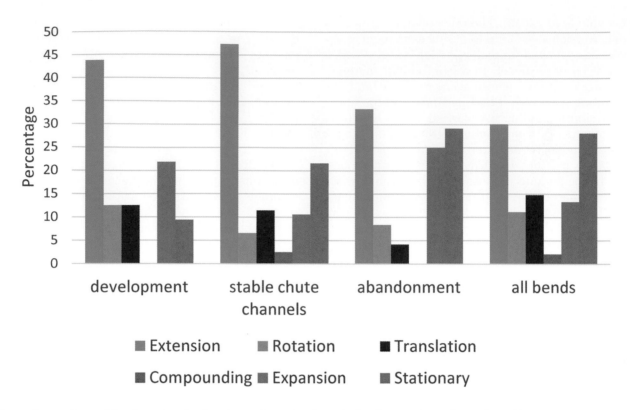

Fig. 5. Comparison of dominant bend migration styles at bends with developing chute channels, stable chutes and during their abandonment as well as all bends in the study reach (Schwendel *et al.*, 2015).

down-valley rotation of the bend, the depositional environment was the point bar tail which is reflected by the coarsening of the deposits from silt to silty sand below a silt dominated surface layer (T2B in Fig. 6B). The latter consist of silt-dominated overbank deposits at considerable but steady accumulation rates of up to $7.5\,cm\,a^{-1}$ (Table 1). Signs of surface erosion such as up to 0.5 m-deep, sharp-edged flood ruts, aligned perpendicular to the transect and free of vegetation and traces of deposition were common in the NW half of the transect (Fig. 7). The undulating boundary between low resistivity and medium resistivity SE of 740 m shows the development of a scroll bar surface with the increase in migration rate after 1975 (Fig. 4B). Subsurface resistivity increases in south-eastern direction, which reflects coarsening due to the relative change in transect location towards the head of the point bar. The ERGI model shows a lens of medium resistivity near the surface at 580 m; however, the raw data reveals that this modelled lens is based only on three level 1 (surface) resistivity measurements and could be an artefact, for example induced by shallow, horizontally spreading buttress roots of bibosi trees (*Ficus insipida*).

In the south-eastern half of the transect the forest vegetation indicates little disturbance and signs of erosion are limited. [210]PB chronology suggests steady accumulation at $1\,cm\,a^{-1}$ at core T2A (Table 1). Between 110 m and a scroll slough at 260 m the ERGI model shows an up to 14 m-deep distinct body of high resistivity (e.g. >200 Ω m) which can be traced on the satellite imagery as a former chute channel (elongated dark green area shown by arrow in Fig. 4B). It is adjoined in the ESE by another shallower high resistivity zone which is probably a wide trough between scroll bars. Lack of satellite imagery does not allow a detailed reconstruction whether these two structures evolved and filled separately or not.

Transect 1

Transect 1 (T1) covers an area that experienced point bar deposition from 1986 to 2009. It has been located downstream of the bend apex during most of that period (Fig. 4). T1 shows a basement of material with a resistivity larger than 50 Ω m

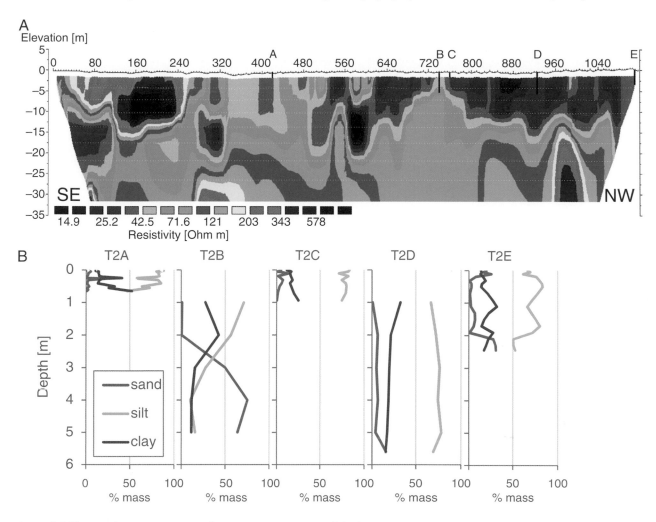

Fig. 6. (A) Electrical resistivity ground imaging inversion model of transect 2 on the point bar side of bend 4. The location and approximate depth of cores are shown. (B) Percentage of clay, silt and sand in floodplain cores along transect 2. The location of bend 4 within the Rio Beni study reach is shown in Fig. 2.

below 15 to 17 m in depth (Fig. 8A), which suggests grain sizes up to sand, if wet (Anomohanran, 2015; Sudha *et al.*, 2009). These are deposits probably older than the recently reworked material above and contain a number of higher resistivity structures. As in T2, the depth of recent point bar deposition is roughly 15 m, which corresponds well with maximum depth from bank top to the deepest part of the channel shown in eight cross-sections at the same bend (Gautier *et al.*, 2010). The more recent low resistivity sediments above the basement are silts, in parts, very wet as indicated by core T1E. This is largely a result of point bar deposition of fines towards the bar tail.

At 1040 m along T1 there is a distinctive rise in the surface elevation of 1.2 m, beyond which the forest was markedly more mature and has estab-

lished after 1989. On the edge of the elevated part of the floodplain the inversion model (Fig. 8A) shows a body of high resistivity of approximately 80 m in width and 5 m in depth. Here, the core T1G did not reach sufficient depth to penetrate the silt-dominated surface layer which, according to ^{210}PB chronology (Table 1), has an average sedimentation rate of 3.7 cm a^{-1} but grain-size characteristics of the high resistivity body can be inferred from core T1F as sand-dominated. This is interpreted as a chute channel that evolved downstream of the apex of the point bar in 1988. It was abandoned in the following year due to aggradation and extension of the point bar to the NE. It must have been filled rapidly, potentially aided by development of a scroll ridge at its downstream end (Fig. 4C) and showed a light vegetation cover

Table 1. ^{210}Pb geochronology of floodplain sediments at the Rio Beni (for locations of the cores see Figs 4F, 6A, 8A and 11A).

Core	Depth [m]	Oldest sediment [a]	Depth to oldest sediment [m]	Penetration rate [10^2 m a^{-1}]	Inventory rate [10^2 m a^{-1}]	Notes
T2A	0.66	69	0.66	1.0	n/a	Probably steady
T2C	0.96	15	0.96	6.4	n/a	1 older deposit
T2E	2.46	33	2.46	7.5	2.9	Probably steady
C6	0.52	1	0.52	52.0	n/a	1 fresh deposit from 2011
T1A	1.46				n/a	Problematic concentrations
T1E	1.37				n/a	Problematic concentrations
T1G	0.84	23	0.84	3.7	n/a	1 older deposit
T3K	2.15	51	0.70	1.4	0.3	Old sediment below, more episodic at top
T3A	1.34	32	0.70	2.2	0.4	Old sediment below, not episodic
T3B	2.52	44	0.70	1.6	0.3	Old sediment below, not episodic
T3C	0.62	17	0.62	3.6	n/a	Episodic
T3D	1.44	19	0.92	4.8	n/a	Episodic
T3E	0.70	46	0.70	1.5	n/a	Episodic, 0.52 m from 2011
T3F	2.25	85	2.25	2.6	1.6	Episodic, 0.25 m from 2011
T3H	0.93	4	0.93	23.3	n/a	0.45 m from 2011
T3J	1.12	1	1.12	112.0	n/a	All from 2011
C4	1.45	19	1.45	7.6	n/a	1 older deposit
C3	0.74	21	0.74	3.5	n/a	1 older deposit
C2	1.16	25	1.16	4.6	n/a	1 older deposit
C1	1.06	9	1.06	11.8	n/a	1 recent deposit
C5	0.74	9	0.74	8.2	n/a	1 recent deposit

in 1990. In 1991 a large chute channel across the point bar had developed with its tail end NE of the filled trough. This chute persisted until 1994 preventing establishment of vegetation to facilitate gradual aggradation of the floodplain (Fig. 4D). The resulting step in the terrain is still visible on T1 at 1050 m (Fig. 8A). This chute channel appears to be filled in 1994 with some vegetation established on a scroll ridge to the NE. It is shown on T1 between 700 m and 1100 m with a modelled depth of 10 m and some internal horizontal zonation. At its centre, grain-size samples are dominated by sand with silt and clay fractions increasing with depth (Fig. 8B). The moisture increased below 2 m but the sediments show no signs of hydromorphy.

The surface in the north-eastern part of the transect shows regular ridges and troughs (Figs 4 and 8A) deposited after 1995 at a fairly constant bend migration rate (22 to 40 m a^{-1}). The vegetation was a mixture of first pioneer species such as *Gynerium sagittatum* (arrow cane), bushes of *Tessaria integrifolia* and *Cecropia membranacea* trees with the former most common in the troughs and *Cecropia* trees on the ridges. The growth forms of these trees showed adaptation

to regular inundation such as stilt roots. The presence of *G. sagittatum* appeared to be more related to ecological disturbance, such as fallen trees or signs of flood deposition, than to surface morphology. Further to the SW the terrain flattens and more mature trees become more common, interspersed in places by *G. sagittatum*, indicating disturbance. The top 5 m of the inversion model between 0 and 640 m shows a number of zones of elevated resistivity (>50 Ω m) which are mainly due to elevated resistivity in level 1 of the raw ERGI data and thus surficial. While the surface layer consists of silt-rich deposits (T1A in Fig. 8B) the content of sand increases with depth (T1C and T1D) or contains distinct layering between deposits of varying colour and grain-size (silt, silty sand and sand) as in T1B (Fig. 8B). Location T1C is situated in a trough while location T1D is only 5 m away on the shoulder of the trough (>1 m higher). With the relief difference taken into account the coarsening of grain-size starts at a similar depth. The silty sand at the bottom of the cores was usually very moist, with signs of hydromorphy. To the north-east of 200 m along the transect the inversion model shows increasing resistivity between

Fig. 7. (A) Evidence of overbank flow deposition behind trees (flow direction away from the observer) and partial burial of buttress roots on transect 2. The handle of the push corer is approximately 0.3 m-long. (B) Evidence of 0.2 m-deep scour ruts on the floodplain at transect 2. The orientation of the ruts in B is aligned with the indicated flow direction in A and is to the NNE across the point bar.

the surface layer and the basement, suggesting an increase in grain-size. These sediments were deposited after 2003 (Fig. 4E) which coincides with the sudden north-eastward shift in migration direction and thus bend-apex processes becoming dominant at this part of the bend with increased coarser sedimentation on the point bar (Nanson & Croke, 1992).

Abandoned meanders

The characteristics of 17 meander bends abandoned following neck cutoff between 1986 and 2015 have been analysed and are summarised in Table 2. Field observations (Fig. 9) and satellite imagery show that the lower limb of the abandoned channel is plugged within the same flood event while the upper end may stay connected to the river. Cutoff bends with high curvature have significantly higher diversions angles on the lower end (Pearson correlation R=0.62, n=17, p=0.008) and a larger difference between upper and lower diversion angles (Pearson correlation R=0.54, n=17, p=0.027). Although this has no immediate effect on the initial differential development of upper and lower plug areas, after two years the difference between lower and upper diversion angle is significantly related to the ratio of the plug areas (Pearson correlation R=-0.55, n=14, p=0.04; Fig. 10). The rapid planform adjustment at the neck cutoff usually results in expansion and down-valley translation of the residual bend. This can lead to a short-term (e.g. 2a) migration into the plug of the lower limb of the abandoned meander and thus influence the measured plug area.

The analysis of the development of the initial fill of a cutoff channel was conducted on bend 53 and 95, both of which experienced a neck cutoff in the 2011 wet season (December 2010 to March 2011, Fig. 3). On March 13th at almost bankfull water level both cutoff meanders were connected showing turbid flow patterns although to a lesser extent than in the new cut channel (Fig. 9). A month later discharge had decreased significantly but at both bends the abandoned channels appeared to be connected with zones of high turbidity. In May, the downstream end of the cutoff was fully blocked while at the upstream end a plug bar had emerged reaching 2/3 (B95) and 4/5 (B53) across the entrance from upstream and resulting in lower turbidity in the cutoff. In June the water level was lowest and thus the plug structure is clearly visible. The lower plugs showed signs of light vegetation and extend towards the old point bar. At the upper end the plug bars had almost closed the entrance with flood channels developing. They emerged up to 2.5 m above the unusually low water level at both sites in September with only a small flood channel at bend 53 and a 25 m-wide, but very shallow, connection at bend 95. The lower plugs reached up to 3 m (bend 95) and more than 6 m (bend 53) above water-level and were colonised by several months old wind-dispersed plants. The lower plug surface at bend 53 was 1 m below the floodplain level and consisted of silty sand throughout the vertical profile at the eroding bank that faced the river (Fig. 9).

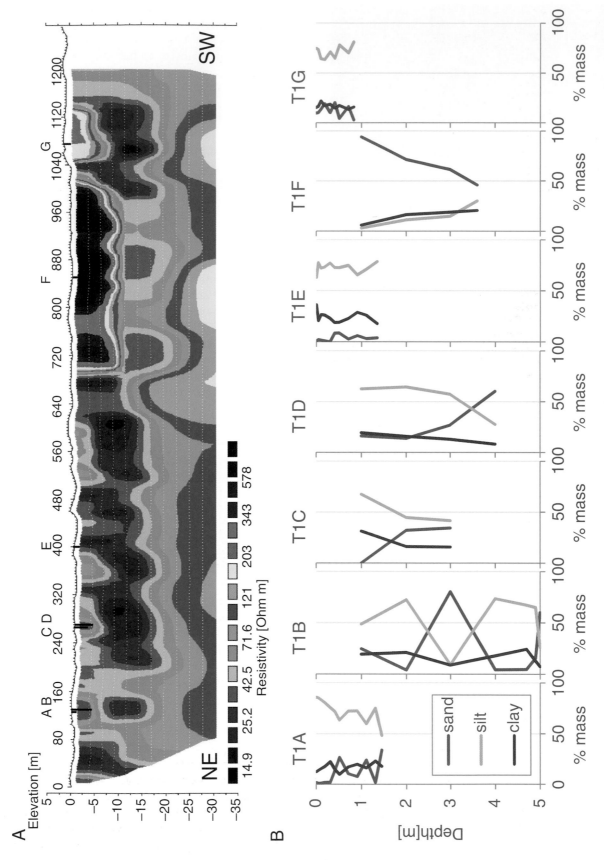

Fig. 8. (A) Electrical resistivity ground imaging inversion model of transect 1 on the point bar side of bend 4. The location and approximate depth of cores are shown. (B) Percentage of clay, silt and sand in floodplain cores along transect 1. The location of bend 4 within the Rio Beni study reach is shown in Fig. 2.

Table 2. Diversion angles between cutoff channel and new channel, ratios between upper and lower plug bar area initially after cutoff, after 1 year and after 2 years and curvature of the cutoff bend at 17 neck cutoffs along the Rio Beni between 1986 and 2015.

Bend	Upper diversion angle [°]	Lower diversion angle [°]	Initial plug area ratio	plug area ratio 1 a	plug area ratio 2 a	Cutoff bend curvature [m⁻¹]
n	17	17	17	15	14	17
Minimum	12	67	0.00	0.00	0.01	0.0006
Mean	47	103	0.62	0.86	1.03	0.0022
Maximum	81	149	2.00	2.77	3.56	0.0042

Transect 3

The sedimentological record of an old infilled neck cutoff channel has been investigated on the cutbank side of bend 4. The abandoned meander bend is at least 10 years older than another meander to the NW which has been cut off before 1960 (Fig. 4A). Transect 3 crosses the meander channel perpendicularly at approximately three quarters of the channel length from its upstream end. The location of the transect corresponds with the lower margin of the oxbow lake until it was more or less filled in 1999 (Fig. 4E). The channel is clearly outlined in the surface morphology with a distinct drop in terrain of 0.9 m at 270 m and a notable rise of 0.8 m around 730 m (Fig. 11A). The surface of the channel fill is relatively flat but interrupted by a muddy 10 m-wide and 0.8 m-deep flood channel that connects to the younger cutoff in the NW. Around the flood channel and towards the cutbank margin the ground was wet with few understorey plants and signs of recent inundation SW of the flood channel and dense stands of *Heliconia rostrata*, a fast-colonising perennial understorey plant (Berry & Kress, 1991), in disturbed but less recently flooded areas. Overbank sedimentation rates of silt-rich sediments in these locations are substantial (around 4 mm a⁻¹) and interlayering of silty and clay-rich sediments reflect episodic deposition processes (Table 1, Fig. 11B). Towards the south-western part of the channel sedimentation rates are lower but with larger contributions from the 2011 flood. The ERGI inversion model shows a horizontally zonated, 15 m-deep channel fill (Fig. 11A) which has a resistivity range typical for wet alluvial silts and clays (Baines *et al.*, 2002; Hoffmann & Dietrich, 2004). This is supported by the cores which show an increase in the clay content on the expense of the silt fraction within the top 2 metres. The channel rises gently towards the point bar where coarse bars appear at depth which have

also been reached by cores T3G and T3I (Fig. 11B). Below the channel electrical resistivity increases which is consistent with a coarsening of the substrate but the inversion model does not resolve any coarse channel lag or plug bar deposits.

The north-eastern cutbank side of the channel is not very well outlined in the inversion model because the cutbank consists here of an older infilled channel (Fig. 4C) and the boundary of the latter align well with the end of the low resistivity zone. Here, the sedimentation rates, calculated from ²¹⁰PB chronology, are lowest (<2.2 mm a⁻¹, Table 1). Towards the SW end of the transect sedimentation rates of silty overbank sediments of up to 1.12 m from the 2011 flood dominate the profiles, corresponding with observations of buried buttress roots and the dying of susceptible trees two years after. The cores C2 to C4 closer to the channel show variable but high sedimentation rates.

Deposition rates along the study reach

Coring along cutbanks, point bars and stretches of straight channel within the study reach showed that deposition rates were highest at the apex of cutbanks (28 cm a⁻¹), followed by cutbanks upstream of the apex (9 cm a⁻¹) while migration rates were highest at the apex and downstream (39 m a⁻¹ and 29 m a⁻¹ respectively, Fig. 12). The latter were also the area with the coarsest soil (>8% sand averaged over the entire core).

DISCUSSION

Abandonment and fill of chute channels

In sand-bed rivers, chute channels tend to form in scroll sloughs (Grenfell *et al.*, 2012), enlarging them or occupying particularly widely spaced scrolls (Fisk, 1947; Hickin & Nanson, 1975). This can be promoted by high bend extension rates

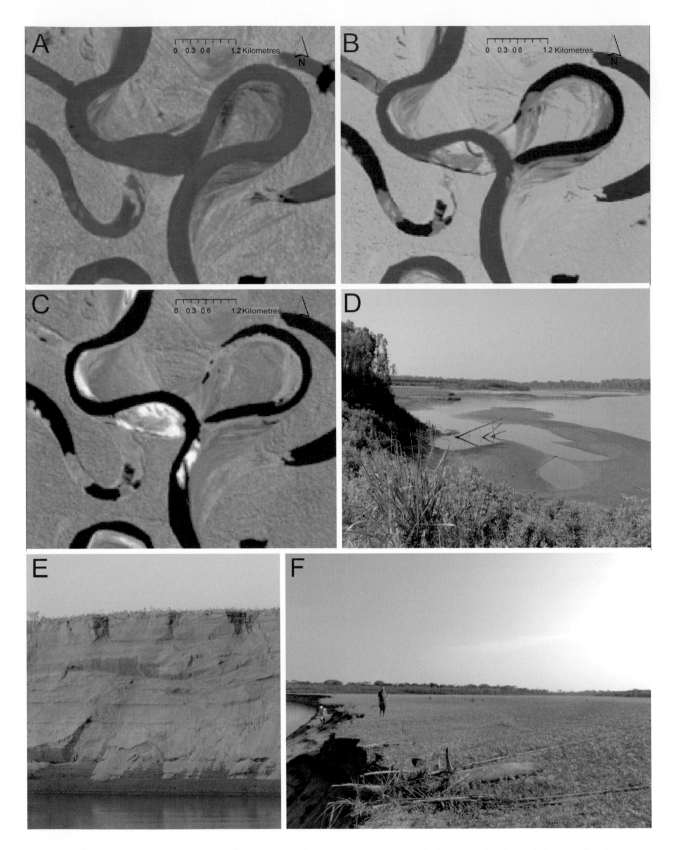

Fig. 9. Evolution of a neck cutoff at bend 53 after the flood season in 2011. Discharge at the date of the satellite imagery is shown in Fig. 3. (A) Landsat 5 image from 13th March 2011 at almost bankfull discharge showing recent cutoff and initial downstream planform adjustment. (B) Landsat 5 image from 16th May 2011 showing deposition of a plug in the lower end of the abandoned channel and plug bar growth at the upper end. (C) Landsat 5 image from 17th June showing continued plug growth sealing off the oxbow lake. Note also substantial sedimentation in oxbow lakes downstream. (D) View from the lower plug upstream to the active channel in September 2011 with the upper plug bar on the far left. Note deposition in front of upper plug due to down-valley translation of the residual bend. (E) View of the erosional face of the lower plug (6 m-high above the water level, silty sand throughout) showing different stages of deposition within no more than 7 months. (F) View across the lower plug surface showing deposition of a large tree trunk on top and several months old vegetation.

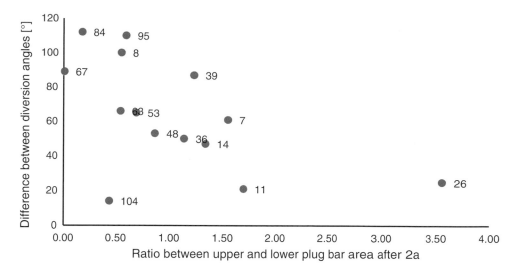

Fig. 10. The difference between lower and upper diversion angle between the new channel and the abandoned channel is significantly correlated (Pearson's R = -0.55, p = 0.04) with the ratio between upper and lower plug area two years after neck cutoff at 14 bends of the Rio Beni. Bend numbers are given in the graph and their location is shown in Fig. 2.

(Grenfell *et al.*, 2012) or short-term phases of rapid cutbank erosion (Parker *et al.*, 2011). The abandonment of chute channels at the Rio Beni is shown to depend more on the dominant migration style of a given bend than the absolute bend migration rate. Chute channel formation is most probable during phases of bend extension. However, this migration style is also related to the overall persistence of stable chutes as well as being a key component in their initiation (Fig. 5; Grenfell *et al.*, 2012). In contrast, bend expansion occurs, in general, at much lower frequencies. However, bends showing expansion have almost twice the probability to experience chute channel abandonment. These chute channels are located near the point bar (Fig. 1) and an increase in bend wavelength will lead to a lengthening of the chute channel, reducing gradient advantage and thus decreasing efficiency compared to the main channel. In the Rio Beni, bend wavelength expansion is typical for the more mobile reaches and is often connected with migration into resistant floodplain sediments (Schwendel *et al.*, 2015). The abandonment of chute channels could not be related to flood peak, cumulative discharge during the wet season or the number of days with discharge above bankfull. After exceptional flood seasons (e.g. 1999, Fig. 3) or after long periods of relatively low flood discharge (Gautier *et al.*, 2010) there appeared a high number of abandoned and newly formed stable chutes but only if sufficient time for the stabilisation of chute channels

since the previous large flood has passed (e.g. not in 2001).

Continuous rotation of bends down-valley seems to result in sufficient coarse sediment delivery to the former bar tail to establish scroll ridges (Fig. 4D) which block off the lower end of chutes (Grenfell *et al.*, 2012; Fig. 1). High lateral bend extension rates have been linked to the formation of stable chutes (Grenfell *et al.*, 2014) but this example shows that a combination of rotation and migration can prevent medium-term chute stability. The apparent low coincidence of purely rotational migration style and the abandonment of stable chutes (8%, Fig. 5) is an underestimation because 18% of the all bends classified as dominantly extending were also affected by rotational planform development (Schwendel *et al.*, 2015).

High bend extension rates appear to be beneficial for the stability of chute channels that have been incorporated into the vegetated point bar (Fig. 5). In contrast, the limited stability of chutes across the unvegetated point bar can be attributed to the low diversion angle with the main channel at their upstream end (Fig. 1), favouring high bed load sedimentation throughout the chute (Constantine *et al.*, 2010; Dieras *et al.*, 2013). This corresponds well with observations in the Rio Beni of even distribution of bed sediment along an active chute channel with the bed surface at the downstream end of many active channels often slightly higher than at the entrance. The composition of active chute channel deposits (e.g. D_{50} 45 µm

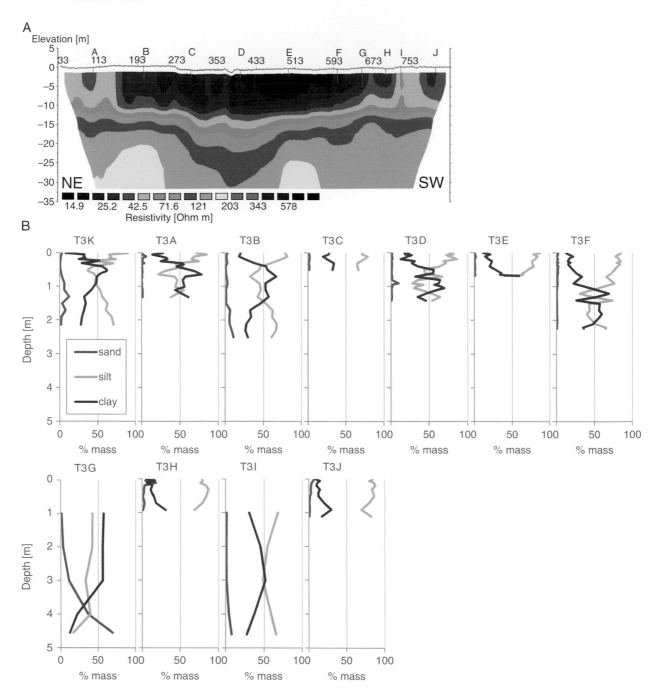

Fig. 11. (A) Electrical resistivity ground imaging inversion model of transect 3 on the cutbank side of bend 4. The location and approximate depth of cores are shown. (B) Percentage of clay, silt and sand in floodplain cores along transect 3. The location of bend 4 within the Rio Beni study reach is shown in Fig. 2.

at bend 4, Fig. 4F) ranged between bedload and suspended load calibre, thus reflecting the mixture of transport mode towards the end of the flood season in these channels, with important implications for the sedimentology of the fill.

According to Grenfell *et al.* (2012) roughly half of the chutes initiated in tropical sand-bed rivers are filled in within a decade due to rapid bend extension and high suspended sediment load. The suspended sediment in the Rio Beni is dominated by fine to medium silt (Aalto *et al.*, 2003; Guyot *et al.*, 1999) while the bed median bed material size is 0.1 to 0.15 mm (fine sand) in the study reach (Guyot *et al.*, 1999). However, the chute channels are only connected to the main channel during floods, when suspended load is probably

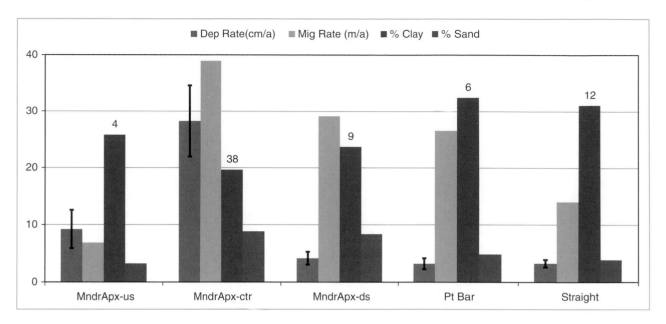

Fig. 12. Floodplain sediment deposition rate at straight reaches and at meander bends of the Rio Beni based on soil cores and ^{210}Pb geo-chronology (error bars show one standard deviation) compared to channel migration and grain-size. Cores from bends were divided according to their position on the point bar (Pt bar) and upstream (MndrApx-us), at (MndrApx-ctr) and downstream of the bend apex (MndrApx-ds) on the cutbank side (number of cores per position given).

higher due to increased stream power and higher upstream sediment supply (Yang, 1979). In addition, diversion of bedload into chute channels during floods will depend on chute channel length and the local flow conditions, specifically the angle between the main channel and the chute (Constantine *et al.*, 2010). The range of resistivity of the filled chute channel at T1 is typical for wet sands (Anomohanran, 2015), which is confirmed by the grain-size characteristics (D_{50}: 148 µm). This grain-size compares well with the median bed-load calibre, suggesting rapid infill within a flood season in 1994, a common mechanism of chute abandonment identified at numerous bends of the Rio Beni during phases of dominant extension in this study; and by others (Grenfell *et al.*, 2012; Grenfell *et al.*, 2014). The chute channels at bend 4 were completely filled to the floodplain level and could not be readily identified from surface topography or satellite imagery, although vegetation structure can be indicative (Fig. 4E).

Fill of abandoned meander bends and formation of plug bars

The abandonment and fill of channels has been explored in numerous studies with the majority focussing on chute cutoffs (e.g. Citterio & Piégay, 2009; Dieras *et al.*, 2013; Grenfell *et al.*, 2012; Micheli & Larsen, 2011; Toonen *et al.*, 2012; van

Dijk *et al.*, 2014; Zinger *et al.*, 2011) while neck cutoffs, in particular at fast-migrating sand-bed rivers, have received less attention (Erskine *et al.*, 1992; Fisk, 1947). Although the processes are similar to chute cutoffs, channel infill in meanders cut off at the neck is also conditioned by the higher length of the abandoned channel. The close proximity of the upper and lower end of the abandoned channel at erosional neck cutoffs results in high gradient advantage of the new channel and this specific geometry creates high diversion angles between the new channel and the old channel at the lower end of the cutoff (Fig. 1). The diversion angles control flow separation with high angles promoting decreased flow and therefore rapid sedimentation in the entrance of the old channel resulting in a plug of bedload material (Constantine *et al.*, 2010). This relationship between the relative diversion angle and plug development after two years is evident at the Rio Beni with the plug bars at the downstream end of the abandoned channel usually of larger area and higher elevation. In the shorter term this relationship can be confounded by plug erosion due to down-valley translation (Fig. 9C and D) and post-cutoff planform adjustment of the new channel (Fuller *et al.*, 2003). The development of a plug that more efficiently seals off the downstream end of the abandoned channel than at the upstream end is an important difference to chute cutoffs (Citterio &

Piégay, 2009; Dieras *et al.*, 2013) which has implications for infill processes. Gautier *et al.* (2007) distinguishes three phases of oxbow infill at the Rio Beni with the initial phase of building up bedload plugs over up to three years (Fig. 1), providing a significant sediment sink to counterbalance the mobilisation of material during cutoff (Lauer & Parker, 2008). This phase is followed by a period of low aggradation with limited hydraulic connectivity to the river via flood channels and finally a phase of infill by suspended sediment when the active channel again migrates towards the abandoned meander.

Unlike gravel bed rivers where plug aggradation may last several years to decades (Hooke, 1995; Kleinhans, 2010), deposition of the bedload plug in sand-bed rivers can happen in the same year as the cutoff, reaching floodplain level within three years (Gautier *et al.*, 2007). The lower end plugs appear to be deposited in several stages mostly within the first year during conditions of substantial flow as indicated by deposition of large tree trunks (Fig. 9E and F). At the upper end of the abandoned channel plug height is less and flood channels across the plug bar are common. This maintains hydraulic connectivity with the main channel during floods and thus the potential of bedload entering the oxbow channel. It is unclear, to what extent these upper end plug bars are reworked during floods and present a threshold for bedload (Dieras *et al.*, 2013). In the few cases, where lower end plugs initially have flood channels, they are usually closed within the first two years and the plug surface is stabilised by vegetation within the first year. As such they present an obstacle to throughflow during floods and promote sedimentation in the oxbow channel.

The second phase is characterised by limited connectivity with the main channel while organic lake deposits and overbank sedimentation constitute the oxbow infill (Gautier *et al.*, 2007). The duration of this phase depends on the distance to the active channel cutbank and the degree of hydraulic connectivity for suspended sediment flux (Citterio & Piégay, 2009; Gautier *et al.*, 2010). This phase applied to the investigated oxbow at bend 4 until approximately 1984 (Fig. 4B) while it filled in progressively from its upper (SW) side until it was almost completely filled in 1999. Thus the surveyed transect beyond the lower end of the lake was expected to consist of early lake deposits with high organic content or flood deposits of the first phase with some evidence of channel lag

deposits and potentially of the alluvial bedload plug at depth (Toonen *et al.*, 2012). The relatively high curvature of the former meander suggests a high difference in upper and lower diversion angle and thus the probability of a significant plug at the lower end as shown at recent cutoffs (Fig. 9C). However, although resistivity increases below the channel, the inversion model does not indicate the presence of distinct coarse alluvial plug deposits within the outline of the channel. The lack of plug deposits could be explained if the meander was originally more elongated but due to the more recent oxbow to the NW the original configuration cannot be inferred from satellite imagery. The internal zonation as shown in the inversion model (Fig. 11A) can probably be attributed to pore moisture content with lowest values near the ephemeral flood channel (Hoffmann & Dietrich, 2004).

The third phase of oxbow channel infill is facilitated by sediment advection via flood channels and overbank sedimentation (Gautier *et al.*, 2007), the latter decreasing exponentially with distance from the channel (Aalto *et al.*, 2003). At the investigated oxbow the approaching channel from the West has from 1984 continuously provided suspended sediment to fill the lake, aided by connecting flood channels at the upstream end. The presence of these channels and their position relatively central at the apex of the approaching meander (Fig. 12) may be the reasons for the higher fill compared to the oxbow further NW (Gautier *et al.*, 2010). The traces of inundation and high deposition rates near the flood channel indicate that these channels play a key role in suspended sediment dispersion to the floodplain, particularly in years with lower than average flood peaks. The delivery of suspended sediment to the floodplain is episodic at the Rio Beni and the spatial variability in sedimentation rates on the cutbank sides (cores C1 to C6) can be attributed to distance from the channel, topography and vegetation structure (Aalto *et al.*, 2003; Gautier *et al.*, 2010). Additionally, the location relative to the bend apex appears to have some influence (Table 1, Fig. 12).

Floodplain heterogeneity

The floodplain inside meander bends consist of a range of bar deposits and overbank sediments (Fig. 1). Coarse grained point bar sediments are deposited around the bend apex with a tendency for fining towards the downstream end of the bar and in vertical direction (Bridge *et al.*, 1995; Willis & Tang, 2010). The process of down-bar fining is well documented at bend 4 with a general coarsening of the point bar deposits when the location of the apex shifted relatively towards the transects (Figs 6A and 8A). Genuine coarsening with depth could be detected at some cores (Fig. 8B), as well as interlayering between silt and sand dominated material, but the spatial resolution of the ERGI model was not sufficient to map these trends for the entire transect.

The lateral extension of bends promotes the formation of regularly spaced convex scroll bars by accretion of finer sediment on transverse in-channel bars or by deposition of suspended load on the point bar or behind obstacles due to flow separation (Nanson & Croke, 1992). The distance between scroll ridges appears to be related to bend migration rate (Hickin & Nanson, 1975). These scroll bars have been evident on the investigated bend (Fig. 8A). However, the surficial zones of elevated resistivity appear not always to correspond to the surface morphology and the vegetation composition, in particular dominance of *Gynerium sagittatum*, was a much better indicator for position of surficial high resistivity zones. *G. sagittatum* is a primary pioneer plant and typically found to first colonise fresh sand deposits via vegetative rhizomes (Kalliola *et al.*, 1992; Kalliola *et al.*, 1991). While most plants on the lower parts of the point bar show some adaptation to the annual flooding (e.g. stilt roots), fewer are tolerant to high rates of surface deposition which can lead to large stands of dead forest on cutbanks after larger floods (e.g. in 2011 at the Rio Beni). Hence, it is possible that the coarser deposits, corresponding to higher resistivity, are sandy overbank deposits which are more recent and have been deposited in older scroll troughs during floods. The strong scroll-trough surface morphology (Fig. 4E and F), traces of significant overbank flow across the point bar (Fig. 7) and potential hydraulic connectivity of the troughs with recent chute channels might provide a pathway for transport of coarser sediments into these areas. Additionally, deposition of fine overbank deposits and scour during floods has probably altered surface morphology.

Downstream of point bars, counter point bar deposits can be found in particular when bends are translating down-valley or rotating in downstream direction while bend wavelength expansion counters their development (Smith *et al.*, 2011; Willis & Tang, 2010). They are of finer grain-size and lower in elevation than point bars (Smith *et al.*, 2009) and the ridges show a concave planform. The north-western part of transect 2 shows a series of successive fill of the migrating channel starting with fine counter point bar sediments over increasingly coarsening point bar deposits, all overlain by overbank sediments. The latter can amount to roughly 3.75 m over 50 years but due to the low elevation of the counter point bar deposits (Smith *et al.*, 2009) sedimentation rate might initially have been higher.

The presence of deep chute channels filled completely with coarser material than the surrounding floodplain and a range of point bar deposits has important implications for future planform evolution of meandering rivers. Schwendel *et al.* (2015) showed that at the Rio Beni the migration rate of bends eroding into a cutbank made out of channel or oxbow lake deposits (58.8 m a^{-1} and 98.8 m a^{-1} respectively) is much higher than that of bends migrating into point bar deposits (43.3 m a^{-1}). Thus, point bars with a high number of deep chute channels will experience a more variable erodibility than predicted by a model assuming uniform erosion rates. Although erosion rates into channel deposits and oxbow lakes are on average higher than for other floodplain units (Schwendel *et al.*, 2015), their high variability indicates a strong influence of their evolutionary stage and the location within the former meander.

CONCLUSIONS

Floodplain evolution at highly dynamic sand-bed rivers is a function of floodplain sedimentation and meander migration. The latter is conditioned by the erodibility of floodplain sediments which reflects a range of floodplain sedimentation processes. While meander cutoffs and erosion of chute channels are responsible for the mobilisation of substantial amounts of sediment (Fuller *et al.*, 2003; Zinger *et al.*, 2011), abandoned channels represent also an important accommodation space which can rapidly offset the downstream effects of cutoffs upstream. Deposition in abandoned channels results in deposits significantly different from the surrounding floodplain. The fill of chute channels and scroll sloughs on the point bar create coarse bedload sediment bodies of varying depth within finer point bar and counter-point bar deposits, which are often not identifiable from aerial imagery but have varying resistance to meander migration. While abandoned meanders are more easily identified, their sedimentology can be complex. The large difference between lower and upper diversion angle of an abandoned channel at neck cutoffs leads to rapid deposition of a bedload plug at the lower end, development of an oxbow lake with sequestration of fine sediments and finally infill via suspended load conditioned by channel migration. However, quantification of the contribution of bedload to the fill of abandoned meanders via flood channels across the plug bars towards the end of the first phase requires more detailed investigation.

Backswamps and oxbow lakes are regarded as the typical environment where clay-rich flood-plain deposits form under the influence of stagnant groundwater levels. The 'clay plugs' are highly resistive to fluvial erosion and are a key influence on meander migration (Schwendel *et al.*, 2015). Although the consequences of this rapid plugging for detailed oxbow stratigraphy and the implications for the development of clay-rich deposits are not fully clear, this study has provided constraints on the time scales of channel fill processes at rapidly migrating rivers dominated by suspended load.

ACKNOWLEDGEMENTS

The authors would like to thank Greg Sambrook-Smith for the provision of the electrical resistivity equipment and Lisa Mol for insightful discussions on this technique. The research was funded by the UK Natural Environment Research Council (grant NE/H009108/1). We would like to thank the two anonymous reviewers whose comments have improved the manuscript.

REFERENCES

Aalto, R., Maurice-Bourgoin, L., Dunne, T., Montgomery, D.R., Nittrouer, C.A. and Guyot, J.L. (2003) Episodic sediment accumulation on Amazonian flood plains influenced by El Nino/Southern Oscillation. *Nature*, **425**, 493–497.

Aalto, R. and Nittrouer, C.A. (2012) Pb-210 geochronology of flood events in large tropical river systems. *Phil. Trans. Roy. Soc. A*, Mathematical Physical and Engineering Sciences, **370**, 2040–2074.

Allen, J.R.L. (1965) A review of the origin and characteristics of recent alluvial sediments. *Sedimentology*, **5**, 89–191.

Allenby, R.J. (1988) Origin of rectangular and aligned lakes in the Beni Basin of Bolivia. *Tectonophysics*, **145**, 1–20.

Anomohanran, O. (2015) Hydrogeophysical and hydrogeological investigations of groundwater resources in Delta Central, *Nigeria. J. Taibah University for Science*, **9**, 57–68.

Baines, D., Smith, D.G., Froese, D.G., Bauman, P. and Nimeck, G. (2002) Electrical resistivity ground imaging (ERGI): a new tool for mapping the lithology and geometry of channel-belts and valley-fills. *Sedimentology*, **49**, 441–449.

Banton, O., Cimon, M.-A. and Seguin, M.-K. (1997) Mapping Field-Scale Physical Properties of Soil with Electrical Resistivity. *J. Soil Sci. Soc. Am.*, **61**, 1010–1017.

Berry, F. and Kress, W.J. (1991) *Heliconia: An identification guide*. Smithsonian Institution Press, Washington.

Bridge, J.S., Alexander, J.A.N., Collier, R.E.L., Gawthorpe, R.L. and Jarvis, J. (1995) Ground-penetrating radar and coring used to study the large-scale structure of point-bar deposits in three dimensions. *Sedimentology*, **42**, 839–852.

Chambers, J.E., Wilkinson, P.B., Wardrop, D., Hameed, A., Hill, I., Jeffrey, C., Loke, M.H., Meldrum, P.I., Kuras, O., Cave, M. and Gunn, D.A. (2012) Bedrock detection beneath river terrace deposits using three-dimensional electrical resistivity tomography. *Geomorphology*, **177–178**, 17–25.

Citterio, A. and Piégay, H. (2009) Overbank sedimentation rates in former channel lakes: characterization and control factors. *Sedimentology*, **56**, 461–482.

Clifford, J. and Binley, A. (2010) Geophysical characterization of riverbed hydrostratigraphy using electrical resistance tomography. *Near Surface Geophysics*, **8**, 493–501.

Constantine, J.A., Dunne, T., Piégay, H. and Mathias Kondolf, G. (2010) Controls on the alluviation of oxbow lakes by bed-material load along the Sacramento River, California. *Sedimentology*, **57**, 389–407.

Crook, N., Binley, A., Knight, R., Robinson, D.A., Zarnetske, J. and Haggerty, R. (2008) Electrical resistivity imaging of the architecture of substream sediments. *Water Resour, Res.*, **44**, W00D13.

Dieras, P.L., Constantine, J.A., Hales, T.C., Piégay, H. and Riquier, J. (2013) The role of oxbow lakes in the off-channel storage of bed material along the Ain River, France. *Geomorphology*, **188**, 110–119.

Dumont, J.F. (1996) Neotectonics of the Subandes-Brazilian craton boundary using geomorphological data: The Maranon and Beni basins. *Tectonophysics*, **259**, 137–151.

Dumont, J.F. and Hannagarth, W. (1993) *River shifting and tectonics in the Beni basin*. In: Third International Conference Geomorphology, Hamilton.

Erskine, W., McFadden, C. and Bishop, P. (1992) Alluvial cutoffs as indicators of former channel conditions. *Earth Surf. Proc. Land.*, **17**, 23–37.

Fisk, H.N. (1947) Fine-grained alluvial deposits and their effects on Mississippi River activity. *USCE Mississippi River Communications*, **1**, 82.

Fisk, H.N. (1944) Geological investigation of the alluvial valley of the lower Mississippi River. Report for U.S. Department of the Army, Mississippi River Commission.

Fuller, I.C., Large, A.R.G. and **Milan, D.J.** (2003) Quantifying channel development and sediment transfer following chute cutoff in a wandering gravel-bed river. *Geomorphology*, **54**, 307–323.

Gautier, E., Brunstein, D., Vauchel, P., Jouanneau, J.-M., Roulet, M., Garcia, C., Guyot, J.-L. and **Castro, M.** (2010) Channel and floodplain sediment dynamics in a reach of the tropical meandering Rio Beni (Bolivian Amazonia). *Earth Surf. Proc. Land.*, **35**, 1838–1853.

Gautier, E., Brunstein, D., Vauchel, P., Roulet, M., Fuertes, O., Guyot, J.L., Darozzes, J. and **Bourrel, L.** (2007) Temporal relations between meander deformation, water discharge and sediment fluxes in the floodplain of the Rio Beni (Bolivian Amazonia). *Earth Surf. Proc. Land.*, **32**, 230–248.

Geotomo (2011) *Manual RES2DINVx64 v4.00*, Geotomo Software, Gelugor.

Gourry, J.-C., Vermeersch, F., Garcin, M. and **Giot, D.** (2003) Contribution of geophysics to the study of alluvial deposits: a case study in the Val d'Avaray area of the River Loire, France. *J. Appl. Geophys.*, **54**, 35–49.

Grenfell, M., Aalto, R. and **Nicholas, A.** (2012) Chute channel dynamics in large, sand-bed meandering rivers. *Earth Surf. Proc. Land.*, **37**, 315–331.

Grenfell, M.C., Nicholas, A.P. and **Aalto, R.** (2014) Mediative adjustment of river dynamics: The role of chute channels in tropical sand-bed meandering rivers. *Sed. Geol.*, **301**, 93–106.

Gueneralp, I. and **Marston, R.A.** (2012) Process-form linkages in meander morphodynamics: Bridging theoretical modeling and real world complexity. *Prog. Phys. Geogr.*, **36**, 718–746.

Guyot, J.L., Jouanneau, J.M., Soares, L., Boaventura, G.R., Maillet, N. and **Lagane, C.** (2007) Clay mineral composition of river sediments in the Amazon Basin. *CATENA*, **71**, 340–356.

Guyot, J.L., Jouanneau, J.M. and **Wasson, J.G.** (1999) Characterisation of river bed and suspended sediments in the Rio Madeira drainage basin (Bolivian Amazonia). *J. S. Am. Earth Sci.*, **12**, 401–410.

Hajek, E. and **Edmonds, D.** (2014) Is river avulsion controlled by floodplain morphodynamics? *Geology*, **42**, 199–202.

Hajek, E., Sheets, B. and **Heller, P.** (2010) Significance of channel-belt clustering in alluvial basins. *Geology*, **38**, 535–538.

Hajek, E. and **Wolinsky, M.** (2012) Simplified process modeling of river avulsion and alluvial architecture: connecting models and field data. *Sed. Geol.*, **257–260**, 1–30.

Hausmann, J., Steinel, H., Kreck, M., Werban, U., Vienken, T. and **Dietrich, P.** (2013) Two-dimensional geomorphological characterization of a filled abandoned meander using geophysical methods and soil sampling. *Geomorphology*, **201**, 335–343.

Hickin, E. and **Nanson, G.** (1975) The character of channel migration on the Beatton River, north-east British Columbia, Canada. *Geol. Soc. Am. Bull.*, **86**, 487–494.

Hoffmann, R. and **Dietrich, P.** (2004) An approach to determine equivalent solutions to the geoelectrical 2D inversion problem. *J. Appl. Geophys.*, **56**, 79–91.

Hooke, J.M. (1995) River channel adjustment to meander cutoffs on the River Bollin and River Dane, northwest England. *Geomorphology*, **14**, 235–253.

Kalliola, R., Puhakka, M. and **Salo, J.** (1992) Intraspecific variation and the distribution and ecology of Gynerium sagittatum (Poaceae) in the western Amazon. *Flora*, **186**, 153–167.

Kalliola, R., Salo, J., Puhakka, M. and **Rajasilta, M.** (1991) New site formation and colonizing vegetation in primary succession on the western Amazon floodplains. *J. Ecol.*, **79**(4), 877–901.

Kleinhans, M.G. (2010) Sorting out river channel patterns. *Prog. Phys. Geogr.*, **34**, 287–326.

Latrubesse, E.M. and **Restrepo, J.D.** (2014) Sediment yield along the Andes: continental budget, regional variations and comparisons with other basins from orogenic mountain belts. *Geomorphology*, **216**, 225–233.

Lauer, J.W. and **Parker, G.** (2008) Modeling framework for sediment deposition, storage and evacuation in the floodplain of a meandering river: Theory. *Water Resour. Res.*, **44**, W04425.

Leopold, M., Völkel, J., Huber, J. and **Dethier, D.** (2013) Subsurface architecture of the Boulder Creek Critical Zone Observatory from electrical resistivity tomography. *Earth Surf. Proc. Land.*, **38**, 1417–1431.

Loke, M.H. (2000) *Electrical Imaging Surveys for Environmental and Engineering Studies: A Practical Guide to 2D and 3D Surveys*, Geotomo Software Ltd., Gelugor.

Loke, M.H., Acworth, I. and **Dahlin, T.** (2003) A comparison of smooth and blocky inversion methods in 2D electrical imaging surveys. *Explor. Geophys.*, **34**(3), 182–187.

Micheli, E.R. and **Larsen, E.W.** (2011) River channel cutoff dynamics, Sacramento River, California, USA. *River Research and Applications*, **27**, 328–344.

Nanson, G.C. and **Croke, J.C.** (1992) A genetic classification of floodplains. *Geomorphology*, **4**, 459–486.

Nanson, G.C. and **Hickin, E.J.** (1986) A statistical analysis of bank erosion and channel migration in western Canada. *Geol. Soc. Am. Bull.*, **97**, 497–504.

Parker, G., Shimizu, Y., Wilkerson, G.V., Eke, E.C., Abad, J.D., Lauer, J.W., Paola, C., Dietrich, W.E. and **Voller, V.R.** (2011) A new framework for modeling the migration of meandering rivers. *Earth Surf. Proc. Land.*, **36**, 70–86.

Plafker, G. (1964) Oriented lakes and lineaments of Northeastern Bolivia. *Geol. Soc. Am. Bull.*, **75**, 503–522.

Reynolds, J.M. (2011) *An Introduction to Applied and Environmental Geophysics.* John Wiley & Sons Ltd, Chichester.

Roche, M.A. and **Fernandez, C.F.** (1988) Water resources, salinity and salt yields of rivers in the Bolivian Amazon. *J. Hydrol.*, **101**, 305–331.

Rowland, J.C., Lepper, K., Dietrich, W.E., Wilson, C.J. and **Sheldon, R.** (2005) Tie channel sedimentation rates, oxbow formation age and channel migration rate from optically stimulated luminescence (OSL) analysis of floodplain deposits. *Earth Surf. Proc. Land.*, **30**, 1161–1179.

Salo, J., Kalliola, R., Häkkinen, I., Mäkinen, Y., Niemelä, P., Puhakka, M. and **Coley, P.D.** (1986) River dynamics and the diversity of Amazon lowland forest. *Nature*, **322**, 254–258.

Schwendel, A.C., Nicholas, A.P., Aalto, R.E., Sambrook Smith, G.H. and **Buckley, S.** (2015) Interaction between meander dynamics and floodplain heterogeneity in a large tropical sand-bed river: the Rio Beni, Bolivian Amazon. *Earth Surf. Proc. Land.*, **40**, 2026–2040.

Smith, D.G., Hubbard, S.M., Lavigne, J., Leckie, D.A. and **Fustic, M.** (2011) Stratigraphy of counter-point-bar and eddy-accretion deposits in low-energy meander belts of the Peace-Athabasca Delta, northeast Alberta, Canada. In: *From River to Rock Record: The preservation of fluvial sediments and their subsequent interpretation* (Eds S.K. Davidson, S. Leleu and C.P. North), *SEPM Spec. Publ.*, **97**, 143–152.

Smith, D.G., Hubbard, S.M., Leckie, D.A. and **Fustic, M.** (2009) Counter point bar deposits: lithofacies and reservoir significance in the meandering modern Peace River and ancient McMurray Formation, Alberta, Canada. *Sedimentology*, **56**, 1655–1669.

Sudha, K., Israil, M., Mittal, S. and **Rai, J.** (2009) Soil characterization using electrical resistivity tomography and geotechnical investigations. *J. Appl. Geophys.*, **67**, 74–79.

Toonen, W.H.J., Kleinhans, M.G. and **Cohen, K.M.** (2012) Sedimentary architecture of abandoned channel fills. *Earth Surf. Proc. Land.*, **37**, 459–472.

van Dijk, W.M., Schuurman, F., van de Lageweg, W.I. and **Kleinhans, M.G.** (2014) Bifurcation instability and chute cutoff development in meandering gravel-bed rivers. *Geomorphology*, **213**, 277–291.

Ward, J.V., Tockner, K., Arscott, D.B. and **Claret, C.** (2002) Riverine landscape diversity. *Freshwater Biology*, **47**, 517–539.

Ward, S.H. (1990) Resistivity and induced polarization methods. In: *Geotechnical and Environmental Geophysics* (Ed. S.H. Ward), *Investigations in Geophysics*, **1**, Society of Exploration Geophysicists, Tulsa, 147–190.

Willis, B.J. and **Tang, H.** (2010) Three-Dimensional Connectivity of Point-Bar Deposits. *J. Sed. Res.*, **80**, 440–454.

Yang, C.T. (1979) Unit stream power equations for total load. *J. Hydrol.*, **40**, 123–138.

Zinger, J.A., Rhoads, B.L. and **Best, J.L.** (2011) Extreme sediment pulses generated by bend cutoffs along a large meandering river. *Nature Geoscience*, **4**, 675–678.

Int. Assoc. Sedimentol. Spec. Publ (2018) **48**, 273–296.

Characterising three-dimensional flow through neck cutoffs with complex planform geometry

DEREK RICHARDS[†], KORY KONSOER[†], CHRISTOPHER TURNIPSEED[‡] and CLINTON WILLSON[§]

[†] *Department of Geography and Anthropology, Louisiana State University College of Humanities and Social Sciences, 227 Howe-Russell-Kniffen Geoscience Complex, Baton Rouge, Louisiana, USA*
[‡] *Coastal Studies Institute, Louisiana State University, Baton Rouge, Louisiana, USA*
[§] *Department of Civil & Environmental Engineering, Louisiana State University, Baton Rouge, Louisiana, USA*

ABSTRACT

The evolution of meanders on alluvial rivers includes periods of accelerated change and strong contrasts in the three-dimensional flow field across the channel, such as when neck cutoffs are present. Our understanding of the three-dimensional flow field through such complex planform configurations remains incomplete. Therefore, the goal of this research is to characterise three-dimensional flow through neck cutoffs with complex planform configurations. An acoustic Doppler current profiler was used to obtain near-bankfull velocity measurements on five neck cutoffs on the White River, Arkansas. The main hydrodynamic characteristics of cutoffs, where the upstream and downstream limbs are oriented subparallel to each other, are 1) tight bend flow resulting from flow redirection of nearly 180° from the upstream to the downstream limb, 2) a zone of flow separation and recirculation adjacent to the cutoff junction corner within the downstream limb; and 3) zones of recirculation at the entrance and exit of the abandoned loop. The neck cutoffs that currently display a planform geometry where the cutoff channel is roughly parallel with the upstream limb and perpendicular to the downstream limb exhibited 1) highly asymmetric flow through the cutoff channel, 2) a zone of stagnation and/or flow recirculation that extends across the entrance and exit of the abandoned loop; and 3) a zone of recirculation along the outer bank within the apex region of the downstream loop. Reversal of helical motion was observed, typically occurring at the entrance to the cutoff channel and within the downstream limb of the loops. The findings from this study are summarised into two different conceptual hydrodynamic models based on planform configuration that explain the complex three-dimensional flow structure observed at these types of neck cutoffs.

Keywords: Meander cutoff, three-dimensional flow, bifurcations, helical flow

INTRODUCTION

As meandering rivers migrate through their floodplain, the channel path may increase in length and sinuosity while decreasing in channel gradient, producing a wide range of geomorphic and habitat complexity, such as channel cutoffs and oxbow lakes. While cutoffs and oxbow lakes are a natural and ubiquitous feature of riverine landscapes, predicting when these events will occur and how they will evolve remains challenging

due to a scarcity of detailed investigations focused on the three-dimensional flow structure through actively evolving cutoffs.

Meander cutoffs on alluvial rivers are typically sudden and extreme events resulting in a change of river course, shortening of the channel length and a local increase in channel gradient (Fisk, 1947; Allen, 1965). Cutoffs on meandering rivers are generally classified as either chute or neck. Chute cutoffs form when overbank flow produces sufficient shear stress to incise a new channel into

Fluvial Meanders and Their Sedimentary Products in the Rock Record, First Edition.
Edited by Massimiliano Ghinassi, Luca Colombera, Nigel P. Mountney and Arnold Jan H. Reesink.

the floodplain, with incision often concentrated within swales or sloughs across the meander neck (Fisk, 1947; Allen, 1965; Erskine *et al.*, 1982; Gay *et al.*, 1998; Constantine *et al.*, 2010b; Micheli & Larsen, 2011). Neck cutoffs occur when a river increases in sinuosity while decreasing in radius of curvature, leading to the migration of the upstream and downstream meander limbs into each other and eventually resulting in bank collapse (Fisk, 1947; Allen, 1965; Ratzlaff, 1981; Gagliano & Howard, 1984; Gay *et al.*, 1998; Micheli & Larsen, 2011).

The current knowledge surrounding the detailed processes involved during and after a cutoff event remains incomplete, partly due to the sporadic and episodic nature of cutoff events, which has limited field-based research as the event occurs (Lewis & Lewin, 1983; Hooke, 1995; Micheli & Larsen, 2011). While some research has focused on the detailed processes involved in chute cutoff and the improvement of chute cutoff models (Shields & Abt, 1989; Constantine & Dunne, 2008; Constantine *et al.*, 2010b; Zinger *et al.*, 2011, 2013), far less similar research has been focused on the processes associated with neck cutoff (Gagliano & Howard, 1984; Fares, 2000; Han & Endreny, 2014; Konsoer *et al.*, 2016b), particularly for rivers with complex meander planform geometry. The lack of detailed investigations of neck cutoffs has led to current conceptual models of neck cutoff and oxbow lake formation being established using relatively simple planform configurations where the upstream and downstream meander limbs are depicted as connected by a relatively straight cutoff channel (Gagliano & Howard, 1984; Konsoer *et al.*, 2016b). However, in natural rivers, the location and planform geometry of cutoff channels in relation to a meander bend can vary substantially, resulting in complex hydrodynamic and morphologic features that are not explicitly accounted for in the current conceptual models.

Previous research on flow through cutoffs has revealed the importance of the diversion angle between the new active channel and the older abandoned channel in determining the rate of disconnection of the abandoned loop from the active channel (Fisk, 1947; Gagliano & Howard, 1984; Shields & Abt, 1989; Constantine *et al.*, 2010a). Fisk (1947) was the first to note that the duration of flow through abandoned bends was controlled by the alignment of the active channel to the upstream and downstream limbs of the bend. The smaller

the diversion angle, the longer the abandoned bend remained active (Fisk, 1947). The diversion angle was found to be generally smaller for chute cutoffs, which explained why flow persisted through abandoned bends formed from chute cutoffs longer than those formed from neck cutoffs (Fisk, 1947). Gagliano & Howard (1984) then produced a model depicting four stages for neck cutoff evolution from active meandering to complete infill and terrestrialisation (Fig. 1A). Similar to Fisk (1947), Galiano & Howard (1984) state that the angle between the abandoned limbs of the bend and the active channel, along with bed load supply and relative gradient advantage of the new cutoff channel compared to the old bend, determined the length of time for complete disconnection of the bend from the active channel. For cases when the diversion angle was slight, flow diverted more easily through the abandoned bend causing prolonged flow (Gagliano & Howard, 1984). Bridge *et al.* (1986) and Hooke (1995) corroborated these results with further field observations and Shields & Abt (1989) found that the diversion angle explained over 90% of variation of bed-load aggradation. Constantine *et al.* (2010a) was the first to describe the physical mechanisms that explain why diversion angle affected plugging and flow duration through abandoned bends. As flow enters the abandoned bend a zone of flow separation and recirculation is produced, where flow velocity and shear stress are reduced (Fig. 1B; Constantine *et al.*, 2010a). The width of this zone of separation is determined by the diversion angle between the new cutoff channel and the abandoned bend, where larger diversion angles result in wider zones of separation and recirculation, leading to deposition of sediment within the entrance to the abandoned bend (Constantine *et al.*, 2010a). Thus, higher diversion angles, lead to faster rates of sediment accumulation and the formation of sediment plugs that disconnect the abandoned bend from the active channel (Constantine *et al.*, 2010a).

Research on chute cutoffs has led to a conceptual model that suggests flow hydrodynamics at cutoffs is analogous to a bifurcation and confluence separated by a relatively short channel (Fig. 1C; Zinger *et al.*, 2013). Field measurements of three-dimensional velocities of a chute cutoff on the Wabash River have shown the presence of a zone of reduced velocity within the original channel opposite the entrance to the cutoff and a zone of separation immediately downstream of the bifurcation within the newly developed cutoff

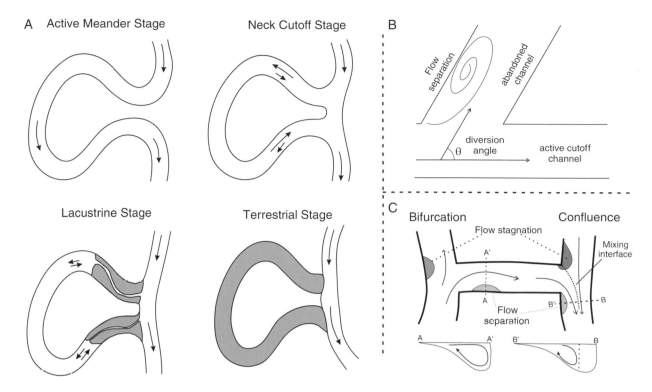

Fig. 1. (A) Image showing the four stages of neck cutoff evolution from active meandering to complete infill, modified from Gagliano & Howard (1984). The grey shaded areas in the Lacustrine and Terrestrial Stages indicate plugging and infill of sediment into the abandoned loop. (B) Zone of flow separation and recirculation occurring within an abandoned channel following cutoff. As diversion angle between cutoff (active) channel and abandoned channel increase, the zone of separation increases. modified from Constantine *et al.*, 2010a. (C) Conceptual model of flow within a chute cutoff, showing similarities with bifurcation and confluence hydrodynamic features. Modified from Zinger *et al.*, 2013.

channel (Zinger *et al.*, 2013). At the exit of the cutoff channel, flow from the cutoff converges with flow from the original channel, producing a zone of stagnation at the upstream junction corner of the confluence while a zone of flow separation forms on the inside bank of the downstream receiving channel (Zinger *et al.*, 2013). These zones of recirculation and stagnation lead to the deposition of sediment and the development of bars that contribute to the abandonment of the original channel and the evolution of the cutoff channel.

Similar hydrodynamic investigations as that by Zinger *et al.* (2013) have not been conducted for neck cutoffs and previous models depict cutoffs with a lack of upstream curvature of the channel entering the cutoff. However, as cutoffs are bifurcations, at least temporarily, research on bifurcations with upstream meander bends can be beneficial for understanding flow structure through cutoffs. Using a combination of 1D, 2D and 3D numerical modelling, it has been shown that in the absence of a gradient advantage between

the bifurcates, the outer bend branch receives more flow and deepens while the inner bend branch receives more bed-load sediment than can be transported and aggrades (Kleinhans *et al.*, 2006, 2008). Direction of sediment transport and flow differs between the bifurcates due to gravitational effects on transverse and longitudinal slopes and helical flow produced from the upstream meander bend (Kleinhans *et al.*, 2006, 2008). The inner bend bifurcate becomes dominant only when the gradient advantage of the bifurcate is enough to counteract the effect of the upstream meander bend flow and sediment division (Kleinhans *et al.*, 2006, 2008). Van Dijk *et al.* (2014) found similar results on chute cutoffs, noting that success or failure of the new cutoff channel was determined by gradient advantage and channel curvature at the bifurcation. The curving flow from the upstream bend resulted in helical motion that steered sediment to the inner bifurcate and flow discharge to the outer bifurcate (van Dijk *et al.*, 2014). Thus, chute cutoff channels that formed on the inner bend received too much sediment and failed while a chute cutoff channel that formed on the outer bend scoured and successfully captured the majority of flow (van Dijk *et al.*, 2014).

The hydrodynamics described by previous research suggest that differences in planform geometry and channel morphology are important factors controlling flow through meander cutoffs. For example, the diversion angle model predicts that a zone of separation and recirculation will form within the entrance to the abandoned channel (Constantine *et al.*, 2010a), whereas the bifurcation-confluence model shows a zone of reduced velocity (or stagnation) within this same region (Zinger *et al.*, 2013). While both of these models do predict the accumulation of sediment within the entrance to the original channel, each model accounts for bar formation through different hydrodynamic mechanisms. The differences between these two models probably reflect the range of planform geometries observed in nature, as well as the morphologic variability produced during the evolution of channel cutoffs.

The purpose of this study is to characterise three-dimensional flow through neck cutoffs with complex planform geometries using detailed field measurements. Multiple neck cutoffs have been documented on the White River, in central Arkansas, each with different planform characteristics. The morphologic evolution of these cutoffs appears relatively slow compared with other observed neck cutoffs, providing an excellent opportunity to investigate in detail the hydrodynamic features associated with these events. By examining five different neck cutoffs on the same river, this paper seeks to address the importance of planform geometry and cutoff location on the three-dimensional flow field and provide a new conceptual model for the three-dimensional flow structure through neck cutoffs.

STUDY SITE

The neck cutoffs presented in this study are located on the lower White River, in central Arkansas (Fig. 2). The lower White River begins near Batesville, Arkansas and runs approximately 745 km before joining the Mississippi River, draining an area of about 8360 km². The channel has a bankfull width of roughly 170 m and bankfull depth of roughly 7 m. Hydrologic data from a United States Geological Survey gaging station at De Valls Bluff, Arkansas (gage: 07077000) indicate a mean annual discharge of 750 m³ s⁻¹ and a peak annual discharge of 2,460 m³ s⁻¹. Generally, higher discharge occurs between March – June and lower discharge occurs between July – December.

Five neck cutoffs have been identified for this study, all located upstream of Clarendon, AR (Fig. 2). All five sites represent cutoffs on elongate meander loops at different stages in morphologic evolution and different planform geometries. Three of the sites are grouped together as having upstream and downstream channels oriented subparallel to each other (Fig. 2). Pumps Bend began cutoff incision in 2014, making it the most recent cutoff event of this study. The width of the narrowest part of the neck in satellite imagery before cutoff was about 30 m and the cutoff channel is located about 900 m from the apex of the inner bank of the meander loop. Franklin Bend is located roughly 9 km upstream of Clarendon, AR and cutoff in 2006 based on aerial photographs. The width of the neck before cutoff was about 15 m and the location of the cutoff channel is about 710 m from the inner bank of the meander loop apex. The third elongate loop cutoff site, Seven Mile Bend, is located ~ 9 km downstream from Pumps Bend, roughly halfway between Pumps and Franklin bends and cutoff of in 2011. The width of the neck before cutoff was about 20 m and the location of the cutoff is about 250 m from the apex of the meander loop along the inner bank. The other two cutoff sites, currently, have a planform geometry where the upstream channel is oriented roughly parallel with the cutoff channel and the downstream channel is roughly perpendicular to the cutoff channel (Fig. 2). Calhoun Bend, located ~ 12 km upstream of Des Arc, AR, appears to have initially cutoff in 1988 and is nearly fully disconnected from the active channel. The furthest upstream cutoff is Devil's Elbow, located ~ 15 km upstream of Georgetown, AR. Based on time-series satellite imagery, Devil's Elbow cutoff sometime between 1983 and 1984.

Previous research on neck cutoffs indicate that the time for plugging and disconnection of the abandoned loop from the active channel occurred over a relatively short time span. Gagliano & Howard (1984) found the time for disconnection ranged between 2 to 10 years for the lower Mississippi River, Hooke (1995) found the time ranged from less than 1 to 7 years for cutoffs in England & Petersen (1963) found the time to

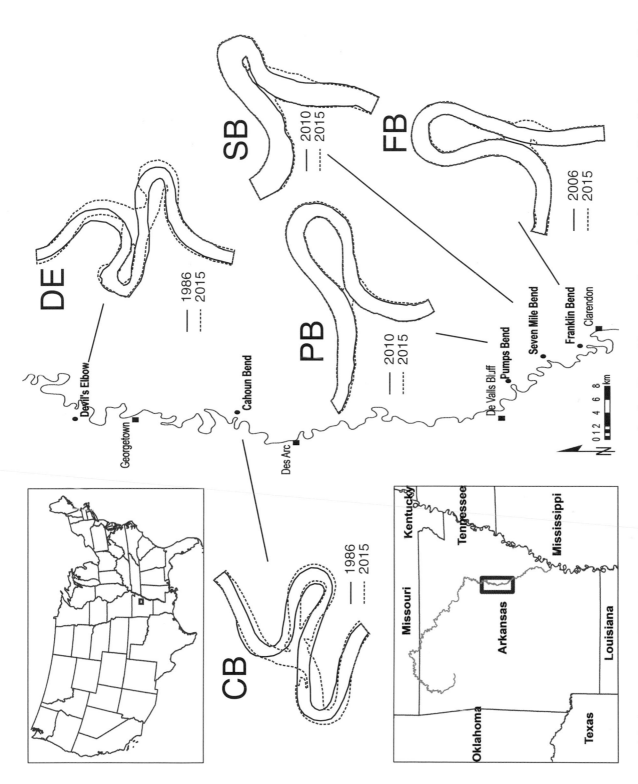

Fig. 2. Map of the location of cutoff sites along the White River, in central Arkansas. Banklines were drawn from for each cutoff site before cutoff and in 2015 to show the overall cutoff evolution. Banklines for Calhoun Bend before cutoff were traced using a 1986 topographic map, while this is after the date of cutoff the data for this map seems have been collected between 1962 and 1981; and the bend was not updated from other earlier versions of topographic maps. Banklines for Franklin Bend before cutoff were traced using the 2006 after cutoff satellite image with the neck being inferred from the image, satellite images from before cutoff were not available. All after cutoff dates were se- at 2015 as this is the last complete satellite imagery that was available for all cutoff sites. Labels: PB: Pumps Bend, SB: Seven Mile Bend, FB: Franklin Bend, DE: Devil's Elbow, CB: Calhoun Bend.

disconnection was around 5 years. Comparatively, the morphologic evolution of the neck cutoffs in this study are relatively slow, as none of them are fully disconnected from the abandoned bend even decades after initiation of the cutoff event.

A recent geomorphic study of the lower White River documented the influence of base level lowering of the Mississippi River and flow regulation (dams) within the upper White River on channel incision, migration rate and sinuosity (Edwards *et al.*, 2016). Findings from this study revealed an incisional knickpoint located near St. Charles, AR, roughly 40 km south-east of Clarendon, AR. Upstream of this knickpoint, where all five cutoff sites are located, migration rates and sinuosity are relatively high compared to downstream of the knickpoint, with sinuosity for the study reach being ~ 2.2 and migration rates being ~ 2 to 4 m yr^{-1}. Upstream of Clarendon, AR the channel slope is about 6×10^{-4}.

Most of the lower White River flows through a forested floodplain, including all five of the cutoff sites in this study, although there are patches where agricultural land meets the river. The planform geometry of the White River exhibits multiple neck cutoffs and numerous other bends with narrow necks suggesting imminent neck cutoff. There also is an apparent lack of chute cutoffs on the White River, despite frequent overbank flood events.

METHODS

The collection of velocity data occurred during 21 to 23 May, 2016, when discharge was ~ 1020 m^3 s^{-1}, corresponding to a river stage roughly 1 m below bankfull stage. A Teledyne-RDI 1,200 kHz Rio Grande acoustic Doppler current profiler (ADCP), mounted to an 18 ft jon boat, was used to obtain three-dimensional velocities for each cutoff site. The ADCP is a four beam system with a resolution of 0.001 m s^{-1} and accuracy of 0.002 m s^{-1}. Boat position was tracked using a Hemisphere A100 integrated differential global positioning system (dGPS), with a 0.6 m accuracy. The velocity measurements were collected along predetermined cross-sections oriented perpendicular to channel centreline, prepared using a modified version of the Matlab script PCS-Curvature (Güneralp & Rhoads, 2008). Cross-sections were spaced approximately every half-channel width ~ 85 m and each cross-section was traversed

two to four times to improve time-averaging of the velocity fields (Szupiany *et al.*, 2007).

WinRiver II was used in the field for data acquisition and later for processing of transect lines. Transect lines were exported from WinRiver II and imported into the Velocity Mapping Toolkit (VMT) where they were combined into the cross-sections. VMT is a suite of Matlab codes developed to process boat-mounted ADCP data and combines transects along similar cross-sections to produce temporally and spatially averaged velocity data (Parsons *et al.*, 2013). Additionally, VMT allows for visualisation of three-dimensional flow fields using various frames of reference. In this paper, the cross-stream frame of reference (orientation parallel and orthogonal to the measured transects) and the Rozovskii (1957) frame of reference (primary *Vp* and secondary *Vs* components of flow oriented to local depth-averaged velocity vector in each vertical profile) into components oriented perpendicular *Vpx* and parallel *Vsy* to the cross-stream plane were used (Fig. 3). The Rozovskii (1957) frame of reference was included in the analyses because it allows for visualisation of flow streamlines where the primary flow direction changes substantially along a cross-section (Rhoads & Kenworthy, 1998). Vertical velocity components are not affected with the Rozovskii frame of reference. Images of the depth-averaged velocity (DAV) and the cross-sectional flow fields were produced using VMT.

RESULTS

Depth-averaged velocity

Maps of depth-averaged velocity vectors through the neck cutoffs with subparallel upstream and downstream channels (i.e. Pumps, Seven Mile and Franklin bends) show many similarities. Flow within the main channel immediately upstream of each cutoff displays a slight asymmetry in the cross-stream pattern of depth-averaged velocity vectors, with the highest velocities ~ 1.0 to 1.2 m s^{-1} along the outer (right) bank and lowest velocities ~ 0.25 m s^{-1} along the inner (left) bank (Figs 4 to 6). As flow continues to travel into the neck cutoff region, the majority of flow discharge is redirected nearly 180° from the upstream limb to the downstream limb of the loop over a relatively short distance, roughly half of the bankfull channel width ~ 100 m. The strong

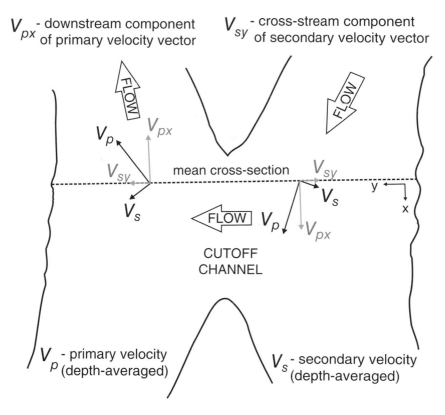

Fig. 3. Rotational definitions for Rozovskii (1957) frame of reference. Block arrows represent general flow direction through the cutoff. V_p and V_s represent the primary and secondary velocities, respectively, of the local depth-averaged velocity. V_{px} represents the downstream component of primary velocity relative to the mean cross-section. V_{sy} represents the cross-stream component of secondary velocity, oriented orthogonal to V_{px}.

redirection and small radius of flow curvature results in flow acceleration (e.g. Franklin Bend, Fig. 5) and advection of high momentum fluid along the bank opposite the exit of the cutoff (Figs 4 to 6). Immediately downstream of the cutoff, the core of high velocity remains along the far bank opposite from the cutoff and displays strong asymmetry in the cross-stream pattern of velocity vectors. At each site roughly 100 metres downstream from the cutoff, the pattern of velocity vectors shows a redirection of the primary flow back towards the centre of the channel (Figs 4 to 6).

At Pumps Bend, the tight bend flow through the cutoff results in a zone of flow separation immediately downstream of the cutoff along the right bank within the downstream limb of the loop and results in a zone of flow recirculation that is ~ 75 m and ~ 200 m in the transverse and streamwise directions, respectively (Fig. 4, zone 1). Velocities within this zone of recirculation have upstream oriented magnitudes of ~ 0.2 m s⁻¹. While the pattern of depth-averaged velocity vectors at Franklin and Seven Mile bends do not dis-

play upstream oriented flow within the zone of separation, flow recirculation is still present at both sites (Figs 5 to 6, zone 1). The lack of data within these two regions was due to shallow water that prevented collection of boat-mounted ADCP measurements. However, previous observations from the field during multiple field campaigns have confirmed the presence of flow recirculation in these regions.

In addition to the zone of separation and recirculation downstream of each cutoff, Pumps and Franklin bend exhibit two zones of recirculation located within the entrance and exit of the abandoned loop (Figs 4 and 5, zones 2 and 3). These zones of recirculation occupy the full width of each channel segment and extend in the streamwise direction for multiple cross-sections. The dimensions and velocities within these hydrodynamic features vary between sites; however, the zones of recirculation at the exit of the abandoned loops show higher velocities (up to ~ 0.6 m s⁻¹) than the zones of recirculation at the entrance (up to ~ 0.1 m s⁻¹; Figs 4 and 5).

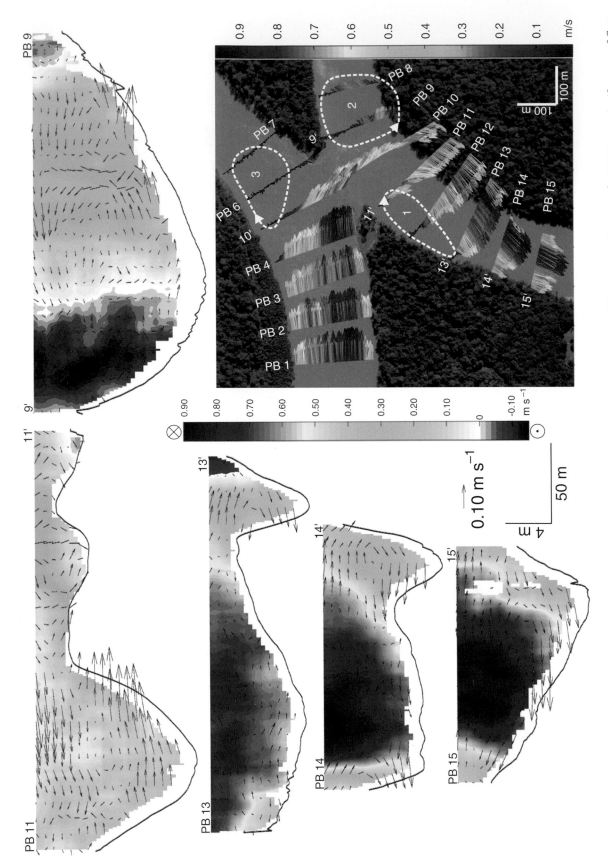

Fig. 4. Map of depth-averaged velocity vectors and selected cross-sectional flow fields at Pumps Bend. White dashed lines in the DAV map signify areas of flow recirculation. (1) zone of flow separation and recirculation, (2 and 3) zones of flow recirculation. For the cross-sectional flow images, primary (colour contours) and secondary (vectors) velocities using the Rozovskii frame of reference as described in methods section. Positive streamwise velocities represent flow into the cross-sectional plane, and negative velocities represent flow out of the cross-sectional plane.

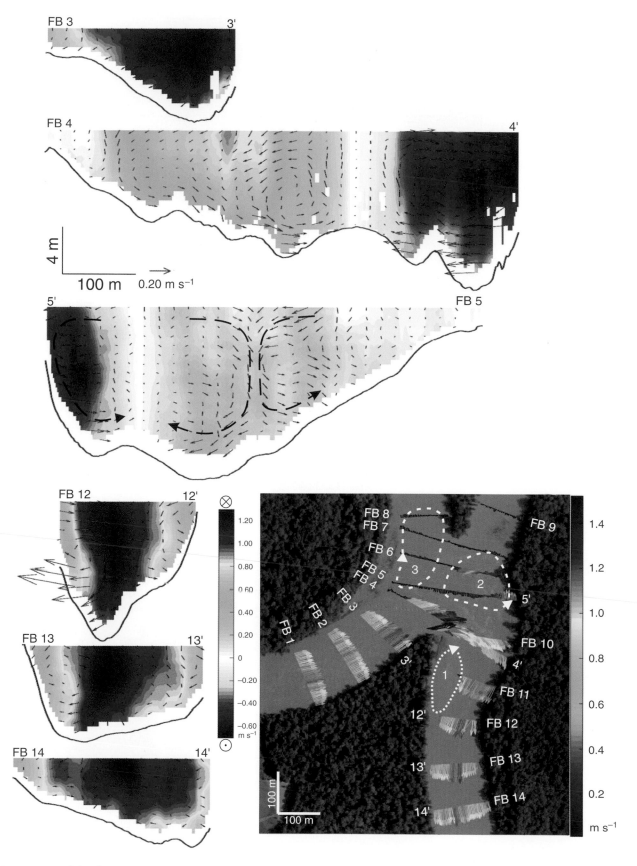

Fig. 5. Map of depth-averaged velocity vectors and cross-sectional flow fields at Franklin Bend. (1) In the DAV image, white dotted line is an area of flow separation and recirculation over shallow bar interpreted from field observations. (2 and 3) White dashed line shows areas of flow recirculation. Primary (colour contours) and secondary (vectors) velocities using the Rozovskii frame of reference as described in methods section are shown in the cross-sectional flow fields. Positive streamwise velocities represent flow into the cross-sectional plane; and negative velocities represent flow out of the cross-sectional plane. Dashed black lines represent areas of helical flow.

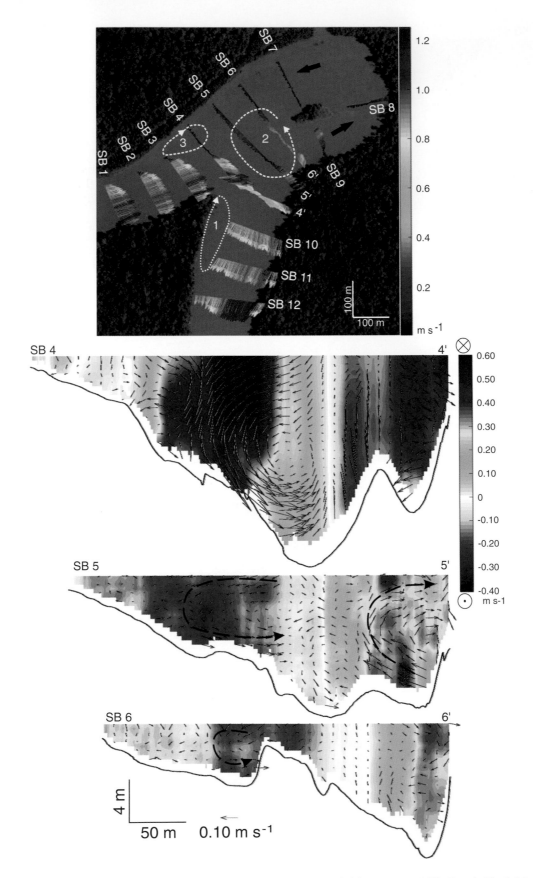

Fig. 6. Map of depth-averaged velocity vectors and cross-sectional flow fields at Seven Mile Bend. Black block arrows signify general flow movement through the loop. (1) White dotted line is an area of flow separation and recirculation over shallow bar interpreted from field observations. (2 and 3) The white dashed lines show zones of flow recirculation. Primary (colour contours) and secondary (vectors) velocities using the Rozovskii frame of reference as described in methods section. Positive streamwise velocities represent flow into the cross-sectional plane; and negative velocities represent flow out of the cross-sectional plane. Dashed black lines represent areas of helical flow.

At Seven Mile Bend, the hydrodynamic features shown by the patterns of depth-averaged velocity vectors are more complex. Within the upstream limb of the loop, a zone of recirculation is located along the bank opposite from the cutoff and has dimensions of ~ 70 m and ~ 140 m in the transverse and streamwise directions, respectively (Fig. 6, zone 3). As flow travels through the cutoff and collides with the far opposite bank in the downstream limb, a portion of the flow (~ 140 m³ s⁻¹, ~13% of discharge) is redirected upstream (Fig. 6). This portion of flow is further divided, with a smaller volume travelling upstream through the abandoned loop (Fig. 6, black arrows) and the majority of flow forming a large zone of recirculation located adjacent to the cutoff island (Fig. 6, zone 2). This zone of recirculation appears to extend the entire width of the neck cutoff and has maximum velocities of ~ 0.6 m s⁻¹ (Fig. 6).

The pattern of velocity vectors at Devil's Elbow shows pronounced flow asymmetry in the cross-stream pattern of depth-averaged velocity through most of the reach, with the core of high velocity apparent throughout (Fig. 7). In the upstream limb of the active channel, the high velocity core is located along the outer (left) bank opposite from the entrance and exit to the abandoned loop. Flow travelling through the cutoff channel has velocities up to ~ 1.25 m s⁻¹ that are initially aligned with flow from the upstream limb but is quickly redirected ~ 90° and collides with the far bank (Fig. 7). The collision of the flow with the bank results in a portion of the flow being redirected in the upstream direction with velocities of ~ 0.25 m s⁻¹, forming a zone of recirculation located across the entrance and exit to the abandoned loop (Fig. 7, zone 1). Flow near the entrance and exit to the abandoned loop shows slight exchange with this zone of recirculation, though velocities are nearly stagnant.

The flow that is redirected ~ 90° out of the cutoff channel immediately enters the apex region of the downstream loop and is strongly redirected ~ 180° over a distance of ~ 100 m (Fig. 7). The tight bend flow through the cutoff channel and downstream receiving limb results in two zones of recirculation. Near the exit of the cutoff channel, a relatively small zone of flow separation forms along the bank opposite the entrance to the abandoned loop and extends ~ 100 m in the downstream direction (Fig. 7, zone 2). Flow travelling through the apex region of the downstream loop collides nearly perpendicular with the far (left) bank, with

a portion of the flow being redirected upstream forming a large zone of recirculation that occupies the majority of the channel within this region (Fig. 7, zone 3).

The overall pattern of depth-averaged velocity vectors at Calhoun Bend is similar to Devil's Elbow. Upstream of the cutoff channel, the core of high velocity is located closer to the left bank and shifts to the right bank as flow travels through the cutoff (Fig. 8). Near the exit of the cutoff channel, the pattern of cross-stream velocity vectors shows a strong asymmetry with the highest velocities ~ 1.0 m s⁻¹ along the bank opposite from the exit of the abandoned loop. Downstream of the cutoff channel, the flow turns ~ 80° over a distance of roughly 600 m into the apex region of the downstream loop and the core of high velocity shifts back to the left bank (Fig. 8). Flow entering the apex region is highly asymmetric with the highest velocities ~ 1.35 m s⁻¹ along the inner bank. The flow is then strongly redirected through the apex, with a ~ 100 m zone of separation and recirculation forming along the inner bank (Fig. 8, zone 2). As flow travels through the apex of the downstream loop and collides with the outer bank, a portion of the flow is redirected upstream with velocities ~ 0.55 m s⁻¹ forming a large zone of recirculation along the outer bank upstream of the loop apex that is ~ 100 m-wide and ~ 400 m-long (Fig. 8, zone 1).

Cross-sectional flow fields

Pumps Bend

Within the upstream limb of Pumps Bend, the three-dimensional flow fields do not exhibit coherent secondary circulation, probably due to a lack of planform curvature upstream of the cutoff. However, the three-dimensional flow field along a cross-section roughly parallel to the neck cutoff at Pumps Bend (PB 10) reveals a complex pattern of secondary circulation and bed topography (Fig. 9). On the left side of the cross-section, the pattern of streamwise velocity contours show a core of high velocity into the cross-sectional plane that occupies the top half of the water column with maximum velocity of ~ 0.6 m s⁻¹ (Fig. 9A). These positive velocities represent flow travelling into the cutoff channel from the upstream limb of the loop. On the right side of cross-section PB 10, streamwise velocities are negative indicating flow moving out of the

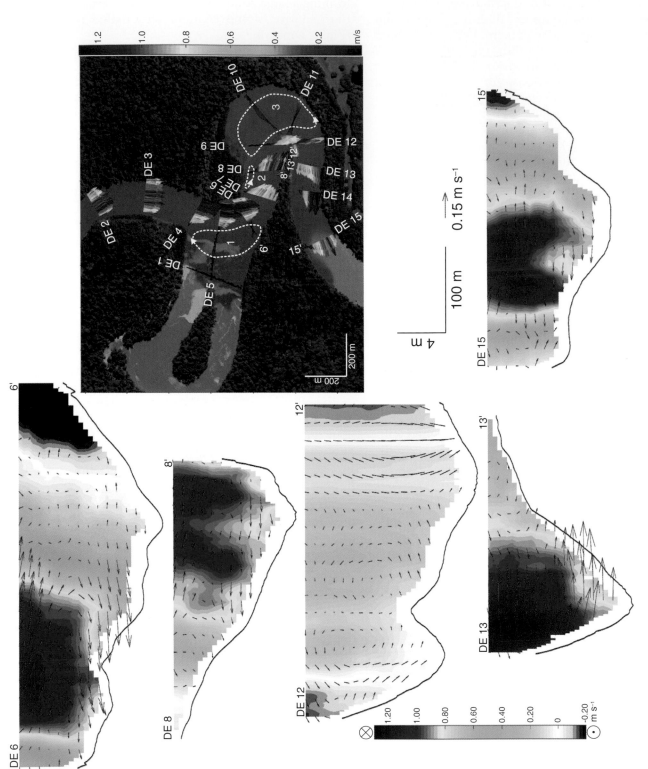

Fig. 7 Map of depth-averaged velocity vectors and cross-sectional flow fields at Devil's Elbow. The DAV image shows large zones of flow recirculation (1 and 3) and a zone of flow separation and recirculation (2) represented by the white dashed lines. For the cross-sectional flow fields, primary (colour contours) and secondary (vectors) velocities using the Rozovskii frame of reference as described in methods section. Positive streamwise velocities represent flow into the cross-sectional plane; and negative velocities represent flow out of the cross-sectional plane.

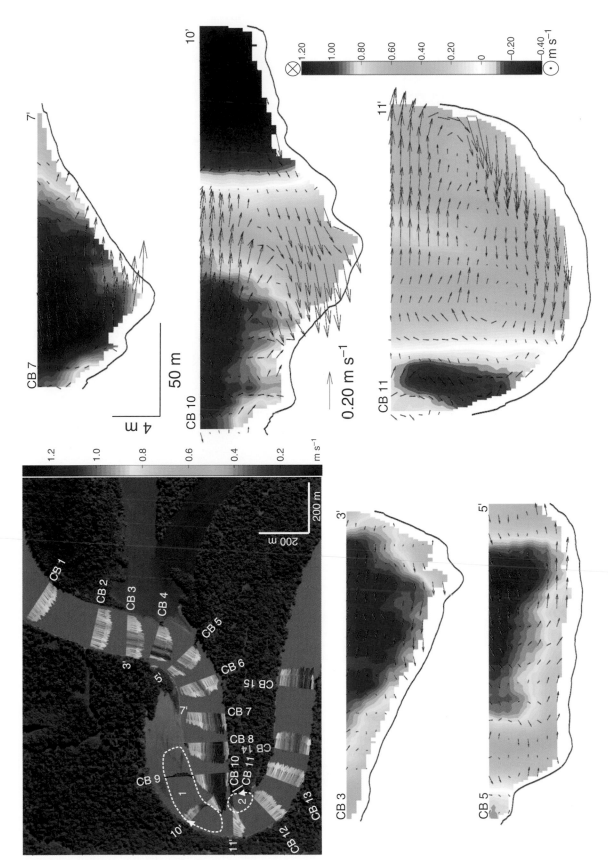

Fig. 8. Map of depth-averaged velocity vectors and cross-sectional flow fields at Calhoun Bend. The white dashed lines indicate zones of flow recirculation (1) and flow separation and recirculation (2) in the DAV image. Primary (colour contours) and secondary (vectors) velocities using the Rozovskii frame of reference as described in methods section in the cross-sectional flow fields. Positive streamwise velocities represent flow into the cross-sectional plane, and negative velocities represent flow out of the cross-sectional plane.

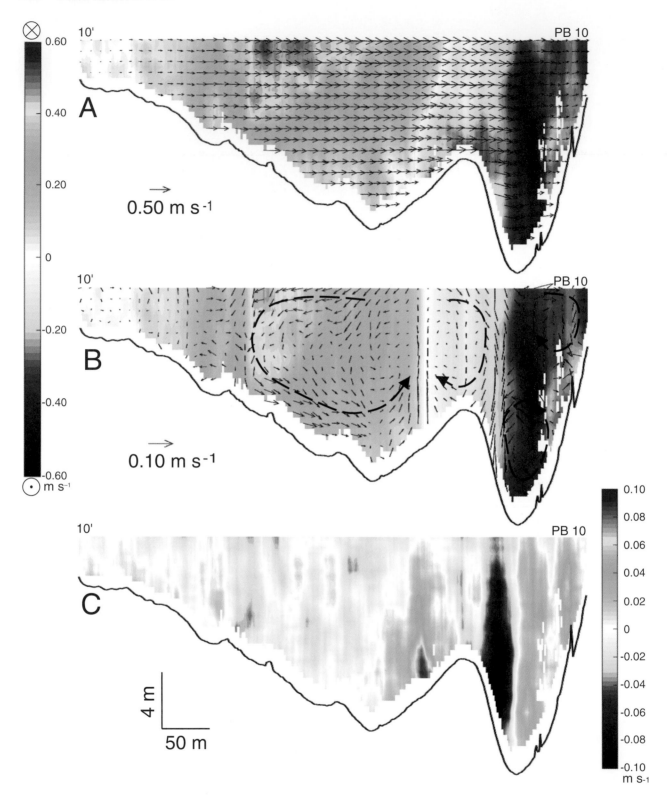

Fig. 9. Cross-sectional flow fields for Pumps Bend cross-section PB10. See Fig. 10 for location of cross-section within cut-off. (A) Streamwise velocities (colour contours) and secondary velocity vectors using cross-stream frame of reference. Positive streamwise velocities represent flow into the cross-sectional plane; and negative velocities represent flow out of the cross-sectional plane. (B) Primary (colour contours) and secondary (vectors) velocities using the Rozovskii frame of reference as described in methods section. Dashed black lines show areas of helical flow. (C) Pattern of vertical velocities within cross-section PB 10.

cross-sectional plane and representing flow leaving the cutoff channel and travelling through the downstream limb of the loop (Fig. 9A). The pattern of streamwise velocities within the right side of the cross-section show a core of fairly uniform velocity ~ -0.6 m s^{-1} that extends from the water surface to the channel bed. Within the right side of the cross-section, the channel bed appears to be scoured to a depth of ~ 17 m below the water surface (Fig. 9A). The pattern of secondary velocity vectors in the cross-sectional plane show nearly uniform lateral flow from the left to right side of the cross-section, with magnitudes up to ~ 0.75 m s^{-1}, representing the general pattern of flow travelling through the cutoff channel.

Using the Rozovskii frame of reference, the pattern of secondary circulation of the flow becomes evident. On the left side of the cross-section, the pattern of secondary velocity vectors show near-surface flow directed toward to the left and near-bed flow directed toward the right (Fig. 9B). This pattern of secondary velocity vectors indicates helical motion of the flow in a counter-clockwise direction as the flow travels into (positive velocities) the cutoff channel. On the right side of the cross-section above the area of higher bed elevation, there is a region of flow where the near-surface secondary velocity vectors are oriented toward the right and near-bed vectors are oriented to the left (Fig. 9B). This area of the cross-section is also bounded by a zones of positive and negative vertical velocities on the left and right side of the region of elevated bed topography, respectively (Fig. 9C). These patterns of secondary velocity vectors and vertical velocity contours also indicate the presence of helical motion of the flow. While the pattern of secondary circulation on the right side of the cross-section appears to be opposite from the pattern on the left side of the cross-section, the primary direction of the flow is out of the cross-sectional plane (negative velocities). Thus, the sense of motion for the helical flow travelling out of the cutoff channel is also counter-clockwise. On the far right side of the cross-section along the bank and above the scour hole, the pattern of secondary velocity vectors show two additional smaller cells with opposite sense of rotation, one near the bed within the scour hole and the other along the top of the bank.

The three-dimensional flow field at the first cross-section downstream from the cutoff (PB 11) shows a ~ 13 m-deep channel on the left side of the cross-section with a pattern of strong counter-clockwise secondary circulation exhibited by the secondary velocity vectors (Fig. 4). At cross-section PB 13, the left side of the channel shallows and widens, with streamwise velocities increasing and patterns of secondary velocity vectors still showing counter-clockwise secondary circulation, though the magnitude of secondary velocities has decreased (Fig. 4). On the right side of cross-section PB 13, a ~ 8 m-deep and ~ 50 m-wide section of the cross-section shows a pattern of clockwise secondary circulation. Continuing downstream, the cross-sectional profile of PB 14 shows more uniform depth across the section and the pattern of clockwise secondary circulation on the right side of the cross-section has increased in size and magnitude (Fig. 4). At the farthest downstream cross-section PB 15, the transverse profile is asymmetric with a thalweg located on the right side of the profile (Fig. 4). The pattern of clockwise secondary circulation now occupies nearly the entire cross-sectional area.

The cross-sectional flow field at the exit of the abandoned loop (PB 9) displays a large cross-sectional area with positive streamwise velocities on the right side indicating flow moving upstream into the abandoned loop and negative velocities on the left side indicating flow moving out of the abandoned loop (Fig. 4). The pattern of secondary velocity vectors within the region of positive streamwise velocity show counter-clockwise secondary circulation, indicating helical motion into the abandoned loop.

Franklin Bend

The cross-sectional flow field just upstream of the cutoff at Franklin Bend FB 3 shows an asymmetric transverse profile with a thalweg near the right side of the section and core of high velocity within the thalweg with velocities up to ~ 1.2 m s^{-1} (Fig. 5). The pattern of secondary velocity vectors show a pattern of clockwise secondary circulation that is present throughout the entire cross-section. Within the cutoff channel, a cross-section oriented parallel to the channel along the upstream junction corner FB 4 shows positive streamwise velocities of the left side of the section indicating flow moving into the cutoff channel and negative streamwise velocities on the right side of the section indicating flow moving out of the cutoff channel (Fig. 5). The pattern of secondary velocity vectors show counter-clockwise secondary circulation within the region of positive streamwise velocities and clockwise secondary circulation within the region of negative streamwise velocities (Fig. 5). Similar to the pattern shown at Pumps Bend, the secondary circulation imposed on the region of negative streamwise velocities indicates a counter-clockwise helical motion of the flow.

Cross-section FB 5 is also located within the cutoff channel but intersects the two zones of recirculation at the entrance and exit of the abandoned loop (Fig. 5). The viewing orientation of FB 5 is from the exit of the abandoned loop looking downstream. Streamwise velocities within this cross-section show a region of negative velocity ~ -0.5 m s^{-1} along the left bank, indicating flow moving upstream into the exit of the abandoned loop (Fig. 5). This region of negative streamwise velocity shows a pattern of counter-clockwise secondary circulation, indicating clockwise helical motion of flow into the abandoned loop. Near the middle of the cross-section, streamwise velocities are positive representing flow from the exit of the loop into the cutoff channel and secondary velocity vectors show a pattern of clockwise circulation (Fig. 5). The right half of cross-section FB 5 also shows positive streamwise velocities; however, the pattern of secondary velocity vectors show a pattern of counter-clockwise circulation.

Downstream of the cutoff channel, the cross-sectional flow fields all display positive streamwise velocities and a progression in the transverse profile from a narrow channel with deep thalweg near the middle of the section (FB 12) to an asymmetric profile with the thalweg near the base of the right bank (Fig. 5; FB 14). At cross-section FB 12, a pattern of strong clockwise secondary circulation is located along the left bank indicating clockwise helical motion of the flow, while the right side of the section shows counter-clockwise secondary circulation (Fig. 5). At the next cross-section downstream FB 13, the pattern of clockwise secondary circulation occupies the entire cross-sectional area and the counter-clockwise secondary circulation is no longer present and by cross-section FB 14 the clockwise secondary circulation has become weak throughout the cross-section (Fig. 5).

Seven Mile Bend

Three cross-sectional flow fields through the cutoff channel at Seven Mile Bend (SB 4 to 6) reveal the highly three-dimensional characteristics of the flow at this site. These three cross-sections are all oriented with a viewing direction from the upstream limb of the loop looking downstream toward the cutoff island. At the first cross-section within the cutoff channel SB 4, the transverse profile shows a ~ 19 m-deep scour hole near the middle of the section (Fig. 6). On the left side of this scour hole, the streamwise velocities are positive with a pronounced pattern of counter-clockwise secondary circulation. On the right side of the deep scour hole, streamwise velocities are negative with a pattern of clockwise secondary circulation, indicating counter-clockwise helical motion of the flow (Fig. 6), similar to Pumps and Franklin bends. Additionally, on the far left side of cross-section SB 4 the pattern of secondary velocity vectors show a third region of secondary circulation which has a clockwise sense of rotation.

At cross-sections SB 5 and 6, the magnitudes of velocity have decreased and the pattern of streamwise velocities has reversed from SB 4, showing negative and positive streamwise velocities on the left and right side of the flow field, respectively (Fig. 6). At SB 5, the pattern of secondary velocity vectors in the region of negative and positive streamwise flow shows counter-clockwise and clockwise secondary circulation, respectively (Fig. 6). These patterns of secondary circulation imposed on the streamwise flow indicate clockwise helical motion into and out of the cross-sectional plane. Within cross-section SB 6, the patterns of secondary circulation have become diminished, with secondary circulation in the region of positive streamwise velocity is no longer apparent, though a weak pattern of counter-clockwise secondary circulation can be seen within the region of negative streamwise velocity (Fig. 6).

Devil's Elbow

As flow travels into the cutoff channel at Devil's Elbow, the patterns of secondary velocity vectors do not exhibit secondary circulation until the end of the cutoff channel at cross-section DE 6, where pronounced clockwise secondary circulation can be seen throughout most of the section where streamwise velocities are positive (Fig. 7). At cross-section DE 8, the transverse profile is asymmetric with the thalweg near the right bank and a pattern of clockwise secondary circulation imposed on positive streamwise velocities. As flow travels through the apex region of the downstream loop, secondary circulation breaks down and does not show a coherent pattern until cross-section DE 12, though the pattern has now reversed to a counter-clockwise sense of rotation (Fig. 7). At cross-section DE 13, the pattern of counter-clockwise secondary circulation intensifies with magnitudes of secondary velocities as high as $\sim 0.8\,\mathrm{m\,s^{-1}}$. Farther downstream at cross-section DE 15, the pattern of secondary circulation reverses again and now shows a clockwise sense of rotation within the middle of the section (Fig. 7).

Calhoun Bend

General patterns of the three-dimensional flow fields at Calhoun Bend are similar to those observed at Devil's Elbow. As flow travels through the cutoff channel from CB 3 to CB 5, the pattern of secondary circulation reverses from a clockwise to counter-clockwise sense of rotation and the counter-clockwise secondary circulation intensifies downstream to cross-section CB 7 (Fig. 8). At cross-section CB 10, the flow turns sharply through the downstream loop apex and positive streamwise velocities now exhibit strong clockwise secondary circulation (Fig. 8). On the right side of cross-section CB 10, streamwise velocities are negative with magnitudes up to $\sim 0.4\,\mathrm{m\,s^{-1}}$ and the pattern of secondary velocity vectors show clockwise sense of rotation, indicating flow travelling upstream with counter-clockwise helical motion. At the apex of the downstream loop CB 11, much of the cross-section displays positive streamwise velocities with a pattern of clockwise secondary circulation, with the exception of a $\sim 30\,\mathrm{m}$-wide region of negative streamwise velocities along the left bank with the top $\sim 2\,\mathrm{m}$ of flow directed to the right and weak counter-clockwise secondary circulation below that extends to the channel bed (Fig. 8).

DISCUSSION

Conceptual model of flow through neck cutoffs

The results from this study detail the complex three-dimensional flow structure through neck cutoffs on meander bends with complex planform geometry and offer insight into the mechanisms responsible for producing the observed hydrodynamic features at these sites. Generally, the pattern of depth-averaged velocity vectors

observed at Franklin, Seven Mile and Pumps bends show a strong redirection of the flow from the upstream to downstream limb by nearly 180° (Fig. 10). As flow travels through the cutoff channel, the core of high velocity is advected against the bank opposite from the upstream junction corner and remains along this bank within the downstream limb of the loop. This pattern of flow is attributed to the location of the cutoff channel within the original elongate loop, where the cutoff channels formed such that flow within the upstream and downstream limbs was oriented roughly subparallel and in opposite directions. This planform configuration is one of the factors contributing to the tight bend flow through the cutoff channel. Other factors responsible for redirecting the flow through the cutoff over such a relatively short distance are the slope advantage resulting from the shortening of the channel length and the water within the entrance and exit of the abandoned loop that provides an adverse pressure gradient along the fluid interface with the flow through the cutoff.

The planform configuration and tight bend flow of these neck cutoffs on elongate loops are also responsible for producing the zones of flow separation and recirculation. Pumps, Franklin and Seven Mile bends all have a zone of separation located along the right bank within the downstream limb of the loop immediately downstream of the upstream cutoff junction corner (Fig. 10). The point of flow separation begins from the junction corner and extents to the point of reattachment ~ 200 to $300\,\mathrm{m}$ along the right bank, resulting in upstream recirculating flow along the bank. At Pumps and Franklin bends, a zone of recirculation is produced within the exit of the abandoned loops, probably from a combination of 1) redirection of flow travelling through the cutoff after it collides with the far bank and resulting super-elevation of the water surface; and 2) exchange of streamwise momentum along the fluid interface between the flow travelling through the cutoff and flow within the exit of the abandoned loop. A third zone of recirculation occurs within the entrance to the abandoned loop and exhibits weaker velocity magnitudes.

In the case of Seven Mile Bend, the location of the cutoff within the loop occurred very close to the apex, resulting in markedly less slope advantage than Pumps and Franklin bends. Thus, high momentum fluid is able to travel upstream through the abandoned loop from the exit to the entrance. Additionally, the zone of recirculation

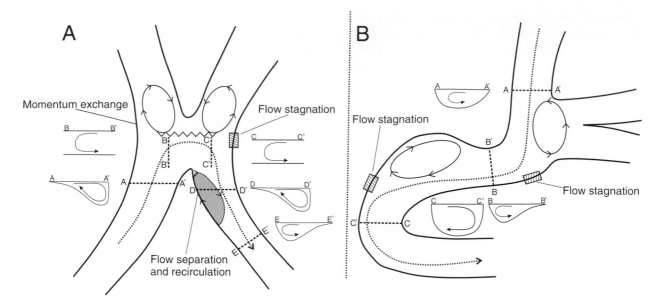

Fig. 10. Model of overall primary and secondary flow patterns for neck cutoffs based off the results found in this research. (A) Hydrodynamics through neck cutoffs that evolve where the upstream and downstream meander limbs are subparallel and oriented in opposite directions from each other. (B) Hydrodynamics through neck cutoffs that evolve where the upstream meander limb forms parallel to the cutoff channel and the downstream limb forms perpendicular. The dotted line represents primary flow movement and direction. Circles with arrows indicate areas of flow recirculation and directionality of the circulation. The circle adjacent to the junction corner in the downstream limb represents an area of flow separation and recirculation. The dashed lines represent the location of cross-sections that have been selected to represent secondary circulation throughout the cutoff. The zig-zag line represents the area of momentum exchange between the primary flow and fluid in the entrance and exit of the abandoned loop. Lastly, the small boxes with lines represent areas of flow stagnation.

that occupies only the exit of the abandoned loop at Pumps and Franklin bends, extends the entire width of the cutoff channel at Seven Mile Bend and displaces the third zone of recirculation upstream and opposite the cutoff junction corner.

The planform geometry and location of the neck cutoffs at Devil's Elbow and Calhoun Bend produced a planform geometry where the upstream channels are oriented roughly parallel with the cutoff channel and the downstream channels are nearly perpendicular to the cutoff channel. In addition, the curvature of the upstream limb results in a core of high velocity that is already positioned along the bank opposite from the entrance to the abandoned loop. The cross-stream pattern of depth-averaged velocity vectors becomes even more asymmetric through the cutoff channel as the spatial pattern of the water surface elevation directs the core of high velocity (Dietrich, 1987) away from the abandoned loop and toward the downstream channel. The zones of flow recirculation located across the entrance and exit of the abandoned loop and within the

apex of the downstream limb appear to be produced by gradients in the water surface elevation redirecting flow upstream and exchange of momentum along the fluid interface between the primary flow and recirculating flow.

The characteristics of tight bend flow and zones of recirculation can be used to explain the patterns of secondary circulation and helical flow observed within the three-dimensional flow fields. Helical motion of flow through the cutoff channel was well-developed once the flow began entering the cutoff. The helical motion of the flow is established by a balance between 1) the pressure gradient force resulting from the super-elevation of the water surface along the fluid interface between the primary flow and flow within the entrance and exit of the abandoned loop; and 2) the centrifugal force resulting from the flow curving through the cutoff channel. The centrifugal forces preferentially act upon the high momentum near-surface fluid causing flow near the water surface to travel outward toward the abandoned loop, while the friction along the channel bed leads to low momentum near-bed fluid being acted upon by the pressure gradient forces leading to flow near the bed being redirected more strongly into the downstream channel.

The spatial pattern of helical flow showed relatively fast rates of development, spatially, and even reversal of helical motion. Reversal of helical motion typically occurred over distances ~ 100 m located at the entrance to the cutoff channels (e.g. Franklin Bend FB 3 and 4), or within the downstream limb of the loops (e.g. Pumps Bend PB 11 to 13). Interestingly, at the entrance to the cutoff channels, the transverse bed profiles still exhibit point bars along the left bank as established prior to cutoff. Thus, the reversal of helical motion through the cutoff results in near-bed flow oriented from the top of the point bar toward the thalweg, a pattern that is opposite from the helical motion typically observed in elongate meander loops with point bars (Whiting & Dietrich, 1993a,b; Konsoer et al., 2016a). Another impressive hydrodynamic feature is the development of helical flow within zones of strong recirculation (e.g. Calhoun Bend CB 10). The helical motion observed within zones of strong recirculation similarly suggest a balance between forces produced by flow curvature and gradients in the water surface elevation.

The development and reversal of helical motion over relatively short streamwise distances observed at the neck cutoffs of this study do not conform to patterns of helical flow structure typically documented within elongate meander loops. In general, the apex region of an elongate loop is connected by relatively straight upstream and downstream limbs, allowing for relatively longer distances for secondary circulation to develop and decay, and commonly only results in a single helical cell throughout the loop (Whiting & Dietrich, 1993a,b; Konsoer et al., 2016a). Compound meander loops, which display two or more offset lobes of curvature, often display a decay in secondary circulation between the two lobes as a result of a reduction in channel curvature (Hooke & Harvey, 1983; Frothingham & Rhoads, 2003; Engel & Rhoads, 2012). As the compound meander loop continues to evolve, a secondary inflection point can develop leading to a reversal in channel curvature and the potential for reversal of helical motion. However, even in the case of compound meander loops, the spatial scales over which secondary circulation develops and decays tends to be on the order of multiple channel widths, rather than the single channel widths observed for the neck cutoffs of this study.

Comparison of results to previous models

The main difference between the model that resulted from this field-based research and previous models is that this model focuses on hydrodynamics instead of morphologic change. Previous models have either focused on change in channel morphology through neck cutoffs or on flow and morphologic change through chute cutoffs. Therefore, this is the first conceptual model based on field data that characterises hydrodynamics through neck cutoffs. However, to aid in further discussion, patterns of flow structure are inferred from morphologic models of neck cutoff (e.g. Gagliano & Howard, 1984) to compare to the results in this research.

The observations of flow structure through neck cutoffs in this study differ from current conceptual models that depict neck cutoffs that form such that flow from the upstream to downstream limb is straight without any flow curvature (Fisk, 1947; Gagliano & Howard, 1984). This classic conceptual model does not provide an explanation for the pattern of complex hydrodynamic features observed in this study. Although the morphologic model of Gagliano & Howard (1984) depicts a simplified planform with a straight cutoff channel, it is important to note that previous works discuss the effect of differences in planform geometry and diversion angle on the rate of sediment plugging (Fisk, 1947; Gagliano & Howard, 1984). Thus, these different models should not be viewed as opposing, but rather capturing different characteristics and controlling factors. One of the main components of the previous conceptual models is that flow travels straight through the cutoff channel, with some of the flow angling into the entrance of the abandoned loop forming a zone of flow separation and recirculation (Fig. 1; Shields & Abt, 1989; Constantine et al., 2010a). Although the planform configurations of Devil's Elbow and Calhoun Bend exhibit an upstream limb that is roughly parallel with the cutoff channel, the downstream limb is oriented perpendicular to the cutoff channel. The pattern of flow through these neck cutoffs was shown to be highly asymmetric with the core of high velocity along the bank opposite from the abandoned loop, thus preventing zones of flow separation at the entrance and/or exit to the abandoned loop and indicating any plugging of sediment at these cutoff sites is by a different mechanism than that described by diversion angle models.

The planform configuration and patterns of flow structure more closely aligns with the bifurcation-confluence model of flow through chute cutoffs (Zinger *et al.*, 2013). In the bifurcation-confluence model, flow from the upstream limb turns sharply into the cutoff channel at the bifurcation producing a zone of separation and recirculation that was also observed in this study. Conversely, the patterns of flow structure through the neck cutoffs on the White River do not agree with the downstream confluence aspect of the model, as the field observations display zone of flow recirculation at the exit of the abandoned loop, while the conceptual models of confluent flow suggest a zone of flow stagnation (Best, 1986; Kenworthy & Rhoads, 1995; Zinger *et al.*, 2013). One possible explanation for this difference is that the bifurcation-confluence model was established with flow still travelling out of the exit of the abandoned loop (Zinger *et al.*, 2013), indicating an early stage of chute cutoff development. However, even in the late stages of chute cutoff evolution, a sediment plug typically forms within the exit of abandoned bend and would also not help to explain the formation of the zone of recirculation observed from these neck cutoffs.

Limitations and future work

The data used in this study was obtained for one flow event (Q = ~ 1020 m^3 s^{-1}) and it is expected that patterns of flow will change significantly with varying discharges, especially during large floods. Thus, obtaining data at multiple discharges is still needed in order to fully understand the hydrodynamics at these cutoff sites. At higher stages, such as the discharge in this study, a sufficient amount of flow moves through the abandoned loops and downstream limbs of the active channel that zones of recirculation can be seen. However, at lower stages plugging of the abandoned loop most probably has a greater impact on the flow pattern and on the steering of flow through the cutoff. Furthermore, for Pumps, Seven Mile and Franklin bends, a bar that forms adjacent to the junction corner in the downstream limb of the active channel emerges at low flow stages and no flow occupies that area. Thus, the

presence, size and strength of the zones of separation and recirculation are clearly dependent on the stage and discharge conditions.

Additionally, the data used in this study was collected at each site in a single day, representing a nearly static morphology, providing little insight into how the hydrodynamics vary as these neck cutoff evolve through time. However, Pumps and Franklin bends display similar planform morphology prior to cutoff and have been shown in this study to have similar hydrodynamic characteristics despite the different stages of cutoff evolution. Therefore, Pumps and Franklin bends could be used as a surrogate for temporal adjustments during cutoff evolution.

The results from this research can provide insight into the spatial distribution of shear stress and sediment transport which results in patterns of erosion and deposition and has implications for the evolution of these cutoffs. While future work on the morphologic evolution of these neck cutoffs still needs to be completed, the expected areas of deposition are within the downstream limb adjacent to the upstream cutoff junction corner (i.e. zone of separation and recirculation) and the entrance and exit of the abandoned loop. These areas of deposition are inferred from the patterns of flow structure documented herein and occur in areas of hydrodynamic contrasts. These areas often being where water surface gradients were locally steepened, redirecting flow upstream and causing momentum exchange along the fluid interface between flow in the cutoff and flow in the abandoned loop (i.e. Pumps and Franklin bends) or between primary and recirculating flow (i.e. Devil's Elbow and Calhoun Bend). Hydrodynamic contrasts also exist in areas of tight bend flow, such as at Pumps, Seven Mile and Franklin bends where channel morphology forced strong flow redirection, leading to the production of zones of flow separation and recirculation in the downstream limb adjacent to the cutoff junction corner. Based on the velocity magnitudes within the zones of separation and recirculation, it can also be expected that the pronounced zones of separation in the downstream limbs would be composed of coarser bedload material, whereas the zones of recirculation within the entrance and exit of the abandoned loops, as well as the large zones of recirculation within the apex region of Devil's Elbow and Calhoun Bend, would be predominately composed of finer bedload and suspended material.

Previous research on oxbow lakes have shown a general pattern of sedimentary architecture with the channel bed of the abandoned bend before cutoff being preserved by coarse-grained deposits (Toonen *et al.*, 2012; Ishii & Hori, 2016). Draped over the coarse-grained deposits is transitional material comprised of mixed-load that represents the time between cutoff and full abandonment of the bend (Toonen *et al.*, 2012; Ishii & Hori, 2016). On top of the transitional deposits are fine-grained laminated channel fill (Toonen *et al.*, 2012; Ishii & Hori, 2016). The relative amounts of transitional and fine-grained deposits depend on the relative rate of plug formation (Ishii & Hori, 2016). Sedimentary deposits in abandoned bends typically show an upward fining sequence (Toonen *et al.*, 2012). However, given that the neck cutoffs of this study appear to be plugging at relatively slow rates and remain hydrologically connected for decades, which is in contrast to previous literature (Peterson, 1963; Gagliano & Howard, 1984; Hooke, 1995), the patterns of oxbow lake infill could be substantially different from what is currently expected for neck cutoffs.

CONCLUSIONS

This paper has documented in detail the three-dimensional flow structure through neck cutoffs with differing planform configurations. The main hydrodynamic characteristics of flow through the neck cutoffs are summarised as follows:

1. Flow through cutoffs with upstream and downstream limbs oriented subparallel to each other exhibited a pattern of strong flow redirection of nearly 180° from the upstream to the downstream limb. The cutoff location within the original elongate loop, forming such that flow in the upstream and downstream limbs is oriented subparallel and in opposite directions, is the main factor contributing to the pattern of tight bend flow. Other factors are the slope advantage from the shortening of the channel length and water within the entrance and exit of the abandoned loops that lead to an adverse pressure gradient along the fluid interface with the flow through the cutoff. In contrast, cutoffs at Devil's Elbow and Calhoun Bend show a planform geometry of a cutoff channel oriented roughly parallel with the upstream channel and a downstream channel that is oriented nearly perpendicular.

2. Zones of separation and recirculation formed at the cutoff sites as a result of planform configuration and tight bend flow. Pumps, Seven Mile and Franklin bends showed a zone of separation in the downstream limb adjacent to the cutoff junction corner that resulted in upstream recirculation of flow. A second recirculation zone was found at the exit of the abandoned loop for Pumps and Franklin bends that resulted from the collision of flow with the far bank, leading to super-elevation of the water surface and the exchange of streamwise momentum along the fluid interface between flow in the cutoff and flow in the abandoned loop exit. A third zone of recirculation was found at the entrance of the abandoned loop. The cutoffs at Devil's Elbow and Calhoun Bend exhibited a zone of flow recirculation across the entrance and exit of the abandoned loop and within the apex of the downstream loop, formed by water surface elevation gradients redirecting flow upstream and the momentum exchange along the fluid interface between primary and recirculating flow.

3. Helical motion was seen at all cutoff sites, formed from the balance between pressure gradient force resulting from water surface super-elevation between primary flow and flow at the entrance and exit of the abandoned loop and from the centrifugal force resulting from curving of the flow through the cutoff channel. The result was a pattern of near-surface flow directed outward toward the abandoned loop and near-bed flow directed inward to the downstream channel. Reversal of helical motion was also seen, usually occurring at the entrance to the cutoff channel or within the downstream limb of the loops.

Results herein led to the production of the first conceptual model of hydrodynamics through neck cutoffs with complex planform configurations, the characteristics of which are not captured fully by prior models. However, the pattern of flow through neck cutoffs on elongate loops does show some agreement with aspects of flow at bifurcations, yet does not conform to patterns observed at channel confluences. While the findings of this study provide some of the first-ever observations of the three-dimensional flow structure through neck cutoffs on large rivers (White River mean annual $Q = 750 \, m^3 s^{-1}$; bankfull width = 170 m, bankfull depth = 7 m), further research on the detailed morphologic evolution of these neck cutoffs is needed in order to further refine a conceptual morphodynamic model that can be used for neck cutoffs. As well, field measurements such as those presented in this study could be used as a foundation for numerical simulations focused on evaluating how variable flow discharge affects patterns of flow and sediment transport through neck cutoffs.

ACKNOWLEDGEMENTS

Research was funded in part by a Louisiana State University—Faculty Research Grant and a 2016 West-Russell Research Award from the LSU Department of Geography and Anthropology. Comments from Arjan Reesink and two anonymous reviewers greatly improved the quality and clarity of this manuscript.

REFERENCES

Allen, J.R.L. (1965) A review of the origin and characteristics of recent alluvial sediments. *Sedimentology*, **5**(2), 91.

Best, J.L. (1986) The morphology of river channel confluences. *Progr. Phys. Geogr.*, **10**(2), 157–174.

Bridge, J.S., Smith, N.D., Trent, F., Gabel, S.L. and Bernstein, P. (1986) Sedimentology and morphology of a low-sinuosity river; Calamus River, Nebraska Sand Hills. *Sedimentology*, **33**(6), 851–870.

Constantine, J.A. and Dunne, T. (2008) Meander cutoff and the controls on the production of oxbow lakes. *Geology [Boulder]*, **36**(1), 23–26. doi:10.1130/G24130A.1

Constantine, J.A., Dunne, T., Piegay, H. and Kondolf, G.M. (2010a) Controls on the alluviation of oxbow lakes by bed-material load along the Sacramento River, California. *Sedimentology*, **57**(2), 389–407. doi:10.1111/j.1365-3091.2009.01084.x

Constantine, J.A., McLean, S.R. and Dunne, T. (2010b) A mechanism of chute cutoff along large meandering rivers with uniform floodplain topography. *Geol. Soc. Am. Bull.*, **122**(5–6), 855–869. doi:10.1130/B26560.1

Dietrich, W.E. (1987) Mechanics of flow and sediment transport in river bends. *Special Publication - Institute of British Geographers*, **18**, 179–227.

Edwards, B.L., Keim, R.F., Johnson, E.L., Hupp, C.R., Marre, S. and King, S.L. (2016) Geomorphic adjustment to hydrologic modifications along a meandering river: Implications for surface flooding on a floodplain. *Geomorphology*, **269**, 149–159. doi:10.1016/j.geomorph.2016.06.037

Engel, F.L. and Rhoads, B.L. (2012) Interaction among mean flow, turbulence, bed morphology, bank failures and channel planform in an evolving compound meander loop. *Geomorphology*, **163–164**, 70–83. doi:10.1016/j.geomorph.2011.05.026

Erskine, W., Melville, M., Page, K.J. and Mowbray, P.D. (1982) *Cutoff and Oxbow Lake. Aust. Geogr.*, **15**(3), 174.

Fares, Y.R. (2000) Changes of bed topography in meandering rivers at a neck cutoff intersection. *J. Environ. Hydrol.*, **8**.

Fisk, H.N. (1947) *Fine-grained alluvial deposits and their effects on Mississippi River activity*: Vicksburg, Miss., Waterways Experiment Station, 1947.

Frothingham, K.M. and Rhoads, B.L. (2003) Three-dimensional flow structure and channel change in an asymmetrical compound meander loop, Embarras River, *Illinois. Earth Surf. Proc. Land.*, **28**(6), 625–644. doi:10.1002/esp.471

Gagliano, S.M. and Howard, P.C. (1984) The neck cutoff oxbow lake cycle along the lower Mississippi River. *Am. Soc. Civ. Eng.*, New York, NY, United States, 147–158.

Gay, G.R., Gay, H.H., Gay, W.H., Martinson, H.A., Meade, R.H. and Moody, J.A. (1998) Evolution of cutoffs across meander necks in Power River, Montana, *USA. Earth Surf. Proc. Land.*, **23**(7), 651–662.

Güneralp, I. and Rhoads, B.L. (2008) Continuous characterization of the planform geometry and curvature of meandering rivers. *Geographical Analysis*, **40**(1), 1–25.

Han, B. and Endreny, T.A. (2014) Detailed river stage mapping and head gradient analysis during meander cutoff in a laboratory river. *Water Resour. Res.*, **50**(2), 1689–1703. doi:10.1002/2013WR013580

Hooke, J.M. (1995) River channel adjustment to meander cutoffs on the River Bollin and River Dane, Northwest England. *Geomorphology*, **14**(3), 235–253.

Hooke, J.M. and Harvey, A.M. (1983) Meander changes in relation to bend morphology and secondary flows. *Int. Assoc. Sedimentol. Spec. Publ.*, **6**, 121–132.

Ishii, Y. and Hori, K. (2016) Formation and infilling of oxbow lakes in the Ishikari lowland, northern Japan. *Quatern. Int.*, **397**, 136–146. doi:10.1016/j.quaint.2015.06.016

Kenworthy, S.T. and Rhoads, B.L. (1995) Hydrologic control of spatial patterns of suspended sediment concentration at a stream confluence. *J. Hydrol.*, **168**(1–4), 251–263.

Kleinhans, M., Jagers, B., Mosselman, E. and Sloff, K. (2006) Effect of upstream meanders on bifurcation stability and sediment division in 1D, 2D and 3D models. Paper presented at the Proc. River Flow.

Kleinhans, M.G., Jagers, H.R.A., Mosselman, E. and Sloff, C.J. (2008) Bifurcation dynamics and avulsion duration in meandering rivers by one-dimensional and three-dimensional models. *Water Resour. Res.*, **44**(8).

Konsoer, K.M., Rhoads, B.L., Best, J.L., Langendoen, E.J., Abad, J.D., Parsons, D.R. and Garcia, M.H. (2016a) Three-dimensional flow structure and bed morphology in large elongate meander loops with different outer bank roughness characteristics. *Water Resour. Res.*, **52**(12), 9621–9641.

Konsoer, K.M., Richards, D. and Edwards, B. (2016b) Planform evolution of neck cutoffs on elongate meander loops, White River, Arkansas, USA *River Flow 2016* (pp. 1730–1735): CRC Press.

Lewis, G.W. and Lewin, J. (1983) Alluvial cutoffs in Wales and the Borderlands. *Int. Assoc. Sedimentol. Spec. Publ.*, **6**, 145–154.

Micheli, E.R. and Larsen, E.W. (2011) River channel cutoff dynamics, Sacramento River, California, USA. *River*

Research and Applications, **27**(3), 328–344. doi:10.1002/rra.1360

Parsons, D.R., Jackson, P.R., Czuba, J.A., Engel, F.L., Rhoads, B.L., Oberg, K.A. and Riley, J.D. (2013) Velocity Mapping Toolbox (VMT): a processing and visualization suite for moving-vessel ADCP measurements. *Earth Surf. Proc. Land.*, **38**(11), 1244–1260. doi:10.1002/esp.3367

Petersen, M.S. (1963) Hydraulic aspects of Arkansas River stabilization. *Journal of the Waterways and Harbors Division*, **89**(4), 29–65.

Ratzlaff, J.R. (1981) Development and cutoff of Big Bend meander, Brazos River, Texas. *Texas J. Sci.*, **33**(2–4), 121–129.

Rhoads, B.L. and Kenworthy, S.T. (1998) Time-averaged flow structure in the central region of a stream confluence. *Earth Surf. Proc. Land.*, **23**(2), 171–191.

Rozovskiĭ, I.L. (1957) *Flow of water in bends of open channels*: Academy of Sciences of the Ukrainian SSR.

Shields, F.D., Jr. and Abt, S.R. (1989) Sediment deposition in cutoff meander bends and implications for effective management. *Regulated Rivers*, **4**(4), 381–396.

Szupiany, R.N., Amsler, M.L., Best, J.L. and Parsons, D.R. (2007) Comparison of fixed- and moving-vessel flow measurements with an aDcp in a large River. *J. Hydraul. Eng.*, **133**(12), 1299–1309. doi:10.1061/(ASCE)0733-9429(2007)133:12(1299)

Toonen, W.H.J., Kleinhans, M.G. and Cohen, K.M. (2012) Sedimentary architecture of abandoned channel fills. *Earth Surf. Proc. Land.*, **37**(4), 459–472. doi:10.1002/esp.3189+

van Dijk, W.M., Schuurman, F., van de Lageweg, W.I. and Kleinhans, M.G. (2014) Bifurcation instability and chute cutoff development in meandering gravel-bed rivers. *Geomorphology*, **213**, 277–291. doi:10.1016/j.geomorph.2014.01.018

Whiting, P.J. and Dietrich, W.E. (1993a) Experimental studies of bed topography and flow patterns in large-amplitude meanders; 1, Observations. *Water Resour. Res.*, **29**(11), 3605–3614. doi:10.1029/93WR01755

Whiting, P.J. and Dietrich, W.E. (1993b) Experimental studies of bed topography and flow patterns in large-amplitude meanders; 2, Mechanisms. *Water Resour. Res.*, **29**(11), 3615–3622. doi:10.1029/93WR01756

Zinger, J.A., Rhoads, B.L. and Best, J.L. (2011) Extreme sediment pulses generated by bend cutoffs along a large meandering river. *Nature Geoscience*, **4**(10), 675–678. doi:10.1038/NGE01260

Zinger, J.A., Rhoads, B.L., Best, J.L. and Johnson, K.K. (2013) Flow structure and channel morphodynamics of meander bend chute cutoffs: A case study of the Wabash River, USA. *J. Geophys. Res. Earth Surf.*, **118**(F4), 2468.

Int. Assoc. Sedimentol. Spec. Publ (2018) **48**, 297–320.

Hydro-sedimentological processes in meandering rivers: A review and some future research directions

KOEN BLANCKAERT

Technische Universität Wien Fakultät für Bauingenieurwesen, Institute of Hydraulic Engineering and Water Resources Management, Wien, Austria

ABSTRACT

In spite of 150 years of research, there are still numerous open questions on the hydro-sedimentological processes occurring in meandering rivers. Recent research on these processes in meandering rivers has been reviewed herein. The dominant processes are first identified and the underlying physics are described in a synoptic way. For each process, the important control parameters are identified and the type of model required for the simulation of the process is considered. Then, the mutual interactions between the different processes are described. Finally, some future research directions are recommended in field investigations, laboratory investigations and numerical modelling.

Keywords: Meandering, curved flow, hydrodynamic processes, sediment transport processes, sedimentology, morphology

INTRODUCTION

Braiding and meandering are the two most common river planforms (Leopold & Wolman, 1957). The present paper focusses on meandering rivers. River planform and bathymetry are shaped by hydro-sedimentological processes, which are defined as interactions between flow processes on the one hand and sediment transport processes at the other. Rozovskii's (1957) book can be considered the seminal contribution with respect to recent research on hydro-sedimentological processes in meandering rivers. Based on a series of laboratory experiments and data from various field experiments, he identified the main flow processes: the curvature-induced secondary flow, the counter-rotating outer-bank cell of secondary flow, the advective redistribution of momentum by the secondary flow, topographic steering, inner-bank flow separation and outer-bank flow separation. Fig. 1 illustrates the main hydro-sedimentological processes in meandering rivers, including the flow processes identified by Rozovskii. Moreover, he established a theoretical model for the curvature-induced secondary flow and its effect on the velocity distribution in wide, shallow and weakly curved meander bends. In the 60 years following Rozovskii's contribution, there has been abundant research on the hydro-morphological processes in meandering rivers by means of field experiments, laboratory experiments and numerical modelling.

The improvement in measurement technology has played a decisive role in experimental research. Earlier contributions were limited to measurements of mean velocity components with low spatial resolution. They were typically performed in laboratory flumes of rectangular cross-section (Rozovskii, 1957; Götz, 1975; de Vriend, 1979). The first measurements of the 3-D flow structure over a bathymetry shaped by hydro-sedimentological processes were made by Dietrich & Smith (1983) in the field and by Odgaard & Bergs (1988) in the laboratory. Nowadays, measurement technology allows measurement of the bathymetry and the three-dimensional flow patterns with high temporal and spatial resolution in the field (Konsoer *et al.*, 2016a) and the laboratory (Blanckaert, 2010). Recent laboratory investigations have typically followed two different strategies. The first strategy consists of designing a laboratory set-up that is

Fluvial Meanders and Their Sedimentary Products in the Rock Record, First Edition.
Edited by Massimiliano Ghinassi, Luca Colombera, Nigel P. Mountney and Arnold Jan H. Reesink.
© 2019 International Association of Sedimentologists. Published 2019 by John Wiley & Sons Ltd.

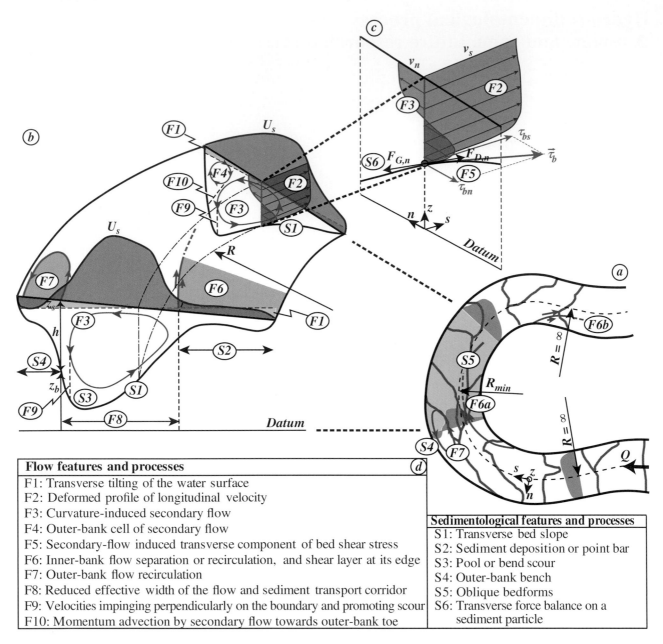

Fig. 1. Conceptualisation of flow and sedimentological features and processes in meandering rivers. (a) Planview of a meander bend; (b) 3-D view of a reach in the meander bend; (c) Profiles of the longitudinal and transverse velocity components, bed shear stress components and forces acting on bed sediments; (d) Legend.

meant to be representative of natural meandering rivers and investigating and explaining patterns of quantities, such as the bathymetry, mean flow, secondary flow, turbulent kinetic energy, turbulent stresses and boundary shear stresses. This approach is often combined with a focus on one specific process. Abad & Garcia (2009), for example, focus on the bedform dynamics, Jamieson *et al.* (2010) on the boundary shear stress and Termini & Piraino (2011) on the secondary flow.

Similar flow and sediment transport processes occur in a wide variety of geometric, hydraulic and sedimentologic settings. However, the relative importance of individual processes and the outcome of their interactions is case-dependent and it is therefore difficult to generalise results from this strategy. The second strategy consists of designing bespoke laboratory flumes and experimental conditions that allow isolating and accentuating individual processes. This second strategy often aims at gaining generic insight into the physics underlying the investigated process and the control parameters, with a broader range of validity. Blanckaert *et al.* (2012), for example, investigate flow processes near the outer bank and Blanckaert (2015) investigate flow separation at the inner bank.

The steady increase in computational capacity has played a decisive role in numerical research. Numerical simulations with high spatial and temporal resolution that resolve the relevant flow processes in meanders have now become computationally feasible (Stoesser *et al.*, 2010; van Balen *et al.*, 2010; Kang & Sotiropoulos, 2011; Constantinescu *et al.* 2011). These high-resolution models often complement data from field and/or laboratory experiments. After validation, they can provide a quasi-complete spatial-temporal description of the flow, information on variables that are difficult to measure, and allow investigating the effects of individual parameters. High-resolution numerical investigations have followed the same two strategies as the experimental investigations. Constantinescu *et al.* (2013) and Kang & Sotiropoulos (2011), for example, have performed complementary numerical research on patterns of mean flow, secondary flow and turbulence in representative configurations. Van Balen *et al.* (2009, 2010) have simulated experiments that were designed to isolate individual processes. These high-resolution investigations have proven very powerful tools for enhancing insight into flow processes. They have shown, for example, that higher-order turbulence closures are required to resolve accurately the flow processes of inner-bank flow separation and recirculation, outer-bank flow separation and the outer-bank cell of secondary flow.

Higher-order numerical models have two major drawbacks. Firstly, they are computationally very expensive, especially when also considering flow-sediment interactions. Secondly, they are black boxes that do not allow identification of the control parameters and do not clearly indicate or quantify the interactions between different processes. Reduced-order models, which are typically integrated over the depth or over the cross-section, are situated at the opposite end of the spectrum with respect to computational requirements. These models typically consist of interacting sub-models for the flow, bathymetry and bank migration. Obviously, the reduced computational requirements are accompanied by a loss of directly resolved processes. Whereas curvature-induced secondary flow can be accounted for by means of parameterisations, these models cannot account for the flow processes near the banks, i.e. the outer-bank cell of secondary flow and flow separation or recirculation near the inner and outer banks. These models are particularly appropriate for investigating the interactions between the different resolved processes. Moreover, when written in nondimensional form, these models clearly identify the control parameters of the different processes.

Ikeda *et al.* (1981) can be considered the seminal reduced-order meander model. This model did not include coupling between the flow and the bathymetry and was limited to weakly curved long meander bends, where all interactions between different processes can be considered as linear and where variables vary slowly and gradually around the bend. Reduced-order models have steadily been improved. De Vriend & Struiksma (1984) included the coupling between the flow and the bathymetry at the linear level. In a series of papers, Seminara and co-workers have focussed on improving the modelling of this coupling and extended it to the nonlinear level (Blondeaux & Seminara, 1985; Zolezzi & Seminara, 2001; Bolla Pittaluga *et al.*, 2009; Bolla Pittaluga & Seminara, 2011). However, the parameterisation of the secondary flow and its effect on the flow distribution remained linear in these models. Blanckaert & de Vriend (2003) have proposed a nonlinear parameterisation of the secondary flow that is valid over the entire curvature range and not limited to configurations with variables that vary slowly around the bend. Blanckaert & de Vriend (2010) have implemented this parameterisation in a meander flow model and Ottevanger *et al.* (2013) have done so in a model for the meander bathymetry. Frascati & Lanzoni (2013) have accommodated meander models for slow variations of the width. Parker and co-workers have focussed on bank accretion and erosion; and the resulting dynamics of the meander width and planform. Their recent contributions (Parker *et al.*, 2011; Eke *et al.*, 2014) are a breakthrough in the modelling of bank migration and meander planform evolution.

The present paper is primarily intended as a review and vision paper for sedimentologists, geomorphologists and ecologists. The paper has the following aims: (i) to review recent research on hydro-sedimentological processes in meandering rivers, highlighting some of the major contributions; (ii) to describe in a synoptic but self-contained way the dominant flow processes, the underlying physics and to provide references for more comprehensive and detailed accounts; (iii) to identify and discuss the dominant control parameters of these processes; (iv) to describe their sedimentological and morphological relevance; (v) to discuss the type of model required for their simulation; (vi) to discuss the interactions between the different processes; and (vii) to identify and discuss challenges for future research.

MATHEMATICAL FRAMEWORK

Fig. 1 defines the terminology used in the present paper and illustrates the main hydro-sedimentological processes. The axis of the orthogonal curvilinear reference system are (s, n, z). The horizontal s axis follows the centreline of the meandering river and is positive in the streamwise direction. The horizontal n axis has its origin at the centreline and is positive to the left. As a consequence, the radius of curvature at the centreline, R, is negative/positive for bends turning to the left/right. By definition, R is infinite at cross-overs and reaches its minimum magnitude R_{min} at apices. The vertical z-axis is positive upward. The bed and water surface elevations are indicated by z_b and z_s, respectively and their difference defines the flow depth $h = z_s - z_b$. The time-averaged local velocity components are (v_s, v_n, v_z) and their depth-averaged values are (U_s, U_n, U_z). The discharge per unit width, also called unit discharge, is $q = U_s h$. The bed shear stress $\vec{\tau}_b$ has longitudinal and transverse components, τ_{bs} and τ_{bn}, respectively; its magnitude is denoted by τ_b.

Secondary flow is defined in the present paper as the flow component perpendicular to the river centreline, i.e. in the cross-section. It can be decomposed into translatory and circulatory parts, which Bradshaw (1987) names 'cross-flow and identifiable downstream vortices'. Bolla Pittaluga & Seminara (2011) name the former 'topographic secondary flow', to highlight that it is mainly induced by topographic steering. The circulatory part includes the curvature-induced secondary flow (Process F3 in Fig. 1B) and the outer-bank cell of secondary flow (Process F4 in Fig. 1B), which will be reviewed further in this paper. This definition of secondary flow is typical for laboratory studies. In field studies, where the definitions of the river centreline and cross-sections are not straightforward, secondary flow is often differently defined as the flow component perpendicular to the local depth-averaged flow.

Most formulae for the sediment transport capacity can be written in the generic form (Mosselman, 2005):

$$\Phi = \alpha \theta^{n-\gamma} \left(\mu\theta - \theta_c \right)^{\gamma} \qquad \text{(Eq. 1)}$$

Here, Φ is the nondimensional sediment transport capacity, also called the Einstein parameter and defined as $\Phi = q_s \left(g\Delta d \right)^{-1/2}$, with q_s the sediment transport capacity per unit width, g the gravitational acceleration, $\Delta = \rho_s/\rho - 1$ the relative submerged density of sediment particles; and d a characteristic sediment diameter. The nondimensional bed shear stress is defined as $\theta = \tau_b \left(\rho g\Delta d \right)^{-1}$, commonly called the Shields parameter, and θ_c is the critical Shields parameter for the initiation of sediment motion (Shields, 1936). Further, α, μ, n and γ are nondimensional coefficients. For the purposes of the present paper, it is sufficient to use a simplified sediment transport equation in the form (Hendersson, 1966):

$$q_s \sim \tau_b^n \qquad \text{(Eq. 2)}$$

In open-channel hydraulics, τ_b and the magnitude of the depth-averaged velocity U are commonly related through a friction coefficient. When using the Chézy-type friction coefficient c_f, for example, this relation is:

$$\tau_b = \rho c_f U^2 \qquad \text{(Eq. 3)}$$

When assuming a constant friction coefficient c_f, Equations 2 and 3 indicate that q_s increases more than linearly with U if the coefficient n is larger than 0.5. This is typically the case because the coefficient n in most sediment transport formulae has values of between 1.5 and 2.5 (Mosselman, 2005). This implies that a transversal redistribution of velocity will lead to a more pronounced transversal redistribution of the sediment transport capacity.

HYDRO-SEDIMENTOLOGICAL PROCESSES

Bends of constant curvature

Although an idealised configuration, an infinite bend of constant curvature and constant cross-sectional shape clearly reveals some flow and

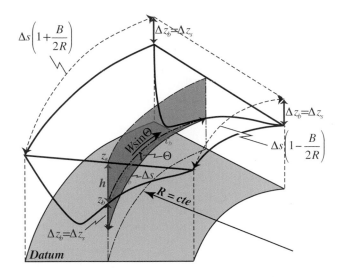

Fig. 2. Schematic representation of a reach in an infinite bend. The blue area represents a control volume of width n.

sedimentological processes. In such an infinite bend, all cross-sections are identical and flow and sedimentologic parameters only vary within the cross-section, not in the longitudinal direction; a condition often denoted as fully developed flow.

Gravitational driving of the flow

Open-channel flow is driven by gravity and opposed by friction on the boundary. When considering uniform flow in a wide and straight open-channel, both forces are in equilibrium in every vertical-longitudinal slice through the water of length Δs, width Δn and flow depth h, resulting in (Fig. 2, for $R = \infty$):

$$W \sin \Theta = \rho g h \Delta s \Delta n \sin \Theta = \tau_b \frac{\Delta s}{\cos \Theta} \Delta n \text{ or } \rho g h S = \tau_b$$

(Eq. 4)

Here W is the weight of the water in the considered slice and the longitudinal water surface slope (which is equal to the longitudinal bed slope in an infinite bed) $\Theta = \operatorname{atan}(-\Delta z_s \Delta s^{-1})$ is represented by:

$$S = \sin \Theta \cos \Theta = \frac{1}{2} \sin 2\Theta = \frac{1}{2} \sin \left(2 \operatorname{atan} \frac{-\Delta z_s}{\Delta S} \right)$$

(Eq. 5)

For the small water surface slopes typically encountered in meandering rivers, $S \sim \Theta \sim \Delta z_s \Delta s^{-1}$. Note that the commonly adopted interpretation of S as the longitudinal surface slope in Equation 4 is

only correct for small slopes. The formulation used here is more general and not limited to small slopes. Note also that the minus sign has been chosen in order to obtain positive values of τ_b and S. Applying Equation 3 then provides relations for U and q:

$$U = c_f^{-1} g^{1/2} h^{1/2} S^{1/2} \text{ and } q = Uh = c_f^{-1} g^{1/2} h^{3/2} S^{1/2}$$

(Eq. 6)

Although the processes in curved open-channels are more complicated, this equation remains very useful for understanding the hydro-sedimentological processes in meandering channels.

Firstly, the relation shows that the flow depth is the dominant factor of influence for the distributions of the flow and sediment transport, which tend towards the deepest part of the cross-section that is typically located in the outer half of the cross-section (Figs 1B and 2). The dominant influence of the flow depth has been confirmed in laboratory and field experiments by Blanckaert (2010) and Konsoer *et al.* (2016), respectively. Dietrich & Whiting (1989) introduced the term topographic steering for this process. It should be noted that topographic steering explains to a large extent why the flow tends towards the deepest part of the cross-section but it does not explain how the flow shapes a bathymetry with deeper and shallower parts.

Secondly, the relation shows that the distributions of flow and sediment transport are driven by S, which is equivalent to the longitudinal water surface slope Θ (Equation 5). The water surface drop Δz_s between two cross-sections perpendicular to the river axis is equal all over the cross-section (Fig. 2) but the longitudinal distance between both cross-sections increases outwards, i.e. from the inner to the outer bank, as (Fig. 2):

$$L = \Delta s \left(1 + nR^{-1} \right)$$

(Eq. 7)

As a result, the longitudinal water surface slope is maximum at the inner bank and decreases outwards (Fig. 2). This planimetric effect impels the highest velocities and sediment transport capacities towards the inside of the bend. This process competes with other processes in determining the final distributions of flow and sediment transport capacity.

Thirdly, the relation shows that the bed friction coefficient c_f is a control parameter for the distributions of the flow and sediment transport. Contrary to straight open-channel flow, the friction factor can considerably vary over the cross-section of a meandering river due to, for example, grain sorting, differences in bedform size, differences in the ratio between grain-size and local flow depth, or hydrodynamic processes.

It can be concluded that the gravitational driving of the flow in a meander is not only controlled by the longitudinal slope and the roughness as in a straight flow but also by the bathymetry and the planform of the river.

The gravitational driving of the flow is directly resolved by 3-D and 2-D depth-integrated models. Its variation over the width of the river due to planimetric effects is accounted for in all recent reduced-order models.

Curvature-induced secondary flow
(Process F3 in Fig. 1B and C)

Flow in a curved open-channel is known to follow a helicoidal path in the central part of the cross-section, with flow in an outwards direction in the upper part of the water column and in an inwards direction in the lower part of the water column. It is often called helical flow or spiral flow and its projection on a cross-section is called curvature-induced secondary flow in the present paper (Process F3 in Fig. 1B and C). Since the first investigation by Boussinesq (1868), it has been abundantly investigated. A detailed review is given by Blanckaert & de Vriend (2004).

Flow in a curved reach is subject to a centrifugal force, which impels the water mass in the outward direction and gives rise to a transverse tilting of the water surface (Process F1 in Fig. 1B). This, in turn, leads to an inward pressure gradient that is about constant over the flow depth. Averaged over the flow depth, the outward centrifugal force and inward pressure gradient nearly balance. It is their local imbalance that gives rise to the formation of secondary flow. Averaged over the flow depth, the curvature-induced secondary flow does not cause any transversal mass redistribution because the water mass that is transported outwards in the upper part of the water column is equal to the water mass that is transported inwards in the lower part, or $\rho U_n = 0$ (Process F3 in Fig. 1C); but, the longitudinal momentum ρv_s that is transported outwards in the upper part of the water column by $v_n > 0$ is typically larger than the longitudinal momentum that is transported

inwards in the lower part of the water column by $v_n < 0$ (Fig. 1C). As a result, averaged over the flow depth, the curvature-induced secondary flow causes transversal momentum distribution because $\rho \langle v_s v_n \rangle > 0$ (the brackets indicate depth-averaged values), leading to a decrease of velocities and sediment transport capacity in the inner part of the cross-section and an increase in the outer part. These physics underlying the curvature-induced secondary flow and its effect on momentum redistribution were already described and modelled by, e.g., Rozovskii (1957) and Johannesson & Parker (1989b), but with a validity range limited to shallow and mildly curved meander bends.

The curvature-induced secondary flow also causes an inwards component of the bed shear stress, τ_{bn}, and of the drag force on bed sediments, $F_{D,n}$ (Process S6 in Fig. 1C). Sediment particles on a transversally inclined bed are also subject to a so-called gravitational pull, $\vec{F}_G$, which has a downslope oriented component $F_{G,n}$ in the cross-section (Process S6 in Fig. 1C). Ottevanger *et al.* (2013) review models for this gravitational pull. As a result, the directions of the bed shear stress, the sediment transport and the depth-averaged flow are typically not concomitant (Fig. 1C; van Bendegom, 1947; Parker & Andrews, 1985; Talmon *et al.*, 1995; Seminara *et al.*, 2002). Sediment is typically transported in the inward direction. Deposition of sediment in the inner-half of the cross-section leads to the formation of a sediment deposition bar attached to the inner bank, often called point bar, and a transverse slope of the bed (Features S1 and S2 in Fig. 1B; Struiksma *et al.*, 1985). The equilibrium transverse bed slope results from the balance between $F_{D,n}$ and $F_{G,n}$ (Process S6 in Fig. 1C; Parker & Andrews, 1985; Talmon *et al.*, 1995). A larger secondary flow magnitude will lead to a larger $F_{D,n}$. In order to maintain $F_{D,n}$ and $F_{G,n}$ balanced, a larger equilibrium transverse bed slope will be required. It can be concluded that a larger secondary flow magnitude will lead to a larger equilibrium transverse bed slope. The upslope drag force is proportional to the projected surface of the sediment particle ($\sim d^2$), whereas the downslope gravitational pull is proportional to the volume of the sediment particle ($\sim d^3$). This causes a transversal sorting of the sediment, with increasing sediment diameters from the inside to the outside (Parker & Andrews, 1985).

The curvature-induced secondary flow also causes an increase in the capacity to transport sediment, both as bedload and as suspended load. The bedload transport capacity is mainly driven by the magnitude of the bed shear stress, τ_b (Equation 2). As compared to straight uniform flow, the secondary flow increases the bed shear stress magnitude by two effects. Firstly, the additional transverse component of the bed shear stress (Process F5 in Fig. 1C) causes an increase by a factor $\left(1 + \tau_{bn}^2 / \tau_{bs}^2\right)^{1/2}$. Secondly, the secondary flow does not only redistribute momentum transversally as discussed before, but also vertically, leading to a modified flow distribution in the water column as compared to a straight uniform flow (Process F2 in Fig. 1C). This typically results in an increased velocity gradient near the riverbed, which causes an increased bed shear stress for the same depth-averaged velocity. This effect was first described by de Vriend & Geldof (1983). Blanckaert (2009) quantified this effect in the central region of the secondary flow cell where $v_n > v_z$, by multiplying the friction factor c_f in Equation 3 by a correction coefficient $\psi_{\text{secondary flow}}$. Hence, the bed shear stress magnitude in a bend can be written as:

$$\tau_b = \left(1 + \frac{\tau_{bn}^2}{\tau_{bs}^2}\right)^{1/2} \psi_{\text{secondary flow}} \rho c_f U^2 \quad \text{(Eq. 8)}$$

This second effect is assumed to be particularly important near the outer bank, where $v_z > v_n$ and where vertical velocities impinge on the bed and promote bend scour (Process F9 in Fig. 1B). The suspended load transport capacity is mainly driven by the turbulence level, which can be quantified by the turbulent kinetic energy, *tke* (Vanoni, 1946). In straight uniform flow, *tke* is directly proportional to τ_b, which reflects the fact that turbulence is generated by friction at the bottom (Nezu & Nakagawa, 1993). The secondary flow enhances in two ways the *tke* and, hence, the suspended load transport capacity. Firstly, the bed shear stress magnitude is enhanced as discussed before. Secondly, the secondary flow induces additional shear in the water column, which leads to additional turbulence production. This effect has been quantified and conceptually modelled by Blanckaert (2009). Both effects can lead to *tke* levels in curved flow that are a multiple of their counterparts in straight uniform flow (Blanckaert (2010) and Termini & Piraino (2011) for laboratory experiments and Sukhodolov (2012) for field measurements).

The hydro-sedimentological implications of the curvature-induced secondary flow are essentially conditioned by its magnitude. In mildly curved flow, its magnitude is known to increase proportionally to the ratio $H R^{-1}$ (e.g. Rozoskii, 1957; Engelund, 1974; de Vriend, 1977; Johannesson & Parker, 1989a), where H is the flow depth in the centre of the secondary flow cell, which is typically confounded with the flow depth at the river centreline and the cross-sectional averaged flow depth (Fig. 1B; Odgaard, 1986). Moreover, the magnitude of the secondary flow in mildly curved flows is known to depend on the bed roughness and to decrease with increasing friction coefficient c_f (e.g. Rozoskii, 1957; Engelund, 1974; de Vriend, 1977; Johannesson & Parker, 1989a). Following de Vriend (1981a), a mildly curved flow is defined in the present paper as a flow where non-linear interactions between different flow processes are negligible and a strongly curved flow one where they are non-negligible. In strongly curved flows, the secondary flow magnitude grows less than linearly with $H R^{-1}$ due to non-linear interactions between the longitudinal flow component v_s, the secondary flow components (v_n, v_z); and the transverse bed slope. Blanckaert (2009) called this effect the saturation of the secondary flow. Blanckaert & de Vriend (2003, 2010) and Blanckaert (2011) have developed a reduced-order model that parameterises the secondary flow magnitude and its hydro-sedimentological effects over the entire curvature range. This model has been validated over a broad range of parameters in the field and in the laboratory (Ottevanger et al., 2012; Wei et al., 2016). This model identifies $c_f^{-1} H B^{-1}$ and BR_{min}^{-1} as additional control parameters; the former being characteristic for river reaches and the latter for individual bends. The saturation of the secondary flow and its effects are favoured in sharp bends (high values of BR_{min}^{-1}) and shallow rough rivers (low values of $c_f^{-1} H B^{-1}$). Wei et al. (2016) propose that a bend is mildly curved when $R B^{-1} > 20 c_f^{0.75} \left(c_f^{-1} H B^{-1}\right)$ and strongly curved when $R B^{-1} < 10 c_f^{0.75} \left(c_f^{-1} H B^{-1}\right)$, where R represents an average value around the bend; but there is no objective delimiting criterion for mildly and strongly curved flows and so other criteria have been proposed in the literature. De Vriend (1981b) proposed a Dean number proportional to $c_f^{-1/2} H^{1/2} R^{-1/2}$, which can be obtained as a combination of both control parameters mentioned before. Bolla Pittaluga & Seminara (2011) proposed the discriminating parameter $c_f^{-1/2} H R^{-1}$ and also adopted different definitions of mildly curved and sharp bends, being bends which generate fairly weak and strong curvature-induced secondary flow, respectively.

Turbulence does not play a dominant role with respect to the generation of the curvature-induced secondary flow (Blanckaert & de Vriend, 2004; van Balen *et al.*, 2010a). Therefore, 3-D flow models with basic turbulence closures can adequately resolve the secondary flow and its effects. As the curvature-induced secondary flow has by definition no depth-averaged transverse velocity, $U_n = 0$, depth-averaged and reduced-order models cannot directly resolve it, but these models can accurately account for the effects of the secondary flow through parameterisations, as discussed previously.

Outer-bank cell of secondary flow (Process F4 in Fig. 1B)

Flow patterns near the outer bank often show peculiar features. Obviously, these near-bank flow features are relevant with respect to the flow's bank erosion potential and the resulting meander planform evolution.

A feature that is often observed in the field (e.g. Bathurst *et al.*, 1977; Bridge & Jarvis, 1977; Thorne & Hey, 1979; de Vriend & Geldof, 1983; Sukhodolov, 2012) and in the laboratory (e.g. Mockmore, 1943; Einstein & Harder, 1954; Rozovskii, 1957) is a so-called outer-bank cell of secondary flow (Process F4 in Fig. 1B) with rotational sense opposite to the curvature-induced secondary flow (Process F5 in Fig. 1B). Blanckaert & de Vriend (2004) and van Balen *et al.* (2010a) have investigated the underlying physics by a combination of laboratory experiments and eddy-resolving numerical simulations. They found that some subtle turbulence processes play an important role in the generation of the outer-bank cell. More precisely, the different ways in which the solid outer bank and the water surface affect the fluxes of kinetic energy between the mean flow and the turbulence play a key role in the generation of the outer-bank cell. They also concluded that the numerical simulation of the outer-bank cell requires a fully 3-D flow model with a turbulence closure that is able to account for fluxes of kinetic energy from the turbulence to the mean flow. This result has been confirmed by successful simulations with eddy-resolving turbulence models (van Balen *et al.*, 2010a; Kang & Sotiropoulos, 2011; Constantinescu *et al.*, 2013).

The analyses of Blanckaert & Graf (2004) and Blanckaert *et al.* (2012) indicate that the outer-bank cell reduces the effective width of the river, that part of the width where most of the discharge and the sediment transport occur (Process F8 in Fig. 1B). This reduced effective width has sedimentological implications, as will be discussed in the next section. There are still open questions on the effect of the outer-bank cell on the stability of the outer bank and on its conditions of occurrence. Bathurst *et al.* (1979) found that the outer-bank cell advects high momentum originating from near the water surface towards the toe of the outer bank (Process F10 in Fig. 1B) and thereby destabilises the bank. Blanckaert & Graf (2004) found that the outer-bank cell had a protective effect on the outer bank, by establishing a buffer region in between the centre-region cell and the outer bank that shelters the outer bank from the effect of the centre-region cell. These observations suggest that the effect of the outer-bank cell on the bank stability is case dependent.

It is also not clear which control parameters affect the outer-bank cell and even under what conditions it occurs. The outer-bank cell does occur in mildly curved bends and even in straight flows (Blanckaert & de Vriend, 2004) and is therefore not a typical feature of sharp bends only. It seems to be promoted by the steepness of the outer bank (Bathurst *et al.*, 1979). Experiments by Blanckaert (2011) and Blanckaert *et al.* (2012) suggest that the magnitude and size of the outer-bank cell grow with increasing curvature ratio $H R^{-1}$, bank roughness and bank steepness.

Interactions between the processes and sedimentological implications

The interactions between the hydro-sedimentologic processes discussed above determine the cross-sectional shape and the flow distribution in an infinite bend. With the exception of the outer-bank cell, reduced-order models for meander dynamics (e.g. Bolla Pittaluga *et al.*, 2009; Parker *et al.*, 2011; Ottevanger *et al.*, 2013) quantify all processes discussed above and their interactions.

The cross-sectional shape in an infinite bend seems to be largely conditioned by bed erodibility. When the bed erodibility is relatively high, such as in most alluvial bends under bankfull conditions, the secondary flow transports sediment inwards and creates a transverse bed slope with flow depths that increase outwards (Fig. 1C). The gravitational driving of the flow than favours velocities that increase outwards as $U{\sim}h^{1/2}$ (Equation 6). Outwards momentum advection by the secondary flow further favours an outwards increase in velocities. This outwards increase in velocities is accompanied by an outwards increase in sediment transport capacity (Equations 2 and 3), which further favours the development of the transverse bed slope. The positive feedback between these different processes is stronger than the planimetric effect (Equation 7) that favours an inwards increase in velocities. The outwards increases in flow depth and velocity have further hydro-sedimentological implications. The deposition of sediment near the inner bank may lead to the formation of a shallow point bar, with negligible contribution to the discharge (Feature S2 in Fig. 1B). The quasi-totality of the discharge and the curvature-induced secondary flow will then be constrained to the deepest outer part of the cross-section, but excluding the region occupied by an outer-bank cell if one occurs. This leads to a quasi flat bed on the point bar and a transverse bed slope that is confined to the region where the secondary flow occurs. A lower erodibility will lead to a less pronounced transverse bed slope and less pronounced topographic steering. Such a situation can occur, for example, in bedrock rivers. In such configurations, an inwards increasing velocity can occur when the planimetric effect becomes dominant. According to the model of Ottevanger *et al.* (2013), this is favoured in bend that are very sharply curved ($BR^{-1} > 1$) and/or very rough and shallow $\left(c_f^{-1}HB^{-1} < 5\right)$. Unfortunately, no experimental data are yet available that would allow assessing these hypotheses based on the model.

The overall sediment transport capacity in an infinite bend differs from that in a corresponding straight uniform flow with the same width, slope and discharge. The overall sediment transport capacity Q_s is obtained by integrating the unit sediment transport capacity q_s from the inner bank, n_i, to the outer bank, n_o:

$$Q_s = \int_{n_i}^{n_o} q_s \, dn \qquad \text{(Eq. 9)}$$

Equations (2), (3) and (8) can be elaborated into:

$$\frac{Q_s}{Q_{s,straight}} = \int_{n_i}^{n_o} \left[\left(1 + \frac{\tau_{bn}^2}{\tau_{bs}^2}\right)^{1/2} \psi_{secondary\ flow} \right.$$
$$\left. \frac{c_f}{c_{f,straight}} \frac{U^2}{U_{straight}^2} \right]^n dn > 1 \quad \text{(Eq. 10)}$$

Values in straight uniform flow are indicated by the subscript straight. The first and second terms in this equation are by definition larger than 1, but the major cause of the modified overall sediment transport capacity in bends is the transverse distribution of $(c_f U^2)^n$. If it is assumed that the width-averaged value of $c_f U^2$ remains unchanged, this is roughly equivalent to the assumption that the cross-sectional area remains unchanged. This is a reasonable assumption in a mildly curved flow (Odgaard, 1986) that is commonly adopted in perturbation analyses of such configurations. The lowest integral value is then obtained with a distribution of $c_f U^2$ that is uniform over the width of the river, such as in an idealised straight river. Any variation of $c_f U^2$ over the river width will then lead to an increased overall sediment transport capacity Q_s.

Although the behaviour in strongly curved bends may be more complicated, these simplified considerations indicate that a curved river can modulate the overall sediment transport capacity by three mechanisms. Firstly, it can reduce the effective width of the river by creating zones with negligible contribution to the discharge and the sediment transport. According to these simplified considerations, this leads to an increase in depth-average velocity and sediment transport that overcompensate for the reduction in effective width. A more detailed discussion on the relationship between the width and the overall sediment transport capacity in straight rivers is given by Carson & Griffiths (1987). Secondly, it can modify the transverse distribution of U, which is closely related to the cross-sectional shape. A linear increase of U from $0.5U$ at the inner bank to $1.5U$ at the outer bank combined with a coefficient $n = 3$ (Equation 2), for example, would lead to a 25% increase in overall sediment transport capacity. Thirdly, it can adapt the transversal distribution of the friction coefficient, for example by grain sorting that affects the skin roughness or by adapting the bedform (Feature S5 in Fig. 1A) characteristics which affect the form roughness.

Bends and meanders of varying curvature

The curvature radius in a natural meander bend gradually evolves from infinity at the cross-over to a minimum at the bend apex and then again to infinity at the next cross-over (Fig. 1A). In compound meanders, several local curvature minima may exist within one bend (Frothingham & Rhoads, 2003). The planform and corresponding longitudinal evolution of the centreline radius of curvature are case-dependent. Due to inertia, the adaptation of the distributions of the flow, sediment transport and bathymetry to the forcing by the local curvature is not instantaneous but requires a finite distance. This leads to modifications to the hydro-sedimentological processes discussed in the framework of infinite bends, as well as additional processes.

Gravitational driving of the flow

The transverse tilting of the water surface (Process F1 in Fig. 1B) is proportional to Fr^2HR^{-1}, where $Fr = U(gH)^{-1/2}$ is the Froude number (Chow, 1959). Experimental data indicate that this transverse tilting adapts quasi-instantaneously to curvature changes (Blanckaert *et al.*, 2013). In the upstream part of a bend (i.e. from the cross-over to the apex), the transverse tilting of the water surface increases because R^{-1} increases. This is accompanied by changes in the downstream water surface slope that scale with $Fr^2H\partial R^{-1}/\partial s$. The downstream water surface slope increases in the inner half of the cross-section and decreases in the outer half, leading to corresponding accelerations and decelerations, respectively, of the flow (Equation 6). Due to mass conservation, these flow accelerations and decelerations are accompanied by an inward transport of mass. The flow accelerations and decelerations are accompanied by an increase and decrease, respectively, of the sediment transport capacity, which contribute to shaping the transverse bed slope. In the downstream part of a bend (i.e. from the bend apex to the next crossover), R^{-1} decreases and the opposite behaviour occurs. Blanckaert *et al.* (2013) quantified and discussed these processes in detail and illustrated them in their fig. 12A.

The flow and the morphology adapt to these changes in the longitudinal water surface slope with different adaptation lengths. Struiksma *et al.* (1985), for example, propose $H(2c_f)^{-1}$ as adaptation length for the flow and $(B/H)^2(\pi^2G)^{-1}H$ as adaptation length for the morphology. As the adaptations of the transverse flow distribution and the transverse bed slope are out of phase, the transverse bed slope typically evolves towards its equilibrium value as a damped oscillation. In the vicinity of an abrupt change in centreline curvature (which induces the change in longitudinal water surface slope), the transverse bed slope is known to overshoot its equilibrium value, leading to the occurrence of the maximum scour depth (Feature S3 in Fig. 1B) and the maximum flow attack on the outer bank (Struiksma *et al.*, 1985; Odgaard, 1988). Zolezzi & Seminara (2001) have called this process overdeepening. They have shown theoretically that the maximum scour depth can occur both downstream and upstream of the change in centreline curvature. Upstream overdeepening occurs in so-called super-resonant meanders, which are typically shallow and characterised by high *Fr*. Downstream overdeepening occurs in so-called sub-resonant meanders, which are typically narrower and characterised by lower subcritical *Fr*. The maximum scour depth in natural meander bends typically seems to occur downstream of a cross-over (Sukhodolov, 2012; Vermeulen *et al.*, 2015), indicating that most natural meander bends are sub-resonant. Surprisingly, these models for the flow and the bathymetry in meander bends have only been validated with observations in laboratory flumes (Struiksma *et al.*, 1985; Odgaard, 1988; Zolezzi *et al.*, 2005; Ottevanger *et al.*, 2013).

Curvature-induced secondary flow (Process F3 in Fig. 1B and C)

As mentioned in the corresponding section on infinite bends, the equilibrium magnitude of the curvature-induced secondary flow depends on c_f and increases linearly with $H\,R^{-1}$ in a mildly curved bend. In a mildly curved bend, the evolution of the secondary flow magnitude around the bend lags spatially behind the evolution of the relative curvature $H\,R^{-1}$ but follows a similar trend (Rozovskii, 1957; Johannesson & Parker, 1989a). The longitudinal evolution of the secondary flow magnitude in strongly curved bends is more complicated, due to the aforementioned nonlinear hydrodynamic feedback mechanisms that limit the magnitude of the secondary flow. Blanckaert & de Vrie nd (2003, 2010) and Ottevanger *et al.* (2013) provide a detailed discussion and analysis. They have shown that these nonlinear feedback mechanisms are typically small in the upstream part of the bend, downstream of the crossover, and increase around the bend. As a result, the secondary flow magnitude can overshoot its equilibrium value in the upstream part of the bend and then continuously decrease further around the bend. Surprisingly, this aspect of the model's behaviour and predictions have only been validated with observations in laboratory experiments over immobile flat (Blanckaert, 2009) and mobile deformed beds (Blanckaert, 2010; Constantinescu *et al.*, 2013; note that the term flat bed is used in the present paper to design a bed that is transversally flat but can be tilted in longitudinal direction). The maximum secondary flow magnitude in natural meander bends typically seems to occur in the upstream part of the bend (Sukhodolov, 2012; Vermeulen *et al.*, 2015), which qualitatively agrees with the modelled behaviour. The overshoot of the secondary flow magnitude is particularly important because it favours the overshoot of the transverse bed slope.

Outer-bank cell of secondary flow (Process F4 in Fig. 1B)

All available data from laboratory experiments (Blanckaert *et al.*, 2012) and numerical simulations (van Balen *et al.*, 2010; Kang & Sotiropoulos, 2011; Constantinescu *et al.*, 2013) indicate that the size and magnitude of the outer-bank cell vary considerably around a bend and that the size and magnitude in an infinite bend may be considerably smaller than the maximum values occurring in a real bend of varying curvature. Unfortunately, no field data is available yet that document the evolution of the outer-bank cell around a bend. However, there is a consensus that these variations in the size and magnitude of the outer-bank cell are sedimentologically relevant because they condition the redistribution of the flow and the sediment transport over the river width and thereby affect the flow attack on the outer bank and the bathymetry. The observed variations around bends with immobile flat bed (Blanckaert *et al.* 2012) and bends with mobile deformed bed (Kang & Sotiropouos, 2011; Constantinescu *et al.*, 2013) are all different, suggesting case-dependency. It is at present not clear what parameters control these variations around the bend.

Flow recirculation at the outer bank (Process F7 in Fig. 1A and B)

In a region of decreasing curvature radius (i.e. in the upstream part of a bend), a zone of horizontal flow recirculation characterised by low velocities

can develop near the outer bank. It is typically associated with the formation of so-called outer-bank benches, which represent a local widening of the river (Feature S4 in Fig. 1A and B). The low-velocity zone of horizontal flow recirculation is thought to act as a buffer between the zone of highest velocities and the outer bank and thereby protects the outer bank from erosion and reduces meander migration (Hickin, 1977; Page & Nanson, 1982). Similar to the outer-bank cell, outer-bank recirculation reduces the effective width of the river (Process F8 in Fig. 1B), which affects the overall sediment transport capacity. Since the first observation of outer-bank recirculation by Hickin (1977), this process has often been observed in natural meander bends. Recent examples include Nanson (2010), Vietz *et al.* (2012), Schnauder & Sukhodolov (2012), Rhoads & Massey (2012); and Vermeulen *et al.* (2015). Reference is made to Blanckaert *et al.* (2013) for a more detailed review on this process.

In spite of the importance of outer-bank recirculation, little is known yet about this flow process, its interaction with sediment transport, its conditions of occurrence and its control parameters.

The occurrence of outer-bank flow recirculation is related to a reduction in the gravitational driving of the flow (Equation 6). As discussed before, the downstream water surface slope is reduced in the outer half of the cross-section in regions of decreasing curvature radius. A sharp decrease in curvature radius can lead to the formation of an adverse water surface slope, which favours flow deceleration and the onset of horizontal flow recirculation. Blanckaert (2010) has postulated that an adverse water surface slope is a necessary condition for the onset of recirculation and proposed the following requirement for its occurrence:

$$\frac{R_{\min}}{B} < \left(\frac{1}{2} \frac{1}{c_f} \frac{H}{B} \right)^{1/2} \qquad \text{(Eq. 11)}$$

Note that the same control parameters were identified for the magnitude of the secondary flow. According to this requirement, horizontal flow separation is favoured by bend sharpness, smoothness and narrowness. Unfortunately, there are no field data available yet to assess this criterion. The occurrence of an adverse water surface slope is thought to be a necessary but not a sufficient condition because other flow processes also play a role.

Blanckaert *et al.* (2013) investigated the role of topographic steering in a bend with an artificial widening at the outer bank, which is reminiscent of an outer-bank bench (Feature S4 in Fig. 1A and B). They compared detailed flow measurements over an immobile flat-bed configuration and a deformed mobile-bed configuration. The latter was characterised by flow depths that increased from the inner to the outer bank and reached maximum values in the outer bank widening. A zone of horizontal flow recirculation that protects the outer bank only occurred over the flat bed. For this configuration, the criterion given by Equation 11 was largely satisfied. Over the mobile bed, velocities remained relatively high in the outer-bank widening, where a complex 3-D flow distribution occurred with return currents only near the bed. The onset of horizontal flow recirculation seemed to be hindered by two effects. Firstly, the increase in flow depth in the outer-bank widening promotes higher velocities due to the gravitational driving of the flow (Equation 6). This increase in flow depth in the outer half of the cross-section is a typical morphological feature of meander bends. Secondly, the deformed mobile bed attenuated the water surface gradient, which opposes the establishment of an adverse water surface gradient. Moreover, the hydro-sedimentological processes also led to a friction factor c_f that was much higher over the mobile flat bed than over the immobile deformed bed. Due to this higher friction factor, the criterion given by Equation 11 was near its critical value.

Inertia also plays an important role with respect to the onset of horizontal flow recirculation at the outer bank. Hodskinson & Ferguson (1998; numerical modelling) and Engel & Rhoads (2012; field measurements) demonstrated the importance of the meander inflow conditions, which are determined by the river planform. An inflow distribution with low velocities near the outer bank favours the onset of horizontal flow recirculation. This is typically the case if the upstream bend turns in the opposite direction, such as in most natural meanders and in Blanckaert *et al.*'s (2013) laboratory experiments.

These previous investigations provide some incomplete and merely qualitative insight into the role of adverse water surface gradients, topographic steering, inertia and hydro-sedimentological processes that affect the bed roughness. More experimental observations in real rivers and laboratory flumes are required to quantify the hydro-sedimentological processes and interactions and to identify the relevant control parameters.

Obviously, a 3-D flow model is required to resolve the process of outer-bank flow recirculation and its sedimentological implications. Due to the lack of insight into the role of turbulence, it is not clear to date what kind of turbulence closure is required.

Flow separation and recirculation at the inner bank (Process F6 in Fig. 1A and B)

Zones of flow separation and horizontal recirculation can also occur at the inner bank. In line with Bagnold (1960), Leeder & Bridges (1975), Simpson (1989) and Blanckaert (2015), flow separation and recirculation are here defined as flow processes that cause the main flow body to separate from the inner bank, whereby a zone of slower moving fluid is situated between the inner bank and the main flow body. A defining characteristic is the occurrence of a shear layer at the separation of the main flow body and the slower moving fluid (Process F6 in Fig. 1B). Two stages of flow separation are considered, which are both schematically illustrated in their fig. 1 of Blanckaert (2015). In the first stage, velocities remain downstream oriented, no recirculation develops and the shear layer does not reattach to the inner bank, implying that the zone of flow separation is not spatially constrained. In the second stage, reverse upstream directed velocities occur, recirculation occurs and the shear layer reattaches to the inner bank, implying that the zone of recirculation is spatially constrained and closed.

Inner-bank flow separation also reduces the effective width of the river (Process F8 in Fig. 1B). The reduced velocities in the zone of inner-bank separation favour the deposition of sediment and the establishment of a deposition bar (Feature S2 in Fig. 1B). This bar plays an important role in processes leading to the accretion of the inner bank and the related meander migration (Pizzuto, 1994; Parker *et al.*, 2011).

In spite of its importance, there are still open questions on the hydro-sedimentological processes related to inner-bank flow separation, the conditions of occurrence and the control parameters.

Blanckaert (2015) has performed a series of 32 laboratory experiments in a very sharp bend with immobile flat-bed configuration, covering a broad range of Froude numbers and $c_f^{-1}HB^{-1}$ values. Inner-bank flow separation occurred in all of these experiments but in none of them was the second stage with recirculation observed. However, inner-bank recirculation did occur in a mobile-bed experiment in the same laboratory flume (Blanckaert, 2010) and is commonly observed in natural meanders (Leeder & Bridges, 1975; Hodskinson & Ferguson, 1998; Ferguson *et al.*, 2003; Frothingham & Rhoads, 2003; Nanson, 2010; Rhoads & Massey, 2012; Schnauder & Sukhodolov, 2012; Vermeulen *et al.*, 2015). These differences point to the importance of hydro-sedimentological processes and interactions.

Zones of inner-bank flow separation are often located in the upstream part of the bend between the cross-over and the apex (i.e. in regions of decreasing curvature radius; Process F6a in Fig. 1A). Based on laboratory experiments, Blanckaert (2015) describes the physics underlying inner-bank flow separation in flat-bed configurations as follows. Flow tends to move along a straight path due to inertia, which favours flow separation. This leads to water mass accumulation at the outer bank and the generation of a transverse tilting of the water surface (Process F1 in Fig. 1B) that provides the inward pressure gradient force required to impose a change of direction on the flow and to guide it around the bend. As discussed before, the development of a transverse water surface tilting is accompanied by increased streamwise water surface slopes at the inner bank that cause flow acceleration and oppose flow separation at the inner bank. Due to mass conservation, this flow acceleration at the inner bank requires mass transport towards the inner bank that further opposes flow separation. Secondary flow promotes flow separation in the upper part of the water column and opposes flow separation in the lower part of the water column. Fig. 12A in Blanckaert *et al.* (2013) illustrates schematically these processes.

Hydro-sedimentological interactions lead to additional processes that favour the onset of inner-bank flow separation and recirculation. A transverse bed slope (Feature S1 in Fig. 1B) and a deposition bar (Feature S2 in Fig. 1B) attached to the inner bank typically develop in the region of decreasing curvature radius. Due to topographic steering, they cause an outwards velocity redistribution that favours inner-bank flow separation. Moreover, the shallowness of the deposition bar causes reduced velocities with further favour flow

separation. Fig. 12B in Blanckaert *et al.* (2013) illustrates schematically these processes. Positive feedbacks between flow and sedimentological processes lead to a widening of the zone of flow separation or recirculation. Blanckaert (2015) observed a reduction of the effective width by about 20% over a flat bed and Blanckaert (2010) a reduction by more than 50% over a mobile bed for a similar discharge in the same flume. Moreover, it favours the onset of the second stage of separation including recirculation.

Whether or not inner-bank flow separation occurs and the location of its occurrence will depend on the magnitude of these flow processes, especially on the spatial lags between them. The fact that the inward pressure gradient force spatially lags behind the outward inertial forces, for example, is supposed to play an important role in the onset of flow separation. Blanckaert (2015) divided the multiple controls on inner-bank flow separation into three groups. First, the geometry of the meander bend, as parameterised by $R_{min}B^{-1}, HB^{-1}$ and especially the streamwise variation of the centreline curvature, dR^{-1}/ds. Inner-bank flow separation should be favoured in shallow wide rivers, for example because establishment of the inward pressure gradient will require a longer distance through the bend than in deep narrow rivers. The second group includes sedimentologic parameters that will determine the friction coefficient and how pronounced the bed morphology gradients are. The third group contains the Froude number that represents inertia. In general, inner-bank flow separation is promoted by rapid changes in curvature, wider rivers and higher Froude numbers.

Inner bank flow separation and recirculation can also occur in the downstream part of the bend between the apex and the next crossover (i.e. in regions of increasing curvature radius) (Process F6b in Fig. 1A). Here, it is favoured by the reduction in the downstream water surface slope in the inner half of the cross-section (Section 3.2.1). Equation 11 remains valid as a condition for the occurrence of an adverse downstream water surface slope. Contrary to outer-bank flow separation, inner-bank flow separation is typically favoured by topographic steering because the inner half of the cross-section is typically becoming shallower in downstream direction from the apex to the next crossover. Inner-bank flow separation in the downstream part of bends has been observed in laboratory experiments with flat bed (Rozovskii, 1957; Blanckaert, 2011) and mobile bed (Blanckaert, 2010; Blanckaert *et al.*, 2013), as well as in natural rivers (Frothingham & Rhoads, 2003; Schnauder & Sukhodolov, 2012). These observations all occurred in long bends that turn over 180° or more. Moreover, a second zone of inner-bank separation also occurred in the upstream part of these long bends.

Very sharp bends, where inner-bank flow separation is most probable, are often rather short. This favours interactions between the potential zones of inner-bank flow separation in the upstream and downstream parts of the bend. There is hardly any insight into these interactions and into the related influence of the bend length.

Similar to outer-bank flow separation, it is obvious that a 3-D model is required for the simulation of inner-bank flow separation but the required sophistication of the turbulence closure remains unknown.

Local effects

A variety of local effects can occur in meandering rivers that redistribute the flow and sediment transport and thereby alter the bathymetry and planform.

Large dead wood (Daniels & Rhoads, 2003, 2004) and macrophytes (Schnauder & Sukhodolov, 2012; Vargas-Luna *et al.*, 2015; Termini, 2016) may occur all over the cross-section. They locally increase the flow resistance and turbulence and cause redistribution of flow and sediment transport. Near the banks, riparian vegetation (Thorne, 1990; Thorne & Furbish, 1995; Darby, 1999; Konsoer *et al.*, 2016) may locally increase the boundary roughness and steer the flow away from the bank. A similar effect may be induced by local bank collapse. If the bank consists of material that is more resistant to erosion than the bed material, so-called slump blocks may deposit at the bank toe and accentuate roughness increase and the steering of the flow away from the bank (Parker *et al.*, 2011; Engel & Rhoads, 2012). The partial coverage of the riverbed by non-erodible bedrock material may also alter the flow distribution (Thorne, 1982; Wood *et al.*, 2001; Parker *et al.*, 2011; Hackney *et al.*, 2016). Engineering interventions often aim at modifying hydro-sedimentological patterns and especially at protection of certain regions from erosion. These interventions can consist of hard structures, such as vanes, baffles or

weirs (Abad *et al.*, 2008; Jamieson *et al.*, 2013; Khosronejad *et al.*, 2014). However, there is a tendency to make more use of soft engineering structures, such as wooden debris (Daniels, 2006; Pilotto *et al.*, 2016).

Interaction between the processes and sedimentologic implications

The interactions between the hydro-sedimentological processes discussed above determine the morphology and the flow distribution in bends of varying curvature. At present, no reduced-order models exist that resolve the near-bank flow processes (i.e. the outer-bank cell, the inner-bank flow separation and the outer-bank flow separation) and their sedimentological effects. The simulation of these hydro-sedimentological processes remains a challenge even for fully 3-D flow models. There is in particular a lack of insight into the interactions between these different processes.

Inner-bank flow separation reduces the effective width and steers the flow outwards. Based on field measurements and numerical simulations, Hodskinson & Ferguson (1998) assumed that this opposes the development of outer-bank flow separation and outer-bank benches, but laboratory experiments by Blanckaert (2010) indicate that the deepening in the pool zone and the related increase in cross-sectional area compensate for this effect. Hickin (1977) indicated that local widening reduces velocities near the outer bank and promotes flow separation there but it is not clear if this is a required condition. Similarly, flow separation at the outer bank also reduces the effective width and steers the flow inwards. It can be presumed to oppose inner-bank flow separation. The mutual influences between flow separation at the inner and outer bank remain largely unexplored. It is reasonable to assume that the degree of interaction between flow processes near both banks decreases with increasing river width.

In a sharply curved meander bend with bankfull aspect ratio of $BH^{-1} \sim 10$, Ferguson *et al.* (2003) observed a zone of inner-bank flow separation that extended most of the way to the outer bank and promoted erosion of the outer bank. In a meander bend with similar aspect ratio, Frothingham & Rhoads (2003) also observed inner-bank flow separation that enhanced the flow attack on the outer bank. In a somewhat narrower meander bend, Rhoads & Massey (2012) observed simultaneously flow separation at the inner and outer banks. They considered this to be an artefact of the development of depositional, vegetated benches within the bottom of an over-widened drainage ditch, which produces a channel planform that is imposed on the flow rather than one that is shaped by the flow. Schnauder & Sukhodolov (2012) observed the simultaneous occurrence of flow recirculation at both banks in a meander bend of aspect ratio $BH^{-1} \sim 10$. The outer-bank recirculation occurred on an outer-bank bench, which represented a considerable local widening of the bend. A similar observation was made by Vermeulen *et al.* (2015) on a meander bend with similar aspect ratio that was an order of magnitude larger.

The main sedimentological importance of the near-bank flow processes is that they keep at a distance from the banks the corridor of dominant discharge, dominant sediment transport and the location of the curvature-induced secondary flow; and that they reduce the effective width of the river (Process F8 in Fig. 1B). This obviously has implications for the flow forcing on the banks and the resulting bank erosion, accretion and meander migration. The confinement of the flow to the reduced width can lead to a deepening of the bed, which increases the cross-sectional area and counteracts the reduction of the effective width.

Vermeulen *et al.* (2015), for example, have observed surprisingly deep scour holes in the central part of a natural meander bend that remain largely unexplained. The occurrence of flow separation at the inner and outer banks in this bend can be assumed to play an important role in the formation of these deep scours.

SOME FUTURE RESEARCH DIRECTIONS

In spite of 150 years of research, there are still numerous open questions on the hydro-sedimentological processes occurring in meandering rivers. An approach that integrates field measurements, laboratory experiments and numerical simulations provides the best guarantee for progress. In recent years, technology for all of these has considerably improved. The next sections will identify some scientific challenges that can be addressed with the presently available technology. The focus is on hydro-sedimentological processes.

Field measurements

As mentioned in the introduction, the hydro-sedimentological patterns in a meandering river depend on the relative importance of individual processes and are therefore case-dependant. The current insight into meandering rivers is largely based on laboratory experiments in configurations with an oversimplified geometry (simplified planform, regular bank shapes, rather uniform bed material, absence of organic matter and biota) and hydrology (constant discharge and sediment supply). These experiments have provided insight into individual processes but they are inappropriate to identify the broad variety of flow and sedimentological patterns that can result from the interaction between these individual processes. Examples are the surprisingly deep scour holes discovered by Vermeulen *et al.* (2015) or the broad range of bedforms including barchan dunes discovered by Konsoer *et al.* (2016a). Therefore, there is a need for field investigations in a variety of different settings. Most previous field investigations concerned alluvial rivers with relatively high bed erodibility. Hydro-morphological processes in rivers with lower erodibility, such as cohesive sediment (Kleinhans *et al.*, 2009; Nanson, 2010) or bedrock rivers (Stark *et al.*, 2010; Inoue *et al.*, 2017), remain largely unexplored. Most previous field investigations focussed on flow patterns and were performed under conditions of quasi-constant discharge, which was typically well-below the morphology-building discharge. Rare examples of field measurements of flow patterns near bankfull flow include Nanson (2010), Engel & Rhoads (2016) and Konsoer *et al.* (2016a). Field investigations that focussed on the temporal dynamics of meandering rivers have mainly been limited to observations on the planform evolution but did not include the temporal dynamics of the bathymetry and the flow patterns (Brice, 1974; Güneralp & Rhoads, 2009). There is a need to design long-term field experiments that incorporate the temporal dynamics of the flow and sedimentologic processes.

Important open questions relate to relevant time scales of the flow and sedimentological processes and their control parameters. It is not clear, for example, to what extent and how fast the morphology adapts to changes in discharge and sediment supply. The transverse bed slope in an alluvial meander bend, for example, is often represented as

$\partial z_b/\partial n = AH\,R^{-1}$, where A is called the scour factor (Engelund, 1974; Odgaard, 1981). According to this simplified parameterisation, the transverse bed slope should vary with the flow depth and thus with the discharge, for conditions above the threshold for sediment transport. The bathymetry is probably more stable than predicted by this simplified parameterisation, due to non-linear hydrodynamic and sedimentological interactions and due to inertia. In a similar way, it is not clear to what extent and how fast bedforms, which condition the roughness, adapt to changes in discharge and sediment supply. More generally, there has been very little research on sediment transport and bedforms in meandering rivers. Notable exceptions are the investigations of Dietrich & Smith (1984) in the field and of Abad & Garcia (2009b) in the laboratory. It is not clear, for example, if the characteristics and control parameters of bed forms in meandering rivers are similar to those in straight channels, or to what extent oblique bedforms (Feature S5 in Fig. 1A) contribute to the redistribution of flow and the sediment transport in meandering rivers. Another open question related to the time scales of relevance concerns the importance of abrupt hydro-sedimentological changes during extreme events and more gradual changes resulting from the normal variability of discharge and sediment supply. This question is particularly important for the formation of cut-off events (Zinger *et al.*, 2011, 2013), for example.

Investigating these questions requires long-term field measuring campaigns. Ideally, they should include measurements of the discharge and sediment supply. Important progress has been made recently in the monitoring of sediment transport with acoustic instruments and this both for bedload (Rennnie *et al.*, 2002; Kostaschuk *et al.*, 2005; Rennie & Church, 2010) and suspended-load (Guerrero *et al.*, 2013; Thorne & Hurther, 2014; Wilson & Hay, 2015a). Bedform-resolving bathymetric surveys and flow measurements should be performed regularly and for a variety of conditions. They are particularly important after extreme events (Leyland *et al.*, 2017). Recent field investigations demonstrate the capabilities of multibeam echosounders for bathymetric measurements and acoustic Doppler current profilers coupled to positioning systems for accurate flow measurements (Konsoer *et al.*, 2016a).

Field measurements most often consist of measurements in sections across the width of the river that aim at resolving the global flow and morphological patterns. There is still a lack of insight into the hydro-sedimentological processes occurring near the banks, in spite of their importance with respect to meander planform stability and evolution. Recently, Engel & Rhoads (2017) reported the first measurements of flow and turbulence near an outer bank in a natural meander bend. Laboratory experiments in simplified configurations have provided incomplete insight into flow separation near both banks and the conditional occurrence of an outer-bank cell of secondary flow. A broader variety of near-bank flow processes can be expected to occur in natural rivers with irregular bank geometry. Investigating these processes requires high-resolution measurements of the bed and bank morphology and turbulence-resolving flow measurements with high spatial resolution near the banks. Lague *et al.* (2013) and Leyland *et al.* (2017) demonstrate terrestrial laser scanners' capability for high-resolution measurements of the bank geometry.

Field investigations are also essential to enhance insight into effects that are difficult to mimic in the laboratory or in numerical investigations. This includes the effect discussed before of large wood, vegetation, or invertebrates on the hydro-sedimentological processes.

Laboratory experiments

As distinct from field investigations, laboratory investigations allow isolating processes and investigating them under controlled conditions with an accuracy that cannot be reached in field investigations. Furthermore, they are repeatable and allow systematic investigation of the effect of individual parameters. However, laboratory investigations are only useful if they mimic processes occurring in the field in a representative way.

A crucial requisite in the design of laboratory experiments is therefore that processes are investigated for the representative range of relevant control parameters. It has been mentioned earlier in this paper that R_{min}/B and $c_f^{-1}H/B$ are important control parameters in meandering rivers. The application of a representative value of the former in the laboratory is straightforward. However, for the latter, many previous laboratory investigations have been performed in flumes that were narrower and smoother than most natural rivers, leading to unrealistic high values of $c_f^{-1}H/B$ (Blanckaert, 2011). This is especially the case in laboratory experiments with flat immobile beds.

This overview has also highlighted the importance of interactions between flow and sedimentological processes, for example with respect to the secondary flow and the near-bank flow separation. Although laboratory experiments over flat immobile beds can provide valuable insight into flow processes, the above considerations suggest that laboratory experiments should preferentially be performed under mobile-bed conditions.

It is notoriously difficult to mimic bank erosion and accretion in laboratory configurations (Parker, 1998) and the majority of previous experiments were performed with fixed planforms. Recent progress in mimicking bank erosion and accretion processes (Peakall *et al.*, 2007; Braudrick *et al.*, 2009) now allows for representative laboratory experiments, that could provide new insight into processes related to bank migration and planform evolution.

Laboratory experiments are particularly appropriate for investigating sediment transport processes in 3-D flows. Current insight into sediment transport processes and sediment transport formulae are largely based on laboratory experiments performed in canonical straight uniform flow configurations. These sediment transport formulae are commonly applied in 3-D flow configurations, neglecting differences between hydro-sedimentological processes in 3-D and 1-D flows. There is a need to enhance insight into the following processes:

i. Turbulent coherent structures, which are largely responsible for peak values in the flow forcing on the sediment particles, are most efficient in picking up sediment (Sumer *et al.*, 2003). Koken *et al.* (2013) have used eddy-resolving numerical simulations to estimate the contribution of turbulent coherent structures to the sediment pick up rate in a curved laboratory flume. However, it has not yet been investigated how the characteristics and sediment pick-up capacity of turbulent coherent structures are affected by curvature. Curvature-induced secondary flow, for example, could be expected to modify the topology of turbulent coherent structures.

ii. The formation of bedforms, their characteristics and their interaction with the flow. Unlike straight uniform flow, the bedform characteristics do vary across the river width in a bend (Abad & Garcia, 2009; Blanckaert, 2010; Jamieson *et al.*, 2010). Abad & Garcia (2009)

observed bedform dimensions in a curved laboratory flume that agreed well with predictors derived for straight flows. However, it remains an open question if the relations between local flow parameters and bedform properties established for straight uniform flow (Garcia, 2008) remain valid in 3-D flows. Bedforms in 3-D flows are also known to be oriented obliquely to the cross-sectional orientation (Feature S5 in Fig. 1A), which induces a transverse near-bed flow component in the lee of the bedform (Dietrich & Smith, 1984). It is an open question how important this transverse flow component is with respect to the redistribution and sorting of the sediment over the river width.

iii. Turbulent sediment fluxes. Sediment is known to be maintained in suspension by upward oriented turbulent fluxes (Vanoni, 1946). In a straight uniform flow, these fluxes are related to the local level of turbulent kinetic energy, *tke* (Nezu & Nakagawa, 1993). It has been discussed before how various processes increase *tke* in meander bends but it remains an open question how this increased *tke* affects the flow's potential to carry sediments in suspension.

iv. The effect of velocities perpendicularly impinging on the boundary (Process F9 in Fig. 1B), which promotes the development of scour. This has been observed in meander bends in the field by Frothingham & Rhoads (2003) and Sukhodolov (2012) and in the laboratory by Jamieson *et al.* (2010) and Blanckaert (2011). It is an open question how these impinging velocities and the related pressure on the boundary affect the flow's capacity to pick up and transport sediment.

Recent progress in acoustic Doppler velocimetry allows non-intrusive simultaneous measurements of entire profiles of the 3D velocity vector and the sediment concentration at high spatial and temporal resolution, even for the highest sediment concentration (Hurther *et al.*, 2011; Naqshband *et al.*, 2014; Wilson & Hay, 2015). Blanckaert *et al.* (2017) have demonstrated that these acoustic techniques are not limited to suspended load transport but also allow the investigation of bedload transport. These acoustic techniques should be particularly appropriate to investigate sediment transport processes in highly 3-D flow configurations.

Laboratory experiments are also appropriate to investigate knowledge gaps on individual hydro-sedimentological processes, to investigate the importance of nonlinear interactions between hydro-sedimentological processes, or to deepen understanding of phenomena discovered in the field. Insight into fundamental processes in meandering rivers is relevant to a broader range of 3-D flow configurations, such as confluences or bifurcations.

Numerical simulations

After validation with experimental data from the field or the laboratory, numerical models can provide additional information in several ways. Firstly, they can provide information on fluid dynamics at high spatial and temporal resolution, contrary to measured data that is typically limited to coarsely spaced cross-sections. Secondly, they can provide full characterisation of spatial-temporal flow structures, such as secondary flows, flow recirculation zones or turbulent coherent structures, contrary to experimental data that infer the existences of such structures from measurements of flow at a few discrete locations. Thirdly, they can provide information on variables that are difficult to measure, such as pressure fluctuations or correlations between different variables. Finally, they allow systematic variation of individual parameters, which is difficult to accommodate in experiments, thereby broadening the domain of investigated conditions. These numerical modelling capabilities have enhanced insight into hydro-sedimentological processes in meandering rivers in the past (e.g. Stoesser *et al.*, 2010; van Balen *et al.*, 2010; Kang & Sotiropoulos, 2011; Constantinescu *et al.*, 2013).

Applications typically have been limited to flow processes in immobile-bed configurations, schematised laboratory configurations and steady flow conditions. The inclusion of sedimentological processes remains a challenge, not only because it increases computational demands but because insight into modelling of the dynamics of sediment transport at high spatial and temporal resolution is at present insufficient. Nabi *et al.* (2013a) have coupled a turbulence-resolving flow model to a physics-based sediment transport model that is based on a description of flow and inter-particular forces on the grain level. They (Nabi *et al.*, 2013b) then validated their model for the case of ripples and dunes in straight steady laboratory flows. Khosrojenad & Sotiropoulos (2014) developed a turbulence-resolving hydro-sedimentological model. Although they adopted a simplified and more conventional sediment transport model as compared to Nabi *et al.* (2013a), they also successfully simulated the development of dunes in straight, steady laboratory flows. The application of such models to real-world configurations under unsteady conditions is now made feasible by two recent advances: 1) Field measuring equipment that allows resolving all relevant details of the boundary form roughness (discussed before) and setting up a computational domain with the required spatial resolution; 2) the increase in computational power. Using the same model as Khosronejad & Sotiropoulos (2014), Khosronejad *et al.* (2016) have recently performed high-resolution turbulence-resolving simulations of hydro-sedimentological processes in a real river during an unsteady flood event. They simulated a 3.2 km-long and 300 m-wide reach of the Mississippi River, including a diffluence-confluence unit. Their grid consisted of 100 million cells and had a resolution of 4 m horizontally and 0.5 m vertically. Their methodology and computational power would be appropriate to investigate hydro-sedimentological processes in most natural meander bends. The appropriateness and capabilities of the sediment transport models adopted by Nabi *et al.* (2013a) and Khosronejad *et al.* (2016) for 3-D flow configurations are unknown at present due to incomplete insight into the sedimentological processes and lacking data from the field and the laboratory for model assessment. A validated high-resolution hydro-sedimentological model would be a powerful tool to provide insight into a variety of processes, including the formation of outer-bank benches, point bars or deep scours.

Contrary to high-resolution models, which are rather black boxes, reduced-order models clearly identify control parameters and the relative importance of individual processes (e.g. Bolla Pittaluga & Seminara, 2011; Parker *et al.*, 2011; Ottevanger *et al.*, 2013). Moreover, they can be applied to investigate much longer spatial and temporal scales, such as the evolution of an entire river reach on geological timescales (e.g. Liverpool & Edwards, 1995; Stølum, 1996; Sun *et al.*, 1996; Lancaster & Bras, 2002; Bogoni *et al.*, 2017). There is a need for better and more detailed field measurements for the assessment and validation of reduced-order models. Remote sensing by satellite can nowadays resolve the natural planform dynamics with high resolution over long reaches (Schwenk *et al.*, 2017). However, it remains a challenge to set up a field measuring site that also provides the required data on the hydrology (flow hydrographs and sediment supply) and bathymetry.

Improving the parameterisation of flow and sedimentological processes is another challenge. Parker *et al.* (2011) and Eke *et al.* (2014) have recently improved the parameterisation of bank accretion and erosion and the resulting planform evolution. Blanckaert & de Vriend (2003, 2010) and Ottevanger *et al.* (2013) have recently improved the parameterisation of the curvature-induced secondary flow and their effect on the flow and bathymetry for a fixed planform. Their parameterisation considerably improved the model performance in the high-curvature range, which is particularly relevant for meanders near cut-off. A logical extension would be the coupling of both approaches to investigate the effect of the improved secondary flow parameterisation on the planform evolution. It is also a challenge to improve the modelling of sediment transport in reduced-order models. Ottevanger *et al.* (2013) have demonstrated that the modelling of the effect of the transverse bed slope on the magnitude and direction of the sediment transport is a key process that requires further attention. Baar *et al.* (2018) summarise recent efforts to improve the modelling of this process.

At present, reduced-order models do not account for flow separation near the inner and outer banks, or the outer-bank cell of secondary flow, although these processes obviously play an important role in the meander planform evolution. It is a challenge to account for the effect of these processes, even in a highly simplified and approximate way. Finally, most reduced-order models are limited to rather academic configurations of quasi-uniform steady flow and a longitudinal profile in equilibrium, i.e. they only consider transverse redistribution in sediment and the resulting morphological changes. There is a need to extend these models towards more realistic configurations, including backwater effects and longitudinal sediment redistribution due to changes in sediment supply. Such extensions would allow the investigation of how the meander planform adapts to changes in the flow regime and sediment supply induced by, for example, climate change or human interventions.

ACKNOWLEDGEMENTS

Arjan Reesink, Bruce Rhoads, John Fenton and an anonymous reviewer are thanked for constructive comments and suggestions that greatly improved the manuscript.

REFERENCES

Abad, J.D. and **Garcia, M.H.** (2009) Experiments in a high-amplitude Kinoshita meandering channel: 2. Implications of bend orientation on bed morphodynamics. *Water Resour. Res.*, **45**, W02402.

Abad, J.D., **Rhoads, B.L.**, **Güneralp, I.** and **García, M.H.** (2008) Flow structure at different stages in a meander-bend with bendway weirs. *J. Hydraul. Eng. ASCE*, **134**(8), 1052–1063.

Baar, A.W., **de Smit, J.**, **Uijttewaal, W.S.J.** and **Kleinhans, M.G.** (2018) Sediment transport of fine sand to fine gravel on transverse bed slopes in rotating annular flume experiments. *Water Resour. Res.*, **54**, 19–45.

Bagnold, R.A. (1960) Some aspects of the shape of river meanders. *US Geol. Surv. Prof. Pap.*, **282-E**, US Geological Survey, Washington, DC.

Bathurst, J.C., **Thorne, C.R.** and **Hey, R.D.** (1977) Direct measurements of secondary currents in river bends. *Nature*, **269**, 504–506.

Bathurst, J.C., **Thorne, C.R.** and **Hey, R.D.** (1979) Secondary flow and shear stress at river bends. *J. Hydraul. Div. ASCE*, **105**(10), 1277–1295.

Blanckaert, K. (2015) Flow separation at convex banks in open channels. *J. Fluid Mech.*, **779**, 432–467.

Blanckaert, K. (2011) Hydrodynamic processes in sharply-curved river bends and their morphological implications. *J. Geoph. Res. Earth Surf.*, **116**, F01003.

Blanckaert, K. (2009) Saturation of curvature induced secondary flow, energy losses and turbulence in sharp open-channel bends. Laboratory experiments, analysis and modelling. *J. Geoph. Res. Earth Surf.*, **114**, F03015.

Blanckaert, K. (2010) Topographic steering, flow recirculation, velocity redistribution and bed topography in sharp meander bends. *Water Resour. Res.*, **46**, W09506.

Blanckaert, K. and **de Vriend, H.J.** (2003) Nonlinear modeling of mean flow redistribution in curved open channels. *Water Resour. Res.*, **39**(12), 1375.

Blanckaert, K. and **de Vriend, H.J.** (2010) Meander dynamics: a nonlinear model without curvature restrictions for flow in open-channel bends. *J. Geoph. Res. Earth Surf.*, **115**, F04011.

Blanckaert, K. and **de Vriend, H.J.** (2004) Secondary flow in sharp open-channel bends. *J. Fluid Mech.*, **498**, 353–380.

Blanckaert, K., **Duarte, A.**, **Chen, Q.** and **Schleiss, A.J.** (2012) Flow processes near smooth and rough (concave) outer banks in curved open channels. *J. Geoph. Res. Earth Surf.*, **117**, F04020.

Blanckaert, K. and **Graf, W.H.** (2004) Momentum transport in sharp open-channel bends. *J. Hydraul. Eng. ASCE*, **130**(3), 186–198.

Blanckaert, K., Heyman, J. and **Rennie, C.D.** (2017) Measurement of bedload sediment transport with an Acoustic Doppler Velocity Profiler. *J. Hydraul. Eng.,* **143**(6), Article number 04017008.

Blanckaert, K., Kleinhans, M.G., McLelland, S.J., Uijttewaal, W.S.J., Murphy, B.J., van de Kruijs, A., Parsons, D.R. and **Chen, Q.** (2013) Flow separation at the inner (convex) and outer (concave) banks of constant-width and widening open-channel bends. *Earth Surf. Proc. Land.,* **38**, 696–716.

Blondeaux, P. and **Seminara, G.** (1985) A unified bar-bend theory of river meanders. *J. Fluid Mech,* **157**, 449–470.

Bogoni, M., Putti, M. and **Lanzoni, S.** (2017) Modeling meander morphodynamics over self-formed heterogeneous floodplains. *Water Resour. Res.,* **53**(6), 5137–5157.

Bolla Pittaluga, M., Nobile, G. and **Seminara, G.** (2009) A nonlinear model for river meandering. *Water Resour. Res.,* **45**, W04432.

Bolla Pittaluga, M. and **Seminara, G.** (2011) Nonlinearity and unsteadiness in river meandering: a review of progress in theory and modelling. *Earth Surf. Proc. Land.,* **36**(1), 20–38.

Boussinesq, J. (1868) Mémoire sur l'influence de frottement dans les mouvements réguliers des fluides; XII – Essai sur le mouvement permanent d'un liquide dans un canal horizontal à axe circulaire. *Journal de Mathémathiques Pures et Appliquées,* **2 XIII**: 413.

Bradshaw, P. (1987) Turbulent secondary flows. *Annu. Rev. Fluid Mech.,* **19**, 53–74.

Braudrick, C.A., Dietrich, W.E., Leverich, G.T. and **Sklar, L.S.** (2009) Experimental evidence for the conditions necessary to sustain meandering in coarse-bedded rivers. *Proc. Nat. Ac. Sc. USA,* **106**(40), 16936–16941.

Brice, J.C. (1974) Evolution of meander loops. *Geol. Soc. Am. Bull.,* **85**, 581–586.

Bridge, J.S. and **Jarvis, J.** (1977) Velocity profiles and bed shear stress over various bed configurations in a river bend. *Earth Surf. Proc. Land.,* **2**, 281–294.

Carson, M.A. and **Griffiths, G.A.** (1987) Influence of channel width on bed load transport capacity. *J. Hydraul. Eng.,* **113**(12), 1489–1509.

Chow, V.T. (1959) Open Channel Hydraulics. McGraw-Hill: New York.

Constantinescu, G., Kashyap, S., Tokyay, T., Rennie, C.D. and **Townsend, R.D.** (2013) Hydrodynamic processes and sediment erosion mechanisms in an open channel bend of strong curvature with deformed bathymetry. *J. Geoph. Res. Earth Surf.,* **118**(2), 480–496.

Constantinescu, G., Koken, M. and **Zeng J.** (2011), The structure of turbulent flow in an open channel bend of strong curvature with deformed bed: Insight provided by detached eddy simulation, *Water Resour. Res.,* **47**, W05515.

Daniels, M.D. (2006) Distribution and dynamics of large woody debris and organic matter in a low-energy meandering stream. *Geomorphology,* **77**(3-4), 286–298.

Daniels, M.D. and **Rhoads, B.L.** (2004) Effect of large woody debris configuration on three-dimensional flow structure in two low-energy meander bends at varying stages. *Water Resour. Res.,* **40**(11), W1130201-W1130214.

Daniels, M.D. and **Rhoads, B.L.** (2003) Influence of a large woody debris obstruction on three-dimensional flow structure in a meander bend. *Geomorphology,* **51**(1-3), 159–173.

Darby, S.E. (1999) Effect of riparian vegetation on flow resistance and flood potential. *J. Hydraul. Eng.,* **125**(5), 443–454.

de Vriend, H.J. (1977) A mathematical model of steady flow in curved shallow channels, *J. Hydraul. Res.,* **15**(1), 37–54.

de Vriend, H.J. (1979) *Flow measurements in a curved rectangular channel.* Rep. No. 9-79, Laboratory of Fluid Mechanics, Department of Civil Engineering, Delft University of Technology, The Netherlands.

de Vriend, H. J. (1981a) *Steady flow in shallow channel bends.* Rep. No. 3-81, Laboratory of Fluid Mechanics, Department of Civil Engineering, Delft University of Technology, The Netherlands.

de Vriend, H.J. (1981b) Velocity redistribution in curved rectangular channels. *J. Fluid Mech.,* **107**, 423–439.

de Vriend, H.J. and **Geldof, H.J.** (1983) Main velocity in short river bends. *J. Hydraul. Eng.,* **109**(7), 991–1011.

de Vriend, H.J. and **Struiksma, N.** (1984) Flow and bed deformation in river bends, in River Meandering, edited by C.M. Elliot, pp. 810–828, ASCE, New Orleans, Louisiana, ISBN 0-87262-393-9.

Dietrich, W.E. and **Smith, J.D.** (1984) Bed load transport in a river meander. *Water Resour. Res.,* **20**(10), 1355–1380.

Dietrich, W.E. and **Smith, J.D.** (1983) Influence of the point bar on flow through curved channels. *Water Resour. Res.,* **19**(5), 1173–1192.

Dietrich, W.E. and **Whiting, P.** (1989) Boundary shear stress and sediment transport in river meanders of sand and gravel. In: River Meandering (Eds S. Ikeda and G. Parker), *Water Resour. Monogr. Ser.,* **12**, 1–50, AGU, Washington, D.C.

Einstein, H.A. and **Harder, J.A.** (1954) Velocity distribution and the boundary layer at channel bends, *Eos Trans. AGU,* **35**(1), 114–120.

Eke, E., Parker, G. and **Shimizu, Y.** (2014) Numerical modeling of erosional and depositional bank processes in migrating river bends with self-formed width: Morphodynamics of bar push and bank pull. *J. Geoph. Res. Earth Surf.,* **119**, 1455–1483.

Engel, F.L. and **Rhoads, B.L.** (2016) Interaction among mean flow, turbulence, bed morphology, bank failures and channel planform in an evolving compound meander loop. *Geomorphology,* **163-164**, 70–83.

Engel, F.L. and **Rhoads, B.L.** (2016) Three-dimensional flow structure and patterns of bed shear stress in an evolving compound meander bend. *Earth Surf. Process. Land.,* **41**, 1211–1226.

Engel, F.L. and **Rhoads, B.L.** (2017) Velocity profiles and the structure of turbulence at the outer bank of a compound meander bend. *Geomorphology,* **295**, 191–201.

Engelund, F. (1974) Flow and bed topography in channel bends, *J. Hydraul. Div.,* **100**(HY11), 1631–1648.

Ferguson, R.I., Parsons, D.R., Lane, S.N. and **Hardy, R.J.** (2003) Flow in meander bends with recirculation at the inner bank. *Water Resour. Res.,* **39**(11), 1322.

Frascati, A. and **Lanzoni, S.** (2013) A mathematical model for meandering rivers with varying width. *J. Geoph. Res. Earth Surf.,* **118**(3), 1641–1657.

Frothingham, K.M. and **Rhoads, B.L.** (2003) Three-dimensional flow structure and channel change in an asymmetrical compound meander loop, Embarras River, Illinois. *Earth Surf. Proc. Land.*, **28**(6), 625–644.

Garcia, M.H. (2008) *Sediment transport and morphodynamics.* In: *Sedimentation Engineering: Processes, Measurements, Modelling and Practice* (Ed. M.H. Garcia), Am. Soc. Civ. Eng., Reston, Va., chap. 2, pp. 21–164,

Götz, W. (1975) Sekundärstromungen in aufeinander folgenden Gerinnekrümmungen. Rep. No. 163, Theodor-Rehbock Flussbaulaboratorium, Karlsruhe, Germany.

Guerrero, M., Szupiany, R.N. and **Latosinski, F.** (2013) Multi-frequency acoustics for suspended sediment studies: An application in the Parana River. *J. Hydraul. Res.*, **51**(6), 696–707.

Güneralp, I. and **Rhoads, B.L.** (2009) Empirical analysis of the planform curvature-migration elation of meandering rivers. *Water Resour. Res.*, **45**(9), Article number W09424.

Hackney, C., Best, J., Leyland, J., Darby, S.E., Parsons, D. and **Aalto, R.** (2015) Modulation of outer bank erosion by slump blocks: Disentangling the protective and destructive role of failed material on the three-dimensional flow structure, *Geophys. Res. Lett.*, **42**, 10, 663–10670.

Henderson, F.M. (1966) *Open-channel Flow.* Macmillan, 522 pp.

Hickin, E.J. (1977) Hydraulic factors controlling channel migration. In: *Research into Fluvial Systems* (Eds R.E. Davidson-Arnott and W. Nickling), Proceedings of the 5th Guelph Geomorphology Symposium. Geobooks: Norwich; 59–72.

Hodskinson, A. and **Ferguson, R.I.** (1998) Numerical modelling of separated flow in river bends: model testing and experimental investigation of geometric controls on the extent of flow separation at the concave bank. *Hydrol. Process.*, **12**, 1323–1338.

Hurther, D., Thorne, P.D., Bricault, M., Lemmin, U. and **Barnoud, J.M.** (2011) A multi-frequency Acoustic Concentration and Velocity Profiler (ACVP) for boundary layer measurements of fine-scale flow and sediment transport processes. *Coast. Eng.*, **58**(7), 594–605.

Ikeda, S., Parker, G. and **Sawai, K.** (1981) Bend theory of river meanders. Part 1: Linear development, *J. Fluid Mech.*, **112**, 363–377.

Inoue, T., Parker, G. and **Stark, C.P.** (2017) Morphodynamics of a bedrock-alluvial meander bend that incises as it migrates outward: approximate solution of permanent form. *Earth Surf. Proc. Land.*, **42**(9), 1342–1354

Jamieson, E., Post, G. and **Rennie, C.D.** (2010) Spatial variability of three-dimensional Reynolds stresses in a developing channel bend, *Earth Surf. Proc. Land.*, **35**(9), 1029–1043.

Jamieson, E.C., Ruta, M.A., Rennie, C.D. and **Townsend, R.D.** (2013) Monitoring stream barb performance in a semi-alluvial meandering channel: Flow field dynamics and morphology. *Ecohydrology*, **6**(4), 611–626.

Johannesson, H. and **Parker, G.** (1989a) Secondary flow in a mildly sinuous channel. *J. Hydraul. Eng.*, **115**(3), 289–308.

Johannesson, H. and **Parker, G.** (1989b) Velocity redistribution in meandering rivers. *J. Hydraul. Eng.*, **115**(8), 1019–1039.

Kang, S. and **Sotiropoulos, F.** (2011) Flow phenomena and mechanisms in a fieldscale experimental meandering stream with a pool-riffle sequence: insight gained via numerical simulation. *J. Geoph. Res. Earth Surf.*, **116**, F03011.

Khosronejad, A., Kozarek, J.L. and **Sotiropoulos, F.** (2014) Simulation-based approach for stream restoration structure design: Model development and validation. *J. Hydraul. Eng.*, **140**(9), 04014042.

Khosronejad, A., Le, T., DeWall, P., Bartelt, N., Woldeamlak, S., Yang, X. and **Sotiropoulos, F.** (2016) High-fidelity numerical modeling of the Upper Mississippi River under extreme flood condition. *Adv. Water Res.*, **98**(1), 97–113.

Khosronejad, A. and **Sotiropoulos, F.** (2014) Numerical simulation of sand waves in a turbulent open channel flow. *J. Fluid Mech.*, **753**(2), 150–216.

Kleinhans, M.G., Schuurman, F., Bakx, W. and **Markies, H.** (2009) Meandering channel dynamics in highly cohesive sediment on an intertidal mud flat in the Westerscheldeestuary, the Netherlands. *Geomorphology*, **105**(3-4), 261–276.

Koken, M., Constantinescu, G. and **Blanckaert, K.** (2013) Hydrodynamic processes, sediment erosion mechanisms and Reynolds-number-induced scale effects in an open channel bend of strong curvature with flat bathymetry. *J. Geophys. Res. Earth Surf.*, **118**, 2308–2324.

Konsoer, K.M., Rhoads, B.L., Best, J.L., Langendoen, E.J., Abad, J.D., Parsons, D.R. and **Garcia, M.H.** (2016a) Three-dimensional flow structure and bed morphology in large elongate meander loops with different outer bank roughness characteristics, *Water Resour. Res.*, **52**, 9621–9641.

Konsoer, K.M., Rhoads, B.L., Langendoen, E.J., Best, J.L., Ursic, M.E., Abad, J.D. and **Garcia, M.H.** (2016b) Spatial variability in bank resistance to erosion on a large meandering, mixed bedrock-alluvial river. *Geomorphology*, **252**, 80–97.

Kostaschuk, R., Best, J., Villard, P., Peakall, J. and **Franklin, M.** (2005) Measuring flow velocity and sediment transport with an acoustic Doppler current profiler. *Geomorphology*, **68**(1-2), 25–37.

Lague, D., Brodu, N. and **Leroux, J.** (2013) Accurate 3D comparison of complex topography with terrestrial laser scanner: Application to the Rangitikei canyon (N-Z). *ISPRS Journal of Photogrammetry and Remote Sensing*, **82**, 10–26.

Lancaster, S.T. and **Bras, R.L.** (2002) A simple model of river meandering and its comparison to natural channels. *Hydrol. Proc.*, **16**, 1–26.

Leeder, M.R. and **Bridges, P.H.** (1975) Flow separation in meander bends. *Nature*, **253** (5490), 338–339.

Leopold, L.B. and **Wolman, M.G.** (1957) River Channel Patterns: Braided, Meandering and Straight. *US Geol. Surv. Prof. Pap.* **282-B**, US Geological Survey, Washington, DC.

Leyland, J., Hackney, C.R., Darby, S.E., Parsons, D.R., Best, J.L., Nicholas, A.P., Aalto, R. and **Lague, D.** (2017) Extreme flood-driven fluvial bank erosion and sediment loads: direct process measurements using integrated Mobile Laser Scanning (MLS) and hydro-acoustic techniques. *Earth Surf. Proc. Land.*, **42**(2), 334–346.

Liverpool, T.B. and Edwards, S.F. (1995) Dynamics of a meandering river. *Phys. Rev. Lett.*, **75**(16), 3016–3019.

Mockmore, C.A. (1943) Flow around bends in stable channels. *Trans. Am. Soc. Civ. Eng.*, **109**, 593–628.

Mosselman, E. (2005) Basic equations for sediment transport in CFD for fluvial morphodynamics. In: *Computational Fluid Dynamics: Applications in Environmental Hydraulics* (Eds P.D. Bates, S.N. Lane and R.I. Ferguson), chap. 4, pp. 71–89, John Wiley, Hoboken, N.J.

Nabi, M., De Vriend, H.J., Mosselman, E., Sloff, C.J. and Shimizu, Y. (2013a) Detailed simulation of morphodynamics: 2. Sediment pickup, transport and deposition. *Water Resourc. Res.*, **49**(8), 4775–4791.

Nabi, M., De Vriend, H.J., Mosselman, E., Sloff, C.J. and Shimizu, Y. (2013b) Detailed simulation of morphodynamics: 3. Ripples and dunes. *Water Resourc. Res.*, **49**(9), 5930–5943.

Nanson, R.A. (2010) Flow fields in tightly curving meander bends of low width–depth ratio. *Earth Surf. Proc. Land.*, **35**(2), 119–135.

Naqshband, S., Ribberink, J.S., Hurther, D., Barraud, P.A. and Hulscher, S.J.M.H. (2014) Experimental evidence for turbulent sediment flux constituting a large portion of the total sediment flux along migrating sand dunes. *Geoph. Res. Lett.*, **41**(24), 8870–8878.

Nezu, I. and Nakagawa, H. (1993) Turbulence in Open-Channel Flows. IAHR-Monograph, Balkema.

Odgaard, A.J. (1986) Meander flow model. I: Development. *J. Hydraul. Eng.*, **112**(12), 1117–1136.

Odgaard, A.J. (1988) River-meander model. I: Development. *J. Hydraul. Eng.*, **115**(11), 1433–1450.

Odgaard, A.J. (1981) Transverse bed slope in alluvial channel bends. *J. Hydraul. Div.*, **107**(HY12), 1677–1693.

Odgaard, A.J. and Bergs, M.A. (1988) Flow processes in a curved alluvial channel. *Water Resour. Res.*, **24**(1), 45–56.

Ottevanger, W., Blanckaert, K. and Uijttewaal, W.S.J. (2012) Processes governing the flow redistribution in sharp river bends. *Geomorphology*, **163-164**, 45–55.

Ottevanger, W., Blanckaert, K., Uijttewaal, W.S.J. and de Vriend, H.J. (2013) Meander dynamics: A reduced-order nonlinear model without curvature restrictions for flow and bed morphology. *J. Geoph. Res. Earth Surf.*, **118**, F20080.

Page, K.J. and Nanson, G.C. (1982) Concave-bank benches and associated floodplain formation. *Earth Surf. Proc. Land.*, **7**, 529–543.

Parker G. (1998) River meanders in a tray. *Nature*, **395**(6698), 111–112.

Parker, G., Shimizu, Y., Wilkerson, G.V., Eke, E.C., Abad, J.D., Lauer, J.W., Paola, C., Dietrich, W.E. and Voller, V.R. (2011) A new framework for modeling the migration of meandering rivers. *Earth Surf. Proc. Land.*, **36**(1), 70–86.

Peakall, J., Ashworth, P.J. and Best J.L. (2007) Meander-Bend Evolution, Alluvial Architecture and the Role of Cohesion in Sinuous River Channels: A Flume Study. *J. Sed. Res.*, **77** (3), 197–212.

Pilotto, F., Harvey, G.L., Wharton, G. and Pusch, M.T. (2016) Simple large wood structures promote hydromorphological heterogeneity and benthic macroinvertebrate diversity in low-gradient rivers. *Aquat. Sci.*, **78**(4), 755–766.

Pizzuto, J.E. (1994) Channel adjustments to changing discharges, Powder River, Montana. *Geol. Soc. Am. Bull.*, **106**(11), 1494–1501.

Rennie, C.D. and Church, M. (2010) Mapping spatial distributions and uncertainty of water and sediment flux in a large gravel bed river reach using an acoustic Doppler current profiler. *J. Geoph. Res. Earth Surf.*, **115**(3), F03035.

Rennie, C.D., Millar, R.G. and Church, M.A. (2002) Measurement of bed load velocity using an acoustic Doppler current profiler. *J. Hydraul. Eng.*, **128**(5), 473–483.

Rhoads, B.L. and Massey, K.D. (2012) Flow structure and channel change in a sinuous grass-lined stream within an agricultural drainage ditch: implications for ditch stability and aquatic habitat. *River Res. Appl.*, **28**(1): 39–52.

Rozovskii, I.L. (1957) Flow of Water in Bends of Open Channels, Acad. of Sci. of the Ukr., Kiev, 1957. (English translation, Isr. Program for Sci. Transl., Jerusalem, 1961).

Schnauder, I. and Sukhodolov, A.N. (2012) Flow in a tightly curving meander bend: effects of seasonal changes in aquatic macrophyte cover. *Earth Surf. Proc. Land.*, **37**(11), 1142–1157.

Schwenk, J., Khandelwal, A., Fratkin, M., Kumar, V. and Foufoula-Georgiou, E. (2017) High spatiotemporal resolution of river planform dynamics from Landsat: The RivMAP toolbox and results from the Ucayali River. *Earth and Space Science*, **4**, 46–75.

Seminara, G., Solari, L. and Parker, G. (2002) Bed load at low Shields stress on arbitrarily sloping beds: Failure of the Bagnold hypothesis. *Water Resour. Res.*, **38**(11), 311–3116.

Shields, A. (1936) Anwendung der Aehnlichkeitsmechanik und der Turbulenzforschung auf die Geschiebebewegung [Application of similarity mechanics and turbulence research on shear flow], Preußische Versuchsanstalt für Wasserbau, Berlin (in German).

Simpson, R.L. (1989) Turbulent boundary-layer separation. *Annu. Rev. Fluid Mech.*, **21**, 205–234.

Stark, C.P., Barbour, J.R., Hayakawa, Y.S., Hattanji, T., Hovius, N., Chen, H., Lin, C.W., Horng, M.J., Xu, K.Q. and Fukahata, Y. (2010) The climatic signature of incised river meanders, *Science*, **327**, 5972, 1497–1501.

Stoesser, T., Ruether, N. and Olsen, N.R.B. (2010) Calculation of primary and secondary flow and boundary shear stresses in a meandering channel. *Adv. Water Resour.*, **33**(2), 158–170.

Stølum, H-H. (1996) River meandering as a self-organization process. *Science*, **271**, 1371–1374.

Struiksma, N., Olesen, K.W., Flokstra, C. and de Vriend, H.J. (1985) Bed deformation in curved alluvial channels. *J. Hydraul. Res.*, **23**(1), 57–79.

Sukhodolov, A.N. (2012) Structure of turbulent flow in a meander bend of a lowland river, *Water Resour. Res.*, **48**, W01516.

Sumer, B.M., Chua, L.H.C., Cheng, N.S. and Fredsoe, J. (2003) Influence of turbulence on bed load sediment transport. *J. Hydraul. Eng.*, **129**(8), 585–596.

Sun, T., Meakin, P., Jøssang, T. and Schwarz, K. (1996) A simulation model for meandering rivers. *Water Resour. Res.*, **32**, 2937–2954.

Talmon, A.M., Struiksma, N. and **Van Mierlo, M.C.L.M.** (1995) Laboratory measurements of the direction of sediment transport on transverse alluvial-bed slopes. *J. Hydraul. Res.*, **33**(4), 495–517.

Termini, D. (2016) Experimental analysis of the effect of vegetation on flow and bed shear stress distribution in high-curvature bends. *Geomorphology*, **274**, 1–10.

Termini, D. and **Piraino, M.** (2011) Experimental analysis of cross-sectional flow motion in a large amplitude meandering bend. *Earth Surf. Proc. Land.*, **36**, 244–256.

Thorne, C.R. (1990) Effects of vegetation on riverbank erosion and stability. In: *Vegetation and Erosion* (Ed. J.B. Thornes), Wiley, Chichester, England, 125–144.

Thorne, C.R. (1982) Processes and mechanisms of river bank erosion. In: *Gravel-Bed Rivers* (Eds R.D. Hey, J.C. Bathurst and C.R. Thorne), 227–271, Wiley, Chichester, U.K.

Thorne, C.R. and **Hey, R.D.** (1979) Direct measurements of secondary currents at a river inflexion point. *Nature*, **280**, 226–228.

Thorne, P.D. and **Hurther, D.** (2014) An overview of the use of backscattered sound for measuring suspended particle size and concentration profiles in non-cohesive inorganic sediment transport studies. *Cont. Shelf Res.*, **73**, 97–118.

Thorne, S.D. and **Furbish, D.J.** (1995) Influences of coarse bank roughness on flow within a sharply curved river bend. *Geomorphology*, **12**(3), 241–257.

Van Balen, W., Blanckaert, K. and **Uijttewaal, W.S.J.** (2010a) Analysis of the role of turbulence in curved open-channel flow at different water depths by means of experiments, LES and RANS. *J. Turb.*, **11**(12), 1–34.

Van Balen, W., Uijttewaal, W.S.J. and **Blanckaert, K.** (2009) Large-eddy simulation of a mildly curved open-channel flow. *J. Fluid Mech.*, **630**, 413–442.

Van Bendegom, L. (1947) Some considerations on river morphology and river improvement. *De Ingenieur*, **59**(4), (in Dutch; English transl.: Nat. Res. Counc. Canada, Technical Translation 1054, 1963).

Vanoni, V.A. (1946) *Transportation of suspended sediment by water*. Am. Soc. Civ. Eng., 67 pp.

Vargas-Luna, A., Crosato, A. and **Uijttewaal, W.S.J.** (2015) Effects of vegetation on flow and sediment transport: Comparative analyses and validation of predicting models. *Earth Surf. Proc. Land.*, **40**(2), 157–176.

Vermeulen, B., Hoitink, A.J.F. and **Labeur, R.J.** (2015) Flow structure caused by a local cross-sectional area increase and curvature in a sharp river bend. *J. Geophys. Res. Earth Surf.*, **120**, 1771–1783.

Vietz, G.J., Stewardson, M.J., Rutherfurd, I.D. and **Finlayson, B.L.** (2012) Hydrodynamics and sedimentation of concave benches in a lowland river. *Geomorphology*, **147–148**, 86–101.

Wei, M., Blanckaert, K., Heyman, J., Li, D. and **Schleiss, A.J.** (2016) A parametrical study on secondary flow in sharp open-channel bends: experiments and theoretical modeling. *J. Hydro-Env. Res.*, **13**(1), 1–13.

Wilson, G.W. and **Hay, A.E.** (2015a) Acoustic backscatter inversion for suspended sediment concentration and size: A new approach using statistical inverse theory. *Cont. Shelf Res.*, **106**, 130–139.

Wilson, G.W. and **Hay, A.E.** (2015b) Measuring two-phase particle flux with a multi-frequency acoustic Doppler profiler, *J. Acoust. Soc. Am.*, **138**(6), 3811–3819.

Wood, A.L., Simon, A., Downs, P.W. and **Thorne C.R.** (2001) Bank-toe processes in incised channels: The role of apparent cohesion in the entrainment of failed bank materials. *Hydrol. Process.*, **15**(1), 39–61.

Zinger, J.A., Rhoads, B.L. and **Best, J.L.** (2011) Extreme sediment pulses generated by bend cutoffs along a large meandering river. *Nat. Geosci.*, **4**, 675–678.

Zinger, J.A., Rhoads, B.L., Best, J.L. and **Johnson, K.K.** (2013) Flow structure and channel morphodynamics of meander bend chute cutoffs: A case study of the Wabash River, USA. *J. Geophys. Res. Earth Surf.*, **118**, 2468–2487.

Zolezzi, G., Guala, M., Termini, D. and **Seminara, G.** (2005) Experimental observations of upstream overdeepening. *J. Fluid Mech.*, **531**, 191–219.

Zolezzi, G. and **Seminara, G.** (2001) Downstream and upstream influence in river meandering. Part 1. General theory and application to overdeepening. *J. Fluid Mech.*, **438**, 183–211.

Int. Assoc. Sedimentol. Spec. Publ (2018) **48**, 321–348.

Unsuccessful cut offs – origin and partial preservation of enigmatic channel fills encased within a large-scale point-bar deposit – The McMurray Formation type section, Alberta, Canada

MILOVAN FUSTIC[†], RUDY STROBL[§], MASSIMILIANO GHINASSI[×] and SHUYU ZHANG[†*]

[†] *Department of Geoscience, University of Calgary, Calgary, Alberta, Canada*
[§] *EnerFox Enterprise, Calgary, Alberta, Canada*
[×] *Department of Geosciences, University of Padova, Padova, Italy*
[*] *Petroleum University of China (East China), Qingdao, Shandong, China*

ABSTRACT

This study describes and attempts to interpret processes leading to the formation and preservation of eight channel-fill deposits encased within an interpreted large scale (40 m-thick) point-bar deposit. The studied channel fills are superbly exposed along the McMurray Formation type section, Alberta, Canada. Exposures were mapped in detail as part of an outcrop study extending over multiple seasons. The apparent emplacement of channel fills into vertically continuous point-bar deposit is difficult to explain by common facies models for meandering rivers. Very deep incisions (15 to 35 m) and sharp erosional contacts of channel bases are indicative of high-magnitude flood events and/or prolonged bifurcation of stream flow that would deepen channel over time. Channel-fill truncation and the presence of younger overlying thick and extensive lateral-accretion deposits suggest point-bar reactivation. The spatial relationship of channel incisions and hosting point-bar reactivation allows for interpreting the order of geologic events for each encased channel-fill: (i) a flooding event and associated increasing discharge causing a channel incision into an existing point-bar; (ii) channel plugging by sediments (due to reduced discharge); and (iii) reactivation of the hosting point bar and/or meander avulsion with associated erosion of the upper part of incised channels, followed by burial (encasement) of channel remnants at the bottom of the incision. The proposed model suggests that all eight encased channel-fills are only minute keels of significantly larger channel incisions that did not ultimately undergo cut off or avulsion. Although these deposits and interpreted sedimentary processes are expected to be common in both modern and ancient meandering river systems and particularly those that have experienced major floods, this phenomenon does not appear to have been described in the scientific literature. Implications are numerous and potentially significant. Proposed depositional models may be useful for distinguishing encased channel deposits from vertically continuous point-bar deposits in subsurface studies. Sand-dominated encased channel deposits may contribute to increased reservoir connectivity.

Keywords: Cut offs, flood events, chute channel, bifurcation, point bar, meandering river

INTRODUCTION

The Lower Cretaceous McMurray Formation, the primary host of the world's largest oil reservoir, the Athabasca Oil Sands Deposit, is formed by an upward transition from fluvial deposits at the base to fully marine deposits at the top and is interpreted to be formed in an overall transgressive setting (Mellon & Wall, 1956; Carrigy, 1959, 1971; Flach & Mossop, 1985; Ranger, 1997, Wightman & Pemberton, 1997, Hein & Cotterill, 2006). Sedimentological, ichnological and palynological

Fluvial Meanders and Their Sedimentary Products in the Rock Record, First Edition.
Edited by Massimiliano Ghinassi, Luca Colombera, Nigel P. Mountney and Arnold Jan H. Reesink.

studies suggest that the majority of the preserved sediments are tidal-fluvial and brackish (Hubbard *et al.*, 2011).

The McMurray Formation type section, located on the eastern bank of the Athabasca River (Fig. 1) has been described by many authors in the past (Bell, 1884; McConnell, 1893; Ells, 1914; McLearn, 1917; Carrigy, 1959; Hein *et al.*, 2001). The outcrop is comprised of three stratigraphic units named Early, Late and the Latest McMurray (Figs 2 and 3). This study is restricted to Late McMurray Formation exposure. The underlying Early McMurray is interpreted as multiple small-scale stacked and interfingered thin (3 to 7 m-thick) meandering-channel deposits and the overlying Latest McMurray as low-sinuosity late-stage channel incisions formed in response to base level drop (Fig. 2).

The Late McMurray Formation is interpreted as a part of a single, large-scale point-bar deposit (up to 40 m-thick and 5 km laterally), comprised of cross-bedded sandstones in lower point bars that transition upwards into alternating finer-grained sandstone and mudstone in the form of inclined heterolithic stratification (IHS; *sensu* Thomas *et al.*, 1987; Fig. 2). The sandstone dominance is caused by a combination of high-energy flow processes that dominate the lower-point-bar environment, while the transition to finer-grained deposits reflects an up-dip waning in flow energy, velocity and stream capacity (Allen, 1970; Hey & Thorne, 1975; Dietrich & Smith, 1983; Dietrich, 1987; Miall, 1996; Bridge, 2003). The unidirectional palaeo-basinward (to the north) orientation of IHS depositional dips along a 1.6 km-long exposure (Fig. 3) is interpreted as due to downstream translation (Ghinassi *et al.*, 2016), a depositional style typical for McMurray Formation (Hubbard *et al.*, 2011; Fustic *et al.*, 2012). The interpreted scale, architecture and depositional processes are analogous to those for subsurface point-bar deposits mapped using dipmeter (Fustic, 2007) and seismic geomorphology (Smith *et al.*, 2009; Hubbard *et al.*, 2011, Labreque *et al.*, 2011, Fustic *et al.*, 2012, Durkin *et al.* 2017 and Martinius *et al.*, 2017).

Although stratigraphy and architecture undoubtedly suggest a large-scale point-bar deposit, a number of unexpected sedimentological observations along the studied exposure (Fig. 3) apparently contradict the point-bar facies model, including occurrences of encased channels within interpreted large-scale point-bar deposit. Channel incisions in point-bars commonly interpreted as syndepositional chute channels (Constantine *et al.*, 2010,

Zinger *et al.* 2011, Ghinassi 2011, Dijk *et al.*, 2012, van Dijk, *et al.*, 2014), rill channels (Choi & Jo, 2015) are commonly shallow and always occur at the top of point-bars. In contrast, the observed encased channels do not occur at the top, are deeply incised and appear emplaced within vertically continuous point-bar deposits. Interpreting their origin in the context of point-bar deposition and the point-bar facies model is not a straightforward process. The objective of this study is to describe and compare channel fills encased in point-bar deposits, interpret the processes leading to their formation and preservation, suggest potential modern analogues and propose implications for subsurface mapping and reservoir developments.

To the best of authors' knowledge, this is the first study that reports partially preserved channel fills as remnants of unsuccessful cut offs and avulsions preserved in the rock record. However, described processes are expected to be common in both modern and ancient meandering river settings and particularly in those that have experienced major floods.

Other peculiar sedimentological observations include: (i) the occurrence of fine-grained strata at the base of the large-scale channel deposit (Fustic *et al.*, 2014); (ii) instead of anticipated large-scale bedforms, the lowermost part (several metres) of the point bars is dominated by thin, decimetre-scale, two-dimensional dunes (planar crossbedding); (iii) the channel base is characterised by an absence of breccia (or other associated channel-lag deposits); (iv) instead of at the base, the cleanest mud-free sand (the best reservoir) occur over 8 m above the channel base; and (v) bidirectional palaeo-currents at the base of the channel *versus* unidirectional ones in the overlying inclined heterolithic strata (IHS).

METHODS

The type section is sub-divided into individual 'bowls' that are exposures separated by natural gullies. Downstream (North) to upstream (South), the bowls are termed A to K, respectively (Fig. 3). Some bowls are further divided into closely spaced exposures, such as bowl B which is subdivided into B1, B2 and B3 (Fig. 3). Eight channel-fill deposits studied are exposed on D, E, I, J2 and K3 to 4 bowls (Fig. 3).

Outcrop data were acquired through detailed two-dimensional outcrop mapping recorded on

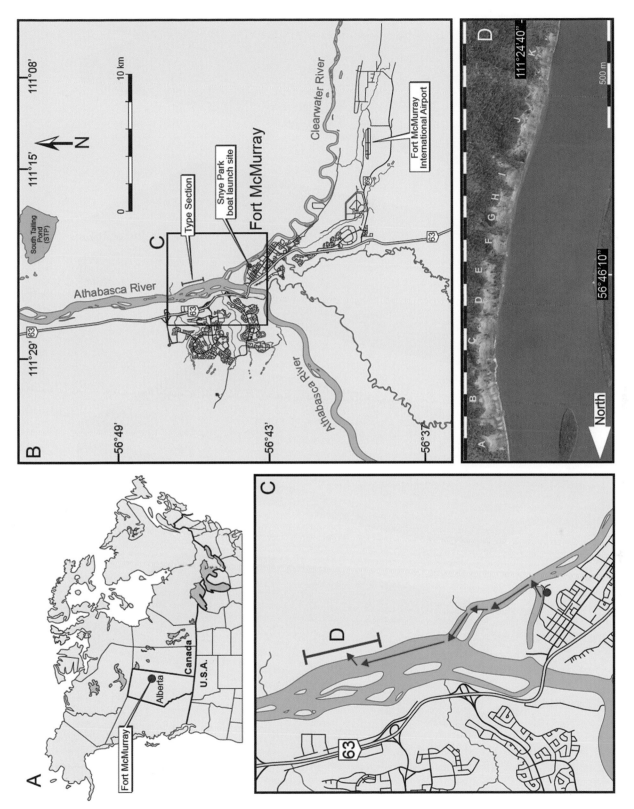

Fig. 1. (A) Location map showing Fort McMurray, Canada. (B) Fort McMurray city map. Annotated are locations of the international airport, boat launch site at Snye Park and location of the outcrop. (C) Zoom in or location of the boat launch site and the McMurray Formation type section outcrop. Red arrows shows boat route. (D) Google Earth image of type section (Image courtesy of Google Earth, ©DigitalGlobe); A to K (yellow letters) is informal geographic subdivision of the outcrop; Studied channel fills are on D to E (channel fill 1), J2 (channel fill 2), I (channel fill 3), K4 (channel fills 4 and 5); and K2 and 3 (channel fills 6 to 8). Key: Ch. – channel; Channel – channel fill.

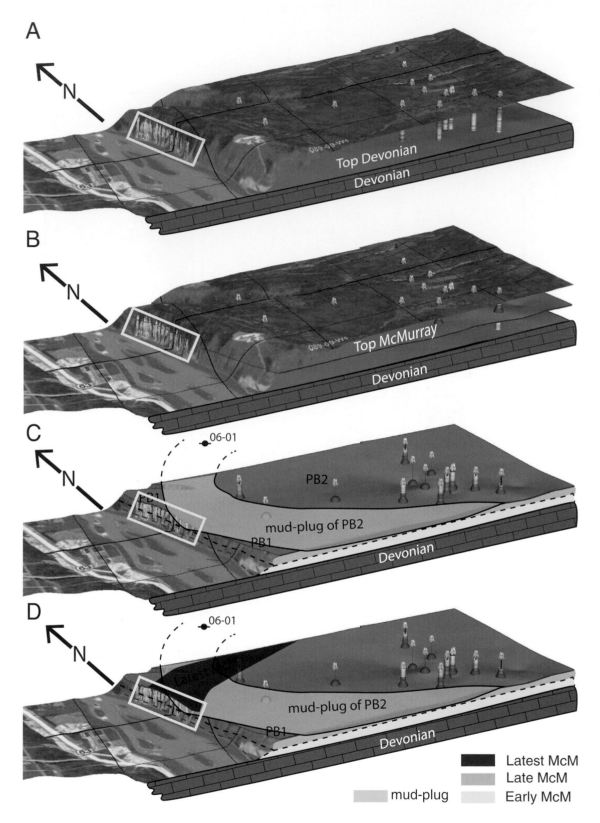

Fig. 2. Three dimensional visualisation of type section outcrop along the right bank of the Athabasca River (flowing northwards), including nearby well locations, stratigraphic subdivision and major architectural elements. (A) Surface based three-dimensional model (vertical exaggeration 5 times) showing the outcrop exposure relative to well locations, topography and underlying Devonian carbonates. Surface image is a combination of (i) Google Maps aerial image, (ii) high-resolution aerial photo and (iii) terrestrial Lidar of outcrop exposure. All three surfaces are snapped on EMD topographic surface. (B) As A, with McMurray Formation top mapped in logs and outcrop. (C) As B, with post-McMurray taken away and McMurray Formation sub-division into 'Early' and 'Late'. NOTE: 'Late McM' is comprised of point-bar 1 [PB1], point-bar 2 [PB2] and mud-plug of PB2. (D) As C with added interpretation of the 'Latest McM' channel incision. Point-bar 2 is younger than (erodes laterally into) PB1. Yellow square shows location of detailed panorama image shown in Fig. 3. Black squares in A and B shows legal sub-division into sections, each section is 402 m by 402 m (0.25 miles by 0.25 miles). Note: studied channel deposits are within interpreted large-scale point-bar 1 [PB1] of 'Late McM'.

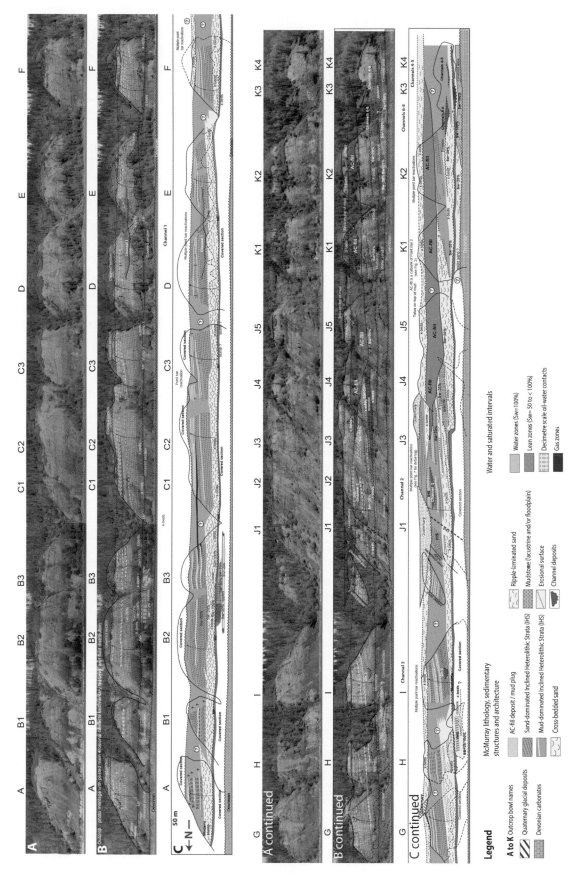

Fig. 3. McMurray Formation type section (A) High-resolution panorama image. (B) and (C) Interpretation of A. Studied channel deposits (coloured orange in B and C) are exposed on D and E, I, J2 and K3 and 4 parts of the outcrop.

high-resolution photographs (created with the use of a Gigapan™; Fig. 3) as well as scaled photo-montages, line-drawings of various geological contacts and features (Fig. 4) and detailed (cm-dm) bed-by-bed logging along some outcrop exposures. Additionally, Pix4D Mapper Software (product of Pix4D Inc.) was used for processing hundreds of georeferenced images taken by unmanned aerial vehicles (UAV) and creating a realistic 3D outcrop model, which was used for visualisation and accurate measurements of individual channel-fill widths, thicknesses and areal extents (Fig. 5A to C).

To understand how channel fills were formed and preserved, detailed outcrop maps of cliff exposures were combined with vertical logs (including lateral extrapolation of geological surfaces over covered parts of the outcrop, i.e. Figs 5B and 6) and cross-cut relationships between channel fills and hosting point-bar deposit were examined in detail. Channel-fill internal architecture and facies descriptions (Table 1) were used to interpret depositional history of each channel fill. At accessible parts of exposure, bed thicknesses and clast-size were estimated using pogo stick (1.5m-long with 0.1m subdivision) and scale (0.2m-long with 0.05m and 0.01m subdivisions); while a 10x magnifying lens and a grain-size chart were used to estimate grain-size and sorting.

Based on the similarities between eight channel-fills with respect to elevation of their base (depth of incision) above the base of the larger-scale point bar deposit, distances from each other and internal architecture and facies, the studied channel fills (named 1 to 8; Fig. 3) are grouped into four distinct types. Specifically, Channel fills 1 and 2 have similar dimensions (Table 1), depth of incision (Fig. 5) and internal architecture (Table 1), Channel fill 3 is isolated, while Channel fills 4 and 5 and 6 to 8 are nested into each other, respectively. Each channel fill and fill type are described in detail and compared. Channel truncation and overlying IHS deposition are discussed in detail in synthesis.

Google-Earth (© DigitalGlobe) and a literature review were used for finding potential modern analogues.

RESULTS

Channel fills 1 and 2: Single course channels with sandy lateral and vertical-accretion fill

Description

Channel fill 1, located on the D and E exposures, is about 16m above the base of the Late McMurray (Figs 3, 5B and 6). This channel fill is 7.5m-thick and 112m-wide and has an aerial exposure of about 640m² (Table 1). It is comprised of 3 distinct architectural elements: (i) breccia 3.73m-thick, up to 107.8m-wide and covering an area of about 650m² (Table 1); (ii) lateral-accretion deposits, comprised of 4 lateral-accretion bed sets, 2.37m-thick, 12.5m-wide and covering an area of about 36.4m² (Table 1); and (iii) two vertical-accretion sets separated by an erosional surface. The first vertical-accretion package, comprised of 4 bed sets, is 12.5m-wide and covers an area of about 36.4m²; and the second vertical-accretion package comprised of seven stacked bed sets is up to 4.26m-thick, 101.25m-wide and covers an area of about 311m² (Table 1). Bioturbation in all beds is absent.

Channel fill 2, located on the J2 exposure about 21m above the base of the Late McMurray (Figs 5C to D and 7), is 51.31m-wide, up to 3.62m-thick and covers an area of about 139m² (Table 1). The Channel-fill is comprised of three architectural elements: (i) cross-beds which are about 1m-thick, up to 19.93m-wide and cover an area of about 13.83m² (Table 1, Fig. 7), (ii) lateral-accretion deposits comprised of 3 major bed-sets: 4m-wide (Table 1, Fig. 7); and (iii) vertical-accretion deposits comprised of 5 vertically stacked bed sets (Fig. 7): 2.88m-thick, 51.31m-wide and covering an area of about 117.72m² (Table 1). Bioturbation in all beds is absent.

Channel fills 1 and 2 have similar, slightly elongated symmetrical concave-up base (Figs 5B to D, 6 and 7) characterised by sharp erosional contacts with underlying IHS. Both Channel fills are truncated by a slightly inclined (northerly dipping) erosional surface that marks the base of the continuous overlying IHS package (Figs 3, 6 and 7).

Differences between Channel fills 1 and 2 include: (i) thickness of 8m, compared to about 4m; (ii) width of 112m, compared to 51m; (iii) presence versus absence of internal erosional surfaces (Fig. 6); and (iv) channel-lag deposit comprised of breccia versus cross-bedded sand.

Interpretation

The bases of channel fills 1 and 2 occur at similar elevation relative to the base of the Late McMurray of 16 and 21m, respectively (Fig. 5D); this suggests that the base of the incision was about 25 and 20m from the bankfull level. Other similarities include (i) the same orientation of lateral-accretion beds, in the opposite (upstream) direction of hosting point-bar (Figs 6 and 7); (ii) limited width of lateral-accretion deposits of 12.5m in Channel-fill 1 and 3.95m in Channel-fill 2 (Figs 6 and 7;

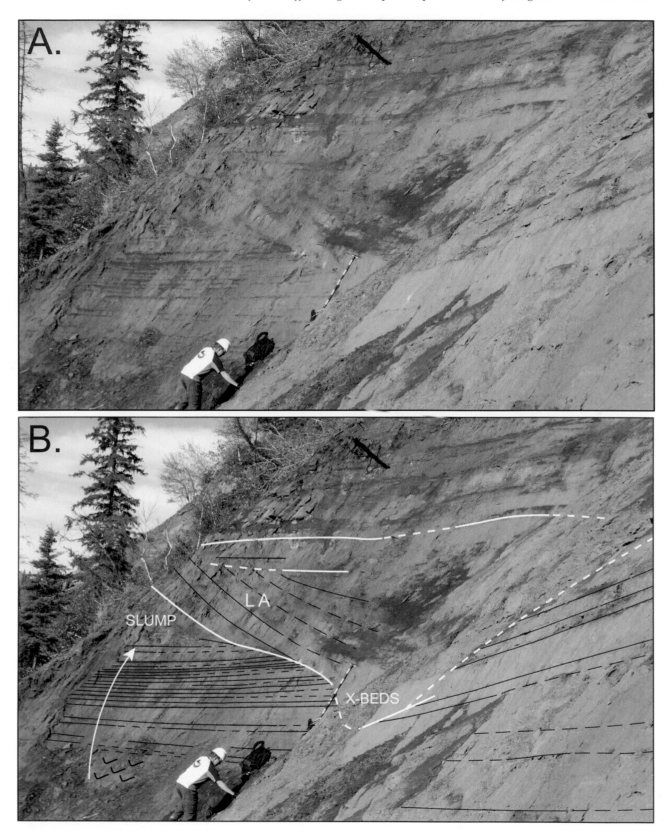

Fig. 4. (A) Close-up view of channel fill 2 exposure on J2 bowl. (B) Interpretation of A showing two-dimensional outcrop mapping, line-drawings of various geological contacts and features and interpretation. Pogo stick for scale is 1.1 m-long with 0.1 m-long black-yellow stripes. Pink dots along channel contact sprayed for easier delineation on photographs taken from distance including those used for gigapan image in Fig. 3. Key: white line – channel; black lines – bedding, LA – lateral-accretion; X-BEDS – cross-beds.

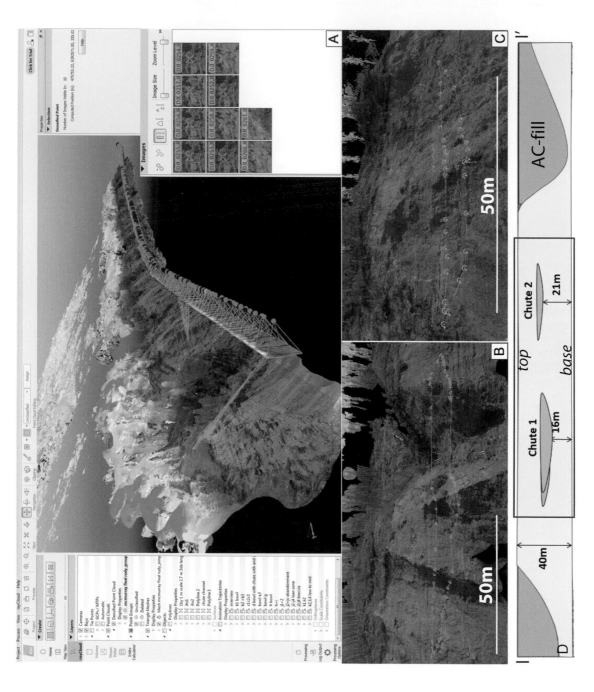

Fig. 5. (A) Three-dimensional outcrop model (Pix4D Mapper) of McMurray Formation type section created from 195 photos at the distance of 70 to 80m from the outcrop. Drone location (green dots) and individual photo coverage (blue rectangles) of each used photo. Blue areas above the outcrop are processing artefacts. Inset shows 10 photographs used to georeference a single point in 3D mesh (multiple merging green lines); (B and C) close-up view of D and E and J2 outcrop exposures with digitised limits of channel fill 1 and 2.; (D) schematic (not to scale) cross-section showing relative positions of channel fill 1 and 2 on the outcrop exposure (red box presenting entire Fig. 3) and within large point-bar (yellow) bounded by abandoned channel fill deposits (AC-fill) illustrated in Fig. 2.

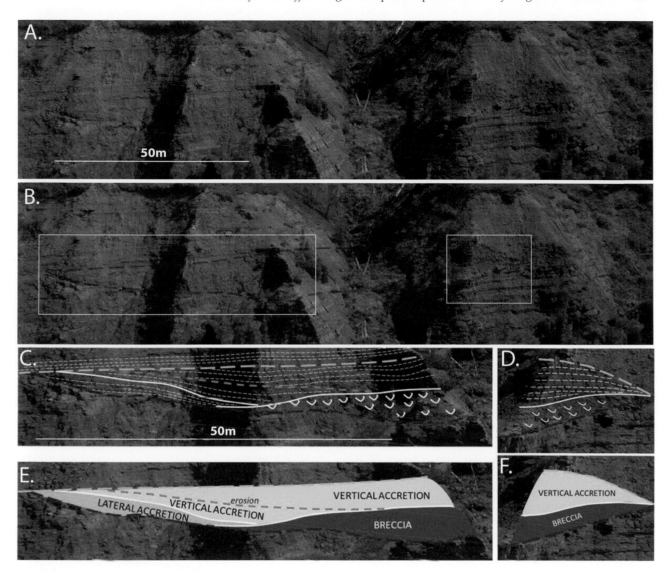

Fig. 6. Channel fill 1 facies and architecture (A) outcrop image (for location see Fig. 3); (B) interpretation of A. Red dashed line shows channel fill 1 limits. White boxes show locations of C to F; (C and D) channel fill 1 facies, (E and F) channel fill 1 architecture. Key: red dashed line – channel fill 1 erosion; green dashed line - channel fill 1 internal erosion; Orange dashed line – truncation surface; White solid lines – facies contacts; White dashed lines – IHS.

Table 1); (iii) significant volumetric predominance of vertical-accretion deposits (Figs 6 and 7; Table 1); and (iv) sand-dominated lithology in both lateral and vertical-accretion deposits (Figs 6 and 7); this suggests the same or very similar processes of incision and infill.

The geometry of observed contacts and geometric relationship of channel fills with the underlying and the overlying sediments allows for interpreting the channel-fill evolution as follows: (i) sharp erosional contacts (Figs 3 and 5 to 7) coupled with significant depth of channel incision and absence of bioturbation interpreted as high-energy subaqueous conditions such as those associated with deep flood scours; channel incision occurred either on the point-bar top or on the point-bar slope; (ii) bed-load transport and deposition of breccia and/or cross-beds at the base of incision during waning flow; (iii) the original course slowly narrows through lateral-accretion (Figs 6 and 7) to accommodate reduced discharge; (iv) as flow further declines, the underfit stream becomes progressively narrower infilling the channel with vertically accreting deposits; (v) complete channel abandonment (plugged with sediments) that prevented the possible return of

Table 1. Channel 1 and 2 dimensions obtained from detailed high-resolution three-dimensional outcrop model.

Channel 1 Entire channel		Thickness(m)	Width(m)	Area(m²)	Base(m)	Top(m)
		7.46	111.91	640.97	112.75	110.3
Facies		**Thickness(m)**	**Width(m)**	**Area(m²)**	**Number of beds**	
	lateral accretion	N/A	12.5	36.4	4	
	breccia	3.73	107.8	221.63	N/A	
	vertical accretion period 1	N/A	8.69	50.81	3	
	period 2	4.26	101.25	310.98	7	

Channel 2 Entire channel		Thickness(m)	Width(m)	Area(m2)	Base(m)	Top(m)
		3.62	51.31	138.7	53.43	51.55
Facies		**Thickness(m)**	**Width(m)**	**Area(m2)**	**Number of beds**	
	lateral accretion	N/A	3.95	7.31	3	
	cross-beds	1.45	19.93	13.83	N/A	
	vertical accretion	2.88	51.31	117.72	5	

streamflow into the channel. The erosional surface between two vertical-accretion sets in Channel fill 1 (Fig. 6) suggests minor channel adjustment and/or shift, possibly caused by a subsequent wave of the same flood. The limited estimated erosion and the fine-grained nature of vertical-accretion set suggests that the second wave was weaker.

The small difference in incision depth of Channel-fills 1 and 2 (Fig. 5D) may suggest these are: (i) two spatially and temporary separate incisions; (ii) two separate contemporaneous incisions; and/or (iii) distal (deeper) and proximal (shallower) parts of the same incision. Fig. 8 illustrates all three scenarios in case incisions occurred from the bar top. An alternative interpretation would envisage a subaqueous incision on the bar slope.

The proposed scenarios are typical for modern chute channel forms associated with meandering river deposits. The fundamental difference is that chute channels occur at the top of the point-bar deposit whilst Channel fills 1 and 2 are truncated by inclined erosional surfaces that mark the base of the overlying IHS (Figs 3, 6 and 7).

Channel fill 3: Single (isolated) channel with clast-dominated fill

Description

Channel fill 3, located on the I bowl exposure, is about 12 m above the base of the Late McMurray (Figs 3, 9 and 10), is 15 m-wide and up to 4 m-thick. It is sharply eroded into decimetre-scale stratified interbeds of fine to medium ripple-laminated sand

and thin (<0.01 m) laterally continuous silt to very-fine grained strata classified as inclined sandstone strata (ISS) and inclined heterolithic strata (IHS; Figs 3, 9 and 10). This channel is filled with very poorly sorted, non-stratified sand and an abundance of scattered, non-imbricated mudstone and heterolithic and sandstone clasts (Fig. 10D to E). Mudstone clasts are commonly grey and range in size from millimetre scale flakes to 0.05 m-thick angular to sub-angular clasts. Heterolithic clasts are commonly light grey or 'rusty' with occasional thin parallel black (due to bitumen saturation) sandstone interbeds. Clast-size ranges from 0.05 to 0.15 m (Fig. 10D to E). Bioturbation is absent. The channel is truncated and overlain by IHS (Figs 9 and 10).

Interpretation

Considering that the hosting point-bar is about 40 m-thick and that the base of Channel fill 3 is about 12 m above the point-bar base, it can be inferred that the base of Channel fill 3 was about 25 to 30 m below bankfull level (Figs 3, 9 and 10).

The sharp erosional contact (Figs 9 to 10), the absence of bioturbation and roots, the heterogeneous nature of the channel fill, comprising of a variety of shapes and sizes of non-imbricated clasts and structureless sand, suggest that incision was followed by confined subaqueous debris transport and rapid deposition. The lithology of the clasts is similar to the hosting meandering-belt facies, suggesting that sediments are locally sourced. The sediment size variation from flakes to the decimetre scale clasts suggests that at the

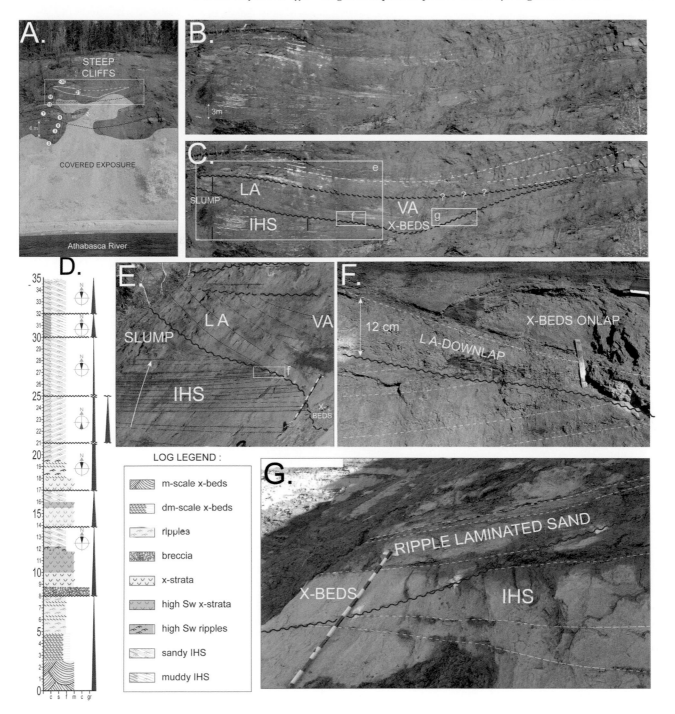

Fig. 7. (A) Bowl B2 (see location in Fig. 3); small white box shows location of channel fill 2. (B and C) Close-up view of channel fill 2. (D) Detailed outcrop log. Channel fill 1 is between 21 and 25 m. Note southward oriented lateral-accretion dips in channel fill 1, versus northward in underlying and overlying IHS. (E) Close-up, oblique view of the left side of channel fill 1. (F and G) Close-up view of facies changes and depositional dip orientation along erosional contact at the base of channel fill 1. Key: LA – lateral-accretion; VA – vertical-accretion; x-beds – cross-beds; HIS – inclined heterolithic strata (same as lateral-accretion, for clarity term is applied to hosting bar-facies only).

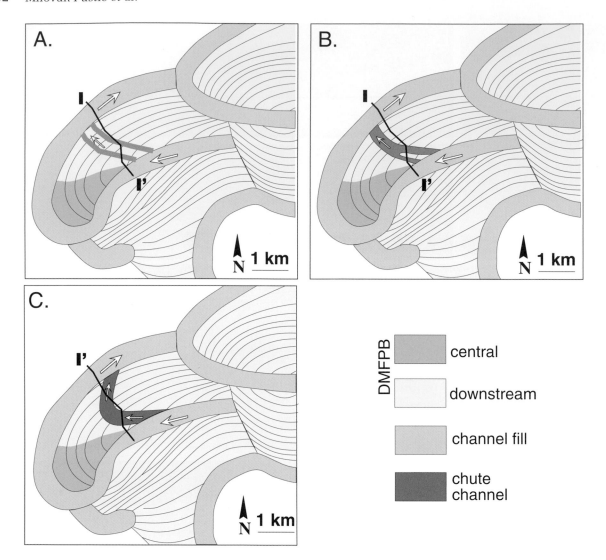

Fig. 8. Three potential scenarios for channel fills 1 and 2 incision into large-scale point-bar deposit: (A) two separate and time independent incisions, (B) two separate contemporaneous channel incisions potentially merging downstream; and (C) parts of the same channel incision. I-I' hypothetical cross-section illustrated in 5D. Key: DMFPB – Downstream Migrating Fluvial Point-bar. AC-fill – abandoned channel fill; central and downstream parts of point-bar deposit (modified after Ghinassi *et al.*, 2016). Note: all three scenarios illustrate period of channel incision. Subsequent meander-bend migration and point-bar reactivation may erode upper parts of incised channels truncating the top of their infill.

time of the event sediments were mostly unconsolidated or semi-consolidated. The rusty colour of some clasts is probably early diagenetic siderite.

Such extremely deep incision into the large-scale point-bar deposit or its slope suggests that incised channel attempted to completely cut off the meander. It was probably caused by a multi-year flow characterised by increasing discharge and progressive channel deepening that at the end was probably reshaped by a major flood and infilled by massflow deposits that caused rapid channel plugging that further prevented the possible re-establishment of streamflow. An alterna-

tive interpretation is that incision was caused by a high magnitude flood through weakly consolidated sediments of the hosting point-bar deposit.

Channel fills 4 and 5: Multiple lateral cut-and-fill channels with aggradational stratified heterolithic fill

Description

Channel fills 4 and 5, located on the K4 bowl exposure, are only 6 to 7 m above the base of the Late McMurray (Figs 3 and 11). Channel fill 4 is

Fig. 9. Channel fill 3 (coloured orange) on bowl 'I' (see location in Fig. 3). Key: McM – McMurray, IHS – inclined heterolithic strata; ISS – inclined sandstone strata; white solid lines – facies contacts; white dashed lines – IHS; red dashed line – base of Late McM; black dashed line – estimated size of incised channel before truncation; curved white line with arrow – fining upward pattern showing 40 m vertical continuity of a single large-scale point-bar; black rectangle – Fig. 10 location. Scale: short vertical pink lines are 1 m apart (at each elevation only the first '0' and the last '10' are labelled).

13.5 m-wide, Channel fill 5 is 12 m-wide and both are up to 2 m-thick. Both channel fills are sharply eroded into fine-grained, cross-bedded sand and IHS (Fig. 11A and B). These channel fills are truncated by a single undulated erosional surface that marks the base of the overlying IHS (Figs 3 and 11A and B). Channel fill 4 (on the left) is partially incised into channel fill 5 (to the right; Fig. 11E to F). Both channel fills have elongated symmetrical concave-up bases (Fig. 11A to C) and are comprised of clean, cross-stratified sand at the base and stratified heterolithic centimetre-scale interbeds of bitumen-saturated (which appear

black due to oil saturation) fine-grained to medium-grained apparently structureless sand and parallel inter-laminated clay-silt and very fine-grained sand (which appear light grey due to lack of oil saturation) strata (Fig. 11C and D). Stratified interbeds follow the concave-up channel bases (Fig. 11C). Bioturbation is absent.

Interpretation

A 30 to 35 m incision is inferred (Fig. 3) suggesting that Channel fills 4 and 5 were almost as deep as the main 40 m-deep channel. This implies that

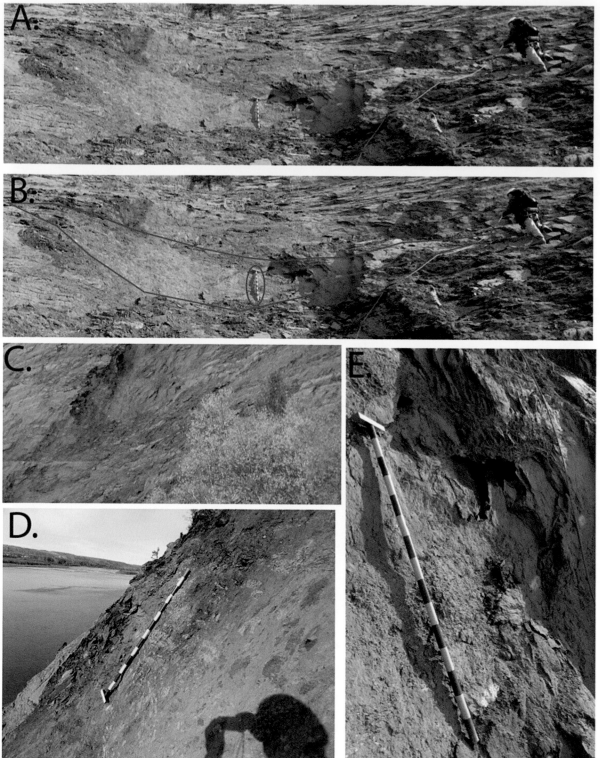

Fig. 10. Channel fill 3. See location in Figs 3 and 9. (A to C) Close-up views and channel delineation with pink dots on outcrop and digital line (red in B) on photo; (D and E) Channel fill comprised of poorly sorted sand (appear black due to oil saturation) and scattered grey and rusty decimetre-scale non-imbricated clasts. Scale: Pogo stick in A, B, D and E) is 1.5 m-long with 0.1 m-long black-yellow stripes.

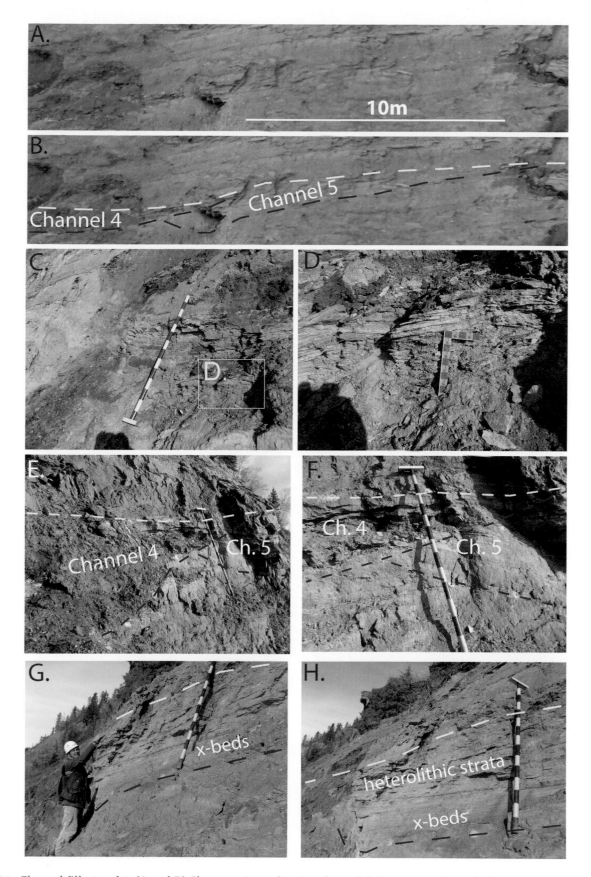

Fig. 11. Channel fills 4 and 5. (A and B) Close-up views showing channel delineation with pink dots on outcrop. (A) and dashed red digital line (B) on photo; (C and D) Close-up view of heterolithic channel 4 fill; yellow box shows location of D.; Scale: 0.1 m by 0.05 m with 0.01 m-long magenta-black stripes. (E and F) Channel fills 4 and 5 cross-cutting. (G and H) Channel fill 5 facies – cross-beds and heterolithic strata. Key: pogo stick in C and E to H is 1.5 m-long with 0.1 m-long black-yellow stripes; x-beds – cross-beds; Ch. – Channel; red dashed line – base of channels 4 and 5 incision; yellow dashed line – erosional truncation line and base of overlying IHS.

Channel fills 4 and/or 5 may have had the potential to divert the entire river flow and become the main channel. The narrow width of incision suggests a stable channel course (without lateral migration) indicative of deepening induced by slope changes and/or increasing discharge. The switch from incision to deposition occurred when flow strength lost its erosive capacity. The absence of breccia and coarse sand, fining upwards trend and the heterolithic nature of fill suggest the overall moderate, yet cyclically varying, strength of decelerating flows.

Although Channel fills 4 and 5 appear as separate entities, the erosional contact with cross-beds and IHS of the Late McMurray, coupled with their cross-cutting relationship and similar widths and characteristics of sediment fills, could indicate two stages of cut-and-fill by the same channel. Firstly, Channel 5 was cut and partially filled; then, with an estimated relative lateral offset of about half-channel width (15 m), the Channel 4 was cut and filled. The offset was probably a result of channel adjustment (i.e. rapid erosion of cutbank or bar slope) to change in discharge and/or channel slope. Strikingly similar channel morphologies and fills suggest the same or very similar depositional processes. Following the maximum incision, during diminishing flow strength, bedload-transported sediments were deposited as cross-beds (Fig. 11G and H). As flow energy, velocity and stream capacity continued to weaken, thin fine-grained strata started blanketing the channel bottom. Alternating heterolithic centimetre scale interbedded sand and fine-grained strata (Fig. 11C, D, G and H) could suggest cyclic variations in flow intensity, which is more difficult to explain by fluvial processes alone. A possible mechanism may include tidal modulation of fluvial discharge. Within the fluvial-tidal transition tidal floods would cause fluvial flow retardation and deposition of fine-grained sediments, while the ebb tides is expected to cause fluvial flow acceleration and thus contribute to the deposition of coarser sediments (sand). Together, the flood and ebb tide may collectively contribute to sediment cyclicity in the upper parts of the channel fills (Fig. 11). Although tidal modulation is not observed in other studied channels, the abundance of brackish-water indicators including skolithos and cylindrichnus within the IHS (Hein *et al.*, 2001) in the upper parts and fluid-mud deposits at the lower parts of the hosting point-bar deposit (Fustic *et al.*, 2014)

supports temporally variable, probably seasonal tidal influence.

Vertical aggradation of channel-fill deposits suggests an overall weakening of the flow strength (Fig. 11).

The weight of overlying sediments caused differential compaction of clay-silt and fine-grained sand strata that dominated central parts of channel fills (Fig. 11A to C).

Channel fills 6 to 8: Multiple vertically stacked cut-and-fill with vertically aggrading breccia and cross-bed deposits

Description

Channel fills 6 to 8, located on K3 and K4 exposure, are about 10 m above the base of the Late McMurray (Figs 3 and 12A). The width of exposure is about 100 m and the thickness is up to 6 m (Figs 3 and 12A to C). All three channel fills have well-exposed concave-up symmetrical bases and sharp erosional contacts with underlying cross-bedded sand and/or IHS (Fig. 12A to C). Channel fill 7 is partially incised into Channel fill 6; Channel fill 8 is partially incised into both Channel fills 6 and 7 (Fig. 12C). The very bottom of each channel is filled by up to 1 m-thick, very poorly sorted, medium to very coarse-grained sand and up to 0.2 m-thick clasts (Fig. 12E to F). Overall, the occurrence and frequency of large clasts is the highest in Channel fill 6 and smaller in Channel fills 7 and 8. The upper parts of each channel fill are comprised of poorly sorted, bitumen-saturated (which appear black), medium to very coarse-grained cross-bedded sand (Fig. 12B to E). Cross-beds are up to 0.5 m-thick in the middle of the channel fills and are systematically thinning upwards and towards the channel edges where thickness is reduced to 0.1 to 0.2 m. Channel fills are truncated by a sharp inclined erosional surface (Figs 3 and 12A to D).

Interpretation

The inferred 30-m incision from the top of the hosting point-bar (Fig. 3) suggests that Channel fills 6 to 8 were almost as deep as the main 40 m-deep channel. This implies that Channel fills 6 to 8 may have had the potential to divert the entire river flow and become the main channel. The sharp erosional contacts of channel bases and vertical stacking (Fig. 12C) indicate three

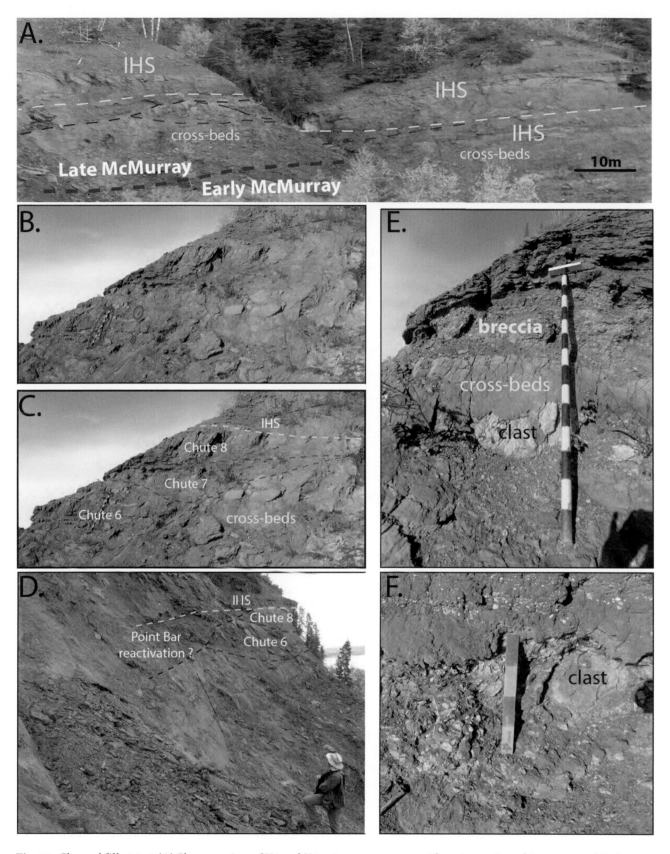

Fig. 12. Channel fills 6 to 8 (A) Close-up view of K3 and K4 outcrop exposures with main stratigraphic contacts; (B) close-up view of channel fills 6 to 8 (West-East facing exposure from gully between K3 and K4); (C) interpretation of B; (D) Channel fills 6 to 8 (North South facing exposure of K3); (E) close-up view of channel fill 7; Note large white coloured blocks; (F) close-up view of clasts (breccia) in channel fill 7. Key: pogo stick in B to E is 1.5 m-long with 0.1 m-long black-yellow stripes and magenta-black stripes in D; x-beds – cross-beds; thick red dashed line – base of Later McMurray; red dashed line – base of channel fills 6 to 8; yellow dashed line – erosional truncation surface. Scale in F: 0.2 m with 0.05 m-long magenta-black stripes.

subsequent episodes of cut-and-fill occurring at the same location. Strikingly similar channel fills suggests the same or very similar depositional processes. Each cut-and-fill cycle included an incision (cut) during the highest discharge, followed by confined subaqueous high-energy flow and deposition of poorly sorted breccia during the initial waning stage. As flow velocity further diminished, cross-bedded sands were deposited on the top of breccias. Cross-bed aggradation and a decrease in the size of bedforms suggest an overall continued decrease in flow energy, velocity and stream capacity.

Slightly shallower incisions of Channel fills 7 and 8, coupled with overall decrease in breccia clasts and cross-beds in each subsequent channel, suggest that the formative flood was characterised by at least two subsequent flood waves. Multiple flood waves are common in rivers associated with large drainage areas, as runoff from individual upstream tributaries arrives (Bridge, 2003).

SYNTHESIS / DISCUSSION

Evolution of encased channel deposits

The described evolution of each of eight channel fills encased in point-bar deposit includes three common stages:

a) Channel incision, caused by major flood event and/or by deepening over time due to increased discharge (Fig. 13B) that in some cases may have been modified and re-shaped by flood events. The absence of the upper part of channel fills does not allow to distinguish whether channels were incised in the bar top such as chute channels (Fig. 13B), or scoured on the point-bar slope.
b) Channel filling with sediments which records a reduction in discharge that may cause complete plugging with sediments and abandonment (Fig. 13C).
c) Erosion of upper parts of channel fills by point-bar reactivation (Fig. 13D). An erosional surface at the top of each channel fill marks the base of 15 (above Channel fill 2) to 25 to 30 (above Channel-fill 3) and potentially up to 35 (above Channel fills 4 and 5) metres of vertically continuous IHS (Fig. 3). These commonly gently inclined erosional surfaces, interpreted as point-bar reactivation surfaces (Fig. 13D), point to frequent relocation events of meandering

channels (Strobl *et al.*, 1997; Martinius *et al.*, 2017). Fustic *et al.* (2013) proposed that some erosional surfaces might be a result of chute-channel cut offs and/or channel avulsion. If an active or abandoned chute-channel bear record of deeper incision than the elevation of a reactivation surface, lower parts of the chute channel are expected to be preserved (Fig. 13D). This is consistent with the notion that the deepest scours have the highest preservation potential as they are below point-bar reactivation surface. New channels continue to migrate and deposit accretion beds (Fig. 13E) until their full abandonment (Fig. 13F).

Described processes are poorly documented in modern and ancient settings. In contrast to many documented cut offs including those seen in the modern Obra (Słowik, 2016) and Rhine (Toonen *et al.*, 2012) rivers where cut off channels are not affected by subsequent point-bar reorientation and are overall filled by finer-grained sediments, the upper parts of the eight channel fills described here are eroded and preserved parts of channel fills are comprised of high-energy deposits. The chute channel cut off and its evolution into the main channel of Allier River at Chatel-de-Lys in France is identified as a more accurate potential modern analogue (Fig. 14; Van Dijk *et al.* 2014). Proposed channel morphology and described channel-fill architecture are analogues to some ancient and/or modern examples. For example lateral accretion within Channel fills 1 and 2 resemble similar processes seen in deposits interpreted as chutes of the Cemalettin Formation of Eocene-Oligocene sedimentary succession in the Boyabat Basin, Turkey (Ghinassi 2011; Ghinassi *et al.*, 2014) and may be inferred from sinuosity of the chute channel on the Missouri River (Fig. 15) and migrated chute channels in a river in Ural Federal District in Russia (Fig. 16). Incised channels migration by cut-and-fill interpreted in Channel fill 1, between Channel fills 4 and 5, and between Channel fills 6 to 8, are similar in nature to those described in Holocene chute channel deposits associated with high-sinuosity river deposition of the Firenze plain, Tuscany, Italy (Ghinassi, 2011).

Evidence for extremely deep incision into the pre-existing large-scale point-bar deposit suggests that a chute channel attempted to cut off the meander. It was possibly caused by temporal increase in discharge and associated channel

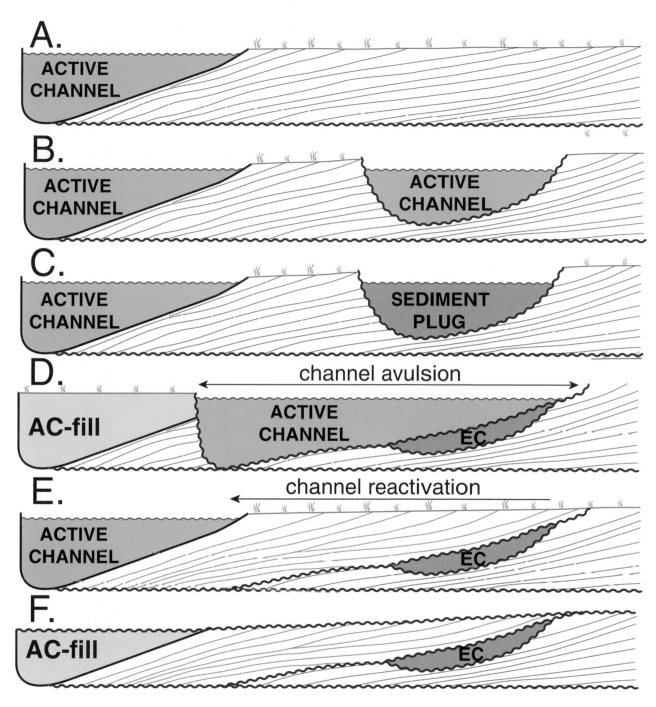

Fig. 13. A schematic diagram of point-bar cross-section, representing key geologic events interpreted from eight channel fills and their relationship with underlying and overlying sediments in the McMurray Formation type section. (A) Large-scale point-bar with typical accretion beds, (B) very deep (due to extreme flood or increasing discharge over long period of time) channel incision into existing point-bar, (C) channel plugging by sediments (due to reduced discharge), (D) point-bar reactivation and/or meander avulsion and erosion of the top part of chute channel by younger sediments, (E) re-establishment of the main river course; and (F) abandonment of meander belt (rock record). Key: red wavy line – major erosional surfaces; black lines – lateral-accretion surfaces; blue lines – lateral-accretion surfaces after point-bar reactivation.

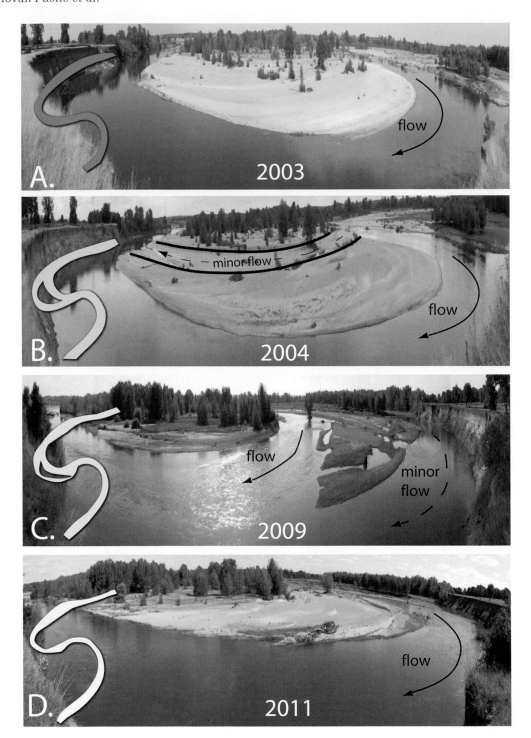

Fig. 14. Oblique images of meander bend at Chateau de Lys showing initiation of a chute cut off and displacement of the chute channel. Taken during discharges less than $50\,m^3\,s^{-1}$ in (A) August 2003; (B) July 2004; (C) September 2009 and (D) September 2011. Figure modified after van Dijk *et al.* (2014).

deepening as documented in modern meandering rivers including Sacramento River (USA) upstream of Colusa, Mackey Bend of the Wabash River (USA) and river Ain (France) near Mollon and near Martinaz (van Denderen *et al.*, 2017). In the case of Channel fill 3 and Channel fills 6 to 8, massflow and breccia deposits suggest that a large flood strongly reshaped a mature incised channel

Fig. 15. Google Earth image of Missouri River stretch and its sinuous chute channel (39° 5′ N, 92° 55′ W).

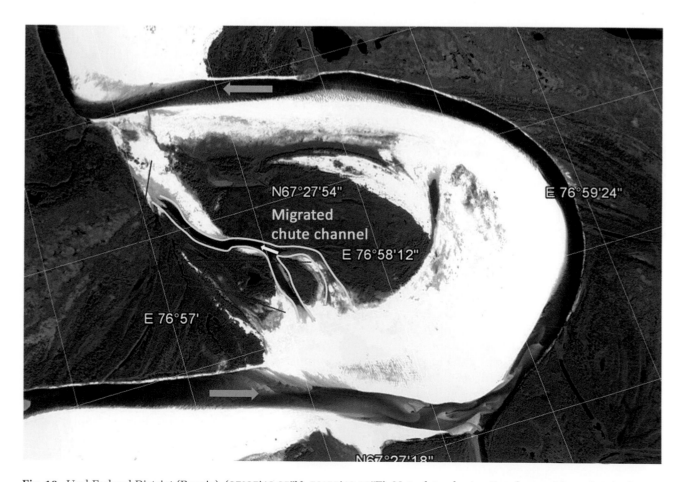

Fig. 16. Ural Federal District (Russia). (67°27′40.85″N, 76°57′46.12″E). Note: lateral migration forms side and point-bars.

and plugged it through collapses of its own banks. In all cases, decelerated flow caused rapid channel plugging with sediments that further prevented the possible channel reactivation.

An alternative interpretation for extremely deep incision is that incision was caused by a high-magnitude flood that might be comparable to catastrophic Quaternary Mississippi floods caused by rapid, yet recurrent discharges of glacial meltwater that have caused numerous avulsions, cut offs and meander straightening along the Mississippi River (Saucier, 1994). Another possibility is that incision was caused by high-magnitude to catastrophic floods characteristic for rivers affected by monsoonal precipitation (Plink-Björklund, 2015). Potentially analogue avulsion, cut offs and meander straightening has been observed on the Situang River, Myanmar, following extremely large storms, that drove major flooding (Stuart 1932).

Whilst strong decadal-scale seasonality and climatic cyclicity have been interpreted from other McMurray Formation outcrops (Jablonski *et al.*, 2016) and subsurface data (Labreque *et al.*, 2011) and whilst channel readjustment has been interpreted as a product of high-energy flood events (Gingras & Leckie, 2017), causes and frequencies of high-magnitude floods in McMurray palaeo-rivers are poorly understood and are the subject of ongoing research.

If deep incisions coupled by frequent point-bar reorientation are caused by high-magnitude floods, then observed shift in frequency of incision events along the type section may indicate climate and/or other allogenic changes. Specifically, the studied channels appear to be more frequent and overall more deeply incised along the upstream than downstream parts of the described 1.6 km-long point-bar exposure (Fig. 3). Specifically, downstream bowls A to H show no encased channel deposits with the exception of Channel fill 1 exposed on D and E, while all other incised channel-fills occur on upstream bowls I to K4 (Fig. 3). Since palaeo-geographic reconstructions suggests that K4 is the oldest and A is the youngest part of a downstream-translating large-scale point-bar deposit (Ghinassi *et al.*, 2016), it can be inferred that the frequency of incisions was higher during the initial stages of point-bar development. This is also evident by the higher number of identified reactivation surfaces along I to K4 than along A to H (i.e. four are mapped just on J2 and only two on H bowl; Figs 3 and 7D). This possibly shows that the studied point bar experienced dramatic and frequent episodes of incisions and reactivation during the early stages (K4 to I) and more stable relatively quiet development during the later stages (H to A) of channel development. Despite described differences, relatively small variation of the thickness of the hosting point-bar deposit (averaging 35 m) suggests relatively stable bankfull discharge throughout the evolution of the point bar.

The absence of bioturbation and roots along channel incisions and within channell-fill deposits suggests that both incision and deposition most likely took place under high-energy and/or in deep subaqueous conditions. Additionally, it is considered unlikely that the river level may have dropped more than a half of its bankfull stage to allow for the development of subaerial exposure.

The observation of cut-and-fill, interpreted in all except Channel fills 2 and 3, suggests frequent discharge fluctuations resulting in shifts between erosional (cut) and depositional (fill) stream capacity. Short distances between cut-and-fill locations within Channel fill 1 and between Channel fills 4 and 5 as well as between 6, 7 and 8, coupled with interpreted rapid sedimentation in all cases, suggests short time (perhaps no more than several days) for subsequent erosional events. These time intervals are typical for multiple flood waves. In large drainage systems, such as that of the continental-scale palaeo-McMurray (Blum & Pecha, 2014; Benyon *et al.*, 2014), most floods consist of several waves in succession, each causing additional water-level rise and discharge increase, as runoff from each individual upstream tributary propagates downstream (Bridge, 2003).

IMPLICATIONS FOR SUBSURFACE MAPPING AND RESERVOIR DEVELOPMENTS

If encountered in subsurface cores, channel deposits of the type described here may contribute to misinterpreting the depositional architecture as a series of stacked small-scale channel deposits, rather than a single vertically continuous large-scale point-bar deposit.

By creating pseudo-wells with gamma ray and dipmeter logs along outcrop exposures (Fig. 17), the authors propose several criteria for recognising described channel-fill deposits from well data in the subsurface: (i) continuity of dip orientation of lateral-accretion deposits below and above

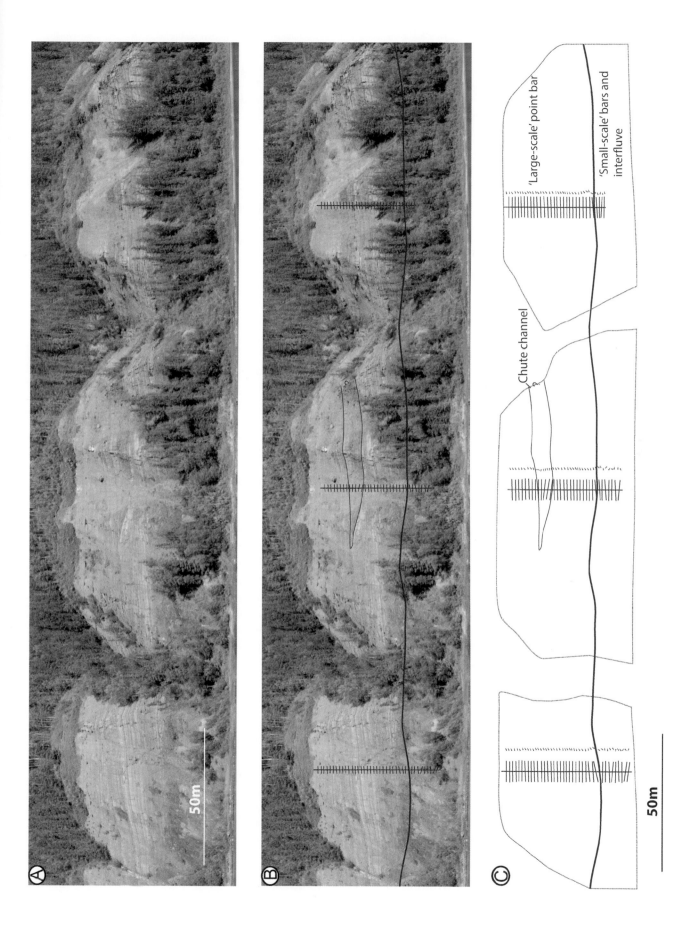

Fig. 17. A chute channel embodied in a large-scale point-bar deposit at the central part of McMurray Formation type section (~2 km north from the confluence of Clearwater and Athabasca; *sensu* Burton, 2011). (A) image of the outcrop, (B) same as *A* with 3 hypothetical wells and the interpreted extent of the chute channel and base of large-scale point-bar as well as hand-picked dips along the lateral-accretion surfaces; (C) stick plot and tad pole along the 3 hypothetical wells.

described channel fills; (ii) absence of clean sand at the same elevation as in lateral-accretion packages in surrounding wells; and (iii) sharp, rather than transitional, contacts between channel-fill sand and overlying lateral-accretion deposits. The presented case study demonstrates the applicability of the proposed recognition criteria (Fig. 17).

The upper parts of point-bar deposits are commonly characterised by decreased reservoir quality due to finer grained sediments including sand-mud interbeds deposited in tidally influenced environments (Thomas *et al.*, 1987: Sisulak & Dashtgard, 2012). Thus, when channels erode portions of the upper point-bar and if their infills contain good quality sand, the overall net-to-gross ratio increases. Additionally, newly deposited sand may contribute to sandbody connectivity in meandering fluvial (and tidal-fluvial) deposits. This is demonstrated in a subsurface study in one of the oil-sands leases in the southern Athabasca region (Fig. 18; modified after Fustic *et al.*, 2013).

As the described channel deposits are formed syndepositionally with the associated point-bar deposits, their preservation in the subsurface complicates point-bar geometry and reservoir architecture. While their recognition in the subsurface is difficult, these deposits may, at least locally, improve reservoir quality and connectivity of sand geobodies. When encountered in the subsurface, their interpretation has to be made with caution because of their specific narrow width (Figs 14 to 18).

SUMMARY AND CONCLUSIONS

The deep incisions on point-bar tops (i.e. sub-aerial floodplain deposits; Fig. 13), and/or subaqueous scours on point-bar slopes, suggest substantial erosive capacity of stream flow indicative of high-magnitude flood events and/or prolonged bifurcation of stream flow typical of rivers undergoing annual discharge fluctuations, as recorded in progressive channel deepening.

Whilst all eight channels have been infilled with sediments during waning and/or overall diminishing flow velocity, variety in the described lithologies (i.e. clasts and grain-size) and in the architecture of fills suggests differences in flow velocities and their dynamics at different locations. Interpreted processes include debris flow (Channel fill 3), cut-and-fill (Channel fills 1, 4, 5 and 6 to 8), tidal modulation (Channel fills 4 to 5),

channel migration by lateral-accretion and channel infilling by vertical accretion (Channel fills 1 and 2). Differences in lithologies and architecture of the eight examined deposits allow for grouping them into four distinct depositional types:

i. Channel fills 1 and 2: Single course channels with sandy lateral- and vertical-accretion deposits.
ii. Channel fill 3: Single (isolated) channel with clast-dominated fill interpreted as subaqueous debris flow suggesting rapid deposition.
iii. Channel fills 4 and 5: Multiple lateral cut-and-fill channels with aggradational stratified heterolithic fill interpreted as tidally modulated deposition.
iv. Channel fills 6 to 8: Multiple vertically stacked cut-and-fill with vertically aggrading breccia and cross-beds suggesting rapid deposition.

All eight channels are interpreted as only minute keels of significantly larger channel incisions that developed as un-successful channel cut-offs and/or avulsions. If any of these channels had happened to evolve to full size, they would have diverted the main flow, become as big as the main channel and eventually developed into the main channel that would have continued migration and erosion including cannibalisation of early 1 and 2 stages (Fig. 13A and B) of their own development.

In fact, each of the described truncation surfaces on the top of encased channel deposits might represent the advanced stage of channel incisions that happened to evolve into the full channel and create cut off. Deeper incision of Channel fills 3 to 8 were more probable to emerge into full cut-offs than shallower incised Channel-fills 1 and 2. The number of interpreted unsuccessful cut offs coupled with a number of mapped and/or inferred reactivation surfaces along 1.6 km-long outcrop exposure suggests the Late McMurray Formation at the type section experienced dramatic history of erosion and deposition. It serves as an ideal analogue for processes inferred from subsurface datasets described by Martinius *et al.* (2017).

The proposed multi-stage development (Fig. 13) is a novel concept, subject to modifications and improvements. Thus, it is expected to stimulate research in analogous modern and ancient settings as well as reproduction in flume and flow simulation studies. In particular, flume experiments focused on understanding the relation between river hydrodynamics and morphodynamics on

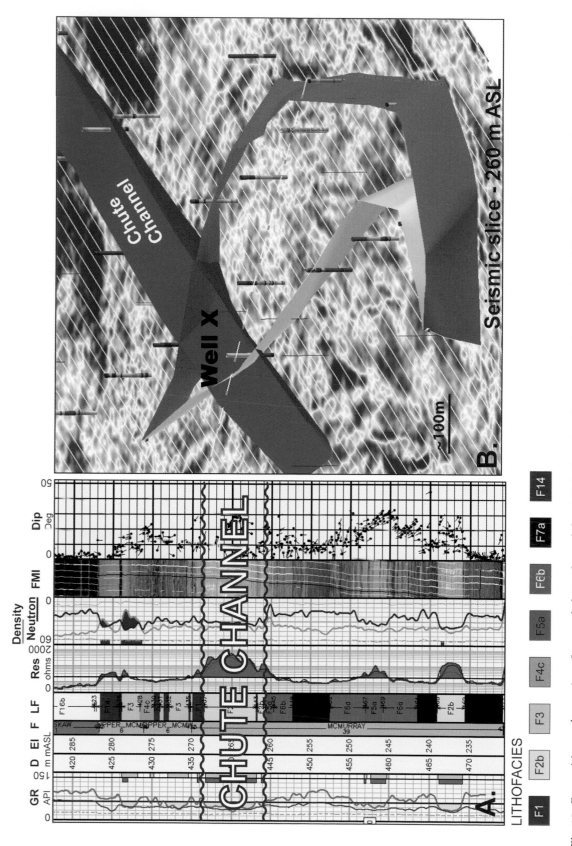

Fig. 18. Recognition and mapping of encased chute-channel deposits in the subsurface, example from an oil-sands lease, Alberta, Canada. (A) A composite log of a selected well showing 1-D recognition and interpretation of a chute channel. The base of the interpreted channel is at 444 m and the channel fill is encased between the overlying and underlying muddy and mixed lithology inclined heterolithic strata. (B) recognition and interpretation in a 3-D spatial model using high-resolution seismic and surrounding well data. Well 'X' is shown in A. (modified after Fustic *et al.*, 2013). Key: orange and grey surfaces show reactivation surfaces within the single point-bar and the red linear feature is the chute channel at the top of the point-bar. GR – gamma ray log; D – Depth; El – elevation (above sea-level); F – formation; LF – lithofacies; Res – resistivity log; FMI – formation imaging log, Dip – dipmeter. Lithofacies: F1 – shale clast breccia, F2 – cross-bedded sand, F3 – compound cross-bedded sand; F4c – sand dominated inlined heterolithic strata with 10 to 30% mudstone beds; F5a – subequal sand-mud inlined heterolithic strata, bioturbated; F6b – mud dominated inlined heterolithic strata, bioturbated; and F7A – laminated mudstone; F14 – rippled to hummocky cross-stratified sand.

evolution of unsuccessful cut offs and conditions leading to a partial preservation of channel fills encased within a point-bar deposit may build on recent works on channel belt architecture and chute cutoffs by van Dijk *et al.* (2012; 2014) and van de Lageweg *et al.* (2013; 2016).

ACKNOWLEDGEMENTS

The authors sincerely thank Automated Aeronautics and VGeoTours for donating original photos of McMurray Formation to construct 3D models, Pix4D Inc. for providing academic license and technical support, Dr. Rudi Meyer for feedback and guidance for interpreting Channel fills 1 and 2 and many colleagues and friends who visited the outcrop with us (us being Fustic and Strobl) and provided useful feedback as well as encouragement to publish this interpretations. We thank Shin Ma for drafting some of figures and Thomas Jerome for help with the three-dimensional visualisation shown in Fig. 2. We particularly thank Bryce Jablonski and Allard Martinius for collecting some early field data with us and for contributing to preliminary ideas that were presented. We sincerely acknowledge support from SID2016 project of Padova University entitled 'From channels to rock record: morphodynamic evolution of tidal meanders and related sedimentary products' (prot. BIRD168939; PI. M. Ghinassi). Dr. Luca Colombera (guest editor) and two anonymous reviewers are most gratefully acknowledged for their insightful reviews.

REFERENCES

Allen, J.R. (1970) *Physical processes of sedimentation.* American Elsevier Pub. Co.

Bell, R. (1885) Report on part of the basin of the Athabasca River, North West Territory. *Geological and Natural History Survey and Museum of Canada Report of Progress for 1882–1884*, p: CC1–CC37.

Benyon, C., Leier, A., Leckie, D.A., Webb, A., Hubbard, S.M. and Gehrels, G. (2014) Provenance of the Cretaceous Athabasca Oil Sands, Canada: Implications for continental-scale sediment transport. *J. Sed. Res.*, **84**, 136–143.

Blum, M. and Pecha, M. (2014) Mid-Cretaceous to Paleocene North American drainage reorganization from detrital zircons. *Geology*, **42**, 607–610.

Bridge, J. (2003) Rivers and Floodplains. Blackwell Publishing, Malden, Mass., 491pp.

Burton, D. (2011) Geologically-based permeability anisotropy estimates for tidally-influenced reservoir analogs using lidar-derived, quantitative shale character data.

Carrigy, M. (1959) *Geology of the Mcmurray Formation-Part III General Geology of the McMurray Area.* Research Council of Alberta, Memoir 1 (available from the Alberta Geological Survey).

Choi, K. and Jo, J.H. (2015) Morphodynamics of tidal channels in the open coast macrotidal flat, southern Ganghwa Island in Gyeonggi Bay, west coast of Korea. *J. Sed. Res.*, **85**, 582–595.

Constantine, J.A., McLean, S.R. and Dunne, T. (2010) A mechanism of chute cutoff along large meandering rivers with uniform floodplain topography. *Geol. Soc. Am. Bull.*, **122**, 855–869.

Dietrich, W.E. (1987) Mechanics of flow and sediment transport in river bends. *River Channels*, **134**, 179–227.

Dietrich, W.E. and J.D. Smith (1983) Influence of the point bar on flow through curved channels. *Water Resour. Res.*, **19**, 1173–1192.

Durkin, P. (2016) The Evolution of Fluvial Meander Belts and Their Product in the Rock Record, University of Calgary, unpublished PhD thesis 2016.

Durkin, P.R., Boyd, R.L., Hubbard, S.M., Shultz, A.W. and Blum, M.D. (2017) Three-Dimensional Reconstruction of Meander-Belt Evolution, Cretaceous McMurray Formation, Alberta Foreland Basin, Canada. *J. Sed. Res.*, **87**, 1075–1099.

Ells, S.C. (1914) *Preliminary report on the bituminous sands of northern Alberta. Government Printing Bureau, Canada Department of Mines, Mines Branch, Report* **281**, p. 92.

Flach, P.D. and Mossop, G.D. (1985) Depositional environments of Lower Cretaceous McMurray Formation, Athabasca Oil Sands, Alberta. *AAPG Bull.*, **69**, 1195–1207.

Fustic, M. (2007) Stratigraphic dip analysis–A novel application for detailed geological modeling of point bars and predicting bitumen grade, McMurray Formation, Muskeg River Mine, northeast Alberta. *Natural Resources Research*, **16**, 31–43.

Fustic, M., Hubbard, S.M., Spencer, R., Smith, D.G., Leckie, D.A., Bennett, B. and Larter, S. (2012) Recognition of down-valley translation in tidally influenced meandering fluvial deposits, Athabasca Oil Sands (Cretaceous), Alberta, Canada. *Mar. Petrol. Geol.*, **29**, 219–232.

Fustic, M., Strobl, R., Jablonski, B., Vik, E., Jacobsen, T., Garner, D. and Martinius, A. (2013) Chute-channel deposits–recognition in outcrop and subsurface with implications for reservoir mapping–examples from Alberta and Utah. *10th International Conference on Fluvial Sedimentology, Leeds, UK, Abstracts.*

Ghinassi, M. (2011) Chute channels in the Holocene high-sinuosity river deposits of the Firenze plain, Tuscany, Italy. *Sedimentology*, **58**, 618–642.

Ghinassi, M., Ielpi, A., Aldinucci, M. and Fustic, M. (2016) Downstream-migrating fluvial point bars in the rock record. *Sed. Geol.*, **334**, 66–96.

Ghinassi, M., Nemec, W., Aldinucci, M., Nehyba, S., Özaksoy, V. and Fidolini, F. (2014) Plan-form evolution of ancient meandering rivers reconstructed from longitudinal outcrop sections. *Sedimentology*, **61**, 952–977.

Gingras M. and Leckie D. (2017) The argument for tidal and brackish water influence in the McMurray Formation reservoirs. *Reservoir*, **2**, 21–24.

Hein, F., Langenberg, C., Kidston, C., Berhane, H., Berezniuk, T. and Cotterill, D. (2001) A comprehensive field guide for facies characterization of the Athabasca Oil Sands, Northeast Alberta. *Alberta Energy and Utilities Board Special Report*, **13**, 415.

Hein, F.J. and Cotterill, D.K. (2006) The Athabasca oil sands—a regional geological perspective, Fort McMurray area, Alberta, Canada. *Natural Resources Research*, **15**, 85–102.

Hey, R.D. and Thorne, C.R. (1975) Secondary flows in river channels. *Area*, 191–195.

Hubbard, S.M., Smith, D.G., Nielsen, H., Leckie, D.A., Fustic, M., Spencer, R.J. and Bloom, L. (2011) Seismic geomorphology and sedimentology of a tidally influenced river deposit, Lower Cretaceous Athabasca oil sands, Alberta, Canada. *AAPG Bull.*, **95**, 1123–1145.

Jablonski, B.V. and Dalrymple, R.W. (2016) Recognition of strong seasonality and climatic cyclicity in an ancient, fluvially dominated, tidally influenced point bar: Middle McMurray Formation, Lower Steepbank River, north-eastern Alberta, Canada. *Sedimentology*, **63**, 552–585.

Labrecque, P.A., Hubbard, S.M., Jensen, J.L. and Nielsen, H. (2011) Sedimentology and stratigraphic architecture of a point bar deposit, Lower Cretaceous McMurray Formation, Alberta, Canada. *Bull. Can. Petrol. Geol.*, **59**, 147–171.

Martinius, A., Fustic, M., Garner, D., Jablonski, B., Strobl, R., MacEachern, J. and Dashtgard, S. (2017) Reservoir characterization and multiscale heterogeneity modeling of inclined heterolithic strata for bitumen-production forecasting, McMurray Formation, Corner, Alberta, Canada. *Mar. Petrol. Geol.*, **82**, 336–361.

McLearn, F. (1917) Athabasca river section. *Geol. Surv. Can. Summary Report*, **1916**, 145–151.

Mellon, G. and Wall, J. (1956) *Geology of the McMurray: Formation. Pt. I. Foraminifera of the upper McMurray and basal Clearwater Formations.*

Miall, A. (1996) *The geology of fluvial deposits.* Berlin: Springer-Verlag.

Miall, A. (2013) *The geology of fluvial deposits: sedimentary facies, basin analysis and petroleum geology.* Springer.

Plink-Björklund, P. (2015) Morphodynamics of rivers strongly affected by monsoon precipitation: Review of depositional style and forcing factors. *Sed. Geol.*, **323**, 110–147.

Ranger, M.J. (1994) *A basin study of the southern Athabasca Oil Sands Deposit.* Unpublished PhD thesis, University of Alberta, p. 290.

Saucier, R.T. (1994) Geomorphology and Quarternary Geologic History of the Lower Mississippi Valley. Volume 2, Army Engineer Waterways Experiment Station Vicksburg MS Geotechnical Lab.

Sisulak, C.F. and Dashtgard, S.E. (2012) Seasonal controls on the development and character of inclined heterolithic stratification in a tide-influenced, fluvially dominated channel: Fraser River, Canada. *J. Sed. Res.*, **82**, 244–257.

Słowik, M. (2016) The influence of meander bend evolution on the formation of multiple cutoffs: findings inferred from floodplain architecture and bend geometry. *Earth Surf. Proc. Land.*, **41**, 626–641.

Smith, D.G., Hubbard, S.M., Leckie, D.A. and Fustic, M. (2009) Counter point bar deposits: lithofacies and reservoir significance in the meandering modern Peace River and ancient McMurray Formation, Alberta, Canada. *Sedimentology*, **56**, 1655–1669.

Strobl, R.S., Muwais, W.K., Wightman, D.M., Cotterill, D.K. and Yuan, L. (1997) Application of outcrop analogues and detailed reservoir characterization to the AOSTRA underground test facility, McMurray Formation, northeastern Alberta. In: *Petroleum Geology of the Cretaceous Mannville Group, Western Canada* (Eds S.G. Pemberton and D.P. James), *Can. Soc. Petrol. Geol. Mem.*, **18**, 375–391.

Thomas, R.G., Smith, D.G., Wood, J.M., Visser, J., Calverley-Range, E.A. and Koster, E.H. (1987) Inclined heterolithic stratification—terminology, description, interpretation and significance. *Sed. Geol.*, **53**, 123–179.

Toonen, W.H., Kleinhans, M.G. and Cohen, K.M. (2012) Sedimentary architecture of abandoned channel fills. *Earth Surf. Proc. Land.*, **37**, 459–472.

van Denderen, R.P., Schielen, R.M., Blom, A., Hulscher, S.J. and Kleinhans, M.G. (2017) Morphodynamic assessment of side channel systems using a simple one-dimensional bifurcation model and a comparison with aerial images. *Earth Surf. Proc. Land.*, DOI: 10.1002/esp.4267

van Dijk, W.M., Schuurman, F., van de Lageweg, W.I. and Kleinhans, M.G. (2014) Bifurcation instability and chute cutoff development in meandering gravel-bed rivers. *Geomorphology*, **213**, 277–291.

van Dijk, W., van de Lageweg, W. and Kleinhans, M. (2012) Experimental meandering river with chute cutoffs. *J. Geophys. Res. Earth Surf.*, **117**, F03023, doi:10.1029/2011JF002314.

van de Lageweg, W.I., van Dijk, W.M., Box, D. and Kleinhans, M.G. (2016) Archimetrics: a quantitative tool to predict three-dimensional meander belt sandbody heterogeneity. *The Depositional Record*, **2**, 22–46.

van de Lageweg, W.I., van Dijk, W.M. and Kleinhans, M.G. (2013) Channel belt architecture formed by a meandering river. *Sedimentology*, **60**, 840–859.

Wightman, D.M. and Pemberton, S.G. (1997) The Lower Cretaceous (Aptian) McMurray Formation: an overview of the Fort McMurray area, northeastern, Alberta.

Zhang, S., Fustic, M. and Strobl, R. (2017) Qualitative and Quantitative Characterization of McMurray Formation Chute Channel Deposits with Implications to Oil Recovery, (abs) Geoconvention, Calgary, May 2017.

Zinger, J.A., Rhoads, B.L. and Best, J.L. (2011) Extreme sediment pulses generated by bend cutoffs along a large meandering river. *Nature Geosci.*, **4**, 675.

Int. Assoc. Sedimentol. Spec. Publ (2018) **48**, 349–384.

Modern and ancient amalgamated sandy meander-belt deposits: recognition and controls on development

ADRIAN J. HARTLEY[†], AMANDA OWEN[‡], GARY S. WEISSMANN[*] and LOUIS SCUDERI[*]

[†] *Department of Geology & Petroleum Geology, University of Aberdeen, Aberdeen, UK*
[‡] *School of Geographical and Earth Sciences, University of Glasgow, Glasgow, UK*
[*] *University of New Mexico, Department of Earth and Planetary Sciences, Albuquerque, New Mexico, USA*

ABSTRACT

Amalgamated sandy meander belts and their deposits are common in modern continental and marine-connected basins yet comprise a minor constituent of the reported fluvial rock record. This suggests that either amalgamated meander-belts are uncommon in the rock record or that the recognition criteria are lacking to identify sandy meandering river deposits. To address this apparent discrepancy, the authors document the range and occurrence of amalgamated sandy meander belts (ASMB) from modern basins and the stratigraphic record. ASMB are widely distributed throughout both present and rock record sedimentary basins occurring in foreland, extensional, cratonic, strike-slip and passive margin basins. They occur in all climatic settings ranging from tundra to hot deserts. Three specific occurrences of ASMB are recognised in modern basins: in the proximal to medial parts of distributive fluvial systems (DFS), as laterally-confined belts that mainly form axial fluvial systems; and as valley-confined meander belts that may infill bedrock, alluvial or coastal plain valleys. From the limited amount of rock record examples of ASMB that are available, it is clear that they occur in similar settings to those observed in modern basins, the recognition of which provides a framework for the further prediction and identification of ASMB in the rock record. The lack of recognition of ASMB in the rock record is considered to be due to an absence of characteristics that allow clear distinction between sandy meandering and braided fluvial deposits. Characteristics considered common to both include: multi-storey, laterally extensive (sheet-like) amalgamated channel belts, dominance of downstream accreting bedforms, no fining upwards grain-size profile and little or no fine-grained sediment and/or soil preservation. In contrast, features considered characteristic of meandering rivers such as inclined heterolithic stratification, high palaeocurrent dispersion, single storey channels and fining upwards grain-size profiles are absent. The authors suggest that no single criterion can be used to definitively identify sandy meander belt deposits in the rock record and that a combination of systematic variations in accretion direction, palaeocurrent dispersal patterns and recognition of storey scale accretion surfaces is necessary to identify clearly this fluvial style. The common occurrence and distribution of sandy meander belts in modern sedimentary basins together with their limited recognition in the rock record suggests that their true stratigraphic distribution has yet to be determined. This has important implications for palaeogeographic reconstructions, understanding the impact of plant colonisation on fluvial planform style and predicting sandstone body dimensions and internal heterogeneity distribution within hydrocarbon reservoirs and aquifers.

Keywords: Meander-belts, rivers, planform, storey, accretion

Fluvial Meanders and Their Sedimentary Products in the Rock Record, First Edition.
Edited by Massimiliano Ghinassi, Luca Colombera, Nigel P. Mountney and Arnold Jan H. Reesink.

INTRODUCTION

Sandy, meandering fluvial channels and their deposits are a common feature of modern continental sedimentary basins. However, despite detailed descriptions and well-established criteria for their recognition being available since the work of Bernard & Major (1963) and developed by Bernard *et al.* (1970), McGowen & Garner (1970), Bluck (1971), Jackson (1976, 1978), Bridge (1985) and Smith (1987) amongst others, they are not widely recognised in the rock record (e.g. Gibling 2006; Hartley *et al.*, 2015). Gibling (2006) in his landmark review paper of fluvial channel deposits noted that the maximum thickness of the studied examples of rock record meandering-river deposits was 38 m, that widths are less than 15 km and typically less than 3 km. He concluded that meandering rivers do not appear to create thick or extensive deposits and probably constitute a relatively minor proportion of the fluvial-channel record. He also pointed out that the lack of recognition of meandering river deposits in the rock record may in part reflect the difficulty of distinguishing coarse grained meander belts from braided-river deposits (see Jackson, 1978), as well as difficulties in recognising lateral accretion deposits in varied outcrop orientations (Willis, 1989, 1993). The discrepancy between the prevalence of meandering rivers in modern basins and their apparent absence in the rock record is significant. It suggests that either the deposits of meandering rivers are uncommon in the rock record and, concomitantly, that the use of modern analogues to interpret the rock record is not appropriate, or, alternatively, it could indicate that the recognition criteria available to identify sandy meandering river deposits are inadequate.

Sandy meander belt deposits host important hydrocarbon reservoirs, e.g. Cretaceous Williams Fork Formation and equivalents, Colorado (Johnson 1989), Cretaceous Fall River Formation, Wyoming and South Dakota (Berg 1968; Willis 1997), McMurrray Formation, Alberta (Mossop & Flach, 1983; Labrecque *et al.*, 2011), Triassic Mungaroo Formation, NW Australia (Jablonski, 1997; Sibley *et al.*, 1999), mineral deposits such as copper (Hitzman *et al.*, 2005) and uranium (e.g. Stokes, 1954; Owen *et al.*, 2016), as well as forming potential geothermal reservoirs (Willems *et al.*, 2015). A key control on the extraction of minerals and fluids from these deposits is the distribution of internal heterogeneities (e.g.

Shepherd, 2009; Owen *et al.*, 2016). Within sandy meander belt deposits, internal heterogeneity distribution occurs at a range of scales which differ significantly from those of sandy braided rivers (e.g. Shepherd, 2009; Hartley *et al.*, 2015; Swan *et al.*, 2018), such that better heterogeneity characterisation will aid resource extraction from these deposits.

The authors aim to examine the discrepancy between the common occurrence of meander belt deposits in modern basins and their apparent under-representation in the rock record by examining the range and distribution of meander-belt deposits in modern continental basins and documenting controls on their spatial distribution. Following this, rock record examples are described and discussed, together with the reasons why sandy meander belt deposits have not been widely recognised in the rock record to date.

Meander-belt definitions and methodology

For modern meandering fluvial systems the authors define a meander belt as the area encompassing the actively meandering channel (Fig. 1), with the average width of the active meander-belt measured over the studied length. An amalgamated meander-belt is the entire area encompassing features indicative of a meandering channel planform (e.g. scroll bars, meander loops, ox-bow lakes etc.) which extends beyond the margin of the active meander-belt (Fig. 1).

Using remotely sensed imagery (including GoogleEarth®, Landsat-7 satellite imagery, topographic maps and aerial photographs where available) the authors have identified 30 amalgamated meander-belts located in actively aggrading sedimentary basins (Fig. 2; Table 1). Herein, this analysis is restricted to aggrading sedimentary basins as they represent the most appropriate analogue for basins preserved in the rock record (Hartley *et al.*, 2010; Weissmann *et al.*, 2010, 2015; Davidson *et al.*, 2013). The authors aim to highlight the range of sandy meander-belts and controls on their development in a variety of different tectonic and climatic settings.

Difficulties in the recognition of modern meander-belts include anthropogenic modification where meander-belts have been utilised for agriculture and/or urban development. Vegetation-abandoned meander-belt deposits may be colonised rapidly by vegetation which thus obscures any evidence for the underlying channel

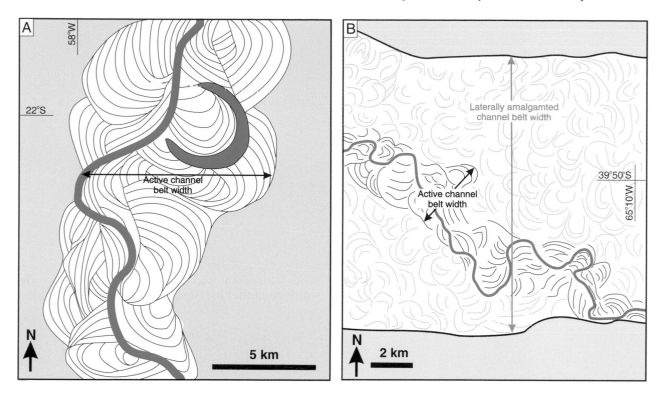

Fig. 1. Interpretive sketch illustrating the definition of active and laterally amalgamated meander-belts from (A) the Rio Paraguay where the active channel-belt width equals the total channel-belt width, pink area represents equivalent flood-plain and distal DFS deposits; and (B) the valley-confined meander belt of the Rio Negro, Argentina, where the active channel-belt width comprises approximately 20% of the laterally amalgamated channel-belt width. Pink area represents valley margins. For location see Fig. 2.

planform and restricts the analysis of lateral channel belt extent. These difficulties mean that many well preserved modern meander belts are restricted to areas of sparse vegetation cover and limited anthropogenic modification such as hot and cold desert areas. Meander belt deposits are recognisable in other climatic settings but may be only partially preserved and/or strongly modified by human activity. This paper focusses primarily on sandy meander-belts; and whilst the grain-size may not be observed directly, the presence of bedforms indicates that sand is the predominant grain-size in most examples and in a number of cases grain-size has been calibrated with ground-based observations and published literature.

Sandy meander-belts – modern examples

Present day observations indicate that three main sandy meander-belt types can be recognised in aggradational settings (Fig. 3): valley-confined, laterally-confined and unconfined. Examples and ranges of each type including tectonic setting, prevailing climate regime, river width and lateral

extent of meander belts are documented for 30 amalgamated meander belts in Table 1 and representative examples are described below.

Valley-confined meander-belts

Valley-confined meander belts are those found in valleys (either bedrock or alluvial) where the fluvial channel and associated overbank flood events are isolated from and have no interaction with adjacent interfluves. Documented examples show significant variations in size and climate (Table 1) with the largest example found in the Lower Mississippi Valley and developed along the Gulf Coast passive margin under a humid, warm temperate climate (Fig. 4A). The valley, which has cut predominantly Quaternary and earlier alluvium, was established in close to its present form around 30,000 years ago (Saucier, 1994). The predominantly sand-dominated amalgamated meander belt (Jordan & Pryor, 1992) has developed mainly in the last 10,000 years (Saucier, 1994). At its widest extent it occupies almost the full valley width in the area between Greenwood

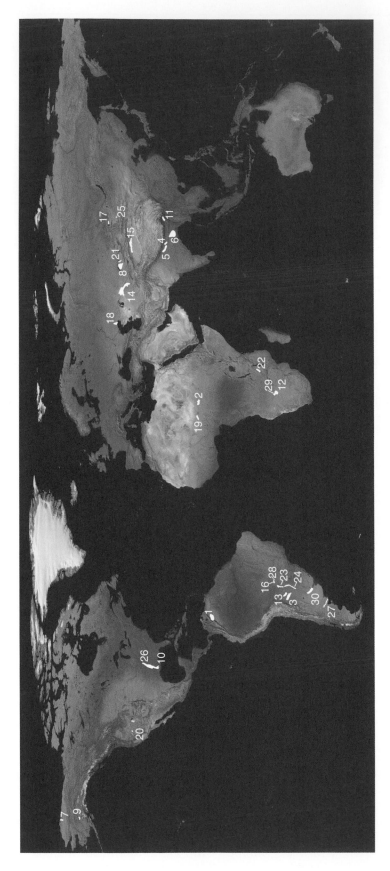

Fig. 2. Location of modern meander-belts used in this study and documented in Table 1. Numbers correspond to numbers in Table 1. Imagery courtesy of Tom Patterson (www.shadedrelief.com).

and Greenville, Mississippi. At this point channel width averages 1500 m, the depth is 30 m, the active channel belt is 24 km wide and the amalgamated channel-belt is 100 km in width, 4 times that of the active channel belt (Fig. 4A). The total area covered by the meander belt is 46,925 km^2 (Table 1).

The Rio Negro in central Argentina drains to the Atlantic passive margin from the east side of the Andes (Fig. 4B). It forms a valley-confined amalgamated meander-belt that extends for 401 km between the towns of Chelforó and Viedma in Rio Negro Province, developed under a cold, arid climate (Fig. 4B). The active meander belt ranges from 2200 to 6000 m in width, bankfull channel width is 450 m and average bankfull depth is 6 m for the study area (Sodano, 1947, Luchsinger, 2006). The valley is cut into Quaternary alluvium and Pliocene fluvial sediments of the Rio Negro Formation and is 20 to 50 m-deep and 3.8 to 17 km-wide (Luchsinger, 2006). The width of the meander-belt varies with the width of the valley. In the upper 100 km of the studied reach, the meander-belt occupies the full width of the valley. For the remaining 300 km, valley width and amalgamated meander belt is twice the width of the active meander-belt but is up to 3 times the width in some areas. The total area covered by the meander belt is 3866 km^2 (Table 1).

The Taquari River in south-west Brazil is located in the back-bulge/cratonic Pantanal Basin and runs westwards through the Pantanal wetlands under an equatorial climate before joining the upper Paraguay River (Fig. 4C and D; Assine *et al.*, 2015). The river exits a dissected tableland plateau and enters the Pantanal Basin west of the town of Coxim, where a valley-confined meander belt occurs within the entrenched upper section of the Taquari distributive fluvial system (DFS; Fig. 4C and D; Assine *et al.*, 2015). The meander belt is entrenched into older Pleistocene DFS lobes, extends for 80 km and occurs between what was the original apex of the Taquari DFS and the current depositional apex (Fig. 4C; Assine, 2005). The depth between the top of the valley and the bankfull elevation of the river decreases downstream from 8 m at the original apex to zero at the end of the valley, from where it passes downstream into an unconfined meander belt on the active depositional surface of the Taquari DFS. The active channel width averages 250 m and the active channel belt is 3.8 km, which approximates to the amalgamated meander belt width of 4 km.

The total area covered by the meander belt is 238 km^2 (Table 1).

Laterally-confined meander-belts

Laterally confined meander-belts are constrained topographically by lateral sediment input or a combination of both lateral sediment and bedrock highs, where the latter may be fault-controlled. A distinguishing feature between laterally-confined and valley-confined meander-belts is that flood events and migrating channel deposits of laterally-confined systems interact with and are interbedded with adjacent floodplain and channel deposits. The majority of laterally-confined examples form axial rivers in sedimentary basins.

The axial meander-belt of the upper Paraguay River, which forms the border between Brazil and Paraguay, is confined by laterally derived sediment input and cratonic basement highs (Fig. 5A and B). The river is developed under an equatorial climate and is located east of the Andean foredeep in a back-bulge setting adjacent to the Brazilian craton. For approximately 590 km between Porto Murtinho in Brazil, where it exits the Pantanal wetlands in the north, downstream to the confluence with the Paraná, the Paraguay is confined to the east by cratonic bedrock highs and small DFS (Fig. 5A and B) and to the west by the large lateral DFS constructed primarily by the Bermejo and Pilcomayo rivers. Over the length of the meander-belt the Paraguay River ranges in width from 300 to 750 m, averaging 650 m, the active meander belt ranges in width from 8 to 13 km, averaging 11 km, and clear scroll-bar topography can be observed (Fig. 5B). Bankfull channel depth is 12 m on average (Drago *et al.*, 2008). For much of the studied reach the amalgamated meander belt and active channel belt width are the same, in places the amalgamated meander belt width may reach 17 km, approximating to 1.5 times the average active channel belt width. The meander belt covers an area of 8093 km^2 (Table 1). There is no valley development along the studied stretch with channel belt sediments passing into adjacent laterally derived sedimentary systems to both the east and west.

The amalgamated meander belt of the Karatal River, Kazakhstan is located in the foreland basin of the Tien Shan mountains from where it is sourced (Fig. 6A and B). It forms a large DFS where it exits the foothills and is developed under a cold, arid climate. The channel belt is confined

Table 1. Sandy meander-belt dimensions from modern basins. Latitude and longitude coordinates indicate the centre of VC – valley-confined. Climate is based on the Köppen-Gegier classification taken from Kottek *et al.* (2006) and refers to these include polar, tropical, which is equivalent to equatorial in some climate classification schemes, continen-lished literature. Notes: For the Gandak and Ghagara channel dimensions are estimates from abandoned channels as Blum *et al.* (2013). For the Okavango DFS widths are taken from the abandoned meander belt (Xugana meander-belt of extent of connected channel belt. Tarim - information from Allen *et al.* (1994). Karatal River – values in parentheses for both an axial and lateral DFS, channel depth from Gilvear *et al.* (2000). Lower Paraguay – average widths and depths Makaske *et al.* (2012) measurement 50 km downstream from terminus. Okavango Panhandle depth from Stanistreet *et al.*

	Type	Location	Tectonic Setting	Fluvial System	Climate	Active Channel Width (m)	Active Channel Belt Width (m)
1	U	Arauca Basin, Colombia/Venez.	Foreland basin (retro-arc)	Unconfined - DFS	Equatorial	400	4000
2	U	Batha River, Chad	Craton	Unconfined DFS	Arid - hot	100	2200
3	U	Bermejo, Argentina	Foreland basin (retro-arc)	Unconfined - Single DFS	Warm - temperate	440	6200
4	U	Gandak, India	Foreland basin	Unconfined - DFS	Warm - temperate	350*	7000*
5	U	Ghagara, India	Foreland basin	Unconfined - DFS	Warm - temperate	350*	6700*
6	U	Hoogly River, India	Foreland/passive margin	Unconfined DFS/Upper delta plain (marine infl.)	Warm - temperate	800	11,500
7	U	Ikpikpuk, Alaska	Foreland basin	Unconfined - Single DFS	Polar – tundra	800	7000
8	U	Ili, Kazakhstan	Foreland basin	Unconfined - Single DFS	Arid - cold	500	5000
9	U	Kuskokwim River, Alaska	Foreland basin	Axial and lateral DFS combined	Snow - cold	300	5000
10	U	Mississippi-Atchafalaya	Passive margin	Unconfined - DFS (marine-influenced)	Warm temp., humid	1200	15,000
11	U	Noa Dihing, India	Foreland basin	Amagamated DFS	Warm - temperate	600	4800
12	U	Okavango, Botswana	Incipient rift	Unconfined - Single DFS	Arid - hot	90	2200
13	U	Pilcomayo, Argentina/Paraguay	Foreland basin (retro-arc)	Unconfined - Single DFS	Warm - temperate	450	4900
14	U	Syr Darya, Kazakhstan	Foreland	Partly confined axial to unconfined DFS	Arid - cold	430	7000
15	U	Tarim, China	Foreland	Partly to unconfined axial DFS	Arid - cold	800	7000
16	U	Taquari (lower section), Brazil	Back-bulge/craton	Unconfined - Single DFS	Equatorial	250	2200
17	U	Tes River, Mongolia	Transpressional	Unconfined - axial DFS (terminates in lake)	Arid - cold	100	1000
18	U	Ural River	Passive margin	Unconfined - DFS (terminates in lake)	Arid - cold	300	6000
19	U	Yobe River, Nigeria	Craton	Unconfined - DFS	Arid - hot	250	5500
20	LC	Humboldt, Nevada	Rift	Laterally confined - axial	Arid - cold	75	850
21	LC	Karatal River, Kazakhstan	Foreland basin	Laterally confined by dunes	Arid - cold	130	2000
22	LC	Luangwa River, Zambia	Incipient rift	Laterally confined by HW and FW DFS - axial	Equatorial	150	4000
23	LC	Paragay, Upper, Paraguay/Brazil	Back-bulge/craton	Laterally confined - axial	Equatorial	600	12,500
24	LC	Paragay, Lower, Argentina/Paraguay	Back-bulge/craton	Laterally confined - axial	Equatorial	650	13,000
25	LC	Zavkhan Gol	Transpressional	Laterally confned - axial	Arid - cold	100	900
26	VC	Mississippi, USA	Passive margin	Valley confined	Warm temp., humid	1500	19,000
27	VC	Rio Negro, Argentina	Passive margin	Valley confined	Arid - cold	375	4100
28	VC	Taquari (upper section) Brazil	Back-bulge/craton	Valley confined	Equatorial	250	3800
29	VC	Okavango, Botswana	Incipient rift	Valley-confined (panhandle)	Arid - hot	100	2200
30	VC	Paraná, Argentina	Craton-Passive margin	Valley confined axial (marine-influenced)	Warm temp., humid	1600	12,000

Paraná data are from the meander-belt. Numbers correspond to those in Fig. 2. U – unconfined, LC – laterally-confined, the climate in the basin in which the meander belts are located. Only the main climate zones were used for classification: tal = snow, drylands = arid and subtropical = warm temperate. Tectonic setting was determined through reference to pub- there are no active channels at present. Depths for the Mississippi-Atchafalaya and Mississippi examples are taken from Stanistreet & McCarthy 1993) as the modern channel is misfit; values in parentheses indicate maximum potential aerial ASMB width are maximum but occur over only 5% of the studied reach. Luangwa River the width of the ASMB includes from Drago et al. (2008). Rio Negro – data from Soldano (1947) and Luchsinger (2006). Taqauari – depth taken from (1993). Paraná data are from Drago & Amsler (1998).

Ratio Active Channel to Channel Belt	Channel Depth (m)	ASMB Max Width (km)	ASMB Max Length (km)	Ratio Channel Belt Width to ASMB Width	Ratio width/ length ASMB	Area (km²)	Location Lat	Location Long
10		75	214	19	3	12,777	7°20'22.49"N	70°55'37.71"W
22		64	224	29	4	9146	13°13'35.99"N	18°52'24.97"E
14		65	212	10	3	9207	24° 4'31.71"S	62°33'17.45"W
20		41	170	6	4	5485	26°35'51.50"N	84°48'14.95"E
19		50	273	7	5	11,543	27°15'39.52"N	81°37'10.42"E
14		130	212	11	2	22,532	23°32'39.90"N	88°47'24.84"E
9		46	118	7	3	2999	70°16'46.58"N	154°50'24.22"W
10		201	214	40	1	21,935	45°40'8.75"N	75°29'44.99"E
17		37	60	7	2	1403	63° 1'7.76"N	154°40'20.72"W
13	20	75	230	5	3	9975	30°51'16.12"N	91°47'26.72"W
8		170	55	35	0.3	6680	27°18'20.82"N	95°11'5.48"E
24		123	119	56	1	8026 (32,055)	19°17'51.24"S	22°43'18.51"E
11	9	37	98	8	3	2657	22°53'29.88"S	62°17'3.78"W
16		126	634	18	5	51,892	44°50'21.92"N	65°11'40.50"E
9		95	388	14	4	22,663	41° 3'13.40"N	83°58'2.41"E
9		38	120	17	3	2910	18°16'57.20"S	55°56'10.90"W
10		27	120	27	4	1920	50°34'46.18"N	93°46'48.42"E
20		60	135	10	2	5199	47°25'3.62"N	51°33'20.16"E
22		64	206	12	3	8794	13°12'59.92"N	12°27'13.92"E
11		6	64	7	11	319	40°38'52.43"N	116°51'19.85"W
15		4 (9)	135	2	34	653	45°48'58.44"N	77°12'1.67"E
27	6	23*	128	6	6	1823	12°56'13.45"S	31°51'53.31"E
21		75	283	6	4	8096	20°21'28.48"S	57°51'23.67"W
20	12	17	590	1	35	7863	25°29'20.31"S	57°39'34.10"W
9		12	64	13	5	616	47° 6'17.74"N	96° 3'39.54"E
13	30	110	580	6	5	46,925	33°21'11.33"N	90°45'0.55"W
11	6	17	401	4	24	3866	39°43'11.77"S	65°28'48.27"W
15	~3.5	4	79	1	20	232	18°21'6.32"S	55° 0'45.80"W
22	5.5	15	100	7	7	1124	18°38'8.37"S	22° 9'12.00"E
8	15	45	180	4	4	8248	33°12'46.35"S	59°57'30.85"W

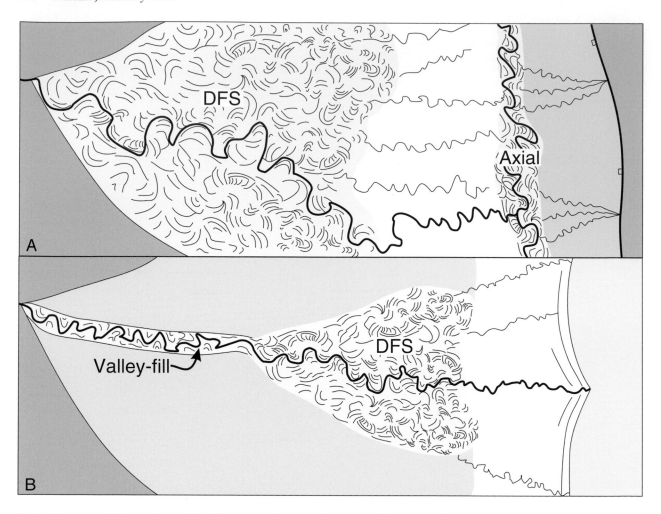

Fig. 3. Conceptual illustration of the different amalgamated meander-belt types. (A) Amalgamated meander belt formed in the proximal and medial parts of a DFS and axial meander-belt. (B) Valley-fill meander-belt feeding a small, marine-connected DFS.

laterally by aeolian dunes, although it is clear that the river breached the surrounding dune field at times in the past with buried channel belts beneath dunes still visible (Fig. 6A). The river is 150 m-wide, the active channel belt is 3 km-wide and the amalgamated meander-belt width averages 4 km but can be up to 9 km. The meander belt forms a relatively narrow strip through the dune field and covers an area of 653 km² (Table 1).

The Zavkhan Gol river in Mongolia (Fig. 7A and B), is developed under a cold arid climate in a transpressional tectonic setting (Bayasgalan *et al.*, 2005). The river enters a small pull-apart basin from the south east in an axial location and expands laterally to form a narrow DFS constrained by a combination of bed rock highs (fault-bounded to the west) some of which source alluvial fans and aeolian dunes, the latter being common particularly in the north of the basin

(Fig. 7A and B). Bankfull width ranges between 50 and 150 m, with an average of 100 m, and the channel belt width is 750 m. Where the river enters the basin in the south east the amalgamated meander belt is 1.5 km-wide and expands to 12 km in the centre of the basin and is dominated by neck cut-off avulsions (Fig. 7B). The total area covered by the meander belt is 616 km², which, although covering a relatively small area, the meander belt does account for over 80% of the sediment within the basin and if preserved would dominate the basin fill.

Unconfined meander-belts

Unconfined meander-belt deposits occur primarily as laterally derived distributive fluvial systems, but may also form axial DFS where lateral sediment input is limited and the axial system is

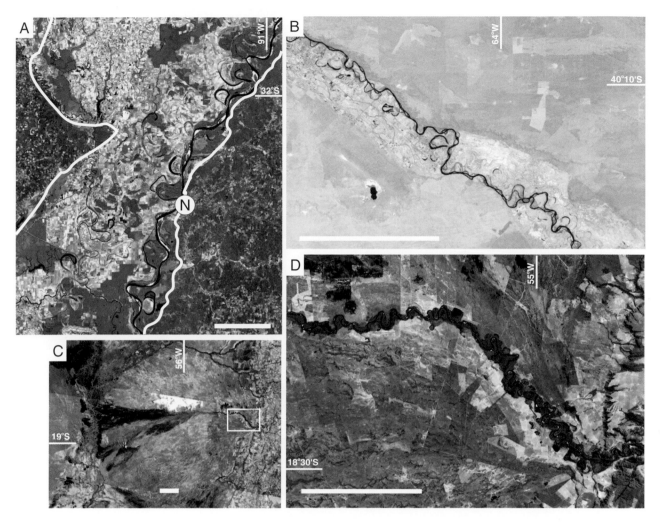

Fig. 4. Examples of valley-confined meander-belts. White horizontal bars in all images represent 20 km. (A) Mississippi Valley (outlined in white). N shows the position of Natchez, Mississippi. (B) Rio Negro, Argentina; (C and D) Taquari River, Pantanal region of Brazil. White box in C outlines area shown in D. For location see Fig. 2.

able to expand laterally. These DFS examples may be composed entirely of amalgamated meander belt deposits (Figs 8 and 9) or transition downstream from a braided to a meandering channel planform (Fig. 10).

In the north-eastern part of the Himalayan foreland basin in eastern Assam Province, India, a number of rivers drain northwards off the peripheral bulge and join the Brahmaputra River under a warm, temperate climate (Fig. 8A). The Noa Dihing is the largest river and together with the Burhi Dihing, Dikhow and the Disang rivers has constructed a 55 × 170 km-wide amalgamated meander belt through lateral amalgamation of individual DFS (Fig. 8A). Average river and active channel belt widths are 600 m and 4.8 km (Noa Dihing), 500 m and 3 km (Buri Dihing), 120 m and 2.2 km (Dikhow) and 70 m and 1.3 km (Disang),

respectively. The rivers on these DFS display sinuosities ranging between 1.37 and 2.06 (Sarma & Basumallick, 1986; Lahiri & Sinha, 2012) with river morphology commonly displaying classic meander belt form with alternating point bars and neck cut-off avulsions, indicated by the presence of oxbow lakes (Fig. 8B; Weissmann *et al.* 2015). Much of the channel belt shifting is considered to be due to lateral meander migration with no apparent preferred migration direction (Lahiri & Sinha, 2012). The total area covered by the laterally amalgamated meander belts is 6680 km² (Table 1).

The sedimentology and development of the Okavango DFS located in north-west Botswana has been widely documented (e.g. McCarthy *et al.*, 1988, 1991, 1992, 2002; Stanistreet & McCarthy, 1993; Stanistreet *et al.*, 1993; Shaw & Nash, 1998;

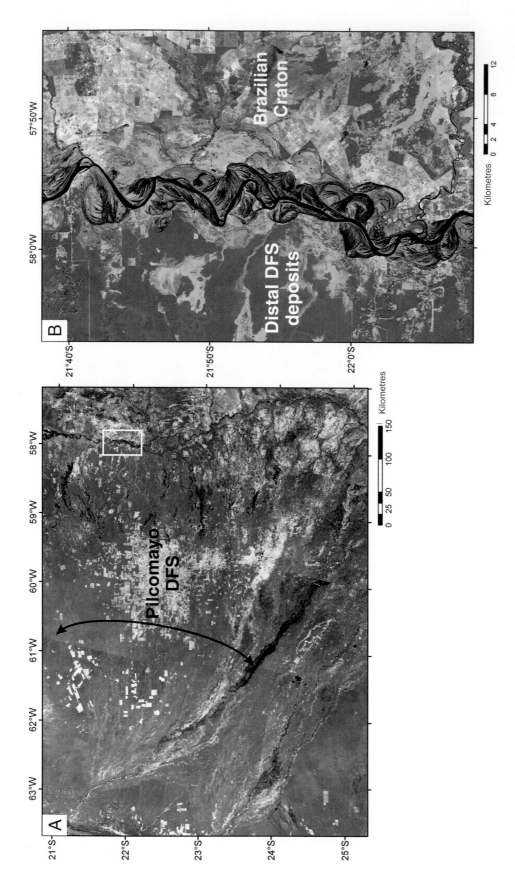

Fig. 5. An example illustrating the relationship between large DFS derived from the Andes and a continental scale axial fluvial system. (A) The large DFS constructed by the Pilcomayo River has pushed the axial Paraguay River eastwards where it abuts against the Brazilian craton. (B) Detail of the Paraguay meander-belt, for location see white box in A. Note that the active and amalgamated channel belt width are approximately the same indicating lateral restriction of the axial system by adjacent depositional systems. For location see Fig. 2.

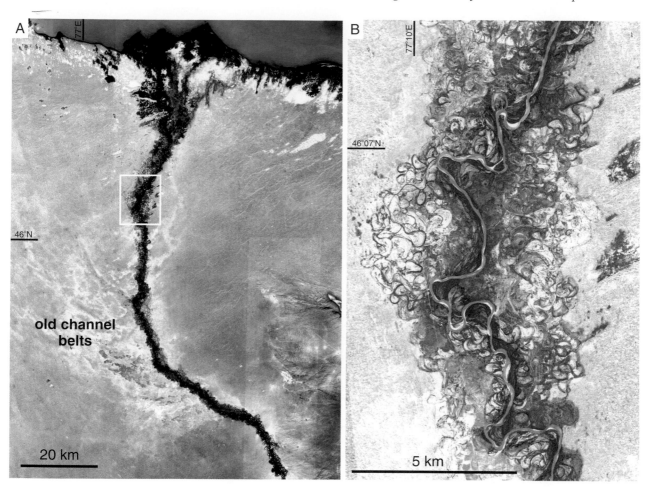

Fig. 6. Lateral confinement of a meander-belt by aeolian dunes, example from the Karatal River, Kazakhstan. (A) Image showing the large scale context of the Karatal meander-belt. The Karatal River drains northwards and forms part of the fill to the Tien Shan foreland basin and terminates in Lake Balkhash to the north. At present the lateral extent of the meander-belt is constrained by aeolian dunes, but it is clear that in the past the meander-belt extended to the west where relict channel-belts can be seen. White box shows location of image in B. (B) Detail of Karatal meander-belt, active meander-belt is approximately half the size of the amalgamated meander-belt in the central part of the image. Imagery from GoogleEarth®. For location see Fig. 2.

Gumbricht *et al.*, 2001, 2004; Ellery *et al.*, 2003; Tooth & McCarthy, 2004) and provides a context for the description of the amalgamated meander belt presented here. The Okavango DFS is developed under a hot, arid climate on the hanging-wall dip slope of a half graben controlled by the Kunyere Fault (Bufford *et al.*, 2012) and occupies the full width of the half-graben (Fig. 9). Meandering channels dominate the fluvial planform with amalgamated channel belts developed in the 4 to 8 m-deep, 100 km-long and 12 km-wide feeder valley (panhandle) and the proximal and medial areas of the DFS, then pass downstream into discrete meander belts separated by floodplain deposits (Fig. 9A and B).

In the proximal apex area of the Okavango DFS where the Okavango River exits the panhandle, the main active channel of the amalgamated channel belt averages 50 m in width with an active channel belt width of 1500 m. Over a downstream distance of 75 km, channel widths and channel belt widths decrease to 25 and 600 m respectively before individual channel belts become separated by floodplain sediment. The sinuosity of the Okavango River decreases downstream from 1.86 in the panhandle to 1.52 on the proximal parts to 1.24 in the medial part of the DFS (Stanistreet *et al.*, 1993). The width of the amalgamated channel belt ranges from 75 to 120 km. The area covered by amalgamated channel belts is difficult to

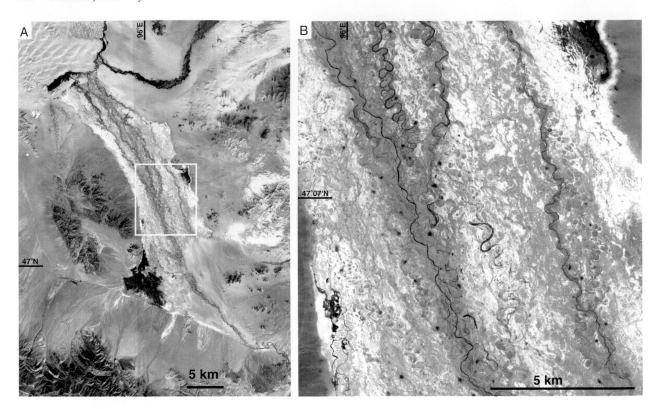

Fig. 7. Laterally confined axial meander-belt which comprises part of a DFS, Zavkhan Gol, Mongolia. (A) Context showing the meander-belt development constrained by alluvial fans sourced from bedrock highs. (B) Detail shown from white box in A, illustrating an active meander-belt width of 1.5 km within an amalgamated meander-belt up to 12 km-wide. Imagery from GoogleEarth®. For location see Fig. 2.

determine precisely due to vegetation cover and burial of older channel belts by recent sediment, but a conservative estimate based on visible connected channel belts is 9062 km². However, if all partially visible fragments of meander belts are considered to be connected across the Okavango area, then the maximum aerial extent of the amalgamated channel belt may be as large as 32,055 km².

The Bermejo River forms one of a number of large DFS developed under a warm, temperate climate that drain eastwards off the Central Andes to the Andean foreland basin in Bolivia, Argentina and Paraguay (Fig. 10) which have been described in detail previously (Iriondo, 2007; Hartley *et al.*, 2010, 2013; Weissmann *et al.*, 2011, 2015; McGlue *et al.*, 2016). The Bermejo River enters the foreland basin as a broad, braided channel 2500 m-wide which diminishes in width and transitions into a meandering river 440 m-wide, 110 km downstream of the DFS apex (Fig. 10A and B). Below the transition the width of the Bermejo gradually decreases over the next 75 km to average 300 m. Similarly, the width of the active channel

belt decreases from 6.2 to 4.3 km over this 75 km reach. Downstream of this point, modern channel belts are discrete and separated by floodplain muds. The amalgamated meander belt developed by the Bermejo is 212 km-long and 65 km-wide and covers an area of 9207 km² (Fig. 10). This area includes relict meandering channel deposits in locations close to the present day apex and immediately adjacent to the current braided Bermejo River channel. This observation suggests that in relatively recent times (late Pleistocene?), changes in discharge/sediment supply in the Bermejo catchment have resulted in the downstream relocation of the braided to meandering transition.

The Tes River marks the border between Mongolia and Russia and is subject to a cold, arid climate (Fig. 11). It forms an axial DFS that drains westwards into Lake Uvs and is sourced from the Tannu-Ola mountain range to the north (Fig. 11). The Uvs basin is a compressional/transpressional basin bounded to the north by a north-dipping thrust fault that is actively uplifting the Tannu-Ola Range and to the west by strike-slip faults

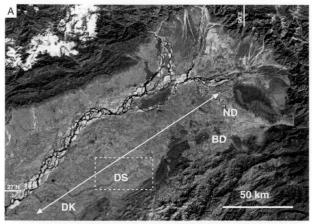

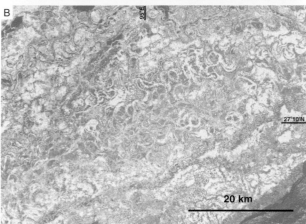

Fig. 8. Example of an unconfined laterally amalgamated meander-belt from the Himalayan foreland basin. (A) Satellite image showing the amalgamated meander belt forming the basin-fill between the mountain ranges to the south-east and the axial braided Brahmaputra River. The meander belt which extends between the arrowed area consists of 4 laterally coalesced DFS which include the Noa Dihing (ND), Burhi Dihing (BD), Dikhow (DK); and Disang (DS) rivers. (B) Digital elevation model of the Disang River meander-belt, the area of which corresponds to the dashed box in A.

(Baljinnyam, 1993; Bayasaglan *et al.*, 2005). The Tes River axial DFS is constrained structurally to the lowest part of the basin. The northern margin of the channel belt abuts alluvial fans derived from the Tannu-Ola range and the southern margin is constrained by a regional northerly dipping stoney desert pediplain onto which the channel belt is onlapping to the south (Fig. 11A, B).

The Tes axial DFS is 120 km in length and comprises a single amalgamated channel belt covering 1920 km^2. Where the river enters the Uvs basin in the east, from the apex it expands from a single channel belt width of 1000 m to a maximum width of 27 km just before the channel belt terminates at

Lake Uvs (Fig. 11A and B). The active channel width decreases downstream over a distance of 100 km from an average of 100 m at the apex to 50 m close to the lake shoreline; and the active channel belt width decreases from 1000 to 700 m over the same distance.

Characteristics of modern sandy meander belts

The examples of amalgamated sandy meander belts (ASMB) documented above and in Table 1 are considered to be representative of the range developed in present day sedimentary basins. The assessment herein is not meant to be a comprehensive list of all ASMB in all sedimentary basins, but rather examples that illustrate the wide range and nature of ASMB that are developed. From the documented examples it is possible to make some general observations regarding present day basins. For example, ASMB occur in all climates ranging from polar tundra to equatorial and in all tectonic settings including foreland, rift, cratonic and passive margin basins. A strong relationship between active channel width and channel-belt width is evident (Fig. 12) with an average ratio of 15 and range of 8 to 23 (Table 1). This relationship is developed irrespective of tectonic or climatic setting and whether the meander belt is confined or not.

A wide range of ASMB dimensions are recorded (Table 1) and these are related primarily to the size of the river and the degree of meander belt confinement. Valley and laterally confined meander belt widths range from 4 to 110 km, but there is significant range in river width (from 75 to 1600 m). To derive a comparison between the different examples, the ratio of active channel belt to amalgamated channel belt is used. Active channel belt width is considered a more robust measurement than active channel width as it shows significantly less variation through the studied reaches. Using this approach the five valley confined ASMB are consistently 1 to 7 times larger than the active channel belt (Table 1). Five laterally confined ASMB show a similar variation being 1 to 7 times larger than the active channel belt, with the remaining example, Zavkhan Gol, proving an exception being 13 times larger than the channel belt (Table 1). In summary, the data suggest a general relationship can be established where laterally and valley confined ASMB are on average 4 to 5 times larger than the width of the active channel belt.

Unconfined ASMB occur as part of DFS and have widths ranging from 27 to 201 km with an

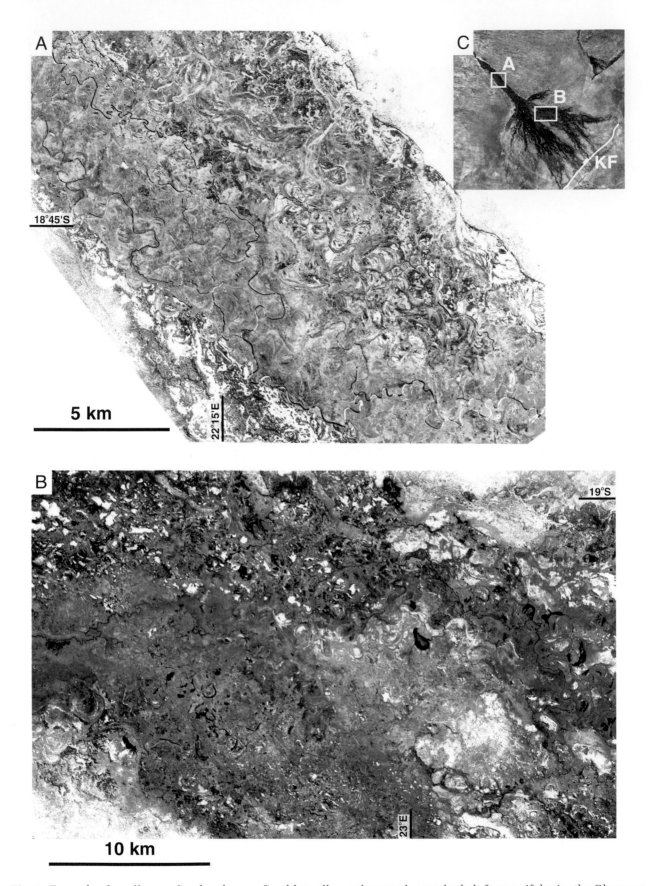

Fig. 9. Example of a valley-confined and unconfined laterally amalgamated meander belt from a rift basin, the Okavango DFS, Botswana. (A) Detail of the panhandle area where the amalgamated meander belt occupies the full width of the valley in the flexural margin to the half-graben. (B) Unconfined laterally amalgamated meander–belt in the proximal-medial area of the Okavango DFS. (C) Image showing location of images in A and B and position of the Kunyere Falut (KF), taken from Bufford *et al.* (2012), which forms the half-graben bounding fault.

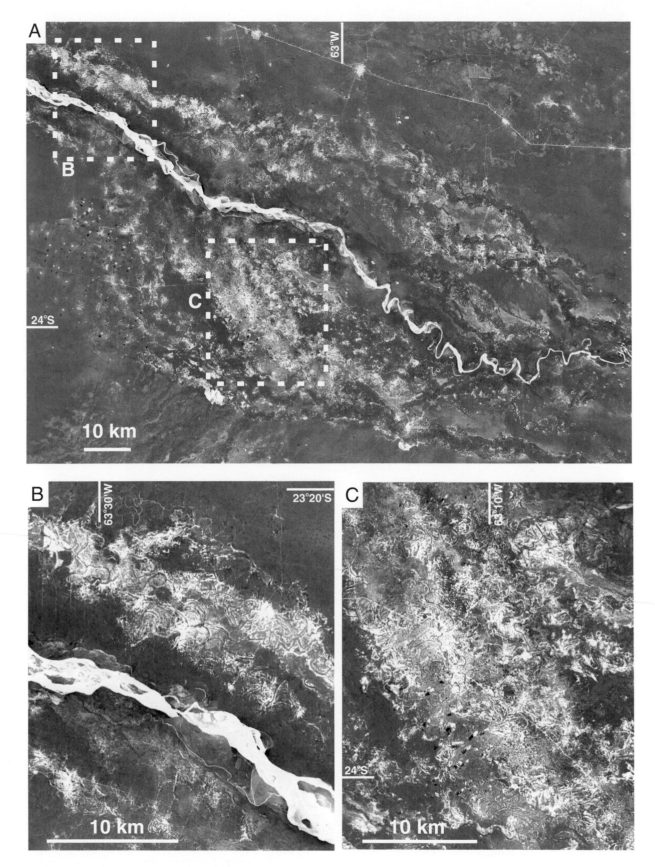

Fig. 10. Unconfined laterally amalgamated meander-belt example form the Rio Bermejo, Argentina. (A) Overall perspective of the Bermejo ASMB. Note that the present day transition from braided to meandering on the modern Bermejo occurs downstream of the main development of the meander belt. (B) Detail showing ASMB deposits adjacent to modern Bermejo braided planform. (C) Detail of ASMB deposits in the main part of the meander-belt. Images modified from GoogleEarth®.

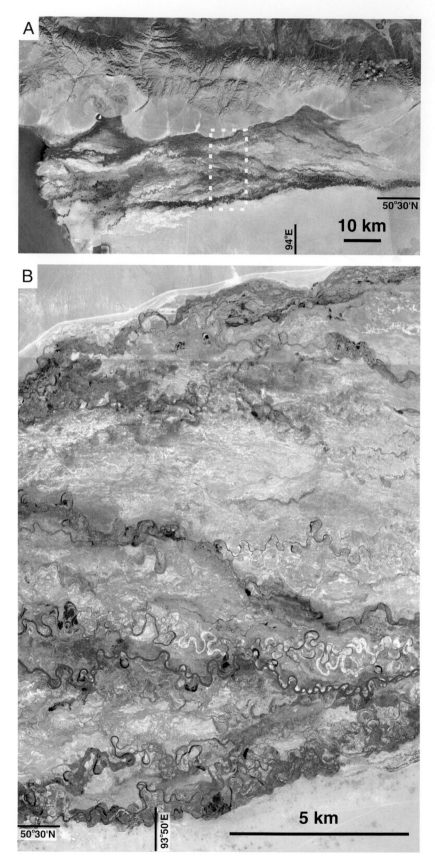

Fig. 11. Laterally-amalgamated meander-belt formed as part of an axial DFS from the Tes River, Uvs Basin, Mongolia. (A) Image illustrating the axial DFS confined to the north by the Tannu-Ola mountain range and draining west into Lake Uvs, constructed entirely of laterally amalgamated meander belt deposits. (B) Detail of area outlined in white dashed box in A. Imagery from GoogleEarth®.

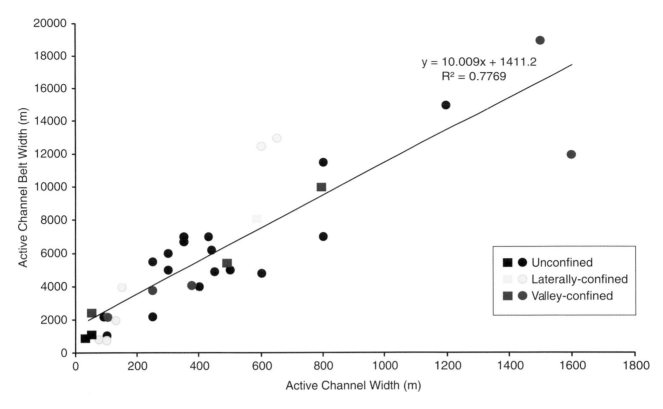

Fig. 12. Graph of active channel width versus active meander-belt width for all different ASMB types: circles indicate modern day examples (Table 1), squares indicate rock record examples (Table 2).

average of 81 km (Table 1). The ratio of channel belt to amalgamated channel belt width ranges between 5 and 56, averaging 18 (Table 1). However, direct comparisons are not straightforward, for example the 170 km-wide Noa Dihing DFS is a composite feature composed of four amalgamated DFS (Fig. 8) and Kuskokwim in Alaska is a composite of lateral DFS and an axial meandering system, consequently only general conclusions can be drawn through comparison of ASMB widths. In addition there is significant variation as to where the ASMB occurs on the DFS. In some examples it transitions downstream from a braided channel planform (e.g. Bermejo & Pilcomayo Fig. 10). In other examples meandering channels are the only planform developed on the DFS and the ASMB is developed immediately downstream of the apex of the DFS (e.g. Okavang, Tes, Ural, Noa Dihing; Figs 8 to 11).

It is important to recognise that the current channel planform may not necessarily be representative of the planform that built much of the DFS (Hartley *et al.*, 2010). For example the transition from braided to meandering on the modern Bermejo River occurs 110 km downstream of the

apex (approximately 15% of the distance between the apex and toe of the DFS), whereas the mapped ASMB is located 70 km upstream of this point. The Okavango shows a similar variation, the modern channel and channel belt are approximately 50% smaller than meander belts commonly visible on the satellite imagery (Fig. 9B), referred to as the Xugana meander belt (Stanistreet & McCarthy, 1993) suggesting that the Okavango ASMB developed at least partly during an earlier wetter period. This interpretation is supported by evidence for progressive desiccation and increased aridity in the Okavango and Kalahari area from the late Pleistocene to present day (Cooke, 1980; Hamandawana *et al.*, 2008).

The width length ratio for unconfined ASMB ranges between 1 and 5 (Table 1). These values are consistently lower than those of valley and laterally confined ASMB where the range for each is 4 to 35 (average 16) and 4 to 24 (average 12), respectively. The difference in ratio values is clearly because of the confinement of the latter two types where the length is significantly greater than the width.

The area covered by ASMB varies considerably between the studied examples (Table 1). The largest ASMB tend to occur as all or component parts of DFS with areas between 5000 and 22,000 km² common. However, only the Syr Darya in Kazakhstan at 51,000 km² is larger than the 47,000 km² valley-confined Mississippi ASMB and it is not uncommon for valley and laterally confined ASMB to cover areas of at least 8000 km².

In summary, in present day basins, large ASMB occur in a range of different climatic and tectonic settings. The extent of ASMB relates primarily to the size of the river which developed the meander belt and the amount of lateral confinement, with ASMB covering 1000s of km² being common features of present day basins.

Sandy meander-belts: rock record examples

The ASMB examples documented above illustrate the wide range of different tectonic, climatic and geomorphic settings in which they form at present. However, these examples represent plan view studies and, despite being located within actively subsiding basins, may not necessarily be preserved in the stratigraphic record. In the following section, published examples of amalgamated meander belts documented in the rock record and summarised in Table 2 are described, discussed and related to different geomorphic, climatic and tectonic settings before discussing both present day and rock record examples.

Valley-confined

Valley-confined ASMB are particularly well known from the Quaternary Gulf of Mexico passive margin in Texas (e.g. Bernard & Le Blanc, 1965; McGowen & Garner, 1970; Aslan & Blum, 1999; Blum & Aslan, 2006; Blum et al., 2013). Coastal plain palaeovalleys were cut and subsequently filled under a warm temperate, humid climate by high sinuosity sandy meander belts referred to as the 'Deweyville' units. The Deweyville ASMB developed in the last 100 kyr during the falling stage and lowstand of the MIS 4-3 and MIS 2 intervals, through episodes of lateral migration and channel belt construction (Blum & Price, 1998; Blum & Aslan, 2006). A composite ASMB typically occupies the lower half the valley and consists of a series of overlapping channel belts. For example the Trinity palaeovalley is 8 km-wide and comprises 3 phases of 'Deweyville' channel-fill composed of 10 to 15 m-thick sand and gravel point bars that amalgamate

to form an 8 km-wide ASMB (Morton et al., 1996; Aslan & Blum, 1999; Blum & Aslan, 2006). The Colorado and Brazos river valleys have similar dimensions. The Colorado River Pleistocene channel belt is 5 to 20 km-wide with Deweyville ASMB making up the lower 10 to 15 m of the 30 m-deep valley-fill (Blum & Price, 1998; Blum & Aslan, 2006). The Brazos Pleistocene valley is 10 to 15 km-wide and contains a lower 10 m-thick ASMB package at the base of the 20 m-deep palaeovalley (Sylvia & Galloway, 2006). Although not constrained precisely, the length of these 'Deweyville' ASMB valley-fills probably extends for at least 50 km upstream of the palaeovalley mouth (e.g. Blum & Price, 1998; Aslan & Blum, 1999; Blum & Aslan, 2006).

The late Triassic Chinle Formation exposed across the Colorado Plateau contains a number of palaeovalleys filled with a range of amalgamated meander belt deposits (e.g. Stewart et al., 1972; Blakey & Gubitosa, 1983, 1984; Demko et al., 1998; Dubiel et al., 1999; Dubiel & Hasiotis, 2011; Hartley & Evenstar, 2018). The Chinle Formation was deposited in a cratonic back-arc basin setting (Blakey & Gubiatosa, 1984; Dubiel & Hasiotis 2011), under a primarily humid/sub-humid climate with a progressive change to an arid/semi-arid climate in the upper part of the formation (Blakey & Gubitosa, 1984; Hasiotis, 2002; Prochnow et al., 2006). Sediment was sourced from the east and transported north-west towards a marine seaway located in present day Nevada (Riggs et al., 1996; Dickinson & Gehrels, 2008). The relatively low subsidence setting (maximum of 600 to 900 m over 21 to 30 Ma) resulted in alternating phases of aggradation and degradation during Chinle deposition (Dubiel & Hasiotis, 2011; Trendall et al., 2012). Degradation is represented by phases of incision with development of the Shinarump Member palaeovalley at the base of the formation, cut into the underlying Moenkopi Formation and the Monitor Butte/Temple Valley and Moss Back Member palaeovalleys within the formation (Dubiel & Hasiotis 2011). An additional mechanism of palaeovalley development in the Chinle Formation is related to salt movement and occurred in east central Utah, near Moab, where ASMB were focussed into palaeovalleys developed between growing salt highs (Hartley & Evenstar, 2017).

The basal Shinarump palaeovalley is developed extensively across the Chinle outcrop area (e.g. Witkind, 1956; Stewart et al., 1972; Blakey &

Table 2. Meander-belt dimensions of documented examples from the rock record. See references for further details of outcrop descriptions and locations. *Channel-belt width estimated from partially preserved scroll bar deposits visible on satellite imagery.

Type	Location	Tectonic Setting	Age	Climate	Estimated Active Channel Width (m)	Estimated Active Channel Belt Width (m)	Ratio Active Channel to Channel Belt	Channel Depth (m)	Preserved ASMB Max Width (km)	Preserved ASMB Max Length (km)	Ratio Channel Belt Width to ASMB Width	Preserved Area (km²)	References
Laterally confined - axial	McMurray Formation, Alberta	Foreland/back-bulge	Aptian	warm temperate, humid	584 (800)	8000	10 to 14	36 (40)	200	400	25	80,000	Labrecque et al. (2011); Bhattacharya et al. (2016)
Unconfined - single DFS	Salt Wash, Utah	Foreland/back-bulge	Kimmeridgian	humid	40 to 75	1000	15	6 to 10	80	140	8	30,000	This paper
Unconfined - two amalgam DFS	Loranca Basin, Spain	thrust-sheet top	Oligo-Miocene	humid to dry	39 to 174			5 to 10	90	20		1800	Díaz Molina & Muñoz-García (2010)
Unconfined - DFS	Cerro Barcino Fmn., Argentina	thermal sag	Cretaceous	seasonal, semi-arid	40	800	20	4.7	12	16	15	99	Foix et al. 2012; Umazano et al. (2017)
Unconfined - DFS	Simsboro Sandstone (Wilcox Group)	Passive margin	Eocene	warm temperate, humid				5	50				McGowen & Garner (1970)
Valley-confined	Brazos Valley, Tx	Passive margin	Pleistocene	warm temperate, humid	792	10,000	13	10	15	50	1.5	750	Epps (1973); Syvia & Galloway
Valley-confined	Trinity Valley, Tx	Passive margin	Pleistocene	warm temperate, humid				15	8	50		400	Blum & Aslan (2006)
Valley-confined	Colorado Valley, Tx	Passive margin	Pleistocene	warm temperate, humid				15	20	50		1000	Blum & Price (1998); Blum & Aslan (2006)
Valley-confined	Scalby Formation, Yorkshire	Extensional basin margin	Bathonian	humid	40 to 70	2250	32 to 56	6					data from upper storey - Ielpi & Ghinassi (2014)
Valley-confined	Shinarump Member, Chinle Fmn.	Back-arc	Carnian	humid				15	10	60		600	Malan (1968); Dubiel et al. (1999)
Valley-confined	Newspaper Rock, Monitor Butte Mbr. Chinle Fmn.	Back-arc	Carnian	humid to semi-arid	500	5000*	10	10					Dubiel et al. (1999); Trendell et al. (2013)

Gubitosa, 1983, 1984; Demko *et al.*, 1998; Dubiel *et al.*, 1999; Dubiel & Hasiotis, 2011) and forms a tributary valley network described in detail by Young (1964), Malan (1968) and Jaireth *et al.* (2010). Palaeovalleys are 1 to 10 km-wide and can be traced for over 50 km. They are 5 to 30 m-deep, although post-depositional erosion prior to deposition of the Monitor Butte Member means thickness values are minima. Valley-fills comprise a series of coarse grained, locally pebbly sandstones commonly with 3 to 4 stories (Fig. 13A). Stories abut against the bounding erosion surface and have no connection to strata outside the palaeovalley. Extensive, mature, well-developed interfluve palaeosols have been widely documented (e.g. Demko *et al.*, 1998; Dubiel *et al.*, 1999; Dubiel & Hasiotis, 2011; Trendell *et al.*, 2012, 2013).

An extensive mud-filled palaeovalley is preserved in the Monitor Butte Member, the stratigraphic unit overlying the Shinarump Member (Demko *et al.* 1998; Dubiel *et al.* 1999). The palaeovalley is up to 100 m-deep and extends laterally for 130 km (Demko *et al.*, 1998; Dubiel *et al.*, 1999;

Dubiel & Hasiotis, 2011). At Petrified Forest National Park, Arizona, the south-western flank of the Monitor Butte Palaeovalley is cut by a subsequent palaeovalley filled by an ASMB referred to as the Newspaper Rock sandstone unit. Here, the palaeovalley is 15 to 20 m-deep and consists of three amalgamated stories 3 to 10 m-thick of fine to medium grained sand-dominated lateral accretion packages that are restricted to the valley-fill and discrete from overbank strata adjacent to the valley (Trendell *et al.* 2013; Fig. 13B). The palaeovalley length cannot be determined due to erosion.

Laterally-confined

Díaz Molina & Muñoz-García (2010) described the architecture of an axial ASMB developed under a relatively humid climate from the Loranca Basin in central Spain. This thrust-sheet top basin developed during the Late Oligocene to early Miocene due to Alpine–related compression (Diaz-Molina *et al.*, 1989; Martinius, 2000). The north-flowing axial system was confined to the east by distal

Fig. 13. Examples of rock record valley-fill ASMB from the Chinle Formation. (A) Shinarump palaeovalley from Capitol Reef National Park. (B) Newspaper Rock palaeovalley from Petrified Forest National Park. In both examples white arrows mark the base and yellow arrows the top of the palaeovalley. Small black arrows follow large-scale lateral accretion surfaces developed within the valley-fill.

edges of DFS and to the west by interbedded alluvial fans derived from the uplifted thrust front (Díaz Molina & Muñoz-García, 2010). The axial systems formed a stacked ASMB over an 80 m-thick stratigraphic interval with channel belts interbedded laterally with floodplain deposits. The width of the meander belt is inferred to be close to 10 km, the length 50 km with channel depths up to 10 m (Díaz Molina & Muñoz-García, 2010).

A laterally confined, axial ASMB is described from the Devonian succession of East Greenland by Olsen & Larsen (1993). These authors describe a 5 to 10 km-wide outcrop belt of predominantly sandy meandering channel deposits of the > 1000 m-thick Andersson Land Formation deposited under a relatively arid climate in an extensional/transtensional basin following collapse of the Caledonian orogen (Larsen & Bengaard, 1991). The formation contains up to 15 m-thick single and mutistorey sandstone bodies which comprise up to 75% of the total formation (Olsen & Larsen, 1993). Lateral constraints to the meander belt are provided by aeolian erg deposits and a transverse terminal meandering fluvial system represented by the Rodstein Formation (Olsen & Larsen, 1993).

The Aptian McMurray Formation comprises extensive ASMB deposits present across eastern Alberta and Saskatchewan and which form the principle heavy oil reservoir unit in the region (e.g. Mossop & Flach, 1983; Flach & Mossop, 1985; Wightman & Pemberton, 1997; Smith *et al.*, 2009; Hubbard *et al.*, 2011; Labrecque *et al.*, 2011). The McMurray is located in the back-bulge area to the Mesozoic West Canada foreland basin and overlies a sub-Cretaceous unconformity which truncates Mesozoic and Palaeozoic strata (Hayes *et al.*, 1994). The meander belt is separated from foredeep strata and confined laterally to the west by a broad, linear palaeotopographic high which represents the forebulge and to the east by cratonic basement (Leckie & Smith, 1992; Blum & Pecha, 2014). The McMurray ASMB was deposited under a humid temperate climate and comprises an orogen parallel, south-east to north-west flowing continental scale fluvial system (Benyon *et al.*, 2014; Blum & Pecha, 2014). Reconstructions from outcrop, seismic reflection and well data indicate active channel widths of 750 to 800 m and channel depths of 30 to 40 m (Lebrecque *et al.*, 2011; Musial *et al.*, 2012; Bhattacharya *et al.*, 2016). Individual fluvial point bars up to ~ 40 m in thickness are predominantly sandy with some siltstone drapes in the upper part of bars associated with lateral accretion sets (Hubbard *et al.*, 2011; Hein *et al.*, 2013; Nardin *et al.*, 2013). The aerial extent of the ASMB is poorly constrained but covers at least an area of 190 x 70 km (Fustic *et al.*, 2012) and may extend for as much as 400 x 200 km based on isopach maps of the McMurray (e.g. Smith, 1994; Nardin *et al.*, 2013), although these latter figures probably represent a minimum due to post-depositional erosion.

Unconfined

A laterally extensive ASMB is preserved as part of the Salt Wash fluvial system which comprises the Salt Wash and Tidwell Members of the Upper Jurassic (Kimmeridgian) Morrison Formation in southern Utah, northern Arizona and western Colorado (Hartley *et al.*, 2015; Fig. 14). The succession is interpreted to represent a large DFS that flowed in a north to north-east direction deposited under a humid climatic regime (Craig *et al.*, 1956; Mullens & Freeman, 1957; Demko *et al.*, 2004; Owen *et al.*, 2015a,b). The system comprises large-scale amalgamated channel belt deposits that can extend tens of kilometres laterally in the proximal region. Downstream, channel belts pass progressively into floodplain facies composed of poorly developed palaeosols, ribbon channels and minor lacustrine units (Owen *et al.*, 2015b).

The ASMB is exposed in what is interpreted to be the proximal to upper medial part of the Salt Wash DFS (Owen *et al.*, 2015a,b, 2016, 2017; Fig. 14). It was mapped over an area of 9000 km^2 using satellite imagery on un-faulted, relatively flat and planar bed surfaces (140 km-long, 80 km-wide, Hartley *et al.*, 2015). To extend this work to areas where strata are folded or buried beneath younger strata, outcrop observations from vertical exposures have been integrated with data on uranium mineralisation within the Salt Wash DFS.

High quality mapping in the 1950s as part of a uranium exploration programme produced detailed maps on the distribution and nature of uranium mineralisation within Salt Wash sandstone bodies (e.g. Fisher & Hilpert, 1952; Stokes 1954). It is clear from these maps (Fig. 14B) that one of the principle controls on uranium mineralisation was the distribution of heterogeneity within point bar deposits of the sandstone bodies (see also Owen *et al.*, 2016). Mineralisation is concentrated where permeability barriers and baffles occur within the ASMB, particularly at

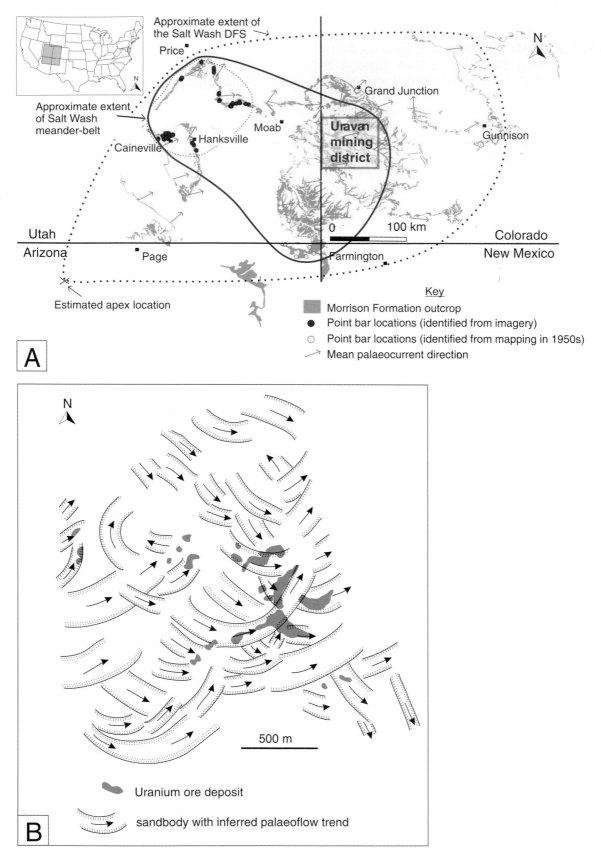

Fig. 14. Salt Wash amalgamated meander belt. (A) Map showing the lateral extent of the meander-belt using the original mapping from Hartley *et al.* (2015) combined with observations from the Uravan mining district in western Colorado where meander-belt deposits are exposed (Owen *et al.*, 2016) and mapping undertaken during uranium exploration (Fisher & Hilpert, 1952; Stokes 1954). The Salt Wash ASMB has an approximate extent of 30,000 km². (B) Map of the relationship between ore mineralisation and sandstone body geometry in the vicinity of Blanding Mines, San Juan County, Utah redrawn and modified from Stokes (1954). The meandering planform can be clearly discerned from the map as can the amalgamated nature of the meander-belt deposits.

boundaries between point bars (Stokes, 1954; Owen *et al.*, 2016). Using the uranium exploration maps from Arizona together with field mapping of ASMB deposits in the uranium mining district of western Colorado (Owen *et al.*, 2016), the ASMB can be extended southwards and eastwards to cover a total area of 30,000 km² (Fig. 14A).

Salt Wash bankfull channel depths based on storey and lateral accretion set thicknesses range from 6 to 10 m and average around 8 m in the more proximal parts of the ASMB (Hartley *et al.* 2015; Owen *et al.*, 2015b,c). Dimensional data derived from satellite imagery and field observations in the Caineville area of SW Utah indicate point bar widths of up 800 m with most between 350 and 400 m in width, similarly based on the measurement of sand-filled channel-plugs, channel widths are up to 75 m but most are between 40 to 50 m wide (Swan *et al.*, 2018).

Two DFS, the Tortola and Villalba de la Sierra, have been described from the Oligo-Miocene succession of the Loranca Basin of central Spain DFS (Díaz-Molina *et al.*, 1989; Díaz-Molina, 1993; Díaz-Molina & Tortosa, 1996; Martinius, 2000; Díaz Molina & Muñoz-García, 2010). DFS were developed under a humid climate and extended laterally for ~60 km with both systems transitioning downstream from braided to meandering channel belt deposits (Díaz-Molina *et al.*, 1989). The transition in both systems occurs approximately 20 to 30 km downstream of the apex and results in the development of an ASMB. Outcrop is not continuous across the two systems however regional facies mapping suggests that an amalgamated meander belt would have extended along strike approximately 90 km in a north-south direction across the basin and extended downstream for 10 to 20 km (Díaz-Molina, 1993; Daams *et al.*, 1996; Martinius, 2000).

The Cretaceous Cerro Barcino Formation of the Cañadón Asfalto Basin of south-central Argentina contains a well-developed amalgamated meander belt (Foix *et al.*, 2012). The formation was deposited in a post-rift thermal sag setting (Umazano *et al.*, 2017). The ASMB flowed west and is interbedded with alluvial fan material on the western edge of the basin forming an unconfined ASMB as part of the proximal to medial part of a DFS. Channels average 1.8 m in depth and 40 m in width but may be up to 4.7 m-deep and 100 m-wide (Foix *et al.*, 2012). Active channel belt widths inferred from point bar dimensions are a minimum of 800 m and the exposed aerial extent of the ASMB is ~99 km².

Discussion

It is clear from the examples documented above that amalgamated sandy meander belt deposits are common in modern continental and marine-connected sedimentary basins under a wide range of climatic and tectonic settings (Table 1). In addition, the documented rock record examples also occur in a similarly wide range of tectonic and climatic settings and overlap at a range of scales (width, length and aerial extent) with those described from present day basins (Tables 1 and 2). The common development and wide aerial extent of modern and ancient ASMBs suggests that their true rock record extent may be greater than the range documented by Gibling (2006). However, it should be noted that the large amalgamated meander belt deposits discussed here have been documented post 2006, e.g. the Salt Wash, Deweyville and McMurray examples (Table 2). These recently described rock record examples together with those from present day basins, suggest that ASMBs are probably under-represented in the stratigraphic record and may be considerably more widespread than has been appreciated previously.

Comparison between modern and ancient ASMB deposits

The data in Tables 1 and 2 detail the main parameters that can be determined for the studied modern and ancient ASMB deposits and allow a comparison between the different examples. Two key issues when comparing the two different deposit types are: 1) the limited control on horizontal dimensions from rock record systems due to lack of exposure through either post-depositional burial or erosion; and 2) the limited information on vertical dimensions in modern systems such as accumulated channel belt thickness. However, despite these issues some comparisons between modern and ancient ASMB can be made. In particular the relationship between the active channel and channel belt width appears to be similar as shown in Fig. 12, with ratios of 15 for modern day and 19 for rock record examples (Tables 1 and 2). These ratios hold firm for all climate regimes, tectonic settings and meander belt type (laterally-confined, valley-confined or unconfined). Other parameters are more difficult to compare due to the limited dataset for rock record examples and the wide range in extent of ASMB deposits related to variations in both lateral confinement and the size of the depositing river(s). In general, it is clear that the largest rivers produce the largest meander belts even when confined within a valley (Mississippi) or laterally confined (McMurray Formation), although it should be noted that even relatively small rivers can produce laterally extensive ASMB deposits covering 100s of km² whether in a confined or unconfined setting (Table 2).

Recognition of ASMB deposits

The lack of recognition of amalgamated meander belt deposits in the rock record is probably due to a combination of strong similarities between the deposits of ASMB and braided rivers and a lack of features normally considered characteristic of meandering river deposits. Common characteristics of both sandy meandering and braided fluvial deposits include the development of multi-storey, laterally extensive (sheet-like) amalgamated channel belt deposits, dominance of downstream accreting bedforms, a lack of well-developed fining upwards grain-size profiles, limited fine grained sediment preservation and limited to no soil development (e.g. Jackson, 1978; Bridge 1985, 1993; Campbell & Hendry, 1987; Hartley *et al.*, 2015; Ghinassi & Ielpi, 2015; Swan *et al.*, 2018). In addition, sandy meandering deposits lack many of the features often considered characteristic of meandering river deposits, particularly inclined heterolithic stratification, high palaeocurrent dispersion (dependent on the scale of investigation), single storey channels, well-developed fining upwards grain-size profiles and well developed soil profiles (e.g. Cant, 1982; Galloway & Hobday, 1996; Willis, 1989; Bridge, 1993; Shepherd, 2009). The lack of clear criteria to identify ASMB deposits, together with the similarity between sandy meandering and braided fluvial deposits, is considered to be the main reason why ASMB successions are not widely recognised in the rock record.

A common criterion that is widely used to distinguish between sandy meandering and braided fluvial systems is the dominance of lateral versus downstream accretion: i.e. whether the orientation of bedforms relative to an accretion surface is dominantly parallel (downstream) or orthogonal (lateral) to that surface. If both the accretion surface and overlying bedforms dip in a similar direction then downstream accretion dominates and a braided planform is normally implied.

Alternatively, if the accretion surface and overlying bedforms show a strong divergence in dip direction (commonly > 45°), then lateral accretion dominates and a meandering channel planform is inferred. A key assumption of this approach is that all braided rivers display predominantly downstream accretion and that all meandering rivers accrete predominantly laterally. This assumption is a generalised oversimplification, for example lateral accretion is the dominant accretion style in braid bars of the Brahmaputra River (Bristow, 1987), whereas downstream accretion is the predominant accretion style in large sandy point bars in the modern Mississippi (Jordan & Pryor 1992) and Ganga (Shukla *et al.*, 1999) rivers. These observations from modern rivers create significant difficulties in using downstream versus lateral accretion as a distinguishing characteristic, a feature also supported by observations from the rock record. In a recent study of an exhumed sandy point bar from the Salt Wash DFS, lateral accretion deposits account for less than 20% of the point bar package (Swan *et al.*, 2018). Also, a detailed description of the Jurassic Scalby Formation of, Yorkshire, northern England, showed that downstream migrating point bar deposits dominate the preserved meander belt succession (Ghinassi & Ielpi, 2015). These latter authors noted specifically that distinguishing between braided and meandering deposits without the aid of excellent planform exposure would be extremely difficult. These observations indicate that the predominance of either downstream or lateral accretion in sandy fluvial successions cannot be used in isolation to determine the planform characteristics of the depositing river.

Large variations in palaeocurrent dispersal (>180°) are commonly used to identify meandering fluvial deposits reflecting the differing orientation of unit bars and dunes when traced around the full extent of a point bar (e.g. Selley, 1968; Jackson, 1978; Bridge, 1985; Le Roux, 1992). However, as discussed above, braided rivers can show significant variability in accretion direction and sandy point bars can show restricted palaeocurrent dispersal patterns or be dominated by downstream accretion (Jordan & Pryor 1982; Bristow, 1987; Shukla *et al.*, 1999; Ghinassi & Ielpi, 2015; Ghinassi *et al.*, 2016). These observations indicate that without additional information, variation in palaeocurrent dispersal is not a useful distinguishing criterion in identification of fluvial planform type.

Whilst careful, detailed analysis can allow recognition of point bar geometries in 2D outcrops (e.g. Ghinassi *et al.*, 2013, 2014), this requires excellent high-quality exposure where the orientation and relationship between accretion surfaces, unit bars and dunes can be established without ambiguity. In cases where exposure quality is poor and where the size of the exposure may not be sufficient for large scale, low angle accretion surfaces to be identified (i.e. exposures are developed at a scale smaller than the meander bend), thus interpretation of channel planform style is difficult (e.g. Swan *et al.*, 2018). This is illustrated in an example for the Salt Wash ASMB which was originally interpreted as a braided river deposit (Petersen, 1984; Robinson & McCabe, 1998), prior to the availability of high resolution satellite imagery which, through the utilisation of plan view exposures, allowed identification of a meandering channel planform (Hartley *et al.*, 2015).

Storey-scale accretion surfaces represent an additional important recognition criterion for the identification of sandy meander belt deposits. These surfaces bound packages of point bar deposits that have aggraded in either lateral, upstream or downstream directions (e.g. Jordan & Pryor, 1992; Ghinassi *et al.*, 2013; Fig. 15) and define phases of scroll-bar development. These surfaces are relatively common in sandy meander belt deposits (e.g. Jordan & Pryor 1992; Ghinassi *et al.*, 2013) but are not as widely recognised in modern or ancient low sinuosity fluvial deposits (Bridge, 1985). Whilst channel scale accretion surfaces do occur in braided fluvial deposits (e.g. Lunt *et al.*, 2004, 2013), they are less regular and more commonly modified by erosion due to fluctuations in flow stage than those developed in sandy meandering rivers.

The points discussed here highlight why ASMB deposits are under-recognised in the rock record and why no single criterion can be used to definitively identify sandy meander belt deposits. A combination of systematic variations in accretion direction, palaeocurrent dispersal patterns and the identification of storey scale accretion surfaces is necessary to clearly identify ASMB deposits.

Meander-belt development and distribution

Determining the controls on amalgamated meander-belt development and distribution can be useful in developing predictive models for their occurrence and preservation within sedimentary basins. The extent of ASMB development in present day basins occurs irrespective of tectonic or climatic setting and is related to whether the meander-belt is either confined laterally by valley walls, bedrock highs or adjacent geomorphic features such as alluvial fans, or is unconfined. However, specific meander belt styles do occur in particular locations within present day basins and it is assumed therefore that similar controls were also operative in rock record examples.

Laterally confined meander belts occur primarily in axial settings within both foreland (e.g. Paraguay-Parana, Brazil-Paraguay-Argentina; Po, Italy; Table 1, Fig. 5) and extensional/transtensional basins (Gawthorpe & Leeder, 2000) where they are confined by lateral sediment input associated with alluvial fans and DFS (e.g. Carson River, Nevada, Peakall, 1998; Luangwa River, Zambia; Table 1), by bedrock highs (e.g. Zavkhan Gol; Fig. 7) of which some may be fault-bounded (e.g. Humboldt; Fig. 16). The rock record examples of axial ASMB described here from both extensional (Andersson Land Formation, Olsen & Larsen, 1993) and compressional settings (Díaz Molina& Muñoz-García, 2010) suggest that a relatively thick vertical accumulation of ASMB deposits can occur (up to 1000m) within a relatively narrow area (up to 10km in width). Stacked meander belt deposits are in general restricted to the lowest point in the basin throughout aggradation of the basin-fill succession (e.g. Gawthorpe & Leeder, 2000).

Large axial ASMB are associated with continental scale fluvial systems and examples include the modern day Paraguay River and the Cretaceous McMurray Formation, both of which occur in a low subsidence, back-bulge foreland basin settings. Differences between these two examples occur because of the amount of lateral confinement, the ratio of channel belt to ASMB width in the Paraguay and the McMurray being 1 and 25, respectively (Tables 1 and 2). For the Paraguay River this is due to lateral confinement by DFS to the west (Pilcomayo and Bermejo) and small craton-derived alluvial fans to the east. In contrast, the McMurray channel belt was able to migrate laterally across the back-bulge area between the peripheral bulge to the west and the craton to the east (Leckie & Smith, 1992; Blum & Pecha, 2014) with no significant lateral confinement. In summary, the main control on the width of axial ASMB deposits is provided by lateral constraints

Fig. 15. Examples of large, storey-scale accretion and accretion surfaces within sandy meandering channel deposits. (A) Unconfined amalgamated meander-belt, showing accretion deposits that scale to the 6 m height of the storey over which they occur, red dashed line marks base of storey, Salt Wash DFS, Caineville, Utah (see also Swan *et al.*, 2018). (B) Storey surface (between black arrows) overlain by a package of storey-scale, parallel laminated accretion deposits lying parallel to the storey surface (between the two dashed black lines) and which is onlapped by a package of point bar (PB) deposits migrating to the right of the photograph, Willwood Formation, South Fork, Wyoming. (C) Large-scale low angle accretion deposits that scale to and overlie storey surfaces. A storey surface is highlighted by the white arrows. The dashed line represents a horizontal bedding plane, Chinle Formation, Long Canyon, near Moab, Utah. (D) Large-scale accretion surface dipping towards the viewer shown by black arrows, valley-confined meander-belt, Shinarump Member, Chinle Formation, Capitol Reef National Park, SUV circled for scale at bottom.

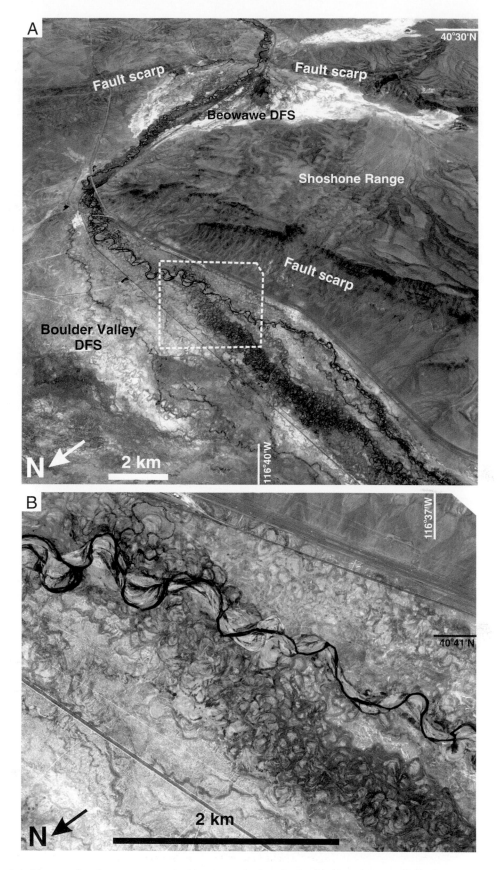

Fig. 16. Example of ASMB development in a half-graben, Humboldt River Nevada. (A) At the top of the image the unconfined meander-belt developed as part of the Beowawe DFS forms where the Humboldt River exits the valley cut through the fault scarp. In the centre of the image the Humboldt again expands to form the meander belt dominated Boulder Valley DFS located in an axial position parallel to the half-graben bounding fault which forms the northern end of the Shoshone Range. The DFS is constrained to the south by small alluvial fans derived from the footwall scarp. (B) Detail of meander-belt deposits highlighted in the white dashed box in A. Imagery from GoogleEarth®.

which vary depending on the amount of lateral sediment input (e.g. alluvial fans and DFS) and basin margin topography (e.g. cratonic bedrock highs, fault scarps).

Valley-fill ASMB are associated with a range of different settings in both the modern and ancient examples. The Okavanago Panhandle example occurs on the flexural margin of a half-graben (Fig. 9), whereas the Taquari example is incised into alluvium of an older DFS (Fig. 4). Both examples show a downstream decrease in valley depth and eventual mergence with active DFS deposition. This suggests that valley development is associated with upstream controls linked to changes in discharge/sediment supply. These valleys develop along the basin flanks, lateral to the basin margin and have been referred to as buffer (Holbrook *et al.*, 2006) or top down valleys (Weissmann *et al.*, 2015).

Many of the ASMB valley-fills documented above are developed on passive margins and are associated with valleys cut during periods of relative sea-level fall and lowstand where the meander-belt is preserved in the lower part of the valley fill, e.g. Morton *et al.*, 1996; Aslan & Blum, 1999; Blum & Aslan, 2006). These include the examples documented for the Texas Gulf Coast but also modern day examples where upstream of the present day shoreline valleys cut during the Pleistocene contain sandy meander belts such as the Mississippi, Paraná and Rio Negro examples (Figs 4 and 5; Table 1). A similar rock record example is known from the Jurassic Scalby Formation of Yorkshire, England, where an amalgamated meander-belt was developed in the lower part of a marine-connected incised valley-fill in the extensional Central North Sea Basin (Ghinassi *et al.*, 2014).

An additional type of valley-fill meander-belt is represented by those recorded in the Chinle Formation. During Chinle deposition alternation between phases of aggradation and widespread incision and degradation resulted in the development of at least 3 phases of widespread tributary drainage network development (Dubiel & Hasiotis, 2011). ASMB deposits are preserved in both lateral and axial palaeovalley fills as part of the linked tributary drainage systems.

In summary, valley-fill ASMB deposits occur in a range of different settings. They are developed as bedrock valley-fills along basins margins and within proximal basin-fill successions, as valleys cut and filled during phases of base-level fall along passive margins and regionally extensive tributary drainage valley-fill networks associated with degradational phases of a basin-fill succession. Valley confined meander belts they can vary from a single active channel belt width (Taquari) to up to 5 times the channel belt width (Mississippi) with preserved thicknesses of 20 to 30 m in rock record examples discussed here (Table 2).

Unconfined ASMB deposits in both the present day and rock record occur primarily as constituent parts of either lateral or axial DFS in extensional, compressional, transpressional, passive margin or cratonic tectonic settings. They may occur either downstream of an upper braided planform reach (e.g. Bermejo, Alaskan examples, Loranca Basin, Salt Wash, – Tables 1 and 2) or the entire DFS may be composed of meander belt deposits with the more proximal, upstream part being amalgamated (e.g. Okavango, Noa Dihing). ASMB on DFS may be formed by an individual fluvial system (e.g. Salt Wash) or represent a composite ASMB formed from laterally adjacent coalesced meander-belts of individual DFS (e.g. Noa Dihing). Most of the examples documented here form lateral to the basin margins although axial DFS are know from modern day basins (e.g. Tes River, Mongolia, Tarim River, China) and it is probable that ASMB will form a constituent part of equivalent basin-fill successions in the rock record. Thicknesses for rock record examples range from 17 m (Loranca Basin, Martinius, 2000) to 160 m (Salt Wash, Owen *et al.*, 2015b, 2017). In summary, unconfined ASMB deposits occur in all tectonic settings as important components of both modern and ancient DFS.

In summary, ASMB deposits occur in a wide range of climatic and tectonic settings with the principal controls on meander belt extent related to the amount of confinement (whether valley or laterally confined or unconfined) and the size of the river. This latter point is illustrated by the contrast in width of the active meander belts of continental scale rivers such as the Mississippi and Paraná which may be two or three times larger than many of the rivers discussed here (Table 1).

Implications for interpretation of fluvial deposits in the stratigraphic record

Considerable attention has been given to the identification of fluvial planform style and how this may have changed in response to plant colonisation of the land surface (e.g. Davies & Gibling

2010a, b; Davies *et al.*, 2011, 2017; Gibling & Davies 2012; Santos *et al.*, 2016). Prior to widespread late Devonian plant colonisation, the primary fluvial style was considered to be braided, whereas through increased floodplain stability related to colonisation, meandering river deposits became increasingly widespread from the late Devonian onwards (Davies *et al.*, 2011). Davies & Gibling (2010a, b) also noted that despite recording an increase in post-late Devonian meandering channel deposits, braided channel deposits still dominated the post-late Devonian continental rock record. Criteria used to recognise meandering fluvial deposits in these studies mainly identified single-storey fining upwards channel bodies displaying well-developed inclined heterolithic stratification (Davies *et al.*, 2011), whereas multi-storey amalgamated sandy meander belts have been documented from the Neoproterozoic of north-west Canada (Long, 1978, 2011), they have not been widely identified in the pre-late Devonian rock record. Whilst it is clear that a substantial proportion of the pre-late Devonian fluvial record is associated with what is termed a 'sheet-braided style' (Cotter, 1977), the observations herein raise the possibility that both the pre-late Devonian and post-late Devonian fluvial rock record may contain significant ASMB deposits that are as yet unrecognised.

The possibility that ASMB deposits have been misidentified as braided river deposits in the rock record raises the possibility that palaeogeographic reconstructions may need to be revised. The widespread development of braided rivers in present day basins is restricted to specific settings normally associated with high bedload sediment supply. In particular, braided rivers are common immediately adjacent to high relief glaciated mountain ranges such as the Himalayan foreland and Canterbury Plains of New Zealand (e.g. Bristow, 1987; Wilson, 1985; Weissmann *et al.*, 2010). However, in basins supplied by catchments that are restricted to relatively low altitudes (<2000 m) or which have not been widely glaciated, braided rivers are restricted either to the flanks of the basin and pass downstream into meander belts (e.g. Bermejo, Pilcomayo, Alaskan foreland basin; Figs 5 and 10; Table 1) or are absent (e.g. Taquari, Okavango; Figs 4 and 9). This observation suggests that braided rivers in post-late Devonian palaeogeographic reconstructions are likely to be restricted to areas of high sediment supply most probably associated with high relief (glaciated) mountainous catchments.

Subsurface application

The difficulties associated with the recognition of ASMB deposits from outcrop studies that have been highlighted above are further compounded when dealing with subsurface datasets (Fralick & Zaniewski, 2012). The recognition of fluvial planform from seismic reflection data can sometimes be distinguished on seismic horizon slice amplitude displays (e.g. Carter, 2003; Gulf of Thailand examples, Hubbard *et al.*, 2011; Reijenstein *et al.*, 2011; Klausen *et al.*, 2015). However, examples of seismic reflection datasets, from which a meandering planform can be identified, are encased within mudstone-dominated floodplain sediments as, unless a sufficient acoustic impedance contrast is present between the channel sandstones and mudstones, it would not be possible to recognise planform from amplitude horizon slice displays (e.g. Brown, 2011). As many of the probable ASMB deposits are likely to be associated with stacked multi-storey channel deposits, recognition of planform from seismic reflection data in most subsurface examples will be difficult.

Other commonly utilised subsurface datasets such as core, wireline and borehole image logs are also unlikely to allow identification of ASMB deposits. The key criteria that can be used to distinguish between amalgamated meandering and braided sandstone bodies such as recognition of storey scale accretion surfaces, palaeocurrent dispersal patterns and downstream *versus* lateral accretion will be extremely difficult to identify confidently with these datasets. The implications of these observations are that ASMB deposits are unlikely to be widely recognised in subsurface studies and by default will most probably be interpreted as braided fluvial deposits. The implications of this misidentification are important, as heterogeneity distribution in braided versus meandering fluvial deposits differs significantly below the channel belt scale. In braided fluvial systems, grain-size variability is restricted to the bedform and unit bar scale due to the relatively shallow flow depth and frequent erosion and redeposition within bars and channels. In contrast, point bars have preferred permeability trends that correspond to storey scale accretion surfaces and the unit bars that comprise these. Accounting for these differences in permeability distribution related to planform type is important when constructing reservoir models of subsurface fluid flow during the appraisal and production of hydrocarbon fields with fluvial reservoirs (Shepherd, 2009a, b).

CONCLUSIONS

An analysis of amalgamated sandy meander belts shows that they are widely distributed throughout both present day and rock record continental and marine connected sedimentary basins. They occur in all climatic and tectonic settings ranging from tundra to desert and including cratonic, foreland, extensional, strike-slip and passive margin basins. ASMB occur in three specific settings: in the proximal to medial areas of DFS, as laterally confined belts that mainly form axial fluvial systems and valley-confined meander belts that may infill bedrock, alluvial or coastal plain valleys. The recognition of these specific settings and the associated sandy meander belt development in present day basins provides a framework for prediction and identification of ASMB in the rock record.

The wide distribution of sandy meander belts in present day basins, but limited recognition in the rock record, suggests that their true stratigraphic distribution has yet to be determined. The lack of recognition of ASMB in the rock record is due to an absence of clear distinguishing characteristics between sandy meandering and braided fluvial deposits. Characteristics such as the development of multi-storey, laterally extensive (sheet-like) amalgamated channel belt deposits, dominance of downstream accreting bedforms, lack of well-developed fining upwards grain-size profiles, limited fine-grained sediment preservation and limited to no soil development are common to both planform types. In contrast, features normally considered characteristic of meandering rivers, such as inclined heterolithic stratification, high palaeocurrent dispersion, single storey channels and well developed fining upwards grain-size profiles, are largely absent. The authors suggest, following previous work (e.g. Jackson, 1978; Bridge 1985), that no single criterion can be used to definitively identify sandy meander belt deposits in the rock record and that a combination of systematic variations in accretion direction, palaeocurrent dispersal patterns and the identification of storey scale accretion surfaces is necessary to identify this fluvial style clearly.

The observation that ASMB deposits are most probably under-recognised in the rock record has important implications for palaeogeographic reconstructions, understanding the impact of plant colonisation on fluvial planform style and predicting the distribution of porosity and permeability in the subsurface for both reservoirs and aquifers.

ACKNOWLEDGEMENTS

We would like to acknowledge the support of the sponsors of the Fluvial Systems Research Group (FSRG) Phase 2, which include BG, BP, Chevron, ConocoPhillips and Total.

REFERENCES

Allen, M.B., Windley, B.F. and Zhang, C. (1992) Cenozoic tectonics in the Urumqi-Korla region of the Chinese Tien Shan. *Geol. Rundsch.*, **83**, 406–416.

Aslan, A. and Blum, M.D. (1999) Contrasting styles of Holocene avulsion, Texas Gulf Coastal Plain. In: *Fluvial Sedimentology VI* (Eds N.D. Smith and J.J. Rogers), *Int. Assoc. Sedimentol. Spec. Publ.*, **28**, 193–209.

Assine M.L. (2005) River avulsions on the Taquari megafan, Pantanal wetland, Brazil. *Geomorphology*, **70**, 357–371.

Assine, M.L., Merino, E.R., Pupim, F.N., Macedo, H.A. and Santos, M.G.M. (2015) The Quaternary alluvial systems tract of the Pantanal Basin, Brazil. *Brazilian Journal of Geology*, **45**, 475–489.

Bernard, H.A. and Major Jr., C.F. (1963) Recent meander-belt deposits of the Brazos River: an alluvial "sand" model (abstract). *AAPG Bull.*, **47**, 350.

Baljinnyam, I. (1993) Ruptures of major earthquakes and active deformation in Mongolia and its surroundings. *Geol. Soc. Am. Mem.*, **181**, P62.

Bayasgalan, A., Jackson, J. and McKenzie, D. (2005) Lithosphere rheology and active tectonics of Mongolia: relations between earthquake source parameters, gravity and GPS measurements. *Geophys. J. Int.*, **63**, 1151–1179.

Benyon, C., Leier, A., Leckie, D.A., Webb, A., Hubbard, S.M. and Gehrels, G.E. (2014) Provenance of the Cretaceous Athabasca Oil Sands, Canada: Implications for continental-scale sediment transport. *J. Sed. Res.*, **84**, 136–143.

Berg, R.R. (1968) Point-bar origin of Fall River Sandstone reservoirs, northeastern Wyoming. *AAPG Bull.*, **52**, 2116–2122.

Bernard, H.A. and LeBlanc. R.J. (1965) Resumé of the Quateranry Geology of the Northwestern Gulf of Mexico Province. In: *The Quaternary Geology of the United States* (Eds H.E. Wright and D.G. Frey), 137–185. Princeton University Press, Princeton.

Bernard, H.A., Major, C.F. Jr., Parrott, B.S. and LeBlanc, R.J. Sr. (1970) Recent sediments of southeast Texas: a field guide to the Brazos alluvial and deltaic plains and the Galveston barrier island complex: Texas Bureau of Economic Geology Guidebook no. 11, The University of Texas at Austin, Austin, Texas, 16 p.

Bhattacharya, J.P., Copeland, P., Lawton, T.F. and **Holbrook, J.** (2016) Estimation of source area, river paleo-discharge, paleoslope and sediment budgets of linked deep-time depositional systems and implications for hydrocarbon potential. *Earth Sci. Rev.*, **153**, 77–110.

Blakey, R.C. and **Gubitosa, R.** (1984) Controls on sandstone body geometry and architecture in the Chinle Formation (Upper Triassic) Colorado Plateau. *Sed. Geol.*, **38**, 51–86.

Blakey, R.C. and **Gubitosa, R.** (1983) Late Triassic Paleogeography and depositional history of the Chinle Formation, southeastern Utah and northern Arizona. In: *Mesozoic Paleogeography of West-Central United States* (Ed. M.W. Reynolds and E.D. Dolly), 57–76. Rocky Mountain Section, SEPM, Denver, CO.

Bluck, B.J. (1971) Sedimentation in the meandering River Endrick. *Scot. J. Geol.*, **7**, 93–138.

Blum, M.D. and **Aslan, A.** (2006) Signatures of climate vs. sea-level change within incised valley-fill successions: Quaternary examples from the Texas Gulf Coast. *Sed. Geol.*, **190**, 177–211.

Blum, M. and **Pecha, M.** (2014) Mid-Cretaceous to Paleocene North American drainage reorganization from detrital zircons. *Geology*, 10.1130/G35513.1.

Blum, M.D. and **Price, D.M.** (1998) Quaternary alluvial plain construction in response to interacting glacio-eustatic and climatic controls, Texas Gulf Coastal Plain. In: *Relative Role of Eustasy, Climate and Tectonism in Continental Rocks* (Eds K.W. Shanley and P.J. McCabe), *SEPM Spec. Publ.*, **59**, 31–48.

Blum, M., Martin, J., Miliken, K. and **Garvin, M.** (2013) Paleovalley systems: insights from Quaternary analogs and experiments. *Earth Sci. Rev.*, **116**, 128–169.

Bridge, J.S. (1985) Paleochannel patterns inferred from alluvial deposits: a critical evaluation. *J. Sed Petrol.*, **55**, 579–589.

Bridge, J.S. (1993) The interaction between channel geometry, water flow, sediment transport and deposition in braided rivers. In: *Braided Rivers* (Eds J.L. Best and C.S. Bristow), *Geol. Soc. London Spec. Publ.*, **75**, 13–71.

Bristow, C.S. (1987) Brahmaputra River: channel migration and deposition. In: *Recent Developments in Fluvial Sedimentology* (Eds F.G. Ethridge, R.M. Flores and M.D. Harvey), *SEPM Spec. Publ.*, **39**, 63–74.

Brown, A.R. (2011) Interpretation of three-dimensional seismic data, 7th ed. *AAPG Mem.*, **42**, SEG Investigations in Geophysics, No. **9**.

Bufford, K.M., Atekwana, E.A., Abdelsalam, M.G. Shemang, E., Atekwana, E.A., Mickus, K., Moidaki, M., Modisi, M.P. and **Molwalefhe, L.** (2012) Geometry and faults tectonic activity of the Okavango rift zone, Botswana: evidence from magnetotelluric and electrical resistivity tomography imaging. *J. Afr. Earth Sci.*, **65**, 61–71.

Campbell, J.E. and **Hendry, H.E.** (1987) Anatomy of a gravelly meander lobe in the Saskatchewan River, near Nipawin, Canada. In: *Recent Developments in Fluvial Sedimentology* (Eds F.G. Ethridge, R.M. Flores and M.D. Harvey), *SEPM Spec. Publ.*, **39**, 180–189.

Cant, D.J. (1982) Fluvial facies models. In: *Sandstone Depositional Environments* (Eds P.A. Scholle and D. Spearing). *AAPG Mem.*, **31**, 115–138.

Carter, D.C. (2003) 3-D seismic geomorphology: Insights into fluvial reservoir deposition and performance, Widuri field, Java Sea. *AAPG Bull.*, **87**, 909–934.

Cooke, H.J. (1980) Landform Evolution in the Context of Climatic Change and Neo-Tectonism in the Middle Kalahari of North-Central Botswana. *Trans. Inst. Brit. Geog.*, **5**, 80–99.

Cotter, E. (1977) The evolution of fluvial style, with special reference to the central Appalachian Paleozoic. In: *Fluvial Sedimentology* (Ed. A.D. Miall), *Can. Soc. Pet. Geol. Mem.*, **5**, 361–383.

Craig, L.C., Holmes, C.N., Cadigan, R.A., Freeman, V.L., Mullens, T.E. and **Weir, G.W.** (1955) Stratigraphy of the Morrison and related formations Colorado Plateau Region. *U.S. Geol. Surv. Bull.*, **1009-E**, 1–52.

Daams, R., Díaz-Molina, M. and **Mas, R.** (1996) Uncertainties in the stratigraphic analysis of fluvial deposits from the Loranca Basin, central Spain. *Sed. Geol.*, **102**, 187–209.

Davidson, S.K., Hartley, A.J., Weissmann, G.S., Nichols. G.J. and **Scuderi, L.A.** (2013) Geomorphic elements on modern distributive fluvial systems. *Geomorphology*, **180–181**, 82–95.

Davies, N.S. and **Gibling, M.R.** (2010b) Cambrian to Devonian evolution of alluvial systems: The sedimentological impact of the earliest land plants. *Earth Sci. Rev.*, **98**, 171–200.

Davies, N.S. and **Gibling, M.R.** (2010a) Paleozoic vegetation and the Siluro-Devonian rise of fluvial lateral accretion sets. *Geology*, **38**, 51–54.

Davies, N.S., Gibling, M.R. and **Rygel, M.C.** (2011) Alluvial facies during the Palaeozoic greening of the land: case studies, conceptual models and modern analogues. *Sedimentology*, **58**, 220–258.

Davies, N.S., Gibling, M.R., McMahon, W.J., Slatre, B.J., Long, D.G.F., Bashforth, A.R., Berry, C.M., Falon-Lang, H.J., Gupta, S., Rygel, M.C. and **Wellman, C.H.** (2017) Discussion on 'Tectonic and environmental controls on Palaeozoic fluvial environments: reassessing the impacts of early land plants on sedimentation'. *J. Geol. Soc. London*, **174**, 947–950.

Demko, T.M., Currie, B.S. and **Nicoll, K.A.** (2004) Regional paleoclimatic and stratigraphic implications of paleosols and fluvial/overbank architecture in the Morrison Formation (Upper Jurassic), Western Interior, USA. *Sed. Geol.*, **167**, 115–135.

Demko, T.M., Dubiel, R.F. and **Parrish, J.T.** (1998) Plant taphonomy in incised valleys: Implications for interpreting paleoclimate from fossil plants. *Geology*, **26**, 1119–1122.

Díaz-Molina, M. (1993) Geometry and lateral accretion patterns in meander loops, examples from the upper Oligocene–early Miocene, Loranca Basin, Spain. In: *Alluvial Sedimentation* (Eds M. Marzo and C. Puidefábregas), *Int. Assoc. Sedimentol. Spec. Publ.*, **17**, 115–131.

Díaz-Molina, M., Arribas-Mocoroa, J. and **Bustillo-Revuelta, A.** (1989) The Tortola and Villalba de la Sierra fluvial fans: Late Oligocene-Early Miocene, Loranca Basin, Central Spain. 4th International Conference on Fluvial Sedimentology, Barcelona-Sitges, Spain. Field Trip 7.

Diaz-Molina, M. and Munoz-Garcia, M.B. (2010), Sedimentary facies and three-dimensional reconstructions of upper Oligocene meander belts from the Loranca Basin, Spain. *AAPG Bull.*, **94**, 241–257.

Díaz-Molina, M. and Tortosa, A. (1996) Fluvial fans of the Loranca Basin, late Oligocene–early Miocene, central Spain. In: *Tertiary basins of Spain, the stratigraphic record of crustal kinematics* (Eds P. Friend and C.J. Dabrio), Cambridge University Press, Cambridge, United Kingdom, 300–307.

Dickinson, W.R. and Gehrels, G.E. (2008) U–Pb ages of detrital zircons in relation to paleogeography: Triassic paleodrainage networks and sediment disperal across southwest Laurentia. *J. Sed. Res.*, **78**, 745–764.

Drago, E.C. and Amsler M. (1998) Bed sediment characteristics in the Paraná and Paraguay Rivers. *Water Internat.*, **23**, 174–183.

Drago, E.C., Paira, A.R. and Wantzen, K.M. (2008) Channel floodplain geomorphology and connectivity of the Lower Paraguay Hydrosystem. *Ecohydrol. Hydrobiol.*, **8**, 31–48.

Dubiel, R.F. and Hasiotis, S.T. (2011) Deposystems, paleosols and climatic variability in a continental system: the Upper Triassic Chinle Formation, Colorado Plateau, U.S.A. In: *From River to Rock Record: The Preservation of Fluvial Sediments and Their Subsequent Interpretation* (Eds S. Davidson, S. Leleu and C.P. North), *SEPM Spec. Publ.*, **97**, 393–421.

Dubiel, R.F., Hasiotis, S.T. and Demko, T.M. (1999) Incised valley fills in the lower part of the Chinle Formation, Petrified Forest National Park, Arizona: Regional stratigraphic implications. In: *National Park Service Paleontological Research* (Eds V. Santucci and L. McClelland), **4**, 78–84.

Ellery, W.N., McCarthy, T.S. and Smith, N.D. (2003) Vegetation, hydrology and sedimentation patterns on the major distributary system of the Okavango Fan, Botswana. *Wetlands*, **23**, 357–375.

Epps, L.W. (1973) Geologic history of the Brazos River. *Baylor Geol. Studies Bull.*, **24**, Baylor University Press, Waco.

Fischer, R.P. and Hilpert, L.S. (1952) Geology of the Uravan mineral belt. *U.S. Geol. Surv. Bull.*, **988-A**, p. 12.

Flach, P.D. and Mossop, G.D. (1985) Depositional environments of lower Cretaceous McMurray Formation, Athabasca Oil Sands, Alberta. *AAPG Bull.*, **69**, 1195–1207.

Foix, N., Allard, J.O., Paredes, J.M. and Giacosa, R.E. (2012) Fluvial styles, paleohydrology and modern analogues of an exhumed, Cretaceoues fluvial system: Cerro Bacino Formation, Canadon Asfalto Basin, Argentina. *Cretaceous Res.*, **34**, 298–307.

Fralick, P. and Zaniewski, K. (2012) Sedimentology of a wet, pre-vegetation floodplain assemblage. *Sedimentology*, **59**, 1030–1049.

Galloway, W.E. and Hobday, D.K. (1996) *Terrigenous Clastic Depositional Systems: Applications to Petroleum, Coal and Uranium Exploration* (2d ed.). New York, Springer-Verlag, 489 p.

Gawthorpe, R.L. and Leeder, M.R. (2000) Tectono-sedimentary evolution of active extensional basins. *Basin Res.*, **12**, 195–218.

Ghinassi, M., Billi, P., Libsekal, Y., Papini, M. and Rook, L. (2013) Inferring fluvial morphodynamics and overbank flow control from 3D outcrop sections of a Pleistocene point bar, Dandiero Basin, Eritrea. *J. Sed. Res.*, **83**, 1066–1084.

Ghinassi, M., Nemec, W., Aldinucci, M., Nehyba, S., Özaksoy, V. and Fidolini, F. (2014) Planform evolution of ancient meandering rivers reconstructed from longitudinal outcrop sections. *Sedimentology*, **61**, 952–977.

Ghinassi, M. and Ielpi, A. (2015) Stratal architecture and morphodynamics of downstream migrating fluvial point bars (Jurassic Scalby Formation, UK). *J. Sed. Res.*, **85**, 1123–1137.

Ghinassi, M., Ielpi, A., Aldinucci, M. and Fustic, M. (2016) Downstream-migrating fluvial point bars in the rock record. *Sed. Geol.*, **334**, 66–96.

Gibling, M.R. (2006) Width and thickness of fluvial channel bodies and valley fills in the geological record: A literature compilation and classification. *J. Sed. Res.*, **76**, 731–770.

Gibling, M.R. and Davies, N.S. (2012) Palaeozoic landscapes shaped by plant evolution. *Nat. Geosci.*, **5**, 99–105.

Gilvear, D.J., Winterbottom, S.J. and Sidingabula, H. (2000) Character of channel planform change and meander development: Luangwa River, Zambia. *Earth Surf. Proc. Land.*, **24**, 1–16.

Gumbricht, T., McCarthy, J. and McCarthy, T.S. (2004) Channels, wetlands and islands in the Okavango delta, Botswana and their relation to hydrological and sedimentological processes. *Earth Surf. Proc. Land.*, **29**, 15–29.

Gumbricht, T.S., McCarthy, T.S. and Merry, C.L. (2001) The topography of the Okavango Delta, Botswana and its tectonic and sedimentological implications. *S. Afr. J. Geol.*, **104**, 243–264.

Hamandawana, H., Chanda, R. and Eckardt, F. (2008) Reappraisal of contemporary perspectives on climate change in southern Africa's Okavango delta sub-region. *J. Arid Environ.*, **72**, 1709–1720.

Hartley A.J. and Evenstar, L.A. (2018) Fluvial architecture in actively deforming salt basins: Chinle Formation, Paradox Basin, Utah. *Basin Res.*, **30** (1), 148–166.

Hartley, A.J., Weissmann, G.S., Nichols, G.J. and Warwick, G.L. (2010) Large distributive fluvial systems: Characteristics, distribution and controls on development. *J. Sed. Res.*, **80**, 167–183.

Hartley, A.J., Weissmann, G.S., Bhattacharya, P., Nichols, G.J., Scuderi, L.A., Davidson, S.K., Leleu, S., Chakraborty, T., Ghosh, P. and Mather, A.E. (2013) Soil development on modern distributive fluvial systems: Preliminary observations with implications for interpretation of paleosols in the rock record. In: *New Frontiers in Paleopedology and Terrestrial Paleoclimatology* (Ed. S. Driese), *SEPM Spec. Publ.*, **104**, 149–158.

Hartley, A.J., Owen, A.E., Swan, A., Weissmann, G.S., Holzweber, B.I., Howell, J., Nichols, G.D. and Scuderi, L.A. (2015) Recognition and importance of amalgamated sandy meander belts in the continental rock record. *Geology*, **43**, 679–682.

Hasiotis, S.T. (2002) Continental Trace Fossils. *SEPM Short Course Notes*, **51**, 132 pp.

Hayes, B.J.R., Christopher, J.E., Rosenthal, L., Los, G., McKercher, B., Minken, D., Tremblay, Y.M. and **Fennel, J.** (1994) Cretaceous Mannville Group of the Western Canada sedimentary basin. In: *Geological Atlas of the Western Canada Sedimentary Basin* (Eds D. Mossop and I. Shetsen), *Can. Soc. Petrol. Geol.*, 317–334.

Hein, F.J., Dolby, G. and **Fairgrieve, B.** (2013) A regional geologic framework for the Athabasca oil sands, northeastern Alberta, Canada. In: *Heavy-oil and Oil-sand Petroleum Systems in Alberta and Beyond* (Eds F.J. Hein, Leckie, D., Larter, S. and J.R. Suter), *AAPG Stud. Geol.*, **64**, 207–250.

Hitzman, M.W., Selley, D. and **Bull, S.** (2010) Formation of sedimentary rock-hosted stratiform copper deposits through Earth history. *Econ. Geol.*, **105**, 627–639.

Holbrook, J., Scott, R.W. and **Oboh-Ikuenobe, F.E.** (2006) Base-level buffers and buttresses: a model for upstream versus downstream control on fluvial geometry and architecture within sequences. *J. Sed. Res.*, **76**, 162–174.

Hubbard, S.M., Smith, D.G., Nielsen, H., Leckie, D.A., Fustic, M., Spencer, R.J. and **Bloom, L.** (2011) Seismic geomorphology and sedimentology of a tidally influenced river deposit, Lower Cretaceous Athabasca oil sands, Alberta, Canada. *AAPG Bull.*, 1123–1145.

Ielpi, A. and **Ghinassi, M.** (2014) Planform architecture, stratigraphic signature and morphodynamics of an exhumed Jurassic meander plain (Scalby Formation, Yorkshire, UK). *Sedimentology*, **61**, 1923–1960.

Iriondo, M.H. (2007) 2. Geomorphology. In: *The Middle Paraná River, Limnology of a subtropical wetland* (Eds M.H. Iriondo, J.C. Paggi and M.J. Parma), Springer, Berlin, pp. 33–52.

Jablonski, D. (1997) Recent advances in the sequence stratigraphy of the Triassic to Lower Cretaceous succession in the northern Carnarvon Basin, Australia. *APPEA Journal*, **37**, 429–454.

Jackson, R.G. (1976) Depositional model of point bars in the lower Wabash River. *J. Sed. Petrol.*, **46**, 579–594.

Jackson, R.G. (1978) Preliminary evaluation of lithofacies models for meandering alluvial streams. In: *Fluvial sedimentology* (Ed. A.D. Miall), *Can. Soc. Pet. Geol. Mem.*, **5**, 543–576.

Jaireth S., Clarke J. and **Cross A.** (2010) Exploring for sandstone hosted uranium deposits in paleovalleys and paleochannels. *AUSGEO News*, 1–5.

Johnson, R.C. (1989) Geologic history and hydrocarbon potential of Late Cretaceous – age low-permeability reservoirs, Piceance Basin, western Colorado. *U.S. Geol. Surv. Bull.*, **1787-E**, 51 p.

Jordan, D.W. and **Pryor, W.A.** (1992) Hierarchical levels of heterogeneity in a Mississippi River meander belt and application to reservoir systems. *AAPG Bull.*, **76**, 1601–1624.

Klausen, T.G., Ryseth, A.E., Helland-Hansen, W. and **Laursen, I.** (2015) Regional development and sequence stratigraphy of the Middle to Late Triassic Snadd Formation, Norwegian Barents Sea. *Mar. Petrol. Geol.*, **62**, 102–122.

Kottek, M., Grieser, J., Beck, C., Rudolf, B. and **Rubel, F.** (2006) World Map of Köppen-Geiger Climate Classification updated. *Meteorol. Z.*, **15**, 259–263.

Labrecque, P.A., Jensen, J.L., Hubbard, S.M. and **Nielsen, H.** (2011) Sedimentology and stratigraphic architecture of a point bar deposit, Lower Cretaceous McMurray Formation, Alberta, Canada. *Bull. Can. Petrol. Geol.*, **59**, 147–171.

Lahiri, S. and **Sinha, R.** (2012) Tectonic controls on the morphodynamics of the Brahmaputra River system in the upper Assam valley, India. *Geomorphology*, **169–170**, 74–85.

Larsen, P.-H. and **Bengaard, H.-J.** (1991) Devonian basin initiation in East Greenland: a result of sinistral wrench faulting and extensional collapse. *J. Geol. Soc. London*, **148**, 355–368.

Le Roux, J.P. (1992) Determining the channel sinuosity of ancient fluvial systems from paleocurrent data. *J. Sed. Petrol.*, **62**, 283–291.

Leckie, D.A. and **Smith, D.G.** (1992) Regional setting, evolution and depositional cycles of the western Canada foreland basin. In: *Foreland Basins and Fold Belts* (Eds D.A. Leckie and R.W. MacQueen), *AAPG Mem.*, **55**, 9–46.

Long, D.G.F. (2011) Architecture and depositional style of fluvial systems before land plants: a comparison of Precambrian, early Paleozoic and modern river deposits. In: *From River to Rock Record* (Eds. S. Davidson, S. Leleu and C.P. North), SEPM, Tulsa. pp 37–61.

Long, D.G.F. (1978) Proterozoic stream deposits: some problems of recognition and interpretation of ancient sandy fluvial systems. In: Fluvial Sedimentology (Ed. A.D. Miall) *Can. Soc. Petrol. Geol. Mem.*, **5**, 313–341.

Luchsinger, H.M. (2006) The late Quaternary landscape history if the middle Rio Negro Valley, northern Patagonia, Argentina: Its impact on preservation of the archaeological record and influence on late Holocene human settlement patterns. PhD Dissertation, Texas A&M University.

Lunt, I.A., Bridge, J.S. and **Tye, R.S.** (2004) A quantitative, three-dimensional depositional model of gravely braided rivers. *Sedimentology*, **51**, 377–414.

Lunt, I.A., Sambrook Smith, G.H., Best, J.L., Ashworth, P.J., Lane, S.N. and **Simpson, C.J.** (2013) Deposits of the sandy braided South Saskatchewan River: implications for the use of modern analogs in reconstructing channel dimensions in reservoir characterization. *AAPG Bull.*, **97**, 553–576.

Makaske, B., Maathuis, B.H.P., Padovani, C.R., Stolker, C., Mosselman, E. and **Jongman, R.H.G.** (2012) Upstream and downstream controls of recent avulsions on the Taquari megafan, Pantanal, South-western Brazil. *Earth Surf. Proc. Land.*, **37**, 1313–1326.

Malan, R.C. (1968) The uranium mining industry and geology of the Monument Valley and White Canyon Districts, Arizona and Utah. In: *Ore Deposits of the United States, 1933–1967*, **1** (Ed. J.D. Ridge), *Am. Inst. Min. Metall. Petrol. Engineers*, New York.

Martinius, A.W. (2000) Labyrinthine facies architecture of the Tórtola fluvial system and controls on deposition (late Oligocene–early Miocene, Loranca Basin, Spain. *J. Sed. Res.*, **70**, 850–867.

McCarthy, T.S., Ellery, W.N. and Stanistreet, I. (1992) Avulsion mechanisms on the Okavango Fan, Botswana: the control of a fluvial system by vegetation. *Sedimentology*, **39**, 779–795.

McCarthy, T.S., Smith, N.D., Ellery, W.N. and Gumbricht, T. (2002) The Okavango Delta – semi arid alluvial-fan sedimentation related to incipient rifting. In: *Sedimentation in Continental Rifts* (Eds R.W. Renaut and G.M. Ashley), *SEPM Spec. Publ.*, **73**, 179–193.

McCarthy, T.S., Stanistreet, I.G. and Cairncross, B. (1991) The sedimentary dynamics of active fluvial channels on the Okavango fan, Botswana. *Sedimentology*, **38**, 471–487.

McCarthy, T.S., Stanistreet, I,G., Cairncross, B., Etlery, W.N. and Ellery K. (1988) Incremental aggradation on the Okavango Delta-fan, Botswana. *Geomorphology*, **1**, 267–278.

McGlue, M.M., Smith, P.H., Zani, H., Silva, A., Carrapa, B., Cohen, A.S. and Pepper, M.B. (2016) An integrated sedimentary systems analysis of the Rio Bermejo (Argentina): Megafan character and overfilled southern Chaco Foreland Basin. *J. Sed. Res.*, **86**, 1359–1377.

McGowen, J.H. and Garner, L.E. (1970) Physiographic features and stratification types of coarse grained point bars: modern and ancient examples. *Sedimentology*, **14**, 77–111.

Morton, R.A., Blum, M.D. and White, W.A. (1996) Valley fills of incised coastal plain rivers, southeastern Texas. Trans. *Gulf Coast Ass. Geol. Soc.*, **46**, 321–331.

Mossop, G.D. and Flach, P.D. (1983) Deep channel sedimentation in the Lower Cretaceous McMurray Formation, Athabasca Oil Sands, Alberta. *Sedimentology*, **30**, 493–509.

Mullens, T.E. and Freeman, V.L. (1957) Lithofacies of the Salt Wash Member of the Morrison Formation, Colorado Plateau. *Geol. Soc. Am. Bull.*, **68**, 505–526.

Musial, G., Reynaud, J.Y., Gingras,M.K., Féniès, H., Labourdette, R. and Parize, O. (2012) Subsurface and outcrop characterization of large tidally influenced point bars of the Cretaceous McMurray Formation (Alberta, Canada). *Sed. Geol.*, **279**, 156–172.

Nardin, T.R., Feldman, H.R. and Carter, B.J. (2013) Stratigraphic architecture of a large-scale point-bar complex in the McMurray Formation: Syncrude's Mildred Lake Mine, Alberta, Canada. In: *Heavy-Oil and Oil-Sand Petroleum Systems in Alberta and Beyond* (Eds F.J. Hein, D. Leckie, S. Larter and J.R. Suter), *AAPG Stud. Geol.*, **64**, 273–311.

Olsen, H. and Larsen, P.-H. (1993) Structural and climatic controls on fluvial depositional systems: Devonian, NorthEast Greenland. In: *Alluvial Sedimentation* (Eds M. Marzo and C. Puigdefábregas), *Int. Assoc. Sedimentol. Spec. Publ.*, **17**, 401–423.

Owen, A, Jupp, P.E., Nichols, G.J., Hartley, A.J., Weissmann, G.S. and Sadykova, D. (2015a) Statistical estimation of the position of an apex: application to the geological record. *J. Sed. Res.*, **85**, 142–152.

Owen, A., Nichols, G.J., Hartley, A.J., Weissmann, G.S. and Scuderi, L.A. (2015b) Quantification of a distributive fluvial system: The salt Wash DFS of the Morrison Formation, SW USA. *J. Sed. Res.*, **85**, 544–561.

Owen, A., Hartley, A.J., Weissmann, G.S. and Nichols, G.J. (2016) Uranium distribution as a proxy for basin-scale fluid flow in distributive fluvial systems. *J. Geol. Soc. London. Spec. Pap.*, **173**, 569–572.

Owen, A., Nichols, G.J., Hartley, A.J. and Weissmann, G.S. (2017) Vertical trends within the prograding Salt Wash distributive fluvial system, SW United States. *Basin Res.*, **29**, 64–80.

Peakall, J. (1998) Axial river evolution in response to half-graben faulting: Carson River, Nevada, USA. *J. Sed. Res.*, **68**, 788–799.

Peterson, F. (1984) Fluvial sedimentology on a quivering craton: Influence of slight crustal movements on fluvial processes, Upper Jurassic Morrison Formation, Western Colorado Plateau. *Sed. Geol.*, **38**, 21–49.

Prochnow, S.J., Atchley, S.C., Boucher, T.E., Nordt, L.C. and Hudec, M.R. (2006) The influence of salt withdrawal subsidence on palaeosol maturity and cyclic fluvial deposition in the Upper Triassic Chinle Formation: Castle Valley, Utah. *Sedimentology*, **53**, 1319–1345.

Reijenstein, H.M., Posamentier, H.W. and Bhattacharya, J.P. (2011) Seismic geomorphology and high resolution seismic stratigraphy of inner-shelf fluvial, estuarine, deltaic and marine sequences, Gulf of Thailand. *AAPG Bull.*, **95**, 1959–1990.

Riggs, N.R., Lehman, T.M., Gehrels, G.E. and Dickinson, W.R. (1996) Detrital zircon link between headwaters and terminus of the Upper Chinle-Dockum paleoriver system. *Science*, **273**, 97–100.

Robinson, J.W. and McCabe, P.J. (1998) Evolution of a braided river system: The Salt Wash Member of the Morrison Formation (Jurassic) in southern Utah. In: *Relative Role of Eustasy, Climate and Tectonism in Continental Rocks* (Eds K.W. Shanley and P.J. McCabe), *SEPM Spec. Publ.*, **59**, 93–107.

Santos, M.G.M., Mountney, N.P. and Peakall, J. (2016) Tectonic and environmental controls on Palaeozoic fluvial environments: reassessing the impacts of early land plants on sedimentation. *J. Geol. Soc. London.*, **174**, 393–404.

Sarma J.N. and Basumallick, S. (1986) Channel form and process of the Burhi Dihing River, India. *Geogr. Ann.*, **68A**, 373–381.

Saucier, R.T. (1994) Geomorphology and Quaternary Geologic History of the Lower Mississippi Valley. Mississippi River Commission, Vicksburg, 205 pp.

Selley, R.C. (1968) A classification of paleocurrent models. *J. Geol.*, **76**, 99–110.

Shaw, P.A. and Nash, D.J. (1998) Dual mechanisms for the formation of fluvial silcretes in the distal reaches of the Okavango Delta fan, Botswana. *Earth Surf. Proc. Land.*, **23**, 705–714.

Shepherd, M. (2009) Braided fluvial reservoirs. In: *Oil Field Production Geology* (Ed. M. Shepherd), *AAPG Mem.*, **91**, 273–277.

Shepherd, M. (2009) Meandering fluvial reservoirs. In: *Oil Field Production Geology* (Ed. M. Shepherd), *AAPG Mem.*, **91**, 261–272.

Shukla, U.K., Singh, I.B., Srivastava, P. and Singh, D.S. (1999) Paleocurrent patterns in braid-bar and point-bar deposits: examples from the Ganga River, *India. J. Sed. Res.*, **69**, 992–1002.

Sibley, D., Herkenhoff, F., Criddle, D. and McLerie, M. (1999) Reducing resource uncertainty using seismic amplitude analysis on the Southern Rankin Trend, North West Australia. *APPEA Journal*, **39**, 128–48.

Smith, D.G. (1994) Paleogeographic evolution of the Western Canada Foreland Basin. In: *Geological Atlas of the Western Canada Sedimentary Basin* (Eds G.D. Mossop and I. Shetson), *Can. Soc. Petrol Geol. and Alberta Research Council*, 277–296.

Smith, D.G., Hubbard, S.M., Leckie, D.A. and Fustic, M. (2009) Counter point bar deposits: lithofacies and reservoir significance in the meandering modern Peace River and ancient McMurray Formation, Alberta, Canada. *Sedimentology*, **56**, 1655–1669.

Smith, R.M.H. (1987) Morphology and depositional history of exhumed Permian point bars in the southwestern Karoo, South Africa. *J. Sed. Petrol.*, **57**, 19–29.

Soldano, F.A. (1947) Regimen y Aprovechamiento de la Red Fluvial Argentina, Parte II: Rios de la Region Arida y de La Meseta Patagonica. Editorial Cimera, Buenos Aires.

Stanistreet, I.G., Cairncross, B. and McCarthy, T.S. (1993) Low sinuosity and meandering bedload rivers of the Okavango Fan: channel confinement by vegetated levees without fine sediment. *Sed. Geol.*, **85**, 135–156.

Stanistreet, I.G. and McCarthy, T.S. (1993) The Okavango Fan and the classification of subaerial fan systems. *Sed. Geol.*, **85**, 115–133.

Stewart, J.H., Poole, F.G. and Wilson, R.F. (1972) Stratigraphy and origin of the Upper Triassic Chinle Formation and related Upper Triassic strata in the Colorado Plateau region. *U.S. Geol. Surv. Prof. Paper*, **690**, 336 p.

Stokes, W.L. (1954) Some stratigraphic, sedimentary and structural relations of uranium deposits in the Salt Wash Sandstone. *US Atomic Energy, Final Report*, RME-3102.

Swan, A., Hartley, A.J., Owen. A. and Howell. J. (2018) Reconstruction of a sandy point bar deposit: implications for fluvial facies analysis. This Volume.

Sylvia, D.A. and Galloway, W.E. (2006) Morphology and stratigraphy of the Late Quaternary lower Brazos valley: implications for paleoclimate, discharge and sediment delivery. *Sed. Geol.*, **190**, 159–175.

Tooth, S. and McCarthy, T.S. (2004) Controls on the transition from meandering to straight channels in the wetlands of the Okavango Delta, Botswana. *Earth Surf. Proc. Land.*, **29**, 1627–1649.

Trendell, A.M., Atchley, S.C. and Nordt, L.C. (2012) Depositional and diagenetic controls on reservoir attributes within a fluvial outcrop analog: Upper Triassic Sonsela Member of the Chinle Formation, Petrified Forest National Park. *AAPG Bull.*, **96**, 679–707.

Trendell, A.M., Atchley, S.C. and Nordt, L.C. (2013) Facies analysis of probable large-fluvial-fan depositional system: the Upper Triassic Chinle Formation at Petrified Forest National Park, Arizona, U.S.A. *J. Sed. Res.*, **83**, 873–895.

Umazano, A.M., Krause, J.M. Bellosi, E.S., Perez, M., Visconti, G. and Melchor, R.N. (2017) Changing fluvial styles in volcaniclastic successions: A Cretaceous example from the Cerro Barcino Formation, *Patagonia. J. S. Am. Earth Sci.*, **77**, 185–205.

Weissmann, G.S., Hartley, A.J., Nichols, G.J., Scuderi, L.A., Olson, M., Buehler, H. and Banteah, R. (2010) Fluvial form in modern continental sedimentary basins. *Geology*, **38**, 39–42.

Weissmann, G.S., Hartley, A.J., Nichols G.J., Scuderi, L.A., Olson, M., Buehler H. and Massengill, L. (2011) Alluvial facies distributions in continental sedimentary basins – Distributive fluvial Systems. In: *From River to Rock Record* (Eds S. Davidson, S. Leleu and C.P.North), *SEPM Spec. Publ.*, **97**, 327–356.

Weissmann, G.S., Hartley, A.J., Scuderi, L.A., Nichols, G.J., Owen, A., Wright, S., Felicia, A.L., Holland, F. and Anaya, F.M.L. (2015) Fluvial geomorphic elements in modern sedimentary basins and their potential preservation in the rock record: A review. *Geomorphology*, **250**, 187–219.

Wightman, D.M. and Pemberton, S.G. (1997) The Lower Cretaceous (Aptian) McMurray Formation: an overview of the Fort McMurray area, northeastern Alberta. In: *Petroleum Geology of the Cretaceous Mannville Group, Western Canada* (Eds S.G. Pemberton and D.P. James), *Can. Soc. Pet. Geol. Mem.*, **18**, 312–344.

Willems, C.J.L., Hamidreza, M.N., Weltje, G.J., Donselaar, M.E. and Bruhn, D.F. (2015) Influence of fluvial sandstone architecture on geothermal energy production. *Proc. World Geothermal Conf., Melbourne Australia*, **2015**, p. 1–12.

Willis, B.J. (1997) Architecture of fluvial-dominated valley-fill deposits in the Cretaceous Fall River Formation. *Sedimentology*, **44**, 735–757.

Willis, B.J. (1993) Bedding geometry of ancient point bar deposits. In: *Alluvial Sedimentation* (Eds M. Marzo and C. Puigdefabregas), *Int. Assoc. Sedimentol. Spec. Publ.*, **17**, 101–114.

Willis, B.J. (1989) Paleochannel reconstructions from point bar deposits: a three dimensional perspective. *Sedimentology*, **36**, 757–766.

Wilson, D.D. (1985) Erosional and depositional trends in rivers of the Canterbury Plains, New Zealand. *J. Hydrology* (New Zealand), **24**, 32–44.

Witkind, I.J. (1956) Uranium deposition at the base of the Shinarump conglomerate, Monument Valley, Arizona. *US Geol. Surv. Bull.*, **1030-C**, 99–130.

Young, R.G. (1964) Distribution of uranium deposits in the White Canyon-Monument Valley district, Utah-Arizona. *Econ. Geol.*, **59**, 850–73.

Int. Assoc. Sedimentol. Spec. Publ (2018) **48**, 385–418.

A novel approach for prediction of lithological heterogeneity in fluvial point-bar deposits from analysis of meander morphology and scroll-bar pattern

CATHERINE E. RUSSELL,[†,‡] NIGEL P. MOUNTNEY,[†] DAVID M. HODGSON[†] and LUCA COLOMBERA[†]

[†] *Fluvial & Eolian Research Group, School of Earth and Environment, University of Leeds, Leeds, UK*
[‡] *School of Geography, Geology and the Environment, University of Leicester, Leicester, UK*

ABSTRACT

Meandering fluvial reaches exhibit a wide range of morphology, yet published interpretations of ancient meander-belt deposits do not reflect the stratigraphic complexity known to be associated with such variability. An improved understanding of processes that generate stratigraphic heterogeneity is important to improve predictions in sedimentary facies distributions in sub-surface settings. Quantification and classification of planform geomorphology of active fluvial point bars and their recently accreted deposits enables determination of spatio-temporal relationships between scroll-bar pattern and resultant meander shape. Scroll-bar deposits describe an overall pattern of lateral accretion that records how a meander has grown incrementally over time. Analysis of 260 active meander bends, from 13 different rivers, classified by a range of parameters including climatic regime, gradient and discharge, has been undertaken. Assessment of scroll-bar morphology and growth trajectory has been undertaken using remotely sensed imagery in Google Earth Pro. Twenty-two distinct styles of meander scroll-bar pattern are recognised within active meander bends. These are grouped into 8 types that reflect growth via combinations of expansion, extension, rotation and translation. A novel technique for predicting the variable distribution of heterogeneity in fluvial point-bar elements integrates meander-shape and meander scroll-bar pattern. The basis for predicting relative lithological heterogeneity is the observation that deposited sediments fine downstream around a meander bend and outwards as a barform grows and tightens due to bend expansion. Observations of these trends are seen in experimental models, modern fluvial systems and in the ancient record at both outcrop and in the sub-surface. These trends permit planform geometries to be compared with distributions of bar-deposit lithology types. The method is applied to predict heterogeneity distribution in both sub-surface and outcrop settings. Seismic-reflection data that image point-bar and related elements of the McMurray Formation (Cretaceous, Alberta, Canada) are used to test the predictive capability of the method by comparing predicted heterogeneity to trends known from analysis of gamma-ray data available from densely distributed well-log records. Outcrop data from a point-bar deposit in the Montanyana Group (Ypresian, southern central Pyrenees, Spain) are used to test the method by comparing heterogeneity predictions with lithologies observed seen in the outcrop. This novel method constrains heterogeneity predictions in fluvial point-bar deposits for which direct lithological observations are not possible or are limited. The method therefore provides the basis of a predictive tool for improving understanding of a fragmentary geological record, including prediction of lithological heterogeneity from outcrops of limited spatial extent, or from subsurface seismic datasets.

Keywords: Fluvial meander, bend morphology, heterogeneity, point bar, scroll bar pattern, river

Fluvial Meanders and Their Sedimentary Products in the Rock Record, First Edition.
Edited by Massimiliano Ghinassi, Luca Colombera, Nigel P. Mountney and Arnold Jan H. Reesink.

INTRODUCTION

Fluvial point bars that develop on river bends accumulate as laterally discontinuous sedimentary architectural elements (e.g. Allen, 1965; Fielding & Crane, 1987; Mackey & Bridge, 1995; Donselaar & Overeem, 2008; Colombera *et al.*, 2017). These elements are internally characterised by complex distributions of lithofacies (e.g. Thomas *et al.*, 1987; Tye, 2004; Miall, 2006; Durkin *et al.*, 2015; 2017; Ghinassi *et al.*, 2016). Architectural elements representative of accumulated and preserved point bars are commonly sand-rich (Allen, 1965), though grain-size may vary from pebbles, through sand, to clay and/or silt (Allen, 1965; Miall, 1988). The lithological variability of point-bar strata depends on the sediment available in the system. The finest sediment may be sand in systems where no clay is present. Yet, more typically, mud fractions are present in the majority of point-bar deposits and the finest sediment fractions referred to in this study are referred to as 'mud-prone' for convenience. Quantifying and understanding the formation and distribution of this lithological variability is important in applied geology because intra-point-bar mud-prone strata can influence permeability pathways by acting as baffles to fluid flow (Fielding & Crane, 1987; Miall, 1988; Tye, 2004) and packages of such strata are known to influence oil and gas production from fluvial hydrocarbon reservoirs (e.g. Brown & Fisher, 1980; Putnam & Oliver 1980; Mossop & Flach, 1983; Hubbard *et al.*, 2011). In point-bar architectural elements, mud-prone deposits commonly alternate with sand-prone deposits forming couplets that are commonly referred to as inclined heterolithic strata (IHS; Thomas *et al.*, 1987: Fig. 1B). Such IHS can take a variety of forms: (i) laterally continuous around the point-bar element and vertically continuous from the top to the bottom of body of the strata; (ii) laterally continuous but vertically discontinuous such that mud drapes are only present in the uppermost or basal parts of the

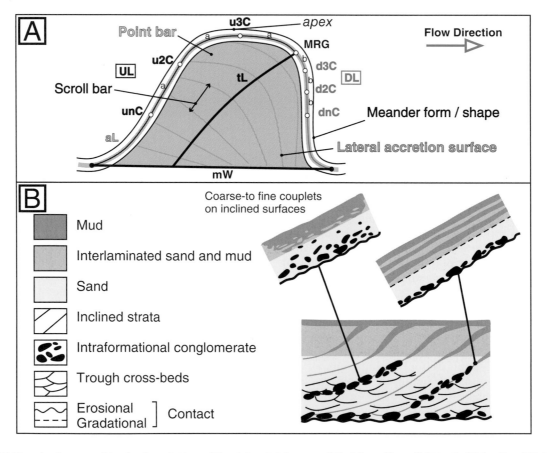

Fig. 1. (A) Terminology used in the description of fluvial point-bars modified from Russell (2017); (B) Inclined Heterolithic Strata (IHS), modified from Thomas (1987). See Methodology section for explanation of terms in A.

preserved architectural element; and (iii) laterally and vertically discontinuous (Thomas *et al.*, 1987; Miall, 1988). IHS is a product of lateral accretion on the inner bank of the fluvial channel (Thomas, *et al.* 1987). Each package of IHS lies between two accretionary or minor erosional surfaces, with the finer-grained (typically mud) fraction usually interpreted to have been deposited in the latter stages of a single major flood event (Bridge & Deimer, 1983). Mud drapes are preserved in the waning flood stage (Thomas *et al.*, 1987); the lower parts of mud drapes may be truncated by the increasing flow energy as the river returns to bank-confined flow. Packages of IHS are especially commonly preserved in tidal settings where tidal 'push' periodically slows flow, allowing deposition of fine-grained sediment fractions from suspension (e.g. Reineck, 1967; Yang & Nio, 1985). Packages of IHS are comparatively less common (though not uncommon) in fluvial settings (Verrien *et al.*, 1967; Geehan *et al.*, 1986) and can reach tens of metres in length (Verrien *et al.*, 1967; Geehan *et al.*, 1986). Where a hydrocarbon reservoir is comprised of stacked point-bar deposits (e.g. Hubbard *et al.*, 2011), the variability in mud-drape orientation within a mosaic of bars will give rise to more complex styles of compartmentalisation. Geological models, which themselves are used to inform reservoir models, have historically simplified or even ignored the variability of heterogeneity within point-bar deposits (Fielding & Crane, 1987; Jordan & Pryor, 1992; Pranter *et al.*, 2009; Hassanpour *et al.*, 2013), typically because the scale of such heterogeneity falls below the resolution of the model being proposed. As such, a disparity exists between the lithological complexity documented from meandering fluvial systems and the relative simplicity of facies models developed for successions of such deposits (Miall, 2006). To improve prediction of this inherent stratigraphic complexity it is important to understand the formative processes that give rise to meanders with particular morphological traits, across a range of physiographic settings (Rosgen, 1985; Gutierrez & Abad, 2014). Studies of modern fluvial systems are useful as a way to inform rock record interpretations (Tye, 2004) because they enable observation of meander-shape variability (Allen, 1965; Hooke, 1984) and the surface expression of meander accretion increments (Daniel, 1971; Thompson, 1986; Ielpi & Ghinassi, 2014; Ghinassi *et al.*, 2016). Scroll bars accrete over time to reflect growth increments of a developing meander (Allen, 1965; Ielpi & Ghinassi, 2014); they form a series of ridges and swales on the point-bar surface (Nanson, 1980; Durkin, 2015; Fig. 1A). Such features arise as a result of primary flow that erodes the outer bank yet simultaneously results in deposition on the inner bank, in turn driving channel migration by bar pull (Van de Lageweg *et al.*, 2014).

The geometry of the meander-bend apex (Fig. 1A) causes the channel to change direction, which itself causes the helical flow structure present within the channel to strengthen (Roberts, 2014). Helical secondary flow causes turbulent bursting, resulting in varying flow velocities within the fluvial channel (Jackson, 1976; Nanson, 1980). Comparatively stronger currents are observed at the outer bank (Leopold & Wolman, 1960; Nanson, 1980; Thompson, 1986; Roberts, 2014) and therefore material is eroded from here to be deposited in the area of weaker flow, on the inner bank (Jackson, 1976; Fustic *et al.*, 2012). Deposited sediment tends to fine downstream in a predictable manner, due to a progressive weakening of flow on the inner bank of the meander bend downstream of a bend apex. Recognition of this process forms the basis for predicting the expected location of mud-prone and sand-prone zones within a meander sandbody. Furthermore, if downstream accretion of the point-bar element is occurring, accommodation space may be created on the concave outer bend of the meander, which may in turn lead to the accretion of a counter point-bar deposit (Smith *et al.*, 2009; Fig. 2A). Downstream from the meander bend apex, erosion occurs where the thalweg meets the outer bank. The distance downstream from the bend apex where such outer bank erosion occurs is dependent on the geometries of adjoining meander shapes in the reach. Extensive data have been collected from meandering fluvial reaches that recognise downstream fining relationships (e.g. Leopold & Wolman, 1960; Jackson, 1975; Bridge & Jarvis, 1982; Thomas *et al.*, 1987; Labrecque *et al.*, 2011; Fustic *et al.*, 2012). The observed relationship between downstream fining and meander geometry may be used alongside analysis of scroll-bar geometry to determine the past shapes of a meander bend. From these relationships, a first-order prediction of the expected relative distribution of heterogeneity can be made from observation of plan-view morphology alone. However, where meander bend dynamics are complex a simplified approach is required in order to develop a predictive model.

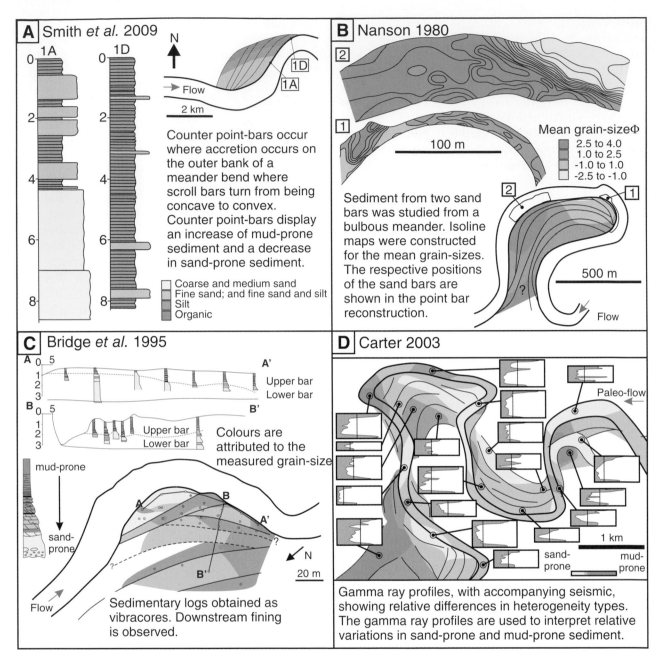

Fig. 2. Literature summary of heterogeneity data in point-bars: (A) An interpreted summary of a counter point bar in a meander from the Peace River, Alberta, Canada (Smith *et al.* 2009). Colours were attributed to defined grain-size and the logs 1A and 1D were used to interpret relative overall planform heterogeneity of the point-bar; (B) An interpreted summary of a point-bar from the Beatton River, British Colombia, Canada (Nanson 1980). Colours were attributed to the defined mean grain-size and sand bars 1 and 2 were coloured accordingly. The relative overall planform heterogeneity of the point-bar was then estimated; (C) Sedimentary logs, obtained as vibracores, were used to interpret the relative overall planform heterogeneity of the point-bar by analysing each log for average grain-size and plotting an appropriately coloured point on the planform map (Bridge *et al.*, 1995). Two lines with their associated logs are depicted (A-A' and B-B') and show an increasing proportion of mud-prone sediment downstream around the point-bar; (D) An interpreted summary of the overall planform heterogeneity of the interpreted seismic from the Widuri field, Java Sea (Carter 2003). The reservoir depth is 1000–1200 m subsea and is an arenite sandstone. The mean relative grain-size, as indicated by vertical black lines, was interpreted from the gamma-ray profiles, where high values are more mud-prone and low values are more sand-prone. These were then used to mark heterogeneity types onto the interpreted seismic, which were in turn used to determine relative sand:mud ratios.

The aim of this study is to develop a method by which the relationship between meander-bend morphology, scroll-bar pattern and the resultant stratigraphic heterogeneity of fluvial meandering reaches can be determined. Specific research objectives are as follows: (i) to document the extensive variability of both meander shape and scroll-bar pattern known from presently active meandering fluvial systems; (ii) to develop a novel scroll-bar classification scheme that is independent of meander shape; (iii) to demonstrate how an understanding of variability of scroll-bar pattern can be used to constrain and predict patterns of relative lithological heterogeneity in accumulated point-bar deposits, in both the sub-surface and at outcrop; and (iv) to discuss the application of this approach to predict the internal lithological organisation of fluvial point-bar elements from reflection-seismic time slices and from exhumed successions of limited lateral or areal extent.

BACKGROUND

Active sedimentary systems are commonly employed as analogues to explain and account for depositional architecture (Bridge & Tye, 2000; Reijenstein *et al.*, 2011; Ghinassi, 2011; Durkin *et al.*, 2015; Ghinassi *et al.*, 2016) and to constrain parameters for geological models used for reservoir modelling (Fielding & Crane 1987; Tye, 2004; Yan *et al.*, 2017). However, current methods for quantifying variability in modern fluvial systems and applying it to ancient successions remain limited, which therefore limits our understanding of morphology and variability of lithological heterogeneity in ancient successions (Tye, 2004; Miall, 2006). Although meandering-river processes can be similar between reaches of different rivers (Leopold & Wolman, 1960), autogenic and allogenic influences (Rosgen, 1985; Blum & Törnqvist, 2002; Ghinassi *et al.*, 2016) result in extensive variability in sedimentary architecture and distributions of heterogeneity (Thomas *et al.*, 1987; Durkin *et al.*, 2015). Two principal morphological descriptors – meander shape and the surface expression of the associated scroll-bar accretion direction – are herein examined to develop an approach for comparing fluvial geometries in order to predict lithological heterogeneity.

Meander shapes

Meander shape refers to the morphology of a bend in a river. In this study, a bend is defined as the shape that lies between the upstream and downstream ends of the tangent that would connect a straight channel if there was no meander (mW in Fig. 1A). Quantification and classification of a static meander shape (i.e. its present form) is difficult because the natural form is highly variable and cannot typically be defined objectively using simple shape parameters (Phillips, 2003; Miall, 2006).

Methods for the investigation of meander shape

Existing approaches for the description and classification of meander forms include shape matching (e.g. Brice, 1974; Allen, 1982; Bridge, 2003; Ielpi & Ghinassi, 2014), measuring sinuosity (Hooke, 2004); and measuring the radius-of-curvature (Nanson & Hickin, 1983; Hudson & Kessel, 2000; Sambrook Smith *et al.*, 2016). However, these approaches are not able to account for complexities in the entire bend shape in a clearly defined, repeatable manner. The approach used in this study (Russell, 2017: see 'Methodology') enables a range of meander shapes to be defined geometrically into 4 main groups (open asymmetric, angular, bulbous and open symmetric), in a repeatable and semi-quantifiable manner (Fig. 3).

Scroll-bar pattern

Episodes of meander growth are recorded within point-bar deposits (Allen, 1965; Thompson, 1986; Ielpi & Ghinassi, 2014) and are expressed as scroll-bar morphologies. Scroll-bar deposits represent the incremental growth (accretion) of point bars through bend expansion, translation, rotation, or combinations thereof (Daniel, 1971). Such growth behaviour results in the development of scroll bars, which are the geomorphic expression of consecutively laterally accreted packages, (Schumm, 1963; Allen, 1965; e.g. Bridge, 2003; Fig. 4C; Ielpi & Ghinassi, 2014; Fig. 4A). Typically, point-bar elements comprise multiple scroll-sets, which are themselves groups of genetically related scroll-bars with a common plan-view trajectory recorded by the progressive shift of the bend apex. Genetically related scroll-bar components are composed of genetically related bedsets in the point-bar stratigraphy (Ghinassi *et al.*, 2016).

		CODE	OUTLINE	DESCRIPTION	APPX. RATIO tL:W	SHAPE AT APEX (Huddleston 1973)
Group 1	OPEN ASYMMETRIC	S1a		One limb perpendicular to width of meander.	1:1	B2 B3
		S1b		One limb perpendicular to width of meander	2:3	C4
		S1c		One limb perpendicular to width of meander. Box-like.	1:2	B2
		S1d		One limb perpendicular to width of meander.	1:2	D2
		S1e		Overturned. Limbs are approximately parallel.	2:1	B4
		S1f		Overturned.	1:1	B3
		S1g		Recumbent.	1:2	A2 B4
		S1h		Overturned. Limbs are approximately parallel.	1:1	B4
		S1i		Recumbent. Rounded apex.	1.5:1	B3 B4
Group 2	ANGULAR	S2a		Top is typically 'pinched'.	1:1	D4
		S2b		Symmetrical. Close to 90°.	1:1	E2
		S2c		Most commonly asymmetrical.	0.5:1	E2
		S2d		Most commonly symmetrical. Pronounced point.	0.5:1	E2
Group 3	BULBOUS	S3a		Not overturned. Rounded apex.	2:1	B3
		S3b		Sometimes overturned.	1:1	B2
		S3c		Not overturned.	2:3	B1 B2
		S3d		Sometimes asymmetric. Rounded shape at apex.	2:1	C3
		S3e		Typically slight asymmetry.	2:1	B2
		S3f		Recumbent. Typically one straight limb.	2:1	B3
		S3g		Overturned. Rounded shape at apex. Irregular.	3:1	C3
Group 4	OPEN SYMMETRIC	S4a		Elongate. Limbs are approximately parallel.	3:1	B4
		S4b		Limbs are approximately parallel.	1:1	B3 B4
		S4c		Typically rounded apex.	1:1	B3 B4
		S4d		Typically skewed.	0.5:1	B3
		S4e		Typically not skewed.	0.25:1	D1

Fig. 3. A classification of meander form and the characteristics of each geometrically defined shape (Russell, 2017).

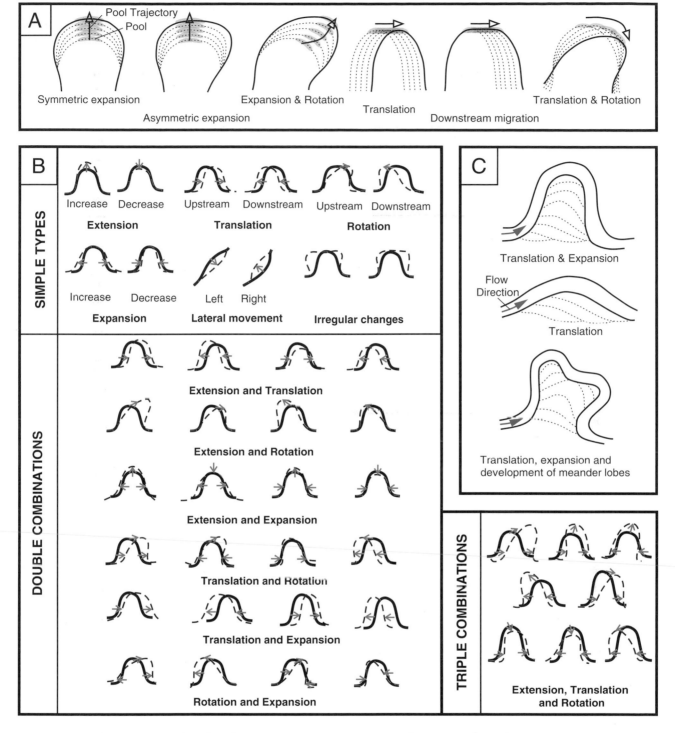

Fig. 4. A figure to show examples of the literature where meander shape change is depicted; (A) Ielpi & Ghinassi, (2014); (B) Hooke (1977); (C) Bridge (2003).

Methods for the investigation of scroll-bar pattern

Widely used models typically depict the most recent episode of meander growth. Many such models commence with an already established sinusoidal shape (e.g. Leopold & Wolman, 1960; Daniel, 1971; Willis & Tang, 2010). Few models depict initiation of point-bar growth from a relatively straight channel reach (Lewin, 1976; Thompson, 1986). Widely applied point-bar facies models (Schumm, 1963; Allen 1964; 1965; 1983; McGowen & Garner, 1970) do not account for the large variability in form observed in modern systems. The freely meandering river reaches selected for examination in this study encompass a broad range of meander shapes and scroll-bar patterns.

Point-bar heterogeneity

Understanding the history of evolution of a meander is essential for improved prediction of lithofacies distribution within its accumulated deposits. Variations in channel geometry, channel orientation and channel position each act as a control on the sedimentology of point-bar deposits (Willis, 1989). Inclined heterolithic strata (IHS) are packages of strata composed of alternating mudstone and sandstone beds, frequently found in fluvial point-bar deposits (Thomas, 1987; Fig. 1B), such as those of the hydrocarbon-bearing Cretaceous McMurray Formation, Alberta (Strobl et al., 1997; Labrecque et al., 2011; Brekke, 2015; Jablonski & Dalrymple, 2016; Durkin et al., 2017). Therefore, prediction of the spatial distribution of occurrence of IHS is important to reservoir geologists and engineers (Lasseter et al., 1986; Miall, 1988; Pranter et al., 2000; Labrecque et al., 2011). The calibre of sediment present commonly fines downstream in a fluvial point-bar deposit (Thomas et al., 1987; Labrecque et al., 2011; Fustic et al., 2012, Nardin et al., 2013). In some cases where sedimentation occurs on the outer bank, the calibre of sediment may fine upstream and produce a counter point-bar deposit (Smith et al., 2009). Counter point bars represent the fine-grained distal end of a point bar where the scroll bars turn from being concave to convex and may be characterised by a transition between sand-prone and mud-prone sediment deposits (Smith et al., 2009). Wightman & Pemberton (1997), Smith (1985), Wood (1989) and Fustic et al. (2012) present models where sediment calibre fines downstream in a point-bar deposit; such trends are observed in many modern meandering rivers (McGowen & Garner, 1970; Jackson, 1975; Bridge & Jarvis, 1982; Smith et al., 2009). The observed scroll-bar patterns in Fig. 2A and C are contrasting, yet both examples exhibit the downstream fining of sediment around a meander bend. Downstream fining has also been recognised extensively in exhumed successions (e.g. Durkin, 2015; Wu et al., 2015; Ghinassi et al., 2016) and in the subsurface (Hubbard et al., 2011; Labrecque et al., 2011).

The downstream-fining of a fluvial point-bar deposit occurs because flow moves at different speeds, and in different directions, around a meander bend (Thompson, 1986; Leopold & Wolman, 1960). The downstream-fining trend has been attributed primarily to a change in flow dynamics at the meander-bend apex (Jackson, 1976) because the meander-bend apex geometry represents a directional change, which itself causes the helical flow present within water flowing in the channel to strengthen (Roberts, 2014). This leads to turbulent bursting, which in turn results in vertical anisotropy of streamflow velocity (Jackson, 1976). Differences in flow velocity are observed laterally across the channel in the region of the meander-bend apex (e.g. Nanson, 1980), whereby the strong helical flow is observed at the outer bank and a weaker secondary helical flow is observed at the inner bank (Leopold & Wolman, 1960; Nanson, 1980; Thompson, 1986; Roberts, 2014). The weaker secondary helical flow, caused by the change in flow direction from the meander-bend apex, encourages the deposition of sediment on the downstream limb of the meander bend. Sediment deposited in this location tends to be relatively finer grained than the material deposited elsewhere on the inner bank of the point-bar deposit (Jackson, 1976; Fustic et al., 2012). This results in a tendency for the relative ratio of 'sand-prone' to 'mud-prone' sediment to decrease downstream (Bridge et al., 1995; Fustic et al., 2012).

METHODOLOGY

The static form (i.e. observed morphology at a moment in time) of 20 consecutive meander bends and their associated scroll bars in a single reach have been studied from each of 13 globally distributed meandering river reaches (Table 1; 260 meanders in total). The rivers chosen for study are from different physiographic locations (Fig. 5; Table 1).

Table 1. A table to show the rivers selected for this study. Data for reach length, gradient and along-reach distance to sea is measured from satellite imagery. Climate data is derived by visual comparison of maps by Peel *et al.* (2007) and the reach location. River discharge data has been collated from the literature; i) Swales *et al.* 2000; ii) Meybeck & Ragu 1997; iv) Bobrovitskaya *et al.* 1996. Where the mean annual discharge data was unavailable, it was calculated using the empirical relationship ($w = 3.15 \, Q^{0.49}$) from Nixon *et al.* (1959), using a channel width measured from satellite imagery (iii).

River	Country	Climate (Peel et al. 2007)		Latitude	Longitude	Reach length (km)	Reach gradient (upstream elevation / downstream elevation)	Mean Annual River discharge (Q=m³s⁻¹)	Along-reach distance to sea (km)
Ok Tedi (Fly)	Papau New Guinea	A	Af	-8.5	143.6833	148.399	6.06473E-05	983[i]	442
Senegal	Senegal	B	BWh	16.0167	-16.5	159.930	1.87582E-05	773[ii]	229
Murray	Australia	B	Bsk	-35.3667	139.3667	42.467	9.41908E-05	250[ii]	522
Yana	Russia	D	Dfd	71.5167	136.5333	102.316	6.84155E-05	1087[ii]	19
Colville tributary	USA	D	Dfc	70.4167	-150.5	22.391	0.00053593	202[iii]	107
Kuskokwim	USA	D	Dfc	60.2833	-162.45	179.915	4.44654E-05	1902[ii]	86
Mississippi	USA	C	Cfa	28.95	-89.4	298.013	4.36223E-05	16769[ii]	421
Purus (Amazon)	Brazil	A	Af	0.1667	-49	304.250	2.95809E-05	28266[iii]	2279
Irtysh (Ob)	Russia	D	Dfd	66.75	69.5	224.830	8.89561E-06	2125[iv]	1495
Kolyma tributary	Russia	D	Dfc	68.5667	160.9667	116.920	8.55286E-06	26368[iii]	144
azos	USA	C	Cfa	29.5833	-95.75	61.690	6.48403E-05	160[ii]	274
Mwenezi (Limpopo)	Mozambique	B	BSh	-25.25	33.5	55.407	0.000631689	409[iii]	481
Appalachicola	USA	C	Cfa	30.7	-85.8667	9.493	0.000210682	679[ii]	108

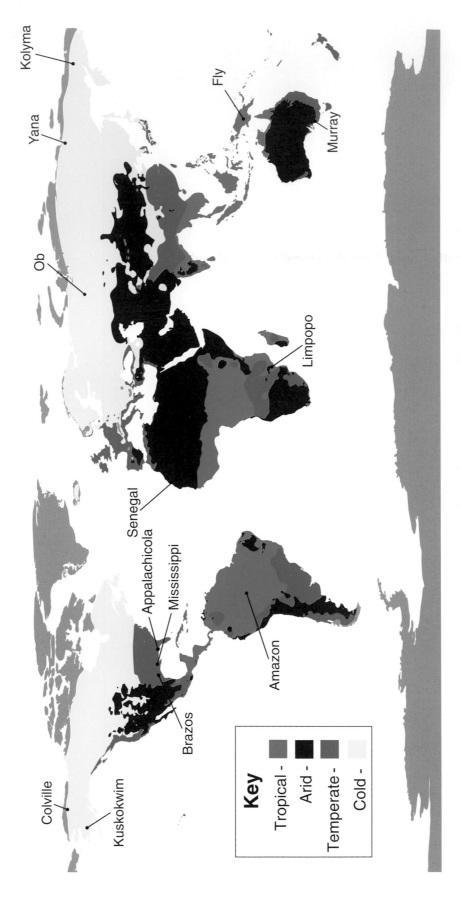

Fig. 5. A map of the simplified climate zones modified from Peel *et al.* (2007). The rivers selected for this study are plotted.

They have been assessed using Landsat imagery from Google Earth Pro (2001). This approach enables the development of a classification that encompasses a wide range of possible meander shapes.

River selection

Climate, gradient and river size each influence how the deposits of meandering rivers accumulate (Shanley, 2004; Schwab, 1976; Ouchi, 1985; Table 1). The rivers investigated in this study are from different climatic settings; they are defined by their regional setting, gradient, discharge and distance of the studied reach from the shoreline (Table 1). The regional climate type for each system has been assigned using the Köppen-Geiger climate classification (Peel *et al.*, 2007; Fig. 5). Gradient has been determined by dividing the elevation difference by the length of the reach along the channel centre line. The elevation uncertainty can be ± 30 m and distance uncertainty up to ± 9 m between points; the elevation measurements recorded were averaged from 10 measurements taken from satellite imagery on GIS software (Google Earth, 2001) in order to increase precision. Additionally, the large-scale nature of this study reduces the impact of these inaccuracies. River size is classified in terms of annual discharge ($m^3 yr^{-1}$) using data from Meybeck & Ragu (1997), Swales *et al.* (2000) and Bobrovitskaya *et al.* (1996). In two cases (The Colville and Kolyma Rivers) where a tributary is examined (as opposed to the main channel), the empirical relationship ($w = 3.15 Q0.49$) from Nixon *et al.* (1959) of channel width (w, in metres) and bankfull discharge (Q in $m^3 s^{-1}$) is used to determine an approximate value for the discharge of the tributary measured in the study. The distance of the reach to the sea was measured along-reach from the most downstream meander measured to the mouth of the channel (m) (Table 1).

Meander-shape classification

The proposed method classifies meander shapes into 4 groups and 25 individual shapes. To define the metrics used in this study, planform images are analysed to characterise individual meanders and the channel centre line is defined. Tangents are then projected into the meander body from the channel centreline and where these tangents connect and form one line, the meander width (mW) is defined

(Fig. 1A). The point of most recent meander growth (MRG) is identified by observing scroll-bar growth direction and a line from the MRG is drawn through the meander body to the line mW, tracing the planform trajectory of the bend apex (tL). Inflection points (or 'mid points'), are identified in the upstream and downstream limbs, respectively. The combination of these geometric definitions identifies the position of intersection lines, which are constructed normal to the channel centreline. Where intersection lines cross within the polygon outlined by mW and the channel centreline, a corner of a polygon is identified. Once all corners have been identified and the polygon is obtained, its area and perimeter is derived. Two metrics are then used to assign a meander to one of 25 categories (Fig. 3): (i) the area of the polygon divided by its perimeter, then normalised by average channel width; and (ii) the length of the meander (tL), divided by the meander width (mW), (Fig. 1A).

Classification of the surface expression of scroll bar pattern

Observations of the surface expression of scroll-bar pattern (Google Earth, 2001), show wide variability in the direction(s) of scroll-bar migration. The surface expression of scroll-bar pattern has been assessed independently of meander shape to enable an objective comparison between channel shape and scroll-bar pattern. The classification scheme developed identifies eight types which are grouped by similar formational processes, i.e. combinations of expansion, extension, translation and rotation. Each type has further sub-divisions, yielding 22 patterns in total (Table 2; Fig. 6). The classification scheme is flexible and encompasses all recognised scroll patterns. The formative processes of the point-bar deposit (such as expansion, or downstream accretion), are similar within each type and differences are highlighted by the subdivisions. If a point-bar deposit is not an accurate visual match it may be categorised into the most appropriate type based on the identified accretion direction(s) (Table 2). The code attributed to each scroll-bar pattern is composed of three parts: the prefix 'T' refers to a scroll bar type (i.e. pattern); the first number (1 to 8), refers to the parent type of scroll bar pattern; the second number (1 to 4), after the decimal subdivides the scroll bar pattern to describe more variability (Table 2; Fig. 6). This scheme accounts for complexity beyond the basic mechanisms of expansion, translation and

Table 2. Descriptions of the classifications of scroll-bar styles identified in this study.

Type	Type sub-division	Description
1	1.1	Extensional (cf. Knighton 1998)
1	1.2	Extensional punctuated rotation (cf. Durkin *et al.*, 2015)
2	2.1	Smooth extension and rotation
2	2.2	Extension and rotation with one major directional change
2	2.3	Extension and rotation with more than one major directional change
3	3.1	Rotation (cf. Daniel 1971) progressing to downstream translation
3	3.2	Extension then downstream translation
3	3.3	Rotation then downstream translation
3	3.4	Extension and rotation then one major directional change, then downstream translation
4	4.1	Expansion
4	4.2	Extension, then expansion is the most recent phase of growth
4	4.3	Rotation, then expansion is the most recent phase of growth
5	5.1	Translation with a straight form (cf. Daniel 1971)
5	5.2	Translation with a curved form (cf. Daniel 1971)
6	6.1	Initial translation then secondary translational movement in a different direction
6	6.2	Initial translation then secondary expansional movement in a different direction
7	7.1	Point-bar composed of recently abandoned point-bar remnants
7	7.2	Only minor movement is visible from surface expression
7	7.3	Point-bar composed of a complex of previously abandoned point-bar remnants
8	8.1	Selection of anomalous remnants of a variety of growth styles
8	8.2	Remnant where the length of the scroll bar is dominantly preserved over migration direction
8	8.3	Remnant where the migration direction is dominantly preserved over the length of the scroll bar

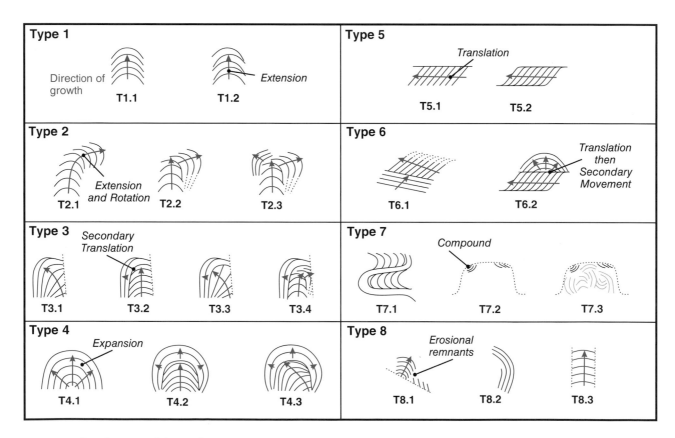

Fig. 6. A classification of the surface expression of scroll-bar pattern. 8 Types have been defined (Table 2); 22 sub-types were defined by identifying variations in the overall process described by the Type category.

rotation; it considers the observed scroll-bar pattern over the whole point-bar deposit so that the accretion history may be derived from the primary observations.

Estimating lithological heterogeneity

In this study, grain-size is considered in terms of relative proportions of sand and mud over the entire thickness of the point-bar deposit at any one point. The degree of heterogeneity has been estimated in a relative manner by using meander shape and the surface expression of scroll-bar pattern through the application of known patterns and trends. In point-bar deposits, heterogeneity can be assessed in the subsurface through observation of IHS in the signatures of gamma-ray logs obtained from well logs, which in turn enables gamma radiation contour mapping (Hubbard *et al.*, 2011; Labrecque *et al.*, 2011). Where this approach is not available, predictive models are produced that typically oversimplify the lithological heterogeneity (Hohn *et al.*, 1997; Ramon & Cross, 1997; Webb & Davis, 1998; Pranter *et al.*, 2000; Tye, 2004).

There is a known variation in grain-size distribution (Labrecque *et al.*, 2011; Fustic, 2012; Durkin *et al.*, 2015), as observed in outcrop whereby there is an increased occurrence of IHS downstream of a bend apex (Pranter *et al.*, 2007; Ghinassi, 2011; Labrecque *et al.*, 2011; Ielpi & Ghinassi, 2014; Durkin, 2015; Fig. 2). Four classes of deposit are defined based on the relative proportion of mud to sand derived from vertical averaging, i.e. by considering proportions over the entire vertical profile of a bar. The classes of deposit are heterogeneity types i–iv, which represent a continuum of coarse lithology through to fine lithology, respectively (Table 3). The end member heterogeneity types (i and iv), are defined by the finest and coarsest sediment in the deposit. In the point-bar deposits, if a specific modal grain-size is not known, then heterogeneity is considered relatively. The relative nature of the zonation in this approach enables systems with different grain-size distributions to be compared. For example, if a point-bar deposit is coarse-grained overall, type iv may represent fine sand, whereas if a point-bar deposit is fine-grained overall, type iv may represent very fine silt or even mud.

The methodology is applied by dividing the upstream and downstream limbs at the meander apex. A cross-over point is identified where the thalweg crosses the channel centreline. The meanders within a reach are ordered by length (tL), which is the distance from the point of most recent growth to the line representing the meander width (mW), (Fig. 1A). This is required because meanders with low tL measurements, relative to the river size, will typically evolve from simpler (so-called relatively more immature) meanders to those with higher tL measurements. The meanders with the lowest tL values are considered first and once the growth history and relative heterogeneity of the simpler forms is constructed, the trends can be carried forwards to assessment of more complex and irregular forms in the reach. Fig. 7 illustrates the analysis and interpretation of the heterogeneity in a complex shape. The first step in attributing the relative lithological heterogeneity to a point-bar deposit is to identify the meander shapes that were developed over different growth stages (Fig. 7A). This may be achieved through observation of scroll-bar geometries, whereby the classified shapes (Fig. 3) are overlain and morphometric matches are identified. Each shape that represents a growth stage is drawn separately and the apex of the shape (the point furthest from the line mW), is constructed (Fig. 7B i–iv). Next, the downstream inflection point is identified where a riffle, or cross-over, probably existed (cf. Leopold & Wolman, 1957; Allen, 1965, 1982). This cross-over identifies occurrence of the most mud-prone sediment (Smith *et al.*, 2009). Where the thalweg is projected to have met the outer bank on the downstream limb, erosion occurred. For individual meander shapes (Figs 7B i–iv), the reach is subdivided into segments between these end members (types i and iv – e.g. Fig. 7B i–iv) and each heterogeneity type should be adjacent to the next one up or down the expected heterogeneity sequence (see key for Fig. 7) because changes in heterogeneity type are modelled here as gradational, though, in practise, changes may be more abrupt. The meander shape groups are now discussed according to the group

Table 3. A table to describe the relative proportions of coarse to fine-grained lithology for each defined heterogeneity type. Coarse and fine refer to the coarsest and finest sediment observed in the point-bar deposit.

Heterogeneity type	Relative proportion of coarse to fine-grained lithology (coarse : fine)
i	75–100% : 0–25%
ii	50–75% : 25–50%
iii	25–50% : 50–75%
iv	0–25% : 75–100%

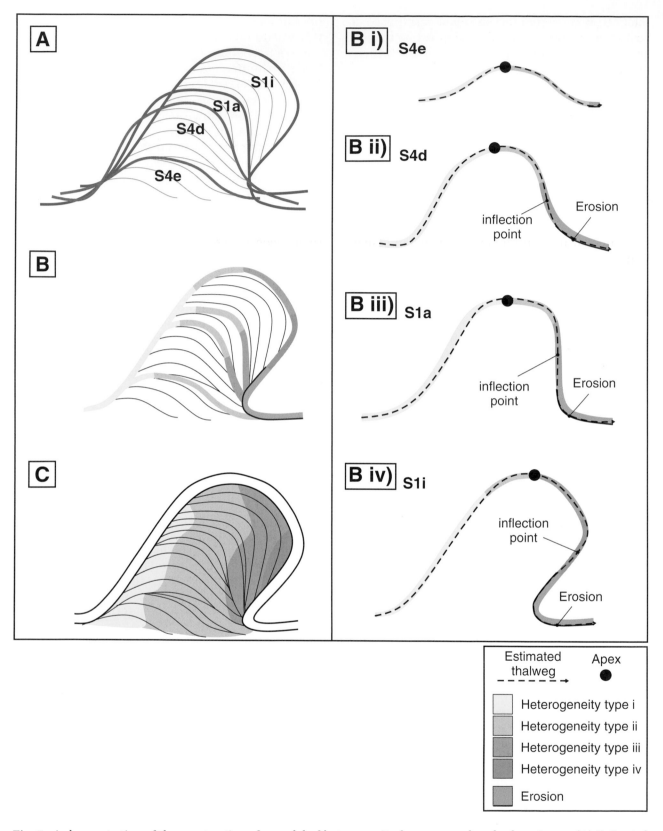

Fig. 7. A demonstration of the construction of a model of heterogeneity from a meander planform image. (A) Estimated meander trajectories through which the most recent shape has transitioned; (B) a compilation of the interpretations of heterogeneity made in B i)–iv); (C) a map of estimated relative grain-sizes for the most recent meander form.

categories (Fig. 3) so that the method may be demonstrated. For an open asymmetric, angular, or open symmetric meander shape (Groups 1, 2 and 4), heterogeneity types i and ii may record accretion on the upstream limb. Heterogeneity type i is not considered on the downstream limb because the energy is typically lower on the inner bank leading to deposition of comparatively finer sediment (Fustic *et al.*, 2012). Heterogeneity type ii will occur past the upstream cross-over and will vary in its downstream extent depending on the geometry of the meander shape and mode of accretion of the point bar. Counter point-bars accrete downstream and may be observed as where the accreting bar turns from being concave to convex (Smith *et al.*, 2009); counter point-bar heterogeneity is constructed using heterogeneity types iii and iv, where type iv is furthest from the inflection point. Bulbous shapes (Group 3), may have more than two inflection points so there may be multiple points of most recent growth, which themselves display grain-size fining around each (Carter, 2003; Fig. 2D). Multiple zones of relative heterogeneity types i–iv should be used if necessary in a complex, multi-phased combination, such as bulbous shapes (Nanson *et al.*, 1980; Fig. 2B).

A map of relative heterogeneity type can be compiled by replacing the lines of the reconstructed meander shapes (Fig. 7A) with heterogeneity types (Fig. 7B i–iv). The plan-view trajectory of the bend apex direction of the scroll bars and other information, such as marked changes in direction of scroll-bar accretion and cross-cutting relationships, are used to predict the distribution of heterogeneity types between the reconstructed lines and thereby to compile the final map of predicted heterogeneity distribution (Fig. 7C). Cross-cutting relationships on the inner bank of the upstream side of meanders are carefully assessed for evidence of upstream erosion. Truncated scroll-bars on the upstream limb indicate erosion and there will probably be a decreased proportion of relative heterogeneity types i and ii in such locations (Jackson, 1976).

Two assumptions are made in applying this methodology: (i) at least one meander in the reach will not be downstream accreting; (ii) at least one meander in the reach will equal or exceed a sinuosity of 2.5. These assumptions, which are justified by observation of modern river forms, enable the rules for the relative distribution of the lithological heterogeneity for the whole reach to be better constrained. These rules have been established because a meander in the reach that is not downstream accreting will probably be characterised by a higher proportion of sand-prone sediment than those that are downstream accreting (Bridge *et al.*, 1995; Smith *et al.*, 2009). Therefore, heterogeneity type i is reserved for use in depiction of scroll-bar deposits that record accretion on the upstream limb or bend apex. A meander in the reach that displays sinuosity equal to or greater than 2.5 will typically be characterised by sediment on the downstream limb that is more mud-prone than sediment on the same limb of a lower sinuosity meander (Jackson, 1976). Therefore, heterogeneity type iv is reserved for use on the downstream limb in shapes with a sinuosity > 2.5. Where an entire river reach is assessed, mud-prone heterogeneity type iv is reserved for the flood plains and abandoned channels. The resolution of the estimation is reliant on the amount to which the fundamental principles of downstream fining can be applied and resolved; volumes of sediment typically span and can cross-cut, multiple unit bars.

RESULTS

Each of the 260 meanders from the 13 river reaches assessed have been classified into one of four Groups (open asymmetric, angular, bulbous, or open symmetric) and subsequently into one of 25 sub-groups (Table 4, Table 1). The surface expression of scroll-bar patterns have been subjectively evaluated and categorised accordingly by Type (1-8) and then into one of 22 sub-classifications. This is achieved via observation of meander morphometry; the surface expression of scroll-bar pattern can be visually compared to a classification (Fig. 6) and the point-bar growth history can be determined to enable identification of the scroll-bar pattern (Table 2; Fig. 6).

Meander shape

Open asymmetric, angular, bulbous and open symmetric shapes (Groups 1, 2, 3 and 4), represent 26, 11, 24 and 39% of the 260 meander shapes assessed, respectively (Table 4; Fig. 8). Open symmetric (Group 4) shapes are the most prevalent with at least 4 in each assessed reach. Occurrence of angular and bulbous shapes (Groups 2 and 3) is more variable. For example, there are no bulbous shapes in the reach of the Kuskokwim River but 10 in the reach of the Purus (Amazon) River. The most common shape in each group is S1g, S2d, S3e and S4d, respectively (Table 4; Fig. 8).

Table 4. Meander shape and scroll-bar pattern observed for the 20 meanders studied in each of the 13 meandering reaches studies. A shape reference in brackets indicates that it is an outlier.

Fly		Senegal		Murray		Yana		Colville		Kuskokwim		Mississippi		Amazon		Ob		Kolyma		Brazos		Limpopo		Appalachicola	
Shape	Scroll	Shape	Scroll	Shape	Scroll	Shape	Scroll	Shape	Scroll	Shape	Scroll	Shape	Scroll	Shape	Scroll	Shape	Scroll	Shape	Scroll	Shape	Scroll	Shape	Scroll	Shape	Scroll
3f	2.2	3c	7.2	3f	2.3	2a	2.2	(3e)	2.3	4d	3.2	4a	2.2	1f	3.1	2d	3.1	4a	3.1	3c	2.3	4d	3.3	(1c)	7.3
1c	3.1	3a	4.2	(4b)	3.1	1g	6.2	1b	3.3	2c	3.4	1h	6.2	2d	6.2	1e	3.3	4a	3.2	3a	2.3	4e	3.1	3d	3.1
4a	2.2	3c	7.3	4e	6.2	1e	3.3	4a	2.2	2c	7.3	4d	3.2	3f	8.2	3f	2.3	1g	3.4	4e	2.2	4e	5.2	4d	3.1
2b	3.3	3c	3.1	4b	3.2	1e	3.4	1f	3.4	4a	7.1	4d	6.2	3f	8.2	1g	3.4	1e	3.1	4e	3.3	4e	3.3	2c	3.1
1e	3.2	(1d)	3.2	3c	8.2	3b	2.3	3f	3.4	1h	3.2	2a	3.4	3a	8.1	3f	2.2	4c	3.3	1h	6.2	4e	1.2	4c	3.1
4c	3.1	2a	5.1	3e	3.1	4e	7.1	3g	2.1	4b	3.4	(1f)	3.4	1f	3.1	3e	2.3	1d	3.2	2d	3.2	3d	3.2	3d	3.2
4e	5.1	1i	6.2	4c	3.2	2d	3.1	4d	6.1	1h	3.4	(4b)	3.2	3e	3.4	3f	4.2	4a	3.1	2d	3.4	4d	3.1	3f	2.2
4d	6.2	4c	3.1	4c	6.2	4c	3.1	4d	3.3	1g	3.2	4d	3.1	1e	3.1	4c	2.2	2d	6.2	4e	5.1	1d	3.1	1g	2.2
2a	4.1	2d	6.2	(4c)	2.2	3e	3.4	1b	3.1	2d	3.3	2d	3.4	3d	8.2	3f	3.4	3d	8.2	4b	6.2	4e	5.2	4c	3.2
1i	2.1	4d	6.1	1e	6.2	4d	6.2	1b	3.3	1b	3.4	4e	6.2	3c	6.2	3g	4.1	3d	3.4	4e	4.2	1c	7.2	4d	3.1
3f	1.1	1d	3.3	1g	3.1	1g	2.1	2a	3.3	1h	2.2	4c	3.1	1e	3.3	4a	8.2	3f	4.2	4e	6.2	4d	3.1	4e	3.1
3a	2.1	1f	6.1	(1f)	7.2	4e	6.2	4e	3.1	4c	3.3	(4d)	3.3	3e	3.1	3f	8.2	4d	6.2	1c	6.2	4d	3.2	2c	3.2
1a	3.1	3b	2.3	3b	2.2	3b	4.2	4a	2.2	4c	2.2	4c	3.3	3g	4.2	1b	3.1	1i	4.2	3d	6.1	4d	3.1	4d	3.1
1e	4.1	4c	3.1	3d	6.1	4c	3.1	3c	7.1	4d	4.1	4d	7.1	4c	2.2	(1d)	4.1	4c	3.1	4b	3.2	4e	5.2	1e	4.3
2a	8.2	4d	3.1	3e	8.2	4d	7.1	2a	2.1	4d	7.1	4e	5.2	3d	3.4	4c	3.4	1i	2.1	4c	3.1	4e	5.2	(1b)	3.1
3g	8.2	4c	6.2	3e	3.3	3b	7.1	1i	2.3	2c	7.1	2b	5.2	4e	5.1	(3e)	2.3	1c	6.2	3e	3.3	4e	5.2	1g	3.1
3c	2.3	(1i)	3.2	1d	3.3	(1g)	3.2	4a	2.1	1g	2.2	2d	7.1	2a	5.1	4d	3.2	3b	2.3	1i	8.1	4e	5.2	4d	1.2
4d	7.1	3c	6.1	4d	7.3	3f	2.1	1i	7.3	1h	3.1	4e	5.2	3f	2.2	2d	6.1	4d	3.2	1e	3.2	1h	3.1	(3f)	3.1
3g	2.3	1g	3.3	1g	3.2	1i	3.1	3d	2.2	4d	5.2	4d	7.3	3a	2.2	4d	6.2	4d	7.3	4a	3.3	2b	3.1	4d	3.1
3g	6.1	1g	6.1	4a	8.2	1i	3.3	2a	5.1	4e	7.1	4e	5.2	1a	3.1	1g	7.3	4d	2.2	(1e)	3.3	2d	1.2	(3d)	3.2

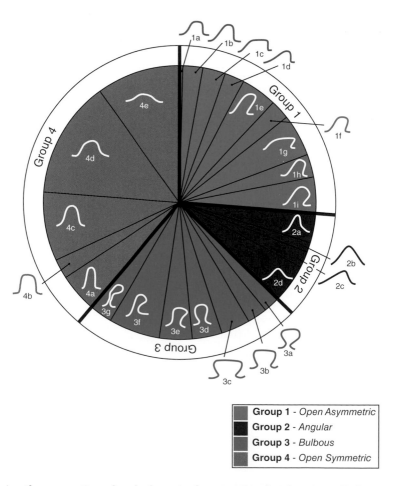

Fig. 8. A pie chart showing the proportion of each shape in the 260 meander shapes studied.

Group 1 - *Open Asymmetric*
Group 2 - *Angular*
Group 3 - *Bulbous*
Group 4 - *Open Symmetric*

Surface expression of scroll-bar pattern

The surface expression of scroll-bar pattern on the 13 river reaches (Table 4; Figs 6 and 9) has been classified into 8 types of migration style, which themselves are further divided into 22 subclassifications. Multi-phased migration (Type 2) and downstream accretion with translation as a secondary phase (Type 3) is present in every river. Downstream accretion with translation as a secondary phase (Type 3) is the most abundant (45.5%) with 6 or more meanders exhibiting this surface expression of scroll-bar pattern in each river. The least consistent scroll-bar pattern identified is translation followed by a secondary movement (Type 6). Seven meanders exhibit this style in the Senegal reach but none in the Kuskokwim reach.

The most common shape for each type is T1.1, T2.2, T3.1, T4.1 and T4.3, T5.1, T6.2, T7.1 and T8.2 (Table 4; Figs 6 and 9). Recording downstream accreting behaviour is important because where downstream accretion occurs there will be no deposition on the upstream limb, so heterogeneity type i will not be observed. Where downstream accretion is not occurring, heterogeneity types i and ii are more widely observed.

Observations from planform morphometric analysis: selected representative examples

Reaches from the Ok Tedi (Fly), Senegal, Murray and Irtysh (Ob) Rivers (Figs 10 and 11) have been selected as key representative examples from the

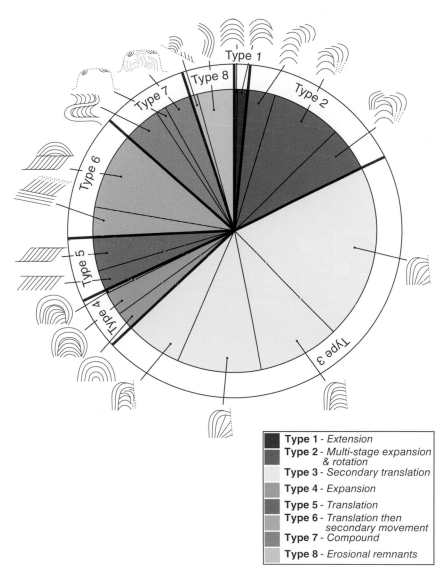

Type 1 - *Extension*
Type 2 - *Multi-stage expansion & rotation*
Type 3 - *Secondary translation*
Type 4 - *Expansion*
Type 5 - *Translation*
Type 6 - *Translation then secondary movement*
Type 7 - *Compound*
Type 8 - *Erosional remnants*

Fig. 9. A pie chart showing the proportion of each scroll bar type in the 260 meander shapes studied.

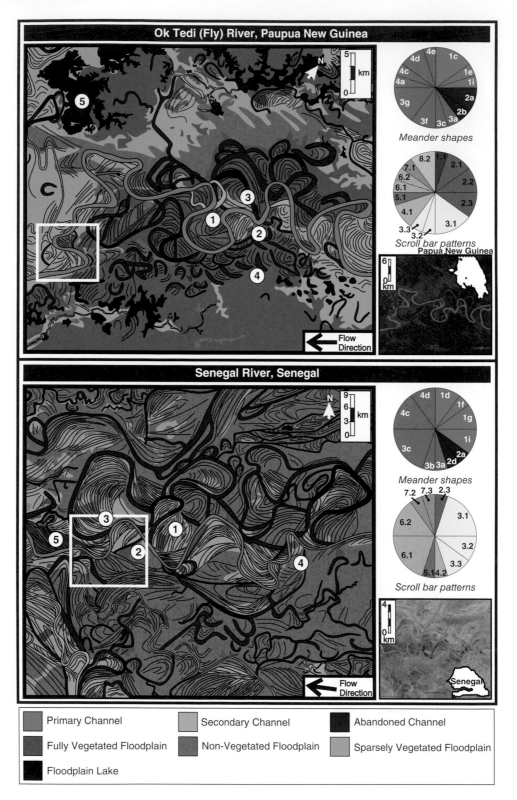

Fig. 10. Pie charts of the proportions of features seen in fluvial meandering systems are shown with planform morphology maps for the Ok Tedi (Fly) River, Papua New Guinea and the Senegal River, Senegal. The white boxes show the sections which are interpreted in Fig. 12. In the Ok Tedi (Fly) River the following features are noted: (1) larger meanders grow through expansion; (2) smaller meanders typically rotate; (3) meanders on the same side of the channel undertake chute cut-off; (4) abandoned loops to the south east overlap; (5) lacustrine environments. In the Senegal River, the following features are noted: (1) meander shapes are commonly both open asymmetric and open symmetric, with occasional bulbous and angular shapes; (2) Meander scroll-bar growth is typically through expansion and translation; (3) clear changes in growth direction are seen in compound forms; (4) vegetation is most dense directly on point bars by the active channel; (5) interestingly, the most closely spaced scroll bars in the studied area are on the most vegetated point bar. A key for the planform images shown in Figs 10 and 11.

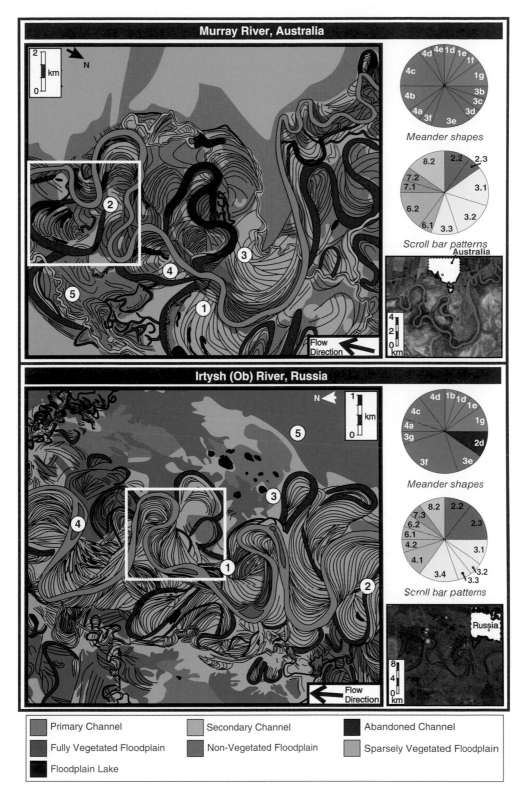

Fig. 11. Pie charts of the proportions of features seen in fluvial meandering systems are shown with planform morphology maps for the Murray River, Australia and the Irtysh (Ob) River Russia. The white boxes show the sections which are interpreted in Fig. 12. In the Murray River, the following features are noted: (1) meander shapes are either small and regular or large and irregular; (2) punctuated growth occurs in co-sets of approximately five; (3) translation and expansion is typical; (4) chute cut-off is indicated by the character of abandoned loops (5) channels connect parts of the main reach. In the Irtysh (Ob) River the following features are noted: (1) meander shape appears to be typically either bulbous, or open symmetric and the principal style of meander growth is translation and rotation; (2) point bars grow in co-sets of 2-6 and undertake incremental rotation; (3) Abandoned loops are sometimes bulbous in nature with scroll-bar pattern orthogonal to the direction of cut-off; (4) Vegetation is most dense on mid-channel bars (5) and on the floodplain beyond the active channel. A key for the planform images shown in Figs 10 and 11.

larger data set to demonstrate different environmental settings (Table 1); the environmental setting of a river may have a controlling influence on the morphology of the reach. By selecting rivers that are morphologically variable, the versatility of the approach is recognised. The planform morphology of reaches from the Ok Tedi (Fly), Senegal, Murray and Irtysh (Ob) Rivers is observed in detail via maps that highlight important morphological features, i.e. abandoned-channel geometry, active-meander shape and the surface expression of scroll-bar forms (Figs 10 and 11). This readily enables visual recognition of trends and features relevant to the investigation.

Ok Tedi (Fly) River, Russia

Meander shapes appear to be mostly open symmetrical (Group 4) with multi-phased growth (Type 2). Analysis of the quantified data reveals that meander shape is equally balanced between open asymmetric and open symmetric; and bulbous shapes make up the majority of meanders (40%). Meander scroll-bar growth shows gradual and punctuated directional change, with major directional changes occurring episodically. Style of meander growth demonstrates a relationship with size (meander area): point-bars > 3 km in length typically expand, whereas point-bars < 3 km in length typically rotate. The quantified data show that there is a wide variety of scroll-bar growth direction. Chute cut-off occurs in sections and abandoned loops on the floodplain overlap each other in a pattern that varies from regular to irregular. Fig. 12A shows a complex and fragmented assemblage of point-bar remnants. There are many sections of scroll-bar that are not obviously related to any active reaches, making it difficult to assign a heterogeneity type. Where heterogeneity can be assigned, it is constructed as fining downstream in point-bar deposits exhibiting scroll-bars that are undertaking rotation (Type 3.1) and punctuated extension (Type 1.2).

Senegal River, Senegal

All four groups of meander shape are represented in this reach, with the open asymmetric type being the most common (35%). Through visual assessment of the reach, scroll-bar growth occurs through expansion and translation in co-sets identifiable by gradual directional changes. Scroll-bar styles that represent downstream accretion (Types

3, 5 and 6), collectively represent 80% of the scroll styles in the reach. More pronounced changes in direction are seen in compound forms and vegetation density increases towards the active reach. Fig. 12B shows the main active reach to be dominantly undertaking rotation and expansion. In the planform map (Fig. 10), there is no evidence of bulbous shapes in the active channel, which could explain why the abandoned reaches are less sinuous than other reaches studied.

Murray River, Australia

Meander shapes increase in irregularity and complexity as they increase in size. Point-bars that are < 1.5 km in length exhibit one scroll pattern (most commonly Type 3 or Type 6) and point-bars that are > 1.5 km in length exhibit punctuated growth in scroll-bar sets. There are no angular shapes in this reach and there is markedly less variation in scroll-bar pattern and meander shape than is typical. Growth by secondary translation (Type 3) is most common (40%). The characters of abandoned reaches indicate chute channel cut-off to be the dominant mode of meander-loop abandonment. Where avulsion occurs, long and narrow channels remain connecting parts of the main reach (Fig. 11 [5]). Fig. 12C shows point-bars that vary in size and complexity. There is an irregular distribution of mud-prone sediment that represents channel remnants abandoned by cut-off mechanisms. Point-bar deposits that are > 3 km in length record incremental growth and punctuated rotation (Type 1.2), whereas point-bar deposits that are < 3 km in length record downstream accretion, dominantly as rotation and translation.

Irtysh (Ob) River, Russia

The most common meander shape group in the studied reach is bulbous (40%), followed by open symmetric and open asymmetric types, which are present in equal proportion (25% each). The principal style of meander growth is translation and rotation (Types 3 and 6), though expansion and multi-phased rotation are both also common (Types 4 and 2). Point bars have grown through incremental rotation and occur in scroll-bar sets of 2–6. Many abandoned loops are bulbous in shape indicating neck cut-off occurred in the past. In these loops, the scroll bars are oriented orthogonal to the point of palaeo cut-off. Fig. 12D shows a

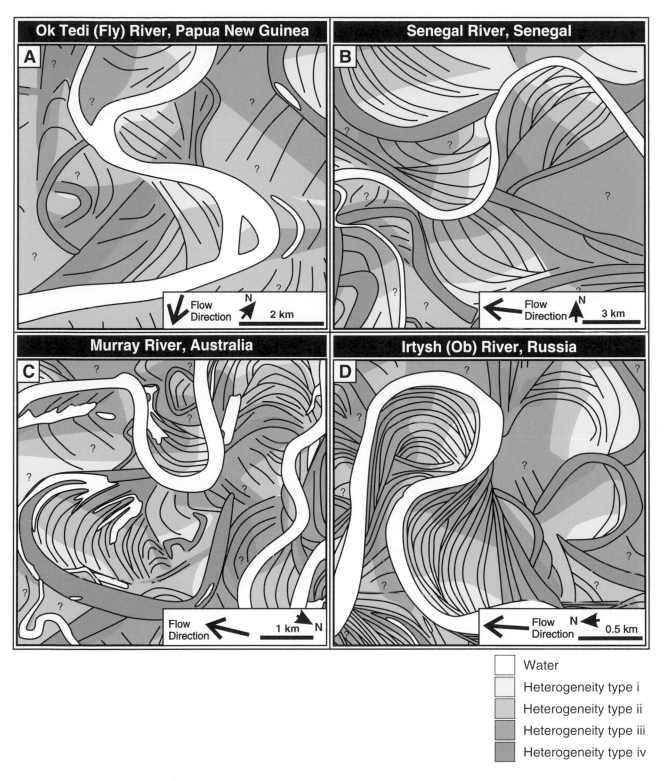

Fig. 12. Interpretations of the heterogeneity types for modern systems in planform; (A) Ok Tedi (Fly) River, Papua New Guinea; (B) Senegal River, Senegal; (C) Murray River, Australia; (D) Irtysh (Ob) River, Russia.

section of the studied area that represents an irregular active reach and regular-shaped abandoned loops. Each point bar is depicted with deposits fining downstream and growth to have occurred through extension then expansion (Type 4.2), or rotation (Type 3.1).

Relationship between meander shape and scroll-bar pattern

Through analysis of the whole data set, each of the four meander shape groups is found to exhibit a wide variety of scroll-bar patterns (Fig. 13). Bulbous shapes (Group 3) have markedly different

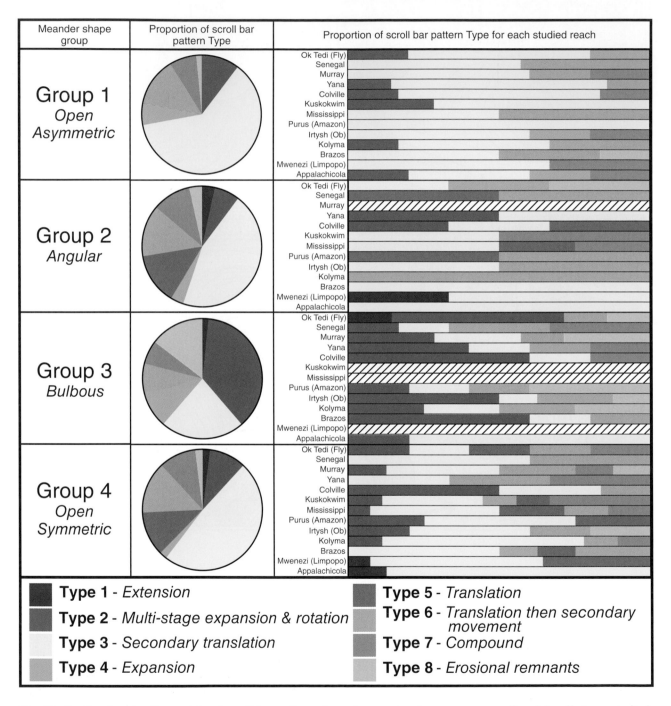

Fig. 13. Graphs showing the variety of scroll bar patterns in each meander shape group overall and for all rivers studied. The pie charts show the proportion of each scroll-bar pattern observed within each meander shape for all 260 meander shapes observed in this study; the bar charts divide these groups into each river reach studied.

Table 5. A table to demonstrate the markedly different trends seen in Group 3 compared with Groups 1, 2 and 4.

	Open Asymmetric (Group 1)	Angular (Group 2)	Bulbous (Group 3)	Open Symmetric (Group 4)
Secondary migration style of downstream translation (Type 3)	62%	45%	23%	49%
Translation (Types 5 and 6)	13%	28%	8%	26%
Downstream accretion (Types 3, 5 and 6)	**75%**	**73%**	**31%**	**75%**
Multi-stage rotation (Type 2)	10%	7%	37%	10%
Fragmented remnants (Type 8)	2%	3%	15%	2%
Sum of multi-stage rotation (Type 2) and fragmented remnants (Type 8)	**12%**	**10%**	**52%**	**12%**

trends in scroll-bar pattern than open asymmetric, open symmetric and angular shapes (Groups 1, 2 and 4; Table 5); they show the least amount of downstream accretion (Types 3, 5 and 6) (31%) compared to the average across the other shapes (74.3%). Bulbous shapes also exhibit the largest proportion of multi-stage rotation (Type 2) and fragmented remnants (Type 8) (52%) compared to the average across the other shapes (11.3%). Angular and open symmetric shapes (Groups 2 and 4), show a similar distribution of scroll-bar pattern. Open asymmetric shapes (Group 1) show a markedly high proportion of downstream accretion via translation as a secondary migration (62%) compared to the average across the other shapes (39%) (Table 5). The diversity of scroll-bar patterns observed within each meander-shape group is notable and highlights the complexity of fluvial point-bar deposits.

DISCUSSION

The novel approach undertaken in this study has shown the relationship between meander shape and scroll-bar pattern to be more complex and varied than previously described (Fig. 13). Different meander shapes exhibit different proportions of scroll-bar pattern type. For example, secondary translation (Type 3; Fig. 6), is the dominantly observed scroll-bar pattern in open asymmetric, angular and open symmetric shapes (Groups 1, 2 and 4 respectively; Fig. 3), whereas multi-stage expansion & rotation (Type 2; Fig. 6) and erosional remnants (Type 8 – Fig. 6), are dom-

inant in bulbous shapes (Group 3; Fig. 3). Hitherto, scroll-bar deposits have typically been modelled with one phase of continuous growth. However, this study has shown that only 33% of the studied point-bar deposits record such a simple evolutionary history. More than one phase of growth is most common at 54% (13% are unknown), (Fig. 9; Table 4).

Assessment of the spatial distribution of lithologic zones with variable grain-size is important to determine sub-seismic architecture in fluvial point-bar deposits and therefore the fluid-flow rates and pathways in reservoir bodies hosted in such deposits (Fielding & Crane, 1987; Miall, 1988). Observation of meander form in seismic datasets enables the geometries of the deposit to be evaluated.

Application of methodologies

The method developed here has been applied to two example case studies: i) a sub-surface study of five fluvial point-bar deposits from the Cretaceous McMurray Formation, Alberta, Canada (Hubbard *et al.*, 2011); and ii) a study of the outcropping Ypresian Pont de Montanyana point-bar deposits, southern central Pyrenees (Cabello *et al.*, 2018).

Heterogeneity prediction in the sub-surface

The point-bar deposits of the Cretaceous McMurray Formation, Alberta, Canada, were deposited in a dominantly fluvial environment (Hubbard *et al.*, 2011; Blum & Pecha, 2014). The area studied is Long Lake in the Athabasca Oil

Sands region of north-eastern Alberta (Fig. 14A), owned by Nexen Inc. (the operator) and Opti Canada Inc. Detrital zircons found in these point-bar deposits identify the sediment as originating from the Appalachian mountains to the south-east of north-eastern Alberta (Blum & Pecha, 2014). During deposition the river system was subject to some tidal influence (Labrecque *et al.*, 2011; Musial *et al.*, 2012; Hein, 2015); although the formative river is not thought to have experienced brackish salinity (Blum & Pecha, 2014; Blum & Jennings, 2016), trace fossils common in marine deposits have been found in these rocks (Musial *et al.*, 2012; Gingras *et al.*, 2016), on the downstream end of a meander bend (Nardin *et al.*, 2013). These marine trace fossils are scattered and few in number, possibly recording rare instances where organisms have been rafted upstream into a dominantly fluvial setting (Blum & Pecha, 2014; Blum & Jennings, 2016). The studied seismic time slice is taken at 275 m depth within the Long Lake area (Labrecque *et al.*, 2011) and is well imaged because of the large size of the preserved point-bar elements and the shallow burial depth (Fig. 14A). High-resolution seismic and gamma-ray data from this succession are used to test the scale-independent methodology developed in this study. The availability of many drill cores that can be tied to seismic imagery for individual architectural elements means that lithofacies can be related directly to point-bar meander shape and scroll-bar pattern for this succession and, for these reasons, the McMurray Formation is considered a very valuable succession on which to test the method developed here.

Labrecque *et al.* (2011) identified seven facies in the point-bar deposit here identified as 'point-bar 1' (Fig. 14). They represent siliciclastic sedimentation and the facies range from massive sandstone with siltstone rip-up clasts, through cross-stratified sandstone, to laminated siltstone (Labrecque *et al.*, 2011). Inclined heterolithic strata IHS (cf. Thomas *et al.*, 1987) is abundant in the succession (Labrecque *et al.*, 2011), making the heterogeneity distribution inherently complex. It is especially important to gain an improved understanding of the lithological heterogeneity of this deposit given that it forms an important oil reservoir (Carrigy, 1962; Mossop & Flach, 1983; Labrecque *et al.*, 2011, Nardin *et al.*, 2013, Moreton & Carter, 2015).

To obtain a prediction of the spatial distribution of lithological heterogeneity present in the McMurray Formation point-bar deposits, the first step requires the interpretation of the morphologies exhibited in the flattened seismic timeslice (Fig. 14A and B; Smith *et al.*, 2009). The point-bar deposits have been numbered as shown in Fig. 4A. Point-bar deposits 1, 2 and 3 exhibit an elongate bulbous shape (S3g of Fig. 3) and a scroll-bar pattern that describes remnants of a point-bar growth history that was strongly erosional (T8 of Fig. 6). Point-bar deposit 4 exhibits an open asymmetric shape (S1d of Fig. 3) and a scroll-bar pattern that describes rotation and translation (T3.1 of Fig. 6). Point-bar deposit 5 exhibits a bulbous shape that is more rounded than point-bar deposits 1, 2 and 3 (S3f of Fig. 3) and describes growth by rotation and translation, followed by expansion at the bend apex (T4.3 of Fig. 6). For each combination of meander shape and scroll-bar pattern (Fig 14B), the heterogeneity of the point-bar deposit is predicted using its migration history (Fig. 7). Examination of the incremental progression of meander growth aids understanding of the spatial distribution of grain-size classes in each point-bar deposit (Fig. 14B); the process is undertaken in order of low to high sinuosity of the final meander form. Studying the simplest point-bar deposits first means that when a meander form is more laterally complex, there is a temporal understanding that may be drawn upon. Point-bar deposit 4 is here considered first since it is characterised by an open asymmetric shape (S1f of Fig. 3), which records a meander with the lowest sinuosity of the point-bar deposits studied.

Point-bar deposit 4 is interpreted to have undergone expansion and rotation. Fig. 14B demonstrates how this may have occurred through the expansion and translation of a low-sinuosity, open symmetric shape, which accreted via expansion and rotation to become more sinuous (Fig. 14B). The meander bend then became asymmetric, due to erosion of the point-bar deposit on the upstream limb, causing the scroll-bars to become truncated. It finally underwent dominantly downstream accretion (Fig. 14B), prior to abandonment via avulsion. Point-bar deposit 5 is interpreted as being a more mature example of the evolution inferred for point-bar deposit 4. The formative meander is interpreted to have evolved to neck cut-off, rather than to have been subject to abandonment via avulsion. Point-bar deposit 5 is interpreted to have undertaken expansion and rotation (T3.1 of Fig. 6) and subsequently further expansion (T4.3 of Fig. 6). This has resulted in a composite, multi-phased

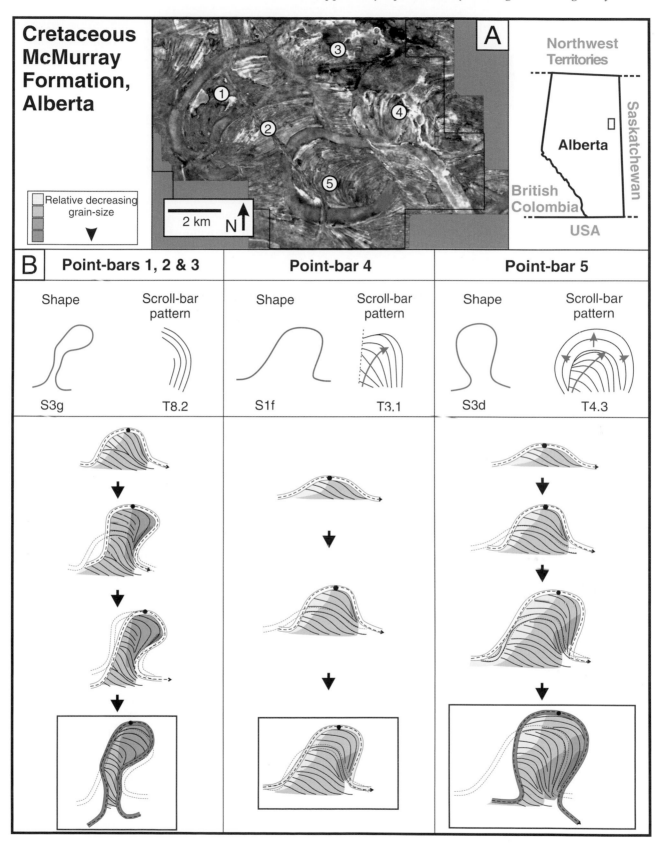

Fig. 14. Interpretations of a seismic time slice from the Cretaceous McMurray Formation, Alberta. The red box on the location map identifies position of the study area within Alberta; (A) Seismic data modified from Hubbard *et al.* (2011) of Nexen's property lease of Lower Cretaceous meandering fluvial-dominated point-bar deposits; (B) the interpretations of the accretion history and resulting heterogeneity for point-bar deposits 1–5 and their associated channel fills.

growth history that concluded in the abandonment of a bulbous shape via neck cut-off. The analysis is more challenging for point-bars 1, 2 and 3 because it is unclear how the point-bar deposits developed. Abandoned point-bar deposits in the Irtysh (Ob) River (Fig. 11, [3]), show a similar scroll-bar pattern to point-bar deposits 1, 2 and 3 of the McMurray Formation (Fig. 14A), whereby the direction of scroll-bar accretion is orthogonal to the direction of neck shortening which caused cut-off. The morphological similarity of the meanders of the Irtysh (Ob) river to point-bar deposits 1, 2 and 3 of the McMurray Formation (Fig. 14A) is here used to inform the interpretation of the mechanism by which migration occurred in point-bar deposits 1, 2 and 3 where it cannot otherwise be derived. In the Irtysh (Ob) River (Fig. 11), the abandoned point-bar deposits that exhibit scroll-bar accretion orthogonal to direction of neck shortening, which caused cut-off (labelled as [3] in Fig. 11), are interpreted to have been formed by irregular, bulbous shapes (S3e, S3f of Fig. 3). The migration histories of point-bar deposits 1, 2 and 3 have therefore been interpreted as consisting of initial growth via rotation and downstream accretion, followed by stages of extension and downstream accretion that generated bulbous shapes. Subsequently, an elongated shape developed due to erosion on both the upstream and downstream limbs until meander loop cut-off occurred (Fig. 14B).

The meander shapes derived were analysed via the methodology outlined in this study (Fig. 7) and used to infer models of lithological heterogeneity (Fig. 14B). These models have been designed to inform the relative grain-size interpretation of the flattened seismic time-slice (Fig. 14A) and not to provide a definitive result as it is only a procedure for estimating relative lithological heterogeneity. By comparing the final models (see red boxes in Fig. 14B), the method predicts that point-bar deposits 1, 2 and 3 should have a higher proportion of fines compared with point-bar deposits 4 and 5. This information is used alongside the estimated spatial distribution of lithological heterogeneities so as to inform the overall interpretation shown in Fig. 15.

Counter point-bar deposits are considered to have arisen from meander translation; the counter-point bars for point bars 3 and 4 have been assigned the finer-grained heterogeneity types iii and iv (Smith *et al.*, 2009 Fig. 2A). Channel fills are also observed in the seismic timeslice; meanders that underwent neck cut-off or chute cut-off are predicted to have experienced a different abandonment process to those that experienced avulsion. A neck cut-off or chute cut-off results in oxbow lake development and the progressive filling of such lakes by fine sediment (heterogeneity types iii and iv); such bodies are modelled as relatively mud prone. A channel that infills due to chute cut-off or avulsion is expected to record an earlier stage of deposition by waning flows, preserved as sand-prone (heterogeneity types i and ii) deposit with asymmetric accretion (Toonen *et al.*, 2012), followed by a phase with limited or no flow represented by concentric filling of mud-prone sediment (heterogeneity type iv; Schumm, 1960; Harms, 1982). As such, these different channel abandonment types have been modelled with different lithologies (Fig. 14B).

In Fig. 14B, the predicted relative lithological heterogeneity is mapped for each point-bar deposit. When the individual point-bar deposits are combined, as in Fig. 15A, the use of the lithological classes is modified so that heterogeneity is expressed as relative to the entire meander belt; the attribution of these relative classes is undertaken with consideration of the occurrence of mud-prone counter-point bars and abandoned-channel deposits. Fig. 15A shows the interpretation of the heterogeneities in each point-bar deposit. Fig. 15B is a gamma ray contour map of the five studied point-bars of the Cretaceous McMurray Formation, Alberta, constructed with data from 213 wells (Hubbard *et al.*, 2011). By comparing the measured gamma-ray data (Fig. 15B), with the predicted lithology distribution (Fig. 15A), the strengths and weaknesses of the technique can be assessed. In the map of predicted heterogeneity (Fig. 15A), a fine-grained margin around the entire coarser-grained point-bar deposits is not present, as inferred from the gamma-ray map for point bars 1 and 3 (Fig. 15B). The contours on the gamma-ray map (Fig. 15B) record small areas of relative lower API values on the upstream and downstream ends of the preserved abandoned channel of point bar 1. This might indicate that mid-channel bars were developed and in turn this might indicate a slow abandonment process, Labrecque *et al.* (2011) recorded a mud-clast facies at the position labelled with 'A' in Fig. 15B, which would have offset the gamma-ray recording so predictions cannot be reliably verified with gamma ray data from deposits that are rich in mud clasts. The predicted model may be able to provide a more generally representative

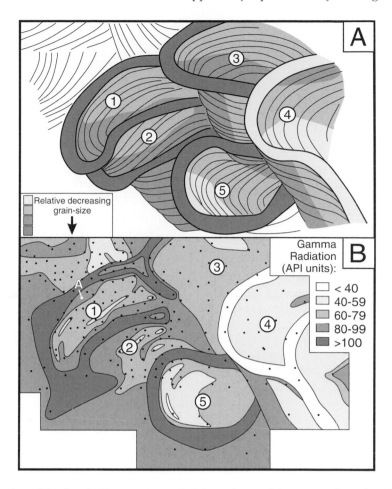

Fig. 15. (A) an interpretation of the data in Fig. 14A compiled through use of the approach outlined in this study (Fig. 14B); (B) a gamma ray contour map constructed with data from 213 wells, modified from Hubbard *et al.* (2011). 'A' marks a location of mud clast facies as defined by Labrecque *et al.* (2011).

relative grain-size distribution in areas with higher-than-normal abundance of mud clasts.

This example application demonstrates the new approach to be effective since it can help mitigate the effects of mud clasts on gamma-ray interpretation and enable the overall heterogeneity to be revealed. The heterogeneity for preserved point-bar deposit 2 was predicted as analogous to preserved point-bar deposit 1 because the morphological data (Fig. 14A) does not enable a more refined analysis; for this reason the prediction for preserved point-bar deposit 2 was not accurate (Fig. 15A). The gamma-ray data for point-bar deposit 4 and its associated counter point-bar (Fig. 15B) indicate that it might be more coarse-grained than predicted (Fig. 15A). Point-bar deposit 4 and its associated counter point-bar deposit are the most recent of the five point-bar deposits depicted in Fig. 14A; the disparity between the prediction and the data

could be indicative of a temporal increase in the calibre of sediment carried thorough this reach during deposition.

Although high-resolution seismic and or gamma-ray data sets are available for the analysis of point-bar heterogeneity in the McMurray Formation (e.g. Hubbard *et al.*, 2011; Nardin *et al.*, 2013; Moreton & Carter, 2015), other subsurface meandering fluvial successions do not benefit from such data sets. In such cases, this assessment method provides a straightforward and cost-effective approach for predicting relative heterogeneity in sub-surface meander-belt deposits, especially when used in conjunction with numerical forward stratigraphic modelling methods (Yan *et al.*, 2017; Yan *et al.* 2018 this volume) and probabilistic methods (Colombera *et al.*, 2012, 2017, 2018).

Where only lower-resolution data sets are available (compared that shown in Fig. 15B; Hubbard *et al.*, 2011), the method developed in this study (Fig. 13; Table 4) may provide insight by enabling prediction of the most probable scroll-bar configuration for an observed meander shape. For example: i) in flattened seismic-reflection data, if the shape of the abandoned river is the only visible feature (e.g. Colombera *et al.*, 2018), the most probable scroll-bar configuration may be predicted for each shape based on known environmental controls; ii) if a flattened seismic image is only available for part of a larger point-bar deposit (or series of deposits), the character of the morphology may be determined and extrapolated beyond the imaged area to aid with prediction up- or down-stream; iii) if no flattened seismic is available, information on the climatic conditions that prevailed during accumulation of the fluvial system may be used to deduce the most probable morphologies for meander shape and scroll-bar pattern using data such as that in Table 4. These

combinations may be used to apply the method to assist in the building of geocellular reservoir models that are more inclusive of morphological variability based on limited data.

Heterogeneity prediction in outcrop

The Montanyana Group is upper Upper Eocene to Oligocene in age and was deposited in the Tremp-Graus Basin as part of a westward-prograding clastic wedge of partly fluvial origin (Marzo *et al.*, 1988). The outcrop studied is approximately 400 m north of the Pont de Montanyana village in the Huesca province (Cabello *et al.*, 2018); the succession is exposed as a cliff that outcrops WNW-NNE and allow the outcrop to be examined from multiple angles. The outcrop face cuts through a fluvial deposit and exposes point-bar deposit 1 (Fig. 16C) along its length. Data presented by Cabello *et al.* (2018) have been summarised

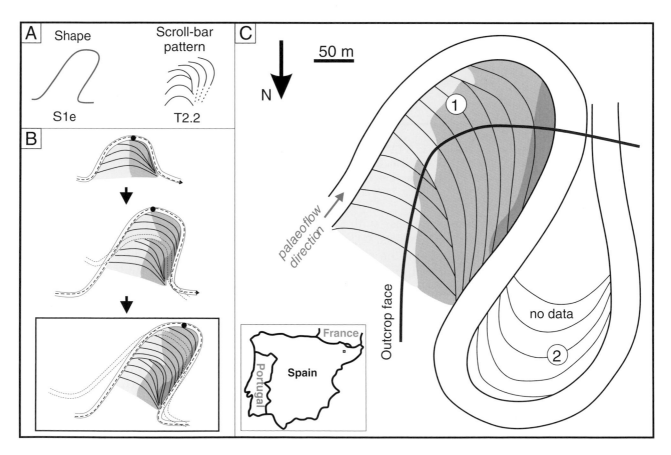

Fig. 16. (A) the morphologies of the meander shape and scroll-bar deposit of the point-bar deposit studied by Cabello *et al.*, (2018); (B) the interpretations of the accretion history and resulting heterogeneity for the studied point-bar deposit; (C) The interpreted palaeogeographic reconstruction of the studied point-bar deposit (modified from Cabello *et al.*, 2018), with summarised lithological heterogeneity overlain. The red box on the location map identifies position of the study area within Spain.

in terms of relative lithologic heterogeneity and overlain onto the interpreted palaeogeographic reconstruction (Fig. 16C). Eleven facies were identified in the outcrop ranging from mudstones to conglomerates and the point-bar deposits observed display lithologic heterogeneity (Cabello *et al.*, 2018). Through the analysis of palaeocurrent indicators, Cabello *et al.* (2018) reconstructed the palaeo-morphology of point-bar deposit 1. It is reconstructed as exhibiting an elongate open asymmetric shape (S1e of Figs 3 and 16A), with a scroll-bar pattern that describes extension and rotation with one major directional change (T2.2 of Figs 6 and 16A).

A prediction of the lithologic heterogeneity is herein obtained (Fig. 16C) through interpretation of the lateral accretion history of the point bar (Fig. 16B). Point-bar deposit 1 is interpreted to have undertaken expansion, rotation and to have dominantly accreted via expansion with some downstream accretion. This may have occurred through the expansion of a low-sinuosity, open shape (S1d of Fig. 3). The meander then rotated causing a directional change in extensional accretion; the point-bar deposit became more elongate prior to abandonment via avulsion. The main difference between the predicted and observed lithologic heterogeneities (Fig. 16), is that heterogeneity type iii extends further towards the apex in the observed data (Fig. 16C) compared to the predicted model (Fig. 16B). This could be a result of the difference in apex shape between the predicted model and the palaeogeographic reconstruction, highlighting a limitation of the predictive models in that the shapes used are inflexible. However, the distributions of lithological heterogeneity predicted by the methodology introduced here (Fig. 16B) are similar to the outcrop observations (Fig. 16C), therefore demonstrating the generally applicable predictive capability of this methodology.

Limitations and future development

The principal limitations of this method are as follows: (i) relative, rather than absolute, grain-size is interpreted; (ii) the scroll-bar classification scheme is largely subjective in nature because scroll-bar morphology is difficult to quantify, though descriptions provided in Table 2 reduce uncertainty by describing the processes of formation of scroll-bar forms and the approach has been shown to be widely applicable; (iii) heterogeneity prediction (Fig. 7) contains a subjective element and although

rules have been clearly defined, uncertainties for predictive models remain in relation to natural variability and limited knowledge thereof; and (iv) because lithological heterogeneity is only considered relatively, classes of heterogeneity might not translate to petrophysical heterogeneity (ranges of porosity and permeability) or net-to-gross ratios univocally. It is anticipated that some of these uncertainties could be reduced through insight gained from future studies.

CONCLUSIONS

Application of the novel methodologies outlined in this study enables a higher level of morphometric detail to be captured and assessed than was possible through use of existing classifications (e.g. Daniel, 1971; Hooke, 1984). The new approach to the assessment of meander morphology provides quantified metrics that enable the past growth of a point-bar deposit to be inferred. Through use of the proposed methodologies, a wider variety of morphometric variability is accounted for that leads to a less biased assessment of fluvial meandering reaches with less subjective comparisons between river reaches, through quantification of system elements. This enables spatial distributions of relative grain-size to be predicted semi-quantitatively, an advantage over classic facies models. The methodology was developed through the analysis of 260 meanders from 13 globally distributed rivers from different environments. Morphometric planform analysis has allowed key features of a meandering reach to be better constrained. Open symmetric (Group 4) shapes are the most abundant (39%) and downstream accretion by translation and rotation is the most typical mode of growth (Type 3.1). Open asymmetric and open symmetric shapes (Groups 1 and 4), are found in every meandering reach studied, whereas angular and bulbous shapes are absent in 8% of rivers studied. There is a very wide distribution of meander scroll type in the 260 studied meander shapes and the interrelationships of these to meander shape exhibit discernible trends, albeit with wide variability. Identification of this morphometric variability can be used to construct a meander model by combining the shape, as identified using the meander classification scheme, with the surface expression of the scroll-bar pattern. The history of point-bar evolution may be reconstructed through understanding of the complex interrelationships of the meander geometries (i.e. meander geometry and scroll-bar patterns) and may be subsequently used to i) provide a template for prediction of the relative heterogeneity of a preserved fluvial deposit; and ii) enable relative heterogeneity maps to be produced. Each measured point bar may be used to aid in the composition of conceptual reservoir models so that statistical limitations may be obtained for features of meandering systems.

The methodology can be applied to interpret lithological heterogeneity in subsurface meander-belt successions for which high-resolution seismic, or gamma-ray data are not available, such as the Triassic Mungaroo and Brigadier Formations and offshore Australia NW Shelf. The approach may also be applied directly to fluvial point-bar elements that are exhumed and exposed in outcrop. Interpretation of meanders in a fluvial reach using both the meander classification scheme and the scroll-bar classification scheme, along with the character of a reach, allows a river to be described in a manner that can aid more insightful interpretation of the rock record. With further study, this approach may be utilised to identify the governing processes for morphological and lithological variability in freely meandering fluvial systems (Rosgen, 1985; Miall, 1988).

ACKNOWLEDGEMENTS

We thank sponsors and partners of the Fluvial & Eolian Research Group (Aker BP, Areva, BHP Billiton, Cairn India [Vedanta], ConocoPhillips, Murphy Oil Corporation, Nexen, Petrotechnical Data Systems (PDS), Saudi Aramco, Shell, Tullow Oil, Woodside and YPF) who are acknowledged for financial support of this research. LC was supported by NERC (Catalyst Fund award NE/M007324/1; Follow-on Fund NE/N017218/1). Jeff Peakall and Robert Thomas are thanked for providing helpful discussions. We thank Guest Editor Massimiliano Ghinassi, Rick Donselaar and an anonymous reviewer for their comments and suggestions, which have significantly improved the paper.

REFERENCES

Allen, J.R.L. (1965) A Review of the Origin and Characteristics of Recent Alluvial Sediments. *Sedimentology*, **5**, 89–191.

Allen, J.R.L. (1982) *Sedimentary Structures, Their Character* and *Physical Basis, Volume 1. Dev. Sedimentol.*, Elsevier Science Publishers.

Allen, J.R.L. (1983) Studies in Fluviatile Sedimentation: Bars, Bar-Complexes and Sandstone Sheets (Low-Sinuosity Braided Streams) in the Brownstones (L. Devonian), Welsh Borders. *Sed. Geol.*, **33**, 237–93.

Allen, J.R.L. (1964) Studies in Fluviatile Sedimentation: Six Cyclothems from the Lower Old Red Sandstone, Anglowelsh Basin. *Sedimentology*, **3**, 163–98.

Blum, M. and Jennings, D. (2016) The McMurray conundrum: Conflicting interpretations of environment of deposition and paleogeography. In: *American Association of Petroleum Geologists, Annual Convention and Exhibition, Calgary, AB, Program with Abstracts*, June.

Blum, M. and Pecha, M. (2014) Mid-Cretaceous to Paleocene North American drainage reorganization from detrital zircons. *Geology*, **42**, 607–610.

Blum, M.D. and Törnqvist, T.E. (2002) Fluvial Responses to Climate and Sea-Level Change: A Review and Look Forward. *Sedimentology*, **47**, 2–48.

Bobrovitskaya, N.N., Zubkova, C. and Meade, R.H. (1996) Discharges and yields of suspended sediment in the Ob and Yenisey Rivers of Siberia. *IAHS Publications-Series of Proceedings and Reports-Intern Assoc Hydrological Sciences*, **236**, 115–124.

Brekke, H. (2015) Cretaceous Forensic Podiatry: Big Game Tracking with a Microresistivity Image Log on a McMurray Formation Scroll Bar. *Bull. Can. Petrol. Geol.*, **63**, 225–42.

Brice, J.C. (1974) Evolution of Meander Loops. *Geol. Soc. Am. Bull.*, **85**, 581–86.

Bridge, J.S. (2003) *Rivers and Floodplains*. Blackwell, Oxford.

Bridge, J.S., Alexander, J.A.N., Collier, R.E., Gawthorpe, R.L. and Jarvis, J. (1995) Ground-penetrating radar and coring used to study the large-scale structure of point-bar deposits in three dimensions. *Sedimentology*, **42**, 839–852.

Bridge, J.S. and Diemer, J.A. (1983) Quantitative interpretation of an evolving ancient river system. *Sedimentology*, **30**, 599–623.

Bridge, J.S. and Jarvis, J. (1982) The Dynamics of a River Bend: A Study in Flow and Sedimentary Processes. *Sedimentology*, **24**, 499–541.

Bridge, J.S. and Tye, R.S. (2000) Interpreting the Dimensions of Ancient Fluvial Channel Bars, Channels and Channel Belts from Wireline-Logs and Cores. *AAPG Bull.*, **84**, 1205–28.

Brown, L.F. and Fisher, W.L. (1980) Principles of Seismic Stratigraphic Interpretation: Interpretation of Depositional Systems and Lithofacies from Seismic Data. *AAPG Department of Education*, **16**.

Cabello, P., Domínguez, D., Murillo-López, M.H., López-Blanco, M., García-Sellés, D., Cuevas, J.L., Marzo, M. and Arbués, P. (2018) From conventional outcrop datasets and digital outcrop models to flow simulation in the Pont de Montanyana point-bar deposits (Ypresian, Southern Pyrenees). *Mar. Petrol. Geol.*, **94**, 19–42.

Carrigy, M.A. (1962) Effect of texture on the distribution of oil in the Athabasca oil sands, Alberta. *J. Sed. Petrol.*, **32**, 312–325.

Carter, D.C. (2003) 3-D seismic geomorphology: Insights into fluvial reservoir deposition and performance, Widuri field, Java Sea. *AAPG Bull.*, **87**, 909–934.

Colombera, L., Mountney, N.P. and McCaffrey, W.D. (2012) A Relational Database for the Digitization of Fluvial Architecture: Concepts and Example Applications. *Petrol. Geosci.*, **18**, 129–140.

Colombera, L., Mountney, N.P., Russell, C.E., Shiers, M.N. and McCaffrey, W.D. (2017) Geometry and compartmentalisation of fluvial meander-belt reservoirs at the barform scale: quantitative insight from outcrop, modern and subsurface analogues. *Mar. Petrol. Geol.*, **82**, 35–55.

Colombera, L., Yan, N., McCormick-Cox, T. and Mountney, N.P. (2018) Seismic-driven geocellular modeling of fluvial meander-belt reservoirs using a rule-based method. *Mar. Petrol. Geol.*, **93**, 553–569.

Daniel, J.F. (1971) Channel Movement of Meandering Indiana Streams. *U.S. Geol. Surv. Prof. Pap.*, **732–A**.

Donselaar, M.E. and Overeem, I. (2008) Connectivity of fluvial point-bar deposits: An example from the Miocene Huesca fluvial fan, Ebro Basin, Spain. *AAPG Bull.*, **92**, 1109–1129.

Durkin, P.R., Hubbard, S.M., Boyd, R.L. and Leckie, D.A. (2015) Stratigraphic Expression of Intra-Point-Bar Erosion and Rotation. *J. Sed. Res.*, **85**, 1238–1257.

Durkin, P.R., Hubbard, S.M., Holbrook, J. and Boyd, R. (2017) Evolution of fluvial meander-belt deposits and implications for the completeness of the stratigraphic record. *Geol. Soc. Am. Bull.*, DOI: 10.1130/B31699.1.

Fielding, C.R. and Crane, R.C. (1987) An application of statistical modelling to the prediction of hydrocarbon recovery factors in fluvial reservoir sequences. In: *Recent Developments in Fluvial Sedimentology* (Eds F.G. Ethridge, R.M. Flores and M.D. Harvey), *SEPM Spec. Publ.*, **39**, 321–327.

Fustic, M., Hubbard, S.M., Spencer, R., Smith D.G., Leckie, D.A., Bennett, B. and Larter, S. (2012) Recognition of down-Valley Translation in Tidally influenced Meandering Fluvial Deposits, Athabasca Oil Sands (Cretaceous), Alberta, Canada. *Mar. Petrol. Geol.*, **29**, 219–232.

Ghinassi, M. (2011) Chute Channels in the Holocene High-Sinuosity River Deposits of the Firenze Plain, Tuscany, Italy. *Sedimentology*, **58**, 618–642.

Ghinassi, M., Ielpi A., Aldinucci, M. and Fustic, M. (2016) Downstream-Migrating Fluvial Point Bars in the Rock Record. *Sed. Geol.*, **334**, 66–96.

Gingras, M.K., MacEachern, J.A., Dashtgard, S.E., Ranger, M.J., Pemberton, S.G. and Hein, F. (2016) The significance of trace fossils in the McMurray Formation, Alberta, Canada. *Bull. Can. Petrol. Geol.*, **64**, 233–250.

Gutierrez, R.R. and Abad, J.D. (2014) On the Analysis of the Medium Term Planform Dynamics of Meandering Rivers. *Water Resour. Res.*, **50**, 3714–3733.

Harms, J.C., Southard, J.B. and Walker, R.G. (1982). Fluvial deposits and facies models. *Structure and Sequence in Clastic Rocks* SEPM Short Course 9, 1–26.

Hassanpour, M.M., Pyrcz, M.J. and Deutsch, C.V. (2013) Improved Geostatistical Models of Inclined Heterolithic Strata for McMurray Formation, Alberta, Canada. *AAPG Bull.*, **97**, 1209–1224.

Hohn, M.E., McDowell, R.R., Matchen, D.L. and Vargo, A.G. (1997) Heterogeneity of Fluvial-Deltaic Reservoirs in the Appalachian Basin: A Case Study from a Lower Mississippian Oil Field in Central West Virginia. *AAPG Bull.*, **81**, 918–936.

Hooke, J.M. (1977) *An Analysis of Changes in River Channel Patterns.* University of Exeter, PhD thesis.

Hooke, J.M. (1984) Changes in River Meanders: Hooke. *Progr. Phys. Geogr.*, **8**, 473–508.

Hooke, J.M. (2004) Cutoffs Galore!: Occurrence and Causes of Multiple Cutoffs on a Meandering River. *Geomorphology*, **61**, 225–238.

Hubbard, S.M., Smith, D.G., Nielsen, H., Leckie, D.A., Fustic, M., Spencer, R.J. and Bloom, L. (2011) Seismic Geomorphology and Sedimentology of a Tidally Influenced River Deposit, Lower Cretaceous Athabasca Oil Sands, Alberta, Canada. *AAPG Bull.*, **95**, 1123–1145.

Hudson, P.F. and Kessel, R.H. (2000) Channel Migration and Meander-Bend Curvature in the Lower Mississippi River prior to Major Human Modification. *Geology*, **28**, 531–534.

Ielpi, A. and Ghinassi, M. (2014) Planform Architecture, Stratigraphic Signature and Morphodynamics of an Exhumed Jurassic Meander Plain (Scalby Formation, Yorkshire, UK). *Sedimentology*, **61**, 1923–1960.

Jablonski, B.V.J. and Dalrymple, R.W. (2016) Recognition of Strong Seasonality and Climatic Cyclicity in an Ancient, Fluvially Dominated, Tidally Influenced Point Bar: Middle McMurray Formation, Lower Steepbank River, North-Eastern Alberta, Canada. *Sedimentology*, **63**, 1–34.

Jackson, R.G. (1976) Depositional model of point bars in the lower Wabash River. *J. Sed. Petrol.*, **46**, 579–594.

Jackson, R.G. (1975) Velocity-Bed-Form-Texture Patterns of Meander Bends in the Lower Wabash River of Illinois and Indiana. *Geol. Soc. Am. Bull.*, **86**, 1511–1522.

Jordan, D.W. and Pryor, W.A. (1992) Hierarchical Levels of Heterogeneity in a Mississippi River Meander Belt and Application to Reservoir Systems. *AAPG Bull.*, **76**, 1601–1624.

Labrecque, P.A., Jensen, J.L., Hubbard, S.M. and Nielsen, H. (2011) Sedimentology and Stratigraphic Architecture of a Point-bar Deposit, Lower Cretaceous McMurray Formation, Alberta, Canada. *Bull. Can. Petrol. Geol.*, **59**, 147–171.

Lasseter, T.J., Waggoner, J.R. and Lake, L.W. (1986) Reservoir heterogeneities and their influence on ultimate recovery. In: *Reservoir characterization* (Eds L.W. Lake and H.B. Jr Carroll), Orlando, Florida, Academic Press, 545–559.

Leopold, L.B. and Wolman, M.G. (1957) *River channel patterns: braided, meandering and straight.* US Government Printing Office.

Leopold, L.B. and Wolman, M.G. (1960) River Meanders. *Geol. Soc. Am. Bull.*, **71**, 769–794.

Lewin, J. (1976) Initiation of Bedforms and Meanders in Coarse-Grained Sediment. *Geol. Soc. Am. Bull.*, **87**, 281–285.

Mackey, S.D. and Bridge, J.S. (1995) Modelling Alluvial Stratigraphy in 3D with Maths. *J. Sed. Res.*, **B65**, 7–31.

Marzo, M., Nijman, W. and Puigdefàbregas, C. (1988) Architecture of the castissent fluvial sheet sandstones, Eocene, South Pyrenees. *Sedimentology*, **35**, 719–738.

McGowen, J.H. and Garner, L.E. (1970) Physiographic Features and Stratification Types of Coarse-Grained

Point bars: Modern and Ancient Examples. *Sedimentology*, **14**, 77–111.

Meybeck, M. and **Ragu, A.** (1997) Presenting the GEMS-GLORI, a Compendium of World River Discharge to the Oceans. *Ass. Hydrol. Sci. Publ.*, **243**, 3–14.

Miall, A.D. (2006) Reconstructing the Architecture and Sequence Stratigraphy of the Preserved Fluvial Record as a Tool for Reservoir Development: A Reality Check. *AAPG Bull.*, **90**, 989–1002.

Miall, A.D. (1988) Reservoir Heterogeneities in Fluvial Sandstones: Lesson from Outcrop Studies. *AAPG Bull.*, **72**, 682–697.

Moreton, D.J. and **Carter, B.J.** (2015) Characterizing alluvial architecture of point bars within the McMurray Formation, Alberta, Canada, for improved bitumen resource prediction and recovery. In: *Developments in Sedimentology* (Eds P.J. Ashworth, J.L. Best and D.J. Parsons), **68**, 529–559.

Mossop, G.D. and **Flach, P.D.** (1983) Deep Channel Sedimentation in the Lower Cretaceous McMurray Formation, Athabasca Oil Sands, Alberta. *Sedimentology*, **30**, 493–509.

Musial, G., **Reynaud, J.Y.**, **Gingras, M.K.**, **Féniès, H.**, **Labourdette, R.** and **Parize, O.** (2012) Subsurface and outcrop characterization of large tidally influenced point bars of the Cretaceous McMurray Formation (Alberta, Canada). *Sed. Geol.*, **279**, 156–172.

Nanson, G.C. (1980) Point Bar and Floodplain Formation of the Meandering Beatton River, Northeastern British Columbia, Canada. *Sedimentology*, **27**, 3–29.

Nanson, G.C. and **Hickin, E.J.** (1983) Channel migration and incision on the Beatton River. *J. Hydraul. Eng.*, **109**, 327–337.

Nardin, T.R., **Howard R.F.** and **Carter B.J.** (2013) Stratigraphic architecture of a large-scale point-bar complex in the McMurray Formation: light detection and ranging and subsurface data integration at Syncrude's Mildred Lake Mine, Alberta, Canada. In: *Heavy Oil and Oil Sand Petroleum Systems in Alberta and Beyond* (Eds F.J. Hein, D.A. Leckie and J.R. Suter), *AAPG Stud. Geol.*, **64**, 273–311.

Nixon, M. (1959) A study of the bank-full discharges of rivers in England and Wales. *Proc. Inst. Civil Eng.*, **12**, 157–174.

Ouchi, S. (1985) Response of Alluvial Rivers to Slow Active Tectonic Movement. *Geol. Soc. Am. Bull.*, **96**, 504–515.

Peel, M.C., **Finlayson, B.L.** and **McMahon, T.A.** (2007) Updated world map of the Köppen-Geiger climate classification. *Hydrology and Earth System Sciences Discussions*, **4**, 439–473.

Phillips, J.D. (2003) Sources of Nonlinearity and Complexity in Geomorphic Systems. *Progr. Phys. Geogr.*, **27**, 1–23.

Pranter, M.J., **Cabrera-Garzón, R.**, **Blaylock, J.J.**, **Davis, T.L.** and **Hurley, N.F.** (2000) Use of multicomponent seismic for the static reservoir characterization of the San Andres formation at Vacuum Field, New Mexico. Society of Exploration Geophysicists. In: *SEG Technical Program Expanded Abstracts 2000*, 1548–1551.

Pranter, M.J., **Cole, R.D.**, **Panjaitan, H.** and **Sommer, N.K.** (2009) Sandstone-Body Dimensions in a Lower Coastal-Plain Depositional Setting: Lower Williams Fork Formation, Coal Canyon, Piceance Basin, Colorado. *AAPG Bull.*, **93**, 1379–1401.

Pranter, M.J., **Ellison, A.I.**, **Cole, R.D.** and **Patterson, P.E.** (2007) Analysis and modelling of intermediate-scale reservoir heterogeneity based on a fluvial point-bar outcrop analog, Williams Fork Formation, Piceance Basin, Colorado. *AAPG Bull.*, **91**, 1025–1051.

Putnam, P.E. and **Oliver, T.A.** (1980) Stratigraphic traps in channel sandstones in the Upper Mannville (Albian) of east-central Alberta. *Bull. Can. Petrol. Geol.*, **28**, 489–508.

Ramón, J.C. and **Cross, T.** (1997) Characterization and Prediction of Reservoir Architecture and Petrophysical Properties in Fluvial Channel Sandstones, Middle Magdalena Basin, Colombia. CT&F – *Ciencia, Tecnologia Y Futuro*, **1**, 19–46.

Reijenstein, H.M., **Posamentier, H.W.** and **Bhattacharya, J.P.** (2011) Seismic Geomorphobgy and High-Resolution Seismic Stratigraphy of Inner-Shelf Fluvial, Estuarine, Deltaic and Marine Sequences, Gulf of Thailand. *AAPG Bull.*, **95**, 1959–1990.

Rosgen, D.L. (1985) *A Stream Classification System. Riparian Ecosystems* and *Their Management: Reconciling Conflicting Uses*, 91–95.

Russell, C.E. (2017) *Prediction of sedimentary architecture and lithological heterogeneity in fluvial point-bar deposits*. PhD thesis, University of Leeds, Leeds, UK.

Sambrook Smith, G.H., **Best, J.L.**, **Leroy, J.Z.** and **Orfeo O.** (2016) The Alluvial Architecture of a Suspended Sediment Dominated Meandering River: The Río Bermejo, Argentina. *Sedimentology*, **63**, 1187–1208.

Schumm, S.A. (1963) Sinuosity of Alluvial Rivers on the Great Plains. *Geol. Soc. Am. Bull.*, **74**, 1089–1100.

Schumm, S.A. (1960) The effect of sediment type on the shape and stratification of some modern fluvial deposits. *Am. J. Sci.*, **258**, 177–184.

Schwab, F.L. (1976) Modern and Ancient Sedimentary Basins: Comparative Accumulation Rates. *Geology*, **4**, 723–727.

Shanley, K.W. (2004) Fluvial Reservoir Description for a Giant, Low-Permeability Gas Field : Jonah Field, Green River Basin, Wyoming, U.S.A., *AAPG Stud. Geol.*, **52**; and Rocky Mountain Association of Geologists 2004 Guidebook, 159–182.

Smith, D.G., **Hubbard, S.M.**, **Leckie, D.A.** and **Fustic, M.** (2009) Counter Point Bar Deposits: Lithofacies and Reservoir Significance in the Meandering Modern Peace River and Ancient McMurray Formation, Alberta, Canada. *Sedimentology*, **56**, 1655–1669.

Smith, N.D. (1985) Proglacial fluvial environment. *Glacial Sedimentary Environments* SC16.

Strobl, R.S., **Muwais W.K.**, **Wightman D.M.**, **Cotterill D.K.** and **Yuan L.P.** (1997) Geological Modelling of McMurray Formation Reservoirs Based on Outcrop and Subsurface Analogues. *Can. Soc. Petrol. Geol. Mem.*, **18**, 292–311.

Swales, S., **Storey, A.W.** and **Bakowa, K.A.** (2000) Temporal and spatial variations in fish catches in the Fly River system in Papua New Guinea and the possible effects of the Ok Tedi copper mine. *Environmental Biology of Fishes*, **57**, 75–95.

Thomas, R.G., **Smith, D.G.**, **Wood, J.M.**, **Visser, J.**, **Calverley-Range E.A.** and **Koster E.H.** (1987) Inclined Heterolithic Stratification-terminology, Description, Interpretation and Significance. *Sed. Geol.*, **53**, 123–179.

Thompson, A. (1986) Secondary Flows and the Pool Riffle Unit: A Case Study of the Processes of Meander Development. *Earth Surf. Proc. Land.*, **11**, 631–641.

Toonen, W.H., Kleinhans, M.G. and **Cohen, K.M.** (2012) Sedimentary architecture of abandoned channel fills. *Earth Surf. Proc. Land.*, **37**, 459–472.

Tye, R.S. (2004) Geomorphology: An Approach to Determining Subsurface Reservoir Dimensions. *AAPG Bull.*, **88**, 1123–1147.

Van de Lageweg, W.I., van Dijk, W.M., Baar, A.W., Rutten, J. and **Kleinhans, M.G.** (2014) Bank pull or bar push: What drives scroll-bar formation in meandering rivers. *Geology*, **42**, 319–322.

Verrien, J.P., Cournad, G. and **Montadert, L.** (1967) Application of production geology methods to reservoir characteristics analysis from outcrop observations. *Proceedings of Seventh World Petroleum Congress, Mexico*, 425–446.

Webb, E.K. and **Davis, J.M.** (1998) Simulation of the Spatial Heterogeneity of Geologic Properties: An Overview. In: Hydraulic Models of Sedimentary Aquifers (Eds G.S. Fraser and J.M. Davis), *SEPM Concepts in Hydrology* and *Environmental Geology*, **1**, 1–24.

Wightman, D.M. and **Pemberton, S.G.** (1997) *The Lower Cretaceous (Aptian) McMurray Formation: an overview of the Fort McMurray area, northeastern, Alberta.* **312–344**.

Willis, B.J. (1989) Palaeochannel reconstructions from point bar deposits: a three-dimensional perspective. *Sedimentology*, **36**, 757–766.

Willis, B.J. and **Tang, H.** (2010) Three-Dimensional Connectivity of Point-Bar Deposits. *J. Sed. Res.*, **80**, 440–454.

Wood, J.M. (1989) Alluvial architecture of the Upper Cretaceous Judith River Formation, Dinosaur Provincial Park, Alberta, Canada. *Bull. Can. Petrol. Geol.*, **37**, 169–181.

Wu, C. and **Bhattacharya, J.P.** (2015) Paleohydrology and 3D Facies Architecture of Ancient Point Bars , Ferron Sandstone, Notom Delta, South-Central Utah. *J. Sed. Res.*, **85**, 399–418.

Yan, N., Colombera, L., Mountney, N.P. and **Dorrell, D.M.** (2018) Three-dimensional modelling of fluvial point-bar architecture and facies heterogeneity using analogue data and associated analysis of intra-bar static connectivity: application to humid coastal-plain and dryland fluvial systems. In: *Fluvial Meanders and their Sedimentary Products in the Rock Record* (Eds L. Colombera, G. Massimiliano, N.P. Mountney and A.J. Reesink), *Int. Assoc. Sedimentol. Spec. Publ.*, **48**, this volume.

Yan, N., Mountney, N.P., Colombera, L. and **Dorrell, R.M.** (2017) A 3D forward stratigraphic model of fluvial meander-bend evolution for prediction of point-bar lithofacies architecture. *Comput. Geosci.*, **105**, 65–80.

Yang, C.S. and **Nio, S.D.** (1985) The estimation of palaeohydrodynamic processes from subtidal deposits using time series analysis methods. *Sedimentology*, **32**, 41–57.

Int. Assoc. Sedimentol. Spec. Publ (2018) **48**, 419–444.

Reconstructing the architecture of ancient meander belts by compiling outcrop and subsurface data: A Triassic example

CÉSAR VISERAS*, SATURNINA HENARES*, LUIS MIGUEL YESTE* and FERNANDO GARCIA-GARCIA*

* *Sedimentary Reservoirs Workgroup, Department of Stratigraphy and Palaeontology, University of Granada, Spain*

ABSTRACT

A multi-approach study is presented by combining outcrop information with data from four fully recovered cores, gamma-ray logs, acoustic and optical tele-viewers and five Ground-Penetrating Radar profiles (GPR) performed behind the outcrop (so-called Outcrop/Behind Outcrop [OBO] characterisation). A detailed analysis of tadpoles corresponding to the surfaces identified at the well walls was carried out by coupling core data with well logging and borehole imaging. GPR information allows correlation among wells in addition to providing three-dimensionality to the architectural elements identified. All these techniques were applied to a 6.3 m-thick and 100 m-wide, asymmetric wedge-shaped sandstone geo-body interpreted as a single-storey meander belt in the distal alluvial plain of the Triassic Red Beds of the Iberian Meseta (TIBEM formation) in south-central Spain. The architectural element analysis on outcrop allowed differentiation of the subenvironments of the main channel, point bars and chute channels. The main-channel element was characterised by a fining and thinning upward succession in which three parts can be distinguished corresponding to the active-channel stage, the progressive abandonment and a final mud plug. The tadpoles on the margin of the channel indicate a consistent dip with the underlying lateral accretion units (LAUs) with a decreasing inclination upward. The heterolithic interval of the progressive abandonment is characterised by radar facies in which mounded and subparallel reflectors alternate. The lateral accretion units of the point bars are characterised by a poorly defined bell-shaped GR curve with anomalously high API values at the lowermost 50 to 60 cm due to the abundance of mud clasts. Other intervals of unexpectedly high API values at the intermediate-upper part of the curve correspond to small, hanging excavated channels. The cross beds, that result from the migration upward of megaripples over the LAUs, appear both in the core and in the GPR to be false horizontal bedding. The erosional scars identified in outcrop and core at the base and within the chute channels correspond to concave-up reflectors in GPR. These channels are filled with rippled sandstones that give rise to very low-amplitude radar facies. The high-resolution characterisation derived from this workflow highlights its great applicability to the study of other non-outcropping ancient meander belts in which more robust sedimentary models are required, such as in geo-steering operations for enhanced oil recovery.

Keywords: Meander belt, outcrop behind outcrop characterisation, well logging, gamma ray, ground-penetrating radar, tadpole, Triassic

INTRODUCTION AND PURPOSE

The importance of three-dimensional geological characterisation in Applied Geology has revitalised interest in a more detailed knowledge of the facies architecture and heterogeneity at different scales of sandstone geo-bodies resulting from meandering river dynamics (e.g. Bjorlykke & Jahren, 2010, Ghinassi *et al.*, 2014; Viseras *et al.*, 2015). In this sense, these main interests are: (1) the precise

Fluvial Meanders and Their Sedimentary Products in the Rock Record, First Edition.
Edited by Massimiliano Ghinassi, Luca Colombera, Nigel P. Mountney and Arnold Jan H. Reesink.

planning of exploitation and the growing emphasis on optimising hydrocarbon recovery in mature reservoirs (Enhanced Oil Recovery, EOR, Richards & Bowman, 1998; Meyer & Krause, 2006); (2) the possibility of using geo-bodies favourable for CO_2 capture (Carbon Capture and Storage, CCS, e.g. Veloso *et al.*, 2016a and 2016b); and (3) the importance placed by hydrogeologists on stratigraphic architecture and sedimentary fabrics as key controlling factors in hydraulic conductivity anisotropy, subsurface flow characteristics and the conditions of the hydraulic limits in detrital aquifers (e.g. Hornung & Aigner, 1999; Calvache *et al.*, 2009; Falivene *et al.*, 2007). Thus, to understand the behaviour of hydrocarbon or CO_2 sandstone reservoirs or aquifers derived from high sinuosity fluvial systems, a detailed 3D knowledge of the architecture is necessary, given their potential high heterogeneity (Corbeanu & Soegaard, 2001; Diaz-Molina & Muñoz-García, 2010; Ghinassi, 2011; Reesink *et al.*, 2014; Viseras *et al.*, 2015).

A substantial volume of quality data is needed to build a precise reservoir model. In this sense, outcrop analogue studies represent a powerful tool, supplementing sparse subsurface data (Miall, 1990; Kocurek *et al.*, 1991; Tyler & Finley, 1991; Wizevich, 1991; Yoshida *et al.*, 2001; Ajdukiewicz & Lander, 2010; Scott *et al.*, 2013; Franke *et al.*, 2015). The selection of appropriate outcrop analogues constitute an indispensable source of information to constrain the reservoir geometry, dimensions and attributes to generate realistic models that include high-resolution facies interpretation and rock heterogeneities (Alpay, 1972; Kocurek, 1988; Ambrose *et al.*, 1991; Miall, 1991; Alexander, 1992; Kostic & Aigner, 2007; Van den Brill *et al.*, 2007; Calvache *et al.*, 2009; Ozkan *et al.*, 2011; Trendell *et al.*, 2012; Ghinassi *et al.*, 2014, Colombera *et al.*, 2014; Klausen & Mørk, 2014; Shimer *et al.*, 2014).

The integrated study of both sources of information – outcrop and subsurface – is so-called outcrop/behind outcrop characterisation (hereafter OBO characterisation) (Browne & Slatt, 2002; Slatt *et al.*, 2011; Viseras *et al.*, 2013, 2015). This workflow has proved to be one of the most effective approaches by providing far more precision in terms of dimensions and characteristics of the geo-body than other studies based on the comparison with recent analogues may do (Szerbiak *et al.*, 2001; Zeng *et al.*, 2004; Donselaar & Schmidt, 2005; Miall, 2006; Colombera *et al.*, 2014).

The aim of this study is to determine key criteria for facies identification and characterisation that can be useful for other non-outcropping meander belts. By applying different imaging techniques, namely GPR, ABI and OBI, together with its validation against outcrop and conventional core analysis, the authors have designed an integrated OBO workflow which has successfully provided a more robust facies model. The studied example corresponds to a meander belt from a Triassic formation that has been already considered as an outcrop analogue for other hydrocarbon-productive reservoirs such as the Algerian TAGI (Trias Argilo-Gréseux Inférieur; Rossi *et al.*, 2002; Dabrio *et al.*, 2005 Viseras *et al.*, 2011b; Henares *et al.*, 2014, 2016). The conclusions reached here can be of significant applied value in mapping the distribution of permeability heterogeneity within this type of fluvial systems improving subsequent derived quantitative models.

GEOLOGICAL SETTING

Triassic red-beds of the Iberian Meseta, south-central Spain (so-called TIBEM by Viseras *et al.*, 2010; Henares *et al.*, 2014, 2016) form part of the continental realm developed during the Tethyan rifting (Sánchez-Moya *et al.*, 2004). In the study area, located north-east of the village of Alcaraz (Albacete province; Fig. 1), the Triassic sedimentary succession, 160 m-thick, consists of siliciclastic alluvial-lacustrine and fluvial stacked facies organised into four subhorizontal depositional sequences (Fernández & Dabrio, 1985; Fernández *et al.*, 1993, Dabrio *et al.*, 2005; Fig. 1). The subdivision and the age of each sequence are based on the correlation with the Tethys level fluctuations and their imprint on the fluvial architecture of the continental realm deposits (Fernández & Dabrio, 1985; Dabrio *et al.*, 2005). According to these authors, the fluvial sandstone of this study is part of Sequence II (Ladinian) together with other sandstone levels corresponding to straight channels and overbank deposits (Fig. 1B). These sedimentary geo-bodies, developed during a general rise in base-level on the basis of the regional correlation with their equivalent coastal sediments (Viseras *et al.*, 2012), are embedded in mud sediments from a distal flood plain which locally show greenish colours as the result of the reducing conditions after pounding/flooding processes (Viseras *et al.*, 2011a). Authors like Arche *et al.*

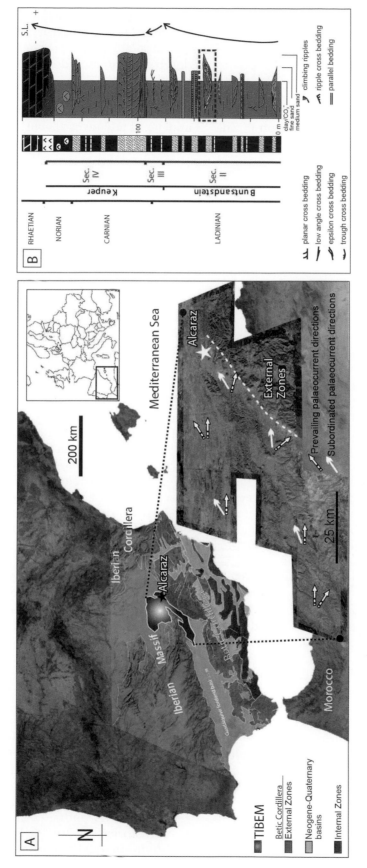

Fig. 1. (A) Geographical and geological location of the TIBEM formation. Arrows indicate the palaeocurrent directions near the study Alcaraz area (after Henares *et al.*, 2011; Viseras *et al.*, 2011a). (B) Simplified TIBEM stratigraphic succession in the Alcaraz area (after Dabrio & Fernández, 1986).

(2002) subdivide the stratigraphic succession according to other criteria such as minor regression-transgression cycles during a general sea-level rise.

At a local scale, a slight tilting to the north (dip < 10°) of the Triassic deposits is recognised, caused by syndepositional or gravitational compaction (Gay, 1989) as sediments adapt to the irregular topography of the Palaeozoic basement.

METHODS AND DATA

The OBO characterisation of the analysed meander belt follows a done-on-purpose workflow consisting of a multi-approach analysis. The data set includes detailed sedimentological descriptions from both surface (outcrop) and subsurface (cores, well data and ground-penetrating radar profiles) sources.

The first step consists of the outcrop-facies analysis for the identification and description of the main architectural elements of the geo-body in terms of geometry, internal structure, sequence trends and vertical and lateral relationship with other elements. After that, a total of four slim-hole (6.25"), behind-outcrop wells target each architectural element for their subsequent subsurface analysis. From south to north, the wells are: MB1 (2.55 m cored), MB2 (1.65 m cored), MB3 (3.8 m cored); and MB4 (6 m cored) with 11 m, 17 m and 32 m of well spacing, respectively (Fig. 2A to C). In the second step, core slabbing was performed to enhance the visibility of the sedimentary features of the core and to determine the simple lithofacies classes (Table 1) according to their predominant grain-size class, texture, fabric and main as well as secondary sedimentary structures (Table 1). Well data include the Gamma Ray log (GR) in addition to borehole imaging from Optical and Acoustic Televiewers (OBI and ABI, respectively). All the surfaces that have been identified in the image logs have been used to produce tadpole plots. In the analysis of these data, the authors have always taken into account the data that correspond to sets bounding surfaces, erosive surfaces or foresets of bedforms, to analyse them separately to be able to carry out a correct environmental interpretation. Tadpole analysis, together with GR patterns, provided information on spatial distribution, orientation and dip of the main sedimentary surfaces and structures. Finally, five ground-penetrating radar (GPR) profiles were carried out behind the outcrop exposure using a shielded 200 MHz antenna and covering a total of 138.1 m of the sandstone geo-body. Penetration, limited to a depth of 10 m, provided high-resolution images of the internal structure of the geo-body. Description and validation of the main Radar Facies with the previous collected data from both the outcrop and wells aided in the 3D interpretation of the sandstone geo-body architecture.

GEO-BODY GEOMETRICAL DESCRIPTION AND FLUVIAL STYLE

The analysed sandstone geo-body is divided into three different sections that extend a total of 57 m. The longest section, oriented approximately NNE-SSW, measures 35 m and is intersected by another two sections of 22 m in length each, both oriented NNW-SSE (Fig. 2). These changes in the orientation of the different exposure sections over the entire outcrop length allow the 3D tracing of the entire geo-body as well of the main elements identified within it.

Based on the architectural elements analysis developed (following the methodology of Miall, 2000), a geo-body with asymmetric wedge geometry can be reconstructed with a concave-up base and flat top at the scale of the entire deposit (Fig. 2D). Some irregularities present at the top are described below. The maximum thickness (3.4 m) is shown near the north-eastern end of the geo-body which pinches out towards the base to the SSW. The maximum lateral extension (measured perpendicular to the main palaeocurrent direction) is 35 m. The aspect ratio (W:T) approximates to 10 and therefore the sandstone geo-body can be classified as a medium-scale, ribbon-like deposit (Friend, 1983; Fernández et al., 1993; Viseras & Fernández, 1994, 1995) completely embedded in fine sediments. In previous studies, this sandstone geo-body has been interpreted as the result of a meander-belt dynamics oriented to the NW-SE in this area (Dabrio et al., 2005, Viseras et al., 2011a, 2016).

The main components within the geo-body, identified according to the bounding surfaces, external geometry, internal structure and position are here called the 'architectural elements' (*sensu* Miall, 1985). They are equivalent to the 'depositional elements' defined by Kostic & Aigner (2007) or the 'stories' of Ford & Pyles (2014). The palaeoenvironmental interpretation will be further

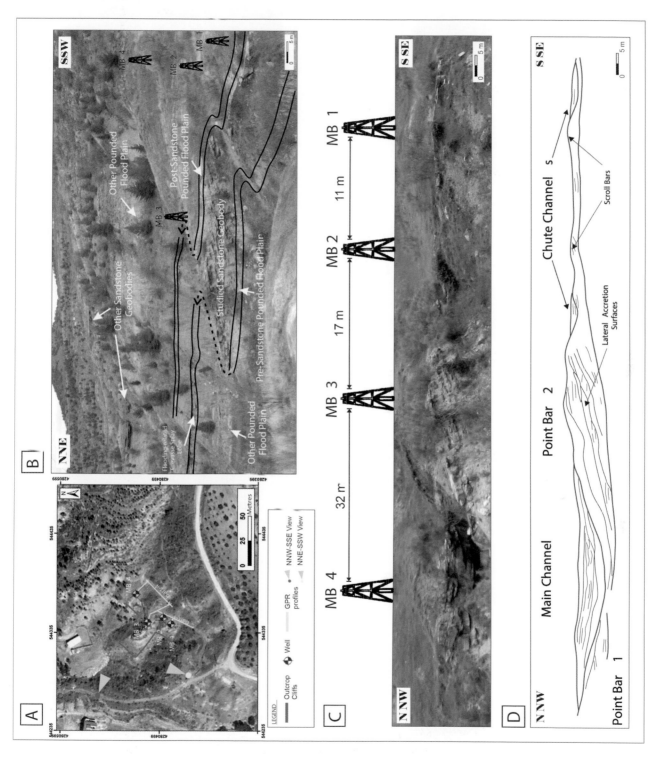

Fig. 2. (A) Orthophotograph showing the outline of the outcrop analysed, the well locations and the GPR profiles (after Henares *et al.*, 2016). (B) Outcrop photo-interpretation from the NNE-SSW view of the sandstone geo-body studied, also showing the well locations. (C) Outcrop photo from the NNW-SSE view of the sandstone geo-body studied, also showing the well locations and inter-well spacing. (D) Facies model of the geo-body analysed, indicating the main architectural elements.

Table 1. Main lithofacies classification from core observations including texture, sedimentary structures and processes, and associated architectural elements.

Major textural class	Code	Texture description	Attribute description	Sedimentary processes	Architectural element	Examples
Gravel	**Gmh**	• Pebble to granule • Clast-supported or matrix-supported	• Diffuse planar bedding • Erosional base • Faint clast imbrication (*) Normal or inverse grading (*) Angular clasts	• High-energy unidirectional current	• Channel bottom • Base of point bar • Chute channel	
Sand	**Sm**	• Very fine-grained to fine-grained sand	• Massive • Faint imbrication of mud clasts (*) Diffuse planar bedding (*) Coal debris	• Channel infilling deposits which result from waning flow	• Lower point bar	
	Sh	• Very fine-grained to fine-grained sand	• Planar lamination	• Upper flow regime under unidirectional flow	• Upper point bar (including scroll bar) • Channel margin • Crevasse lobe	
	Sp	• Fine-grained sand	• Planar cross-bedding	• Ripple migration	• Main channel • Lower point bar	
	St	• Very fine-grained to fine-grained sand	• Trough cross-bedding (*) Mud chips lining the cross beds	• 3-D megaripples and 3-D dune migration • Channel infilling	• Main channel	

	Grain size	Sedimentary structures	Process	Environment
Sr	Very fine-grained and fine-grained sand to silty sand	Current ripple-lamination (*) Flaser bedding	Ripple migration under unidirectional flow	Main channel Lower and upper point bar (including scroll bar) Chute channel
Src	Very fine-grained to fine-grained sand	Asymmetric climbing ripple cross-stratification	Ripple migration and fallout from suspension rate control the angle of the ripple climb	Upper point bar (including scroll bar)
Srw	Very fine-grained sands with silty layers	Wavy lamination	Oscillatory flow under low energy conditions	Main channel (margin) 
Slp	Very fine-grained to fine-grained sand	Low-angle cross-stratification	Down flow ripple migration into the channel and up-dip ripple migration in lateral accretion units	Mid point bar 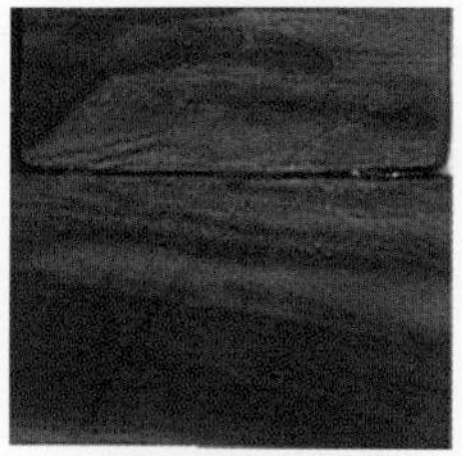
Fr	Clay (silt rare)	Massive Root traces Root-mottled (*) Scattered linsen bedding	Soil formation in overbank deposits	Flood plain 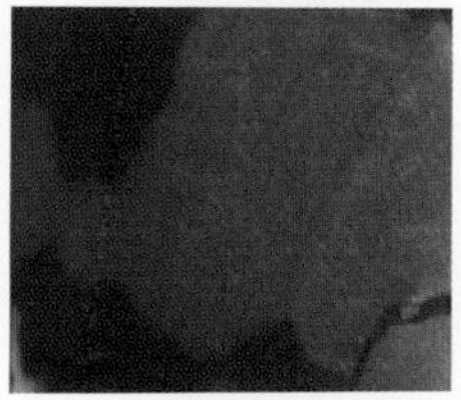
Fl	Clay and silt	Planar bedding under low flow regime	Suspension settling	Point bar (mud drape over lateral accretion units) Main channel (abandoned, clay plug) Flood plain 

Sand

Mud

explained in the next sections but an initial classification of the main architectural elements identified includes: the main channel fill, two successive point bars (called point bar 1 and point bar 2) and two subsidiary channels interpreted as chute channels on the point bar 2 (Fig. 3). In addition, the subsurface data lead to the identification of a third non-outcropping point bar which is aligned to the south with the two previous ones (point bar 3).

According to the geophysical data, the channel shows a typical asymmetric cross-section with an internal gently sloping accretionary bank and a steep external erosion bank (Fig. 3). The apparent symmetry of the main channel in outcrop, which is atypical of meandering fluvial systems, is just an artefact triggered by the oblique orientation of the exposure respect to the main palaeodrainage direction.

In the flood plain in outcrop, two greenish levels enriched in carbonates are identified at the base and the top of the entire sandstone geo-body (Fig. 2B). These levels (here referred to as the pre-sandstone and post-sandstone ponded flood plain) provide information of the timing of the formation of this geo-body. Such deposition might take place in between two generalised flood events on the flood plain, when the water table rose above the topographic surface.

MULTI-APPROACH DESCRIPTION OF ARCHITECTURAL ELEMENTS

The 3 architectural elements identified in the studied sandstone geo-body, which are main channel, point bar and chute channel, are characterised on the basis of their main features in outcrop, core, well and GPR (Fig. 4). Therefore, this characterisation includes: (i) outcrop-derived observations and measurements; (ii) definition of 11 simple lithofacies in the cores (1 gravelly, 8 sandy and 2 lutitic; Table 1); (iii) GR pattern analysis according to its geometry; (iv) tadpole trend analysis for surfaces and sedimentary structures characterisation; and (v) main radar facies description using Table 2 as a reference.

Main channel fill

Outcrop features

This architectural element consists of a sedimentary geo-body of 4 m maximum thickness and 15 to 20 m lateral extent (measured perpendicular to the palaeoflow direction). The thickness of the sandstone beds shows a thinning-upward trend with 3 different intervals. The lowermost interval corresponds to a continuous sandstone bed that occupies up to 75% of the total channel width and presents a biconcave-up lens geometry with a thickness ranging from 0 to 70 cm. The next overlying interval is characterised by heterolithic facies with an alternation of decimetric layers of sandstones and siltstones covering the entire channel width (Fig. 5). Locally, the stratification is dissected by internal shallow erosional scars (30 cm deep; 1.5 m wide; Fig. 5A). The last interval corresponds to a mudstone layer with a flat convex-down lens shape that occupies the most central part of the channel. This mudstone connects laterally with the aforementioned layer of post-sandstone ponded flood plain capping the entire sandstone geo-body (Fig. 2B).

Core features

In the central part of the channel (Well MB4, Fig. 5) a fining and thinning upward succession can be identified (FTU). At the base of this succession, a centimetric bed with abundant weakly-imbricated mud intraclasts (lithofacies Gmh – Table I) occurs. On top of it, a set of 60 cm-thick trough cross beds (St) is overlain by an alternation of ripple sandstones (Sr) and siltstones with horizontal lamination or ripples (Fl). An erosional scar can be discerned upward in the succession. Finally, a 2 m-thick, horizontally laminated (Fl) claystone bed caps the succession, containing abundant rhizoliths and decimetric rippled siltstone intercalated.

Towards the marginal part of the channel (well MB3, Fig. 6), the succession shows 2 erosional scars lined by a sandstone bed with abundant mud intraclasts (Gmh). On top of this, sandy facies with planar cross lamination (Sp) and wavy lamination (Srw) are deposited.

Well data

The GR curve is quite different between the centre and the margin of the channel (Figs 5 and 6). In the central part, the curve shows a clear bell-shaped trend but with a blunt base due to anomalous high API values. On the channel margin, the lower part of the curve shows a saw-tooth-shaped trend and only after the

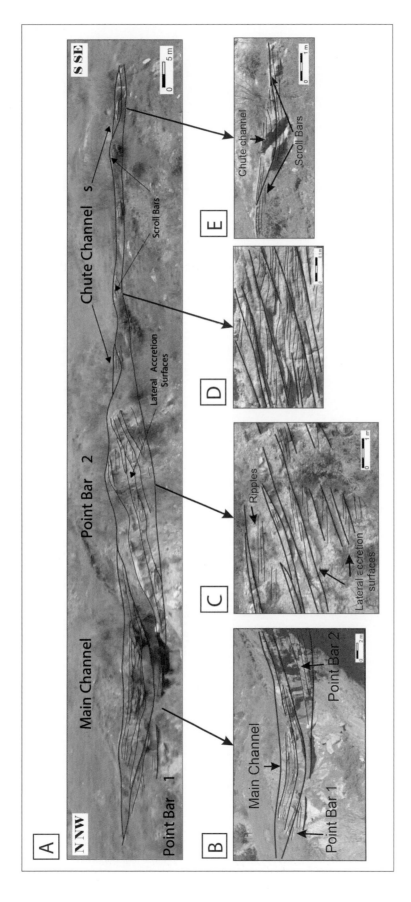

Fig. 3. (A) Outcrop photo-interpretation from the NNW-SSE view of the sandstone geo-body studied. Arrows indicate the position of some outcrop close-view photo interpretations. (B) Internal organisation of the main channel infill as well as the base of the two identified point bars. (C) Internal mega cross-bedding in the intermediate part of point bar 2. (D) Current and climbing ripples in the upper part (scroll bar) of point bar 2. (E) Internal sedimentary structure of one of the chute channels. Note the internal bedding parallel to the erosive base although unconformable with respect to the bedding in the upper part of the point bar 2.

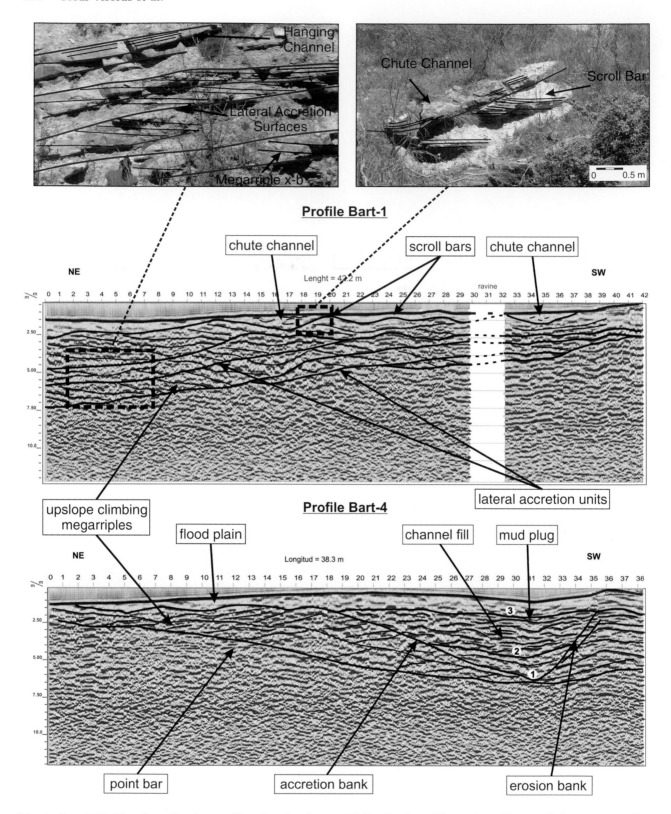

Fig. 4. Two 200 MHz radar reflection profiles showing the spatial distribution of the main architectural elements as well as other recognisable sedimentary features and their appearance in the outcrop.

Table 2. Description of the main radar facies and surfaces identified in the radar reflection profiles based on geometrical criteria and their potential associated lithofacies and depositional environments.

Type	Radar facies code	Facies description	Potential lithofacies	Environmental interpretation	GPR image
Parallel / subparallel	PR	Parallel even to subparallel (sometimes low angle dipping) with high amplitudes	Fl, Sh, Sm	Floodplain main channel	
Mounded	MD	Wavy subparallel, discontinuous reflections with low to high amplitudes	Gmh, Sp, Srw, Slp, Sh, St	Main channel and chute channel	
Short slightly dipping	SD	Short slightly dipping (5-9°) reflectors lying against major and continuous surfaces that dip in the opposite direction. Medium amplitudes	Sr, Src, Slp, Sp, Sh	Point bar	
Chaotic	CH	Irregular, discontinuous, chaotic reflections with very low to medium amplitudes	Fl	Floodplain, chute channel and clay plug	

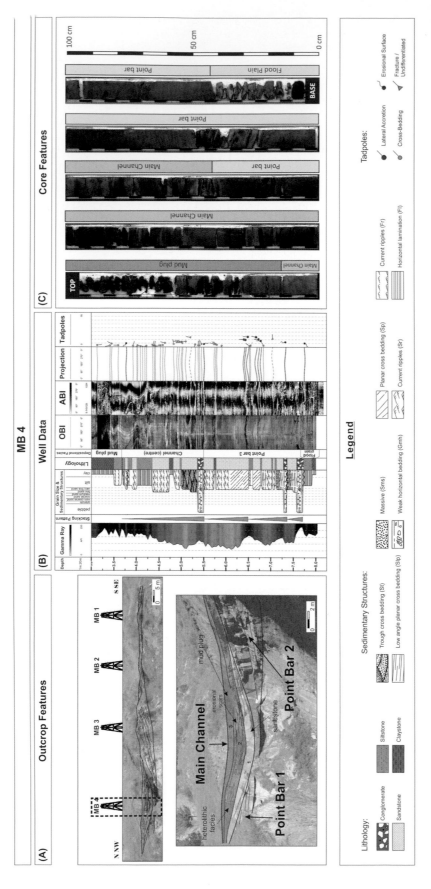

Fig. 5. (A) Outcrop photo-interpretation on the main architectural elements intersected by well MB4. 1,2,3 indicate the different stages of channel infill: 1 – active channel, 2 – progressive abandonment, 3 – ox-bow lake. (B) Well data, lithostratigraphic log, borehole imaging and tadpole analysis in well MB4. (C) Core details from well MB4.

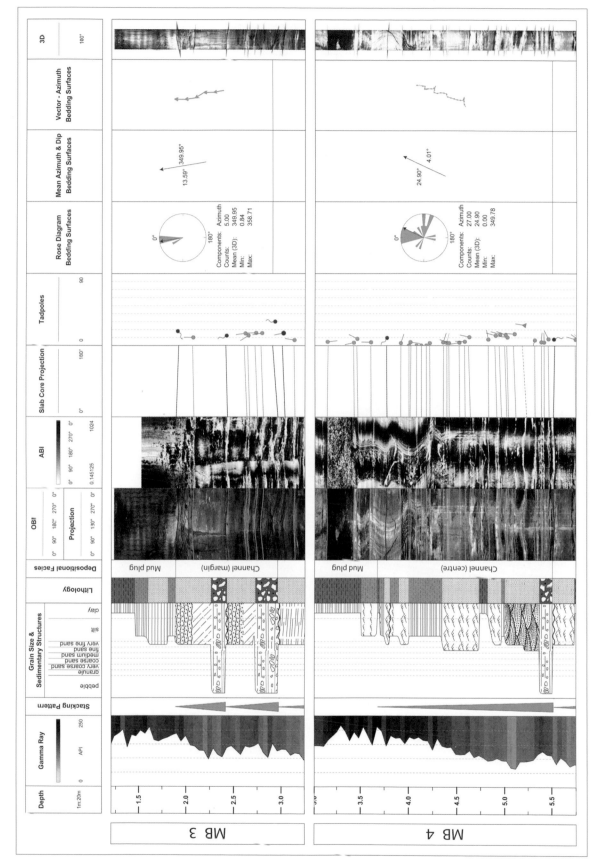

Fig. 6. Tadpole analysis for the spatial characterisation of the main surfaces and sedimentary structures recognised in the channel centre and margin (wells MB4 and MB3, respectively), indicating their orientation also in 3D. See Fig. 5 for legend.

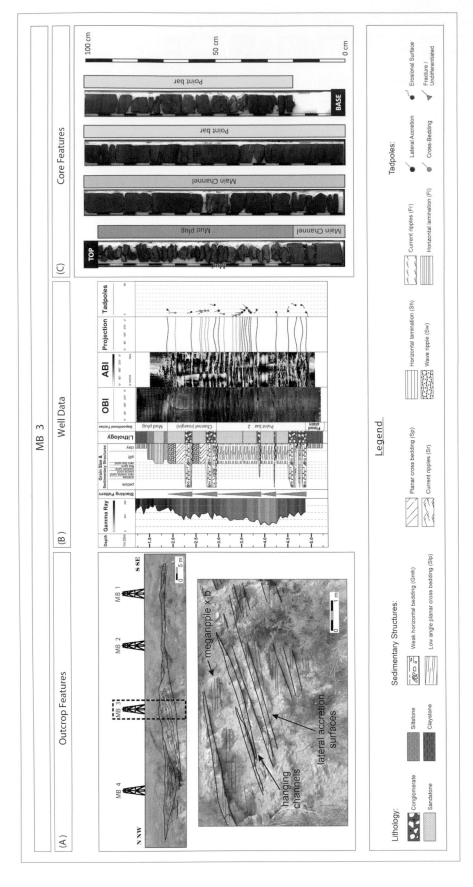

Fig. 7. (A) Outcrop photo-interpretation on the main architectural elements intersected by well MB3. Some internal structures are highlighted. (B) Well data, lithostratigraphic log, borehole imaging and tadpole analysis in well MB3. (C) Core details from well MB3.

internal erosional scar, a relatively irregular bell-shaped trend becomes discernible.

The tadpoles in the central part of the channel indicate horizontal surfaces with a northward, less than 10° dip which suggests palaeohorizontality in coherence with the local tilting. In this central part (well MB4), the tadpoles corresponding to cross beds indicate the palaeoflow directions towards the S and W. However, on the margin of the channel (well MB3), the main surfaces appear to be inclined consistently with the underlying lateral accretion units, whose dip values decrease upwards. In this marginal sector, the cross beds indicate the dip direction towards the N and NNW.

GPR features

The predominant radar facies is the mounded-type (MD, Table 2) although certain subparallel types are also recognised, with a length ranging between 3 and 5 m (radar facies PR, Fig.4). The difference between the various infilling channel phases is highlighted by the contrast between the radar facies MD and PR. In the lower part, the radar reflectors present intermediate amplitudes with mounded patterns whereas in the intermediate part of the channel these amplitudes increase. Towards the top, these amplitudes decrease again resembling the MD radar facies of the lower part.

Point-bar deposits

Outcrop features

The second architectural element occupies the greatest extent of the outcrop. Two examples are recognised, herein referred as point bar 1 and point bar 2 (Fig. 3), being the last one which the description is based on due to its better exposure. It shapes an asymmetric sigmoid with the greatest thickness (close to 6 m) towards the northern part and wedges out at the base until disappearing among the fine flood plains. The top shows several convex-up irregularities spaced between 5 and 7 m apart.

The inner structure is dominated by a mega cross-bedding, tilted between 13 and 15° towards the NW, which becomes asymptotic at the base and the top of the element (Fig. 3). This feature highlights the occurrence of diverse sandstone units with sigmoidal geometry within it. These successive sigmoidal units are delimited by clayey siltstone layers of centimetric to decimetric thickness parallel to the main bedding. Towards the middle part of the sigmoids, internal planar cross-bedding indicates upslope palaeoflow, separated between 30 and 50° from the main flow direction, highlighted by the direction of stretch of the adjacent channel (Fig. 7; rose diagram of Fig. 8). In the uppermost part, small scale structures result of current and climbing ripples abound. They evidence a local upslope palaeocurrent of up to 75° respect to the main channel direction (Fig. 8).

Locally, and systematically towards the upper third of the sigmoidal units, the cross bedding may be interrupted by a channelised-base and flat-top deposit. It measures approximately 50 cm-thick and reaches 5 m laterally with a consistent general inclination with that of the mega cross-beds (Fig. 7).

Core features

In the core, from base to top, repetitive cycles can be discerned, consisting of (Fig. 7): (a) a slightly erosive base floored by a centimetric-decimetric-thick, mud intraclast-rich bed (Gmh); (b) a low-angle planar cross-bedded sandstone (Slp) with a thickness ranging from 10 to 60 cm; (c) a centimetric claystone-siltstone with horizontal lamination (Fl). The dip between the uppermost claystone layer and the cross-bedded sandstone sets is opposite.

Well data

The GR signal of point bar 2 shows a bell-shaped trend both at the margin (Wells MB 1 and 2) and the central part (Well MB 3, Fig. 7). However, at the base of the central part, this bell shape is not so clear due to high API values. In the third non-outcropping point bar, situated to the south of the two exposed ones and inferred by geophysics (point bar 3), the GR pattern is similar. This GR curve shows great irregularity towards the base, although from 4.5 m depth upwards it shows the typical bell-shaped trend described in the previous ones (Fig. 7B).

The tadpoles corresponding to the main accretion surfaces in point bar 2 show a counter-clockwise turn. Thus, in MB1 and MB2, being the latter situated towards the N but both in the southern sector of the architectural element, these surfaces tilt towards the SE-ESE and more

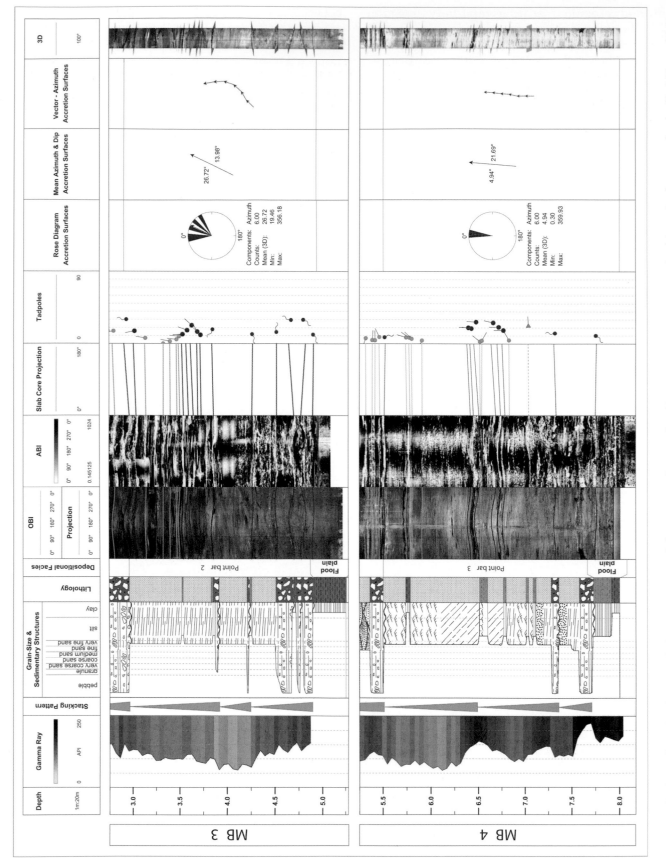

Fig. 8. Tadpole analysis for the spatial characterisation of the main surfaces and sedimentary structures recognised in point bars 2 and 3 (wells MB3 and MB4, respectively), also indicating their orientation in 3D. See Fig. 6 for legend.

clearly towards the ESE, respectively. However, in MB3 they tilt towards the NE (Figs 7, 8 and 9). On the other hand, the tadpoles corresponding to minor cross beds indicate movement directions almost oblique and ascendant respect to the inclination of the main surfaces. The tadpole analysis corresponding to the main accretion surfaces of the non-outcropping point bar 3 (targeted in well MB4) indicates a clockwise rotation (Fig. 5).

GPR features

The radar facies characteristic of this element is composed of short slightly dipping (5 to 9°) reflectors (radar facies SD, Table 2) lying against major, continuous surfaces that dip in the opposite direction (Fig. 4). Furthermore, the main reflectors show a convex geometry locally and towards the high part (Fig. 4).

Chute channel fill

Outcrop features

At the top of point bar 2, two erosional scars are recognised on which symmetrical channelised deposits were formed, featuring a lateral extension of 3 m and a maximum thickness of 70 to 80 cm (Fig. 3). In these channelised bodies, an internal bedding parallel to the erosive base is recognised which is unconformable in respect to the upper point bar bedding (Fig. 9).

Core features

The base of this architectural element is systematically erosive, with local internal erosional scars (Fig. 9). The predominant lithofacies in both studied examples (point bars 1 and 2) is sandstone with ripples (Sr). Locally, a decametric-thick, mud intraclast-rich layer may develop with an erosive base (chute channel in well MB2).

Well data

This element has insufficient thickness to identify a genetically significant GR trend.

GPR features

The radar facies of this element has a very low amplitude (almost reflection free). Locally, poorly marked chaotic (CH) or mounded elements (MD) of scarce continuity can be recognised (Table 2, Fig. 4).

DISCUSSION: ENVIRONMENTAL RECONSTRUCTION INTEGRATING OUTCROP AND SUBSURFACE DATA

The identified architectural elements, the palaeocurrent dispersion and the occurrence of lateral accretion units deduced from the outcrop and subsurface data indicate that the analysed sandstone geo-body formed from high-sinuosity channel dynamics. This interpretation is also consistent with previous studies on the same area (Dabrio *et al.*, 2005, Viseras *et al.*, 2011a, 2016). In the main channel, three different infill stages are recognised (Fig. 5). The lowermost stage is associated to a sandstone bed that occupies the most depressed part and belongs to the active-channel phase. The intermediate heterolithic interval corresponds to a progressive abandonment, where sandstone layers deposited during the tractive transport stages, whereas the siltstone layers do so during the decrease in the flow. The discharge fluctuations are also manifested by erosive scars. The morphology and composition of the last interval suggests that it may be considered as a mud plug developed when the channel was definitively abandoned and transformed into an oxbow lake (Viseras *et al.*, 2006; Ghinassi *et al.*, 2014). This fining- and thinning-upward trend recognised in the channel fill can be also identified in the GR curve by a typical bell-shape (Fig. 5). In the intermediate part of the channel, GPR shows an alternation of reflectors with high and low electromagnetic impedance (Fig. 4). This feature can be correlated with the heterolithic unit represented by the alternation of sandstones and siltstones from the progressive channel abandonment.

Further evidences of the high sinuosity of the fluvial system are the strong dip differences between the lateral accretion surfaces identified in the wells MB3 and MB4. These are rather close to each other (32 m), indicating their position in successive point bars attached to opposite channel margins.

The vertical lithofacies association (Gmh-Sp-Srw; Table 1) points out that the channel abandonment took place initially in an progressive way, probably by a chute cut-off process as described in other examples such as the Mississippi River valley (Guccione *et al.*, 2001). This process might have taken place at a moment of flow diversion towards the subsidiary chute channels identified on top of point bar 2. However, there are also evidences that suggest a more rapid avulsion process as the final abandonment event as it has been

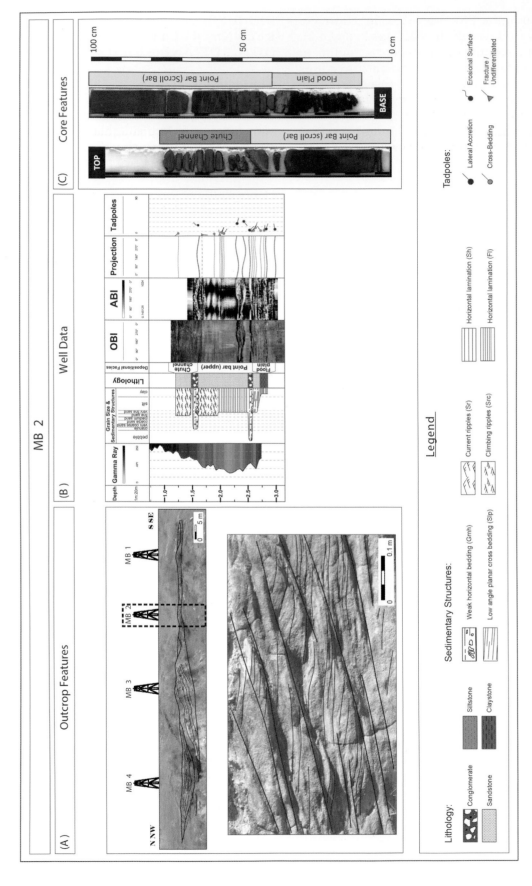

Fig. 9. (A) Outcrop photo-interpretation of the main architectural elements intersected by well MB2. Some internal structures are highlighted. (B) Well data, lithostratigraphic log, borehole imaging and tadpole analysis in well MB2. (C) Core details from well MB2.

described in other examples in the literature (Kraus & Wells, 1999). These evidences are: 1) the great difference in dimension between the main channel and the two chute channels (5 times higher and 6 times wider); and 2) the occurrence of the post-sandstone long-term ponded flood-plain layer in continuity with the mud plug (Fig. 2). Thus, this avulsion may be related with a general flood of the plain which caused the subsequent migration of the meander belt to a new and more depressed position in the flood plain.

The occurrence of low-energy facies on top of a shallow erosional surface in the upper part of the channel margin (lithofacies Srw, Table 1) may be caused by the settling down of the suspended load in small channels excavated in the intermediate-to-upper part of the lateral accretion unit adjacent to the channel. This type of smaller channels may correspond to the high channels described by Harms *et al.* (1963) in point bars with two topographic levels. Nevertheless, they show greater similarity with the channels described as hanging channels by Burge & Smith (1999). The presence of this hanging channel can be related with the break in the bell-shaped GR trend in MB3, which towards the upper third presents a decimetric level with anomalous high API values (Fig. 7).

Large point bars were deposited in the inner banks of the sinuous channel, characterised by the development of typical sigmoidal units (epsilon cross bedding *sensu* Allen, 1970; Figs 3 and 4). The helicoidal flow pattern developed in the meander curve is evidenced by the occurrence of ripples migrating upslope on the epsilon-type cross bedding surfaces (Allen 1970). This type of flow lead to the formation of ripples and megaripples (large sand waves or dunes according to Reineck & Singh, 1973) that ascended obliquely over the lateral accretion surfaces to create a positive relief in the higher part of the point bar, resulting in an uneven top (accretional topography, Puigdefabregas & Van Vliet, 1978). Flow deceleration caused the suspended load to settle which generated a mud drape on top of the lateral accretion surfaces. During drought periods, partial erosion of the upper part of these lateral accretion surfaces may occur, resulting in shallow erosional scars with a local stepped profile similar to that described by Harms *et al.* (1963) in the Red River. Nevertheless, the silt-rich, low-energy facies that fill these depressions suggest that they could have also been the result of slack-water type sedimentation. Similar situations have been described in meandering systems as a consequence of the flow-separation process that can occur near the downstream end of the bar (Nanson, 1980). Burge & Smith (1999) described high-sinuosity systems in which a secondary hanging channel can be generated in the higher part of the point bar due to eddies forming as the flow strikes the channel margin perpendicularly. This local alteration of the flow may cause that large part of the suspension load deposit on top of the erosional surfaces that it has partially excavated on the lateral accretion unit. This facies is interpreted as slackwater fluvial deposits where flow reversals have been linked to eddy evolution (e.g. passage of small eddies or instabilities in the main eddy or eddy pulsations) during high-discharge stage changes in the margin of modern river channels (Kochel & Baker, 1982; Rubin & McDonald, 1995). Different mechanisms (bedrock spurs or protected sites, distributary mouths) in channel margin areas produce local energy dissipation that results in ineffective flow areas, slow-moving water, flow separation from the main eddy shear lines and local recirculation zones (Leeder & Bridge, 1975; Baker, 1984; Benito *et al.*, 2003).

In the GR record, the stacking of the lateral accretion units resulted in a bell-shaped curve, which is more evident from about 50 to 60 cm above the base of the sandstone geo-body. This effect at the base of the GR pattern is caused by the occurrence of an intra-formational mud-pebble conglomerate (lithofacies Gmh − Table 1) at the point bar base, which indicates an apparent upward-coarsening grain-size motif (Selley, 2004). The proper identification of this artefact in the GR signal is of paramount importance because the point bar base may be misinterpreted as floodplain deposits prior to the channel installation. The small hanging channels identified correspond to decimetric intervals of high API value that appear towards the middle-upper part of the curve (Fig. 7).

At the lower and higher parts of the bar, the megaripples are clearly identified as sandstones with planar cross-bedding (lithofacies Sp - Table 1) caused by their migration upward over the lateral accretion surfaces as the result of the helicoidal flow (Fig. 7A and C). However, at the intermediate part of the bar, these megaripples seem to be almost horizontal or low-angle cross-stratified (lithofacies Slp, Table 1; Fig. 7A and C), both in outcrop and core, due to the pronounced dip of the lateral accretion surfaces.

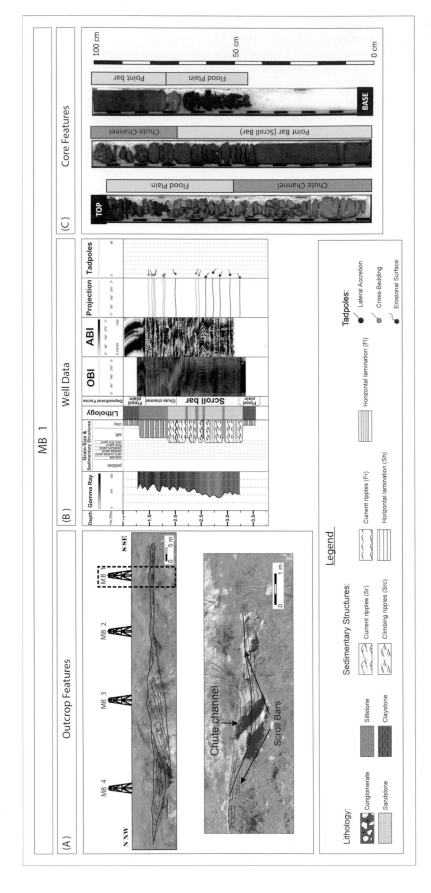

Fig. 10. (A) Outcrop photo-interpretation of the main architectural elements intersected by well MB1. Some internal structures are highlighted. (B) Well data, lithostratigraphic log, borehole imaging and tadpole analysis in well MB1. (C) Core details from well MB1.

Towards the upper part of the point bar, the GPR reflectors may show a concave geometry corresponding to the irregular relief generated by the scroll bars. Similar situation was described by Van de Lageweg *et al.* (2014) in experimentally formed meandering rivers (Fig. 4). When discharge is maximum, part of the water may be channelled through the associated swells at the top of the point bar and excavate the two identified chute channels. In these subsidiary channels, part of the sediment of the point bar top may be reworked forming a lag pavement of mud clasts (Figs 9 and 10). However, these channels were fed mainly by the main channel bedload. Their reduced dimensions with respect to the main channel and the presence of the internal erosion scars suggest that these are equivalent to the type-2 chute channels of Ghinassi (2011) in evolutionary stage III. Evidence of this is also their activity constrained to major flow peaks. This indicates that they were not responsible for the final channel abandonment represented by the mud plug although may have actively participated in the formation of the heterolithic intermediate part of the main channel. This interval shows flow fluctuations but systematically lower energy than at the initial stage.

According to Ghinassi *et al.* (2014), the clockwise rotation of the tadpoles corresponding to the

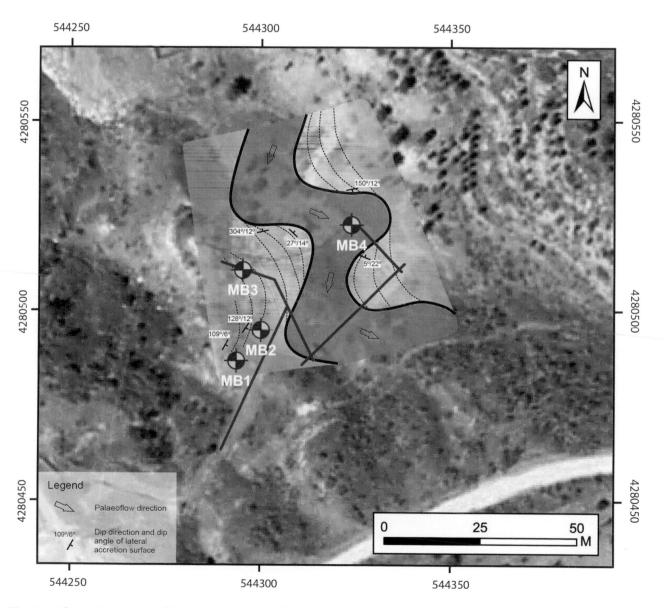

Fig. 11. Schematic location of the main elements of the meander belt.

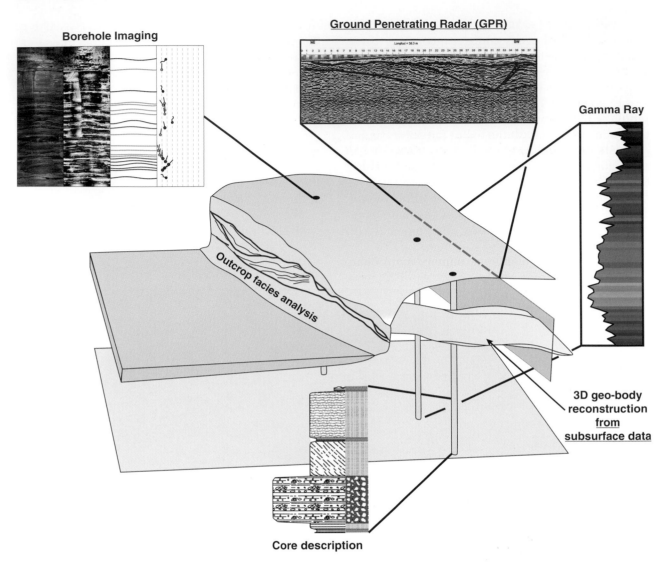

Fig. 12. Illustration of the OBO characterization workflow designed and applied in this study together with the potential results from its application.

lateral accretion units inclination between wells MB3 and MB1 (Figs 7 and 10) would suggest that MB3 would be located upstream whereas MB1 would be placed in the downstream part of the same point bar on the right bank of the main channel (Fig. 11). Similarly, the vertical evolution of the tadpoles in the lateral accretion units of wells MB2 and MB4 shows an anti-clockwise turn in the first one contrary to what happens with MB4 (Figs 5 and 6). According to previous work (e.g. Brekke *et al.*, 2017) this change in orientation indicates that successive point bars have been targeted in both wells, one on the right bank and the other one on the left bank of the same main channel. The latter point bar (the one on the left bank)

is only inferred from GPR (point bar 3, Fig. 11). In that way, tadpole analysis when coupled with core-derived and outcrop-derived observations can be a powerful tool to reconstruct palaeocurrent directions and depositional dips, which are of paramount importance for facies interpretations (Fig. 12).

CONCLUSIONS

Fluvial reservoirs are notoriously complex because of their intrinsic permeability heterogeneity within and between the reservoir sandstone bodies. In the present study, the integrated application of advanced visualisation techniques, validated against outcrop and conventional core and gamma-ray (GR) log data was of vital importance to characterise the spatial distribution of the fluvial facies and inherent permeability baffles to a centimetre-scale vertical resolution. The Outcrop/Behind Outcrop (OBO) workflow employed here combined the sedimentological analysis of a single-storey Triassic meander belt outcrop with Ground-Penetrating Radar (GPR) profiles and behind-outcrop Optical and Acoustic Televiewer (OBI and ABI, respectively) imaging techniques; and with the analyses of dip tadpoles, conventional core and GR-logs. The integrated OBO workflow allowed for the recognition of three architectural elements, namely main channel, point bars and chute channels. The main channel fill sandstone comprises a fining-upward and thinning-upward depositional trend. The point bar element is built up by the stacking of several depositional cycles with an erosional base lined with a lag of mud intraclasts which constitute a permeability baffle. Up the point-bar slope, a low-angle cross-bedded sandstone is present which is capped by a siltstone layer with a depositional dip in the opposite direction of the point-bar dip. The chute channel element is composed of climbing-ripple laminated sandstone with basal and internal erosional surfaces locally lined with mud intraclasts.

This study directly links sedimentological information with petrophysical and image-log response, and highlights that the application of the OBO characterisation workflow presented here provides a solid database for the characterisation of the spatial distribution of reservoir rock properties from outcrop analogues.

ACKNOWLEDGEMENTS

Funding was provided by the research project CGL2013-43013-R (MEC–FEDER) by the Repsol-University of Granada agreement and by the research group RNM369 (JA). The authors are indebted to Advanced Logic Technology (ALT) for providing the academic WellCAD license as well as to the Consejería de Agricultura of Castilla-La Mancha (JCCM), the city hall of Alcaraz and Bart and María Alcazar (landowners) for the drilling licenses. The authors also thank the REPSOL, CEPSA E.P and Crimidesa companies for their support. Thanks to David Nesbitt for the final language revision of the manuscript. The quality of work has improved significantly in response to the suggestions of N. Mountney and two anonymous reviewers. We also want to thank the teachings of Prof. Juan Fernández, the first one who showed us this outcrop.

REFERENCES

Ajdukievwick, J.M. and **Lander, R.H.** (2010) Sandstone reservoir quality prediction: the state of the art. *AAPG Bull.*, **94**, 1083–1091.

Alexander, J. (1992) A discussion on the use of analogues for reservoir geology. In: *Advances in Reservoir Geology* (Ed. M. Ashton), *Geol. Soc. London, Spec. Publ.*, **69**, 175–194.

Allen, J.R.L. (1970) Studies in fluviatile sedimentation: a comparison of fining-upward cyclothems, with special reference to coarse-member composition and interpretation. *J. Sed. Petrol.*, **40**, 293–323.

Alpay, O.A. (1972) A practical approach to defining reservoir heterogeneity. *J. Petrol. Tech.*, **24**, 841–848.

Ambrose, W.A., **Tyler, N.** and **Parsley, M.J.** (1991) Facies heterogeneity, pay continuity and infill potential in barrier-island, fluvial and submarine-fan reservoirs: examples from the Texas Gulf Coast and Midland Basin. In: *The three-dimensional facies architecture of terrigenous clastic sediments and its implications for hydrocarbon discovery and recovery* (Eds A.D. Miall and N. Tyler), *SEPM Concepts in Sedimentology and Paleontology* **3**, Tulsa, p 7–13.

Arche, A., **López-Gómez, J.** and **García-Hidalgo, J.F.** (2002) Control climático, tectónico y eustático en depósitos del Carniense (Triásico Superior) del SE de la Península Ibérica. *J. Ib. Geol.* **28**, 13–30.

Baker, V.R. (1984) Flood sedimentation in bedrocks fluvial systems. *Mem. Can. Soc. Petrol. Geol.*, **10**, 87–98.

Benito, G., **Sánchez-Moya, Y.** and **Sopeña, A.** (2003) Sedimentology of high-stage flood deposits of the Tagus River, Central Spain. *Sed. Geol.*, **157**, 107–132.

Bjorlykke, K. and **Jahren, J.** (2010) Sandstones and sandstones reservoir. In: *Petroleum Geoscience - From Sedimentary Environments to Rock Physics* (Ed. H. Bjorlykke), Springer, Berlin, 119–150.

Brekke, H., **MacEachem, J.A.**, **Roenitz, T.** and **Dastgard, S.E.** (2017) The use of microresistivity image logs for facies interpretations: An example in point-bar deposits of the McMurray Formation, Alberta, Canada. *AAPG Bull.*, **101**, 655–682.

Browne, G.H. and **Slatt, R.M.** (2002) Outcrop and behind-outcrop characterisation of a late Miocene slope fan system, Mt. Messenger Formation, New Zealand: *AAPG Bull.*, **86**, 841–862.

Burge, L.M. and **Smith, D.G.** (1999) Confined meandering river eddy accretions: sedimentology, channel geometry and depositional processes. *Int. Assoc. Sedimentol. Spec. Publ.*, **28**, 113–130.

Calvache, M. L., **Ibáñez, S.**, **Duque, C.**, **Martín Rosales, W.**, **López Chicano, M.**, **Rubio, J. C.**, **González, A.** and **Viseras, C.** (2009) Numerical modelling of the potential effects of a dam on a coastal aquifer in southern Spain. *Hydrol. Process.*, **23**, 1268–1281.

Colombera, L., Mountney, N.P., Felletti, F. and McCaffrey, W.D. (2014) Models for guiding and ranking well-to-well correlations of channel bodies in fluvial reservoirs, *AAPG Bull.*, **98**, 1943–1965.

Corbeanu, R.M. and Soegaard, K. (2001) Detailed internal architecture of a fluvial channel sandstone determined from outcrop cores and 3-D ground-penetrating radar: Examples from the Middle Cretaceous Ferron Sandstone, east-central Utah. *AAPG Bull.*, **85**, 1583–1608.

Dabrio, C.J. and Fernández, J. (1986) Depositos de rios trenzados conglomeraticos Plio-Pleistocenicos de la Depresión de Granada. *Cuadernos de Geologia Iberica*, **10**, 31–53.

Dabrio, C., Fernández, J. and Viseras, C. (2005) Triassic Fluvial Sandstones (Central South Spain) - An excellent analogue for the TAGI Reservoir of Algeria, European Association of Geoscientists and Engineers. 67th EAGE Conference and Exhibition, *Field Trip Guides*, **1**. Van Houten (Holanda), 35 pp.

Díaz-Molina, M. and Muñoz-García, M.B. (2010) Sedimentary facies and three-dimensional reconstructions of upper Oligocene meander belts from the Loranca Basin, Spain. *AAPG Bull.*, **94**, 241–257.

Donselaar, M.E. and Schmidt, J.M. (2005) Integration of outcrop and borehole image logs for high-resolution facies interpretation: example from a fluvial fan in the Ebro Basin, Spain, *Sedimentology*, **52**, 1021–1042.

Falivene, O., Cabrera, L., Muñoz, J.A., Arbués, P., Fernández, O. and Sáez, A. (2007) Statistical grid-based facies reconstruction and modelling for sedimentary bodies. Alluvial-palustrine and turbiditic examples. *Acta Geol.*, **5**, 199–230.

Fernández, J., Bluck, B.J. and Viseras, C. (1993) The effects of fluctuating base level on the structure of alluvial fans and associated fan delta deposits: an example from the Tertiary of the Betic Cordillera, Spain. *Sedimentology*, **40**, 879–893.

Fernández, J. and Dabrio, C. (1985) Fluvial Architecture of the Buntsandstein-facies Redbeds in the Middle to Upper Triassic (Ldinian-Norian) of the Southeastern Edge of the Iberian Meseta (Southern Spain). In: *Aspects of fluvial sedimentation in the Lower Triassic Buntsandstein of Europe. Lecture Notes in Earth Sciences* (Ed. D. Mader), Springer-Verlag, Berlin, 411–435.

Ford, G.L. and Pyles, D.R. (2014) A hierarchical approach for evaluating fluvial systems: Architectural analysis and sequential evolution of the high net-sand content, middle Wasatch Formation, Uinta Basin, Utah, *AAPG Bull.*, **98**, 1273–1304.

Franke, D., Hornung, J. and Hinderer, M. (2015) A combined study of radar facies, lithofacies an three-dimensional architecture of an alpine alluvial fan (Illgraben fan, Switzerland). *Sedimentology*, **62**, 57–86.

Friend, P. (1983) Towards the field classification of alluvial architecture or sequence. In: *Modern and Ancient Fluvial Systems* (Eds J.C. Collinson and J. Lewin), *Int. Assoc. Spec. Publ.*, **6**, 337–344.

Gay, S.P. (1989) Gravitationla compaction, a neglected mechanism in structural and stratigraphic studies: New evidence from Mid-Continent, *USA. AAPG Bull.*, **73**, 641–657.

Ghinassi, M (2011) Chute channels in the Holocene high-sinuosity river deposits of the FIrenze plain, Tuscany, Italy. *Sedimentology*, **58**, 618–642.

Ghinassi, M, Nemec, W., Aldinucci, M., Nehyba, S., Özaksoy, V. and Fidolini, F. (2014) Plan-form evolution of ancient meandering rivers reconstructed from longitudinal outcrop sections. *Sedimentology*, **61**, 952–977.

Guccione, M.J., Burford, M., Kendall, J., Curtis, N., Odhiambo, B., Porter, D. and Shepherd, S. (2001) Channel fills of all scales from the Mississippi River alluvial valley. *Geol. Soc. Am.*, **33**, 356.

Harms, J.C., Mackenzie, D.B. and Mc Cubbing, D.G. (1963) Stratification of modern sands of the Red River, Lousiana. *J. Geol.*, **71**, 566–580.

Henares, S, Caracciolo, L., Cultrone, G., Fernández, J. and Viseras, C. (2014) The role of diagenesis and depositional facies on pore system evolution in a Triassic outcrop analogue (SE Spain). *Mar. Petrol. Geol.*, **51**, 136–151.

Henares, S., Caracciolo, L., Viseras, C., Fernández, J. and Yeste, L.M. (2016) Diagenetic constraints on heterogeneous reservoir quality assessment: a Triassic outcrop analogue of meandering fluvial reservoirs, *AAPG Bull.*, **100**, 1377–1398.

Henares, S., Viseras, C., Fernández, J., Pla-Pueyo, S. and Cultrone, G. (2011) Triassic Red Beds in SE Spain: evaluation as potential reservoir rocks based on a preliminary petrological study. *AAPG. Search and Discovery Article*, #50541.

Hornung, J. and Aigner, T. (1999) Reservoir and Aquifer Characterisation of Fluvial Architectural Elements: Stubensandstein, Upper Triassic, Southwest Germany. *Sed. Geol.*, **129**, 215–280.

Klausen, T.G. and Mørk, A. (2014) The Upper Triassic paralic deposits of the De Geerdalen Formation on Hopen: Outcrop analog to the subsurface Snadd Formation in the Barents Sea. *AAPG Bull.*, **98**, 1911–1941.

Kocurek, G. (1988) First order and superbounding surfaces in eolian sequences – eolian sediments revisited. *Sed. Geol.*, **56**, 193–206.

Kocurek, G., Knight, J. and Havholm, K. (1991) Outcrop of semi-regional three-dimensional architecture and reconstruction of a portion of the eolian Page Sandstone (Jurassic). In: *The three-dimensional facies architecture of terrigenous clastic sediments and its implications for hydrocarbon discovery and recovery* (Eds A.D. Miall and N. Tyler), *SEPM Concepts in Sedimentology and Paleontology* **3**, Tulsa, p 25–43.

Kostic, B. and Aigner, T. (2007) Sedimentary architecture and 3D ground-penetrating radar analysis of gravelly meandering river deposits (Neckar Valley, SW Germany). *Sedimentology*, **54** (4), 789–808.

Kraus, M.J. and Wells, M. (1999) Recognizing avulsion deposits in the ancient stratigraphical record. In: *Fluvial Sedimentology VI* (Eds N.D. Smith and J. Rogers), *Int. Assoc. Sedimentol. Spec. Publ.*, Blackwell Science, **28**, 251–268.

Leeder, M.R. and Bridge, P.H. (1975) Flow separation in meander beds. *Nature*, **253**, 338–339.

Meyer, R. and Krause, F.F. (2006) Permeability anisotropy and heterogeneity of a sandstone reservoir analogue: An estuarine to shoreface depositional system in the Virgelle Member, Milk River Formation, Writing-on-Stone

Provincial Park, southern Alberta. *Bull. Can. Petrol. Geol.*, **54** 301–318.

Miall, A.D. (1985) Architectural-element analysis: A new method of facies analysis applied to fluvial deposits. *Earth-Sci. Rev.*, **22**, 261–308.

Miall, A.D. (1991) Hierarchies of architectural units in Terrigenous clastic rocks, and their relationship to sedimentation rate. In: *The three-dimensional facies architecture of terrigenous clastic sediments and its implications for hydrocarbon discovery and recovery* (Eds A.D. Miall and N. Tyler), *SEPM Concepts in Sedimentology and Paleontology 3*, Tulsa, p 6–12.

Miall, A.D. (1990) *Principles of Sedimentary Basin analysis*, 2nd edition: Springer Verlag Inc., New York, 668 p.

Miall, A.D. (2000) *Principles of Sedimentary Basin analysis*, 3rd edition, updated and enlarged Edition: Springer Verlag Inc., Berlin Heidelberg, 616 p.

Miall, A.D. (2006) Reconstructing the architecture and sequence stratigraphy of the preserved fluvial record as a tool for reservoir development: A reality check, *AAPG Bull.*, **90** (7), 989–1002.

Nanson, G.C. (1980) Point bar and floodplain formation of the meandering Beatton River, northeastern British Columbia, Canada. *Sedimentology*, **27**, 3–29.

Ozkan, E., Brown, M.L., Raghavan, R. and **Kazemi, H.** (2011) Comparison of fractured-horizontal-well performance in tight sand and shale reservoirs. *Soc. Petrol. Eng.*, **14** (2), 1–12.

Puigdefabregas, C. and **van Vliet, A.** (1978) Meandering stream deposits from the Tertiary of the southern Pyrenees. In: *Fluvial sedimentology* (Ed. A.D. Miall). *Bull. Can. Soc. Petrol. Geol.*, **5**, 459–469.

Reesink, A.J.H., Ashworth, P.J., Sambrook Smith, G.H., Best, J.L., Parsons, D., Amsler, M. L., Hardy, R.J., Lane, S.N, Nicholas, A.P., Orfeo, O., Sandbach, S.D. Simpson, C.J. and **Szupiany, R.N.** (2014) Scales and causes of heterogeneity in bars in a large multi-channel river: Río Paraná, Argentina. *Sedimentology*, **61**, 1055–1085.

Reineck, H.E. and **Singh, I.B.** (1973) *Depositional Sedimentary Environments*. Springer-Verlag. Berlin, 550 pp

Richards, M. and **Bowman, M.** (1998) Submarine fans and related depositional systems II: variability in reservoir architecture and wireline log character. *Mar. Petrol. Geol.*, **15**, 821–839.

Rossi, C., Kälin, O., Arribas, J. and **Tortosa, A.** (2002) Diagenesis, provenance and reservoir quality of Triassic TAGI sandstones from Ourhoud field, Berkine (Ghadames) Basin, Algeria. *Mar. Petrol. Geol.* **19**, 117–142.

Sánchez-Moya, Y., Arribas, J., Gómez-Gras, D., Marzo, M., Pérez-Arlucea, M. and **Sopeña, A.** (2004) Inicio del rifting. El comienzo del relleno continental. In: *Geología de España* (Ed. J.A. Vera), Madrid. SGE-IGME. 485–487.

Scott, A., Hurst, A. and **Vigorito, M.** (2013) Outcrop-based reservoir characterisation of a kilometer-scale sand-injectite complex. *AAPG Bull.*, **97**, 309–343.

Selley, R.C. (2004) *Ancient Sedimentary Environments; and their sub-surface diagnosis*. Taylor and Francis Group, London, 297 pp

Shimer, G.T, McCarthy, P.J. and **Hanks, C.L.** (2014) Sedimentology, stratigraphy and reservoir properties of an unconventional, shallow, frozen petroleum reservoir in the Cretaceous Nanushuk Formation at Umiat field, North Slope, Alaska, *AAPG Bull.*, **98** (4), 631–661.

Slatt, R.M., Buckner, N., Abousleiman, Y., Sierra, R., Philp, P., Miceli-Romero, A., Portas, R., O'Brien, N., Tran, M., Davis, R. and **Wawrzyniec, T.** (2011) Outcrop/behind outcrop (quarry), multiscale characterisation of the Woodford Gas Shale, Oklahoma. In: *Shale reservoirs—Giant resources for the 21st century* (Ed. J. Breyer), AAPG Mem., **97**, 1–21.

Szerbiak, R.B., McMechan, G.A., Corbeanu, R., Forster, C. and **Snelgrove, S.H.** (2001) 3-D characterisation of a clastic reservoir analog: from 3-D GPR data to a 3-D fluid permeability model. *Geophysics*, **66**, 1026–1037.

Trendell, A.M., Atchley, S.C. and **Nordt, L.C.** (2012) Depositional and diagenetic controls on reservoir attributes within a fluvial outcrop analog: Upper Triassic Sonsela Member of the Chinle Formation, Petrified Forest National Park, Arizona: *AAPG Bull.*, **96**, 679–707.

Tyler, N. and **Finley, R.J.** (1991) Architectural controls on the recovery of hydrocarbons from sandstone reservoirs. In: *The three-dimensional facies architecture of terrigenous clastic sediments and its implications for hydrocarbon discovery and recovery* (Eds A.D. Miall and N. Tyler), *SEPM Concepts in Sedimentology and Paleontology 3*, Tulsa, p 1–5.

Van de Lageweg, W.I., Van Dijk, W.M., Baar, A.W., Rutten, J. and **Kleinhans, M.G.** (2014) Bank pull or bar push: What drives scroll-bar formation in meandering rivers? *Geology*, **42** (4), 319–322.

Van den Bril, K., Gregoire, C., Swennen, R. and **Lambot, S.** (2007) Ground-penetrating radar as a tool to detect rock heterogeneities (channels, cemented layers and fractures) in the Luxembourg Sandstone Formation (Grand-Duchy of Luxembourg), *Sedimentology*, **54**, 949–967.

Veloso, F.M.L., Liesa, C.L., Soria, A.R., Meléndez, M.N. and **Frykman, P.** (2016a) Outcrop scale reservoir characterisation and flow modelling of CO_2 injection in the tsunami and the barrier island-Tidal inlet reservoirs of the Camarillas Fm. (Galve sub-basin, Teruel, NE Spain). *Sed. Geol.*, **343**, 38–55.

Veloso, F.M.L., Navarrete, R., Soria, A.R. and **Meléndez, M.N.** (2016b) Sedimentary heterogeneity and petrophysical characterisation of Barremian tsunami and barrier island/inlet deposits: the Aliaga outcrop as a reservoir analogue (Galve sub-basin, Teruel, NE Spain). *Mar. Petrol. Geol.*, **73**, 188–211.

Viseras, C. and **Fernández, J.** (1994) Channel migration patterns and related sequences in some alluvial fan systems. *Sed. Geol.*, **88**, 201–17.

Viseras, C. and **Fernández, J.** (1995) The role of erosion and deposition in the construction of alluvial fan sequences in the Guadix Formation (SE Spain). *Geol. Mijnbouw* **74**, 21–33.

Viseras, C., Fernández, J. and **Henares, S.** (2011a) Anatomy of straight *vs* meandering fluvial channels in Triassic red beds of S Spain and Morocco. Implications as reservoir rocks. 28th IAS Meeting of Sedimentology, Zaragoza (Spain), Abstract Book, 107.

Viseras, C., Fernández, J. and **Henares, S.** (2011b) Facies architecture in outcropping analogues for the TAGI

Reservoir. Exploratory Interest. AAPG International Conference and Exhibition, Milan (Italy), *Search and Discovery Article*, #90135.

Viseras, C., Fernández, J. and **Henares, S.** (2012) Processes of meandering channel abandonment as generators of reservoir heterogeneities. Examples from the Triassic red beds of the High Atlas (Morocco). 29th IAS Meeting of Sedimentology, Schladming (Austria), *Abstract Book*, 304.

Viseras, C., Henares, S., Fernández, J. and **Jaimez, J.** (2013) Outcrop/Behind outcrop characterisation in onshore Western Mediterranean basins of Southern Iberia. AAPG European Regional Conference and Exhibition, Barcelona (Spain), *Search and Discovery Article*, 41175.

Viseras, C., Henares, S., Fernández, J., Yeste, L.M., Pla-Pueyo, S. and **Calvache, M.L.** (2015) Towards 3D reservoir modeling of outcrop analogs through integrated outcrop/behind outcrop characterisation. A Triassic example. 31th IAS Meeting of Sedimentology, Krakov (Poland), *Abstract Book*, **564**.

Viseras, C., Henares, S., Yeste, L.M. and **García-García, F.** (2016) Subsurface VS outcrop key features to reconstruct ancient point bar deposits with examples from Spain and Morocco. 32nd IAS Meeting of Sedimentology, Marrakech (Morocco), *Abstract Book*.

Viseras, C., Soria, J.M., Durán, J.J., Pla, S., Garrido, G., García-García,F. and **Arribas, A.** (2006) A large-mammal site in a meandering fluvial context (Fonelas P-1, Late Pliocene, Guadix Basin, Spain). Sedimentological keys for its paleoenvironmental reconstruction. *Palaeogr. Palaeoclimatol. Palaeoecol.*, **242**, 139–168.

Wizevich, M.C. (1991) Photomosaics of outcrops: useful photographic techniques. In: *The three-dimensional facies architecture of terrigenous clastic sediments and its implications for hydrocarbon discovery and recovery* (Eds A.D. Miall and N. Tyler), *SEPM Concepts in Sedimentology and Paleontology* **3**, Tulsa, p 22–24.

Yoshida, S., Jackson, M.D., Johnson, H.D., Muggeridge, A.H. and **Martinius, A.W.** (2001) Outcrop studies of tidal sandstones for reservoir characterization (Lower Cretaceous Vectis Formation, Isle of Wight, Southern England). In: *Sedimentary Environments Offshore Norway – Palaeozoic to Recent* (Eds O.J. Martinsen and T. Dreyer), *Norwegian Petroleum Society Special Publication*, **10**, 233–257.

Zeng, H. and **Hentz, T.F.** (2004) High-frequency sequence stratigraphy from seismic sedimentology: Applied to Miocene, Vermilion Block 50, Tiger Shoal area, offshore Lousiana, *AAPG Bull.*, **88**, 153–174.

Int. Assoc. Sedimentol. Spec. Publ (2018) **48**, 445–474.

Reconstruction of a sandy point-bar deposit: implications for fluvial facies analysis

ALISTAIR SWAN[†], ADRIAN J. HARTLEY[†], AMANDA OWEN[‡] and JOHN HOWELL[†]

[†] *Department of Geology and Petroleum Geology, University of Aberdeen, Aberdeen, UK*
[‡] *School of Geographical and Earth Sciences, University of Glasgow, Glasgow, UK*

ABSTRACT

The Salt Wash distributive fluvial system (DFS) of the Upper Jurassic Morrison Formation consists of stacked fluvial channel bodies interbedded with overbank deposits. The Salt Wash DFS has previously been recognised as a braided fluvial fan covering an area of over 100,000 km². However, the addition of high-resolution satellite imagery means planform exposures are now recognisable and point bar deposits can be identified in the sand-rich proximal to medial reaches of the system, requiring a reassessment in this area of the DFS. An individual exhumed point bar deposit has been identified within a 30,000 km² plan view area of meander belt in central Utah, which offers a unique perspective into the preserved internal distribution of facies and architecture. Field techniques have been utilised in conjunction with LiDAR and heli-drone three-dimensional outcrop datasets to compare measurements of the bedform and barform architectural elements within two contrasting outcrop exposure styles. One outcrop (Caineville Wash) has extensive plan view and vertical exposures, whereas the other (Caineville Road) is semi-restricted to 2D canyon exposures with limited planform exposure. The contrasting exposure styles have been used to develop criteria for the interpretation of sandy meandering river deposits in 2D exposures, where plan view characteristics are not available. Internally, the point bar body exposed at Caineville Wash outcrop can be subdivided into three portions from the plan view imagery: upstream, central and downstream portions of a point bar. Mapping of the internal architecture of the fluvial bar allows recognition of downstream and laterally accreting components. The upstream and downstream portions of the point bar are predominately composed of downstream migrating barform deposits; and the central portion of the point bar consists of laterally accreting elements. Sedimentary logs taken from around the outcrop display vertical profiles commonly considered characteristic of a braided fluvial system. Through understanding the internal architecture of the Caineville Wash point bar deposit, it is possible to create a planform reconstruction of a stacked multi-storey channel body from a two-dimensional outcrop. Results indicate that point bar deposits in sand-rich fluvial systems may have been incorrectly interpreted as braided deposits due to: 1) the presence of a significant proportion of downstream accreting elements within the point bar deposits; and 2) 2D sedimentary logs considered characteristic of a braided fluvial system (such as vertical grain-size trends and repeating conglomerate lags). Subsequently, sandy meander-belt deposits may be under-represented within the proximal – medial portion of the Salt Wash DFS.

Keywords: Distributive fluvial system, fluvial architecture, fluvial facies analysis, Salt Wash Member, point bar, Morrison Formation, Meander-bend deposits

Fluvial Meanders and Their Sedimentary Products in the Rock Record, First Edition.
Edited by Massimiliano Ghinassi, Luca Colombera, Nigel P. Mountney and Arnold Jan H. Reesink.
© 2019 International Association of Sedimentologists. Published 2019 by John Wiley & Sons Ltd.

INTRODUCTION

Understanding the planform depositional style of a fluvial system is an integral part of interpreting fluvial deposits, as the interpreted fluvial planform architecture (braided *vs.* meandering) is commonly considered to have an effect on the larger scale shape, dimensions and internal heterogeneity of preserved fluvial sandstone bodies (Miall, 1977; Gibling 2006; Hartley *et al.*, 2015). As fluvial deposits form hydrocarbon reservoirs, host mineral deposits comprise important aquifers, understanding and reconstructing fluvial planform can be important in determining the characteristics and heterogeneity apparent within fluvial succession, particularly when maximising production from hydrocarbon reservoirs (Campbell, 1976; Gibling, 2006; Smith *et al.*, 2009; McKie *et al.*, 2010).

Commonly, the interpretation of fluvial planform has been conducted using sedimentary logs, net:gross, width:thicknesses and facies classification analysis (Miall, 1988; Bridge, 1993; Miall, 1995). Braided deposits are interpreted to have a high volume of gravel facies, dominated by trough cross-bedding forming an overall sheet-like geometry (Miall, 1977; Allen, 1983; Bridge 1993; Miall, 1996; Gibling, 2006). In addition, these deposits are dominated by the downstream migration of unit bars during flood flow stage, resulting in bed-form palaeoflow direction being equal or similar to the barform accretion direction (Miall, 1977). In a vertical profile (Fig. 1), meandering fluvial deposits are traditionally recognised as a fining upward package from a coarse-grained pebble lag passing upwards into fine sand with possible preservation of overbank fines at the top of the package (Jackson 1976; Donselaar & Overeem, 2008; Ghazi & Mountney, 2009). Common features present within point bar deposits include gravel lags, trough cross-stratification and inclined heterolithic stratification (Jackson, 1976; Jackson, 1981; Plint, 1983; Smith, 1987; Donselaar & Overeem, 2008; Ghazi & Mountney, 2009). Floodplain deposits may be present at the top of the point bar succession (Fig. 1), although in some cases the point bar deposit may be truncated, such that the full profile of the classic vertical succession may not be present (Plint, 1983). This will result in a similar vertical sedimentary succession to that of a braided river (Fig. 1) raising the issue of how to recognise point bar deposits from a simple 1D sedimentary log – or in subsurface core – or whether confident recognition is possible at all.

Ground penetrating radar, together with shallow coring, has been used to try and image internal features within modern point bars (Bridge *et al.*, 1995; Neal, 2004). However, these studies use examples with significant internal heterolithic

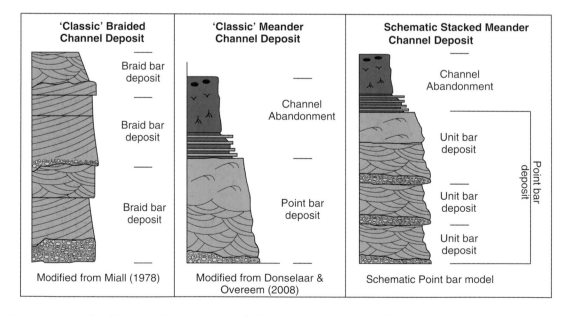

Fig. 1. A comparison of sedimentary logs (see Fig. 6 for key). The classic braided channel sedimentary log (Modified from Miall, 1978) is notably different to the log profile of the meandering channel deposit (Modified from Donselaar & Overeem, 2008); however, stacked meandering channel deposits may appear similar to those of braided channels deposits.

packages, which have good acoustic contrast and allow increased resolution. The identification of point bar deposits in the rock record is commonly reliant on the identification of heterolithic lateral accretion packages (Muisal *et al.*, 2012; Nardin *et al.*, 2013). Only a few examples of point bar deposits occur in the rock record, which are preserved solely within sand dominated fluvial systems (Ori, 1982; Arche, 1983; Campbell & Hendry, 1987; Schirmer, 1995; Kostic & Aigner, 2007; Hartley *et al.*, 2015, 2018).

Hartley *et al.* (2015, 2018) used satellite imagery and field studies to document a meander belt in the Salt Wash DFS (Fig. 2A), Utah, where 63 point bars were observed in planform. This meander belt is interpreted to cover a minimum planform area of 30,000 km^2 and is the largest exposed meander belt document in the rock record. The meander belt covers a significant area within the sand-dominated proximal to medial portion of a well documented distributive fluvial system (DFS) (Craig *et al.* 1955; Owen *et al.*, 2015a,b; Owen *et al.*, 2017a). Previous studies have interpreted the Salt Wash DFS as representing a braided fluvial fan (e.g. Peterson, 1978; Robinson & McCabe, 1997; Robinson & McCabe, 1998). However, the interpretation of Robinson & McCabe (1998) was based on the relationship of intra-channel deposits to overbank deposits, a lack of lateral accretion and the sheet-like nature of the sandstone bodies.

This study builds upon the initial findings of Hartley *et al.* (2015) and highlights the potential under-documentation and recognition of sand-dominated point bars in the Salt Wash DFS, and in the fluvial rock record in general, and provides well-exposed examples that can be used as a basis for increased understanding and recognition of ancient sandy point bar deposits.

Aims

This study documents two outcrops (Caineville Wash and Caineville Road) within the Salt Wash DFS meander belt, which have contrasting exposure style. The internal architecture and facies from the Caineville Wash outcrop has vertical and plan view exposures, allowing a 3D representation of the deposit to be constructed. This outcrop has been compared with a two-dimensional canyon exposure at Caineville Road, which only has a minor planform exposure to aid intepretation. The two outcrops are situated 15 km

apart, but as the Salt Wash DFS is 550 km in length (Owen *et al.*, 2015a) it is inferred that no substantial changes in fluvial style should be expected over such a relatively short distance. An assessment will be made from these two outctops on how features documented in the vertical and planform sections can be recognised and aid planform interpretation.

GEOLOGICAL BACKGROUND OF THE SALT WASH DFS

The Salt Wash DFS, which is stratigraphically part of the Upper Jurassic Morrison Formation, has been documented in detail since the 1950s (Craig *et al.*, 1955; Mullens & Freeman, 1957; Peterson, 1978; Tyler & Ethridge, 1983; Robinson & McCabe, 1998; Owen *et al.*, 2015a,b; Owen *et al.* 2017a). It was deposited in a back-bulge area during the early stages of foreland basin development in the Late Jurassic when the palaeo-Pacific oceanic plate subducted beneath the North American Plate (DeCelles, 2004; Owen *et al.*, 2017a). The Salt Wash DFS extends from northern Arizona into north-eastern Utah and western Colorado (Fig. 2A; Craig *et al.*, 1955; Owen *et al.*, 2015a,b) and was deposited in an arid to semi-arid climate with seasonal precipitation (Mullens & Freeman, 1957; Tyler & Ethridge, 1983; Peterson, 1994; Robinson & McCabe, 1998; Dunagan & Turner, 2004; Hasiotis, 2004; Owen *et al.*, 2017a). The source area for the Salt Wash DFS is located to the south-west of the basin, (Craig *et al.*, 1955; Mullens & Freeman, 1957). Recent statistical analysis of palaeocurrent data predict the apex to be located in northern Arizona (Owen *et al.*, 2015a,b; Fig. 2A), placing it at the Mogollon-Sevier syntaxis, or solely within the western portion of the Mogollon Highlands (Peterson, 1980; Peterson, 1994; Kjemperud *et al.*, 2008; Weissmann *et al.*, 2013; Owen *et al.*, 2017a).

In southern Utah, the Morrison Formation is stratigraphically divided into the Tidwell, Salt Wash and Brushy Basin members (Fig. 2B). Lithostratigraphically, the Salt Wash DFS comprises the two basal lithostratigraphic units; the Tidwell and Salt Wash members. The basal Tidwell Member, composed of overbank, minor fluvial and lacustrine deposits, represents distal deposits of the system (Hasiotis, 2004; Weissmann *et al.*, 2013; Owen *et al.*, 2015b). The Salt Wash Member represents a relatively more proximal

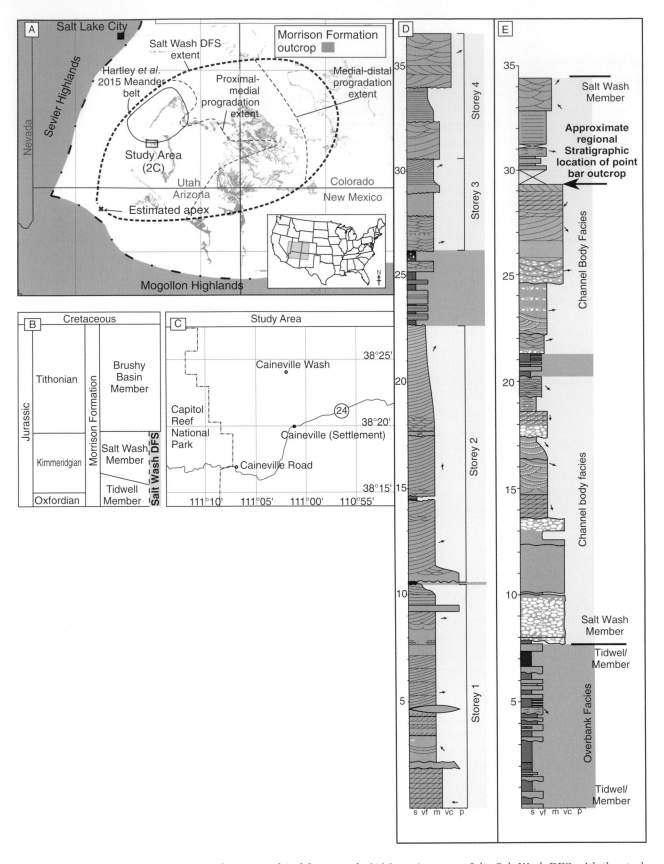

Fig. 2. Salt Wash DFS location map and stratigraphical framework. (A) Location map of the Salt Wash DFS with the study area placed within the map and the Meander Belt identified by Hartley *et al.* (2015). Modified from Owen *et al.* (2017a). (B) Stratigraphy of the Salt Wash DFS (based on Owen *et al.*, 2015a). (C) Detailed outcrop location map. Caineville Wash, point bar outcrop located approximately 15 km to the north, north-east of the Caineville Road. (D) Sedimentary log through the Caineville Road section. (E) Sedimentary log through the Caineville Wash section (See Fig. 6 for key). The section of interest is noted towards the top of the sedimentary log for context.

part of the DFS and is composed of large amalgamated fluvial channel belts and minor overbank deposits, which together with the Tidwell Member form an overall progradational succession (Peterson, 1980; Peterson, 1994; Kjemperud *et al.*, 2008; Weissmann *et al.*, 2013; Owen *et al.*, 2015b; Owen *et al.* 2017a).

Study area

Owen *et al.* (2015b) produced a quantified system scale map of the Salt Wash fluvial system, allowing context to be given to the two studied outcrops. These outcrops are located within the extensive meander belt established by Hartley *et al.* (2015) and are situated in the proximal to medial portion of the DFS. The Caineville Wash outcrop is located 8 km north of Caineville settlement while the Caineville Road outcrop is located 10 km to the south-west on the eastern border of Capitol Reef National park (Fig. 2C).

The exposure at Caineville Wash (Fig. 2C) is a laterally extensive rock plinth with low relief. Erosion of the overlying, mud-dominated, Brushy Basin Member, has resulted in excellent exposure of planform elements across the outcrop as well as full vertical exposure. Caineville Road (Fig. 2C) is situated within a canyon and forms a cliff section that extends laterally for 1 km and is approximately 30 metres in height. A planform exposure located 500 m south-west of the main cliff has been used to aid interpretation of the cliff section.

METHODOLOGY

To compare the two different outcrops, a total of 11 sedimentary logs were recorded, together with multiple photopanels at each location and 3D LiDAR and heli-drone data. At Caineville Wash, palaeocurrent measurements were taken from the planform surface of the outcrop using a global positioning system (GPS) (accuracy < 5 m) and a compass-clinometer. Seven sedimentary logs were taken around the perimeter of the outcrop to capture the variability in deposit characteristics. Photopanels were taken and interpreted to illustrate the lateral variability in facies architecture. The perimeter of the outcrop comprises an 'inside' and 'outside' cliff face. Both the inside and outside of the outcrop have full lateral exposure. A heli-drone was remotely flown over the outcrop and multiple photographs taken of the outcrop from various angles. Approximately 120 photographs were taken from the heli-drone, then processed through Agisoft PhotoScan to create a 3D outcrop model (Fig. 3). Once processed, interpretations of the 3D outcrop model were made using the LIME virtual outcrop suite (see Enge *et al.*, (2007) for a detailed description of the workflow). At Caineville Road, four sedimentary logs were taken around the outcrop, two logs covered the full vertical extent and one captured approximately 70% of the total exposed section. A high-resolution ground-based LiDAR scan combined with photomosaic was conducted to produce a virtual outcrop model. The LiDAR scan covers a total outcrop length of approximately 1.06 km in an area of 0.1 km². Palaeocurrent data were collected from cross bedded strata on the planform exposure to the south-west of the cliff section using a GPS and compass clinometer.

Facies and facies association

Table 1 contains a description and interpretation of the lithofacies at both study sites, which form two distinct facies associations: fluvial channel and overbank. A sedimentary log taken through the full Salt Wash DFS at Caineville Wash had a fluvial channel percentage of 65% (Fig. 2). The fluvial channel facies association is mainly composed of conglomerate lag deposits (Cm), cross-bedded sandstone (St), laminated sandstones (Sh) and ripple laminated sandstone (Sr). Infrequent heterolithic sections are present within the fluvial channel in the form of structureless mudstone (Fs). Overbank deposits are composed of structureless mudstone (Fs), laminated mudstone (Fl); and horizontally laminated sandstone (Sh).

Architectural elements

Surfaces identified within a fluvial channel body characterise the original channel fill or bar migration pattern (Fig. 4). Here, the authors identify four distinct surface-bounded architectural elements (further details in Table 2) which include channel body, storey, barform and unit bar (lateral and downstream) elements. A channel body represents a portion of a single channel belt or multiple amalgamated channel belts, encompasses all storey and internal accretion surfaces (Owen *et al.* 2017b) and is encased within floodplain deposits. A storey surface bounds an individual storey and

Fig. 3. Oblique views of the two digital outcrop models. (A) Caineville Wash digital outcrop model acquired by drone photogrammetry. (B) Caineville Road digital outcrop model acquired by LiDAR scanning.

represents the erosional phase of a river channel which has cut into underlying fluvial deposits within a channel belt (Friend, 1978; 1979). A bar-bounding surface represents the outer limit of a barform within a channel and records the overall accretion direction of a bar. Surfaces often display an offset dip direction to subsequent bar bounding surfaces (Fig. 4C). Accretion surfaces represent the palaeo-avalanching fronts of unit bars which migrated within either a downstream or laterally accreting barform. Internally, between accretion surfaces, bedforms such as cross-stratification that represent individual dunes within unit bars are developed (Allen 1963; Miall 1977; Bridge 1993, 2006; Ghazi & Mountney 2009) (Fig. 4).

RESULTS

Caineville Wash – Planform, facies distribution and architecture

The Caineville Wash outcrop benefits from having a significant amount of planform exposure (Fig. 5), which can be described and interpreted independently of, then related to, the vertical exposures that bound the outcrop. The northern and southern parts of the outcrop (Fig. 5) display clear arcuate scroll bar morphology, but lack any vertical exposure. The central part of the outcrop occurs within the same meander complex and forms a minor (3 to 8 m) vertical cliff face. The basic architecture was mapped initially onto high-resolution satellite imagery allowing interpretation of the deposit as the partially preserved section of a point bar (Fig. 5). Mapping of palaeocurrent data from bedforms on the planform surface and comparison to accretion surface dip directions allowed identification of the bar accretion direction as well as point bar development through time.

Four photopanel sections (Fig. 6 and Fig. 7) illustrate the vertical architecture of the outcrop. Sections 1 and 2 (Fig. 6) describe the upstream and central parts of the outcrop and are divided due to a significant change in architecture. Sections 3 and 4 (Fig. 7) are located on the inside cliff face of the outcrop and were also separated due to a change in architecture.

Table 1. Description and interpretation of the identified facies.

	Lithofacies	Description	Interpretation
Cm.	Conglomerate	Matrix and clast-supported. Clasts are well rounded and are either igneous, sandstone or mudstone. Max clast size 30 cm diameter. Matrix medium-coarse sand. Occasionally stratification can be seen but most commonly structureless gravel. Conglomerate interbeds can be up to 0.5 m in thickness and 10 m in width.	Deposition of sediment in the proximal area by high flow stage in channel. Clasts transported by traction in the deepest section of the channel. Crude bedding related to bar formation and migration.
St.	Cross-bedded sandstone	Sandstone lithofacies which has a grain-size distribution from fine sand to pebbles. Includes trough and planar cross bedding. Pebbles are commonly present on foresets of cross beds especially in more proximal portions of the system. Cross bed sets can be up to 3 m in thickness and laterally extensive for 15 m.	Cross-beds are interpreted as bedload deposits of a river system which has formed 3D and straight crested dunes. Dunes formed under medium-upper flow conditions.
Sh.	Horizontal laminated sandstone	Well-sorted horizontal lamination often with cross bedding above and below. Inclusions of pebbles and mud clasts on laminations. Up to 3 m in thickness and up to 20 m in lateral width.	A product of upper flow regime within a fluvial channel or by dropping out of suspension in a lower flow regime channel (splay – crevasse or terminal). Internal variations in grain-size can be attributed to fluctuations in flow velocity.
Sr.	Ripple laminated sandstone	Asymmetrical ripples are commonly present at the tops of sand bodies or within thin sheet-like sands in the floodplain. Sediment size is no greater than coarse sand. Ripple lamination thickness is up to 0.5 m; and up to 5 m in width.	A product of lower-flow regime. Formed under low water conditions when the sediment size is less than medium and flow velocity is low. Most commonly found in the top sections of a proximal channel – related to channel abandonment or in sheet channels where deposition is thinly dispersed over a large area.
Fl	Laminated mudstone - siltstone	Finely laminated siltstone and mudstone extend up to 300 m across outcrops. Infrequent horizons may show traces of roots. Colour varies from red/purple - grey. Commonly 0.2 to 0.4 m thickness and laterally extensive for up to 100 m.	Fine grain-size and lamination indicates deposition from suspension and associated with overbank flood events. Colour is related to floodplain drainage conditions: grey poorly drained, red well drained and oxidised.
Fs	Structureless Mudstone - Siltstone (Fs)	Structureless mud-siltstone commonly tops multistorey channel sandstone bodies. Usually there is a gradational contact at the base and erosional boundary at the top where a fluvial channel has cut into the deposit. Desiccation cracks & rootlets are common. Purple/red - grey in colour. Commonly 0.05 to 0.15 m in thickness and 20 to 30 m in width.	Formed in floodplain environment. Sheet deposits of mud/silt over a broad expansive floodplain. Little or no erosion as sediment drapes succession below. Desiccation cracks, rootlets and nodules record subaerial exposure.

Section 1

Section 1 displays the upstream portion of the point bar and includes logs 1 and 2 (Fig. 6). The base of log 1 shows a basal erosion surface truncating underlying floodplain or channel abandonment deposits. The point bar deposit comprises poorly sorted pebbly sandstone with grain-size varying from pebble-grade to fine sand, with pebbles commonly present above the basal erosion surface, as well as punctuated lags further up the section.

Throughout section 1, a total of 6 unit bar accretion surfaces are recognised with an average dip direction towards the north-west (339°). The average palaeoflow obtained from trough cross bedding on the vertical section is towards the north-west

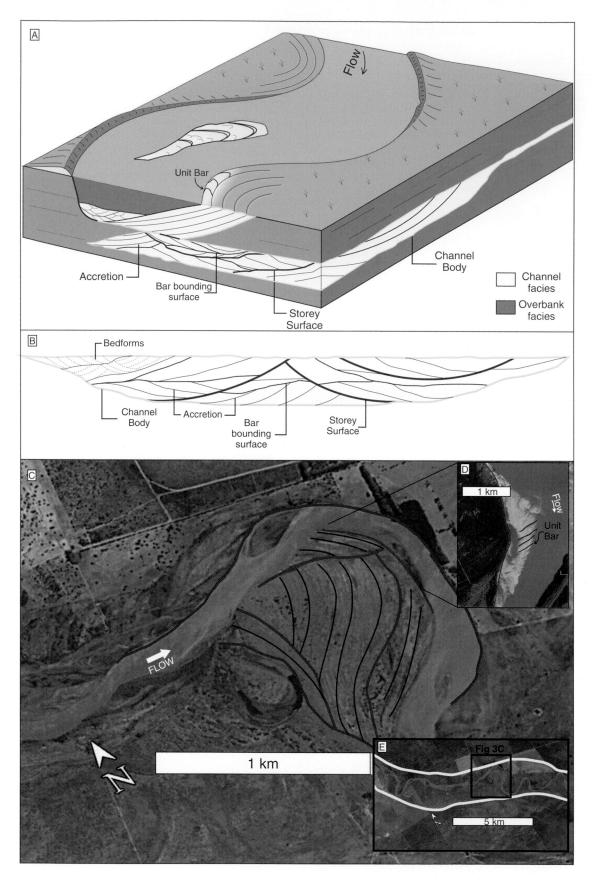

Fig. 4. Architecture scheme (A) A schematic interpretation illustrates the relationship of fluvial architecture within a laterally extensive channel belt. (B) Channel body cross section illustrates the depositional interpretation of each architectural surface. (C) Modern day example from a Rio Colorado (39°12'31.63″S 63°40'28.86″W) point bar deposit displaying the relationship between surfaces identified in the rock record – Image from Google Earth©. (D) A high resolution Google Earth© image displays the unit bars within a point bar deposit. (E) A planform view of the active Rio Colorado channel belt. (Note image is from Mississippi River, USA - 32°42'38.47″N 91° 3'52.20″W – and has been used due to the lack of high resolution imagery at the Rio Colorado example).

Table 2. Description, interpretation and dimensions of the observed fluvial architecture in the Salt Wash DFS.

Surface	Description	Interpretation	Dimensions
Channel body margin	The basal surface of the channel margin is typically represented by an erosional surface cutting into underlying strata with a conglomerate lag directly above the surface. The upper margin of the channel body may show an abrupt contact with overbank facies or a gradational transition from to floodplain or channel abandonment deposits.	Channel body surfaces represents the outer limit of a fluvial channel belt.	Individual channel bodies can be up to 10 km in width in proximal reaches. Up to 20 m in thickness.
Storey surface	Storeys are represented by an erosional surface within a channel body. A storey can extend from the upper channel margin down to the base of the channel margin Storey surfaces commonly preserve a thin layer of mud above the erosional surface. A storey surface is only truncated by an earlier storey surface or the outer channel limits.	Storey surfaces represent the erosional scour left by a once active river channel which has eroded into the underlying channel deposits.	Up to 200 m in length. Up to 10 m in height.
Barform bounding surface	Barform bounding surfaces can be erosional or abrupt contacts internally of a storey. The surface can often be recognised where it truncates accretion surfaces. Barform bounding surfaces are truncated by storey surfaces and channel body margins.	Barform boundaries represent the outer limit of a barform deposited within a fluvial channel.	Up to 100 m in length. Up to 10 m in height.
Accretion surface	Inclined surfaces which often preserve a thin layer of mud (up to 10 cm). Accretion surfaces with a dip direction parallel to the local palaeoflow are interpreted as downstream accretion surfaces whereas accretion surfaces with a dip direction more than 60° to the local palaeoflow are interpreted as lateral accretion surfaces. Accretion surfaces may be truncated by storey surfaces and channel body margins.	Accretion surfaces represent the migration of a bar deposit within a fluvial channel. Lateral accretion represents the perpendicular or orthogonal migration of a point bar deposit relative to the palaeoflow. Downstream accretion relates to mid channel bar migration or the upper and lower portions of a downstream migrating point bar.	Up to 50 m in length. Up to 8 m in height.

(332°). Internal unit bar dip directions show a comparable azimuth towards the north and north-west.

Section 2

Section 2 is an extension of section 1 towards the point bar centre (Fig. 5B) and includes sedimentary logs 3 and 4 (Fig. 6D). Unit bar accretion surfaces continue to dip northwards. However, mid-way through section 2 the accretion direction of the surface relative to the direction of palaeoflow becomes increasingly oblique. The blue surface highlighted in Figs 6C and 8B illustrates a barform bounding surface that separates the two distinctly different packages. The package of

sediment above the blue line dips towards the north (005°) while palaeoflow, derived from trough cross bedding, is offset towards the east (087°). The sediment below the surface dips towards the north-east (060°).

Logs 3 and 4 (Fig. 6D) both contain the barform bounding surface, denoted in blue on the sedimentary logs, as an erosional surface with a thin (10 cm) pebble lag directly above the surface. The profile of log 3 fines upwards from a pebble lag base into a fine, clean sandstone. Internally there are multiple gravel lags associated with unit bar accretion surfaces within this overall fining upwards sequence.

Transitioning north around the outcrop (Fig. 6C) the prominent dipping unit bar accretion surfaces

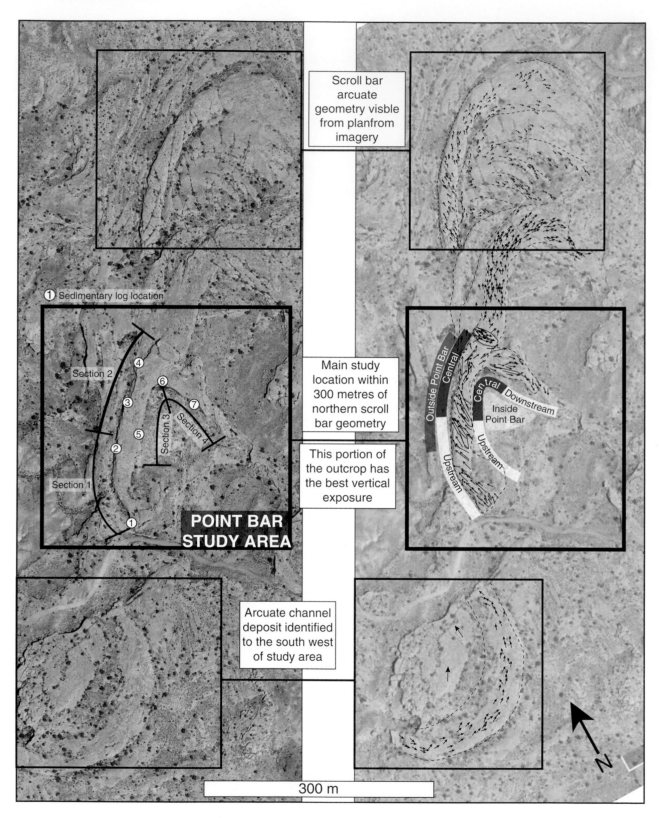

Fig. 5. Caineville Wash planform imagery displays the arcuate geometries commonly associated with point bar deposits. The planform geometries are interpreted (dashed line) with palaeoflow data (black arrows) and palaeoflow 'trains' (red line arrows) collected in the field. The main study area is located in the centre box. Images from Google Earth©.

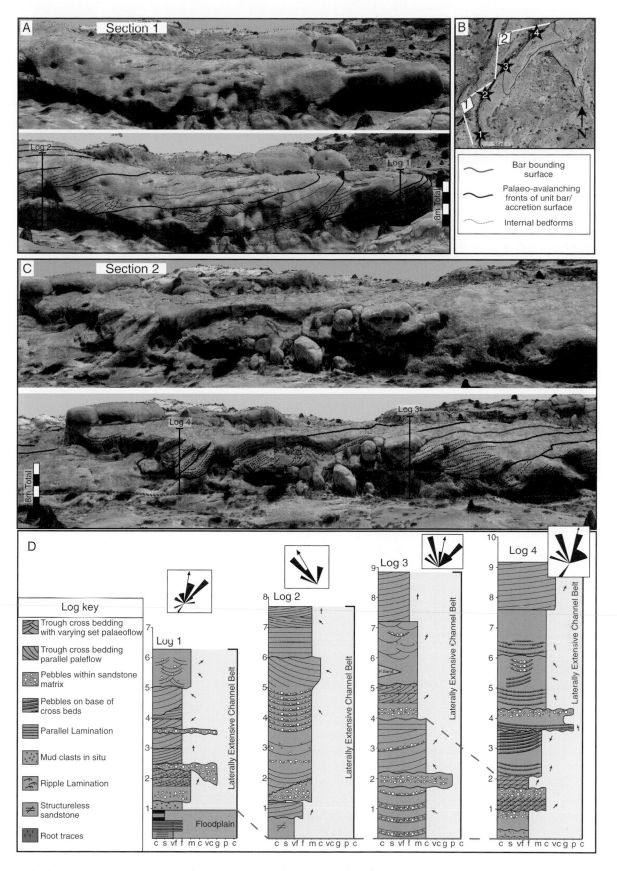

Fig. 6. The depositional architecture of the external outcrop perimeter. (A) The upstream portion of the point bar. Downstream migrating elements dominate. (B) Location map showing the vertical section images in context. The blue surface is the laterally accreting element of the point bar. (C) The top end of the point bar nearest the apex with blue surface showing lateral accretion as in (B). The four sedimentary logs were taken around the perimeter of the outcrop with the location of each noted in both lateral and planform images.

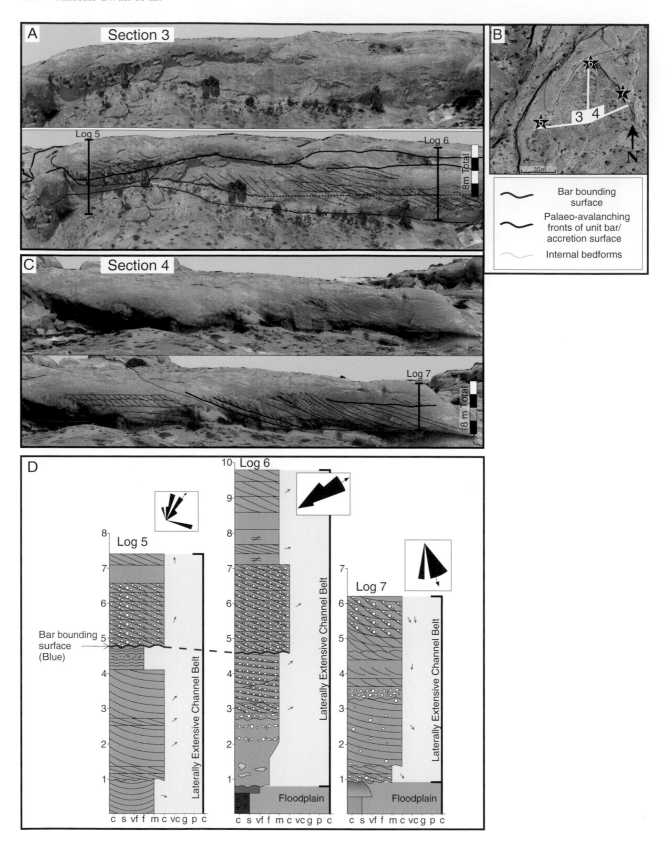

Fig. 7. (A) Upstream portion of the bar on the inside of the point bar outcrop. B) Aerial view of the outcrop showing photograph locations. (C) The downstream portion of the outcrop. (D) Sedimentary logs from the internal portion of the outcrop. See Fig. 6 for key.

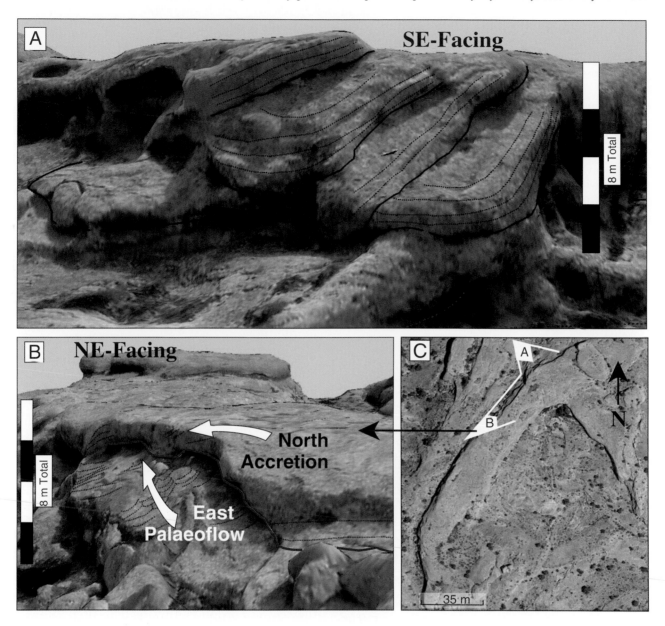

Fig. 8. View of the middle portion of the point bar. (A) The location of the cascading bars can be seen at the apex section of the point bar deposit. (B) An image taken from the heli-drone 3D data shows the package of sediment migrating towards the north-west (left) with the underlying sediment migrating towards the north-east (into the image). (C) Location map showing the angle and direction of the image in A and B. (Legend of symbols as in Fig. 5).

are no longer visible. The bar bounding surface dominates the central portion of the deposit and the uppermost part of the deposit is dominated by small scale accretion surfaces that represent small isolated bar features, referred to as cascading bars. The cascading bar features are unit bars which measure 1.5 metres in thickness, up to 5 metres in width and dip towards 010°. Contacts between the cascading barforms are conformable and parallel to the lateral accretion surfaces. Some of the

contacts are represented by low angle erosional surfaces. The cascading bars are exclusively located in the central portion of the point bar and appear as a barform migrating from the top to the base point bar (Fig. 6A). Log 4 (Fig. 6D) is situated at the northernmost section of the point bar and transects through the cascading barforms. At the base of the sedimentary log, multiple erosion surfaces are present with associated gravel lags interpreted as the lower extension of lateral accretion surfaces. Between lateral accretion surfaces, pebbles are noted on the base of trough cross beds. Up to 25% of the log displays pebbles on the trough cross bedding. The cascading bars are only noted directly above the bar bounding surface and are associated directly with the northwards migration of the point bar as sediment was deposited over the underlying point bar deposit (Fig. 8B). Similar bedforms were recorded by Jordan & Pryor (1992) from the upper portion of a modern Mississippi River point bar. In their example, bedforms migrating through a bar top chute channel formed downslope to oblique downstream accreting bedforms where the channel intersected the lateral accretion surface. Here, chute channels are not preserved on the planform side due to post-depositional erosion of the top of the outcrop.

Section 3

Section 3 is located on the east (inside) face of the outcrop, extends from the upstream portion to the central portion of the point bar and represents the inside equivalent to section 1 and section 2 on the external outcrop perimeter (Fig. 7A). Logs 5 and 6 are located within section 3. This section displays unit bar accretion surfaces that dip in the same direction (towards the north-west) as those present in section 1 and which represent the downstream accretion of beds within the point bar. The blue surface, identified as a lateral accretion surface in section 2, can also be traced around the outcrop to section 3. At the base of Log 5 (Fig. 7D), average palaeoflow orientated towards 095° has been recorded in the bedforms, which suggests the lower portion of this section of the outcrop represents the central part of the point bar. The upper portion of Log 5 (Fig. 7D) above the accretion surface displays a change in palaeoflow direction (345°) which is strongly oblique to that in the lower portion of the sedimentary log (110°), suggesting the upper part of the deposit records a second phase of point bar development. Sedimentary log 5 (Fig. 7D) transects through the accretion surface on the inside portion of the outcrop. The profile of the sedimentary log differs from the sedimentary log on the outside of the outcrop as there is no clear fining upwards profile. The bar bounding surface is noted in log 5 and 6 on the inside portion of the point bar due to the erosional base and the influx of coarse-grained sediment recorded midway up the sedimentary logs above the blue surface.

Section 4

Section four covers the only downstream portion of the exposed point bar and is located on the inside portion of the outcrop with one sedimentary log (Log 7) through the section (Fig. 7). The section comprises stacked trough cross-bed sets with palaeoflow towards the south-east (144°). The direction of palaeoflow at this location is directly opposite to that of the palaeoflow recorded in section 1 (Fig. 9). With the significant change in palaeoflow, it is possible to associate this section with the downstream portion of the point bar. There are two unit bar accretion surfaces identified that migrated parallel to palaeoflow with an average dip direction towards 128°. The downstream portion of the outcrop documents migration of the unit bars towards the south (170°).

Planform

The planform section of the outcrop allows trough cross-bed palaeoflow directions to be mapped on the planform exposure together with the architectural surfaces identified on the vertical sections (Fig. 9A and B), which can be correlated across the outcrop (Fig. 9C and D). Palaeoflow directions from section 1 (upstream section) of the point bar are directed towards the north-west (334°), parallel to unit bar accretion surface dip directions, which represent downstream point bar migration (Fig. 9A). Section 2 (central section) of the point bar displays the largest disparity in palaeoflow direction, with over 30% of the palaeoflow directed towards the north-west and over 30% directed towards the south-east (Fig. 9A). The average palaeoflow direction from bedforms in the central portion is towards 046° (Fig. 9B) with unit bar accretion dip direction towards the north, indicating lateral migration of the unit bars (Fig. 9B). Section 4 of the outcrop displays palaeoflow towards the south-east (144°) and unit bar accretion surface dip direction towards 128°, suggesting predominantly downstream-directed bar migration. The planform data compliment those measured from the vertical exposures, as a change of palaeoflow (up to 180°) is observed as a full transition around the outcrop is completed.

Summary

The Caineville Wash outcrop illustrates the complexity of a sand-dominated point bar. The internal architecture is changeable over a short

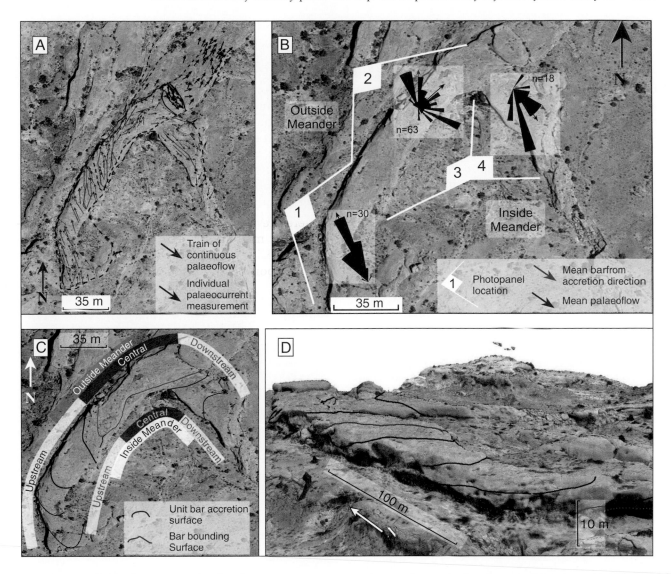

Fig. 9. Caineville Wash outcrop. (A) Palaeoflow measurements superimposed on point bar deposit. (B) A three-dimensional outcrop has exposure on the planform section and approximately 8 m-high perimeter of the outcrop. (C) Barform palaeo-avalanche fronts (black lines) and lateral accretion surface (blue) plotted on the planform imagery. (D) A virtual outcrop image of the point bar deposit. Palaeoflow was recorded using a handheld GPS and individually mapping the location of the bedform palaeoflow readings.

distance and the sedimentary logs are typically medium-coarse grained, poorly sorted and generally display no systematic grain-size profile. The lithofacies of the sedimentary logs do not always display a simple fining upwards trend (Bridge *et al.* 1995) but rather can resemble a classic braided log signature which displays multiple erosional surfaces with associated gravel lags and trough cross bedding encased within a storey channel body. The erosional surfaces represent the base of individual unit bars or accretion surfaces which form the overall point bar deposit.

The Caineville Wash outcrop represents a single storey channel body. Accretion surfaces indicate migration of unit bars which form the overall point bar deposit. Here, it is recognised that specific portions of the point bar deposit comprise either downstream migrating unit bars or laterally accreting unit bars. Measurement of trough cross-bedding allows palaeocurrent directions to be established with comparison to unit bar migration direction allowing reconstruction of point bar development through time. This point bar outcrop is considered to be a combination of translational (50%) and expansional (50%) components based on descriptions by Jackson (1976), as the preserved planform displays migration lateral to palaeoflow (expansion towards the north) and downstream migration (translation towards the north-east) (Fig. 5). The point bar deposit displays an arcuate architecture with a slightly oblique direction of migration relative to palaeoflow. By identifying the unit bar migration direction relative to palaeoflow, it is clear that the point bar comprises a significant amount of downstream accreting elements; consequently, downstream accreting unit bars within a fluvial succession are not restricted solely to braided river deposits but may also constitute a smaller fragment of a larger point bar (e.g. Ghinassi & Ielpi, 2015).

From the architectural descriptions and sedimentary logs, it is possible to identify at least 2 phases of point bar development. Phase one is illustrated in Fig. 10, where the upstream and downstream portions of the outcrop are considered time equivalent. The upstream portion (seen in section blue, Fig. 10A) developed initially outwards towards the north-west (339°) and then rotated to migrate towards the south-west (144°) in the downstream section. The second phase of point bar development can be seen above the blue line (red section Fig. 10) recording point bar migration towards the north. Following the first phase of point bar development, the point bar expanded to the north-east (040) (Fig. 11B) and on

reaching sinuosity equilibrium it evolved through a combination of expansion (towards the north-east, 040) and downstream migration (towards the east, 080) (Fig. 11B).

The point bar bounding surface (separating phase 1 and 2 of the point bar) can be traced on all studied outcrops, extends north of the study site and may be related to channel avulsion. However, available outcrop towards the north is poor and offers only a limited planform section without any vertical exposure. A planform reconstruction (Fig. 11) illustrates the development of the point bar deposit through time.

Caineville Road

The Caineville Road study site consists of a fluvial channel body exposed over 800 m continuously and up to 30 m in height (Fig. 12). To the south-west of the cliff section (500 m) a 1.5 km² area offers partially exposed planform outcrop which is used to aid interpretation of the cliff section (Fig. 13). The road section has an element of three-dimensionality due to an active river channel creating a sinuous canyon exposure (Fig. 12). Vertically, the section can be subdivided into four major channel storeys (from 1 at the base to 4 at the top) and LiDAR panels that cover the lateral variations. The LiDAR imagery allows centimetre scale measurements of the internal architecture

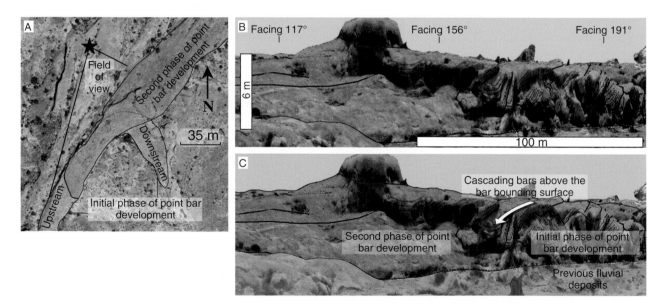

Fig. 10. (A) Planform image illustrates the 2 phases of point bar development. (B) Vertically exaggerated virtual outcrop image illustrates the point bar deposit with the architecture mapped on. (C) Annotated virtual outcrop image illustrates the two phases of point bar development.

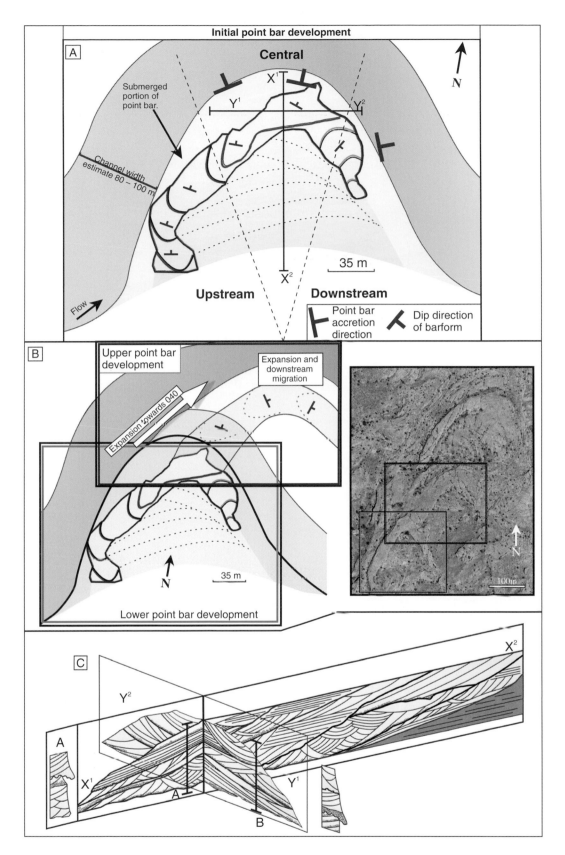

Fig. 11. A schematic reconstruction of the Caineville Wash outcrop. (A) Planform reconstruction with the vertical exposure overlaid to guide the perspective of the internal architecture. Dotted lines represent potential scroll bar morphology placing the outcrop in context with the depositional setting. (B) A continued planform reconstruction of the point bar identifies the progressive development of the bar towards the north-east (030°). (C) A three-dimensional schematic cross section illustrates the change of architecture through the point bar deposit and relationship to associated sedimentary logs.

Fig. 12. Caineville Road study site. Planform satellite imagery displays the extent of the vertical exposure (marked in red) and the lack of planform exposure due to the coverage of vegetation (Image courtesy of Google Earth©).

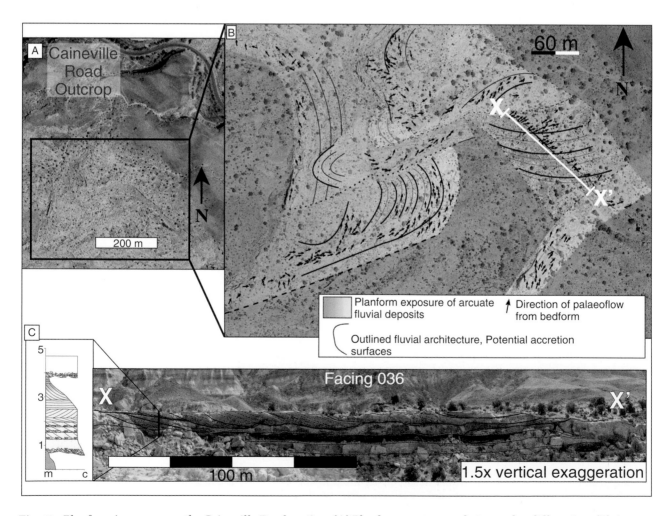

Fig. 13. Planform imagery over the Caineville Road section. (A) Planform exposure relative to the cliff section. (B) Arcuate geometries with palaeoflow mapped onto the planform exposure. (C) Interpreted cross section from the area described in A and B (sedimentary log - Legend of log symbols as in Fig. 6).

calibrated by sedimentary logs. The outcrop has been subdivided into 4 lateral sections. The subdivisions are directly connected around the perimeter of the outcrop and have been divided based on outcrop architecture (Fig. 12).

Storey 1

The base of storey 1 is not exposed, but the top of the storey exposed in sections 1 and 4. Storey 1 (sedimentary log 1, Fig. 14D) fines upwards over-all from a coarse sand at the base to very fine sand at the top, punctuated by coarse-grained lags. Trough cross-bedding comprises the majority of the deposits with pebbles and mud clasts commonly present on the base of troughs. Towards the top of the storey, a cascading bar structure is present between 9 and 11 m on log 1 (Fig. 14D). The top of storey 1 is defined by a thin (10 to 20 cm) heterolithic package.

In section 1 (Fig. 14) the top portion of storey 1 comprises bars with large scale (5 metres in thickness), low angle (approximately 5° dip of strata) trough cross beds, probably associated with large-scale dunes or barforms within the channel. The mean accretion direction of the bar set is towards 060° with palaeoflow towards 052°. The low disparity between palaeoflow and bar migration suggests bedform/unit bars at the top of storey 1 migrated downstream towards the north-east.

Storey 2

Storey 2 is exposed around the full perimeter of the outcrop. It has an average thickness of 10 m and extends for 400 m laterally, but due to exposure limitations, the margins were not observed. In section 1 (Fig. 14), erosion at the base of storey 2 occurs into the underlying heterolithic package and the upper portion of storey 1. The storey comprises a bar bounding surface dipping towards 060°, with accretion surfaces displaying varying dip directions. Beneath the bar bounding surface, accretion surfaces are horizontal, above the bounding surface accretion surfaces dip towards the north-east. Trough cross-bedding within the bars displays palaeoflow towards 060° indicating downstream bar migration. Storey 2 fines upwards into an interpreted channel abandonment fill which has inclined heterolithic surfaces.

In section 2 (Fig. 15), only a minor portion of the storey is exposed as it dips beneath the road. The outcrop displays a set of bedforms with palaeoflow towards 005°. Log 2 (Fig. 15C), which crosses storey 2 in section 2, displays a fining upwards profile from medium-grained sand into channel abandonment mud deposits. Internally the channel-fill is dominated by parallel bedding and scattered pebbles throughout.

Throughout section 3 (Fig. 16), storey 2 displays multiple unit bar accretion surfaces dipping northwards (approximately 025°). Bedforms in the northern portion of section 3 display palaeoflow towards 140°. The disparity between palaeoflow and accretion direction indicates lateral accretion dominates in this portion of the section. The southern portion of the section displays unit bar accretion and bedforms dipping towards (043°) and accretion surfaces towards (051°) suggesting that section 3 has a transition from downstream accretion into lateral accretion.

The outcrop angle at section 4 (Fig. 17) is perpendicular to sections 1 and 3, offering a cross-sectional viewpoint of the storey. Internally, storey 2 shows a fining upward profile from medium sand into channel abandonment deposits. More predominant heterolithic bedding is visible towards the top of the storey.

Storey 2 is dominated by accretion surfaces that dip towards the north and north-east (025–060°) and palaeoflow from the bedforms is towards the north and east (005–145°) resulting in downstream barform accretion. In section 1 and 2 storey 2 is dominated by accretion surfaces which are interpreted as downstream accreting. These sections together are interpreted as the upstream portion of a point bar (Fig. 18B). In section 3, the same storey displays accretion surfaces that migrated perpendicular to palaeoflow, indicating lateral accretion. The accretion surfaces that dip towards the north and north-east in section 3 may represent the transition from the upstream portion to the central portion of the point bar where lateral accretion is more probably situated.

Storey 3

In section 1 (Fig. 14), storey 3 comprises a 3 m-thick storey body which increases in thickness towards the west. Storey 3 is included in 3 sedimentary logs around the outcrop (Figs 14D, 15C and 16C) and ranges from medium sand with pebbles to very fine sand. The storey is dominated by trough cross bed sets and accreting barforms. There is no evidence for any accretion surfaces within section 1, although erosion does occur at

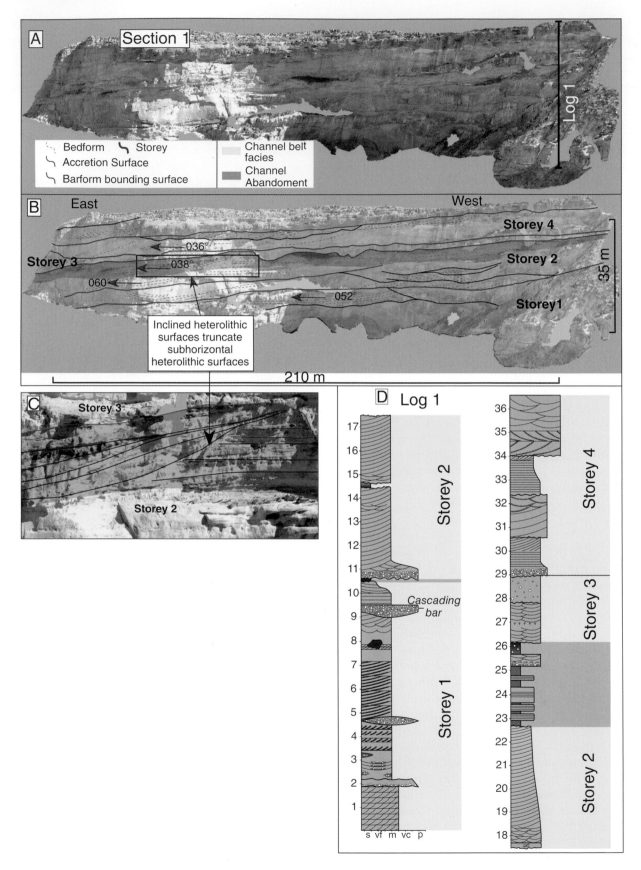

Fig. 14. Section 1 of Caineville Road outcrop – (A) LiDAR image shows the section in 3D. (B) Schematic interpretation of section 1. (C) An annotated representation of the inclined heterolithic strata within the channel abandonment. (D) Sedimentary log taken from the western section of the outcrop. (See Fig. 6 for log key).

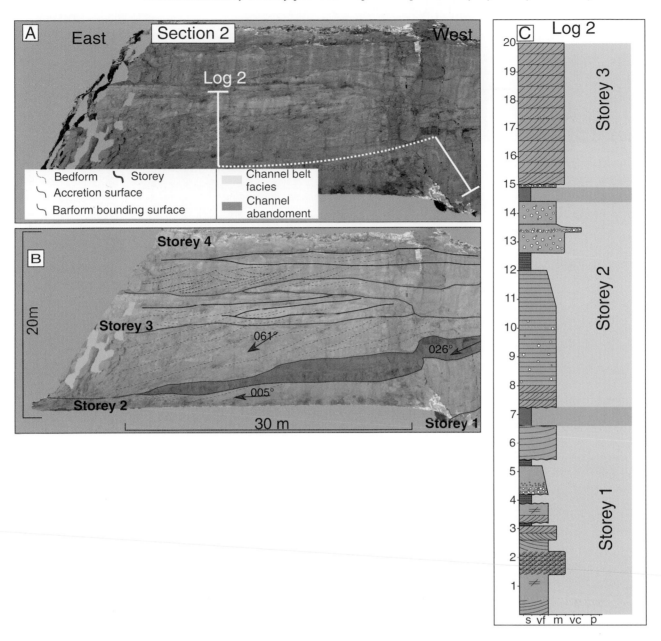

Fig. 15. Section 2 of Caineville Road outcrop. (A) Lidar image of the northern portion of the outcrop. (B) A schematic interpretation of section 2. (C) The sedimentary log 2 covers two-thirds of the outcrop, but the top third is inaccessible at this location. (See Fig. 6 for log key).

the base of the body into underlying channel abandonment deposits. Similar to the Caineville Wash outcrop, a thicker mud package is preserved in the upstream portion point bar. Minor bedforms are noted; however, no accurate palaeoflow can be recorded due to inaccessibility. Eastwards, an additional 1 m of erosion at the base of the channel results in complete removal of underlying channel abandonment deposits.

In section 2 (Fig. 15) the storeys increase in thickness from 3 to 10 m. A bar bounding surface can be traced from the upper limit of the storey and around the perimeter of the outcrop (into section 3) for approximately 100 m. Accretion surfaces occur beneath the bar bounding surface and dip towards 061°. These are superseded by smaller scale trough cross beds towards the top of the storey which indicate palaeoflow towards

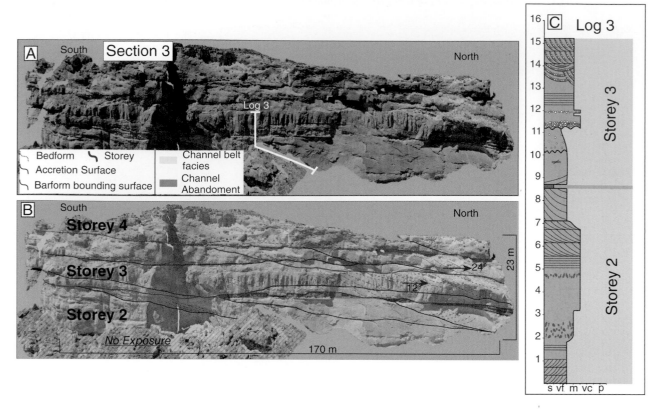

Fig. 16. Section 3 of Caineville Road outcrop. (A) LiDAR Image of the eastern face. (B) A schematic interpretation of the LiDAR image. (C) Log 3 only captures half of the outcrop due to limited accessibility. (See Fig. 6 for log key).

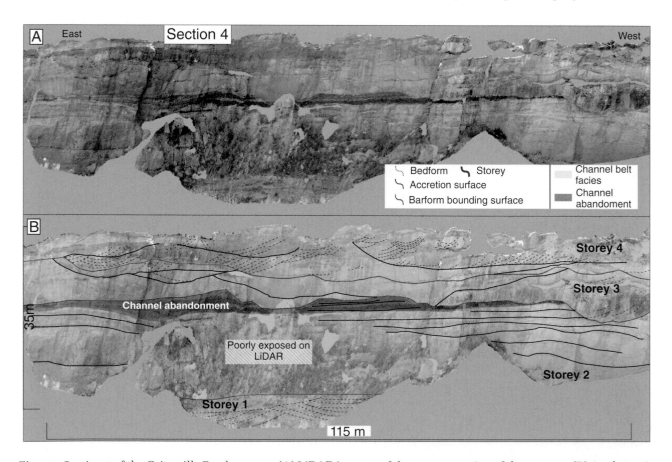

Fig. 17. Section 4 of the Caineville Road outcrop. (A) LiDAR imagery of the eastern portion of the outcrop. (B) A schematic interpretation of section 4 from the LiDAR imagery, Sauropod footprints visible at the base of storey 3.

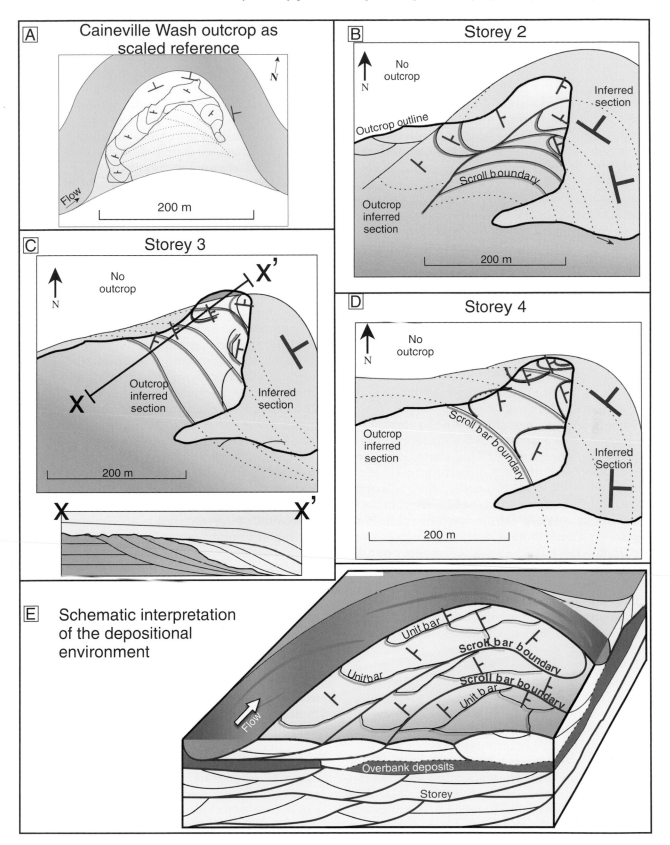

Fig. 18. Schematic reconstruction of the Caineville Road outcrop. (A) The Caineville Wash outcrop can be used as a scaled planform comparison. (B to D) The storeys that have been reconstructed all show barform elements associated with point bar development. (E) A schematic three-dimensional reconstruction of the outcrop displays the interaction of barforms, accretion surfaces and storey surfaces within one outcrop.

081°, suggesting this section of the storey is downstream accreting. The grain-size of the storey is medium-grained sandstone throughout (Fig. 15C).

In section 3 the northward dipping erosional surface (blue line Fig. 16) continues towards the base of the body before changing trajectory and climbing up through the body indicating that the outcrop bisects the barform bounding surface at an oblique angle at section 3 (Fig. 16). The accretion surfaces and large-scale bedforms within the channel indicate palaeoflow towards the north/ north-east (012–024°). Bedforms in section 3 have a palaeoflow parallel to the accretion surfaces, which is a product of downstream migration. Log 3 captures the lower portion of storey 3 with a consistent grain-size of fine sand, with pebbles present on trough cross-bed sets.

Storey 3 decreases in thickness in section 4 (Fig. 17), averaging 5 m. The basal surface shows a significant amount of erosion, cutting downwards up to 3 m into underlying storey and channel abandonment deposits. Internally, barforms dip towards the east. The mean palaeoflow is also towards the east.

The interpretation of storey 3 in the vertical sections is that it is a transect of the upstream portion of point bar. The majority of the accretion surfaces are downstream dipping, generally towards the north and north-east. Internally, the palaeoflow displays a direction similar to the adjacent accretion surfaces. When using the Caineville Wash outcrop as a scaled reference, it is possible to suggest that section 1, 2 and 3 represent a transect through the upstream portion of the point bar. This is due to the significant portion of downstream dipping accretion surfaces and an increase in storey thickness towards the north-east, which was also noted in the Caineville Wash outcrop (Fig. 18C).

Storey 4

Storey 4 comprises the upper section of the outcrop, therefore the true thickness of the body cannot be defined and is only observed in 1 of the 3 sedimentary logs (log 1 Fig. 14D). In section 1 (Fig. 14) storey 4 displays accretion surfaces dipping towards the north-east with a palaeoflow direction parallel to the direction of bar migration (036°). Internally, bedforms are large-scale (up to 5 m in thickness). The lower barform which is recorded in log 1 is approximately 5 m-thick and fines upwards from a conglomerate lag to a fine sand displaying parallel lamination and trough cross bedding probably associated with low angle bedform migration in the river channel. The upper portion of storey 4 is a coarse grained, clean sandstone composed of trough cross bedding.

In section 2 (Fig. 15), storey 4 displays barforms that pinch out onto the underlying storey channel base. The erosional bases of the barforms in the centre of the channel do not display any clear dip direction or accretion, although a general palaeoflow to the east can be determined. Section 3 (Fig. 16) displays the lower portion of storey 4. Internally, two minor barforms can be identified which have a shallow dip towards the north/ north-east and with internal bedforms also showing a similar dip direction suggesting the unit bars in this portion of storey 4 migrated downstream.

Interpretation of storey 4 is challenging due to vertical exposure limitations. By mapping the dip direction of the barforms and estimating the palaeoflow direction form the LiDAR dataset, it is possible to reconstruct a point bar deposit which has multiple stacked unit bars dipping towards the east (Fig. 18D). In section 1 (Fig. 14) the barforms are migrating parallel to the direction of palaeoflow, suggesting this section could be the upstream portion of the point bar. Palaeoflow and barform dip direction in sections 3 and 4 (Figs 16 and 17) display a change in flow direction and are more probably associated with the downstream portion of a point bar.

Planform

The outcrop has no planform exposure directly associated with the cliff face due to vegetation coverage (Fig. 12). However, 500 m to the south-west, a 1.5 km² area of planform exposure is visible (Fig. 13A). The Caineville Road deposits dip gently northwards, which results in this planform exposure correlating approximately to Story 4 from the top of the road cliff exposure. Interpretation of the planform imagery allows identification of large, arcuate geometries similar in scale to those seen at Caineville Wash. With the addition of palaeoflow data superimposed on planform imagery, radial patterns throughout the planform region are clearly identified (Fig. 13B). A small 10 x 200 m-high cliff section displays internal accretion surfaces that dip towards the southeast (155°), perpendicular to the north-east-directed

(051°) palaeoflow within the cliff face and on the planform surface (Fig. 13C). Divergence in dip direction between internal accretion surfaces and palaeoflow indicators suggests that the small cliff face (Fig. 13C) contains the upstream/central portion of a point bar deposit. Throughout the planform section, multiple scroll bar surfaces are visible which are often truncated beneath overlying scroll bars.

Summary

The planform section to the south-west of the Caineville Road section provides an additional level of confidence to the interpretation of the cliff section as a fluvial meander belt. Arcuate geometries are clearly visible from the satellite imagery, with radial palaeoflow and scroll bar surfaces migrating lateral to the regional palaeoflow. Correlation of the planform exposure to the cliff section at the same stratigraphical level over a distance of 500 m supports the interpretation of the cliff section as a series of stacked point bar deposits.

The cliff section comprises a multi-storey channel body that consists of four storeys and associated channel abandonment facies. Within each storey, the barform accretion direction has been identified and compared to the palaeoflow to determine whether downstream or lateral migration dominates the fluvial style. A schematic planform reconstruction of the outcrop has been developed (Fig. 18), comparing the internal architecture scales of the Caineville Road outcrop with the Caineville Wash outcrop. The diagram displays the individual accretion surfaces within the storey, but the lack of complete outcrop means the full extent of the fluvial deposit cannot be fully determined.

By identifying the direction of barform accretion direction and bedform palaeoflow, it is possible to create a scaleable reconstruction of the outcrop including the area where outcrop has been eroded. By overlaying the Caineville Wash outcrop on top of the Caineville Road outcrop, it is possible to identify the scale of the point bar architecture and infer a reasonable interpretation (Fig. 18). Due to the proximity of the two outcrops and the relatively minor downstream distance between them, coupled with the planform architecture noted in Fig. 13, the scales and sizes of the channel barforms are assumed to be comparable.

DISCUSSION

The information presented here recognises the difficulty of interpreting the deposits of sandy meander belts due to the complexity of the internal architecture of point bars. The Caineville Wash outcrop represents a sand rich point bar deposit within a proximal-medial portion of a DFS. The outcrop comprises lateral and downstream dipping accretion surfaces (barforms), trough cross strata (bedforms) and lacks inter-channel mud deposits; which are commonly recognised as a characteristic of coarse-grained meandering river deposits (Jackson, 1976; Ori, 1982; Arche, 1983; Schirmer, 1995; Kostic & Aigner, 2007; Hartley *et al.*, 2015) or braided river deposits (Miall, 1977; Cant & Walker, 1978; Lunt *et al.*, 2004). Six of the seven sedimentary logs at the Caineville Wash section display features which are observed also in braided fluvial systems (Ore, 1964; Rust, 1972; Smith, 1972; Rust, 1977; Miall, 1978; Ori, 1982; Allen, 1983; Miall, 1993; Lunt *et al.*, 2004). The logs commonly include coarse-grained pebbly and gravel deposits, dominated by trough cross bedding. Internally, multiple downstream dipping surfaces are present which dip parallel to the palaeoflow directions and bound downstream migrating unit bars. Frequently, point bar deposits are recognised by lateral accretion surfaces (Miall, 1977; Thomas *et al.*, 1987; Kostic & Aigner, 2007; Ghazi & Mountney, 2009; Smith *et al.*, 2011). However, in this study laterally accreting barforms represent only one-third of the total planform of the outcrop. The remaining two-thirds are dominated by downstream accretion which is more commonly associated with low sinuosity braid bar deposits. A superficial analysis of the downstream accreting barforms in the point bar deposit and in the 2D cliff section at Caineville Road could lead to the misinterpretation of the sections as representing downstream migrating mid-channel braid bars. The comparison of the two outcrops illustrates how challenging it is to interpret a sand rich fluvial outcrop from a vertical cliff section even with prior understanding of the fluvial depositional style. Caineville Road does show evidence of both downstream and laterally accreting surfaces, but as a stand-alone outcrop without any planform knowledge, it could be difficult to conclude that it represents a series of amalgamated point bars. However, at the Caineville Wash site, the planform imagery displays large, arcuate bedding geometries associated with scroll bars of a meandering fluvial system. It is clear that care must be taken when trying to interpret fluvial deposits and this example highlights that sand rich outcrops with downstream migrating bar-scale architecture may not always have been deposited by braided rivers.

The direct application of the Caineville Wash outcrop is to use the measured size and scales of the internal point bar architecture as a contextual framework for the neighbouring outcrop at Caineville Road. Although the exact stratigraphic relationship of the sections cannot be accurately determined, it is assumed that the Caineville Wash outcrop is broadly time equivalent to the Caineville road deposit. The point bar deposit at Caineville Wash is located approximately 30 metres above the base of the Salt Wash Member (Fig. 2), placing it stratigraphically in the centre of the Salt Wash member at approximately the same level as Stories 1 and 2 in the Caineville Road outcrop. The exact size and dimensions of the Caineville Wash deposit can be overlain directly on to the Caineville Road outcrop to better understand the probable scale of the meander deposit. This places the 2D cliff section storey channels into a planform perspective (Fig. 18) and shows that the Caineville Road outcrop is, although considered to be a large extensive outcrop, 30 metres in vertical height and 600 metres of lateral exposure; not sufficiently extensive to allow the confident identification of accretion or bar bounding surfaces to reconstruct accurately the fluvial deposit to the level that can be determined at the Caineville Wash outcrop. The Caineville Road outcrop also does not display a direct perpendicular transect of the channel deposit, which may result in accretion surfaces being near horizontal (Bridge & Tye, 2000). Without a detailed analysis of architecture and the direction of accretion, no assumptions should be made of the planform character of the deposit without taking the wider fluvial system understanding into account when making a depositional reconstruction.

The Salt Wash DFS has been widely cited as representing a braided fluvial system (Peterson, 1978; Tyler & Ethridge, 1983; Robinson & McCabe, 1997, 1998). However, the results published here strengthen the more recent interpretations made by Weissmann *et al.* (2013), Hartley *et al.* (2015, 2018) and Owen *et al.* (2015a,b) that significant portions of the Salt Wash DFS were formed by a predominantly meandering fluvial system. The meander belt (Hartley *et al.* 2015, 2018) includes a significant portion of the Salt Wash DFS and therefore the medial portion of the DFS cannot be regarded as predominantly braided. An improved knowledge of the internal sedimentary architecture of sand dominated point bar deposits in the rock record will contribute to an increased understanding of the nature of the fluvial system that formed the Salt Wash DFS.

McGowen & Garner (1970), Donselaar & Overeem, (2008); Smith *et al.*, (2011); Ielpi & Ghinassi, (2014), Ghinassi *et al.* (2015) and Ghinassi *et al.* (2016) all provide examples of sandy point bar deposits in the rock record. McGowen & Garner (1970) have investigated the contrasts between fine-grained and coarse-grained point bar deposits and show that a coarse-grained point bar deposits show no trend in grain-size compared with a well developed fining upwards trend seen in fine grained point bar deposit – a comparable finding to this study with respect to the use of 1D sedimentary log profiles. The example given by Donselaar & Overeem (2008) is located in the medial-distal portion of the Ebro DFS. When compared with the Salt Wash DFS it is a greater distance from the apex when the scale is normalised and subsequently, contains as much as 70% overbank/heterolithic facies in the vertical profile. The examples given by Ghinassi *et al.* (2014) Ielpi & Ghinassi (2015), Ghinassi *et al.* (2016) were confined within valleys and are therefore not directly comparable to the unconfined Salt Wash DFS. For example, meander bend migration is predominantly downstream in confined systems (Nicoll & Hickin, 2010; Ghinassi *et al.* 2016), whereas the unconfined Salt Wash DFS examples show examples of both meander belt migration lateral and downstream relative to local palaeoflow. The point bar models noted here are still relevant for the associated depositional environment. However, studies which have identified laterally migrating sand dominated point bars are particularly limited.

The knowledge acquired from the outcrop study presented here can also be applied to subsurface datasets. Subsurface studies of fluvial systems using cored sections and borehole images will have difficulty in recognising the large-scale unit bar and lateral accretion surfaces needed to identify point bar deposits unless seismic data are available with sufficient resolution to image the planform geometry of the fluvial system. An example from the Gouvca survey, Barents Sea

(Fig. 19), illustrates the difficulty of interpreting fluvial strata solely from core derived-sedimentary logs (Klausen *et al.* 2014). The profile of the sedimentary log (Fig. 19B) could potentially be interpreted to represent a braided fluvial channel package, due to the coarse-grained stacked sequences within the storey, if the interpretation was conducted without any planform informa-tion. However, a Z-slice through the seismic sur-vey reveals a planform view of the fluvial deposit displaying a large arcuate architecture, character-istic of point bar development. Without the plan-form view, a braided fluvial interpretation is probable, with significant implications for the interpretation of barform geometries, internal architecture, sandstone body dimensions, lateral

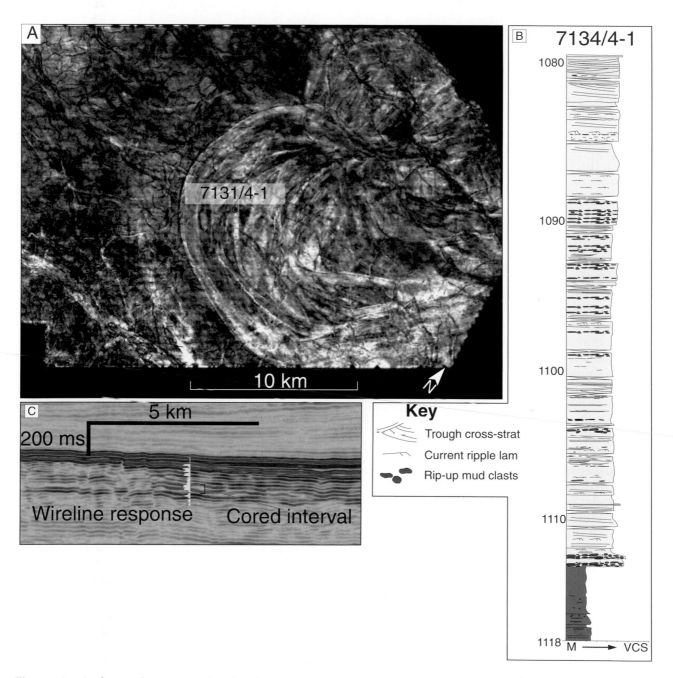

Fig. 19. A point bar can be recognised in the planform time-slice of the Gouvca survey, Barents Sea. A well drilled through the centre of this structure displayed a thick (25 m) section of core. Initially, the core was identified as a braided channel sandstone until further analysis of the planform architecture was conducted. (Modified from Klausen *et al.,* 2014).

extent and potential production strategies in economically viable fluvial deposits.

CONCLUSION

In this study, the combined usage of satellite imagery, drone photospectrometry, 3D digital outcrop data from LiDAR and sedimentological field-based methods have been utilised to describe and interpret point bar deposits within a sand-dominated DFS. The data presented cover a range of scales, the outcrop scale descriptions cover the detailed sedimentology, the photospectrometry from the drone and LiDAR datasets provide detailed images for the intermediate (barform) scale and the satellite imagery provides resolution to cover the planform fluvial character.

Key architectural elements were identified from the Caineville Wash outcrop where the planform exposure offered an additional dimension to assist with interpretation. Subsequently, at the neighbouring Caineville Road outcrop, the same workflow was conducted to interpret the fluvial character without the use of planform imagery.

Key learning from this study shows that multiple techniques must be utilised in order to identify correctly the depositional character of the fluvial system. Sedimentary logs alone will not capture the three-dimensionality of a fluvial outcrop or the larger scale barform architecture. Recent studies have developed our understanding of criteria used in the recognition of meandering channel deposits at outcrop (e.g. Cain & Mountney 2009; Ielpi & Ghinassi 2014; Ghinassi & Ielpi 2015; Owen *et al.* 2017b); however, in subsurface studies where fluvial planform cannot be inferred from seismic reflection data, models are commonly still derived from 1D sedimentary logs.

It is important to recognise that if a 2D cliff section is dominated by downstream accreting surfaces then interpretation as a meandering fluvial system is equally as viable as that of a braided system as, the Caineville Wash outcrop demonstrates that considerable portions of a point bar deposit may be composed of downstream accreting elements. It is suggested that vertical logs can no longer be the sole way in which planform is interpreted in fluvial deposits, but rather a close analysis of palaeocurrent directions from bedform and accretion surfaces, in conjunction with architectural panels and an understanding of the fluvial system (and therefore barform-size) and outcrop-size is needed.

ACKNOWLEDGEMENTS

The authors would like to thank BG, BP, Chevron, ConocoPhillips and Total for sponsoring Phase 2 of the Fluvial Systems Research Group consortium, which funded the PhD project of the senior author. We also appreciate the input of Gary Weissmann and his students at UNM, which helped with the understanding of the Caineville Wash outcrop. The authors would like to thank the Virtual Outcrop Group, Bergen, for supplying the Virtual Outcrops which have been used in the study and for the use and support of the free LIME software which is available as an academic download at http://virtualoutcrop.com/lime.

REFERENCES

Allen, **J.R.L.** (1983) Studies in fluviatile sedimentation: Bars, bar-complexes and sandstone sheets (low-sinuosity braided streams) in the brownstones (L. Devonian), Welsh borders. *Sed. Geol.*, **33**, 237–293. doi:10.1016/0037-0738(83)90076-3

Allen, **J.R.L.** (1963) The classification of cross-stratified units, with notes on their origin. *Sedimentology*, **2**, 93–114. doi:10.1111/j.1365-3091.1965.tb02116.x

Arche, **A.** (1983) Coarse-Grained Meander Lobe Deposits in the Jarama River, Madrid, Spain. *Mod. Anc. Fluv. Syst.*, **6**, 313–321, doi:10.1002/9781444303773.ch25

Bridge, **J.** (1993) Description and interpretation of fluvial deposits: a critical perspective. *Sedimentology*, **40**, 801–810. doi:10.1111/j.1365-3091.1993.tb01361.x

Bridge, **J.S.** (2006) Fluvial facies models: recent developments. *Facies Models Revisited*, **84**, 85–170. 10.2110/pec.06.84.0085.

Bridge, **J.S.**, **Alexander**, **J.**, **Collier**, **R.E.L.**, **Gawthorpe**, **R.L.** and **Jarvis**, **J.** (1995) Ground-penetrating radar and coring used to study the large-scale structure of point-bar deposits in three dimensions. *Sedimentology*, **42**, 839–852. doi:10.1111/j.1365-3091.1995.tb00413.x

Bridge, **J.S.** and **Tye**, **R.S.** (2000) Interpreting the dimensions of ancient fluvial channel bars, channels and channel belts from wireline-logs and cores. *AAPG Bull.*, **84**, 1205–1228. doi:10.1306/E4FD4B07-1732-11D7C1865D

Cain, **S.A.** and **Mountney**, **N.P.** (2009) Spatial and temporal evolution of a terminal fluvial fan system: The Permian organ rock formation, South-east Utah, USA. *Sedimentology*, **56**, 6, 1774–1800, doi:10.1111/j.1365-3091.2009.01057.x

Campbell, **C.V.** (1976) Reservoir Geometry of a Fluvial Sheet Sandstone. *AAPG Bull.*, **60**. doi:10.1306/C1EA3609-16C9-11D7C1865D

Campbell, **J.** and **Hendry**, **H.** (1987) Anatomy of a gravelly meander lobe in the Saskatchewan River, near Nipawin, Canada. In: *Recent Developments in Fluvial Sedimentology* (Eds **F.G. Ethridge**, **R.M. Flores** and **M.D. Harvey**). SEPM, **39**, 179–189.

Cant, **D.J.** and **Walker**, **R.G.** (1978) Fluvial processes and facies sequences in the sandy braided South Saskatchewan River, Canada. *Sedimentology*, **25**, 625–648. doi:10.1111/j.1365-3091.1978.tb00323.x

Craig, L.C., Holmes, C.N., Cadigan, R. A., Freeman, V.L., Mullens, T.E. and Wier, G.W. (1955) Stratigraphy of the Morrison and related formations Colorado Plateau Region. A Preliminary Report. *Geol. Surv. Bull.*, **1009-E**, 125–168.

DeCelles, P.G. (2004) Late Jurassic to Eocene evolution of the cordilleran thrust belt and foreland basin system, Western U.S.A. *Am. J. Sci.*, **304**, 105–168. doi:10.2475/ajs.304.2.105

Donselaar, M.E. and Overeem, I. (2008) Connectivity of fluvial point-bar deposit: An example from the Miocene Huesca fluvial fan, Ebro Basin, Spain. *AAPG Bull.*, **92**, 1109–1129. doi:10.1306/04180807079

Dunagan, S.P. and Turner, C.E. (2004) Regional paleohydrologic and paleoclimatic settings of wetland/lacustrine depositional systems in the Morrison Formation (Upper Jurassic), Western Interior, USA. *Sed. Geol.*, **167**, 269–296. doi:10.1016/j.sedgeo.2004.01.007

Enge, H.D., Buckley, S.J., Rotevatn, A. and Howell, J.A. (2007) From outcrop to reservoir simulation model: Workflow and procedures. *Geosphere*, **3**, 469–490. doi:10.1130/GES00099.1

Friend, F., Slater, M.J. and Williams, R.C. (1979) Vertical and lateral building of river sandstone bodies, Ebro Basin, Spain. *Geol. Soc. London*, **136**, 39–46.

Friend, P.F. (1978) Distinctive features of some ancient river systems. Fluvial Sedimentology: *Mem. Can. Soc. Pet. Geol.*, **5**, 541–542.

Ghazi, S. and Mountney, N.P. (2009) Facies and architectural element analysis of a meandering fluvial succession: The Permian Warchha Sandstone, Salt Range, Pakistan. *Sed. Geol.*, **221**, 99–126. doi:10.1016/j.sedgeo.2009.08.002

Ghinassi, M. and Ielpi, A. (2015) Stratal Architecture and Morphodynamics of Downstream-Migrating Fluvial Point Bars (Jurassic Scalby Formation, U.K.). *J. Sed. Res.*, **85**, 1123–1137. doi:10.2110/jsr.2015.74

Ghinassi, M., Ielpi, A., Aldinucci, M. and Fustic, M. (2016) Downstream-migrating fluvial point bars in the rock record. *Sed. Geol.*, **334**, 66–96. doi:10.1016/j.sedgeo.2016.01.005

Ghinassi, M., Nemec, W., Aldinucci, M., Nehyba, S., Özaksoy, V. and Fidolini, F. (2014) Plan-form evolution of ancient meandering rivers reconstructed from longitudinal outcrop sections. *Sedimentology*, **61**(4), 952–977.

Gibling, M.R. (2006) Width and Thickness of Fluvial Channel Bodies and Valley Fills in the Geological Record: A Literature Compilation and Classification. *J. Sed. Res.*, **76**, 731–770. doi:10.2110/jsr.2006.060

Hartley, A.J., Owen, A., Swan, A., Weissmann, G.S., Holzweber, B.I., Howell, J., Nichols, G. and Scuderi, L. (2015) Recognition and importance of amalgamated sandy meander belts in the continental rock record. *Geology*, **43**, 679–682. doi:10.1130/G36743.1

Hartley, A.J., Owen, A., Weissmann, G.S. and Scuderi, L. (2018) Modern and Ancient Amalgamated Sandy Meander-Belt Deposits: Recognition and Controls on Development. In: *Fluvial Meanders and Their Sedimentary Products in the Rock Record* (Eds M. Ghinassi, L. Colombera, N. Mountney, A. J. Reesink and M. Bateman), *Int. Assoc. Sedimentol, Spec. Publ.*, **48**, 349–384.

Hartley, A.J., Weissmann, G.S., Nichols, G.J. and Warwick, G.L. (2010) Large Distributive Fluvial Systems: Characteristics, Distribution and Controls on Development. *J. Sed. Res.*, **80**, 167–183. doi:10.2110/jsr.2010.016

Hasiotis, S.T. (2004) Reconnaissance of Upper Jurassic Morrison Formation ichnofossils, Rocky Mountain Region, USA: paleoenvironmental, stratigraphic and paleoclimatic significance of terrestrial and freshwater ichnocoenoses. *Sed. Geol.*, **167**, 177–268. doi:10.1016/j.sedgeo.2004.01.006

Ielpi, A. and Ghinassi, M. (2014) Planform architecture, stratigraphic signature and morphodynamics of an exhumed Jurassic meander plain (Scalby Formation, Yorkshire, UK). *Sedimentology*, 1923–1960. doi:10.1111/sed.12122

Jackson, R.G. (1976) Depositional model of point bars in the Lower Wabash river. *J. Sed. Petrol.*, **46**, 579–594. doi:10.1306/212F6FF5-2B24-11D7-C1865D

Jackson, R.G. (1981) Sedimentology of muddy fine-grained channel deposits in meandering streams of the American Middle West. *J. Sed. Petrol.*, **51**, 1169–1192.

Kjemperud, A.V., Schomacker, E.R. and Cross, T.A. (2008) Architecture and stratigraphy of alluvial deposits, Morrison Formation (Upper Jurassic), *Utah. AAPG Bull.*, **92**, 1055–1076. doi:10.1306/03250807115

Klausen, T.G., A.E. Ryseth, W. Helland-Hansen, R. Gawthorpe and I. Laursen (2014) Spatial and Temporal Changes In Geometries of Fluvial Channel Bodies From the Triassic Snadd Formation of Offshore Norway: *J. Sed. Res.*, **84**(7), 567–585, doi:10.2110/jsr.2014.47.

Kostic, B. and Aigner, T. (2007) Sedimentary architecture and 3D ground-penetrating radar analysis of gravelly meandering river deposits (Neckar Valley, SW Germany). *Sedimentology*, **54**, 789–808. doi:10.1111/j.1365-3091.2007.00860.x

Lunt, I.A., Bridge, J.S. and Tye, R.S. (2004) A quantitative, three-dimensional depositional model of gravelly braided rivers. *Sedimentology*, **51**, 377–414. doi:10.1111/j.1365-3091.2004.00627.x

McGowen, J.H. and Garner, L.E. (1970) Physiographic Features and Stratification Types of Coarse-Grained Point Bars: Modern and Ancient examples. *Bur. Econ. Geol.*, **14**.

McKie, T., Jolley, S.J. and Kristensen, M.B. (2010) Stratigraphic and structural compartmentalization of dryland fluvial reservoirs: Triassic Heron Cluster, Central North Sea. In: *Reservoir Compartmentalization* (Eds S.J. Jolley, Q.J. Fisher, R.B. Ainsworth, P.J. Vrolijk and S. Delisle), *Geol. Soc. London, Spec. Publ.*, **347**, 165–198. doi:10.1144/SP347.11

Miall, A.D. (1977) A review of the braided-river depositional environment. *Earth-Sci. Rev.*, **13**, 1–62. doi:10.1016/0012-8252(77)90055-1

Miall, A.D. (1993) Description and interpretation of fluvial deposits: a critical perspective. *Sedimentology*, **40**, 801–810.

Miall, A.D. (1995) Description and interpretation of fluvial deposits: a critical perspective. *Sedimentology*, **42**, 379–389. doi:10.1111/j.1365-3091.1993.tb01361.x

Miall, A.D. (1978) Lithofacies types and vertical profile models in braided river deposits: a summary. *Fluv. Sedimentol.*, **5**, 597–600.

Miall, A.D. (1988) Reservoir Heterogeneities in Fluvial Sandstones: Lesson From Outcrop Studies. *AAPG Bull.*, **72**(6), 682–697. doi:10.1306/703C8F01-1707-11D7-C1865D

Miall, A.D. (1996) *The geology of fluvial deposits: sedimentary facies, basin analysis and petroleum geology.* Springer, Berlin; New York.

Mullens, **T.E.** and **Freeman**, **V.L.** (1957) Lithofacies of the Salt Wash member of the Morrison formation, Colorado plateau. *Geol. Soc. Am. Bull.*, **68**, 505. doi:10.1130/0016-7606(1957)68[505:LOTSWM]2.0.CO;2

Musial, **G.**, **Reynaud**, **J.Y.**, **Gingras**, **M.K.**, **Fenies**, **H.**, **Labourdette**, **R.** and **Parize**, **O.** (2012) Subsurface and outcrop characterization of large tidally influenced point bars of the Cretaceous McMurray Formation (Alberta, Canada). *Sed. Geol.*, **279**, 156–172. doi:10.1016/j.sedgeo.2011.04.020

Nardin, **T.**, **Feldman**, **H.R.** and **Carter**, **B.J.** (2013) Stratigraphic architecture of a large-scale point-bar complex in the McMurray Formation: Syncrude's Mildred Lake Mine, Alberta, Canada. Heavy-oil and oil-sand petroleum systems in Alberta and beyond: *AAPG Stud. Geol.*, **64**, 273–311. doi:10.1306/13371583St643555

Neal, **A.** (2004) Ground-penetrating radar and its use in sedimentology: Principles, problems and progress. *Earth-Science Rev.*, **66**, 261–330. doi:10.1016/j.earscirev.2004.01.004

Nicoll, **T.J.** and **Hickin**, **E.J.** (2010) Planform geometry and channel migration of confined meandering rivers on the Canadian prairies. *Geomorphology*, **116**, 37–47. doi:10.1016/j.geomorph.2009.10.005

Ore, **T.H.** (1964) Some criteria for recognition of braided stream deposits. *Contrib. to Geol. Univ. Wyoming*, **3**, 1–14.

Ori, **G.G.** (1982) Braided to meandering channel patterns in humid-region alluvial fan deposits, River Reno, Po Plain (northern Italy). *Sed. Geol.*, **31**, 231–248. doi:10.1016/0037-0738(82)90059-8

Owen, **A.**, **Ebinghaus**, **A.**, **Hartley**, **A.J.**, **Santos**, **M.G.M.** and **Weissmann**, **G.S.** (2017b) Multi-scale classification of fluvial architecture: An example from the Palaeocene-Eocene Bighorn Basin, Wyoming. *Sedimentology*, **64**(6), 1572–1596, doi:10.1111/sed.12364

Owen, **A.**, **Jupp**, **P.**, **Nichols**, **G.J.**, **Hartley**, **A.J.**, **Weissmann**, **G.S.** and **Sadykova**, **D.** (2015a) Statistical estimation of the position of an apex : Application to the geological record. *J. Sed. Res.*, **85**, 142–152.

Owen, **A.**, **Nichols**, **G.J.**, **Hartley**, **A.J.** and **Weissmann**, **G.S.** (2017b) Vertical trends within the prograding Salt Wash distributive fluvial system, *SW United States. Basin Res.*, **29**, 64–80. doi:10.1111/bre.12165

Owen, **A.**, **Nichols**, **G.J.**, **Hartley**, **A.J.**, **Weissmann**, **G.S.** and **Scuderi**, **L.A.** (2015b) Quantification of A distributive fluvial system: The Salt Wash DFS of the Morrison. *J. Sed. Res.*, **85**, 544–561.

Peterson, **F.** (1978) Measured sections of the lower Member and Salt Wash Member of the Morrison Formation (Upper Jurassic) in the Henry Mountains mineral belt of southern Utah. *US Geol. Surv. Report*, **78**-1094, 97 pp.

Peterson, **F.** (1994) Sand dunes, sabkhas, streams and shallow seas: Jurassic paleogeography in the southern part of the western interior basin. *Mesozoic Systems of the Rocky Mountain Region, USA*, 233–272.

Peterson, **F.** (1980) Sedimentology as a strategy for uranium exploration: concepts gained from analysis of a uranium bearing depositional sequence in the Morrison Formation of south-central Utah. Uranium in Sedimentary Rocks: Application of the Facies Concept to Exploration, *SEPM*, **6**, 65–126.

Plint, **A.G.** (1983) Sandy Fluvial Point-Bar Sediments from the Middle Eocene of Dorset, England. In: *Modern and Ancient Fluvial Systems*. Blackwell Publishing Ltd., Oxford, UK, **6**, 355–368. doi:10.1002/9781444303773.ch29

Robinson, **J.W.** and **McCabe**, **P.J.** (1998) Evolution of a braided river system: the Salt Wash member of the Morrison Formation (Jurassic) in southern Utah. In: *Earth Surface Processes, Landforms and Sediment Deposits* (Eds J. Bridge and R. Demicco), *SEPM Spec. Publ.*, **59**, 93–108.

Robinson, **J.W.** and **McCabe**, **P.J.** (1997) Sandstone-body and shale-body dimensions in a braided fluvial system: Salt wash sandstone member (Morrison formation), Garfield County, *Utah. AAPG Bull.*, **81**, 1267–1291. doi:10.1306/522B4DD9-1727-11D7-C1865D

Rust, **B.R.** (1972) Structure and process in a braided river: *Sedimentology*, **18**(3 to 4), 221–245, doi:10.1111/j.1365-3091.1972.tb00013.x.

Schirmer, **W.** (1995) Valley Bottoms in the late Quaternary. Z. f. *Geomorphology N.F.* **100**, 27–53.

Smith, **D.G.** (1987) Meandering river point bar lithofacies models: modern and ancient examples compared. In: *Recent Developments in Fluvial Sedimentology* (Eds F.G. Ethridge, R.M. Flores and M.D. Harvey), *SEPM Spec. Publ.*, **39**, 83–91.

Smith, **D.G.**, **Hubbard**, **S.M.**, **Lavigne**, **J.R.**, **Leckie**, **D.A.** and **Fustic**, **M.** (2011) Stratigraphy of Counter-Point-Bar and Eddy-Accretion Deposits in Low-Energy Meander Belts of the Peace-Athabasca Delta, Northeast Alberta, Canada. In: *From River to Rock Record: The preservation of fluvial sediments and their subsequent interpretation* (Eds S.K. Davidson, S. Leleu and C.P. North), *SEPM Spec. Publ.*, **97** 143–152. doi:10.2110/sepmsp.097.143

Smith, **D.G.**, **Hubbard**, **S.M.**, **Leckie**, **D.A.** and **Fustic**, **M.** (2009) Counter point bar deposits: Lithofacies and reservoir significance in the meandering modern Peace River and ancient McMurray formation, Alberta, Canada. *Sedimentology*, **56**, 1655–1669. doi:10.1111/j.1365-3091.2009.01050.x

Smith, **N.D.** (1972) Some sedimentological aspects of planar cross-stratification in a sandy braided river. *J. Sed. Petrol.*, **42**, 624–634.

Thomas, **R.G.**, **Smith**, **D.G.**, **Wood**, **J.M.**, **Visser**, **M.J.**, **Calverley-Range**, **E.A.** and **Koster**, **E.H.** (1987) Inclined heterolithic stratification – terminology, description, interpretation and significance. *Sed. Geology*, **53** (September), 123–179, doi:10.1016/S0037-0738(87)80006-4.

Tyler, **N.** and **Ethridge**, **F.G.** (1983) Depositional Setting of the Salt Wash Member of the Morrison Formation, Southwest Colorado. SEPM, *J. Sed. Res.*, **53**, 67–82. doi:10.1306/212F8157-2B24-11D7-C1865D

Tyler, **N.** and **Ethridge**, **F.G.** (1983) Fluvial architecture of Jurassic Uranium-bearing sandstones, Colorado Plateau, Western United States. *Mod. Anc. Fluv. Syst.*, 533–547. doi:10.1002/9781444303773

Weissmann, **G.S.**, **Hartley**, **A.J.**, **Scuderi**, **L.A.**, **Nichols**, **G.J.**, **Davidson**, **S.K.**, **Owen**, **A.**, **Atchley**, **S.C.**, **Bhattacharyya**, **P.**, **Michel**, **L.** and **Tabor**, **N.J.** (2013) Prograding distributive fluvial systems - geomorphic models and ancient examples. In: *New Frontiers in Paleopedology and Terrestrial Paleoclimatology* (Eds S.G. Driese, L.C. Nordt and P.J. McCarthy), *SEPM Spec. Publ.* 131–147. doi:10.2110/sepmsp.104.16

Zeng, **X.**, **McMechan**, **G.A.**, **Bhattacharya**, **J.P.**, **Aiken**, **C.L. V**, **Xu**, **X.**, **Hammon**, **W.S.** and **Corbeanu**, **R.M.** (2004) 3D imaging of a reservoir analogue in point bar deposits in the Ferron Sandstone, Utah, using ground-penetrating radar. *Geophys. Prospect.*, **52**, 151–163. doi:10.1046/j.1365-2478.2003.00410.x

Int. Assoc. Sedimentol. Spec. Publ (2018) **48**, 475–508.

Fluvial point-bar architecture and facies heterogeneity and their influence on intra-bar static connectivity in humid coastal-plain and dryland fan systems

NA YAN[†], LUCA COLOMBERA[†], NIGEL P. MOUNTNEY[†] and ROBERT M. DORRELL[‡]

[†] *Fluvial & Eolian Research Group, School of Earth and Environment, University of Leeds, Leeds, UK*
[‡] *Institute of Energy and Environment, University of Hull, Hull, UK*

ABSTRACT

Many published studies detail the sedimentology and stratigraphic architecture of meandering fluvial systems and their preserved successions based on data from modern systems, outcrops or subsurface. However, the broader understanding of the behaviour of depositional systems and its role in determining the nature of the resultant depositional units remains limited due to difficulties associated with the collection of appropriate data on three-dimensional sedimentary-facies distributions and on the temporal evolution of the form of architectural elements. To overcome such limitations, numerical lithofacies and stratigraphic modelling approaches provide a valuable suite of tools to examine the sensitivity of intrinsic system behaviour to different controls that operate at varying spatial and temporal scales. This study utilises a three-dimensional forward stratigraphic model, the 'Point-Bar Sedimentary Architecture Numerical Deduction' (PB-SAND), informed by data and relationships extracted from a sedimentological database, to model the facies architecture and heterogeneity of fluvial point-bar elements in two contrasting environmental settings: humid coastal plains *vs.* dryland fluvial fans. This study demonstrates a workflow that uses high-resolution seismic imagery to constrain the planform evolution of preserved point-bar elements, in combination with data from appropriate geological analogues (five humid coastal-plain and eight dryland fluvial-fan systems) to constrain modelling parameters (i.e. bar thickness, facies proportion and mud-drape geometry). The method applies a statistical analytical approach to constrain the ranges and types of sedimentary architectures and facies heterogeneity known for humid and dryland meandering fluvial depositional systems. Modelling results demonstrate the effects of increasing compartmentalisation of sand geo-bodies by mud drapes, whose density increases vertically towards the bar top of point-bar architectural elements. Modelling results also demonstrate how the compartmentalisation of point-bar elements in the models constrained on humid coastal-plain analogues, compared to those based on dryland-fan analogues, exhibits a larger variation in response to the spatial distribution of mud drapes and their discontinuity in three dimensions. The modelling outputs are able to represent realistic architectural geometries, capture the three-dimensional complexity of sedimentary architecture and incorporate styles of facies heterogeneity that can be employed in the quantitative analysis of connectivity. The PB-SAND simulations can, therefore, be used to enhance conventional reservoir models and thereby to improve the realism of fluid-flow models.

Keywords: Forward stratigraphic model, numerical model, PB-SAND, FAKTS, meandering river, facies heterogeneity, connectivity, reservoir compartmentalisation

Fluvial Meanders and Their Sedimentary Products in the Rock Record, First Edition.
Edited by Massimiliano Ghinassi, Luca Colombera, Nigel P. Mountney and Arnold Jan H. Reesink.

INTRODUCTION

As meandering rivers migrate laterally over flood-plains, sediment is eroded from the cut-bank on the outer bend and deposited on the inner bank, developing point bars (Leopold & Wolman, 1960; Nanson & Croke, 1992). Point-bar deposits are present in a wide range of fluvial sedimentary environments: they evolve under the influence of both humid and dryland climatic regimes; they develop in both coastal and inland settings; and they are present in both fluvial fans and alluvial valleys (Hartley *et al.*, 2015; Maynard & Murray, 2003; Reijenstein *et al.*, 2011; Smith, 1987; Trendell *et al.*, 2013; Wu *et al.*, 2015). Considerable initial research effort focused on the analysis of the plan-view shape and geometric regularity of meanders, their relation to flow patterns and behaviour (Carlston, 1965; Leopold & Langbein, 1966) and also to geometry-based classifications of meander shape (Brice, 1974). Statistical relationships between geometric measures of three-dimensional bodies (e.g. the ratio of channel depth to width and the ratio of channel width to meander length) have also been documented, although established relationships vary greatly between different systems, or even within the same system but using different datasets (Fielding & Crane, 1987; Leeder, 1973; Leopold & Wolman, 1960; Lorenz *et al.*, 1985; Smith, 1987). The internal facies architecture and heterogeneity of point-bar deposits have been recognised to relate to the behaviour of meander-bend transformations including expansion, translation, rotation and combinations thereof (Ghinassi *et al.*, 2014; Jackson, 1976; Makaske & Weerts, 2005). However, detailed research on this topic has, to date, largely been restricted to individual case studies based on analysis of outcrop successions or seismic slices (Bridge, 2003; Labrecque *et al.*, 2011; Miall, 1996).

Computer simulations using forward stratigraphic modelling approaches enable different hypotheses to be tested relating to the mechanisms for the generation of point-bar stratigraphy and lithofacies distributions at various spatio-temporal scales. Process-based models that apply hydro-morphological relations and empirical equations can reproduce quantitative relationships between fluid flow and sediment transport but require large computational resources to resolve differential equations (Bridge, 1982; Brownlie, 1983; Sun *et al.*, 2001a; Sun *et al.*, 2001b; Yan *et al.*, 2017). Meanwhile, the geometry of meanders can be complex; their form and associated point-bar deposits evolve in a complex manner over time in response to processes that are commonly nonlinear and difficult to replicate using such modelling approaches (Sun *et al.*, 1996; Willis & Tang, 2010). By contrast, although geometric-based numerical models do not explicitly account for complicated hydraulic processes, they do benefit from high computational efficiency, which competitively enables rapid visualisation of large phase-spaces, stochastic analysis of modelling results and examinations of controlling parameters. In particular, models of this type can replicate the complicated shape of meanders and associated point-bar architectural elements. Furthermore, they possess the flexibility required to honour field data. For example, geometric-based models can account for bedding geometries and facies distribution patterns in both plan-view and cross sections. Substantiation of modelled stratigraphy of this type relies on the availability of appropriate data for the purposes of comparison to natural systems (Hassanpour *et al.*, 2013; Pyrcz *et al.*, 2015; Rongier *et al.*, 2017; Yan *et al.*, 2017). One important application of geometric-based modelling approaches is in the prediction of the detailed geometries and facies composition of architectural elements within point-bar bodies. For example, the spatial arrangement (density of occurrence, lateral extent, spacing and continuity) of mud drapes, which are common features of fluvial point-bar elements, can be modelled to investigate their influence on the stratigraphic compartmentalisation of hydrocarbon reservoirs. Supported by appropriate datasets, geometric modelling approaches can facilitate understanding of heterogeneities of preserved architectures controlled by different environmental factors, which can help constrain parameter settings for static and dynamic reservoir models in cases where required primary data at a particular locality are not available.

This study uses a three-dimensional forward stratigraphic model, the Point-Bar Sedimentary Architecture Numerical Deduction (PB-SAND; Yan *et al.*, 2017), which can be informed by data derived from geological analogues. In this work, PB-SAND simulations are constrained using outputs from a relational database that stores quantified sedimentological data from modern classified fluvial systems and their ancient equivalents preserved as fluvial successions, the Fluvial Architecture Knowledge Transfer System

(FAKTS; Colombera *et al.*, 2012a, b; 2013). A systematic comparison of facies architecture and heterogeneity of point-bar elements is undertaken for two selected types of fluvial environments: humid coastal plains *vs.* dryland fluvial fans.

The aim of this study is to demonstrate how the PB-SAND model in combination with the FAKTS database can generate realisations of 3D point-bar architectures that can improve our generic understanding of the form and distribution of architectural elements and facies heterogeneity in particular types of fluvial environmental settings. Specific research objectives are as follows: (1) to introduce a geometric-based approach to modelling the spatial arrangement and discontinuity of mud drapes commonly present in fluvial point-bar elements; (2) to demonstrate a workflow that uses a relational database of sedimentological data (FAKTS) in combination with statistical analysis to parameterise PB-SAND geometric models; (3) to generate 3D architectures that facilitate our understanding of facies heterogeneity, connectivity of geo-bodies and net-reservoir compartmentalisation within point-bar elements; and (4) to explore generic differences between two types of depositional systems, selected on the basis of available datasets.

DATA AND METHODS

The PB-SAND stratigraphic modelling software primarily uses a geometric-based approach combined with process-based and stochastic-based methods. The model is able to predict the complex 3D geometry and internal lithofacies distribution of point-bar elements in response to different types of channel migration under theoretically unrestricted horizontal and vertical resolutions by virtue of the vector-based modelling technique employed. However, the model resolutions may be restricted, in practice, by the resolutions of available input data. The FAKTS database stores case examples of fluvial systems in a standardised manner from both literature and field studies, thereby enabling the quantitative comparison of sedimentary architectures between different systems (Colombera *et al.*, 2012b; 2013). Parameterised by FAKTS, the PB-SAND model is herein used to reconstruct sedimentary architectures of point-bar accumulations in three dimensions and to provide insight into the differences in facies heterogeneity of these types of deposits in humid coastal-plain systems compared with dryland fluvial-fan systems. A detailed explanation of the algorithms used in the PB-SAND model is described in Yan *et al.* (2017).

Analogue data

Table 1 summarises the geological analogues that have been considered in this study. These analogues are from humid coastal-plain settings (5 depositional systems) and dryland fan settings (8 depositional systems). Typical examples of the sedimentary architecture and geometry of point-bar deposits formed in humid coastal-plain settings are illustrated in Fig. 1. The Middle Miocene Upper

Table 1. Depositional systems extracted from the FAKTS database for informing the PB-SAND model.

	Lithostratigraphy units
Humid coastal plain systems	1. Ferron Sandstone Mb. of the Mancos Shale, Utah, USA (Corbeanu *et al.*, 2004; Wu *et al.*, 2015)
	2. Iles Formation, Mesaverde Group, Colorado, USA (Anderson, 2005)
	3. Lower Williams Fork Formation, Piceance Basin, Colorado, USA (Pranter *et al.*, 2007)
	4. McMurray Formation, Alberta, Canada (Hubbard *et al.*, 2011; Jablonski, 2012)
	5. Neslen Formation, Mesaverde Group, Utah, USA (FRG in-house data)
Dryland fan systems	1. Balfour Formation, Karoo Basin, South Africa (Catuneanu & Elango, 2001)
	2. Brushy Basin Member, Morrison Formation, Colorado, USA (Miall & Turner-Peterson, 1989)
	3. Bunter Sandstone Formation, Bacton Group, UK (Olsen, 1987)
	4. Chinle Formation, Arizona, USA (Trendell *et al.*, 2013)
	5. Huesca fluvial fan, Ebro Basin, Spain (Donselaar and Overeem, 2008)
	6. Kayenta Formation, Colorado, USA (FRG in-house data)
	7. Trentishoe Formation, Hangman Sandstone Group, Devon, UK (Tunbridge, 1984)
	8. Westwater Canyon Member, Morrison Formation, Colorado, USA (Miall & Turner-Peterson, 1989)

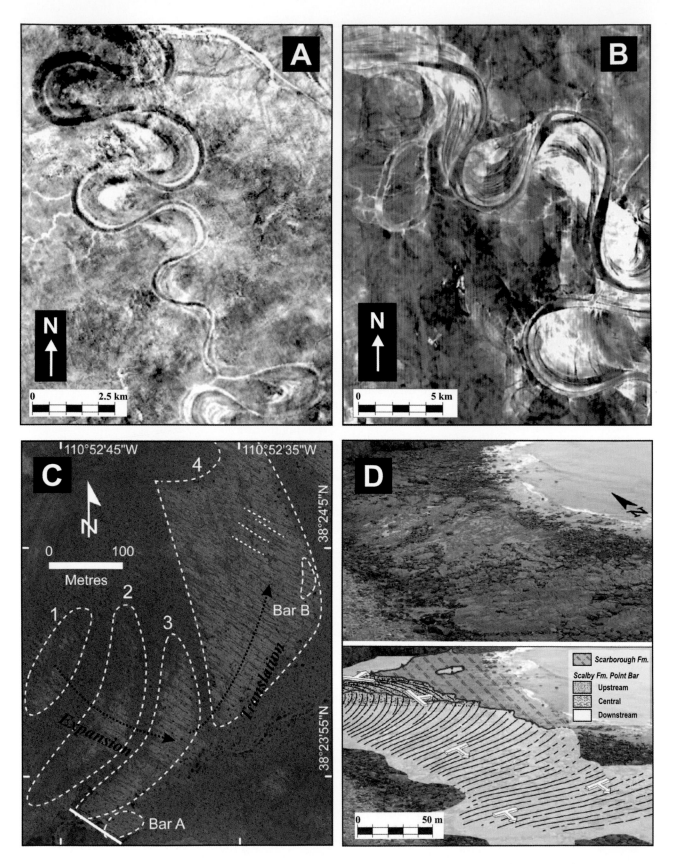

Fig. 1. Examples of fluvial point-bar elements in humid coastal-plain systems. (A) Point-bar elements in the Upper Arang Formation, Middle Miocene, West Natuna Basin, off-shore Indonesia, imaged in high-resolution 3D seismic data (adapted from Maynard & Murray, 2003). (B) Point-bar elements in the Tertiary Pattani rift basin, Gulf of Thailand, imaged in high-resolution 3D seismic data (adapted from Reijenstein *et al.*, 2011). (C) An ancient point-bar element in the Ferron Sandstone of Cretaceous Notom Delta, Utah, USA; note the channel migration directions indicated on the satellite image (adapted from Wu *et al.*, 2016). (D) Point-bar elements in the Jurassic Scalby Formation, UK; note the interpretations of different parts of point-bar elements (adapted from Ghinassi & Ielpi, 2015).

Arang Formation (West Natuna Basin, offshore Indonesia) was deposited under the influence of a humid climate (Maynard & Murray, 2003). The high-resolution seismic image in Fig. 1A shows point-bar elements formed by lateral migration of a meandering channel during part of a Highstand Systems Tract (Maynard & Murray, 2003; Morley, 2000). The seismic plan-view image from the Tertiary Pattani rift basin, Gulf of Thailand (Fig. 1B), by contrast, reveals meander development by both expansion and downstream-translation (Reijenstein *et al.*, 2011). An ancient point bar in the Cretaceous Ferron Sandstone of the Notom Delta succession, Utah, USA (Fig. 1C) has been interpreted to have experienced an early stage of expansion followed by translation (Wu *et al.*, 2015). Outcrop exposures of the Jurassic Scalby Formation, UK (Fig. 1D) show evidence of a multi-storey arrangement of bar deposits: point-bar elements grew dominantly via downstream translation in the more confined, lower part of the succession and via bend expansion in the less confined upper part (Ghinassi & Ielpi, 2015).

Analogue examples of point-bar deposits and their sedimentary architecture in dryland-fan systems are presented in Fig. 2. The aerial photograph in Fig. 2A shows meander-bend accretion packages exposed in plan-view from the Upper Triassic Chinle Formation at Petrified Forest National Park, Arizona, USA. Sandstone facies are dominant and each inclined cross-stratified set consists of a fining-upward package (Trendell *et al.*, 2013). The plan-view of meanders of Reiersvlei in the Permian Beaufort Group of the South-western Karoo, South Africa is shown in Fig. 2B. An exposed outcrop cross-section reveals thin sandstone wedges that interdigitate with top-stratum mudrocks (Smith, 1987). Exhumed sandy point-bar deposits are further exemplified in Fig. 2C and D, which depict examples from the Upper Jurassic Morrison Formation, Utah, USA (Hartley *et al.*, 2015). The outcrops of point-bar deposits from the Miocene Huesca fluvial fan, Ebro Basin, Spain (Fig. 2E) comprise steeply inclined (<20°) siltstone-draped lateral-accretion surfaces; some of the silty drapes extend down through the uppermost two-thirds of the sandstone body thickness (Donselaar & Overeem, 2008). Preserved mud drapes are relatively abundant and are laterally more extensive than those preserved in most other documented point-bar successions accumulated in arid-climate settings.

FAKTS database

The Fluvial Architecture Knowledge Transfer System (FAKTS; Colombera *et al.*, 2012a, 2013) is a relational database that contains data on sedimentary units occurring at different scales of observation and that are assigned to depositional systems classified on a number of attributes describing their depositional context and geological controls. The FAKTS includes data on the geometry, spatial relationships and lithological organisation of sedimentary bodies recognised in modern and ancient fluvial systems, data that are derived from the published scientific literature and from original unpublished field studies. This study uses data from modern point bars and from architectural elements interpreted as the preserved product of point-bar deposition or as barforms recording lateral accretion. For this study, the database has been interrogated and filtered by depositional-environment type to enable the examination of quantified information relating to (i) humid coastal-plain; and (ii) arid or semiarid fluvial-fan systems. Information on the geometry and lithological organisation of point-bar elements can then be applied to condition PB-SAND simulations (Table 1).

PB-SAND model

The Point-Bar Sedimentary Architecture Numerical Deduction model (PB-SAND; Yan *et al.*, 2017) is able to simulate the sedimentary architecture of fluvial point bars developed by different types of meander-bend transformations. This is achieved by identifying and defining plan-view channel trajectories at multiple stages of bar evolution as the sinuosity of a meander bend increases or the bend apex changes in migration direction, for example, due to a component of growth via rotation or downstream translation. In this study, three consecutive point-bar elements interpreted from a 3D seismic image from the Upper Arang Formation of Middle Miocene, West Natuna Basin (Fig. 3A, Maynard & Murray, 2003) are selected as a representative example of modelled planform morphology. The point-bar elements from this system have been chosen for the modelling exercise because of the high quality of the seismic imagery in which the channel fill and the bar accretion packages can be clearly identified (Fig. 3A). The same planform morphology is used for all simulations so that

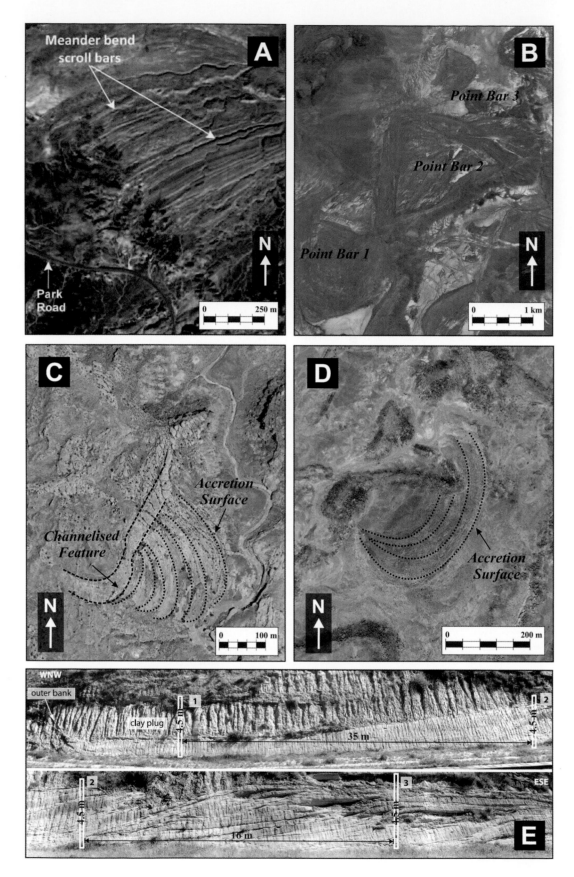

Fig. 2. Examples of fluvial point-bar elements in interpreted dryland fluvial-fan systems. (A) Meander-bend scroll bars in the Upper Triassic Chinle Formation at Petrified Forest National Park, Arizona, USA (adapted from Trendell *et al.*, 2013). (B) Reiersvlei meanders (32° 02′ S, 22° 03′ E) in the Permian Beaufort Group of the South-western Karoo, South Africa (image from Google Earth; *cf.* Smith, 1987). (C) Point-bar elements (38° 24′ N, 111° 01′ W) in the Upper Jurassic Morrison Formation, Utah, USA (image from Google Earth; *cf.* Hartley *et al.*, 2015). (D) Point-bar elements (39° 10′ N, 110° 52′ W) in the Upper Jurassic Morrison Formation, Utah, USA (image from Google Earth; *cf.* Hartley *et al.*, 2015). (E) Outcrop of a point-bar sandstone body with well-preserved clay plug and lateral-accretion surfaces from the Miocene Huesca fluvial fan, Ebro Basin, Spain (adapted from Donselaar & Overeem, 2008).

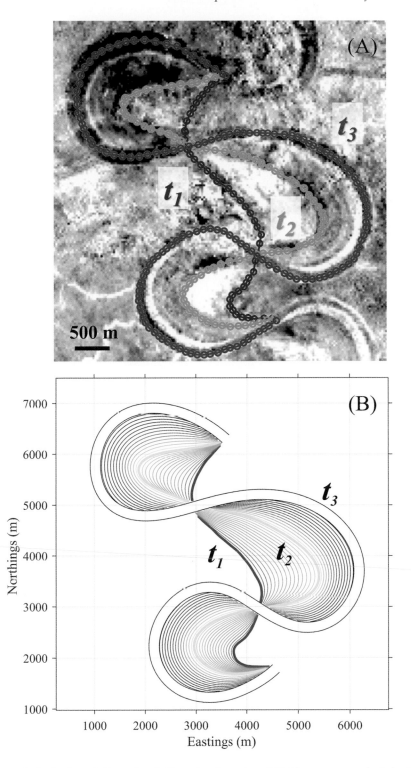

Fig. 3. Modelling the morphological evolution of point bars in plan-view. (A) Blue, green and pink circles represent input coordinates of channel trajectories at t_1, t_2 and t_3 respectively, which are digitised from the seismic image in the Upper Arang Formation of Middle Miocene, West Natuna Basin, off-shore Indonesia (Fig. 1A; Maynard & Murray, 2003). (B) Morphology of point bars modelled by PB-SAND with highlighted channel trajectories at t_1, t_2 and t_3.

comparisons of facies heterogeneity and reservoir connectivity between different simulation outputs are possible. The point-bar elements from this system are expansional forms that are the predominant meander-bend transformations in many fluvial systems including in those of both humid and dryland settings (Bridge, 2003).

To model point-bar morphology in plan-view, three key channel positions are identified from the high-resolution seismic imagery (t_1, t_2 and t_3 in Fig. 3A). Channel positions between the three controlling times are then interpolated and generated (rendered) within PB-SAND by specifying the number of modelled channel trajectories between each of these three channel positions. Via this method, the planform morphology of modelled point bars (Fig. 3B) can be efficiently set to mimic closely those seen in a natural dataset, in this case, the seismic image (Fig. 3A).

The PB-SAND model allows simulation of the geometry and facies distribution of lateral-accretion packages of point bars developed through different growth behaviours and generating vertical cross sections of any orientation by virtue of its vector-based modelling approach. In particular, it is possible to model inclined mud-draped lateral-accretion packages that commonly form permeability baffles in such deposits and act to partition otherwise sand-dominated bar deposits. Significantly, thin mud drapes (relative to the overall thickness of a point-bar element) can be modelled as a discrete lithology type; the vector-based modelling approach of PB-SAND does not suffer from issues related to resolution, as would be the case for raster-based modelling approaches that attempt to incorporate thin packages of strata. This capability is especially important in the assessment of lithofacies arrangements which might impair hydrocarbon production due to compartmentalisation (Jolley *et al.*, 2010).

The PB-SAND model enables the generation of multiple families of mud drapes with different thickness, frequency of occurrence and pinch-out extent along a bar slope. A detailed explanation of the modelling algorithms used in the PB-SAND model is provided by Yan *et al.* (2017). The extent down a bar slope to which mud drapes extend is expressed as a width measured in plan-view (Fig. 4). Two additional parameters defined in the

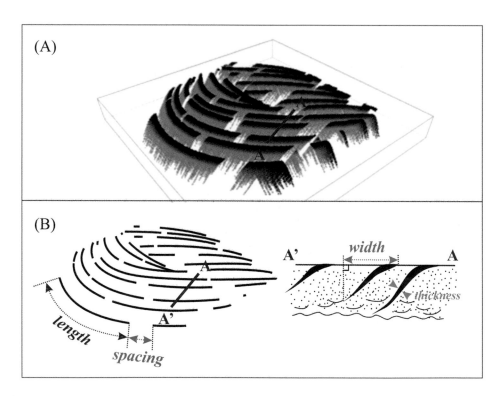

Fig. 4. Examples of the length, lateral spacing, width (the projected extent in the horizontal surface) and thickness (the average thickness perpendicular to the bar-accretion slope) of mud drapes shown in a 3D modelling output of a modelled point bar generated using PB-SAND. (A) Mud drapes shown in a 3D-view of a modelled point-bar example. (B) Schematic illustration of the length, spacing, width and thickness of mud drapes. Cross section A-A' is depicted with a pink line in both the perspective and cross-sectional views.

model are used to account for spatial heterogeneity and discontinuity of mud draping on the same lateral-accretion surface in plan-view: the length of continuous mud drapes around the curved lateral-accretion surface and the spacing between two laterally discontinuous mud drapes along the same curved surface (Fig. 4). By default within the model, these parameters follow truncated Gaussian distributions defined by their respective minimum and maximum values and standard deviations, though customised distributions can optionally be defined. Different families of mud drapes can be modelled either independently or simultaneously with their respective length and spacing distributions (Fig. 5).

PB-SAND allows simulation of variations in lithofacies associations both temporally (as a meander evolves) and spatially (throughout a meander). Facies proportions and trends in facies distributions can be flexibly defined based on analogue data. Data collected from 13 depositional systems (Table 1) have herein been used to constrain and parameterise the model inputs, for example, by defining the bar thickness, the thickness and width of mud drapes and the facies proportions.

Table 2 reports measures of point-bar thickness, muddy facies thickness and facies (grain-size) proportions in humid coastal-plain and dryland fluvial-fan systems. The data were extracted from the FAKTS database and are based on literature studies considered representative of the two types of systems (Table 1). Mean point-bar thickness is 10.3 m for humid coastal plains (35 point-bar elements, $s = 8.86$ m, where s is the sample standard deviation) and 5.1 m for dryland fluvial fans (16 point-bar elements, $s = 4.51$ m). A bar thickness of 8 m is used in the model as a representative value to generate point-bar examples in both systems. This is also consistent with the general relationship between point-bar thickness and point-bar width from 175 fluvial point-bar elements in FAKTS (Colombera *et al.*, 2017), as applied to modelled point-bar elements with bar widths in the order of magnitude of 10^3 metres (Fig. 3). Consideration of the influence of the variations in the bar thickness itself is beyond the scope of this study. Modelled mud drapes are constrained by mudstone beds in analogue point-bar elements due to the scarcity of available data pertaining to mud drapes, although mudstone beds refer to muddy deposits that may not necessarily form drapes on bar slopes. The mean thicknesses of

mud drapes used in the models are 0.1 m for humid coastal plains and 0.4 m for dryland fans, based on the mean thickness of mudstone beds from the chosen analogues (Fig. 6), summarised in Table 2. Given a mud-drape thickness, a mud-drape width has been deduced using an equation from a fitted power-law relationship between the mud-bed thickness and the mud-bed width based on complete measurements of field-derived data in FAKTS (Fig. 7). For a determined mud drape width, the relative position where mud drapes pinch out downslope on accretion surfaces (expressed as a percentage of bar thickness) is projected using simple trigonometric functions for a maximum slope of 15° (which is typical for these types of deposits); results are summarised in Table 3. The input describing the length of the mud drapes along the accretion surfaces has been chosen to reflect positive scaling between the width and the length: a thicker mud drape tends to exhibit a greater length (i.e. lateral continuity). Fig. 8 presents the proportions of different grain-size classes that comprise point-bar elements in humid coastal-plain settings and in dryland-fan settings, respectively. Given that the modelled point-bar elements are principally developed by meander expansion (Fig. 3), a fining-upward grain-size trend or facies association typically produced by such a mode of growth is used (Miall, 1996). The proportions of mud facies are set to increase with the sinuosity of meander-bends from t_2 to t_3. This mimics scenarios observed in rivers and outcrop analogues (Durkin *et al.*, 2015; Miall, 1996). Facies associations in this study are defined by different proportions of mud, sand and gravel lithofacies. Specifically, five facies associations used in simulations are randomly sampled between their respective minimum and maximum proportions (excluding outliers) in a way that ensures proportions of mud, sand and gravel lithofacies sum to 1. This sampling method allows a relatively small number of simulations to cover a wide range of the parameter space. Results from Mann-Whitney U tests show that significant differences are present between sampled proportions of the two systems regarding mud, sand and gravel lithofacies with the p value of 0.093, 0.021 and 0.005, respectively. The resulting facies associations sampled are shown in Table 4. Furthermore, the proportion of mud facies gradually increases from t_2 (Fig. 3) over 10 accretion packages until twice its standard value, whereas the associated proportion of sand facies

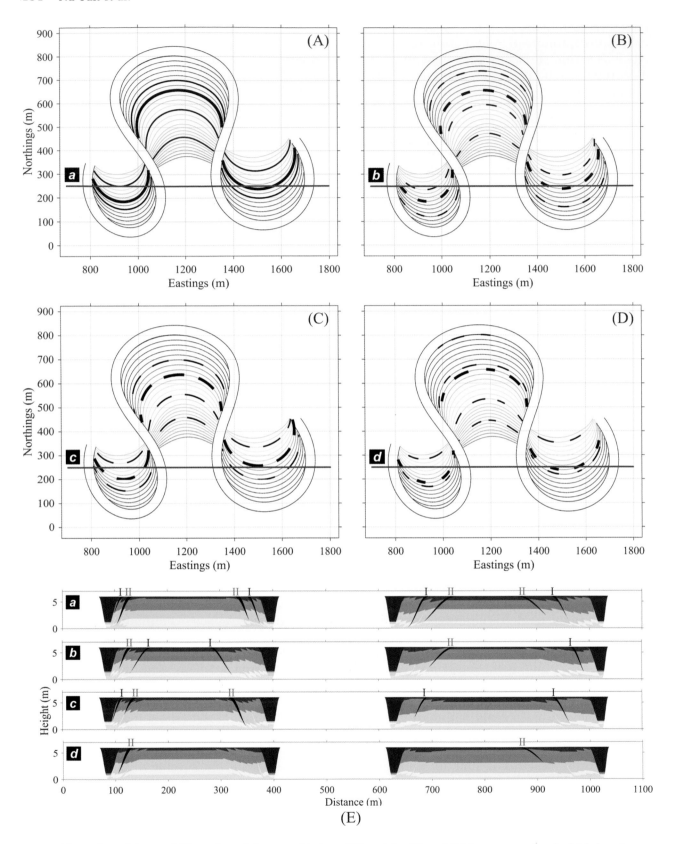

Fig. 5. Examples of the modelling of mud drapes with two different families of thicknesses, each of which has its own respective frequency and length and spacing distributions. (A) Continuous mud drapes. (B) The length and spacing of mud drapes are equal. (C) The length of mud drapes is larger than the spacing between them. (D) The length of mud drapes is smaller than the spacing between them. (E) Examples of cross sections showing two families of mud drapes with different thicknesses (mud drape 'family' I and 'family' II).

Table 2. Statistics of point bars in humid coastal plains *vs.* dryland fans from FAKTS database. $\bar{X}$ and *s* denote the sample mean and the sample standard deviation, respectively.

		$\bar{X}$	*s*	Min.	Max	Percentiles 25	Percentiles 50	Percentiles 75
Humid coastal plain	Point-bar Thickness (m)	10.27	8.86	3.00	17.00	5.60	7.00	11.80
	Mud Facies Thickness (m)	0.12	0.14	0.01	0.24	0.04	0.06	0.14
	Mud Facies Proportion (%)	12	14	0	22	3	11	17
	Sand Facies Proportion (%)	81	14	67	96	72	86	90
	Gravel Facies Proportion (%)	3	0	0	8	0	1	04
Dryland fan	Point-bar Thickness (m)	5.13	4.51	1.00	6.30	1.90	3.65	6.15
	Mud Facies Thickness (m)	0.39	0.41	0.06	0.68	0.10	0.25	0.59
	Mud Facies Proportion (%)	4	10	0	10	0	0	7
	Sand Facies Proportion (%)	95	10	88	100	91	100	100
	Gravel Facies Proportion (%)	1	0	0	0	0	0	0

decreases; the proportion of gravel facies is kept constant. This mimics the sedimentary response to a progressive decline of flow energy induced by a higher degree of energy dissipation and decreased gradient around a meander-bend (van de Lageweg *et al.*, 2014), which causes a larger proportion of finer facies to be deposited.

To mimic inherent variability in facies arrangements commonly seen in natural systems, PB-SAND allows modelling periodic changes in the proportions of facies and the inclinations of bounding surfaces between facies. In this study, proportions of each facies are set to change periodically and systematically ± 10% of the bar thickness, over the duration of a period. The period is set to follow a normal distribution curve with a minimum of 5 accretion packages and a maximum of 10 accretion packages. Moreover, the inclinations of facies bounding surfaces vary periodically across accretion packages from horizontal (0°) to gently inclined (*ca.* 0.5°) and back (0°). The period of inclination change is also set to follow a normal distribution curve with a range of 8 to 12 accretion packages. The facies variations and inter-digitation induced through this modelling approach permit generation of more realistic facies architectures, as documented from naturally occurring point-bar deposits (e.g. Miall, 1996; Labrecque *et al.*, 2011).

Given that the distribution of mud drapes is known to exert a critical control on the compartmentalisation and connectivity of hydrocarbon reservoirs of fluvial point-bar origin, two groups of parameter settings regarding mud drapes are simulated so that a wide range of possible scenarios may be considered; the parameter groups I and II are defined in Tables 5 and 6, respectively. The

mud drapes in parameter group I encompass a larger range and hence a wider variability in terms of occurring period, length and spacing of lateral extent than those of parameter group II. Five sets of parameters are randomly sampled within parameter group I or II for both types of systems (humid and dryland), resulting in 20 simulations in total. The detailed parameter settings for simulations sampled from each group are summarised in Tables 7 and 8, respectively. By testing a range of parameter settings (e.g. the thickness, length and spatial discontinuity of mud drapes and facies proportions), the modelling outcomes can be analysed to explore potential generic differences between fluvial systems developed in humid coastal-plain *vs.* dryland fluvial-fan settings and to examine uncertainty and sensitivity of system components.

Connectivity analysis

To determine the connectivity of sand-bodies within point-bar elements, the tortuosity of modelled geo-bodies and the connectivity changes along and across the modelled channel-belt direction are analysed for simulations of both humid coastal-plain and dryland-fan point-bar deposits. As these analyses require gridded facies models, vector modelling outputs from PB-SAND are rasterised by converting generated cross-section images into ASCII (American Standard Code for Information Interchange) files with facies codes that can be imported into geo-gridded software packages (e.g. Schlumberger Petrel). The resolution of gridded models is controlled by the spacing between cross sections and the pixel size of rasterised cross-section images; the resolution

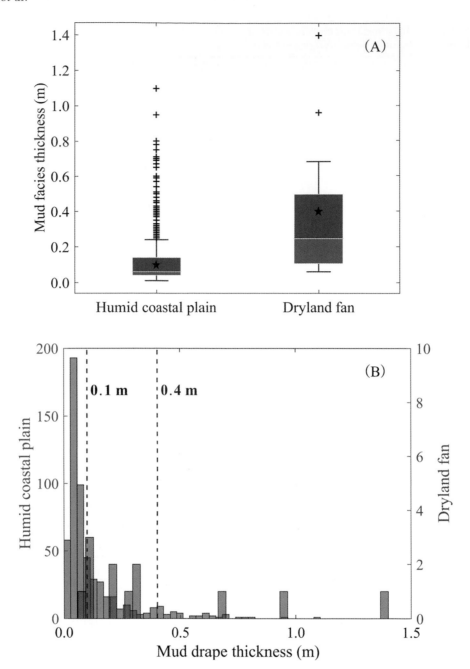

Fig. 6. Mud-facies thickness in humid coastal-plain (561 facies units) *vs.* dryland fluvial-fan (12 facies units) systems; data are extracted from FAKTS (Table 2). (A) Apparent thickness of mud facies in humid coastal-plain *vs.* dryland fluvial-fan systems. The bottom and top of a box denote the first and third quartiles; the white band inside of the box (between the blue and the red) denotes the median; the whiskers on two sides denote the minimum and maximum values excluding outliers (outside of 1.5 interquartile range); and the black star denotes the mean value. (B) Histogram of mud drape thickness in humid coastal-plain *vs.* dryland fluvial-fan systems with a mean of 0.1 m and 0.4 m, respectively, shown as dashed lines; the vertical axis denotes the number of counts.

used here is 10 m, 10 m and 0.1 m along *x* (across channel belt), *y* (along channel belt) and *z* (vertical) directions, respectively. The high resolutions enable retention of detailed expressions of sedimentary geometries and mud drapes in cross sections. The gridded facies models can then be adapted to the format required for connectivity analysis. Here, the tortuosity of sand geo-bodies and the volumetric distribution of sand in the largest connected clusters are estimated using the

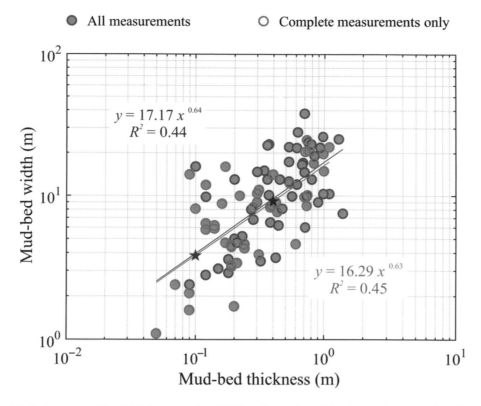

Fig. 7. Relationship between mud-bed thickness and width (i.e. the projected horizontal extent of mud drapes orthogonal to the channel centreline); data are extracted from point-bar case studies in FAKTS (five fluvial systems). Incomplete measurements include widths that have not been fully measured due to partial exposure in outcrops. There are 45 complete and 50 incomplete measurements in total. The blue and red stars denote the representative mud drapes used in the model for humid coastal-plain and dryland fan systems, respectively (Fig. 6 and Table 2).

Table 3. Properties of mud drapes in the model (Figs 6 and 7).

Mud drape	Thickness (m)	Width (m)	Frontal position (% of bar thickness)
Humid coastal plain	0.1	3.8	10
Dryland fan	0.4	9.1	30

algorithm proposed by Deutsch (1998). The tortuosity of porous media is generally defined by the ratio of the convoluted flow pathway to the straight-line distance between two points. In the approach by Deutsch (1998) a proxy measure of tortuosity of three-dimensional geo-bodies is instead obtained as the ratio of their surface areas to volumes. To investigate how mud drapes influence compartmentalisation of sand within point-bar elements, changes of cumulative sand proportions with the number of largest connected clusters are also explored in horizontal slices of the upper portions of the bars (i.e. 40%, 30% and 20% of bar thickness from the bar top).

Furthermore, the connectivity functions of sandy facies are analysed with CONNEC3D (Pardo-Igúzquiza & Dowd, 2003). The connectivity function is defined as the probability that two points separated by a distance along a certain direction are connected through a path contained in the same phase (e.g. sand lithofacies) (Allard, 1993; Allard, 1994). The connectivity functions along x (across channel belt) and y (along channel belt) are examined, respectively, within three horizontal slices of upper point-bar deposits, including 40%, 30% and 20% of bar thickness from the bar top.

RESULTS

Fig. 9 shows the distributions of lithological facies proportions within modelled point-bar elements from randomly positioned one-dimensional vertical samples (1000 samples for each modelling scenario). In consistency with analogue data (Fig. 8 and Table 2), the proportion of mud facies representing humid coastal-plain systems is higher

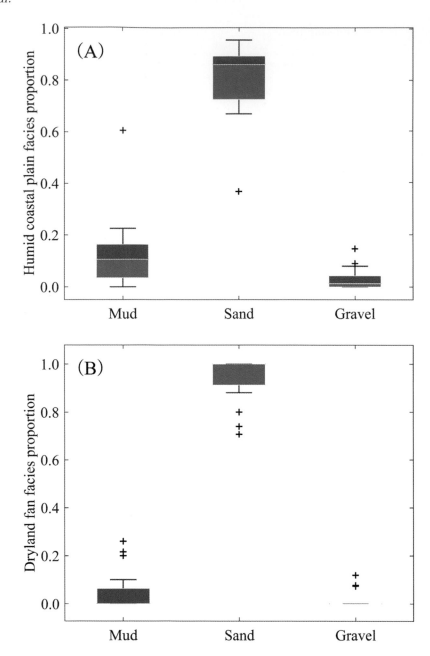

Fig. 8. Lithological facies proportions of mud, sand and gravel in (A) humid coastal-plain systems (25 point-bar elements) *vs.* (B) dryland fluvial-fan systems (26 point-bar elements) (Table 2).

Table 4. Randomly sampled facies associations. Proportions of gravel, sand and mud facies are sampled within their respective minimum and maximum limits drawn from FAKTS and in a way that ensures the overall proportions sum to 1.

	Facies association	FA_1	FA_2	FA_3	FA_4	FA_5
Humid Coastal Plain	Mud Facies Proportion (%)	4	10	22	18	22
	Sand Facies Proportion (%)	91	85	71	79	75
	Gravel Facies Proportion (%)	5	5	7	3	3
Dryland Fan	Mud Facies Proportion (%)	8	3	9	10	6
	Sand Facies Proportion (%)	92	97	91	90	94
	Gravel Facies Proportion (%)	0	0	0	0	0

Table 5. Parameter group I of mud drapes in humid coastal plain *vs.* dryland fan systems. The length and spacing of mud drapes are exemplified in Fig. 4.

Parameters	Frontal position (% of bar thickness)		Period (accretion increment #)		Length (m)		Spacing (m)	
	Mean	Range	Mean	Range	Mean	Range	Mean	Range
Humid coastal plain	10	0–10	1–5	6–10	200–2000	200–1000	1000–3000	200–1000
Dryland fan	30	0–30	3–7	8–12	1000–3000	200–1000	200–2000	200–1000

Table 6. Parameter group II of mud drapes in humid coastal plain *vs.* dryland fan systems. The length and spacing of mud drapes are exemplified in Fig. 4.

Parameters	Frontal position (% of bar thickness)		Period (accretion increment #)		Length (m)		Spacing (m)	
	Mean	Range	Mean	Range	Mean	Range	Mean	Range
Humid coastal plain	10	0–10	1–3	4–6	100–1000	100–400	200–2000	100–400
Dryland fan	30	0–30	3–5	6–8	200–2000	100–400	100–1000	100–400

Table 7. Parameter settings randomly sampled from parameter group I. The 'G' and 'S' are the abbreviation of 'group' and 'simulation', respectively.

	ID #	Facies association #	Frontal position (% of bar thickness)		Period (accretion increment #)		Length (m)		Spacing (m)	
			Min.	Max.	Min.	Max.	Min.	Max.	Min.	Max.
Humid coastal plain	G1 S1	FA_4	5	15	2	10	100	300	800	1600
	G1 S2	FA_5	9	11	4	6	1100	2100	2600	3000
	G1 S3	FA_1	7	13	1	8	0	800	1700	1900
	G1 S4	FA_3	8	12	3	9	400	800	1100	1700
	G1 S5	FA_2	6	14	5	7	700	1300	2100	3100
Dryland fan	G1 S6	FA_4	26	34	5	12	2200	2600	800	1200
	G1 S7	FA_1	15	45	6	8	1700	1900	1300	2300
	G1 S8	FA_5	18	42	7	10	2300	3300	1000	1800
	G1 S9	FA_2	22	38	4	11	1000	1800	1500	1700
	G1 S10	FA_3	23	37	3	9	1300	1900	100	700

Table 8. Parameter settings randomly sampled from parameter group II. The 'G' and 'S' are the abbreviation of 'group' and 'simulation', respectively.

	ID #	Facies association #	Frontal position (% of bar thickness)		Period (accretion increment #)		Length (m)		Spacing (m)	
			Min.	Max.	Min.	Max.	Min.	Max.	Min.	Max.
Humid coastal plain	G2 S1	FA_5	6	14	2	5	500	900	600	1000
	G2 S2	FA_2	7	13	1	3	750	1050	1300	1500
	G2 S3	FA_1	9	11	3	4	50	150	350	650
	G2 S4	FA_3	5	15	1	5	950	1050	1300	1700
	G2 S5	FA_4	8	12	2	4	300	500	950	1050
Dryland fan	G2 S6	FA_1	18	42	4	6	350	450	150	250
	G2 S7	FA_5	25	35	5	6	150	450	150	450
	G2 S8	FA_2	16	44	4	8	1550	1650	0	400
	G2 S9	FA_3	22	38	3	8	1150	1450	450	750
	G2 S10	FA_4	15	45	3	7	700	1100	500	900

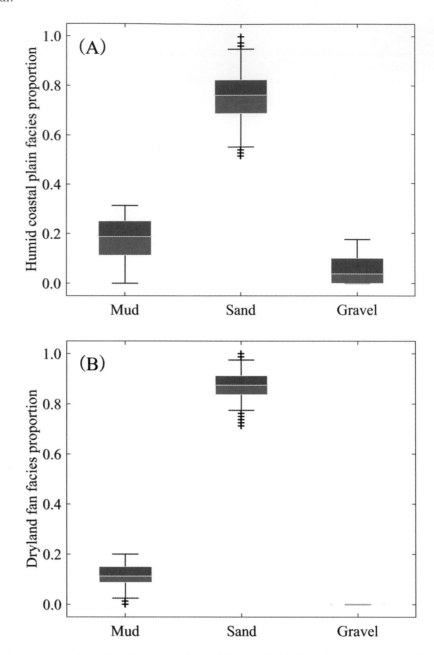

Fig. 9. Lithological facies proportions of mud, sand and gravel from PB-SAND modelling outputs in (A) humid coastal-plain systems *vs.* (B) dryland fluvial-fan systems; which are closely comparable to their counterparts in the input data (Fig. 8). 1000 vertical sections are randomly sampled for each modelling scenario, which leads to 10000 vertical samples overall for each system.

than that of dryland-fan systems; sandstone facies are dominant in both system types and gravel facies are absent from the dryland-fan systems (Fig. 9). Due to the scarcity of analogue data, the inputs are probably biased. However, PB-SAND modelling allows exploration of a wider range of possible scenarios by introducing various degrees of randomness. The plan-form realisations of the modelled point bars are shown in Fig. 10. As

expected, there are clear differences between the outputs for humid coastal plains and dryland fans in terms of the spatial arrangement of mud drapes, specifically, with respect to the frequency of occurrence of mud drapes and the extent to which mud drapes extend laterally. Compared with the parameter group I, mud drapes in parameter group II occur more frequently, have generally smaller lengths in plan-view and are spaced more closely

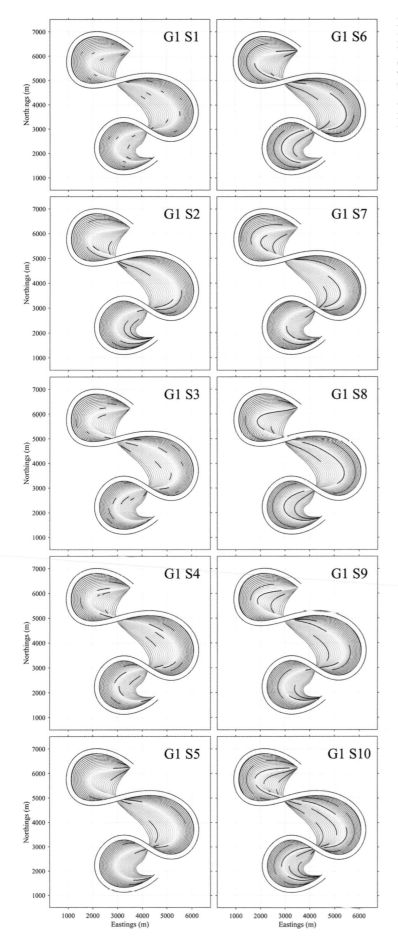

Fig. 10. Part 1 (*To be continued*) The plan-view point-bar morphology and mud-drape distributions modelled by PB-SAND. The detailed parameter settings for each simulation can be found in Tables 5 and 6 along with their respective ID number. The 'G' and 'S' are the abbreviation of 'group' and 'simulation', respectively. Mud drapes are represented as black lines distributed along accretion surfaces (coloured lines).

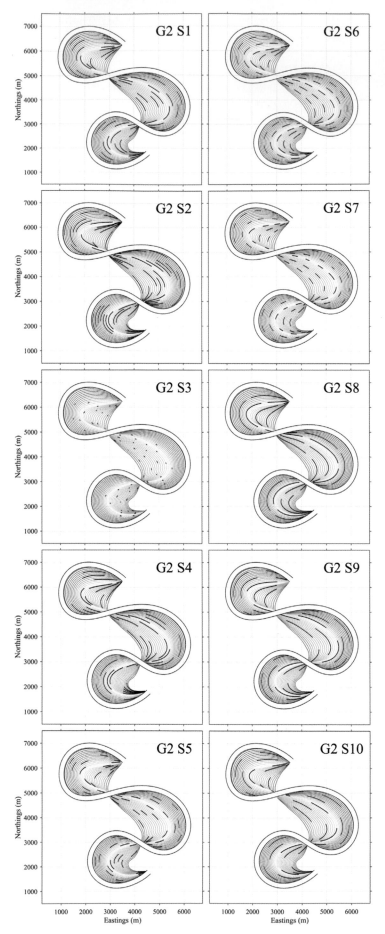

Fig. 10. Part 2 (*Continuation*) The plan-view point-bar morphology and mud-drape distributions modelled by PB-SAND. The detailed parameter settings for each simulation can be found in Tables 5 and 6 along with their respective ID number. The 'G' and 'S' are the abbreviation of 'group' and 'simulation', respectively. Mud drapes are represented as black lines distributed along accretion surfaces (coloured lines).

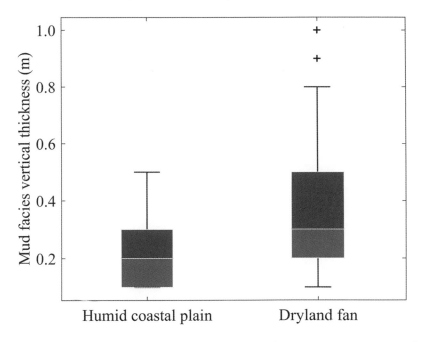

Fig. 11. The vertical thickness of mud drapes sampled from PB-SAND simulation outputs. 1000 vertical sections are randomly sampled for each 3D modelling scenario, which leads to 10000 vertical samples overall for each system. In these vertical sections, 659 and 1273 mud drapes are intersected in humid coastal-plain and dryland fan systems, respectively. The vertical thickness of mud drapes resembles closely between the modelling outputs and the input data (Fig. 6A).

along a lateral-accretion surface. Moreover, in both groups, mud drapes in humid coastal plains are present more frequently (smaller period) but are relatively less continuous (greater average spacing relative to length) than in dryland fluvial fans (greater average length relative to spacing). The vertical thickness of mud drapes sampled from PB-SAND simulation outputs for humid coastal plains is smaller, on average, than that of outputs for dryland fans (Fig. 11), in agreement with the findings from analogue data (Fig. 6 and Table 2). However, the vertical thickness of mud drapes from the modelling outputs (Fig. 11) is slightly larger, on average, than the thickness of mud drapes from the analogues (Fig. 6). This probably reflects the fact that 1D vertical sections provide apparent measures of thickness because: (i) the thickness from sampled 1D sections is oblique to the orientation of the mud drapes; and (ii) sampled sections could intersect the mud drapes at any position down the bar slope and the drapes themselves tend to thin downslope. Fig. 12 presents three-dimensional fence diagrams of representative modelled examples for humid-coastal-plain and dryland-fan systems. The planform graphs and fence diagrams for all simulations can be found in Appendix 1. Examples of cross sections modelled are shown in Fig. 13.

Fig. 14 shows the tortuosity of sand across the modelled segment of meander belt in the humid coastal-plain systems and for three disconnected point-bar elements in the dryland fan systems, computed using face connections only (i.e. cell-to-cell connections can only be established through the six faces of a cell) (Deutsch, 1998). For bars modelled using data from humid coastal-plain analogues, gravel facies preserved on the bases of channel bodies connect the adjacent point-bar elements together forming a single connected geo-body in situation where gravel facies act as permeable media (*cf.* 'string-of-beads' *sensu* Donselaar & Overeem, 2008), whereas modelled point-bar elements based on dryland fluvial-fan analogues are completely isolated by entirely muddy channel deposits (*cf.* Donselaar & Overeem, 2008). The sand compartmentalisation for both systems increases progressively in the upper parts of point-bar elements where an increased presence of mud drapes act to partition the geo-bodies (Fig. 15). The sand compartmentalisation in bars that reflects humid coastal-plain analogues generally has a greater sensitivity to the spatial distribution of mud drapes as compared with those based on dryland-fan analogues. It can be seen from Figs 16 and 17 that there is an evident anisotropy in the connectivity function across and along

(A) Humid Coastal Plain G2 S2

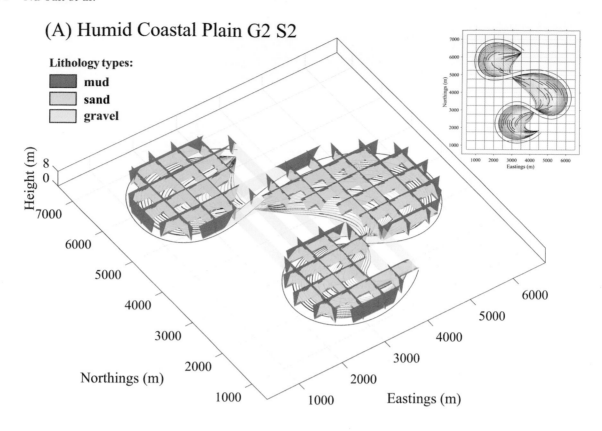

(B) Dryland Fan G2 S9

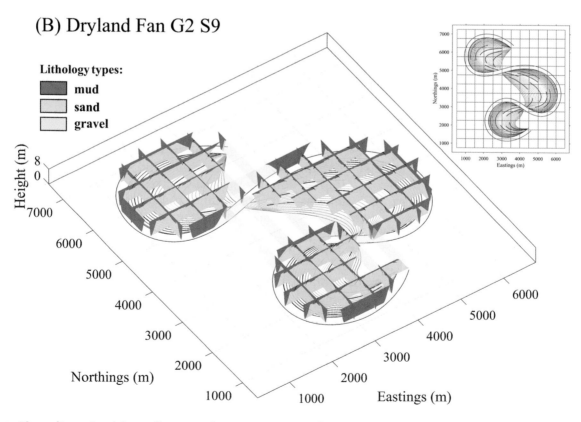

Fig. 12. Three-dimensional fence diagrams of representative simulations in (A) humid coastal-plain systems and (B) dry-land fluvial-fan systems. Complete planform graphs and 3D fence diagrams for all simulations can be found in Appendix 1.

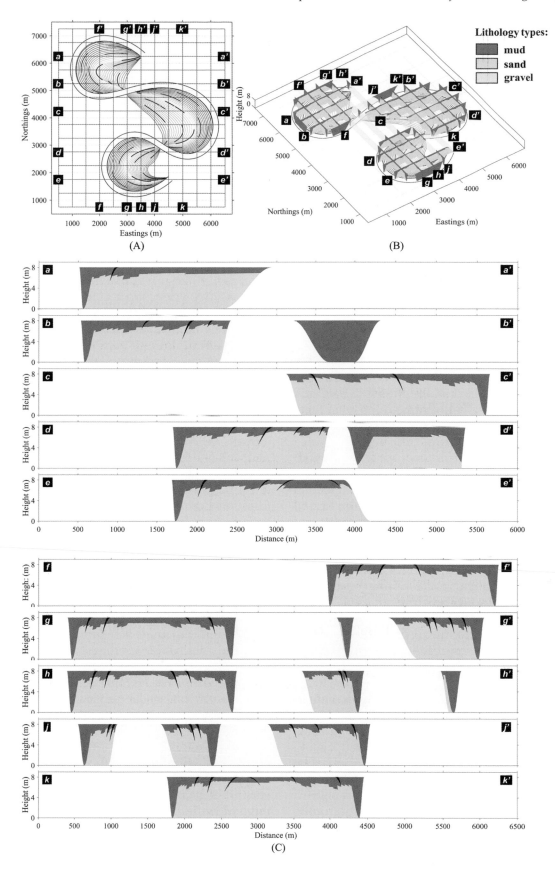

Fig. 13. Examples of cross sections of point-bar elements modelled in simulation G2 S10.

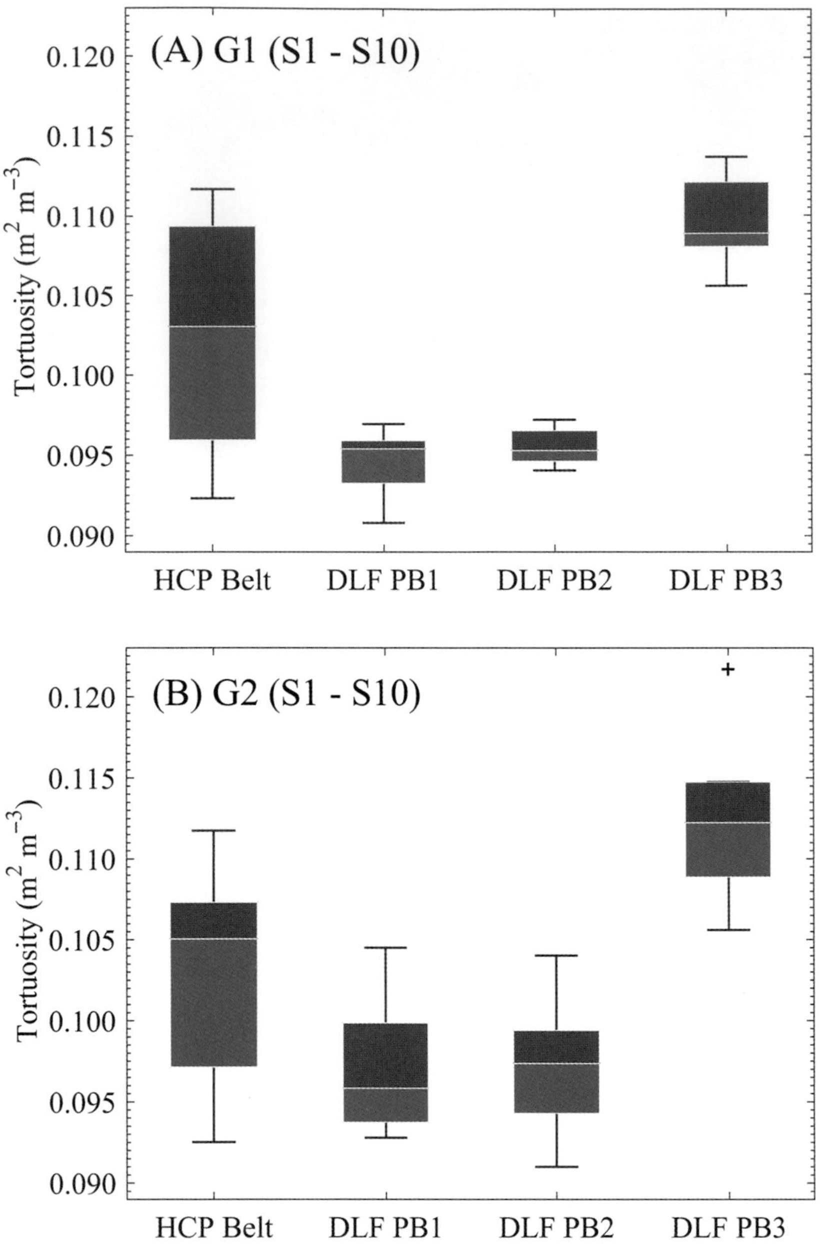

Fig. 14. Tortuosity (estimated as the ratio of surface area to volume; Deutsch, 1998) of connected sand geo-bodies. A single connected deposit is recognised in the meander belt modelled using the humid coastal-plain analogues (HCP Belt), whereas three isolated point-bar elements (DLF PB1, DLF PB2 and DLF PB3) are modelled using the dryland fluvial-fan analogues. The three point-bar elements are connected by gravels at the base of the abandoned-channel fill in the simulation examples based on humid coastal-plain analogues because it is assumed that gravels act as permeable media; the bar deposits are instead isolated by the muddy channel fills in the simulation examples based on dryland fluvial-fan analogues, where gravels are not present.

channel belt directions for both humid coastal plain and dryland fan systems. As the proportion of upper-bar deposits decreases from 40% to 20% of bar thickness, so connectivity functions in both directions vary to a greater extent between simulations with different configurations of mud drapes.

Moreover, for the direction orthogonal to the channel-belt axis, connectivity functions reduce more quickly at a smaller lag of distance but less quickly at a larger lag of distance than they do along the channel-belt axis. Furthermore, the connectivity functions in point-bar models based on

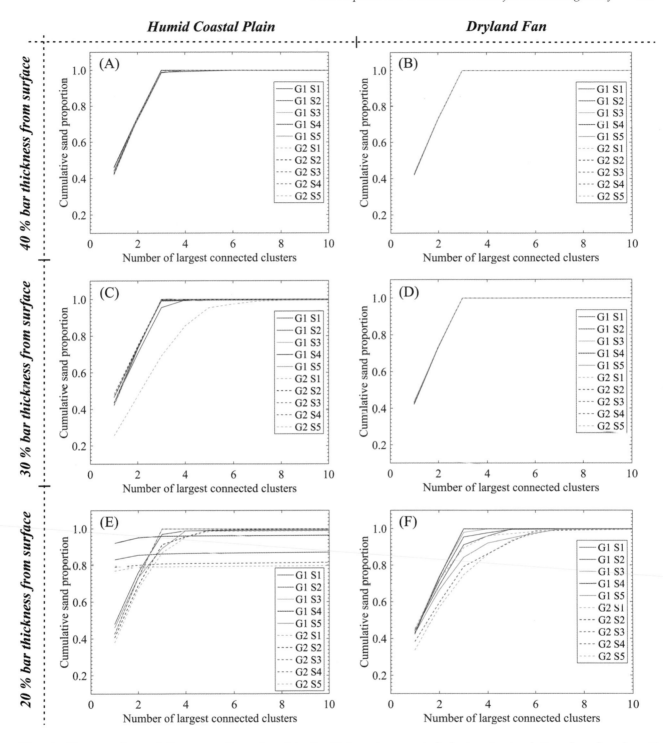

Fig. 15. Changes of cumulative proportion of sand in the 10 largest connected clusters in simulations based on humid coastal-plain and dryland fluvial-fan analogues, analysed in three-dimensional grids. The size of upper-bar deposits sampled is 6020 m (301 cells) in *x*-axis, 6500 m (650 cells) in *y*-axis and 3.6 m, 2.7 m and 1.8 m (36, 27 and 18 cells) in *z*-axis for 40%, 30% and 20% bar thickness, respectively. The connected cluster decreases in size along *x*-axis.

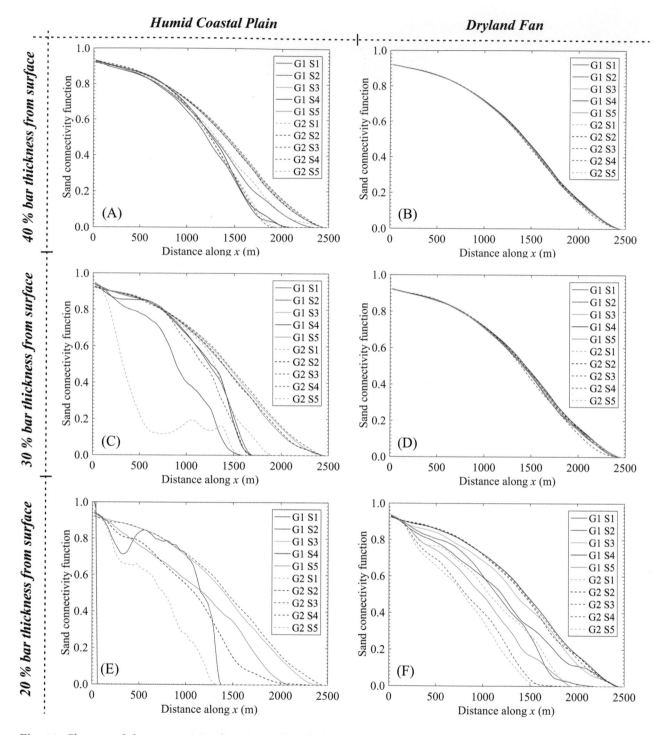

Fig. 16. Changes of the connectivity functions of sand along *x* (across channel belt) direction in simulations based on humid coastal-plain and dryland fluvial-fan analogues, analysed in three-dimensional grids. The size of upper-bar deposits sampled is 6020 m (301 cells) in *x*-axis, 6500 m (650 cells) in *y*-axis and 3.6 m, 2.7 m and 1.8 m (36, 27 and 18 cells) in *z*-axis for 40%, 30% and 20% bar thickness, respectively.

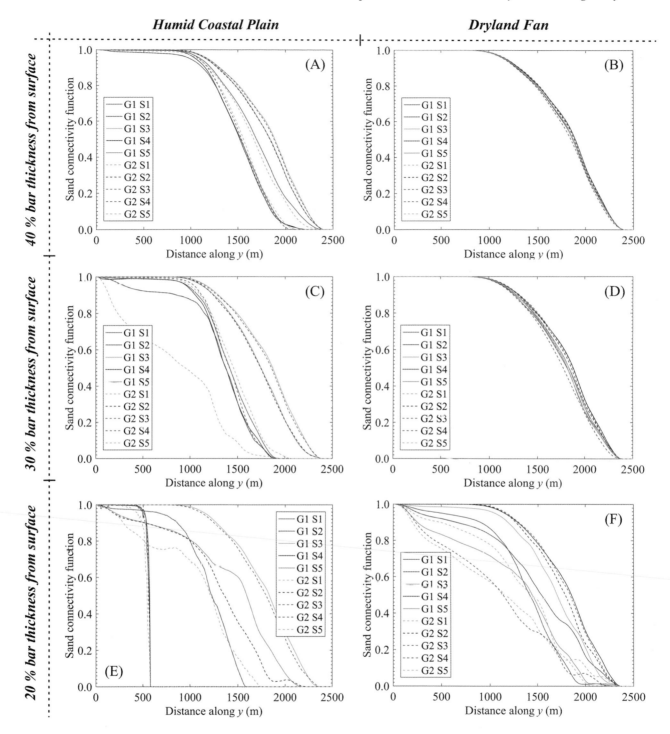

Fig. 17. Changes of the connectivity functions of sand along *y* (along channel belt) direction in simulations based on humid coastal-plain and dryland fluvial-fan analogues, analysed in three-dimensional grids. The size of upper-bar deposits sampled is 6020 m (301 cells) in *x*-axis, 6500 m (650 cells) in *y*-axis and 3.6 m, 2.7 m and 1.8 m (36, 27 and 18 cells) in *z*-axis for 40%, 30% and 20% bar thickness, respectively.

humid coastal-plain analogues, in general, exhibit a larger variability and are more sensitive to the spatial distribution of mud drapes than those of dryland fans.

DISCUSSION

This study utilises the PB-SAND model in combination with appropriate analogue datasets and empirical relationships drawn from the FAKTS database to explore the sedimentary architecture of point-bar elements in humid coastal-plain systems as compared with dryland fluvial-fan systems. Rather than modelling specific examples, this study is the first attempt to develop general insights into the compartmentalisation styles of fluvial point-bar elements in these two different depositional contexts. Parameter settings used in the numerical modelling are restricted by the limited availability of analogue data and the comparisons between the systems should not be generalised to represent general behaviours of these types of systems more widely. Nonetheless, the workflow – which entails using the PB-SAND modelling approach to characterise fluvial point-bar architecture in a particular depositional environment, constrained by real-world case examples from appropriate geological analogues – allows a quantitative examination of uncertainties of different parameters pertaining to geometries of deposits that affect reservoir connectivity at different scales. More rigorous evaluation of the system sensitivities to individual parameters can be achieved by varying systematically parameter values within their respective known or predefined ranges.

A fining-upward trend is applied in this study to model the expected facies arrangement of lateral-accretion packages of expansional point-bar deposits and muddy facies increase as a point-bar element evolves and increases in sinuosity. More complicated spatial changes in facies associations could be incorporated into future modelling exercises, for instance, downstream-fining that is associated with the development of counter-point-bar deposits (Ghinassi *et al.*, 2016; Smith *et al.*, 2009). Comparisons between point-bars formed by different styles of meander-bend transformations are also the subject of related on-going research. Although the analysis of connectivity may be undertaken in terms of different indicators of reservoir properties, mainly categorised into static-connectivity (e.g. lithological connectivity of

sandbodies) and dynamic-connectivity (e.g. flow and transport connectivity) metrics, a geological model that can accurately quantify the geometries of geo-bodies and the heterogeneity of lithofacies plays a crucial role in any assessment of this type (Knudby & Carrera, 2005; Larue & Legarre, 2004; Renard & Allard, 2013). The resulting permeability heterogeneity can cause flow fingering and may impact sweep efficiency in hydrocarbon reservoir production. Understanding the uncertainty of impermeable lithofacies distribution is hence critical to assess compartmentalisation and the appraisal of hydrocarbon reservoirs within point-bar deposits of meandering river systems. This study demonstrates that reservoir compartmentalisation can be assessed using the PB-SAND modelling approach with multiple realisations by varying a range of parameter settings based on deterministic components with varying degree of stochastic controls. Informed by available data, deterministic decisions could be made before exploring facies distribution and sedimentary architecture that probably control reservoir connectivity, for example, point-bar transformation styles and their planform evolutions that may be interpreted from high-resolution seismic imagery for subsurface ancient fluvial deposits or from satellite imagery for modern river systems. To mimic the natural variability in the deposits, PB-SAND has the capability to allow specification of stochastic components (e.g. a range of fluctuation) on the deterministic elements, for example, the random proportions of different facies and variations in the inclination of bounding surfaces. For parameters that cannot be fully determined, multi-dimensional parameter space that includes combinations of wide-ranging parameter values derived from geological analogues or estimated from empirical relationships can be used to encompass possible scenarios and the associated uncertainties. PB-SAND is able to honour available datasets through deterministic components whilst maintaining sufficient flexibility in controlling desirable ranges of uncertainties through stochastic components. The resulting modelling outcome can further guide the collation and assimilation of field data, which in turn can improve modelling algorithms.

Informed by real-world data (e.g. analogues, seismic and core data) and with the advantage of unrestricted resolution, PB-SAND can be employed in the construction of geo-cellular reservoir models that can realistically and accurately represent flow barriers or baffles, especially inclined mud-draped surfaces, as well as high-permeability flow paths. In particular, due to the lack of data and/or restrictions on resolution of geo-cellular models, conventional reservoir modelling workflows commonly oversimplify or overlook the geometries of mud drapes and their spatial heterogeneities, which may significantly impact the compartmentalisation of otherwise relatively permeable geo-bodies, the net-reservoir tortuosity and the measures of static and dynamic connectivity (Alpak & van der Vlugt, 2014; Hovadik & Larue, 2010; Larue & Legarre, 2004). This is schematically illustrated in Fig. 18, albeit for a simple 2D scenario. An increased presence of mud drapes in the upper parts of point-bar elements partitions geo-bodies significantly. The workflow used in this study can be integrated with conventional reservoir-modelling practice to assess sedimentological scenarios and associated uncertainties and to investigate interactions and relative sensitivities of system parameters, which can further assist in decision making and production forecasts. The unrestricted resolution of the vector-based modelling approach of PB-SAND permits flexibility in the way modelling outputs are gridded (grid orientation and resolution;

unstructured grids). The fine-scale details of modelling outputs can be preserved through multiphase flow-based upscaling and improve clastic reservoir models (Alpak & Barton, 2014; Nordahl *et al.*, 2014; Ringrose *et al.*, 2008). Furthermore, the 3D modelling outputs generated from PB-SAND can be used to develop a library of training images for Multiple-Point Statistics simulations (Guardiano & Srivastava, 1993; Strebelle, 2002; Strebelle & Journel, 2001), for application to different depositional environments (e.g. fluvial and estuarine).

CONCLUSIONS

Based on analogue data from the Fluvial Architecture Knowledge Transfer System (FAKTS), this study utilises the Point-Bar Sedimentary Architecture Numerical Deduction (PB-SAND) to model the sedimentary architecture and facies heterogeneity of point-bar elements associated with two classes of depositional environments: humid coastal plains and dryland fluvial fans. Mud drapes in both system types extend from the bar top, down the bar front, to pinch-out in the lower part of the bar bodies.

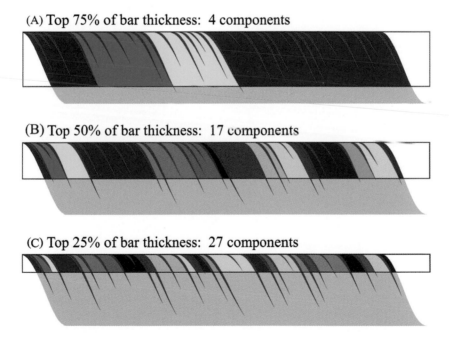

Fig. 18. Idealised 2D cross-section of a point-bar element that illustrates the effect of sample size (shown as black frames) on the number of connected components due to compartmentalisation by mud drapes, in situation where mud drapes are continuous along lateral-accretion surfaces in plan-view. Laterally discontinuous mud drapes enhance connectivity within point-bar elements (Figs 16 and 17) and reduce the number of connected clusters (Fig. 15). The different components (compartments) are denoted by various colours.

Taking progressively thinner horizontal slices of uppermost point-bar deposits (from 40% to 20% of bar thickness from the bar top), mud drapes tend to occur more frequently thereby increasing compartmentalisation with proximity to the bar top. The connectivity functions both across and along channel-belt directions, meanwhile, vary more greatly as a progressively thinner section of the uppermost parts of point-bar geo-bodies is considered. Proxies for sand compartmentalisation in point bars modelled using analogue data from humid coastal-plain systems are generally more variable and are more sensitive to the spatial distribution of mud drapes, compared to the point bars modelled using analogue data from dryland fluvial-fan systems. Given that the results shown in this study are restricted by limited analogue data available for the chosen depositional-system types, the outcome should be regarded as a preliminary attempt at a comparison. However, this study demonstrates a workflow that can use data and relationships from appropriate analogues to constrain geological modelling efforts and to improve resulting three-dimensional facies models by generating realistic architectural geometries and sophisticated spatial distributions of facies associations. The grid-free facies models generated by PB-SAND can further be integrated into conventional reservoir-modelling practices, particularly, to aid in the assessment of the compartmentalisation and connectivity of permeable geo-bodies. This modelling approach has implications in the analysis of hydrocarbon reservoir performance, notably, in the prediction of the fraction of residual oil that is trapped in attic compartments and whether water-injection techniques may enhance production.

ACKNOWLEDGEMENTS

The authors thank Nexen Energy ULC, Canada for provision of financial support for development of PB-SAND and FRG-ERG sponsors and partners AkerBP, Areva, BHPBilliton, Cairn India (Vedanta), ConocoPhillips, Murphy Oil, Nexen Energy, Petrotechnical Data Systems, Saudi Aramco, Shell, Tullow Oil, Woodside and YPF for financial support of the development of the FAKTS database. Luca Colombera has been supported by NERC (Catalyst Fund award NE/M007324/1; Follow-on Fund NE/N017218/1). Paul Durkin, one anonymous reviewer and the editors are also thanked for their valuable comments and suggestions, which have led to an improved manuscript with greater clarity.

APPENDIX 1

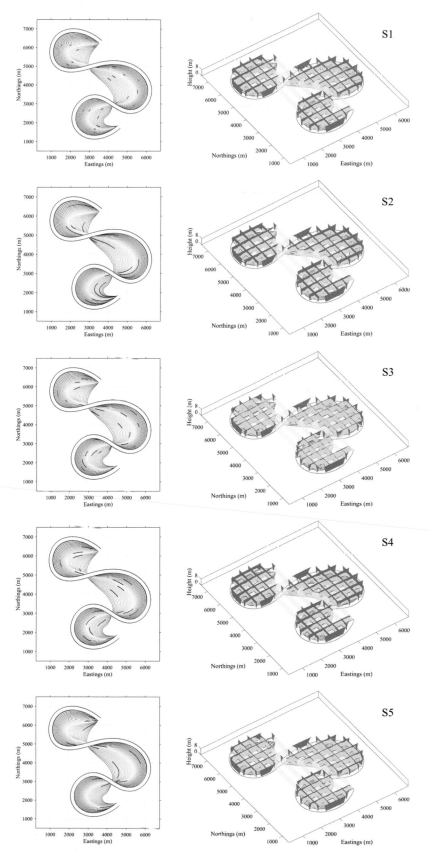

Fig. A.1. Simulation examples of point bars in humid coastal plain systems, for parameter group I.

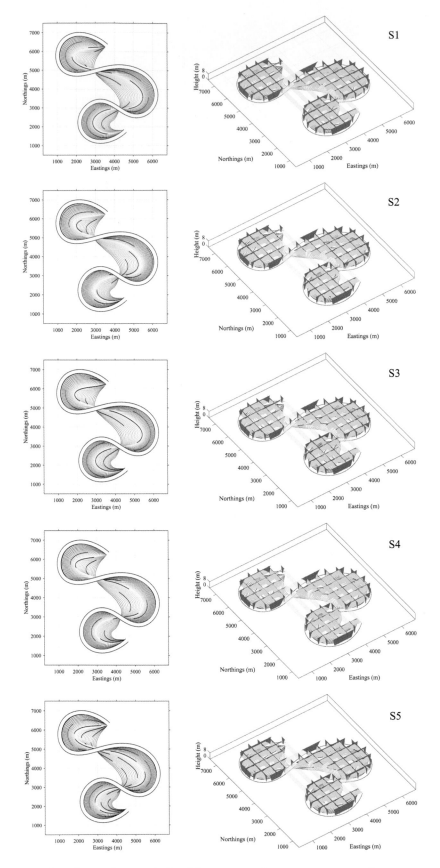

Fig. A.2. Simulation examples of point bars in dryland fan systems, for parameter group I.

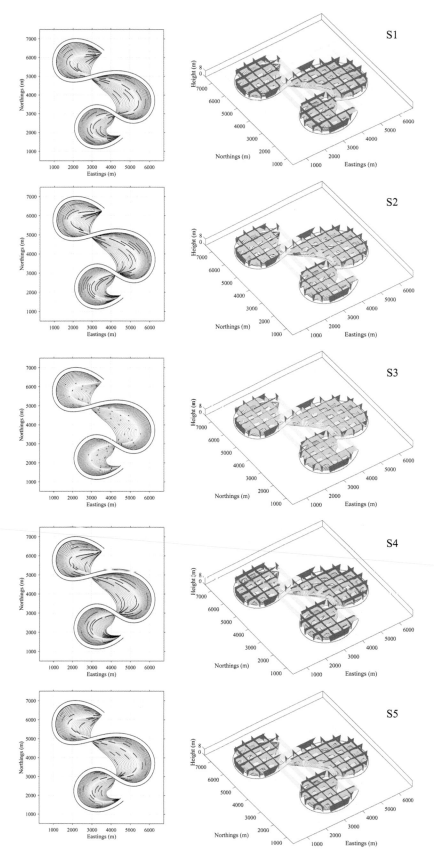

Fig. A.3. Simulation examples of point bars in humid coastal plain systems, for parameter group II.

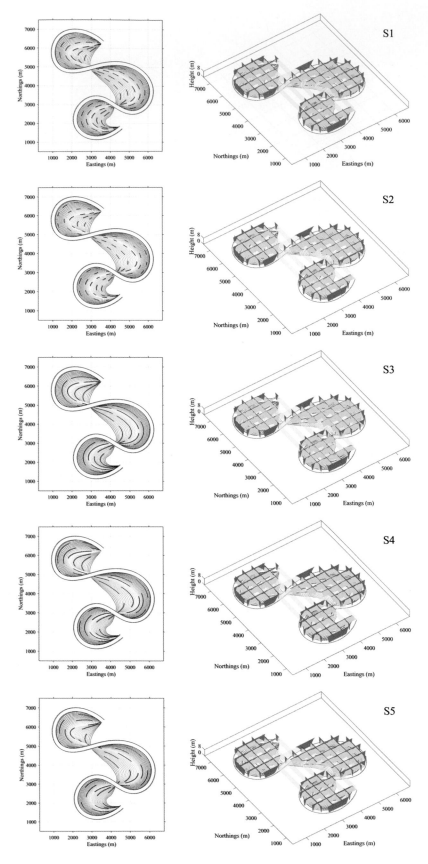

Fig. A.4. Simulation examples of point bars in dryland fan systems, for parameter group II.

REFERENCES

Allard, D. (1993) On the connectivity of two random set models: the truncated Gaussian and the Boolean. In: *Geostatistics Tróia '92: Volume 1* (Ed. A. Soares), 467–478. Springer Netherlands, Dordrecht.

Allard, D. (1994) Simulating a geological lithofacies with respect to connectivity information using the truncated Gaussian model. In: *Geostatistical Simulations: Proceedings of the Geostatistical Simulation Workshop, Fontainebleau, France, 27–28 May 1993* (Eds M. Armstrong and P.A. Dowd), 197–211. Springer Netherlands, Dordrecht.

Alpak, F.O. and **Barton, M.D.** (2014) Dynamic impact and flow-based upscaling of the estuarine point-bar stratigraphic architecture. *J. Petrol. Sci. Eng.*, **120**, 18–38.

Alpak, F.O. and **van der Vlugt, F.F.** (2014) Shale-drape modeling for the geologically consistent simulation of clastic reservoirs. *J. Soc. Petrol. Eng.*, **19**, https://doi.org/10.2118/169820-PA

Anderson, D.S. (2005) Architecture of crevasse splay and point-bar bodies of the nonmarine Iles Formation north of Rangely, Colorado: implications for reservoir description. *The Mountain Geologist*, **42**, 85–107.

Brice, J.C. (1974) Evolution of meander loops. *Geol. Soc. Am. Bull.*, **85**, 581–586.

Bridge, J.S. (1982) A revised mathematical model and FORTRAN IV program to predict flow, bed topography and grain size in open-channel bends. *Comput. Geosci.*, **8**, 91–95.

Bridge, J.S. (2003) *Rivers and Floodplains: Forms, Processes and Sedimentary Record.* Blackwell Publishing, Malden.

Brownlie, W.R. (1983) Flow depth in sand-bed channels. *J. Hydraul. Eng.-Asce*, **109**, 959–990.

Carlston, C.W. (1965) The relationship of free meander geometry to stream discharge and its geomorphic implications. *Am. J. Sci.*, **263**, 864–885.

Catuneanu, O. and **Elango, H.N.** (2001) Tectonic control on fluvial styles: the Balfour Formation of the Karoo Basin, South Africa. *Sed. Geol.*, **140**, 291–313.

Colombera, L., Felletti, F., Mountney, N.P. and **McCaffrey, W.D.** (2012a) A database approach for constraining stochastic simulations of the sedimentary heterogeneity of fluvial reservoirs. *AAPG Bull.*, **96**, 2143–2166.

Colombera, L., Mountney, N.P. and **McCaffrey, W.D.** (2013) A quantitative approach to fluvial facies models: Methods and example results. *Sedimentology*, **60**, 1526–1558.

Colombera, L., Mountney, N.P. and **McCaffrey, W.D.** (2012b) A relational database for the digitization of fluvial architecture: concepts and example applications. *Petrol. Geosci.*, **18**, 129–140.

Colombera, L., Mountney, N.P., Russell, C.E., Shiers, M.N. and **McCaffrey, W.D.** (2017) Geometry and compartmentalization of fluvial meander-belt reservoirs at the barform scale: Quantitative insight from outcrop, modern and subsurface analogues. *Mar. Petrol. Geol.*, **82**, 35–55.

Corbeanu, R.M., Wizevich, M.C., Bhattacharya, J.P., Zeng, X. and **McMechan, G.A.** (2004) Three-dimensional architecture of ancient lower delta-plain point bars using ground-penetrating radar, Cretaceous Ferron Sandstone, Utah. *Regional to Wellbore Analog for Fluvial–Deltaic Reservoir Modeling, The Ferron of Utah* (Eds T.C. Chidsey Jr., R.D. Adams and T.H. Morris), *AAPG Studies in Geology*, **50**, 427–449.

Deutsch, C.V. (1998) Fortran programs for calculating connectivity of three-dimensional numerical models and for ranking multiple realizations. *Comput. Geosci.*, **24**, 69–76.

Donselaar, M.E. and **Overeem, I.** (2008) Connectivity of fluvial point-bar deposits: An example from the Miocene Huesca fluvial fan, Ebro Basin, Spain. *AAPG Bull.*, **92**, 1109–1129.

Durkin, P.R., Hubbard, S.M., Boyd, R.L. and **Leckie, D.A.** (2015) Stratigraphic expression of intra-point-bar erosion and rotation. *J. Sed. Res.*, **85**, 1238–1257.

Fielding, C.R. and **Crane, R.C.** (1987) An application of statistical modelling to the prediction of hydrocarbon recovery factors in fluvial reservoir sequences. In: *Recent Developments in Fluvial Sedimentology* (Eds R.M. Flores and M.D. Harvey), *SEPM Society for Sedimentary Geology*, **39**, 321–327.

Ghinassi, M. and **Ielpi, A.** (2015) Stratal architecture and morphodynamics of downstream-migrating fluvial point bars (Jurassic Scalby Formation, U.K.). *J. Sed. Res.*, **85**, 1123–1137.

Ghinassi, M., Ielpi, A., Aldinucci, M. and **Fustic, M.** (2016) Downstream-migrating fluvial point bars in the rock record. *Sed. Geol.*, **334**, 66–96.

Ghinassi, M., Nemec, W., Aldinucci, M., Nehyba, S., Özaksoy, V. and **Fidolini, F.** (2014) Plan-form evolution of ancient meandering rivers reconstructed from longitudinal outcrop sections. *Sedimentology*, **61**, 952–977.

Guardiano, F.B. and **Srivastava, R.M.** (1993) Multivariate geostatistics: beyond bivariate moments. In: *Geostatistics Tróia '92: Volume 1* (Ed. A. Soares), 133–144. Springer Netherlands, Dordrecht.

Hartley, A.J., Owen, A., Swan, A., Weissmann, G.S., Holzweber, B.I., Howell, J., Nichols, G. and **Scuderi, L.** (2015) Recognition and importance of amalgamated sandy meander belts in the continental rock record. *Geology*, **43**, 679–682.

Hassanpour, M.M., Pyrcz, M.J. and **Deutsch, C.V.** (2013) Improved geostatistical models of inclined heterolithic strata for McMurray Formation, Alberta, Canada. *AAPG Bull.*, **97**, 1209–1224.

Hovadik, J.M. and **Larue, D.K.** (2010) Stratigraphic and structural connectivity. In: Reservoir Compartmentalization (Eds S.J. Jolley, Q.J. Fisher, R.B. Ainsworth, P.J. Vrolijk and S. Delisle). *Geol. Soc. London Spec. Publ.*, **347**, 219–242.

Hubbard, S.M., Smith, D.G., Nielsen, H., Leckie, D.A., Fustic, M., Spencer, R.J. and **Bloom, L.** (2011) Seismic geomorphology and sedimentology of a tidally influenced river deposit, Lower Cretaceous Athabasca oil sands, Alberta, Canada. *AAPG Bull.*, **95**, 1123–1145.

Jablonski, B.V. (2012) *Process sedimentology and three-dimensional facies architecture of a fluvially dominated, tidally influenced point bar: Middle McMurray Formation, Lower Steepbank River area, northeastern Alberta, Canada.* Queen's University (Canada).

Jackson, R.G. (1976) Depositional model of point bars in the lower Wabash River. *J. Sed. Res.*, **46**, 579–594.

Jolley, S.J., Fisher, Q.J. and **Ainsworth, R.B.** (2010) Reservoir compartmentalization: an introduction. In: *Reservoir Compartmentalization* (Eds S.J. Jolley, Q.J. Fisher, R.B. Ainsworth, P.J. Vrolijk and S. Delisle), *Geol. Soc. London Spec. Publ.*, **347**, 1–8.

Knudby, C. and **Carrera, J.** (2005) On the relationship between indicators of geostatistical, flow and transport connectivity. *Adv. Water Res.*, **28**, 405–421.

Labrecque, P.A., Hubbard, S.M., Jensen, J.L. and Nielsen, H. (2011) Sedimentology and stratigraphic architecture of a point bar deposit, Lower Cretaceous McMurray Formation, Alberta, Canada. *Bull. Can. Petrol. Geol.*, **59**, 147–171.

Larue, D.K. and Legarre, H. (2004) Flow units, connectivity and reservoir characterization in a wave-dominated deltaic reservoir: Meren reservoir, Nigeria. *AAPG Bull.*, **88**, 303–324.

Leeder, M.R. (1973) Fluviatile fining-upwards cycles and the magnitude of palaeochannels. *Geol. Mag.*, **110**, 265–276.

Leopold, L.B. and Langbein, W.B. (1966) River meanders. In: *Sci. Am.*, June issue, 60–70.

Leopold, L.B. and Wolman, M.G. (1960) River meanders. *Geol. Soc. Am. Bull.*, **71**, 789–794.

Lorenz, J.C., Heinze, D.M., Clark, J.A. and Searls, C.A. (1985) Determination of widths of meander-belt sandstone reservoirs from vertical downhole data, Mesaverde Group, Piceance Creek Basin, Colorado. *AAPG Bull.*, **69**, 710–721.

Makaske, B. and Weerts, H.J.T. (2005) Muddy lateral accretion and low stream power in a sub-recent confined channel belt, Rhine-Meuse delta, central Netherlands. *Sedimentology*, **52**, 651–668.

Maynard, K. and Murray, I. (2003) One million years from the Upper Arang Formation, West Natuna Basin, implications for reservoir distribution and facies variation in fluvial deltaic deposits. In: *Proceedings of the Annual Convention – Indonesian Petroleum Association*, **29**, 267–276.

Miall, A.D. (1996) *The Geology of Fluvial Deposits.* Springer-Verlag, New York.

Miall, A.D. and Turner-Peterson, C.E. (1989) Variations in fluvial style in the Westwater Canyon Member, Morrison Formation (Jurassic), San Juan Basin, Colorado Plateau. *Sed. Geol.*, **63**, 21–60.

Morley, R.J. (2000) *Origin and evolution of tropical rain forests.* John Wiley & Sons.

Nanson, G.C. and Croke, J.C. (1992) A genetic classification of floodplains. *Geomorphology*, **4**, 459–486.

Nordahl, K., Messina, C., Berland, H., Rustad, A.B. and Rimstad, E. (2014) Impact of multiscale modelling on predicted porosity and permeability distributions in the fluvial deposits of the Upper Lunde Member (Snorre Field, Norwegian Continental Shelf). In: *Sediment-Body Geometry and Heterogeneity: Analogue Studies for Modelling the Subsurface* (Eds A.W. Martinius, J.A. Howell and T.R. Good), *Geol. Soc. London Spec. Publ.*, **387**, 85–109.

Olsen, H. (1987) Ancient ephemeral stream deposits: a local terminal fan model from the Bunter Sandstone Formation (L. Triassic) in the Tønder-3,-4 and -5 wells, Denmark. In: Desert Sediments: Ancient and Modern (Eds L.E Frostick and I. Reid), *Geol. Soc. London Spec. Publ.*, **35**, 69–86.

Pardo-Igúzquiza, E. and Dowd, P.A. (2003) CONNEC3D: a computer program for connectivity analysis of 3D random set models. *Comput. Geosci.*, **29**, 775–785.

Pranter, M.J., Ellison, A.I., Cole, R.D. and Patterson, P.E. (2007) Analysis and modeling of intermediate-scale reservoir heterogeneity based on a fluvial point-bar outcrop analog, Williams Fork Formation, Piceance Basin, Colorado. *AAPG Bull.*, **91**, 1025–1051.

Pyrcz, M.J., Sech, R.P., Covault, J.A., Willis, B.J., Sylvester, Z. and Sun, T. (2015) Stratigraphic rule-based reservoir modeling. *Bull. Can. Petrol. Geol.*, **63**, 287–303.

Reijenstein, H.M., Posamentier, H.W. and Bhattacharya, J.P. (2011) Seismic geomorphology and high-resolution seismic stratigraphy of inner-shelf fluvial, estuarine, deltaic and marine sequences, Gulf of Thailand. *AAPG Bull.*, **95**, 1959–1990.

Renard, P. and Allard, D. (2013) Connectivity metrics for subsurface flow and transport. *Adv. Water Res.*, **51**, 168–196.

Ringrose, P.S., Martinius, A.W. and Alvestad, J. (2008) Multiscale geological reservoir modelling in practice. In: *The Future of Geological Modelling in Hydrocarbon Development* (Eds A. Robinson, P. Griffiths, S. Price, J. Hegre and A. Muggeridge), *Geol. Soc. London Spec. Publ.*, **309**, 123–134.

Rongier, G., Collon, P. and Renard, P. (2017) A geostatistical approach to the simulation of stacked channels. *Mar. Petrol. Geol.*, **82**, 318–335.

Smith, D.G., Hubbard, S.M., Leckie, D.A. and Fustic, M. (2009) Counter point bar deposits: lithofacies and reservoir significance in the meandering modern Peace River and ancient McMurray Formation, Alberta, Canada. *Sedimentology*, **56**, 1655–1669.

Smith, R.M.H. (1987) Morphology and depositional history of exhumed Permian point bars in the southwestern Karoo, South Africa. *J. Sed. Res.*, **57**, 19–29.

Strebelle, S. (2002) Conditional simulation of complex geological structures using multiple-point statistics. *Math. Geol.*, **34**, 1–21.

Strebelle, S.B. and Journel, A.G. (2001) Reservoir modeling using multiple-point statistics. In: Soc. Petrol. Eng. Annual Technical Conference and Exhibition, 30 September–3 October, New Orleans, Louisiana, https://doi.org/10.2118/71324-MS

Sun, T., Meakin, P. and Jøssang, T. (2001a) A computer model for meandering rivers with multiple bed load sediment sizes: 1. Theory. *Water Resour. Res.*, **37**, 2227–2241.

Sun, T., Meakin, P. and Jøssang, T. (2001b) A computer model for meandering rivers with multiple bed load sediment sizes: 2. Computer simulations. *Water Resour. Res.*, **37**, 2243–2258.

Sun, T., Meakin, P., Jøssang, T. and Schwarz, K. (1996) A simulation model for meandering rivers. *Water Resour. Res.*, **32**, 2937–2954.

Trendell, A.M., Atchley, S.C. and Nordt, L.C. (2013) Facies analysis of a probable large-fluvial-fan depositional system: the Upper Triassic Chinle Formation at Petrified Forest National Park, Arizona, USA. *J. Sed. Res.*, **83**, 873–895.

Tunbridge, I.P. (1984) Facies model for a sandy ephemeral stream and clay playa complex; the Middle Devonian Trentishoe Formation of North Devon, UK. *Sedimentology*, **31**, 697–715.

van de Lageweg, W.I., van Dijk, W.M., Baar, A.W., Rutten, J. and Kleinhans, M.G. (2014) Bank pull or bar push: What drives scroll-bar formation in meandering rivers? *Geology*, **42**, 319–322.

Willis, B.J. and Tang, H. (2010) Three-dimensional connectivity of point-bar deposits. *J. Sed. Res.*, **80**, 440–454.

Wu, C., Ullah, M.S., Lu, J. and Bhattacharya, J.P. (2016) Formation of point bars through rising and falling flood stages: Evidence from bar morphology, sediment transport and bed shear stress. *Sedimentology*, **63**, 1458–1473.

Yan, N., Mountney, N.P., Colombera, L. and Dorrell, R.M. (2017) A 3D forward stratigraphic model of fluvial meander-bend evolution for prediction of point-bar lithofacies architecture. *Comput. Geosci.*, **105**, 65–80.

Int. Assoc. Sedimentol. Spec. Publ (2018) **48**, 509–542.

Emergent facies patterns within fluvial channel belts

BRIAN J. WILLIS* and RICHARD P. SECH[†]

*Clastic Stratigraphy Team, Earth Science Department, Chevron Energy Technology Company, Houston, TX, USA

[†]*Reservoir Management Research Team, Chevron Energy Technology Company, Houston, TX, US*

ABSTRACT

The geometry of fluvial channel belts and their internal facies are predicted by considering patterns of erosion and deposition during river channel migration and bend cutoff. Along-belt heterogeneities are defined by the geometry and internal facies of one or more rows of storeys, each divided into five lithic sub-units that formed adjacent to specific areas along a river channel: 1) Bar head, 2) Bar tail, 3) Concave bank, 4) Thalweg scour pool lags; and 5) Abandonment fill. These substoreys are characterised by contrasting vertical facies trends; and their relative abundance within a channel belt depends on how the river changed position over time. Channel bend expansion leads to preservation of more inner bank substoreys (bar head and bar tail), whereas downstream translation favours preservation of deposits in bar-tail and concave-bank areas of the channel. Channel belts change thickness laterally, reflecting shallower maximum erosion depths when channel segments are straighter and locally deeper erosion as they become more sinuous. Bolt mean thickness and lateral facies variations reflect the gradual erosional sweep of channels as they change shape and deposit storeys. The distribution of belt sand thickness values is only partially related to river channel depth at any one time. Similarly, the average character of vertical grain-size trends (bell-shaped *versus* blocky-shaped well log trends) does not reflect a static measure of a river but rather records the relative preservation of deposits from different depositional sub-environments along migrating river segments over time. The resulting facies patterns preserved within fluvial channel belts are generally not channel shaped but rather reflect emergent bodies formed in a 'Walther's Law' sense along the lateral sweep path of specific areas within the channel over time (before channel switching during bend cutoff). To a significant degree, the geometry of these facies patterns are preserved even when belts are composed of multiple rows of storeys formed by a long period of channel migration and cutoff. Facies models for fluvial channel belts need to focus on this emergent organisation of lithic bodies formed as river channels evolve in shape and position, rather than contrasting static views of river pattern.

Keywords: Fluvial, heterogeneity, connectivity, reservoir, design simulation

INTRODUCTION

Fluvial channel belt reservoirs can have complicated production behaviours; and a rich variety of internal heterogeneities have been identified by a long history of outcrop and modern analogue studies (Bridge 2003; Miall 1996, 2013). Despite a general appreciation of these complexities for reservoir heterogeneity prediction (Jordan & Pryor, 1992; Jones *et al.*, 1995; Keogh *et al.*, 2007; Pranter *et al.*, 2007; Willis & Tang, 2010; and many others) typical reservoir models used in field assessments focus on channel belt connectivity rather than details of internal facies patterns. To a significant extent this shortcoming reflects the complex 3D geometry of facies variations within these

Fluvial Meanders and Their Sedimentary Products in the Rock Record, First Edition.
Edited by Massimiliano Ghinassi, Luca Colombera, Nigel P. Mountney and Arnold Jan H. Reesink.
© 2019 International Association of Sedimentologists. Published 2019 by John Wiley & Sons Ltd.

reservoirs, which are difficult to extrapolate from the 2D cross sections exposed in outcrop, or modern river plan view maps. Without accurate 3D models of a range of channel belt reservoir types it is difficult to define variations in subsurface flow patterns caused by specific types of heterogeneities or to assess which types of facies variability have the greatest influence on reservoir performance.

The influence of fluvial reservoir heterogeneities on reservoir performance was previously examined by Jones *et al.* (1995) by constructing fluvial channel belt models with a hierarchy of internal variations. Different scales of geologic features (simulation design factors) were systematically varied between high and low values to construct reservoir models with a range of internal heterogeneity patterns. Small-scale variations were defined by upscaling the influence of grain-size variations between different laminae within cross stratified sandstones (their 'level 1'). Medium-scale variations were defined by changing relative abundance and vertical proportion of lithofacies within channel belts (their 'level 2'). Larger-scale variations were defined by the number of channel belts and their connectivity (their 'levels 3 and 4'). The resulting reservoir models were used in a design simulation study of reservoir performance under water-flood displacement to define the relative impact of varying these different scale features.

Jones *et al.* (1995) concluded that changing large-scale connectivity patterns had the greatest impact on production forecasts. Small-scale grain-size variations between laminae within sedimentary structures also had an important impact on production by changing fluid−fluid−rock interactions in multi-phase flows. They suggested that intermediate-scale factors that "govern the spatial arrangement of lithofacies within sandstone bodies do not have as strong an impact on performance". This conclusion is commonly cited in reservoir assessments as justification to focus reservoir simulation scenarios for production forecasts on large-scale channel belt geometry and connectivity. Internal facies patterns are typically simplified to geostatistically defined regionalised pixelated patterns. The intermediate-scale geologic features defined in the Jones *et al.* (1995) models were relatively coarse and diffuse compared with the geologic details they defined at both the finer (laminae) and coarser (channel-belt) scales. A key motivation for this current study has

been to assess if these intermediate-scale facies variations have a more important influence on reservoir performance when these scales of geologic variation are defined more explicitly.

For this study a relatively simple process-based river channel model (Bridge 1977, 1982, 1992) is used to construct a range of facies patterns within channel belts following the methods used by Willis & Tang (2010). This channel model has been shown to provide reasonable predictions of grain-size patterns and depth changes along moderately sinuous rivers. Although more sophisticated river flow and sediment transport models are available, constructing deposits with the Bridge model is computationally fast and allows explicit control of input channel geometry and position over time so as to allow direct cause and effect comparisons between river migration and preserved facies patterns. The proportion, arrangement and shape of reservoir facies preserved within channel belts vary as a function of river evolution. The focus of this study is on abstracting general facies patterns and trends preserved within the resulting channel belt deposits that might be compared with subsurface observations. An analysis of deposit thickness variations and the spatial distribution of vertical grain-size profiles across a range of different channel-belt models show how subsurface log data might be used to diagnose reservoir architecture. The goal is not to predict variations within all types of fluvial channel belts but rather to explore in more detail variations in intermediate scale 3D facies distribution that might be preserved by the same river channel with differences in migration pattern and river bend cutoff frequency. In the companion paper (Willis & Sech, this volume) the authors vary the intra-channel-belt facies patterns described here across a range of possible configurations to understand the potential impact that facies organisation might have on reservoir performance.

CHANNEL BELT DEFINITION

Reservoirs in most fluvial deposits are defined by the proportion, size and spatial distribution of channel belts. Channel belts form because a river typically deposits sediments for an extended period along a relatively fixed route across its floodplain, before a comparatively abrupt avulsion to a new path. Although the resulting sandstone belt is broadly channel shaped, it is important to emphasise these belts do not have the same dimensions as river channels. During formation of a channel belt different river channel segments meander, deposit sands along their banks as they expand in sinuosity and migrate downstream and split or become abandoned following cutoff. The cumulative sweep of channel segments and growth of sand bars typically results in the formation of a channel belt an order of magnitude wider than the river channel. Specifics of river channel belts will be discussed in more detail later, after terminology is reviewed to distinguish river channel and channel belt characteristics (Fig. 1).

River channels

A river channel is typically characterised by width, depth, sinuosity and braiding. Along stream changes in channel flow direction, channel splits and junctions, alternations between deeper pools and shallower riffles along the river and the spacing of river sand bars are all generally related by a characteristic spacing that scales with bankfull river discharge. There is extensive literature on river hydraulic geometry and controls on river migration behaviour, so this will not be discussed in detail here (Fig. 2; Leopold & Wolman, 1957, 1960; Collinson, 1978, Lorenz *et al.*, 1985; Fielding & Crane, 1987; Gibling, 2006; Bridge & Mackey, 1993; Singh, 2003; Eaton, 2013; and Gleason, 2015). As an approximate specification, single-thread river channel bankfull width is on the order of 10 times the mean bankfull depth (however, width-to-depth ratio tends to increase with river size); and the channel bend wavelength tends to be 10 to 20 times the channel bankfull width. Although braided rivers, with multiple coeval channel threads, can have much higher width-to-depth ratios than rivers with a single active channel thread, it is less obvious that the mean width-to-depth ratio of individual channel threads within a braided river is different from those within single-thread rivers. This is important because in highly braided rivers depositional features are inferred to scale with local channel thread dimensions, rather than necessarily to the full river discharge. Maximum channel scour depth generally increases with the sinuosity of an individual channel segment within both single-thread and multi-thread (braided) rivers. Braided rivers can also display deeper channel junction scours. As a general estimate, maximum depth is 2–3 times mean bankfull, but scour pools may be locally deeper (e.g. Best & Ashworth, 1997).

Storeys

A deposit formed as an individual channel segment which underwent an increase in sinuosity, migrated in position and had then been cut off and abandoned is termed a 'storey' (following Allen, 1965, 1979). The downstream length of a storey scales in a general way with river discharge and its width relative to length reflects the final sinuosity of the associated channel thread before abandonment (Fig. 2A and B). The number of storeys within a cross section of a channel belt reflects the speed of lateral channel migration and switching relative to the longevity of the channel belt (i.e. rate of river avulsion). The average number of storeys in cross sections perpendicular to the channel belt centreline defines the number of 'storey rows'. A narrow channel belt might be composed of a single row of storeys, each separated along the belt by a continuous channel abandonment fill formed during river avulsion (e.g. Donselaar & Overeem, 2008, "string of beads"). A wider channel belt may contain many rows of storeys formed as multiple channel segments migrated and were cutoff over an extended period of time (Fig. 1D; See also Collinson, 1978).

It is important to distinguish the number of coeval channel threads ('braiding') from the number of preserved 'storey rows'. Although braided rivers commonly display frequent channel switching, which leads to preservation of multiple storey rows, single-thread rivers can also preserve many storey rows given a long period of channel migration and cutoff within the belt. Higher-sinuosity rivers generally preserve storeys with more circular planform shape and lower-sinuosity rivers preserve storeys with downstream-elongate planform shape (Fig. 2A). The common inference that single-thread rivers have high-sinuosity channels and braided rivers have low-sinuosity channels is overly simplistic (see also discussions in Bridge 1985, 1993a; Brierley & Hickin 1991; and Ethridge, 2010). Single-thread rivers can be low-sinuosity and preserve downstream elongate storeys (e.g. a typical delta distributary channel) and braided rivers can be dominated in some cases by moderately high-sinuosity channel segments (e.g. Fig. 1 shows a river that has multiple, moderately sinuous, active channel threads at flood discharge).

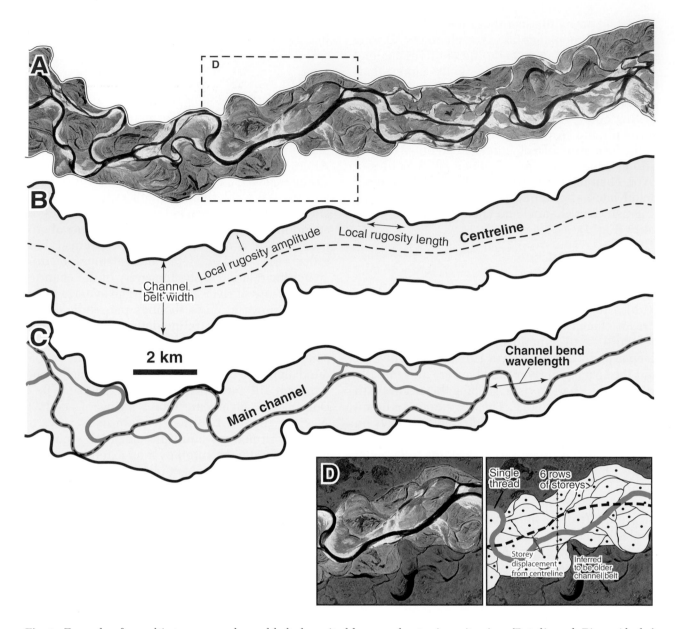

Fig. 1. Example of a multi-storey-row channel belt deposited by a moderate-sinuosity river (Putuligayuk River, Alaska). Imaged reprocessed from USGS LandsatLook (https://landsatlook.usgs.gov). (A) Photo of channel belt and location of insert in part D. (B) Outline of channel belt. Length of bend centreline relative to a straight line connecting centreline ends defines channel belt 'wandering' (as with channel sinuosity, a number always greater than 1). Length of the edges of the channel belt relative to the centreline defines rugosity. The amplitude and wavelength of local edge rugosities provide an indication of the dimensions and shape of the internal storeys. (C) Channel sinuosity, defined by a trace of the main channel centreline, is very different from channel-belt wandering. (D) Trace of plan view dimensions of preserved storeys along one segment of the channel belt. Dots define the centroid of each preserved storey and average deviation of these points from the belt centreline trace divided by mean storey width defines storey swing (e.g. red arrow, labelled storey displacement, shows the deviation of one storey centre from the belt centreline), which is also a measure of the number of preserved storey rows. This river has: 0.001 slope, 1.4 sinuosity, 1.52 braiding, 156 m channel width, 1375 m channel bend wavelength, 2485 m mean bankfull channel belt width (±128 m, 1 standard deviation), 1.05 wandering, 1.21 rugosity, 2 storey length-to-width ratio, 5.7 storey rows and 1.13 storey swing.

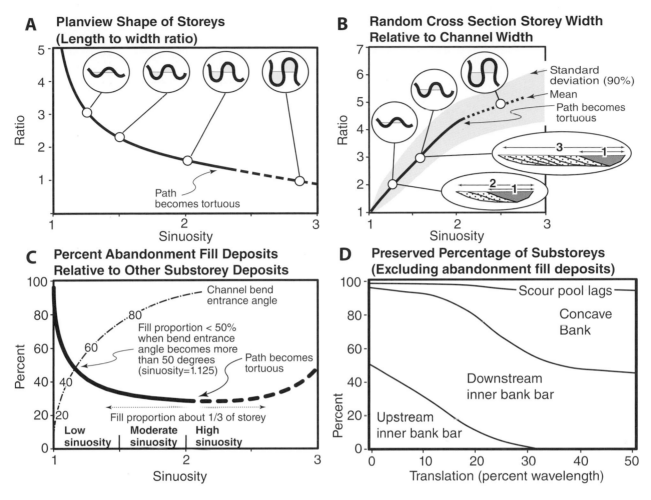

Fig. 2. Dimensions and shape of channels, storeys and channel belts. (A) Plan view shape of storeys (length to width ratio) if the channel planform geometry follows the Langbein & Leopold (1966) theory of minimum variance and channel-width-to-bend-wavelength ratio is 16. (B) The proportion of laterally accreted bar to channel-abandonment fill deposits increases with sinuosity (defined by numerical model described in Willis & Tang, 2010). Measuring the ratio of channel bar to channel abandonment fill in outcrop-exposed cross section provides a minimum estimate of channel-segment sinuosity. Solid line shows the mean ratio observed in random cross sections of simulated channel deposits with different sinuosity (shaded area shows 90% standard deviation of measurements along cross sections with varying orientation). (C) Ratio of channel abandonment fill relative to other types of substorey deposits within a channel belt depends on sinuosity (defined by numerical model described in Willis & Tang, 2010). When the bend geometry becomes tortuous (sinuosity greater than about two) Langbein & Leopold's (1966) sine-generated planform curve is probably an inappropriate description and this graph thus becomes less predictive. For lower-sinuosity cases the preserved channel abandonment fill proportion tends to decrease as a linear function of the deviation of mean flow direction. (D) Preserved proportion of substoreys (other than channel abandonment fill) depends mostly on relative amounts of downstream channel migration (translation), rather than the final channel sinuosity (graph defined by numerical model described in Willis & Tang, 2010). (E) Empirical relationship between river channel width and depth with meander bend radius of curvature, compiled from many sources by Williams (1986). (F) Empirical relationships between channel width and depth (following Leeder, 1973) and channel width and channel belt width (after Leopold & Wolman 1960). Combined function defines relationship between channel depth and channel belt width. (G) Empirical relationship between mean bankfull and channel belt width (after Fielding & Crane 1987). Lines define specific cases. 1a, upper limit defines deposits of straight, non-migrating rivers. 1b, Upper limit of rivers that show clear lateral accretion; WCB = 0.95 DM2.07. 2a, Best fit line to data; WCB = 12.1 DM1.85. 2b, Relationship defined by Collinson, 1978; WCB = 64.6 CM1.54. 3, Laterally restricted braided rivers; WCB = 513 DM1.35. (H) Empirical Relationship between channel mean bankfull and preserved channel belt thickness (after Fielding & Crane 1987). Combined with (G), this relationship can be used to estimate relationships between channel belt width and thickness. (I) Channel belt thickness and width compiled from modern rivers by Blum *et al.*, 2013, (blue circles), Bridge & Mackey, 1993, (red circles); and Fielding & Crane, 1987, (orange triangles). A range of river systems is represented in this data, including many higher-sinuosity and multi-storey-row cases. As Fielding & Crane suggested, their mean bankfull values were divided by 0.55 to get the belt thicknesses that were plotted. (J) Change in single-storey-row width (WCB) to meander bend wavelength (λ) with increasing channel sinuosity (SN) using the theoretical minimum variance relationship defined by Langbein & Leopold (1966).

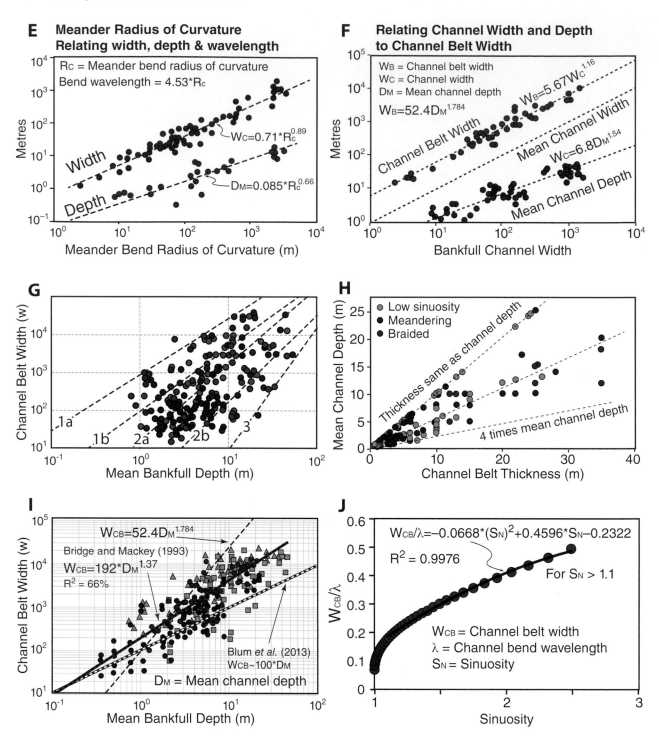

Fig. 2. (Continued)

Storey stacking

Stacking of storeys within belts can be characterised by the average distance by which storey centre points deviate from the belt centreline measured as a function of mean storey width; a measure here termed 'storey swing' (Fig. 1D). High storey swing reflects the migration and switching of the channel over a wide area lateral to the belt centreline, whereas low swing reflects restriction of the channel path to a narrower belt; and probably strong cannibalisation of any earlier formed storeys by those preserved just before belt abandonment. The lateral displacement of storeys (and the width of the channel belt) generally increase with river size, but the dimensionless swing parameter (displacement divided by mean storey width) focuses less on river size than on changes in storey shape and lateral amalgamation. The term 'multi-lateral' has been used in reference to channel belts with high storey swing and has also been used to define a contrast with vertical stacking in a 'multi-storey' channel belt (Potter, 1967; Marzo *et al.*, 2006; Bristow & Best, 1993; and many others). Neither term is very useful in the context of describing 3D variations within channel belts because all belts contain multiple storeys that are laterally juxtaposed horizontally and locally vertically stacked in some pattern (i.e. even belts that contain a single row of storeys contain multiple storeys laterally offset down depositional dip). Although the term 'multi-storey channel belt' can be usefully applied in 1D (core and well log), too often the term is used without clear distinction between storeys vertically stacked within a belt and vertically connected belts within a larger sandstone body (Bridge & Tye, 2000). The term multi-lateral is seldom informative without additional qualifications to indicate the lateral direction of offset of the storeys (dip *vs* strike) and the orientation of the exposed cross section relative to the regional floodplain slope. Storey swing, as defined here, focuses on just a single dimension of storey stacking (i.e. displacement of storey centres perpendicular to the belt centreline in a plane inclined with the mean floodplain slope).

Channel belt shape

Most fluvial channel belts are fairly straight even where the internal channel(s) are highly sinuous (Fig. 1). The sinuous character of a channel-belt centreline is termed here channel belt 'wandering' (Fig. 1B), so as to be distinct from channel sinuosity. Belt wandering generally reflects subtle changes in floodplain depositional topography caused by, for example, differential compaction and subsidence across the floodplain, rather than by any regular oscillation intrinsic to the river flow patterns (cf. channel sinuosity; Armstrong *et al.*, 2014; Alexander & Leeder 1987; Peakall *et al.*, 2000; Taha & Anderson, 2008; Tornqvist, 1994; Smith *et al.*, 1989; Ethridge *et al.*, 1999;

Stouthamer, 2001; Stouthamer & Berendsen, 2001). The width of a channel belt reflects the size and shape of the internal storeys, the number of storey rows and overall storey swing. Variations in channel belt width related to irregularity of the edge of the channel belt (termed 'rugosity'; Payenberg *et al.*, 2014) can provide an estimate of the along-stream length and aspect ratio of the internal storeys. Where a channel belt contains a single row of storeys, the sinuous character of the channel belt edge (i.e. the definition of belt 'rugosity' preferred by Payenberg *et al.* 2014) can be related in some cases directly to the sinuosity of the formative channels. When adjacent storeys become more highly amalgamated within multi-storey-row belts and when channels migrate downstream as storeys form, channel belt rugosity is generally significantly less than the sinuosity of the channel fills within the belt (e.g. Fig. 1).

Quantifying scales of internal heterogeneities within channel belts based on their external shape is difficult because multiple variables control belt mean edge rugosity, including: 1) river discharge influences on scale, 2) channel sinuosity and migration influences on storey shape; and 3) storey swing and their lateral amalgamation within the belt. Due to this intrinsic ambiguity, interpretation of specific belt edge features can be more enlightening than gross measures of rugosity along an entire belt. Where rugosity is low relative to the wavelength of the edge variations one can infer the storeys have low aspect ratio (i.e. internal storeys have low width-to-length plan view shape, characteristic of storeys formed by a low-sinuosity river channel). Where the wavelength of edge rugosities is long relative to channel-belt width one can infer there are few storey rows and low storey swing. As the length of storeys is expected to scale with river discharge (see river hydraulic geometry summaries referenced previously), the wavelength of the edge rugosities may provide some indication of river size in cases where only the margins of a belt is imaged (e.g. in seismic horizon images).

Although the thickness of a channel belt generally reflects maximum channel scour depth, which scales to river discharge, it also depends on local channel segment sinuosity, braid-channel-junction scour frequency and net aggradation within a belt over time (discussed in more detail below). Thus, channel-belt thickness can be highly variable. This means that there is inherent uncertainty in estimating channel belt reservoir volumes (using width to depth ratio) based on just a thickness measure typically observed in an individual well log or by a width measure typically observed in horizon slices of seismic volumes.

VARIATIONS WITHIN STOREYS

Grain-size variations within storeys reflect flow processes and sediment sorting within a river channel and the relative preservation of different areas of the river bed as the channel migrates through time. Geometric relationships between channels, their migration and preserved storey deposits are discussed in detail by Willis (1989, 1993a) and Willis & Tang (2010), based on a consideration of the migration of a simple numerical model of a river channel developed by Bridge (1976, 1982, 1992). As this paper also analyses variations defined by the Bridge model, that work is summarised here with emphasis on highlighting variations within preserved intra-storey facies. Key considerations are the topography of the channel bed around a river bend, lateral sorting of grain sizes across the channel and how the channel migrates over time (Fig. 3). Other useful generalised discussions of process variations along individual river channel bends that control preserved facies variations within channel belts can be found in Bluck (1971), Bridge & Jarvis (1976), Jackson (1976), Dietrich (1987) and Ghinassi *et al.* (2016).

Predicting facies patterns within channel belts can be defined geometrically as an application of Walther's Law applied to a 3D landform. If a channel is static in position and the deposit forms by simple incision and subsequent vertical aggradation during abandonment, all areas of the channel bed would be preserved. If a channel migrates by pure expansion (i.e. an increase in sinuosity with little net vertical aggradation before abandonment), both upstream and downstream inner bank areas will be preserved, but not the outer bank areas along channel bends. If the channel migrates by pure downstream translation, only downstream facing surfaces of the channel bed will be preserved (downstream inner bank and upstream concave bank areas). Although river-bend rotation (i.e. development of channel-bend planform asymmetry) can also influence which parts of the channel bed are preserved (Willis & Tang, 2010), in this study those influences are assumed to be secondary because river channels with extreme tortuosity are not considered (See, however, Smith *et al.*, 2009, 2011). Similarly, very high sinuosity channels with significant bend neck tightening before cut off are not considered explicitly in this discussion (i.e. inner-bank deposits formed closer to the bend apex may be preserved in higher proportion when inner bank areas adjacent to bend entrances and exits formed during earlier stages of bend expansion are preferentially eroded during later stages of channel migration). Although any combination of vertical aggradation and lateral migration is hypothetically possible, bars on river bends seldom migrate upstream and they almost always shift laterally at faster rates than they vertically aggrade (an exception to the latter is during final stages of channel abandonment). Examination of accretion topography on modern point bars suggest lower sinuously channel bends are typically more prone to downstream translation than higher sinuosity bends. Thus, earlier in the evolution of a storey the channel may translate more rapidly downstream and during later stages become mostly expansional. Slow vertical aggradation relative to lateral migration of fluvial channels suggests that the main control on which areas of a channel bed will be preserved is the relative rate of bend expansion to downstream translation (Willis & Tang, 2010). This contrasts with deep water channels, which have been shown to vertically aggrade more rapidly as they are displaced laterally during depositional events (Posamentier & Kolla, 2003; Wynn *et al.*, 2007; Deptuck *et al.*, 2007).

In the simple case considered here, river depth along an individual channel thread alternates such that a deep pool eroded along one side of the channel shoals downstream onto a sand bar (Fig. 3). Deposition of the sand bar is in turn associated with the lateral shift of flows toward the opposite channel bank, where the next thalweg pool is scoured (Fig. 3A). Deeper flows in pools are faster on average and decelerate with expansion and shoaling downstream. Alternating thalweg pools are associated with erosional retreat of adjacent river banks and enlargement of bars on opposite sides of the channel. Thus, thalweg pools are observed along outer banks of river bends and bars build along inner banks, as channel sinuosity gradually increases (Leopold & Wolman, 1960; Engelund, 1974; Ikeda *et al.*, 1989).

Channel curvature induces a centrifugal force that moves faster surface flows outward and flows along the channel bed inward along the channel

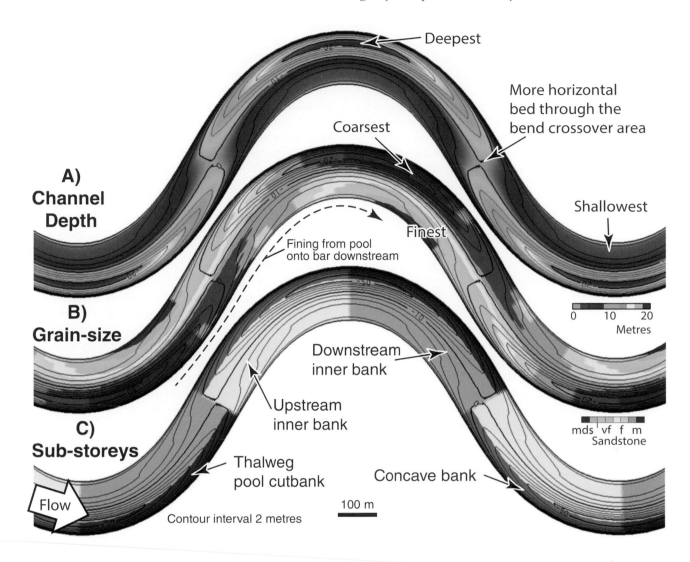

Fig. 3. Predicted variations of the river bed along a curved channel segment defined by the model. Geometry of the channel bed is shown by two-metre depth contours. (A) Colours highlight depth contours. The channel is deepest along the outer bank where channel plain-view curvature is at a maximum near the bend apex and is shallowest near the bend apex along the inner bank. In bend cross-over areas, where the plain-view curvature is at a minimum, the channel bed is relatively flat. In natural rivers the maximum depth is generally shifted somewhat downstream of the bend apex due to flow inertial-lag effects. (B) Colours show lateral variations in predicted mean grain-size superimposed on the channel-depth contours. Mean grain-size is coarsest in the deeper thalweg flows along the outer bank near the bend apex and finer-grained as flows shoal into the bar along the inner bank of the adjacent downstream bend. Uniform cross stream mean grain-size values along the upstream end of the bend reflect poor sorting under expanding flows that move coarser bed load from a thalweg pool through the bend crossover area. Finer grain sizes along the inner bank in the downstream end of a bend reflect preferential inward movement of finer grains under the secondary flows. (C) Distinct areas within the channel shown on depth contours are defined to classify deposits with different spatial grain-size patterns in the resulting storey. These lithostratigraphic 'substoreys' are: 1) upstream inner bank (bar head), 2) downstream inner bank (bar tail), 3) upstream outer bank (concave bank); and 4) downstream outer bank (thalweg pool). Vertical aggradation of deposits within the final abandonment fill defines a fifth substorey type.

bend. This causes a net inward movement of sediment until accumulations along the inner bend build a transverse slope that counterbalances the resulting inward movement of sediment (Engelund, 1974). The influence of the inward-directed

secondary (helical) flow on bed topography is greatest along the downstream end of a channel bend. Although the helical nature of flow is commonly exaggerated as a corkscrew flow pattern, secondary flows are typically an order of magnitude slower than downstream current speeds, such that they move cross stream a distance less than channel width as water moves the length of a channel bend (Bridge & Jarvis, 1982). Thus, the threshold of grain transport depends mostly on downstream directed currents, with secondary components of flow having more subtle influence on lateral cross channel grain-size sorting (Bridge & Dominic, 1984; Dietrich & Smith, 1984).

Flow enters a channel bend with the secondary current moving in the opposite direction (because the bend upstream has the opposite curvature) and strong cross-stream flow expansion related to fast thalweg-pool flows decelerating onto the upstream end of the inner bend bar (Dietrich & Smith, 1983). In downstream ends of a bend, flows deepen and accelerate along the outer bank as centrifugal forces driven by the local bend curvature overcome any residual transverse flow momentum impacted by the bend upstream and flow expansion associated with thalweg crossover occurring near the bend entrance (Bridge & Jarvis, 1982; Dietrich & Smith, 1983; Dietrich, 1987).

Grain-size variations within fluvial channel deposits reflect lateral sorting patterns formed as flows expand from deeper outer bend thalweg pools, through a bend crossover and onto the inner bank bar at the head of the next channel bend downstream (Fig. 3B). Grain-size distributions preserved in the final deposits depend on which areas of the channel bed become preserved (Fig. 3C). Fast flows along the outer bank thalweg pool carry the coarsest grains. Grains fine on average away from these pools as flows expand and decelerate onto the downstream adjacent bar. This results in coarser grains extending topographically higher on the upstream end of a bar than on the downstream end (Bridge & Jarvis, 1982; Dietrich & Smith, 1984).

Bar-head deposits

Coarser-grained deposits topographically high on the upstream end of a bar that are exposed during low river stage are commonly referred to as 'bar-head gravels' (Bluck 1976). Downstream migration can preserve these coarser-grained bar-head deposits on top of finer-grained deposits formed on the bar farther downstream if the channel bed aggrades during river stage changes. Upward-coarsening bar-head deposits are typically only observed in relatively straight rivers with flashy discharge (Cant & Walker, 1978; Lunt *et al.* 2004).

For the more sinuous channel cases that are the focus here, preserved bar-head deposits are generally characterised by poorly vertically sorted deposits that formed under rapidly expanding flows where coarser sediments were moving cross stream from the inner to outer bank of the bend. Such deposits, with weak vertical grain-size trends, are commonly characterised by a 'blocky' well log pattern (Willis, 1989; Bridge & Tye, 2000)

Bar-tail deposits

Deposits preserved along downstream inner banks of bends (on the 'bar tail') typically fine upward defining a 'bell-shaped' log pattern. This reflects the preferential inward movement of finer grains under fully developed helical secondary flows, rather than continued downstream deceleration of flows out of the thalweg pool at the bend entrance. The relative strength of the inward directed secondary current and thus the prominence of the inward grain fining, increases with channel bend sinuosity. These inward fining trends on the channel bar can be related to upward fining in the deposits formed as the bar grows and migrates downstream (Willis 1989; Bridge & Tye, 2000).

Concave-bank deposits

Directly downstream of the bar-tail channel area, across the relatively flat bed of the bend crossover area, is the upstream end of the outer bank of the next bend downstream (labelled the 'concave bank' area in Fig. 3C). In this area, curvature-induced centrifugal forces reverse but inertia and pressure gradients generated by rapid expansion of flows from the thalweg pool on the opposite channel bank hinder helical secondary flow development. In most geomorphic descriptions, this area is part of the 'cutbank' (an area of net erosion) but, as indicated by Willis & Tang (2010), a significant proportion of storeys must be deposited within this area in cases where channels migrate downstream (up to 40%). Proposed depositional mechanisms and associated predictions of the character of the deposits that accumulate in the concave bank area vary widely (Carey 1969; Nanson & Page, 1983; Alexander 1992; Burge & Smith 1999; Makaske & Weerts 2005; and Smith *et al.* 2009).

In relatively straight, high-energy rivers a component of the rapid thalweg pool flow may have enough inertia to continue across the top of the bar downstream and through an inner bend chute channel, before rapidly decelerating as it re-joins the main channel flow in the bar lee area. In this case the concave-bank area can be filled by 'chute bar' deposits, which in some cases comprise angle-of-repose bedsets that coarsen upward. Such deposits are expected to be somewhat finer grained on average than those preserved under expanding flows in the bar-head area but they can be sandier and better sorted than those formed under areas with stronger helical secondary flows along the bar tail (Bluck, 1976).

In more sinuous rivers, concave bank areas tend to accumulate suspended sediment in sluggish waters slowed under the reversed centrifugal force at the bend entrance (relative to that of the bend upstream) and rapid expansion of thalweg flows from the opposite bank. Smith *et al.* (2009) called these deposits 'counter point bars' because they are observed at low river stage as accumulations on the opposite side of the channel from the inner-bank point bar. Like the term 'chute bar' (discussed previously), the term 'counter point bar' is somewhat misleading as these deposits represent a morphodynamic downstream extension of a larger channel bar along the inner bank of the bend (rather than deposits distinct from the inner-bank sediment accumulations formed by flow patterns around a channel bend). In some cases where bends have very high curvature a separate circulation eddy develops in the concave bank area (Burge & Smith, 1999; Smith *et al.*, 2011, Ferguson & Parsons, 2012). In other cases sediments preserved in this area may be deposited during low flow stage forming a 'concave bank bench' (Nanson & Page, 1983) and preserved as the channel begins to migrate downstream during the rising stage of a subsequent flood.

In this study, focusing on storeys formed by moderately sinuous channels, the authors describe the concave-bank deposits as fine-grained overall and strongly upward fining. Bridge's point bar model (Bridge 1982, 1992) predicts progressively greater inward fining of deposits with distance downstream along the bar tail and here the authors assume this trend continues through the concave-bank area. This assumption is reasonable for cases with fine-grained concave-bank deposits but might not be characteristic of storeys formed by all types of rivers (i.e. particularly low-sinuosity higher-slope systems, as indicated above).

Thalweg scour pool lag deposits

Downstream outer bank areas of a river bend, comprising the thalweg pool, are generally erosional. Deposits can be preserved in this area due to net vertical bed aggradation, episodic channel depth changes during river stage changes and stepped migration patterns. Deeper areas along these pools, formed where flows impinged upon cutbanks more strongly or near braid-junction scours, may also preserve deposits in this area of a channel bend. The deposits tend to be coarse-grained lags locally preserved directly above the most deeply incised area under a storey. Where channels migrate by expansion over time, these deposits will occur as coarser-grained lags at the base of upward-fining bar-tail deposits. The deposits can also contain thick accumulations of mud chip conglomerate and local large floodplain slump blocks.

Channel abandonment fills

Deposits within an abandonment fill formed during channel bend cutoff or river avulsion generally fine upward due to longer-term declining discharge (reflecting the switching of river flows to a different channel segment). Sandier fills are generally inferred to reflect gradual channel abandonment and muddier fills more rapid abandonment. It is also expected that higher-sinuosity channel-bend segments will be cut off faster than lower-sinuosity segments because the cutoff path of a higher-sinuosity channel segment defines a greater relative slope advantage.

Bridge (1993) suggested that when a lower-sinuosity channel bend is cut off, abandonment fills generally fine upward due to progressively weaker flows and fine downstream due to progressive blockage of the upstream bend entrance (Bridge *et al.* 1986). For higher-sinuosity channel segments, he suggested that both ends of an abandoned channel would be quickly blocked so that most of the channel fill is relatively fine-grained (mud plug 'oxbow lake'; Fisk 1952; Allen, 1963). Sun *et al.* (1996) suggested the mud plug facies preferentially preserved along the margins of channel belts, difficult to subsequently erode, would tend to restrict subsequent channel segments from migrating much beyond the lateral limits of the earliest-formed channel-bend loops (tending to restrict storey swing).

Willis & Tang (2010) modelled abandonment fills by gradually decreasing river discharge whilst keeping channel width constant. They predicted that bend axis pool scours would fill with sand faster than areas with flat beds near bend crossovers. The result was generally thicker, sandier, more gradually upward-fining abandonment fills along meander-bend axes and overall muddier fills in crossover areas. Another scenario for such gradual discharge decline might be the development of smaller channels and bars that continue to migrate along the path of the larger, partially abandoned channel. Bridge's (1993) abandonment-fill model, predicting progressive lateral fining of deposits from abandoned bend entrance to apex, suggests better continuity between adjacent storeys on opposite sides of the channel-belt axis. The gradual discharge decline model suggested relatively sandy fills under the apex of a cut off bend oxbow (increasing along belt connectivity parallel to the channel-belt axis) and poor connectivity between adjacent bar deposits across the belt axis (see Willis & Tang, 2010).

Substorey type

The discussion above indicates that a storey can be subdivided into deposits with different character based on which areas of the channel bed were locally preserved. Five substorey deposits are defined (Figs 4 to 6): 1) Bar-head deposition in the upstream inner bank area, 2) Bar-tail deposition in the downstream inner bank area, 3) Concave-bank deposition in the upstream outer bank area, 4) Thalweg scour pool lag deposition in the downstream outer bank area; and 5) Abandonment channel fill. The relative proportion of substoreys and details of grain-size patterns within substoreys defines large-scale heterogeneity patterns within fluvial channel belts (Figs 4 and 5).

The proportion of channel abandonment fill, relative to other substorey types formed during active channel migration, is defined mostly by the sinuosity of the channel at bend cutoff (Figs 2C and 4). For storeys formed by lower-sinuosity channels, the proportion of channel abandonment fill is an inverse linear function of channel bend entrance angle. This trend toward decreased proportion of abandonment fills in higher-sinuosity rivers weakens for rivers with highly torturous plan view pattern. The relative proportion of substoreys formed during channel migration depends mostly on the relative rates of channel expansion to downstream translation (Figs 2 and 4).

CHANNEL BELT DIMENSIONS

Defining the dimensions of channel belt sandstone bodies is critical to reservoir volume estimates, particularly where individual belts are isolated within lower net-to-gross (low channel-belt proportion) depositional successions. It is common practice to use geometric relationships between channel width, depth and bend (bar) wavelength observed in modern rivers to predict dimensions of channel belts and internal heterogeneities from sparse subsurface data (e.g. from storey thickness observed in a well log or belt width observed in a seismic stratigraphic horizon slice; Fig. 2; see also summaries in Bridge & Mackey, 1993; Gibling, 2006; Bridge & Tye, 2000; Tye, 2013; Blum et al., 2013; and Colombera et al., 2017). As it is well known that maximum channel depth increases with channel sinuosity as well as river discharge, geomorphic relationships are most commonly defined using mean river depth measured in bend crossover areas where the channel floor tends to be relatively flat (e.g. Leopold & Wolman, 1957, 1960). Although the size of a channel belt generally increases with mean bankfull channel depth of the formative river at flood discharge, the mean thickness of the deposit tends to be significantly greater than mean bankfull channel flow depth (Fig. 2; Fielding & Crane, 1987). The basal relief of a channel belt is related to elevation variations along the floor of the river channel, in that it is the lateral sweep of the active channel over time that forms the channel belt. The distribution of channel depth along modern river analogues cannot be used directly to predict the distribution of belt thickness without considering how channel shape evolved with sinuosity during deposition of each storey. Similarly, the width of a channel belt increases with river discharge and varies with the final sinuosity of the channel that formed each storey and the number and swing of preserved storey rows. As there are multiple controls on both channel belt thickness and width, empirical equations relating channel belt width and thickness are expected to show wide variance from mean values (Fig. 2I).

Channel-belt thickness

The top of the sand within a channel belt is generally more horizontal than its base. Modern river bars extend upward to about 80 to 90% of flood water depth. The elevation of the top of the sand within the belt is expected to vary between

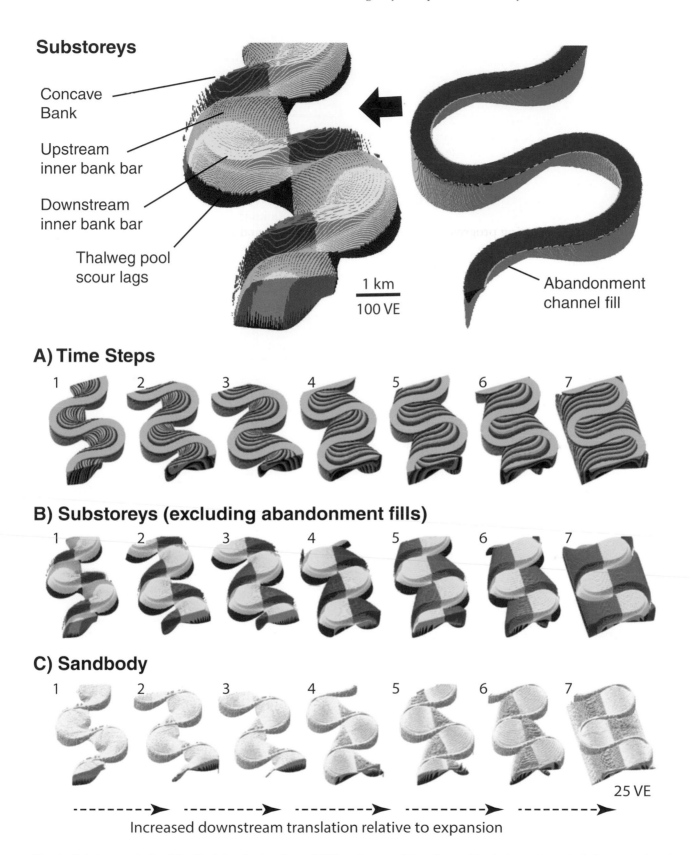

Substoreys

Concave Bank

Upstream inner bank bar

Downstream inner bank bar

Thalweg pool scour lags

1 km
100 VE

Abandonment channel fill

A) Time Steps

1   2   3   4   5   6   7

B) Substoreys (excluding abandonment fills)

1   2   3   4   5   6   7

C) Sandbody

1   2   3   4   5   6   7

25 VE

Increased downstream translation relative to expansion

Fig. 4. Substoreys, defined by the lateral migration of different areas of the channel bed, show characteristic vertical and lateral grain-size patterns. Grain-size variation within storeys of fluvial channel belts reflect patterns of grain sorting on the bed of the river and preservation patterns of different areas within the channel during migration. Examples arranged from left to right show changes in substorey geometry with progressively more downstream translation superimposed on a constant rate of bend expansion (sinuosity increase).

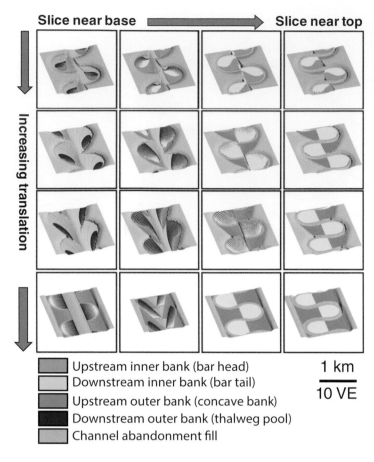

Slice near base ➡ **Slice near top**

Increasing translation

Legend
Upstream inner bank (bar head)
Downstream inner bank (bar tail)
Upstream outer bank (concave bank)
Downstream outer bank (thalweg pool)
Channel abandonment fill

1 km / 10 VE

Fig. 5. Substorey geometry shown in relation to topography of the channel belt basal erosion surface for a single row of storeys. Each example shows a slice through the channel belt at varying depth superimposed on the fixed surface geometry of the basal erosion surface. Coarse-grained outer bank deposits (e.g. thalweg scour pool lags) deposited within deep pools are preferentially preserved topographically lower within the storey and finer grain inner-bank and concave-bank deposits are preserved higher in the storey. The shape of the coarsest-grained deposits is more circular for the case where the storey grows only by expansion and is downstream elongate where it grows by downstream translation. The character of deposits within the different substoreys controls storey-scale heterogeneity patterns within the channel belt.

different substoreys. Sands tend to be preserved to a higher elevation over bar-head deposits and the vertical transition to mud tends to be at lower elevation where deposits grade upward more rapidly in concave-bank and abandonment-fill substoreys. In ancient deposits, it can be difficult to distinguish muds deposited in the upper part of a channel belt from overlying floodplain deposits. As typically abandoned alluvial ridges are initially topographically elevated relative to the surrounding floodplain, their top surface tends to be a depositional hiatus defined by a palaeosol. In such cases the thickness of muddy deposits preserved in the upper part of the channel belt (i.e. the difference between the top of a channel belt and its internal sandbody) can be estimated. It is common to observe separation between the top of a channel belt sand and the top of the overlying

palaeosol to be between 0% and 20% of the belt thickness.

Relief along the basal surface of the channel belt sandstone reflects deeper scour in thalweg pools relative to crossover areas (Fig. 5). Thicker areas within individual storeys tend to have circular plan view shape where a channel migrates by expansion and have more along stream elongate planform when channel bend translation is relatively faster (Fig. 5). In a single-storey-row channel belt in which channels migrated by expansion only (i.e. migrate only by increasing sinuosity), each storey will be separated by an abandonment channel fill. In cases with downstream translation, storeys are probably connected by a continuous basal sand layer that contains coarser-grained thalweg scour pool lags. In all cases the upward-fining character of the final channel abandonment

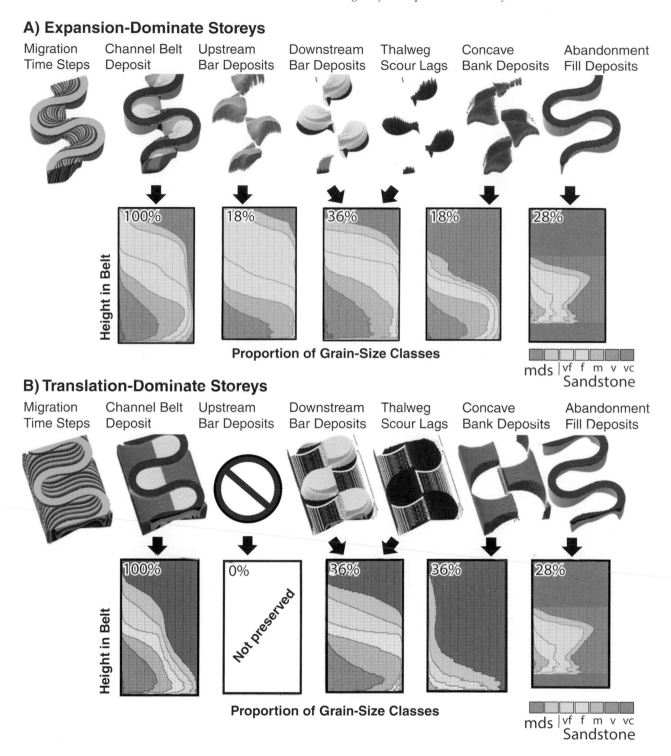

Fig. 6. Averaged vertical grain-size trends within different types of substoreys and relative preservation with (A) and without (B) downstream translation superimposed of bend expansion. Preservation of more bar-head substorey deposits in the case without translation leads to deposits with less vertical fining on average ('blocky' well log pattern). Downstream transition preserves more bar-tail and concave-bank deposits with upward-fining grain-size trends ('bell' well log pattern). The apparent upward-coarsening grain-size trend within the channel abandonment substorey reflects a lateral averaging of grain-size along the bend (including bend crossover areas with more uniform grain-size above mean depth and bend apex areas with thicker basal coarse-grained deposits). At any one location, these fills uniformly fine upward.

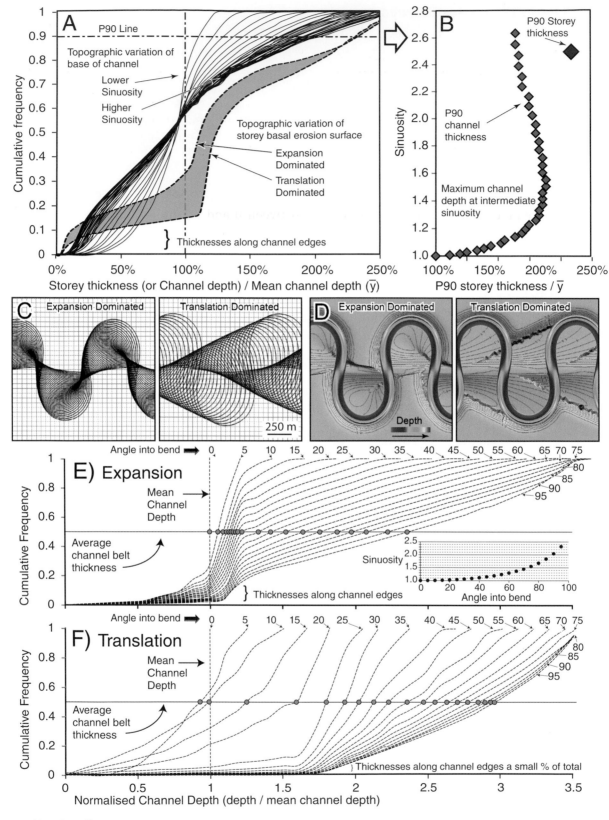

Fig. 7. (Continued)

fill surrounding each storey can be a critical control on inter-storey connectivity. See Willis & Tang (2010) for a broader discussion.

Lateral sweep of a migrating river tends to carve out a basal erosion surface with a mean incision depth significantly greater than the mean bankfull depth of the channel. Defining relationships between mean bankfull channel depth, spatial changes in maximum channel depth along a river bend, temporal changes in the distribution of depth as a river increases in sinuosity during channel migration and the resulting mean and spatial variance in channel belt thickness is not simple. Although precise predictions of the distribution of spatial variations in channel belt thickness are difficult because of the range of potential variables, some general guidance is provided based on modelling single-storey-row channel belts (Fig. 7). Channel belts were constructed by changing the input angle of flow to a channel bend in 0.5 degree increments, assuming the channel centreline planform follows the minimum variance path of Langbein & Leopold (1966) and predicting the channel depth distribution using the model of Bridge (1992). In this construct the input angle (the angle defined in plan view between that at which river flow enters the channel bend relative to that defined by the regional dip of the flood plain) is directly related to channel sinuosity.

The model of a straight channel is predicted to have a flat bed, which everywhere has the same depth. A slightly more sinuous channel is predicted to have subtly deeper areas along cutbank margins and shallower areas where sediments have accumulated along the inner bank of the bend. As river sinuosity increases further, differences between channel depth along cut bank areas and shoaling inner bank areas increases. The predicted distribution of local channel depth for rivers of different sinuosity (entrance angles between 0 and 95 degrees) is shown normalised by the predicted mean bankfull depth in Fig. 7A. As channel belt thickness is expected to be more sensitive to changes in local maximum channel depth, the p90 channel depth (i.e. a value greater than 90% of the measured depths in that channel but less than the deepest 10%) is used to define channel depth variance. The p90 depth value is predicted to increase as channel sinuosity increases from 1 to 1.5, but then to decline somewhat as river channels become highly sinuous (Fig. 7B). This reflects increased cross channel secondary flows as a river progresses from low to moderately sinuous but then an overall gradual

Fig. 7. Channel belt thickness depends on both formative channel depth and patterns of channel migration during storey deposition. (A) Cumulative frequency distribution plots show variations in local channel depth (thin black lines) relative to mean bankfull channel depth ($\bar{y}$) for channel segments with varying sinuosity. A straight channel with vertical side walls is assumed to have constant mean bankfull depth (vertical dash-dot line at 100% normalized mean bankfull depth). As a channel becomes more sinuous, outer bend areas of the channel bed deepen and inner bend areas shoal (i.e. widening the spread of the cumulative frequency distribution curve; compare curves for lower and higher-sinuosity cases). The range of channel belt thickness formed by migration of an evolving channel is the same as the observed range of channel depth (minimum and maximum values of thin black lines and dashed lines) but the distribution of belt thickness values is very different (i.e. compare distributions channel depth distributions indicated by the thin black lines and those defined by the thick dashed lines bounding the orange polygon). Channel migration preferentially preserves a record of areas most deeply eroded at each time step. Channel bend translation superimposed on expansion increases the average channel belt thickness by increasing the area swept by the migrating channel (change in distribution indicated by orange polygon). (B) The greatest channel erosional depth is predicted to increase from low-sinuosity to moderate-sinuosity channels (sinuosity from 1 to 1.5). Very high-sinuosity channels are predicted to have slightly lower maximum erosion depth due to a weakening of secondary currents related to an effective decrease in downstream slope along the longer channel centreline path. Lower maximum depth in the most sinuous cases results in overlapping channel depth distributions for higher-sinuosity cases shown on the graph in part A. (C) Plan view of temporally evolving channel centrelines for analysed expansion-dominated and translation-dominated cases. (D) Depth of the final channel (colours showing depth normalized by mean bankfull) shown on the channel belt basal erosion surface following migration expansion-dominated and translation-dominated cases. (E) Distribution of channel belt thickness formed by a channel with 100 m width, 10 m depth and 1250 m bend wavelength. In each case the channel begins straight and migrates by increasing sinuosity (i.e. expansion) to the indicated bend entrance angle, which is a direct function of sinuosity applying the minimum variance curve of Langbein & Leopold (1966). Above an entrance angle of about 70 degrees (channel segment sinuosity ~ 1.5) the maximum channel depth begins to decrease but mean channel-belt thickness continues to increase because more area is swept by deeper areas of the migrating channel. (F) Similar to part E, except channel begins at the specified bend entrance angle of 90 degrees (i.e. sinuosity ~ 2.3) and translates downstream a distance greater than one wavelength (i.e. no expansion during migration).

decline in flow rates as the lengthening path of the river reduces effective local downstream slope.

Although maximum channel belt thickness will be the same as maximum channel depth (given no net vertical aggradation), the distribution of channel belt thicknesses cannot be related directly to the distribution of channel depths (compare probability distribution functions of channel depth and channel belt thickness in Fig. 7A). Derivation of a channel belt thickness distribution from channel depth variations requires assumptions about river migration patterns. Two end member cases where considered for which channel migration was either by expansion (sinuosity increase) or downstream translation. In the expansion case (Fig. 7E) the channel started straight and gradually increased to a specified input angle (the highest input angle modelled generated a channel with maximum sinuosity of 2.4). In the translation case (Fig. 7F) a channel with fixed sinuosity migrated down floodplain slope a distance greater than one wavelength (sinuosity was 2.4). In both cases the surface eroded during channel migration was recorded and topography of this basal erosion surface was used to define a cumulative frequency distribution of channel-belt thickness (Fig. 7E and F).

Ignoring the tapered edges of belts along individual storey margins, the basal surface of a channel belt can range from mean bankfull channel depth to 2 to 3 times mean bankfull channel depth (depending on the final sinuosity of channel segments before cutoff). In the moderately high-sinuosity cases shown in Fig. 7, mean channel-belt thickness is about 125% mean bankfull channel depth and the p90 thickness is about 225% mean bankfull channel depth, even though the maximum erosional depth of the channel and belt thickness are the same.

Although the methods used to construct Fig. 7 can be used to estimate variability in channel belt thickness (given assumptions about final river sinuosity and migration pattern), at least three other factors potentially influence channel belt thickness distribution: 1) Changes in channel floor topography associated with variations of in other hydraulic geometry parameters; 2) Smaller-scale erosion patterns along the channel; and 3) The stacking of storeys within multi-row belts.

The influence of variations in channel geometry is shown by a comparison between the one example in Fig. 7A with those in Fig. 7E and F. The channel used to generate Fig. 7A had a width to depth ratio of 20 and width to bend wavelength ratio of 16, whereas those used to generate Fig. 7E

and F had a width to depth ratio of 10 and width to bend wavelength ratio of 12. The maximum scour depth predicted in Fig. 7A is 250% of mean bankfull channel depth, compared with 350% in Fig. 7E and F. The predicted thickness distributions (including mean values) vary between these two examples. In the channel depth model used here, increasing channel width and depth relative to bend wavelength tends to increase the strength of the secondary current and the predicted maximum scour depth along the bend apex. River hydraulic geometry data provides broad constraints to define reasonable channel depth, width and wavelength ratios, but specific cases can vary significantly from the mean. Different types of river channel models (incorporating different assumptions about flow and sediment interactions) also vary in their prediction of river channel geometry. Thus, details of input geometry and river model assumptions can significantly influence the predicted channel belt thickness distribution.

Erosion initiated around smaller-scale bedforms within a channel, like dunes or unit ('free') bars (Struiksma *et al.*, 1985; Seminara & Tubino, 1989; Lunt *et al.*, 2004; Ashworth *et al.*, 2011; Reesink & Bridge, 2011), or more local bed irregularities (e.g. variations in substrate cohesiveness) can also change local channel shape and therefore impose spatial variations in the distribution of channel-belt thickness. In many cases these irregularities are probably small enough to be ignored in channel belt thickness estimates but they can be particularly important in straighter channels. The model used here defines channel basal relief as a function of the strength of curvature induced secondary flows, such that the floor of straight channels are predicted to be horizontal and the floor of sinuous channels slope toward deeper outer bank scours. In natural rivers the floors of straight channels tend to develop significant relief defined by scours around smaller-scale unit bars and dunes. Such erosional patterns suggest that the modelled channel-belt thickness values defined in Fig. 7E and F are probably significant underestimates for straight channels. A channel belt formed by braided rivers can also have junction scours where erosion can be as much as five times mean bankfull (Best & Ashworth, 1997).

Increases in mean channel-belt thickness relative to mean bankfull channel depth become more pronounced where more storeys are stacked as different channel segments grow in sinuosity, migrate downstream and are cut off during the life of the channel belt. Thus, it is expected that ratios of mean belt thickness to mean bankfull channel depth will be higher in belts with multiple storey rows relative to single-storey-row belts. In the hypothetical model limit, a belt built by an infinite number of storeys that are randomly placed in position will everywhere have a thickness of the maximum channel depth.

Relating channel belt width and thickness

The width of a channel belt can be reliably predicted from its thickness only by specifying, at a minimum, the channel sinuosity at story cutoff and the number and swing of internal storey rows. Very general guidance provided by geomorphic relationships observed in modern channel belts (e.g. Blum *et al.*, 2013, suggestion that belt width to thickness is typically 100; see also Fig. 2E) are useful but do not account for the wide uncertainty inherent in merging dimensions of belts formed by rivers with widely different discharge, sinuosity, channel switching frequency and belt avulsion frequency. Williams (1986), Bridge & Mackey (1993), Bridge (2003) and Colombera *et al.* (2017) list equations to estimate channel belt width from belt thickness or channel geometry (e.g. width, depth and sinuosity). However, these equations suffer from high data variance from the empirical functions, or theoretical mixing of the multiple non-normally distributed variables that control channel belt width. Most importantly, a clear distinction is required between the plan view shape of the storeys (defined by channel sinuosity) and the number of storey rows (controlled by rates of channel cutoff relative to river avulsion).

Any estimate of belt width from thickness will be highly uncertain without additional constraints. If one assumes an average channel segment sinuosity at bend cutoff, which could be based on belt rugosity observations in seismic slices, variance of belt thickness in different penetrations, or vertical log character, then mean bankfull channel depth can be inferred from belt thickness using Fig. 7. Channel bend wavelength is typically 100 to 125 times mean bankfull channel depth (although there is considerable variation in this relationship). With mean bankfull depth, one can estimate channel width. Estimates of channel width, wavelength and sinuosity can be combined with the minimum variance curves of Langbein & Leopold (1966) to estimate cross stream width of an individual story row (Fig. 2F). Although this method can provide an estimate of

single-row width, it still leaves the number of storey rows and their swing to be inferred.

The suggestion that braided (multi-thread) rivers form wider channel belts than meandering (higher-sinuosity, single-thread) rivers is generally wrong, or at least wrong as often as correct. References supporting this common suggestion in the interpretation of ancient channel belts are commonly circular and not based on specific observations that constrain channel pattern (Ethridge, 2011). This error has propagated through published channel belt dimension databases that uncritically accept previous authors' interpretations of channel pattern (Gibling, 2006). Although belts formed by braided rivers may generally contain more storey rows formed by rapid cutoff of individual channel threads, storeys formed by high-sinuosity rivers are individually much wider. For example, braided river channel belts on the modern Ganga plains aggrading on megafans in front of the Himalayan uplifts in India are generally fairly narrow relative to their depth due to rapid river avulsion. In contrast, the multi-storey-row channel belts of the Mississippi River just south of Memphis (USA) have much higher width-to-thickness ratios.

In the best-case scenario, where subsurface channel belts can be observed both locally in core and more broadly in seismic stratigraphic slice, one might be able to use thickness-to-width ratios to estimate the number of storey rows (i.e. single *vs* multiple) when channel segment sinuosity and bend wavelength can be estimated by reference to internal seismic amplitude variations or belt edge rugosity patterns (Fig. 2F). There is some evidence from modern rivers that channel belts might generally decrease in width-to-depth ratio through a backwater zone near the coast (Blum *et al.*, 2013) or might decrease in width-to-depth ratio down basin more broadly with preferential extraction of bed-load relative to suspended-load (Strong *et al.*, 2005). Estimates of channel-belt width-to-thickness ratios based on larger-scale stratigraphic patterns are an active area of research.

RELATING LOG PATTERN TO CHANNEL-BELT CHARACTER

Although it is popular to relate vertical well log pattern (e.g. 'blocky' *versus* 'bell') to river channel pattern (e.g. 'meandering' *versus* 'braided') this oversimplification leads to naive assumptions about subsurface heterogeneity patterns. Channel belts formed by even the most braided (multi-thread) river systems can locally preserve upward-fining substorey deposits (e.g. within some parts of channel abandonment fills). As indicated previously, even high-sinuosity rivers tend to locally preserve successions with little upward fining (i.e. a 'blocky' pattern) within bar-head and, potentially, chute bar substoreys. Any inferences about channel-belt character based just on well log observations thus requires at a minimum comparison of enough channel belt penetrations such that together they can be inferred to provide a representative sample of substorey character and proportion. Interpretations of channel character from well logs becomes particularly problematic in higher net-to-gross successions where it is more difficult to distinguish isolated from amalgamated channel belt sandstones. The first step in well log interpretation is always identification of individual storeys and the recognition that top-truncated storeys within a succession of amalgamated channel belts may not provide a complete record of the amount of upward fining within the observed storeys.

Variability in vertical log pattern observed within individual channel belts is shown in Fig. 8. Three cases were developed using the Bridge point-bar model (Bridge 1976, 1977, 1982) following the methods described in Willis & Tang (2010). In each case the river had 15 m mean bankfull depth, 175 m channel bankfull width, 4 km bend wavelength and mean grain-size of medium sand (predicted grain sizes varied from mud to very coarse sand). The river was initially straight and migrated to a maximum sinuosity of 1.4 by incrementing the bend entrance angle from 1 to 65 degrees in 2 degree steps. The final channel abandonment fill was created by reducing channel discharge by successive factors of 2 over 5 time steps on the fixed final channel planform, which decrease channel depth and produced a fill style similar to the coarser-grained fill type modelled by Willis & Tang (2010). The segment of each model shown is slightly longer than one channel bend wavelength, displaying a bit more than two storeys that expanded in opposite direction away from the channel-belt centreline. As these modelled channel belts contain a single row of storeys, storey centres are defined easily by locations where the belt is wider and crossover areas dominated by abandonment fill deposits separating storeys by narrow segments of the channel belt.

For the first case (Fig. 8A), migration was only by bend expansion (sinuosity increase), which preserved approximately equal amounts of bar-head and bar-tail substorey deposits and almost no concave-bank deposits. For the other cases (Fig. 8B and C) progressively more downstream translation was added to the migration pattern (0.5 and 1 wavelength downstream migration distance, respectively). For the latter case (Fig. 8C) this migration pattern resulted in preservation of only bar-tail, concave-bank and abandonment-fill substoreys, characteristic of translation-dominated storey deposits. It is important to emphasise that although the river bed evolved during channel migration, the channel used to generate the deposits was the same at the same migration step of the different cases, so that variations in deposits reflect only preservation differences under different migration patterns rather than any differences in the river at any one time step.

Vertical grain-size trends defined at locations spaced 100 m apart along each channel belt are shown by graphs within Fig. 8 (graph boxes are centred on the associated sample position). The grain-size logs were smoothed by a 3 metre moving average to highlight vertical trends. The horizontal axis of each box shows grain-size increasing to the right from mud to coarse sand and the vertical axis shows the depth within the deposits (i.e. values for deposits higher within graph are shallower). The colour within each box defines local mean grain-size. The vertical distance spanned by each grain-size log indicates local deposit thickness. All the channel belts display a wide range of local vertical grain-size variations (i.e. well-log patterns) and local deposit thickness variations, which can be related to the intra-storey depositional trends described in detail by Willis & Tang (2010). These patterns will be summarised only briefly here.

Along specific cross sections, deposits are thin near one margin of the channel belt and thicken toward the opposite margin, reflecting that maximum channel erosion was less when the river was straight and increased as the channel became more sinuous. This cross belt thickening trend alternates direction for successive cross sections spanning different storeys, reflecting that successive river bends expanded in sinuosity in opposite directions. For the expansion dominated case (Fig. 8A), relatively fine-grained channel abandonment deposits dominate bend crossover areas and the centres of the storeys are sandier.

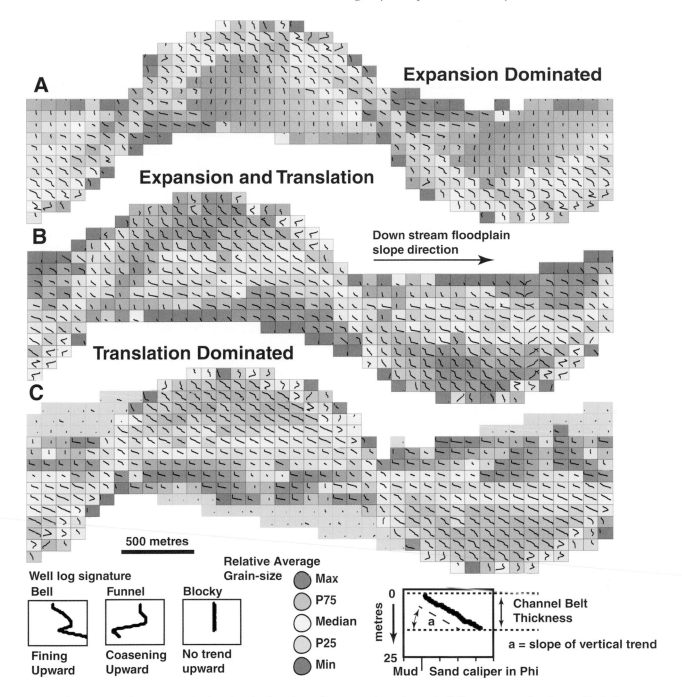

Fig. 8. Shape maps showing vertical grain-size logs in well penetrations through different areas of a channel belt for cases: (A) where the belt formed by a channel that migrated mostly by expansion (sinuosity increase), (B) by both expansion and downstream transition; and (C) by mostly downstream translation. Colours show the local mean grain-size of the deposit. Graphs, centred on the sample location, show grain-size on the horizontal axis and depth on the vertical axis.

The deposits are coarser-grained on average within the bar-head deposits preserved in upstream parts of storeys and the well logs there are characterised as having blocky patterns (little vertical change in mean grain-size). Downstream parts of storeys are finer-grained on average and

display more bell shaped (upward-fining) log patterns characteristic of bar-tail deposits. The most pronounced upward-fining logs occur in thicker deposits toward the outer margins of the belt, where abandonment fill deposits overlie the bar-tail or thalweg-pool deposits. Directly along final cutbank areas local upward-coarsening logs are preserved due to episodic incision during successive time steps.

With downstream translation (Fig. 8B and C), fewer bar-head deposits are preserved and thus there are fewer well logs that can be characterised as blocky pattern. Across-belt lateral grain-size trends become more pronounced than along depositional dip trends within storeys, reflecting preservation of concave-bank deposits. The concave-bank deposits can have coarse-grain basal deposits where thalweg pool sands are preserved but tend to be dominated by mud. In the case dominated by downstream translation (Fig. 8C) very thin coarse-grained deposits preserved along the edge of the channel belt adjacent to areas with concave-bank deposits with no overlying muds at shallower depths are a gridding artefact (average value colour shown as grey).

Models of multiple-row channel belts were developed from the three cases just described by adding a wandering transform to the geometry of the modelled storeys (a sine curve with 4000 to 6000 m wavelength along the belt axis and with amplitudes of a few 100 m was added to plan-view grid-block positions). Three belts with different degrees of internal storey swing were constructed for each case (Fig. 9). Summary plots of the vertical log patterns show the preserved substorey distribution, local average grain-size of wells, local standard deviation in grain-size and local vertical grain-size trend. Stacking multiple storey rows within a belt makes lateral grain-size patterns more complicated but to a significant degree many of the same local patterns remain. For example, in the expansion-dominated case the downstream-fining trends across storeys can still be observed locally despite the added discontinuities. In the translation-dominated case cross stream trends from coarser-grained inner bank to concave-bank deposits remain prominent. Although facies-body continuity lengths do decrease somewhat, the general scaling of these heterogeneity patterns is largely retained. Although the abandonment fills separating bar deposits remain relatively continuous, locally they intersect in an anastomosing pattern. In these models the abandonment fills were relatively sandy and variations in their distribution might have had more impact on heterogeneities patterns if they had been muddier. It is commonly assumed that multi-storey belts will be more homogeneous because amalgamation of storeys would tend to break up the continuity of contrasting facies. The qualitative inference from these models is that the character of facies patterns within individual storeys may have a more primary control on reservoir heterogeneity patterns than the number of storey rows within channel belts. Emplacement of new storeys tends to add similar facies patterns as those removed.

The distribution of local average grain-size, the vertical standard deviation of grain-size and vertical grain-size trend observed in the models shown in Fig. 9 demonstrate pronounced differences for expansion-dominated and translation-dominated migration cases (Fig. 10). The expansion-dominated case is coarser-grained overall and contains a smaller proportion of deposits that would be characterised by distinct upward-fining grain-size trend (i.e. 40% bell-shaped well-log patterns *vs.* 70% for the translation dominated case). This highlights the difficulty of relating log character to river pattern, since the river that generates the one deposit was exactly the same as the other at each time step. It was the relative preservation of deposits formed in different areas within these rivers that defined these contrasts. The relative proportion of well-log patterns also appears relatively insensitive to the number of storey rows within belts (Fig. 10). This was somewhat unexpected as it was assumed that adding more channel-fill deposits, potentially overlying coarser-grained older storey deposits, would tend to increase the number logs with upward fining tends. This conclusion might be different for cases with high rates of net channel-belt aggradation.

DEFINING CHANNEL-BELT CHARACTER FROM CROSS SECTION

Cross sections of channel belts exposed in outcrop provide a richer record of channel belt character than individual core or well log because one can observe lateral facies trends rather than simply isolated vertical changes (Fig. 11). A review of the extensive literature describing the character of fluvial deposits in outcrop is beyond the scope of this paper (however, see Bridge, 2003; Miall 1996, 2013). Lateral facies trends and the geometry of bedding predicted by the channel model used here have been presented previously (Willis, 1989, 1993a; Willis & Tang 2010), so only a brief discussion is included here. As with vertical logs, the first step in interpreting channel belt cross sections is to distinguish deposits of different

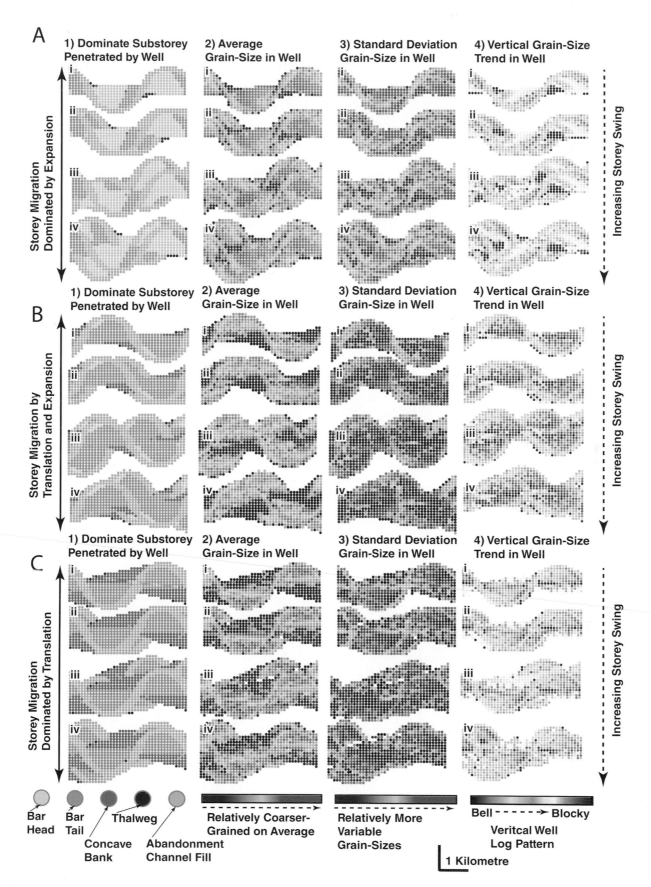

Fig. 9. Summary of vertical grain-size logs predicted for different locations within channel belts for cases where: (A) the belt formed by a channel that migrated mostly by expansion (sinuosity increase), (B) by a channel that migrated by both expansion and downstream transition; and (C) by a channel that moved mostly by downstream translation. For each case: 1) shows the distribution of preserved sub-story deposit types, 2) local vertical averaged grain-size, 3) local vertical standard deviation in grain-size; and 4) strength of the vertical grain-size trend (defined by the slope of a line fit to the data by least squares method). Additional cases show examples where storeys shifted laterally as they migrated, with progressively higher storey swing for cases i, ii, iii and iv, respectively. Cases A1i, B1i and C1i are shown in more detail in Fig. 8A to C, respectively.

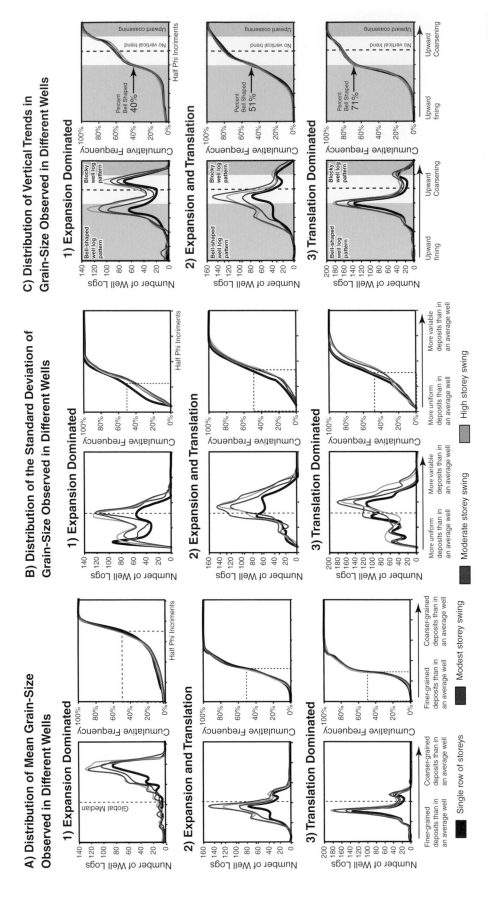

Fig. 10. Summary of vertical grain-size logs predicted for different locations within channel belts: (A) Mean grain-size; (B) Standard deviation of grain-size; and (C) Vertical trend in grain-size for cases of (1) Channel migration by expansion. (2) Expansion and downstream transition. (3) Downstream translation. Black curve on graphs shows character of a single storey row channel belt, whereas coloured curves show the character of channel belts with similar inter-storey facies patterns (i.e. preserved by the same channel migration path) but with different amounts of lateral storey stacking (storey swing). Local vertical grain-size variations within each of these cases are shown in Fig. 9. Vertical grain-size patterns are interpreted to define a 'blocky' well log pattern when the slope of the regression line fit to local vertical grain-size values is steeper than a threshold and either funnel-shaped (upward coarsening) or bell-shaped (upward-fining) when less steep that the chosen threshold.

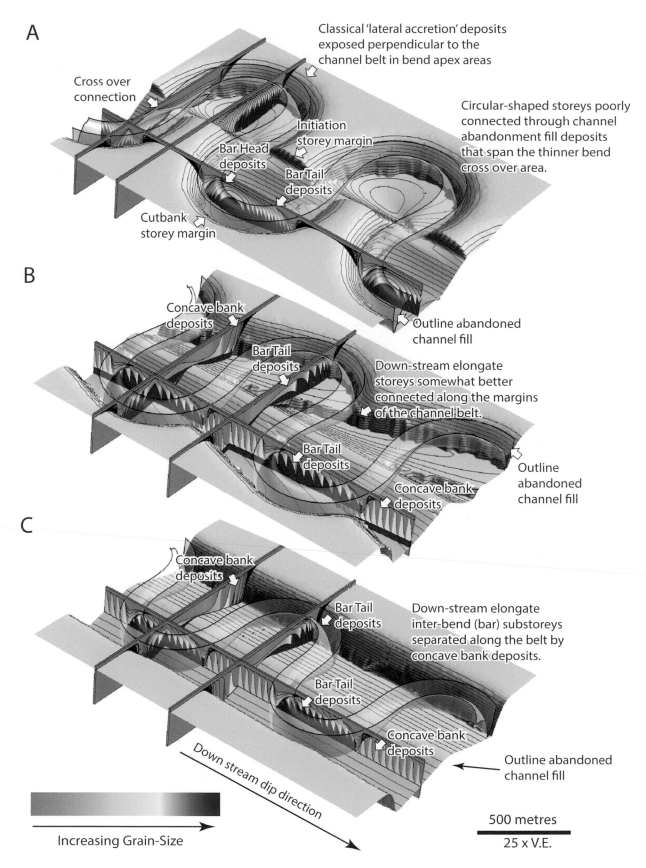

Fig. 11. Selected cross sections through single-storey-row channel belts that were deposited by: (A) dominantly channel bend expansion, (B) both expansion and downstream translation, (C) dominantly downstream translation. Facies variability patterns across different substoreys can be related to topography of the basal erosion surface of the channel belt (shown with half metre contours).

storeys within individual belts (typically separated by erosion surfaces). This is seldom a trivial process because there can be a hierarchy of erosion surfaces and facies changes related to river stage variations or local erosion at scales smaller than the channel bend cut-off that delineates a storey (e.g. Willis 1993b; Bridge & Tye, 2000). Individual storeys can be characterised in terms of their length, lateral change in thickness, internal facies trends and changes in bedding geometry relative to internal palaeoflow indicators. Facies and thickness changes are related to variations across different substorey deposits (as described in the previous section). The number of storeys stacked laterally along an exposed channel belt cross section provides an indication of the number of storeys rows and their swing. The difficulty in interpretation of these patterns is that all cross sections provide an oblique view relative to local palaeoflow orientation and there is typically uncertainty in separating those variations related to spatial changes across the storey and from those related to temporal variations in river deposition (e.g. related to stage change, changes in plan view channel orientation or bend migration path).

Selected cross sections parallel and perpendicular to modelled channel belt deposits are shown in Fig. 11. The models, including cases for which migration was by expansion, translation, or a combination of the two, were described by Willis & Tang (2010). These models are similar to those described in previous sections of this paper, except that discharge was fluctuated between each migration step to highlight the development of river flood bedding as well as substorey facies patterns. As relationships between belt thickness variations, bedding geometry and vertical facies trends are difficult to show in 2D figures, Willis & Tang (2010) provided a collection of movies showing serial cross section slices though single row channel belts with different proportion of substoreys (see Journal of Sedimentary Research data archive @ http://www.sepm.org/pages.aspx?pageid=246).

In the classical case of a cross section perpendicular to the belt centreline, that everywhere transects the apex of a symmetrical river bend, across a deposit generated by expansion only (Fig. 11A), beds initially steepen as the storey becomes thicker away from the initiation storey margin toward the abandonment fill and associated cut bank storey margin. The inclined beds are uniformly perpendicular to internal flow indica-

tors and the distance spanned by an inclined bed is about 2/3 that of the final channel fill (Allen 1979). The ratio between the length of the storey and the distance spanned by an internal inclined bed provides an indication of channel bend sinuosity (Fig. 2B) and by inference the plan view shape of the storey (Fig. 2A; Willis & Tang, 2010). A more general strike cross section (somewhat oblique to bent axis normal or not passing uniformly though apex positions of each successive bend position) will show lateral changes in facies, storey thickness and palaeocurrent orientation related to both changing position within the channel bend and changes in channel shape with sinuosity increase (Willis, 1989, 1993a).

In dip-oriented cross sections, beds within bar head deposits dip up slope (oblique to local palaeoflow within the meandering channel) and those within bar tail deposits dip down slope (Fig. 11A). Although bed inclinations are typically less steep in dip-oriented cross sections (relative to strike-oriented cuts), they can be obvious in outcrop (Willis 1993b). With increased migration by translation (relative to expansion) proportionally fewer bar head deposits are preserved and more concave bank deposits. Concave bank deposits have only recently been widely recognized in outcrop (despite what is expected to be a common occurrence based on modern river bar accretion patterns), presumably because 'chute bar' deposits were generally inferred to be downstream accreting bars and 'Counter point bar' deposits were generally associated with adjacent channel abandonment fill deposits (however, see Smith *et al.*, 2009; Labrecque *et al.*, 2011; and Durkin *et al.*, 2015). Bedding within the concave bank deposits shown in this model reflects the migration of a steep outer bank cut bank channel margin (Fig. 11B, C). In nature, the bedding within concave band deposits can vary significantly between examples, from steep downstream dipping cross sets formed on the avalanche facies of chute bars to horizontal bedded low flow laminated muds within concave bank benches (Bluck, 1976; Carey 1969; Nanson & Page, 1983; Alexander 1992; Burge & Smith 1999; Makaske & Weerts 2005; and Smith *et al.* 2009). Although strike-oriented cross sections tend to be less useful for channel width estimation (typically used in palaeo-discharge estimates), they provide better constraints on storey length (which can also be related to river discharge), channel migration pattern (expansion *vs.* translation) and mechanisms controlling storey stacking pattern (frequency of loop neck *vs.* chute cut off) (Willis 1993a, 1993b and Willis & Tang, 2010).

FACIES PATTERNS WITHIN SINGLE AND MULTI-ROW STOREY BELTS

An example of a single-storey-row and a multi-storey-row channel belt with similar internal storeys is shown in Fig. 12. In both cases the storeys formed during channel migration by modest expansion and downstream translation (bends cut off at lower sinuosity than in the previous examples). Although facies variations (and the predicted mean and variance of permeability and porosity values) within the two belts are similar, the multi-row belt has about 5% fewer abandonment-fill deposits than the single row case. Storey stacking within channel belts deposited by lower-sinuosity channels generally increases channel abandonment fill proportion (sinuosity < 1.125), whereas in those formed by higher-sinuosity rivers added storeys generally decrease channel abandonment fill proportion. Other sub-storey proportions are similar, indicating that adding storeys generally removes these facies by erosion in similar proportion to that in which they are added.

Larger-scale facies patterns within channel belts can be simplified into two classes of rock bodies: 1) Those that have significantly lower permeability than the mean (typically abandonment fills and in some cases concave-bank deposits); and 2) Those that are significantly more permeable than the mean value (commonly thalweg outer bend or junction-scour lags but also in some cases bar-head and chute-channel deposits). Heterogeneity patterns within a channel belt depend on lateral associations of these contrasting facies bodies across belt thickness variations defined by the topography of the basal erosion surface (Fig. 12A).

In the multi-storey-row channel belt the various channel-abandonment fills link up in anastomosing map-view patterns that result from local migration and cut-offs of different curved channel segments over time (Fig. 12B2). In three-dimensions the role of these channel-abandonment fills in separating coarser-grained flow-unit continuity is complicated. Higher-sinuosity channel segments cut deeper on average and all channels incise deeper along the bend apex than along crossover areas. The lateral swing of storeys in

multi-storey-row channel belts locally superimpose deposits of shallower channel segments over deposits that were previously formed along deeper channel segments. Therefore, although the abandonment fills cut all the way through the belt in some places, at other locations they cut only part way through. This lateral variability in channel incision depth is superimposed on complicated lithic patterns within the abandonment-fill deposits (discussed previously in Willis & Tang, 2010). In multi-storey-row channel belts the topography of a basal erosion surface reflects the cumulative effect of the deepest incision that occurred in specific areas of the channel course as different storeys formed and thus the distribution of channel belt thickness in different locations tends to be skewed towards the maximum channel depth (c.f. single-storey row belts that tend to have dominant mode closer in value to mean bankfull channel depth; Fig. 12C).

Coarser-grained (i.e. high-permeability) deposits occur preferentially in areas where the basal erosion surface of the belt is more deeply incised (Fig. 12A3 and B3). In the single-storey-row example these thicker areas tend to be isolated along alternating sides of the belt. In multi-storey-row channel belts it is more probable for these coarser-grained bodies to locally amalgamate, as highlighted by the distribution and size of high flux areas within the single-storey-row and multi-storey-row examples in Fig. 12A4 and B4. Predicting the continuity of preferential flow pathways is very important in some fluvial channel-belt reservoirs because they can lead to irregular sweep and early water breakthrough during production. However, such a prediction is complicated because the distribution of connected lags can change for rivers with different sediment load and in storeys with different migration and aggradation patterns, as well as different storey stacking arrangements.

Streamline flow simulation was used to test if the single and multi-row cases displayed different reservoir behaviour. Simulations on the belts were run to sample two scales of facies variations; 1) A five-spot pattern of injectors on a half kilometre grid spacing (Fig. 12A4 and B4) with producers at grid centres (60 acre spacing of injectors and thus 15 acre spacing for injector-producer pairs; and a simulated line drive flood between a row of injectors at one end of the channel belt and producers at the other [Fig. 12A5 and B5], spaced about 7.5 km apart, i.e. about 2.5 times channel

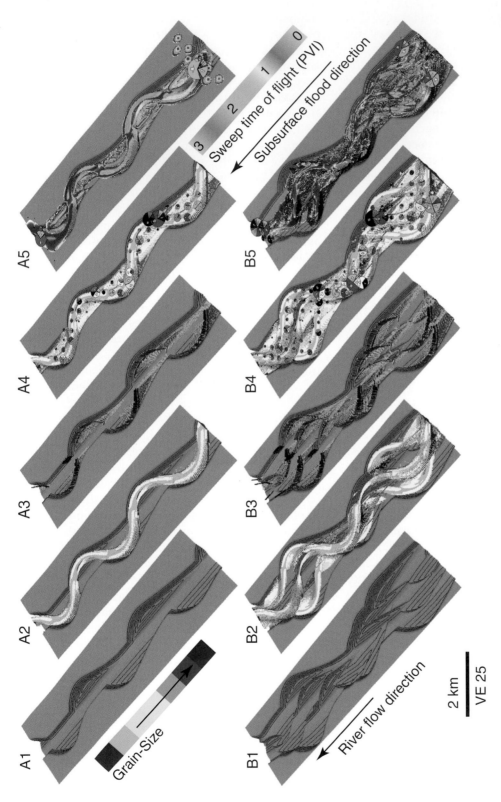

Fig. 12. Single-storey-row (A) and multi-storey-row (B) channel belts. Channels used to generate the models had 175 m width, 15 m mean depth, 3750 m wavelength and sinuosity that increased from 1 to 1.175 as the channel migrated 1 km downstream. The mean grain-size transported was upper fine-grained sandstone and grain-size varied between clay and very-coarse-grained sandstone. Grain sizes predicted within the channel belts were used to define permeability and porosity using empirical functions of core measured from a conventional hydrocarbon reservoir of interest to Chevron, which calculated permeability to vary between 0.01mD and 3.5D; and porosities between 14 and 24%. Small-scale permeability anisotropy (grid block Kh/Kv) was set to 10. (A1 and B1) Oblique view of basal erosion surface with 5 m contour interval. (A2 and B2) Channel-abandonment-fill deposits superimposed on the basal erosion surface are coloured by grain-size. (A3 and B3) Coarsest 20% of the channel belt deposits superimposed on the basal erosion surface are coloured by grain-size (coarse sand to granule). Note the concentration of these coarser grain sizes in lows on the basal erosion surface and the lateral streaking of these deposits along the channel belt due to downstream storey migration. (A4 and B4) Sandy deposits within channel belts superimposed on the basal erosion surface are coloured by grain-size (mudstones are invisible). The diameter of pie charts, centred on each well, show relative flux between injectors and producers spaced about 350 m apart. Areas of sectors in pie charts show the relative contribution of injectors to producers (and vice versa). Arrows point toward injectors and away from producers. A5 and B5) Sandy deposits within channel belts superimposed on the basal erosion surface are coloured by the dimensionless time they were swept (dimensionless time defined in terms of pore volumes injected). The diameter of pie charts centred on each well show flux from the different injectors at one end of the channel belt model to the producers at the opposite end.

bend wavelength). Streamline simulations compared recovery as defined by sweep efficiency at 1 pore volume injected (PVI), dimensionless flux (defined in terms of pore volumes injected) and Dynamic Lorenz coefficient (defined by Snook, 2009) for the single-storey-row and multi-storey-row cases (Fig. 13). The two production scenarios were designed to investigate scale differences in dynamic reservoir performance: the five-spot well distribution measures 'local' substorey variations and a line-drive flood along the belt through several adjacent storeys measures inter-storey heterogeneity.

The cumulative frequency distribution of the sweep efficiency of the different wells at one pore volume injected (Fig. 13A) indicates better recovery overall by line-drive floods compared to the spot-pattern production. Although the multi-storey-row belt had lower average recovery (at 1 PVI) than the single-storey-row belt under five-spot production, differences were subtle and there was wide overlap in values of individual wells developed within the single-row and multi-row belts. Visual inspection of the streamlines in time sequence suggested that this difference in mean recovery reflects greater direct connection of injector-producer well pairs in local areas through basal coarse-grained lags in the multi-storey-row belt, which could leave finer-grained deposits unswept. The best wells in the line-drive single-storey-row and multi-storey-row belt floods had similar recovery and flux. The apparent increase in the number of poor preforming wells in the multi-storey-row case might not be significant owing to the low total number of wells being compared (i.e. only 3 were active at each end of the single-row case).

Flux between injectors and producers in the local 5 spot patterns was significantly greater on average than for line-drive flood production, which reflects broader averaging of higher and lower reservoir quality deposits over longer distances. Higher flux for the multi-storey-row case reflects alignment of fast streaks over longer distances (Fig. 12). Higher dynamic Lorenz coefficients measured for five-spot production wells relative to those under line drive production (indicating greater heterogeneity) also reflects permeability averaging over the longer distances travelled by the line-drive streamlines.

The simulation experiments suggested there are some differences in reservoir behaviour for the single and multi-storey belt cases related to local alignment of either finer-grained or coarser-grained facies bodies. Although these differences do not appear to be dramatic, relative to differences caused by different inter-storey facies patterns, observed contrasts motivated a more extensive design simulation study to predict the relative influence of intra-storey facies patterns, storey amalgamation and inter-belt connectivity (see Willis & Sech, this volume).

CONCLUSIONS

1. Channel-belt internal heterogeneities are defined by the geometry and facies changes across one or more rows of storeys. Each storey is formed by the migration and cutoff of a curved channel segment. Storeys can have circular or down-slope-elongate planform geometry, depending on how the channel migrates. They tend to thicken as the associated channel becomes more sinuous.

2. Storeys can be divided into five lithic sub-units that formed adjacent to specific areas along a river channel: 1) Bar head, 2) Bar tail, 3) Concave bank, 4) Thalweg scour pool lags; and 5) Abandonment fill. These substoreys are characterised by contrasting vertical facies trends; and their relative abundance within a channel belt depends on how the river migrated and switched positon over time.

3. Predicting facies patterns within channel belts can be defined geometrically as an application of Walther's Law applied to a 3D landform and thus requires an appreciation of how channels change position and shape as they migrate. Channel migration by expansion tends to preserve inner-bank deposits (bar head and bar tail), whereas migration by downstream translation favours preservation of downstream facing areas within the channel (bar-tail and concave-bank deposits). The proportion of preserved channel-abandonment fill relative to other types of substoreys (bar and concave bank) generally decreases with river sinuosity. Multiple rows lead to modest decreases in preserved channel-fill proportion in cases such as those modelled where the formative channel has moderate sinuosity (cf. cases formed by very low-sinuosity channels, where additional amalgamation of storeys can increase channel-abandonment fill proportion).

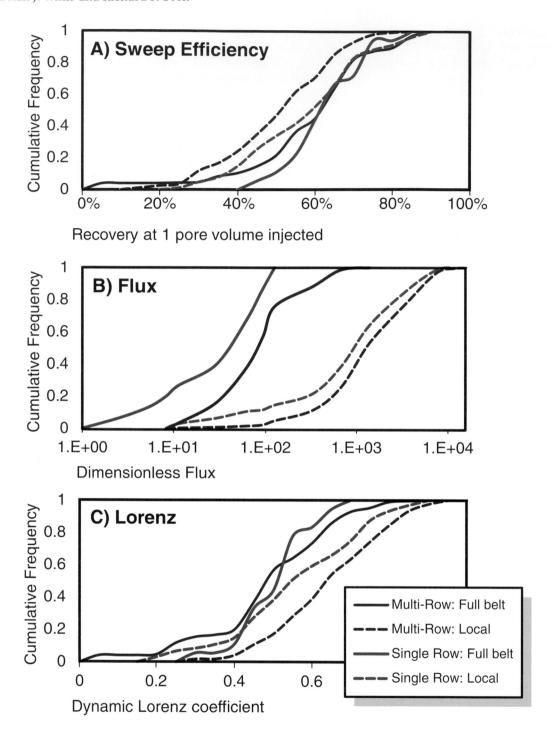

Fig. 13. Comparison of streamline flow simulations of a single-storey-row belt (blue lines) and multi-storey-row belt (red lines) for: 1) Injector – producer wells in a half-kilometre grid five-spot pattern ('local', dashed lines) and 2) A long distance flood along the dip direction of the channel belt ('full', solid lines). (A) The distribution of sweep efficiency at 1 PVI compiled for all injectors and producers. (B) The distribution of flux compiled for all injectors and producers. (C) Dynamic Lorenz coefficient calculated from streamlines traveling from all injectors to all producers.

4. Channel-belt width, typically an order of magnitude greater than channel width, will increase with river discharge, sinuosity and the number and swing of storey rows. Channel belt thickness varies with channel sinuosity and is typically 2 to 3 times mean bankfull channel depth.

5. Although it is generally true that larger rivers form thicker channel belt deposits, the magnitude of variations in local belt thickness change with river sinuosity, channel bend migration pattern and lateral storey stacking patterns. As the preserved distribution of deposit thicknesses along a belt is not a simple function of channel depth at any one time, it should come as no surprise that empirical relationships between channel geometry and preserved channel belt dimensions show wide variations from predicted average trends.

6. The proportion of vertical logs penetrating a channel belt with strongly upward-fining trend (bell-shaped log pattern) and those with more subtle vertical grain-size variations (blocky-shaped log pattern) depends of the proportion of preserved substorey deposits. Channel migration dominated by expansion, preserving more bar-head deposits, would be characterised by more blocky well patterns; and those that migrated more by downstream translation (preserving more bar-tail and concave-bank facies) are predicted to be characterised by more bell-shaped well log penetrations. These conclusions are also sensitive to processes of concave-bank deposition, as chute bar growth into concave-bank areas would also tend to increase the predicted proportion of well penetrations with blocky well log pattern (a process more common in straighter channels than the higher-sinuosity cases modelled here). In any case, individual well penetrations are not expected to be diagnostic of channel-belt types and subsurface interpretation requires comparison of the relative proportion of well log patterns sampled in multiple penetrations.

7. Single-row and multi-row belts may have some differences in reservoir behaviour related to how facies link up across stacked storeys. Stacking of storeys within channel belts does not dramatically change facies proportion because those removed during the emplacement of successive storeys occur in similar proportion to those added.

ACKNOWLEDGEMENTS

Thanks to members of Chevron's Clastic Stratigraphy R&D team for discussions on fluvial reservoir characterization: Bryan Bracken, Sean Connell, Kristy Milliken, Tobi Payenberg, Pete Sixsmith and Morgan Sullivan (team lead). Dennis Dull, Michael Pyrcz and Fabien Laugier provided advice on reservoir modelling. Sean Connell and Michael Pyrcz provided peer reviews of an internal company report from which this contribution was extracted. Matt Pranter, Milovan Fustic and associate editor Luca Colombera are thanked for comments on the initial manuscript that helped clarify the presentation.

REFERENCES

Alexander, J. (1992) Nature and origin of a laterally extensive alluvial sandstone body in the Middle Jurassic Scalby Formation. *J. Geol. Soc. London*, **149**, 431–441.

Alexander, J. and **Leeder, M.R.** (1987) Active tectonic control on alluvial architecture. In: *Recent Developments in Fluvial Sedimentology: Contributions from the Third International Fluvial Sedimentology Conference* (Eds F.G. Ethridge, R.M. Flores and M.D. Harvey). *SEPM Spec. Publ.*, **39**, 243–252.

Allen, J.R.L. (1965) A review of the origin and characteristics of recent alluvial sediments. *Sedimentology*, **5**, 88–191.

Allen, J.R.L. (1979) Studies in fluviatile sedimentation: an elementary geometrical model for the connectedness of avulsion-related channel sand bodies. *Sed. Geol.*, **24**, 253–267.

Allen, J.R.L. (1963) The classification of cross-stratified units with notes on their origin. *Sedimentology*, **2**, 93–114.

Armstrong, C., Mohrig, D., Thomas, H., Terra, G. and **Straub, K.** (2014) Influence of growth faults on coastal fluvial systems: Examples from the late Miocene to Recent Mississippi River Delta. *Sed. Geol.*, **301**, 120–132.

Ashworth, P.J., Sambrook Smith, G.H., Best, J.L., Bridge, J.S., Lane, S.N., Lunt, I.A., Reesink, A.J.H., Simpson, C.J. and **Thomas, R.E.** (2011) Evolution and sedimentology of a channel fill in the sandy braided South Saskatchewan River and its comparison to the deposits of an adjacent compound bar. *Sedimentology*, **58**, 1860–1883.

Best, J.L. and **Ashworth, P.J.** (1997) Scour in large braided rivers and the recognition of sequence stratigraphic boundaries. *Nature*, **387**, 275–277.

Bluck, B.J. (1976) Sedimentation in some Scottish rivers of low sinuosity. *Trans. Roy. Soc. Ed.*, **69**, 425–456.

Bluck, B.J. (1971) Sedimentation in the meandering River Endrick. *Scot. J. Geol.*, **7**, 93–138.

Blum, M., **Martin, J.**, **Milliken, K.** and **Garvin, B.M.** (2013) Paleovalley systems: Insights from Quaternary analogs and experiments. *Earth-Sci. Rev.*, **116**, 128–169.

Bridge, J.S. (1982) A revised mathematical model and FORTRAN IV program to predict flow in open-channel bends. *Comput. Geosci.*, **8**, 91–95.

Bridge, J.S. (1992) A revised model for water flow, sediment transport, bed topography and grain-size sorting in natural river bends. *Water Resour. Res.*, **28**, 999–1013.

Bridge, J.S. (1976) Bed topography and grain-size in open channel bends: *Sedimentology*, **23**, 407–414.

Bridge, J.S. (1993a) Description and interpretation of fluvial deposits: A critical perspective. *Sedimentology*, **40**, 801–810.

Bridge, J.S. (1977) Flow, bed topography, grain-size and sedimentary structure in open channel bends: A three-dimensional model. *Earth Surf. Proc. Land.*, **2**, 401–416.

Bridge, J.S. (1985) Paleochannel patterns inferred from alluvial deposits: A critical evaluation. *J. Sed. Petrol.*, **55**, 579–589.

Bridge, J.S. (2003) *Rivers and Floodplains; Forms, Processes and Sedimentary Record.* Blackwell Publishing, Oxford, 491 pp.

Bridge, J.S. (1993b) The interaction between channel geometry, water flow, sediment transport and deposition in braided rivers. In: *Braided Rivers* (Eds J.L. Best and C.S. Bristow), *Geol. Soc. Spec. Publ.*, **75**, 13–71.

Bridge, J.S. and **Dominic, D.F.** (1984) Bed load grain velocites and sediment transport rates. *Water Resour. Res.*, **20**, 476–490.

Bridge, J.S. and **Jarvis, J.** (1976) Flow and sedimentary processes in the meandering river South Esk, Glen Clova, Scotland. *Earth Surf. Proc. Land.*, **1**, 303–336.

Bridge, J.S. and **Jarvis, J.** (1982) The dynamics of a river bend: A study in flow and sedimentary processes. *Sedimentology*, **29**, 499–541.

Bridge, J.S. and **Mackey, S.D.** (1993) A theoretical study of fluvial sandstone body dimensions. In: *The Geological Modeling of Hydrocarbon Reservoirs* (Eds S. Flint and I.D. Bryant), *Int. Assoc. Sedimentol. Spec. Publ.*, **15**, 213–236.

Bridge, J.S., **Smith, N.D.**, **Trent, F.**, **Gable, S.L.** and **Bernstein, P.** (1986) Sedimentology and morphology of a low-sinuosity river: Calamus River, Nebraska Sandhills. *Sedimentology*, **33**, 851–870.

Bridge, J.S. and **Tye, R.S.** (2000) Interpreting the dimensions of ancient fluvial channel bars, channels and channel belts from wire-line logs and cores. *AAPG Bull.*, **84**, 1205–1228.

Brierley, G.J. and **Hickin, E.J.** (1991) Channel planform as a non-controlling factor in fluvial sedimentology: the case of the Squamish River floodplain, British Columbia. *Sed. Geol.*, **75**, 67–83.

Bristow, C.S. and **Best, J.L.** (1993) Braided rivers: perspectives and problems. In: *Braided Rivers* (Eds J.L. Best and C.S. Bristow), *Geol. Soc. Spec. Publ.*, **75**, 1-H.

Burge, L.M. and **Smith, D.G.** (1999) Confined meandering river eddy accretions: sedimentology channel geometry and depositional processes. In: *Fluvial Sedimentology VI* (Eds N.D. Smith and J. Rogers), *Int. Assoc. Sedimentol. Spec. Publ.*, **28**, 113–130.

Cant, D.J. and **Walker, R.G.** (1978) Fluvial processes and facies sequences in the sandy braided South Saskatchewan River, Canada. *Sedimentology*, **25**, 625–648.

Carey, W.C. (1969) Formation of floodplain lands. Proceedings, *J. Hydraul. Div. Am. Soc. Civ. Eng.*, **95**, 981–994.

Collinson, D. (1978) Vertical sequence and sand body shape in alluvial sequences. In: *Fluvial Sedimentology. Mem. Can. Soc. Petrol. Geol.*, **5**, (Ed. A.D. Miall), 577–588.

Colombera, L., **Mountney, N.P.**, **Russell, C.E.**, **Shiers, M.N.** and **McCaffrey, W.D.** (2017) Geometry and compartmentalization of fluvial meander-belt reservoirs at the bar-form scale: Quantitative insight from outcrop, modern and subsurface analogues. *Mar. Petrol. Geol.*, **82**, 35–55.

Deptuck, M.E., **Sylvester, Z.**, **Pirmez, C.**, **O'Byrne, C.** (2007) Migration–aggradation history and 3-D seismic geomorphology of submarine channels in the Pleistocene Benin-major Canyon, western Niger Delta slope. *Mar. Petrol. Geol.*, **24**, 406–433.

Dietrich, W.E. (1987) Mechanics of flow and sediment transport in river bends. In: *River Channels: Environment and Process* (Ed. K.S. Richards), *Inst. British Geogr. Spec. Publ.*, **18**, 179–227.

Dietrich, W.E. and **Smith, J.D.** (1984) Bedload transport in a river meander. *Water Resour. Res.*, **20**, 1355–1380.

Dietrich, W.E. and **Smith, J.D.** (1983) Influence of the point bar on flow in curved channels. *Water Resour. Res.*, **19**, 1173–1192.

Donselaar, M.E. and **Overeem, I.** (2008) Connectivity of fluvial point-bar deposits: An example from the Miocene Huesca fluvial fan, Ebro Basin, Spain. *AAPG Bull.*, **92**, 1109–1129.

Eaton, B.C. (2013) Hydraulic geometry: empirical investigations and theoretical approaches. In: *Treatise on Geomorphology* (Eds J. Schroder and E.E. Wohl), *Fluvial Geomorphology*, **9**, Academic Press, San Diego, 313–329.

Engelund, F. (1974) Flow and bed totography in channel bends. *J. Hydraul. Div. ASCE*, **100**, 1631–1648.

Ethridge, F.G. (2010) Interpretation of ancient fluvial channel deposits: review and recommendations. In: *From River to Rock Record* (Eds K. Stephanie, S.K. Davidson, S. Leleu and C.P. North), *SEPM Spec. Publ.*, **97**, 9–35.

Ethridge, F.G., **Skelly, R.L.** and **Bristow, C.S.** (1999) Avulsion and crevassing in the sandy, braided Niobrara Rivr: Complex responce to base-level rise and aggradation. In: *Fluvial Sedimentology VI* (Eds N.D. Smith and J. Rodgers). *Int. Assoc. Sediment. Spec. Pub.* **28**, 179–191.

Ferguson, R.I. and **Parsons, D.R.** (2012) Flow in meander bends with recirculation at the inner bank. *Water Resour. Res.*, **39**, 1322–1235.

Fielding, C.R. and **Crane, R.C.** (1987) An application of statistical modeling to the prediction of hydrocarbon recovery factors in fluvial reservoir sequences. In: *Recent Developments in Fluvial Sedimentology* (Eds F.G. Ethridge, R.M. Flores and M.D. Harvey), *SEPM Spec. Publ.*, **39**, 321–327.

Fisk, H.N. (1952) Geological investigation of the Atchafalaya Basin and the problem of the Mississippi River diversion: Vicksburg, Mississippi, U.S. Army Corps of Engineers, Waterways Experiment Station, pp. 145.

Ghinassi, M., Ielpi, A., Mauro, A. Aldinucci, M. and **Fustic, M.** (2016) Downstream-migrating fluvial point bars in the rock record. *Sed. Geol.*, **334**, 66–96.

Gibling, M.R. (2006) Width and thickness of fluvial channel bodies and valley fills in the geological record: A literature compilation and classification. *J. Sed. Res.*, **76**, 731–770.

Gleason, C. (2015) Hydraulic geometry of natural rivers: A review and future directions. *Prog. Phys. Geogr.*, **1**, 1–24.

Ikeda, H. (1989) Sedimentary controls on channel migration and origin of point bars in sand-bedded meandering rivers. In: River Meandering (Eds S. Ikeda and G. Parker). *AGU Wat. Res. Mon.* **12**, 51–68.

Jackson, R.G. (1976) Depositional model of point bars in the lower Wabash River. *J. Sed. Petrol.*, **46**, 579–594.

Jones, A., Doyle, J., Jacobsen, T. and **Kjonsvik, D.** (1995) Which sub-seismic heterogeneities influence waterflood performance? A case study of a low net-to-gross fluvial reservoir. In: *New Developments in Improved Oil Recovery* (Ed. H.J. De Haan), *Geol. Soc. Spec. Publ.*, **84**, 5–18.

Jordan, D.W. and **Pryor, W.A.** (1992) Hierarchical levels of heterogeneity in the Mississippi River Meander belt and application to reservoir systems. *AAPG Bull.*, **76**, 1601–1624.

Keogh, K.J., Martinius, A.W. and **Osland, R.** (2007) The development of fluvial stochastic modelling in the Norwegian oil industry: A historical review, subsurface implementation and future directions. *Sed. Geol.*, **202**, 249–268.

Langbein, W.B. and **Leopold, L.B.** (1966) River Meanders: Theory of Minimum Variance. *U.S. Geol. Surv. Prof. Pap.*, **422-H**, 15 pp.

Leopold, L.B. and **Wolman, M.G.** (1957) River Channel Patterns: Braided, Meandering and Straight. *U.S. Geol. Surv. Prof. Pap.*, **28B**, 39–85.

Leopold, L.B. and **Wolman, M.G.** (1960) River Meanders. *Geol. Soc. Am. Bull.*, **71**, 769–793.

Lorenz, J.C., Heinze, D.M., Clark, J.A. and **Searls, C.A.** (1985) Determination of widths of meander-belt sandstone reservoirs from vertical downhole data, Mesaverde Group, Piceance Creek Basin, Colorado. *AAPG Bull.*, **69**, 710–721.

Lunt, I.A., Bridge, J.S. and **Tye, R.S.** (2004) A quantitative, three-dimensional depositional model of gravelly braided rivers. *Sedimentology*, **51**, 377–414.

Mackey, S.D. and **Bridge, J.S.** (1995) Three-dimensional model of alluvial stratigraphy: theory and application. *J. Sed. Res.*, **65**, 7–31.

Makaske, B. and **Weerts, H.J.T.** (2005) Muddy lateral accretion and low stream power in a sub-recent confined channel belt, Rhine–Meuse delta, central Netherlands. *Sedimentology*, **52**, 651–668.

Marzo, M., Nijman, W. and **Puigdefabregas, C.** (2006) Architecture of the Castissent fluvial sheet sandstones, Eocene, South Pyrenees, Spain. *Sedimentology*, **35**, 719–738.

Miall, A.D. (2013) *Fluvial Depositional Systems*, Springer-Verlag, Inc., Heidelberg, 316 pp.

Miall, A.D. (1996) *The Geology of Fluvial Deposits: Sedimentary Facies, Basin Analysis and Petroleum Geology*, Springer-Verlag, Inc., Heidelberg, 582 pp.

Nanson, G.C. and **Croke, L.C.** (1992) A generic classification of floodplains. *Geomorphology*, **4**, 459–486.

Nanson, G.C. and **Page, K.** (1983) Lateral accretion of fine-grained concave benches associated with meandering rivers. In: *Modern and Ancient Fluvial Systems* (Eds J.D. Collinson and J. Lewin), *Int. Assoc. Sedimentol. Spec. Publ.*, **6**, 133–143.

Payenberg, T., Willis, B., Pusca, V., Sixsmith, P., Bracken, B., Posamentier, H., Pyrcz, M., Sech, R., Connell, S., Milliken, K. and **Sullivan, M.** (2014) Channel Belt Rugosity in Reservoir Characterization. AAPG, Annual Convention and Exhibition, Houston, Texas, April 6–9. Search and Discovery Article #41420.

Peakall, J., Leeder, M.R., Best, J. and **Ashworth, P.** (2000) River response to lateral ground tilting: a synthesis and some implications for the modeling of alluvial architecture in extensional basins. *Basin Res.*, **12**, 413–424.

Posamentier, H.W. and **Kolla, V.** (2003) Seismic geomorphology and stratigraphy of depositional elements in deep-water settings. *J. Sed. Res.*, **73**, 367–388.

Potter, P.E. (1967) Sand bodies and sedimentary environments: a review. *AAPG Bull.*, **51**, 337–365.

Pranter, M.J., Ellison, A.I., Cole, R.D. and **Patterson, P.E.** (2007) Analysis and modeling of intermediate-scale reservoir heterogeneity based on a fluvial point-bar outcrop analog, Williams Fork Formation, Piceance Basin, Colorado. *AAPG Bull.*, **91**, 1025–1051.

Reesink, A.J.H. and **Bridge, J.S.** (2011) Evidence of bedform superimposition and flow unsteadiness in unit-bar deposits, South Saskatchewan River: Canada. *J. Sed. Res.*, **81**, 814–840.

Seminara, G. and **Tubino, M.** (1989) Alternate bars and meandering: free, forced and mixed interactions. In: *River Meandering, No. 12* (Eds S. Ikeda and G. Parker), *Am. Geophys. Union*, Washington, DC, p. 153–180.

Shook, M. (2009) A Robust Measure of Heterogeneity for Ranking Earth Models: The F PHI Curve and Dynamic Lorenz Coefficient. *Soc. Petrol. Eng.* Annual Technical Conference and Exhibition, 4–7 October, New Orleans, Louisiana, SPE **124625**, p. 1–13.

Singh, V.P. (2003) On the theories of hydraulic geometry. *Int. J. Sed. Res.*, **18**, 196–218.

Smith, D.G., Hubbard, S.M., Lavigne, J.R., Leckie, D.A. and **Fustic, M.** (2011) Stratigraphy of counter point bar and eddy-accretion deposits in low energy meander belts of the Peace-Athabasca delta, northeast Alberta, Canada. In: *From River to Rock Record: The Preservation of Fluvial Sediments and Their Subsequent Interpretation* (Eds S.K. Davidson, S. Leleu and C.P. North), *SEPM Spec. Publ.*, **97**, 143–152.

Smith, D.G., Hubbard, S.M., Leckie, D.A. and **Fustic, M.** (2009) Counter point bar deposits: lithofacies and reservoir significance in the meandering modern Peace River and ancient McMurray Formation, Alberta, Canada. *Sedimentology*, **56**, 1655–1669.

Smith, N.D., Cross, T.C., Dufficy, J.P. and **Clough, S.R.** (1989) Anatomy of an avulsion. *Sedimentology*, **36**, 1–23.

Stouthamer, E. (2001) Sedimentary products of avulsions in the Holocene Rhine-Meuse Delta, The Netherlands. *Sed. Geology*, **145**, 73–92.

Stouthamer, E. and **Berendsen, H.J.A.** (2001) Avulsion frequency, avulsion duration and interavulsion period of holocene channel belts in the Rhine-Meuse delta, The Netherlands. *J. Sed. Res.*, **71**, 589–598.

Strong, N., **Sheets, B.A.**, **Hickson, T.A.** and **Paola, C.** (2005) A mass-balance framework for quantifying downstream changes in fluvial architecture. In: *Fluvial Sedimentology VII* (Eds M. Blum, S. Marriott and S. Leclair), *Int. Assoc. Sedimentol. Spec. Publ.*, **35**, 243–253.

Struiksma, N., **Olesen, K.**, **Flokstra, C.** and **De Vriend, H.** (1985) Bed deformation in curved alluvial channels. *J. Hydraul. Res.*, **23**, 57–79.

Sun, T., **Jøssang, T.**, **Meakin, P.** and **Schwarz, K.** (1996) A simulation model for meandering rivers. *Water Resour. Res.*, **32**, 2937–2954.

Taha, Z.P. and **Anderson, J.B.** (2008) The influence of valley aggradation and listric normal faulting on styles of river avulsion: a case study of the Brazos River, Texas, USA. *Geomorphology* **95**, 429–448.

Tornqvist, T.E. (1994) Middle and late Holocene avulsion history of the River Rhine (Rhine–Meuse delta, Netherlands). *Geology*, **22**, 711–714.

Tye, R.S. (2013) Quantitatively modeling alluvial strata for reservoir development with examples from Krasnoleninskoye field, Russia. In: *Proceedings, Symposium in Applied Coastal Geomorphology to Honor Miles O. Hayes* (Eds T. Kana, J. Michel and G. Voulgaris), *J. Coastal Res.*, 129–152.

Williams, G.P. (1986) River meanders and channel size. *J. Hydrol.*, **88**, 147–164.

Willis, B.J., 1993b, Ancient river systems in the Himalayan foredeep, Chinji Village area, northern Pakistan, *Sed. Geol.*, **88**, 1–76.

Willis, B.J. (1993a) Bedding geometry of ancient point bar deposits. In: *Alluvial Sedimentation* (Eds M. Marzo and C. Puigdefabregas), *Int. Assoc. Sedimentol. Spec. Publ.*, **17**, 101–114.

Willis, B.J. (1989) Paleochannel reconstructions from point bar deposits: A three-dimensional perspective. *Sedimentology*, **36**, 757–766.

Willis, B.J. and **Sech, R.** (this volume) Quantifying impacts of fluvial intra-channel-belt heterogeneity on reservoir behavior. Meandering Rivers (Eds xxxx) *Int. Assoc. Sedimentol. Spec. Publ.*, **48**, xxx–xxx.

Willis B.J. and **Tang, H.** (2010) Three-dimensional connectivity of point-bar deposits. *J. Sed. Res.*, **80**, 440–454.

Wynn, R.B., **Cronin, B.T.** and **Peakall, J.** (2007) Sinuous deep-water channels: Genesis, geometry and architecture. *Mar. Petrol. Geol.*, **24**, 341–387.

Int. Assoc. Sedimentol. Spec. Publ (2018) **48**, 543–572.

Quantifying impacts of fluvial intra-channel-belt heterogeneity on reservoir behaviour

BRIAN J. WILLIS* and RICHARD P. SECH[†]

* *Clastic Stratigraphy Team, Earth Science Department, Chevron Energy Technology Company, Houston, TX, USA*
[†] *Reservoir Management Research Team, Chevron Energy Technology Company, Houston, TX, US*

ABSTRACT

Fluvial channel belt deposits formed by a meandering river are modelled to predict variations in fluid flow patterns through subsurface hydrocarbon reservoirs and aquifers. A series of numerical experiments examine the impact of channel belt connectivity and intra-channel belt facies patterns on reservoir behaviour during tracer flood displacement. Channel belts with either connected or disconnected adjacent storeys are modelled to contrast the static connectivity and dynamic flow behaviour of fluvial deposits with progressively higher net to gross (channel belt proportion). The results suggest intra-storey connectivity is a critical factor in estimating reservoir performance in successions with low channel belt proportion. A larger design simulation experiment compared cases with different intra-storey facies patterns and storey stacking arrangements within belts. Although the 'river' in these models stayed the same (constant discharge, hydraulic geometry and mean grain-size), this experiment predicted a wide range of average channel belt heterogeneity for cases with contrasting channel migration, storey stacking and channel abandonment style. A line-drive water flood through these channel belts modelled with a subsurface streamline flow simulator had sweep efficiency between 24 and 54% and average flux varied by a factor of 4. This experiment was repeated twice more to demonstrate large increases in recovery and flux with increasing net channel belt aggradation. Further experiments demonstrate changes in development behaviour where two channel belts are connected. Recovery increased from parallel connected belts where they were offset enough to separate vertically their respective basal lags, as this geometry led to a more stable displacement. Where channel belts cross cut at higher angle, recovery was less sensitive to the amount of vertical offset. These experimental results indicate that flux variations across facies with different permeability can significantly impact reservoir production behaviour. Improvements in reservoir characterisation and models for production forecasts, need to be focused on better definition of 3D facies heterogeneity patterns within channel belts rather than just predicting channel belt geometry and static connectivity.

Keywords: Fluvial, heterogeneity, connectivity, reservoir, design simulation

INTRODUCTION

Preserved 3D architecture and spatial distribution of reservoir properties within fluvial deposits vary over a wide range of length scales, from millimetre-thick mud drapes to variable connectivity patterns between adjacent channel belts over kilometres (Bridge & Diemer, 1983, Jordon & Pryor, 1992; Dreyer *et al.*, 1993; Willis 1993a, b; Jones *et al.* 1995; Keogh *et al.*, 2007; and many others). The representation of heterogeneity across these length scales within reservoir models is limited by several factors, including (1) an understanding of the geometry and spatial relationships of fluvial sandbodies, (2) a reliable appreciation of which types of heterogeneity influence fluid flow behaviour under different development scenarios; and (3) the ability to specify geological features using

Fluvial Meanders and Their Sedimentary Products in the Rock Record, First Edition.
Edited by Massimiliano Ghinassi, Luca Colombera, Nigel P. Mountney and Arnold Jan H. Reesink.

conventional modelling approaches. In most fluvial deposits channel belts comprise the main reservoir elements and the discussion here is restricted to understanding the connectivity and internal heterogeneity patterns within these bodies.

Typical field models of fluvial deposits focus on the abundance, geometry and variable orientation of channel belts and define depositional heterogeneities within these reservoir elements using statistical techniques that represent a stationary arrangement of short correlation heterogeneity features. Although such models are adequate for reservoir volumetric calculations, they typically lead to overestimates of reservoir performance under a variety of development scenarios. This is because they tend to be less useful in defining abrupt changes in reservoir properties that occur across a range of bedding scales. Advanced reservoir modelling approaches that seek to improve the representation of reservoir heterogeneity patterns within channel belts by evolving deposit descriptions as a sequence of events (e.g. Lopez *et al.*, 2009; Pyrcz *et al.*, 2009) or embedding intrabelt geologic objects (Hassanpour *et al.*, 2013; Alpak & Barton, 2015) are limited by the narrow range of heterogeneity they can produce and their ability to remain geologically realistic under the requirements of conditioning to wells, seismic-derived trends and production data (Pyrcz *et al.*, 2015). Depositional models of fluvial channel belt deposits have become increasing sophisticated and complex (e.g. c.f. Bridge, 1977; Parker & Andrews, 1986; Sun *et al.*, 1996; Nicholas, 2013; Schuurman *et al.*, 2013, Van de Lageweg *et al.*, 2015) but the ability to specify depositional model input parameters based on sparse subsurface reservoir data and to incorporate this predicted complexity into field scale models for reservoir behaviour forecasting remains limited. Given the broad range of heterogeneity scales possible within fluvial channel belts and pronounced differences in the character of heterogeneities of specific scale in difference types of fluvial deposits, there is a clear need to triage which scales and types of heterogeneity will have the greatest impact on reservoir performance prediction. The combinatorics of assessing all scales and types of heterogeneities possible in any type of fluvial deposits for reservoir performance uncertainly prediction will probably always be intractable. This study focuses on facies patterns that occur across the deposits of individual river channel bends (termed 'storeys'), rather than smaller scale features like grain-size variations across lamina within sedimentary structures or bed-scale variations imparted by seasonal river stage changes.

For this study, a relatively simple process-based river channel model (Bridge 1977, 1982, 1992) is used to construct a range of channel belt internal facies patterns and lateral thickness variations using the methods of Willis (1989, 1993a) and Willis & Tang (2010). The Bridge model was used rather than a more robust physical description because the authors wished to control details of channel plan view geometry, surface grain-size patterns and migration explicitly so that the influence of different factors on subsurface facies patterns could be viewed separately. The goal of this paper is to assess the impact of different intra-channel-belt heterogeneity patterns on reservoir behaviour.

FLUVIAL CHANNEL BELT RESERVOIR HETEROGENEITY

Facies patterns and reservoir property trends preserved within the modelled channel belt deposits are described in the companion paper (Willis & Sech, this volume) and therefore will be only briefly summarised here. Channel belts form because a river is generally restricted to a specific course for a period of time before avulsing elsewhere on the floodplain. Channel belts are composed of multiple storeys that form due to channel bend migration and become stacked along the channel belt during local channel switching associated with river bend cutoff (Fig. 1). All channel belts contain multiple storeys along their length. The number of storey 'rows' is defined by the lateral stacking of storeys in depositional strike cross section. For single row channel belts, storeys are separated by a continuous channel abandonment fill along the belt, whereas for multi-storey-row belts they are stacked in more complex patterns both downstream and laterally adjacent to each other (Fig. 1).

Storeys can be subdivided into lithic sub-units that formed adjacent to specific areas along a river channel: 1) Bar head, 2) Bar tail, 3) Concave bank, 4) Thalweg scour pool lags; and 5) Abandonment fill. These 'substoreys' are characterised by contrasting vertical facies trends and their relative abundance within a channel belt depends on how the river changed position over

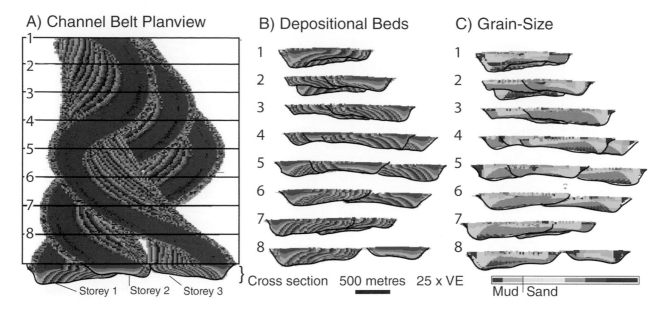

Fig. 1. Model of multi-storey-row channel belt showing: (A) Plan view of depositional bedsets within laterally stacked storeys; (B) Cross sections of depositional bedsets; and (C) Cross sections showing grain-size distribution (blue is mudstone and sand sizes are defined by Phi increment colours). In-cross sections the base of storeys are marked by black lines.

time. The lateral offset of adjacent storeys relative to the channel belt centreline is termed 'storey swing'. Individual storeys tend to thicken and lateral facies trends become more pronounced as the river becomes more sinuous. When a channel migrates dominantly by expansion (sinuosity increase) storeys are typically characterised by a strong downstream fining trend. Migration by translation leaves more downstream elongate coarse-grained basal lags. The character of channel abandonment fill deposits can vary depending on rates that the river is diverted during bend cutoff. Slow river diversion can leave fills nearly as coarse-grained as the deposits preserved during river migration, whereas when diversion is rapid abandonment deposits can be dominated by mud.

Although the vertical stacking of storeys within a channel belt is typically related to aggradation rate, it also becomes more common as the number of storey rows and magnitude of storey swing increases. Given expected thickness variations within storeys, the many close laterally offset storeys in a multi-row belt tend to locally vertically overlap more frequently than those in a single-row belt. For example, a single-row channel belt formed by channels that did not migrate downstream will be everywhere a single storey thick, even when the belt aggraded vertically over time, but multi-row belts with storeys formed by similar

channel migration pattern will locally contain vertically superimposed storeys even for a case with no net vertical aggradation. Fig. 1 shows a modelled channel belt with 2 to 3 storey rows depending on the strike-oriented cross section (obviously in oblique cross sections the belt could appear to have more storey rows). This modelled channel belt had no net vertical aggradation. The base of individual storeys cut less deeply on average when the migrating channel was relatively straight and cut deeper as the channel became more sinuous (as with the general case discussed previously). Due to these storey thickness changes, lateral offset of subsequent storeys during channel switching produces local vertical storey overlap (1D 'multi-storey' character; e.g. in Fig. 1 this is particularly clear in cross sections 2, 3 and 7). Changes in reservoir behaviour related to interactions between intra-storey facies patterns (substoreys) and storey stacking patterns within belts are the focus of this paper.

RESERVOIR CONNECTIVITY AND NET TO GROSS

It is commonly inferred that fluvial reservoir production performance will 'improve' with increasing channel belt proportion relative to overbank deposits. This is based on the general

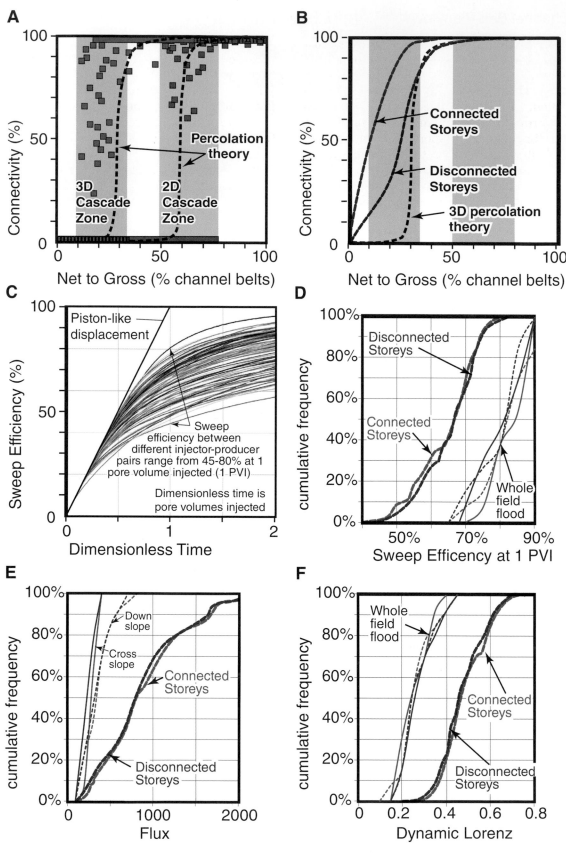

Fig. 2. (Continued)

idea that the amalgamation of different channel belts will tend to provide flow pathways around local reservoir barriers or baffles and thus allow more uniform subsurface flow patterns during production.

Alluvial architecture models (LAB: Leeder, 1978; Allen, 1979; and Bridge & Leeder, 1979) predict the proportion of channel belt sandstones from the dimensionless depositional variables: 1) channel-belt width relative to floodplain width; and 2) channel belt thickness relative to the average thickness the floodplain aggrades between periods of river avulsion. These models were based on the idea that floodplain aggradation declines systematically away from the edge of a vertically aggrading channel belt down an alluvial ridge that becomes progressively more elevated over time. This elevated ridge in turn leads to strong compensational stacking of successive belts. Bridge & Mackey (1993a, b) and Mackey & Bridge (1995) used derivations of LAB models to examine changes in the proportion and connectivity of channel belts with variations in alluvial architecture parameters. They predicted a threshold increase in channel belt connectivity with increasing net to gross, such that channel belts were mostly isolated reservoirs where channel-belt proportion was less than about 50% and became completely connected where channel-belt proportions were greater than about 70%.

A broader series of numerical experiments documenting how variations in channel (belt) geometry influenced relationships between channel proportion and reservoir connectivity was reported by Larue & Friedmann (2005), Larue & Hovadik (2006, 2008) and Hovadik & Larue (2007).

In these models, channel bodies of a specified range of geometry and orientation were randomly placed within a modelled volume. They also observed threshold increases in reservoir connectivity with increasing channel-belt proportion (Fig. 2A, 'cascade zone'), but indicated that this threshold occurred when channel belts occupied 20 to 30% of the succession. They related this threshold behaviour to 'Percolation Theory' (following King, 1990), which describes the behaviour of connected clusters of points randomly positioned in space.

King (1990) showed that a consideration of geobody connectivity in 2D defined a threshold increase in connectivity at about 55 to 65% channel-belt proportion. When applied to 3D objects and volumes, Percolation Theory predicts a threshold increase in connectivity at much lower net to gross values (20 to 30%; King 1990). Differences in the net to gross of the connectivity threshold predicted by Mackey & Bridge (1995; 55 to 65%) and Larue & Hovadik (2006, 2008; 20 to 30%) can be understood in terms of their channel belt placement rules. The compensational stacking inferred by Mackey & Bridge (1995) resulted in straight parallel channel belts that had a connectivity threshold within increasing channel belt abundance more similar to that predicted by a 2D application of Percolation Theory), whereas randomly placed belts are characterised by 3D percolation threshold values (Fig. 2A).

Hovadik & Larue (2007) examined two other issues that hinder prediction of geobody connectivity from a direct application of the Percolation Theory of randomly connected points: 1) volume support; and 2) spatial correlation of permeable

---

Fig. 2. Comparison between static and dynamic measures of model connectivity. (A) Larue & Hovadik (2006) predicted relationships between net to gross and static connectivity for 2D and 3D analysis. Vertical coloured bands highlight their 'cascade zones' defining the narrow range of net to gross values across which reservoir bodies transition from unconnected channel belts to amalgamated volume-wide sandstones. Dashed lines define the threshold increase in connectivity predicted by Percolation Theory for 3D (left line) and 2D (right line) equant-shaped reservoir bodies that are small relative to the volume considered (King 1990). (B) Calculated relationships between net to gross and static connectivity for a reservoir volume populated by single-row channel belts with connected (blue line) and disconnected (red line) adjacent storeys (e.g. Fig. 2). The cascade zones defined in Larue & Hovadik (2006) and the threshold line defined by 3D Percolation Theory (black line) are shown for reference. (C) Sweep efficiency of wells shown in Fig. 3 for the case of a channel belt with unconnected adjacent storeys. At one pore volume injected, recovery as defined by tracer displacement varies from 45 to 80% for different producing wells. Comparison of the cumulative frequency distribution of (D) Sweep effectiveness, (E) Flux; and (F) Dynamic Lorenz coefficient for wells in the model shown in Fig. 3 and a similar model with disconnected adjacent storeys. Thicker dashed lines define values calculated with the well distribution shown in Fig. 3. Blue lines denote the connected-storey case and red lines indicate for the disconnected-storey case. Thinner lines show values calculated with a row of injector and producer wells in separate rows on opposite sides of the models. For the thinner lines, blue represents the connected-storey case and red for the disconnected-storey case; dash lines represent floods down depositional slope and solid lines floods along depositional strike.

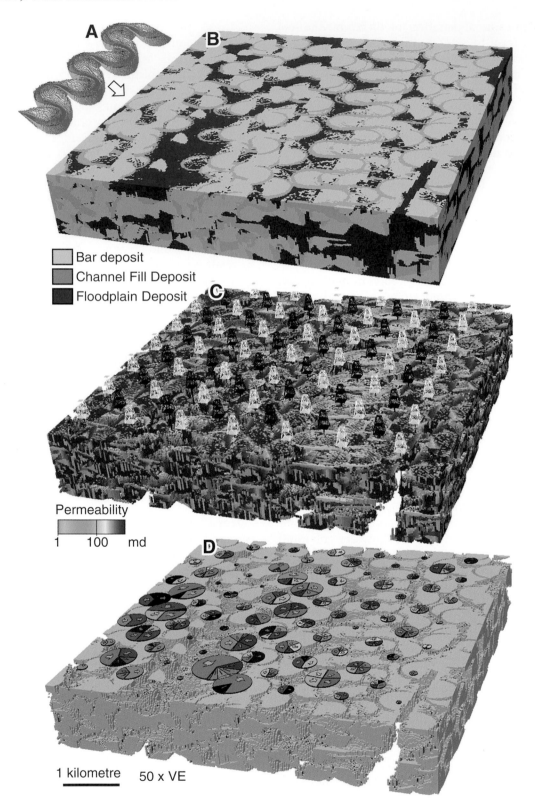

Fig. 3. Model result used to examine differences between channel belts with connected *vs* disconnected adjacent storeys. (A) Segment of the single-row channel-belt model. (B) Static model showing the distribution of 100 channel belts within a volume of impermeable overbank deposits (channel-belt proportion is 55%). (C) Distribution of permeability based on model predicted grain-size variations within the accumulated channel belts. Floodplain deposits are not shown. (D) Results of flow simulation of the model shown by relative flow rates between injector and producer wells in different areas of the reservoir volume. The diameter of pie charts, centred on each well, show relative injector or producer flow rates. Arrows, pointing away from or toward the well distinguish injectors from producers (respectively). Pie slices indicate the relative contribution of adjacent wells to flow at the specific well.

facies. The cascade zone broadens from that predicted by Percolation Theory where the reservoir bodies are the same order of magnitude in size as the model volume in at least one direction (e.g. channel belts are commonly modelled with long axis spanning the model volume; see also King 1990). This effect can be clearly seen in Fig. 2A, where the centre of the cascade zone for channel connectivity defined by numerical experiment is shifted to the left (10 to 20%; i.e. higher connectivity values at a given net to gross) relative to that predicted by 3D Percolation Theory (20 to 30%). Geologic deposits generally display a spatial correlation structure in permeability values imparted by the depositional setting (e.g. sandbodies connected *a priori* along the axis of channel belts are in addition to those that become connected randomly with increasing net to gross). Local spatial correlation of points, whether defined by the placement of continuous geologic bodies of specified shape or geostatistical methods, broadens the cascade zone by increasing connectivity in the lower net to gross range.

Despite potential corrections required for 1) pseudo-2D effects (for very parallel channel belts), 2) the volume support of measurements (along the channel belt axis), 3) local spatial correlation patterns of rock properties within belts; and 4) non-random positions of belts (e.g. compensational stacking or clustering), Percolation Theory provides a reasonable method to define an initial estimate of static reservoir connectivity across channel belts (King, 1990; King *et al.* 1999; Hovadik & Larue, 2007; Sadeghnejad *et al.*, 2011a, b).

Larue & Friedmann (2005) and Larue & Hovadik (2006, 2008) used reservoir simulation to estimate the influence of channelised reservoir architecture on recovery efficiency. They showed that channelised reservoir models with similar channel proportion but a wide range of different geometry (e.g. channel width, thickness, width/thickness ratio, sinuosity, orientation) had similar dynamic recovery behaviour. They concluded that the architecture of the channel bodies was an "insignificant uncertainty" relative to other variables important in reservoir development assessments. Although the channels modelled in most of their experiments had relatively simple internal facies variations, they presented a few examples of models with more organised internal permeability contrasts in patterns that might occur in deep water settings. These models with more explicit facies variations showed more pro-nounced differences in reservoir performance. These results suggested a need to predict facies patterns within fluvial channel belts more deterministically to improve forecasts of production performance.

CONNECTED STOREYS AND CONNECTED BELTS

To look at potential interactions of channel belt internal heterogeneity patterns and channel belt amalgamation, a numerical experiment compared two sets of field scale reservoir models constructed by placing a succession of single-row channel belt models into a reservoir volume with evenly spaced wells (Fig. 3). The channel belts were constructed by the methods described by Willis & Tang (2010). In the first set, storeys within a single-row channel belt were the same as those shown in the expansion-dominated example in Willis & Sech (this volume, their Fig. 4; see model parameters in Table 1A). In this case, storeys defined thicker plan view circular sandstone bodies connected by thin sands along basal parts of the abandonment fills. For the second set of models, the channel belt contained similar storeys but the connections between adjacent storeys were broken by a one-grid-block wide vertical impermeable barrier along each channel belt axis. Although adjacent storeys within a channel belt tend to become disconnected as abandonment channel fills become muddier, a less natural disconnection is imposed here to focus specifically on the impacts of connected *versus* disconnected storeys along the channel belt without any associated changes in net sand volume. This experiment was designed to define the extent to which channel belts with similar internal facies patterns would display different patterns of connectivity with increasing net to gross for cases where adjacent storeys were either connected or disconnected and to understand how these two reservoir architecture scenarios would lead to contrasting reservoir production behaviour.

Field-scale reservoir models (Fig. 3) were constructed by placing a succession of channel belts at a random location on the northern wall of the simulation block, rotating that channel belt in a horizontal plane by a random angle between 0 and 30 degrees either east or west. Each channel belt extended the full length of the simulation block. Each time a new channel belt was added, the total

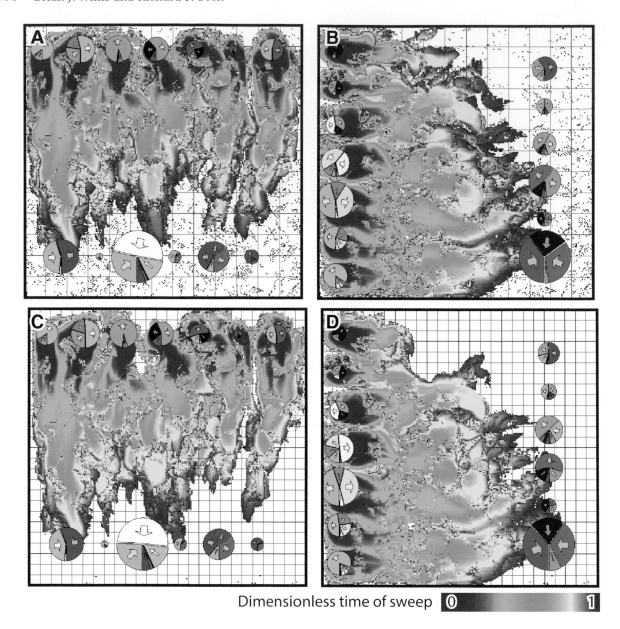

Dimensionless time of sweep 0 ▬▬▬▬▬▬ 1

Fig. 4. (A) Flood front at 1 pore volume injected (1 PVI) is coloured by dimensionless time for streamlines through the connected-storey model swept down depositional dip from a row of injectors to a row of producers. (B) Flood front at 1 PVI through the connected-storey model swept across depositional dip. (C) Flood front at 1 PVI through the connected-storey model swept across depositional dip. (D) Flood front at 1 PVI through the disconnected-storey model swept across depositional dip. In all cases areas swept after one pore volume injected are transparent.

channel belt proportion in the model and reservoir volume (amalgamated channel belt volume) in contact with each well was calculated.

The placement of channel belts within the volume produced a complex crisscrossing of reservoir bodies in three dimensions. As the proportion of channel belts increased, the average well connected to larger volumes in the same general way as that predicted by Percolation Theory. The

channel belts became completely connected for both the connected-storey and disconnected-storey cases when channel belt proportion exceeded 50%.

The connected-storey channel belt models showed a pronounced leftward shift in the cascade zone relative to that predicted by Percolation Theory (i.e. more connected volume at lower net to gross). This reflects the expected *a priori*

Table 1. Model parameters.

(A) Experiment set 1: Channel Belt Connectivity with increased net-to-gross

Channel dimensions: width is 175 m and mean depth is 15 m.
Planform: Bend wavelength is 3800 m and maximum sinuosity is 2.6.
Migration by expansion with no net vertical channel belt aggradation.
Mean grain-size is medium sand (varied from clay to very coarse sand).
Permeability is 0.001 to 350 mD and porosity is 10 to 22%[*1].
Small scale (grid block) anisotropy (Kh/Kv) is 10.
Wells: 500 m (62 acre) spacing. Five spot well pattern is 259 acre.
Chanel belt model grid is 7000 m by 300 m by 25 m
Chanel belt model grid block size is 5 m by 5 m by 45 layers[*2]
Simulation grid (field scale) is 6.5 km by 4.5 km by 35 m.
Simulation grid block size is 10 m by 10 m by 0.2 m.

(B) Experimental set 2: Intra-storey heterogeneity and belt development behavior

Channel dimensions: width is 60 m and mean depth is 3 m.
Planform: Bend wavelength (1000 m), maximum sinuosity of 1.32.
Translation cases 1 to 7 shifted $^{-1}/_8$, $0,^{+1}/_8,^{+1}/_4,^{+3}/_8,^{+1}/_2,^{+3}/_4,^{+1}$ channel bend wavelengths[*3].
Translation cases 8 and 9 shifted $^{+1}$ wavelength with exponential increasing expansion rate.
Net vertical channel belt aggradation cases are 0 m, 0.75 m and 1.5 m)
Mean grain-size is medium sand and varied from clay to very coarse sand.
Permeability is 0.001 to 2900 mD and porosity is 12 to 23%.[*1]
Small scale (grid block) anisotropy (Kh/Kv) is 10.
Wells: 300 m spacing in rows along simulation block entrance and exit.
Line-drive tracer flood over 1.3 km (about one channel bend wavelength).
Channel belt flow simulation model grid is 1.5 km by 1 km by 12 m.
Channel belt flow simulation model grid block size is 5 m by 5 m by 0.2 m.

[*1]. Relationships between grain size and permeability and porosity were defined by empirical fits to core plug data from specific fields of interest to Chevron. Experiment set 1 and 2 referenced different field data.
[*2]. Depositional models were GoCad™ Sgrids with layer thickness that varied depending on the local thickness of aggradation during a migration time step. These grids were resampled onto a rectangular grid for subsurface production simulations
[*3]. Migration by expansion was by increasing the bend input flow angle in equal steps and applying the minimum variance curve of Langbein & Leopold (1966) to define the channel planform.

continuity of permeable grid blocks along the placed channel belt objects. This shift becomes less pronounced for higher net to gross values with the disconnected-storey channel belt model because individual storeys are much smaller than the modelled volume. Put another way, placing a succession of spatially-linked disconnected storeys ('chains of beads'; Donselaar, & Overeem, 2008) into the volume increases connectivity faster than would be expected from the placement of random storeys but not as fast as placing channel sands that are continuous across the whole model.

Both connected-storey and disconnected-storey channel-belt reservoir models were used in reservoir simulation to estimate differences in reservoir production behaviour. In both models used in simulations the channel belt proportion was 55% and all the channel belts and their internal storeys were 100% connected. The connected-storey channel belt model is shown in Fig. 3 (the disconnected-storey channel belt model is not shown

because it looks similar). In this experiment half of the wells used in the static connectivity studies were defined as water injectors and half as producers, such that producers were at the centre of each square kilometre area defined by the injector pattern (i.e. defining a 'five-spot' well production pattern). Injector and producer flow rates were calculated using a Chevron internal streamline fluid-flow simulator that predicted tracer displacement, which is equivalent to water displacing water, using a method based on a finite volume discretisation and fast solver for stationary transport equations (like that described in Natvig *et al.*, 2007; Natvig & Lie, 2008; Shahvali *et al.*, 2012; Moyner *et al.*, 2014).

Different injector-producer well pairs had remarkably different recovery efficiencies (Fig. 2C; 45 to 85%) considering that at channel belt proportion of 55% reservoir bodies are highly connected and flow pathways within the kilometre-scale, production pattern seldom appeared tortuous. Differences in recovery efficiency mostly reflect contrasting permeability within the deposits along flow paths between injectors and producers rather than breaks in static connectivity. Differences in development behaviour between the connected-storey and disconnected-storey cases were compared using three metrics: 1) Effective recovery, 2) Dimensionless flux; and 3) Dynamic Lorenz coefficient.

Effective recovery

Effective recovery was defined as the volume fraction of the model that was swept by injection fluid at 1 pore volume injected (PVI). Recovery was defined for individual injectors and producers and then averaged for all wells. Sweep efficiency between injector and producer wells was remarkably similar for the connected-storey and disconnected-storey models (Fig. 2D) suggesting production behaviour of the two models would be similar. To test if production similarities among models were artefacts of the specific scale measured (i.e. injector-producer pattern map area like that of the storeys), additional flow tests of the models placed separate rows of injectors and producers along opposite ends of the models. This was repeated in a north-south arrangement to flood along depositional dip and an east-west arrangement to flood along depositional strike. Field-scale line-drive flood recovery was higher than for the 5 spot well pattern, probably because a much larger volume of water passed through grid blocks closer to the injectors before sweeping 1 pore volume of the whole model. As with the 5 spot production pattern, there was little difference between the connected-storey and disconnected-storey models under full field line sweep. The measure of recovery used here is not sensitive to injector breakthrough at the producing wells. In many hydrocarbon reservoirs breakthrough is the defining metric of useful production before a well gets shut in or is switched into an injector.

Dimensionless flux

Flux is defined by the movement of fluid out of injectors or into producers over a dimensionless time value defined in terms of pore volumes injected (values are illustrated by the diameter of pie charts in Fig. 3D). In relatively homogeneous deposits producers will be dominantly connected to fluids swept by the four adjacent injectors.

There will be higher flux between an injector and producer connected by a high permeability body. There is no flux between an injector and producer separated by a continuous mudstone barrier. A distinct clustering of channel belts, separated by larger continuous patches of overbank mudstone, is observed in the front face of the field-scale model (Fig. 3B) and is also shown by the along slope alignment of injectors and producers with greater flux (larger area of pie charts in Fig. 3D). Channel belt clustering is observed in outcropping fluvial successions and can define 'sweet spot' fairways in the subsurface. These production fairways, which occur in this case despite the random placement of the channel belts along the upstream edge of the model, hint that the delineation of sweet spots within fluvial reservoirs may reflect the geometry of belt intersections rather than the character of specific belts. Systems with randomly stacked channel belts might produce more continuous channel belt connection 'sweet spots' where channel belts tend to be more parallel and wander little across the floodplain, relative to examples where channel belts tend to crisscross at higher angles (c.f., channel belt patterns modelled by Mackey & Bridge, 1995, vs. Larue & Friedmann, 2005).

The flux from injectors to producers varies widely for different wells but the average and variance of flux in wells is remarkably similar for the connected-storey and disconnected-storey models (Fig. 2). Flux between rows of injectors and producers on opposite sides of the field-scale models is on average lower than between wells in 5 spot injector-producer patterns, reflecting that longer flow paths tend to average out travel time through higher and lower permeability deposits. Distinct fast streaks in flood fronts progressing down depositional dip were observed in both the connected-storey (Fig. 4A) and disconnected-storey models (Fig. 4C). The similarity of these simulated floods confirms that local barriers to flow between storeys along belts have little impact on sweep between wells when the overall model is well connected. The finger-like flood front contrasts with the more uniform depositional strike flood front (Fig. 4B and D) and suggests that for this model a depositional dip-oriented flood will break through faster than a strike-oriented flood. Flux differences between the connected and disconnected storey cases is expected to be greater for lower net to gross cases (i.e. net to gross values below percolation threshold).

Dynamic Lorenz coefficient

The dynamic Lorenz coefficient (Shook, 2009) is a popular method to characterise heterogeneity from steady-state streamline simulation utilising calculations of total travel time, flux and the intersected pore volume along streamlines. A dynamic Lorenz coefficient of one indicates infinitely strong heterogeneity and a value of zero indicates completely homogeneous deposits. Like the other metrics discussed above, the distribution of dynamic Lorenz coefficient for different wells does not vary much between the connected-storey and disconnected-storey models. Field-scale streamline paths have lower dynamic Lorenz coefficient values (relative to those for the 5 spot production), which reflects that high permeability streaks generally do not extend the full length of the models. Variations within the injector-producer 5 spot well pattern mostly sample intra-storey variations and the field-wide flood patterns reflect connections among facies across different channel belts.

The channel belts described in this section were simplified as much as possible. Further understanding channel belt connectivity effects on reservoir performance requires a better appreciation of channel belts with different internal storey stacking arrangement and belts than contain storeys with contrasting internal facies patterns (as discussed below).

LATERAL STOREY STACKING

A set of channel belt models was constructed to examine the influence of storey stacking on subsurface fluid flow. For this experiment three model parameters were varied to construct a design simulation set: 1) Changing rates of downstream translation relative to expansion varied the proportion of preserved substoreys; 2) Single-storey-row and multi-storey-row channel belts with varying lateral swing changed the horizontal stacking of the storeys; and 3) Varying grain-size of the channel abandonment fill, relative to the bar deposits, changed facies contrasts along the belt. Variation in each of these parameters is defined below, after details of the base model are described (see model parameter specifications in Table 1B).

The variety of channel belt storey models used in this experiment are broadly like those presented in Willis & Sech (this volume, their fig. 4). Nine storey model cases differed only in the amount and rate of downstream translation relative to sinuosity increase. Relative rates of downstream translation changed the ratio of preserved substoreys, such that Case 1 had slightly more bar head than bar tail deposits and the higher downstream translation cases had progressively more bar tail and concave bank substorey deposits. The proportion of channel abandonment fills, dependent on final channel sinuosity rather than amount of translation, remained constant at about ~25% of the modelled deposits. Abandonment fill proportion decreased slightly for multi-storey-row relative to single-row cases and cases with less lateral storey swing.

Storey swing within multi-storey-row channel belts was defined by adding a wandering transform to the plan view of successive storey rows in the amalgamated composite depositional model (4 to 6 km wavelength sine curve shift in plain view position along the belt axis with amplitudes of a few 100 m). A high swing example of a multi-storey-row channel belt constructed in this way is shown in Fig. 1. The experiment included: 1) single-row case (no wandering centerline transform), 2) a case with three storey rows that where each simply shifted horizontally in positon along the channel belt centreline (the 'translation only' case); and 3) Cases generated by wandering transforms with progressively higher amplitude (the low, mid and high swing cases). Two realisations of each multi-row case were generated to insure flow simulation results were representative, rather than having differences caused by some fortuitous alignment of lithic features. Fig. 5 shows the resulting plan view sandstone body geometries of the different channel belt cases used in this experiment.

The final parameter varied was the grain-size of the abandonment fill relative to other substoreys. Three cases were generated (Fig. 6). The best reservoir quality abandonment fill is dominated by sand deposited during gradually decreasing discharge with constant mean grain-size. Subtle upward fining trends resulted from the decreased competency of the shallowing flows. A finer-grained sandy fill was defined by gradually decreasing mean grain-size and discharge within the channel, which resulted in a fill on average finer-grained than the average inner bank bar deposits and more strongly upward fining than for the first channel fill case (Fig. 6B). The final fill type, dominated by mudstone, was defined by rapidly dropping the channel discharge and mean grain-size (Fig. 6C). In this case a thin accumulation of finer-grained sand accumulated at the base of the fill along the bend apex and silt accumulated at the base of the fill along bend crossover areas.

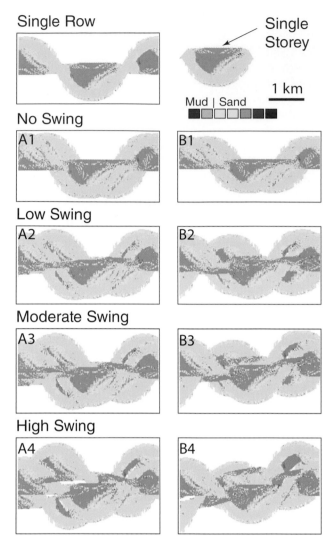

Fig. 5. Plan view of sandstone body for channel belts with different storey swing values used in simulation study. Storeys within each case are the same but their stacking within the belt varies. Cases numbered 1 to 4 have progressively higher swing and cases labelled A and B show different realisations of belt models with the same swing.

Streamline simulation results

A full factorial simulation design compared production differences for the nine cases of downstream translation (i.e. each with different substorey ratios and associated internal facies patterns), nine cases of storey stacking (single row and two each of translation only, low swing, mid swing and high swing) and three types of channel abandonment fill for a total of 243 channel belt models. It should be emphasised that the river channel itself was not changed; i.e. constant discharge (except during channel abandonment), mean channel dimensions, floodplain downstream slope, mean grain-size during channel

migration and final sinuosity before bend cutoff. Differences between the models were defined by comparing three simulation results: 1) Recovery at one pore volume injected, 2) Dimensionless flux; and 3) Dynamic Lorenz coefficient (Fig. 7). A subset of the models in Figs 8, 9 and 10 allow visual comparison of the geometry of the tracer flood front at one pore volume injected.

Average recovery for the different modelled cases varied from 27 to 54% at 1 PVI. Recovery calculated for individual wells had a much greater variance (<10% to >90%; pie charts in Figs 8 to 10), but much of this deviation was caused by flux interactions among adjacent injectors (or adjacent producers) related to very local differences in the contacted facies, rather than being an obvious reflection of overall model heterogeneity. Recovery was best for cases where storeys grew by expansion only and declined with increased rates of downstream translation. Examination of the flood front path through these models suggested the more disconnected the high permeability basal lags in adjacent storeys are, the more stable the sweep down the channel belt. It is not obvious why this trend in recovery reversed for the two cases during which bends expanded at a decelerating rate (8 and 9). For higher translation cases, basal lags became connected along the whole length of the belts considered, such that the flood front advanced rapidly along fingers through the model (Figs 8 to 10, compare cases A5 to E5 within in each figure respectively). In actual reservoirs, recovery for cases with connected coarse-grained lags would probably be lower than predicted here because early water breakthrough would negatively impact production. As basal lags begin to link up, average flux rates though the model increased by more than a factor of 2 (Fig. 7B). Thus, faster delivery generally comes at the cost of lower recovery.

Recovery was lowest for the no-storey-swing cases (single-row and translation-only cases) and improved with higher swing (Fig. 7A). In cases where channels translated further downstream while forming the storeys, the influence of storey swing on recovery was less pronounced. Differences between cases with no storey swing and those with some swing where greater than that between the low and high swing cases. Thus, any amount of swing tends to significantly increase the frequency of flow pathways around lower quality facies. Cases with finer-grained abandonment channel fills generally had lower recovery. This reflects bypass of some compartments in upper parts of bar deposits that are isolated by the finer-grained abandonment fills. Model dynamic Lorenz coefficients indicate the

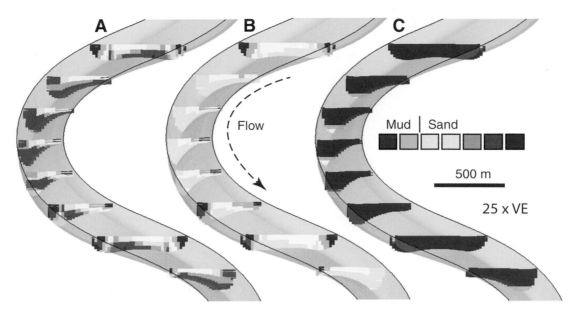

Fig. 6. Grain-size variations of channel-abandonment fills for successive cross sections spaced along depositional dip: (A) Coarser-grain sandy fill, (B) Finer-grained sandy fill; and (C) Mudstone-dominated fill.

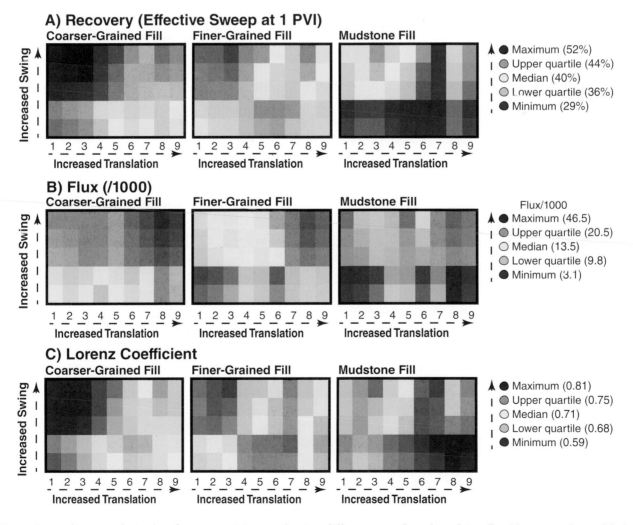

Fig. 7. Streamline simulation results summarising production differences under a line-drive flood between channel belts with different substorey proportion (cases labelled 1 to 9), different storey swing patterns (vertical axis of each case) and different abandonment fill mean grain-size (cases in separate graphs). (A) Average recovery after one pore fluid injected. (B). Average dimensionless flux of line-drive flood between injector and producer wells (dimensionless time defined in terms of pore volumes injected). (C) Mean Lorenz coefficient calculated from variations of streamlines between injector and producer wells. Colour scale grades between quartile classes of results.

Channel Fill Grain-Size Similar to Bar Deposit
1) No Net Aggradation

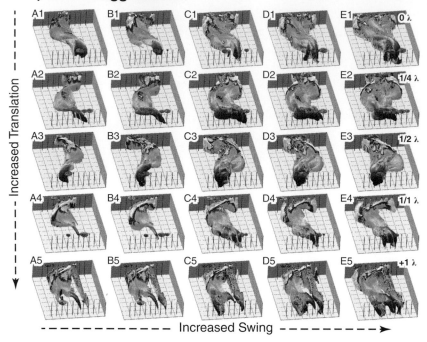

2) Net Aggradation 1/2 Mean Channel Depth

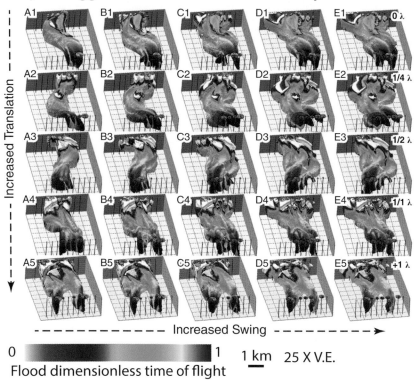

0 ▭▭▭ 1   1 km   25 X V.E.
Flood dimensionless time of flight

Fig. 8. Visualisation of line-drive flood front at one pore volume injected defined by time of flight of streamlines between injector and producer wells. Streamlines from injector wells (upper edge of block) reach areas coloured white first and blue, yellow, red progressively later. Pie charts centred on injector and producer well show relative flux. This figure shows cases with coarser-sand-dominated abandonment fill deposits, in contrast to those in Fig. 9 and 10 which show cases with finer-sand-dominated and mud-dominated channel abandonment fills, respectively). (1) No aggradation cases are a subset of those shown in Fig. 7. (2) Net aggradation cases are a subset of those shown in Fig. 12. In each set, cases A1 to A5 are related to translation cases 1, 3, 5, 7 and 9 on the corresponding graph figures. Cases A1 to E1 are related to swing cases: single row, translation only (1), low swing (2), medium swing (2) and high swing (2), respectively (Fig. 5).

Channel Fill Finer-Grained Than Bar Deposits
1) No Net Aggradation

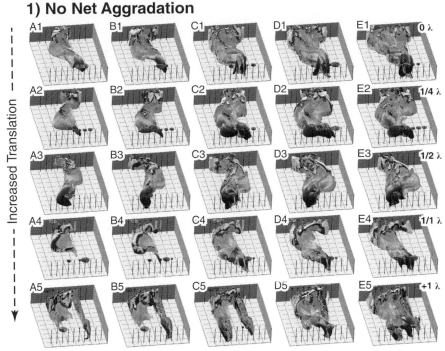

2) Net Aggradation 1/2 Mean Channel Depth

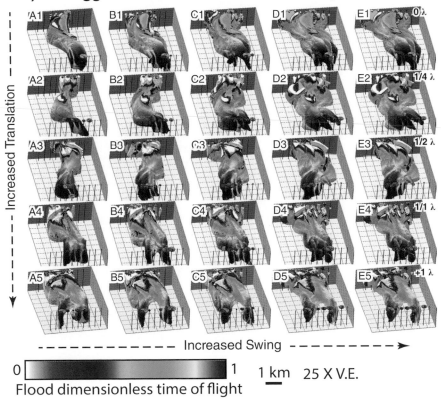

Fig. 9. Visualisation of line-drive flood front at one pore volume injected defined by time of flight of streamlines between injector and producer wells for cases with finer-sand-dominated channel abandonment fills. Refer to caption of Fig. 8 for explanation of the different cases shown.

Channel Fill Dominated by Mudstone
1) No Net Aggradation

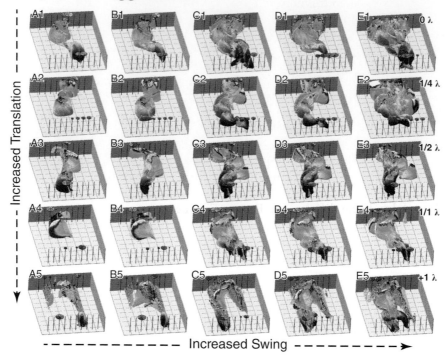

2) Net Aggradation 1/2 Mean Channel Depth

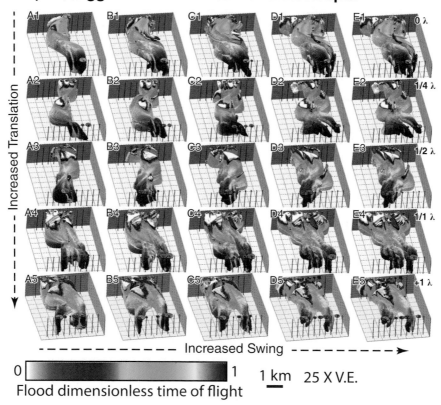

Fig. 10. Visualisation of line-drive flood front at one pore volume injected defined by time of flight of streamlines between injector and producer wells for cases with mud-dominated channel abandonment fills. Refer to caption of Fig. 8 for explanation of the different cases shown.

most heterolithic channel belts are those with moderate to high downstream translation, no storey sweep and muddy abandonment fills (Fig. 7C). The least heterolithic were the high-swing cases containing sandy abandonment fills.

It was not possible to hold the geometry of storeys and the pore volume of models the same for cases with different swing. Volume differences between models were only about 20% and such differences are unlikely to change relative ranks of recovery and flux of the different models. An advantage of the dynamic Lorenz coefficient for comparing models is that this parameter is defined by both the flux and intersected volume along the streamlines and thus comparisons are little affected by overall model volume differences. For the channel belts modelled in this experiment, the relative ranking based on recovery and dynamic Lorenz coefficient were nearly identical.

A box plot summarising values of individual wells measuring each model provide a qualitative measure of the relative impact of channel migration, storey swing and channel abandonment fill grain-size on the reservoir performance metrics (recovery, flux and Lorenz coefficient; Fig. 11). In general, channel migration pattern (and associated variations in preserved sub-storey proportion) has greater impact on median values than does the amount of storey swing. That said, single row belts have distinctly lower recovery and flux than multi-row belts with similar internal storeys. The highest translation cases show marked increases in flux because the basal lags begin to link continuously along the belt. Although the amount of swing impacts the variability between different wells penetrating the same belt, some of this difference reflects that high swing channel belts are wider and thus are penetrated by more active wells. Channel fill character (sandier *vs* muddier) reduces recovery and flux in ways remarkably independent of either migration or swing. Muddier fills also led to more variability between different well penetrations.

VERTICAL STOREY STACKING

The design simulation study reported in the previous section was repeated twice more to test the influence of net vertical aggradation within the channel belt. For one case the modelled channel belt had 1/2 of the mean channel depth net aggradation as a three-row channel belt developed and the other case net aggradation was 1 mean chan-

nel depth (for comparison, channel belt average thickness varied from 1.5 to 2.5 mean channel depth, depending on the channel migration pattern). Aggradation occurred as each storey grew, bed by bed, such that for the multi-row cases individual storeys aggraded 1/3 of the total net thickness as they migrated (i.e. 1/6 and 1/3 mean channel depth, respectively). All other inputs were the same as in the previous example: constant discharge during channel migration, 9 migration cases, 9 swing cases and three types of channel abandonment fill. Adding two levels of net aggradation required an additional 486 channel belt simulation models. Results from streamline flow simulations of these models are in Fig. 12 and a sub-set of the models in Figs 8 to 10 allow visual comparison of the geometry of the tracer flood front at 1 PVI.

Streamline simulation results

Vertical aggradation significantly increased average recovery (Fig. 12A) and flux (Fig. 12B) and decreased overall model heterogeneity (Lorenz coefficient). Flood fronts became broader and more uniform (Figs 8 to 10, c.f. parts 1 and 2 in each figure, respectively). A key influence was thickening of the coarser-grained basal lag deposits and the rise in finer-grained abandonment fill deposits above the base of the channel belt. The basal lags climb to higher levels within storeys, which tended to broaden the flood vertically into upper intervals of the belt. The zero-storey-swing belts (e.g. single-row and translation-only cases), uniformly the worst performers in the no aggradation case, can display the best overall recovery where there is net vertical aggradation. Storey swing and grain-size of the abandonment fill has less relative influence on overall recovery because floods move directly down the channel belt. Cases with both high aggradation and downstream channel migration have somewhat lower recovery because extremely high flux rates through thick coarse-grained basal lag deposits continuously connected along the channel belt tended to completely bypass lower permeability facies. Initial water breakthrough from injectors to producers was very rapid.

High net channel belt aggradation rates were chosen for this experiment to test the magnitude of potential impacts on channel belt reservoir production parameters. Although some have postulated that channel belts can aggrade enough to double their non-aggraded thickness before

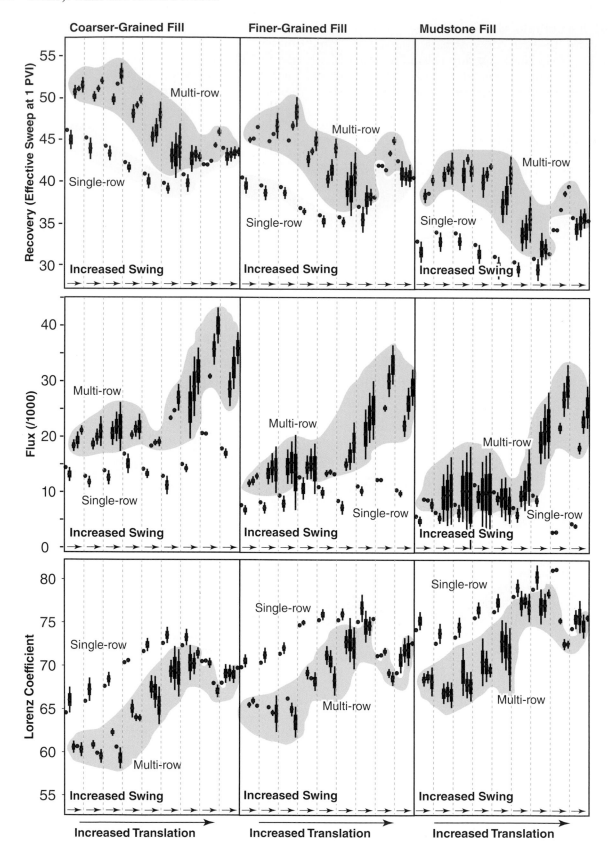

Fig. 11. Summary of simulation results showing statistics for individual wells for all cases with varying channel migration behaviour (sub-storey proportion), number of storeys and amount of storey swing. Red dots show the medium result and box and whisker plots show the range of p50 and p90 results, respectively. Wells in a model that did not penetrate the channel belt (i.e. intersected only overbank deposits) were not included in this compilation.

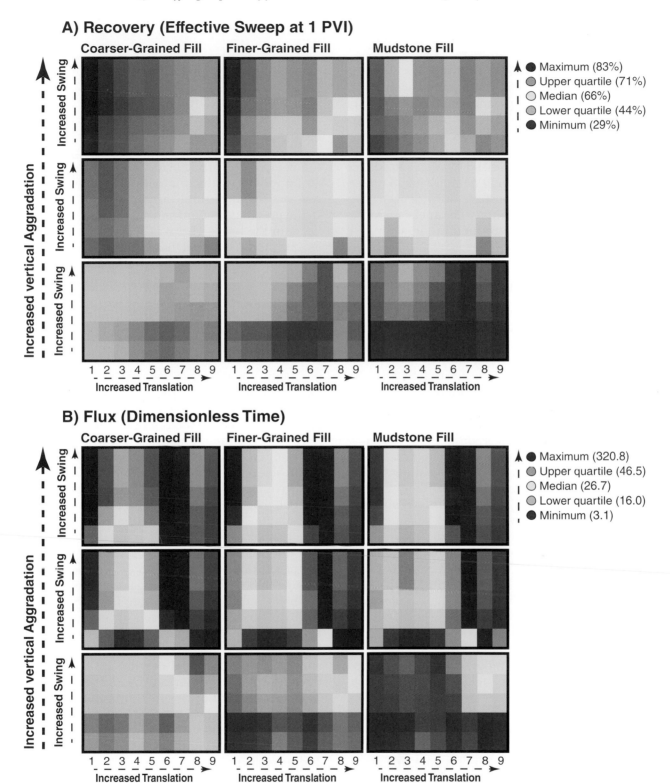

Fig. 12. Streamline simulation results summarising production differences under a line-drive flood between channel belts with different substorey proportion (cases labelled 1 to 9), different storey swing patterns (vertical axis on each plot, in the order shown on Fig. 6: single row, A1, B1, etc.) and different abandonment fill mean grain-size (the lower row of boxes are the no vertical aggradation case, the middle row cases vertically aggraded one half mean depth and the top row cases aggraded one mean depth). The no aggradation cases repeat the graphs shown in Fig. 7 but at a different colour scale that matches the others presented on this figure. (A) Average recovery after one pore fluid injected. (B) Average dimensionless flux of line-drive flood between injector and producer wells (dimensionless time defined in terms of pore volumes injected).

avulsion (Hajek & Heller, 2012; Mohrig *et al.*, 2000), the authors consider these to be extremely rapid vertical aggradation rates. One important result highlighted here is that single-storey-row belt reservoirs tend to have sub-par recovery when there is no net belt aggradation but better than average recovery where there is high net aggradation (Fig. 12). Single-storey-row belts are typical of distributaries on large deltas, which are also areas commonly characterised by high subsidence rate (either due to sediment compaction of underlying deltaic mud or isostatic subsidence). This may explain why distributary channels are recognised to be favourable production targets within deltaic successions.

Impact of more modest river channel aggradation

Although it is clear that high rates of river channel vertical aggradation can have a pronounced impact on heterogeneity patterns within the resulting deposits, it is less clear if these influences increase gradually with increasing aggradation rate. Van de Lageweg *et al.* (2016) recently showed that mean bedset thicknesses increased abruptly at a threshold aggradation rate (based on a river morphodynamic model). The implication is that reservoir behaviour may also show threshold changes with aggradation rate. To define the linearity of relationships between reservoir properties and channel aggradation rate, selected cases from the experiment described above were repeated for finer increments of net aggradation (Fig. 13). This revised experiment included only single row cases, with three amounts of expansion relative to translation and three channel abandonment fill types, for cases with between 1/32 and 1/2 net vertical aggradation. As expected from the previous experiments, channel fill character had a significant influence on recovery, flux and Lorenz coefficient but here the focus is on the rate of change in values with net aggradation for each given migration pattern and channel fill example. In all cases there appears to be a gradual linear increase in recovery with increases aggradation. Flux increases with aggradation rate are somewhat less linear, rising rapidly in the range of 1/8 mean channel depth aggradation. This acceleration in average flux is more pronounced for the finer channel abandonment fill cases. This nonlinear rate of average flux change is also shown by increases in heterogeneity as measured by the Lorenz coefficient.

VERTICAL CHANNEL BELT STACKING

The term 'multi-storey' is commonly used in reference to stacked channel belts, rather than simply stacked storeys within an individual channel belt (Willis, 1993b; Bridge & Tye, 2000). It can be difficult to tell these two situations apart, particularly in the subsurface where the lower (top-truncated) storeys are less than half the thickness of the complete overlying storey. However, there is an important distinction between belts with storeys stacked due to rapid aggradation (as described in the previous section) and amalgamated channel belts. In the former, basal lags climb in the storey as individual channel bars aggrade, whereas in the latter storey facies distributions are changed only due to storey top truncation within the lower belt(s). With amalgamated channel belts, basal relief of the upper channel belt cuts to varying depth, potentially superimposing coarser-grained lag facies against different facies within the underlying channel belt. Even when the stacked channel belts have similar internal facies patterns, variations in the location of specific substoreys are expected to increase flow paths through the amalgamated sandstone reservoir.

To examine the influence of connected channel belts on reservoir performance a numerical experiment stacked two channel belts with varying vertical overlap (Fig. 14). The along stream axes of the belts were parallel. There was no net vertical aggradation of storeys within belts. The stacked channel belts were similar, but not the same, such that the two connected belts had the same substorey proportion, number of storey rows and storey swing (Fig. 5). For each amalgamated set of channel belts, subsurface streamline flow simulations were run on cases with 8 levels of vertical overlap, from 100% (i.e. belts cut down from the same horizon) to no overlap (i.e. one belt is everywhere above the other), so that they do not touch (even though they are penetrated by the same wells). The subsurface flow simulation setup was the same as described in the previous section. Nine storey migration cases, three variations in storey swing, three types of channel abandonment fill and 8 vertical channel belt offsets resulted in 648 flow simulation models.

Streamline simulation results

As in the previous examples, the simulation results compared are recovery at 1 PVI (Fig. 15A), dimensionless flux (Fig. 15B) and overall model

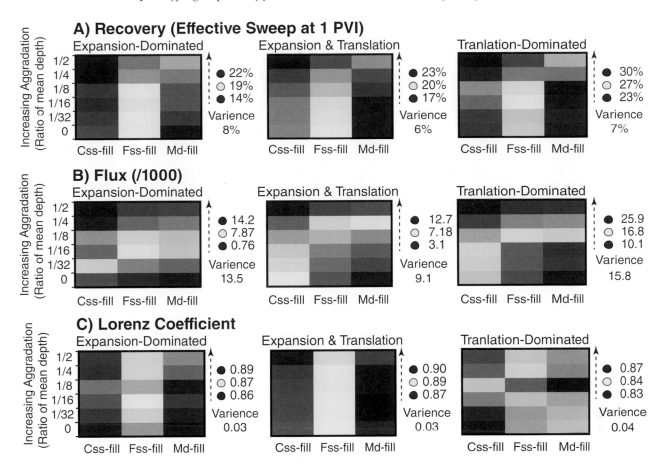

Fig. 13. Streamline simulation results summarising production differences under a line-drive flood for modest differences in vertical aggradation rate. Cases are based on single storey row channel belts with migration dominated by expansion, combined expansion and translation; and dominantly translation (as indicated by label above each graph). For each of these migration cases examples with coarser-grained sandy abandonment fill (C-fill), finer-grained sandy abandonment fil (F-fill) and mudstone dominated fill (Md-fill) are arranged in different columns. The vertical axis of each graph shows values for these cases with varying vertical aggradation rate during migration (defined as a fraction of mean channel depth). A) Average recovery after one pore fluid injected. B) Average dimensionless flux of line-drive flood between injector and producer wells (dimensionless time defined in terms of pore volumes injected). Colours grade across quartile classes averaged for each case (minimum, lower quartile, median, upper quartile, maximum) so as to highlight relative rate of change with aggradation rate.

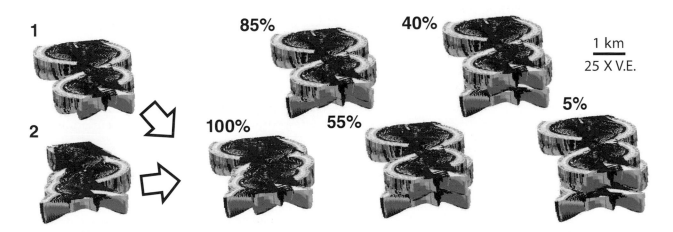

Fig. 14. Stacking of vertically amalgamated channel belts. Oblique view of grain-size variations within two channel belts with a similar number of storey rows and storey swing (the case shown is a high swing example). Hotter colours are coarser grained. Similar channel belts were amalgamated within a combined reservoir grid with variable vertical offset, ranging for 100% overlap to only 5%. Flow simulation results of the amalgamated belts are shown in Fig. 15.

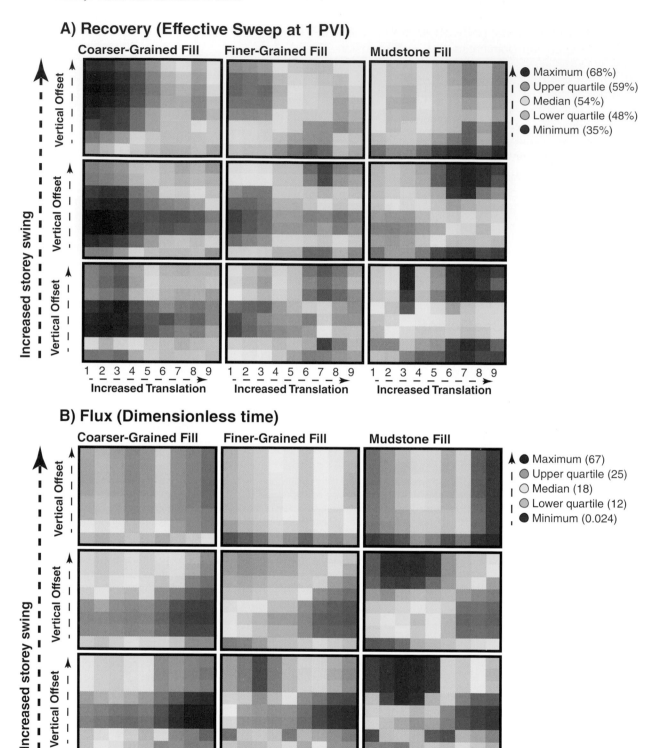

Fig. 15. Streamline simulation results of aligned, vertically-amalgamated, channel belts summarising production differences under a line-drive flood between a row of injectors and producers. Injectors and producers were arranged in the same way as the examples shown in Figs 8 to 10. Simulation results shown for: (A) Average recovery after one pore fluid injected and (B) Average dimensionless flux of line-drive flood between injector and producer wells (dimensionless time defined in terms of pore volumes injected). Graphs show simulation results arranged by channel belts with the same abandonment channel fill character (columns) and the same storey swing (rows: lower swing cases along the lower row and high swing cases along the upper row). Each graph axis shows substorey proportion defined by the amount of downstream translation (numbered 1 to 9) along the horizontal axis and vertical axis the vertical offset between connected channel belts (from 100 overlap for the basal value and 5% connected at the top; as in Fig. 16 and intermediate cases).

heterogeneity (dynamic Lorenz coefficient). When amalgamated belts cut down from the same horizon (100% overlap) the result was a smoother belt basal erosion surface, more continuous basal lags and lower rugosity margins (Fig. 14). Recovery increased by about 10% relative to the single channel belt case, which is similar to the effect caused by greater storey swing. This is not surprising as amalgamation of co-located channel belts had a similar influence on facies patterns as simply adding more storey rows within an individual belt. As was observed in the isolated channel belt experiments, translation generally led to lower overall recovery and higher average flux. Downstream elongation of basal coarse grained lags allowed more bypass and generated a flood front with more elongate fingers along the base of the channel belt (Fig. 15A). This effect was greater when the two channel belts were completely separated because differences in overall recovery of two separate belts intersected by the same well is addictive. Belts that vertically overlap by 50% had significantly higher recovery and flux than either the completely vertically overlapping cases or the vertically separated cases. In this case coarser-grained basal deposits of the upper belt most effectively cross cut finer facies within the lower belt. This change in recovery with overlap (i.e. first increased recovery with increasing vertical offset and then decreased recovery as the belt vertical separation because greater than about 50%) becomes more pronounced for cases with

higher storey swing and finer-grained abandonment fills. Flux differences with variations in the amount of belt vertical overlap have been more pronounced for storeys formed by more downstream channel translation and those with coarser-grained abandonment fills (Fig. 15B). The most heterolithic channel belts had complete vertical overlap, fine-grained abandonment fills and storeys formed by channels with strong downstream transition. In these cases the abundance of fine-grained-fills and downstream elongate basal lags led to very uneven flood fronts. The most homogeneous reservoirs were those with moderate vertical overlap, relatively coarse-grained fills and storey growth dominated by expansion because in these cases spatial heterogeneity patterns tend to be well distributed within the belt and therefore have more limited extent.

CROSS-CUTTING CHANNEL BELTS

The experiment described in the section directly above was repeated for the case where channel belts cross cut at 90 degrees (Fig. 16). The subsurface flow simulation model had the same setup as the previous section, except that two sets of injector wells were each located in a row at opposite ends of the east-west-oriented channel belt and two sets of producers were similarly located in rows at each end of the north-south-oriented channel belt. Thus the line-drive flood moved

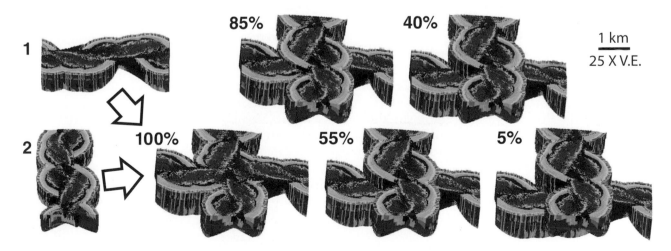

Fig. 16. Stacking of intersecting channel belts. Oblique view of grain-size variations within two channel belts with a similar number of storey rows and storey swing (the case shown is the low swing examples in Fig. 5 and associated graphs of flow simulation results in Fig. 7). Hotter colours are coarser grained. The similar channel belts were amalgamated within a combined reservoir grid with variable vertical offset, ranging for 100% overlap to only 5%. Flow simulation results of the amalgamated belts are shown in Fig. 17.

both directions toward the central segment of the first belt before diverting sideways toward the rows of producer wells across the area of belt intersection. The connected belts had higher overall recovery than isolated belts (Fig. 17A). As before, downstream channel translation during storey deposition and finer-grained channel abandonment fills increased overall reservoir heterogeneity and reduced recovery. The degree of vertical channel belt overlap had significantly less influence on recovery than for the case where the superimposed belts were parallel. A pronounced decrease in flux (and, surprisingly, an increase in recovery) occurred where a belt just touched an overlying belt (5% overlap). Although presumably recovery was increased by forcing flow upward though the lower-quality facies capping the lower channel belt, slow flow rates in this case may make the production less economical (Fig. 17B). Very low vertical overlap would presumably also increase uncertainty in lateral reservoir continuity because if belts locally become unconnected no flow would pass from one to the other.

DISCUSSION

A large number of modern river and outcrop analogue studies have been completed over the last several decades with the purpose of better defining reservoir heterogeneities within fluvial channel belts. These studies demonstrate that this reservoir class has a complicated variety of internal facies architecture formed by river systems with different depositional style (i.e. channel pattern, sediment load and rate of channel migration and switching relative to river avulsion rate). Much of this work is summarised in the compilations of Bridge (2003) and Miall (1996, 2013). Despite this broad literature, an understanding of how subsurface variability will influence the development of these types of reservoirs, or how this variability can be characterised using sparse subsurface data, is still lacking. There is no clear consensus as to which types of variations will have first order influences on production behaviour and without such guidance it is difficult to develop modelling workflows to define uncertainty forecasts (e.g. compare conclusions of Larue & Hovadik, 2008; Willis & Tang 2010; Issautier *et al.* 2014; Alpak & Barton, 2015; and Willems *et al.*, 2017).

Channel belt organisation

There is renewed interest in documenting the organisation of channel belts within larger-scale alluvial successions and, in particular, for differentiating compensational, random and clustered channel belt arrangements and associated relationships between channel belt proportion and connectivity (e.g. Strong, *et al.*, 2005; Straub *et al.*, 2009; Hajek *et al.*, 2010; Wang *et al.*, 2011; Flood & Hampson, 2015; and Villamizar *et al.*, 2015). Despite the definition of new metrics to quantify channel belt arrangements, the 3D geometry of bodies defined by channel belt intersections and the impact of channel belt connectivity patterns on reservoir performance remains poorly understood. By reference to Percolation Theory it can be inferred that parallel channel belts will tend to have lower average static connectivity, at a given net to gross, compared to systems where channel belts crisscross at high angles. Channel belt stacking modelled in this study suggests that when parallel belts in these successions do connect they tend to produce more down-stream continuous sweet spots. Although fluvial successions with belts that crisscross at higher angles tend to be better connected overall, any sweet spots defined by connected belts are expected to be more localised.

Channel belts formed by river systems with more prominent alluvial ridges can be inferred to be more strongly compensational stacked and to remain parallel for longer distances down the floodplain (Bridge & Mackey, 1993a; Karssenberg & Bridge, 2008). Those with overbank deposits dominated by avulsion deposits may have more subtle alluvial ridges and channel belts with a wider range of orientations.

Based on the results of the authors' numerical experiments, dense amalgamation of channel belts within a single stratigraphic interval may not significantly increase reservoir performance (relative to isolated belts). This will be true where deposits eroded by subsequent channel belts are similar to those that each successive belt adds, such that the intra-storey facies variation patterns remain the same. As described previously, connected belts with the greatest boost in reservoir performance over isolated belts are those that are vertically offset by about one half belt thickness, as this arrangement is most effective in breaking up more continuous heterogeneities. The situation obviously becomes more complicated where the river

A) Recovery (Effective Sweep at 1 PVI)

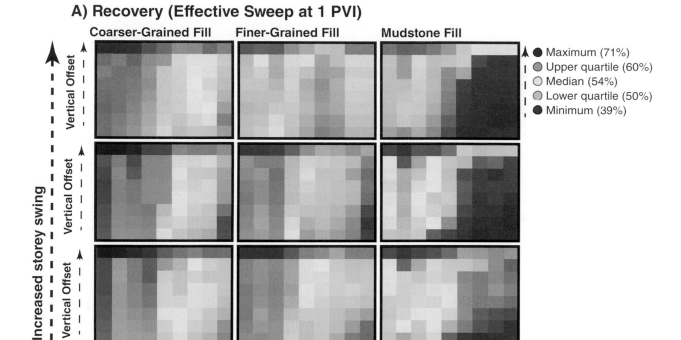

B) Flux (Dimensionless time)

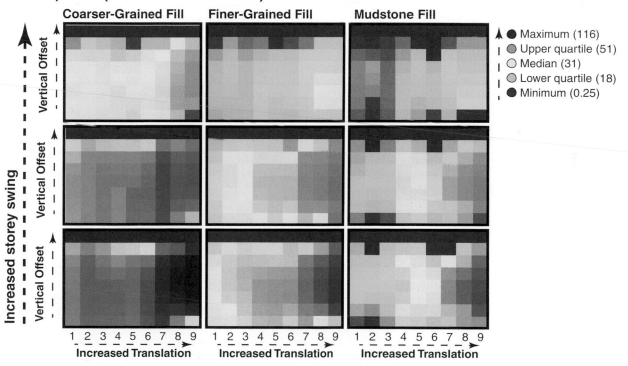

Fig. 17. Streamline simulation results of cross-cutting, vertically-amalgamated, channel belts summarising production differences under a line-drive flood between rows of injectors and rows of producers. Simulation results shown for: (A) Average recovery after one pore fluid injected; and (B) Average dimensionless flux of line-drive flood between injector and producer wells (dimensionless time defined in terms of pore volumes injected). Graphs show simulation results are arranged by channel belts with the same abandoned channel fill character (columns as labelled) and the same storey swing (rows: with no swing cases at the bottom and high swing cases at the top). On each box graph the horizontal axis shows substorey proportion defined by the amount of downstream translation and vertical axis the vertical connection between 100% and 5% (as in those shown in Fig. 16 and intermediate cases).

system depositing the subsequent channel belt changes over time. In such cases the geometry of sweet spots (plan view definition) or favoured completion intervals (vertical definition) may be controlled by contrasting facies within the different belts rather than channel belt arrangement.

Depositional mechanisms that produce clustered channel belts (e.g. avulsion node lock, fault related subsidence variations or channel belt reoccupation) can produce composite connected bodies of varying shape (e.g. Mackay & Bridge, 1995; Dalman *et al.*, 2015). For example consider two systems in which successive channel belts fan away from a common 'major' avulsion node (Fig. 18). If the avulsion node is locked in place for long periods by, for example, structural confinement into the basin or an elevation drop across a local fault, then belts below this confined node may stay relatively parallel. In systems where major avulsion nodes are free to shift laterally in position, belts fanning downstream from successive node positons may crisscross at higher angles. The implication is that for reservoir performance prediction it is not enough to characterise how channel belt geometry and internal facies patterns change down basin (and, by reference to Walther's Law, within stratigraphic successions) but there is also a need to characterise channel belt intersection patterns.

Reservoir models of heterogeneity

Modelling workflows for channelised reservoirs have been added in recent years to standard reservoir modelling software (e.g. GoCAD™ and Petrel™). For the most part these methods were developed to investigate deep water channels with heterogeneity definitions focused on axis to margin rock property trends and edge bounding mudstones (Clark & Pickering, 1996; Larue & Hovadik, 2008; McHargue *et al.* 2011; Meirovitz *et al.*, 2016). These deep water intra-channel trends do not sufficiently characterise natural variations within fluvial channel belts because they cannot define the fundamental links between grain-size and thickness patterns within storeys. Similarly, fluvial channel belt internal heterogeneities are not characterised well by models that define belts as being composed of amalgamated intertwined sandy spaghetti-like threads. The Larue and others numerical experiments (Larue & Friedmann, 2005; Larue & Hovadik, 2006, 2008; and Hovadik & Larue, 2007) made a convincing argument that it is not the variation in channelised reservoir geometry that imparts major uncertainties in reservoir assessments (or even static connectivity, where net to gross is known) but rather variations in facies patterns that can change flow tortuosity. The experiments discussed herein also

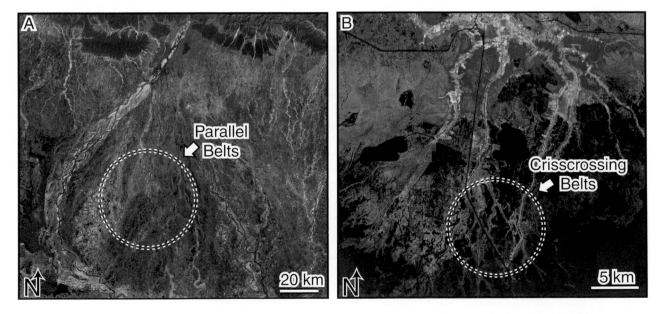

Fig. 18. Channel belt patterns. (A) Channel belts tend to be parallel basinward of a tectonically defined entrance into the basin. Kosi River 'megafan' in northern India. (B) Where major avulsion nodes shift position in time channel belts with crisscrossing patterns can be preserved. Lower Mississippi River Delta, USA. Images reprocessed from USGS LandsatLook (https://landsatlook.usgs.gov).

highlight the importance of predicting the continuity of facies with varying permeability. Where fluvial channel belt internal heterogeneities are modelled naively (e.g. as homogeneous sand or with rock properties defined as weakly-structured pixelated bodies) it should not be surprising that the models tend to underestimate field development uncertainties.

The geologic details required in reservoir models change with the questions being asked and variations of the overall reservoir system. For example simple channel belt models may be adequate for volumetric assessments of gas reservoirs, whereas geologically rich models are needed to accurately predict steam floods of heavy oil reservoirs. In either case, the geologic details that are added to a reservoir model need to reflect accurately the scales and types of heterogeneities that will most probably impact development forecasts. Applying incorrect internal heterogeneity models to fluvial channel belts (e.g. a channel-shaped body with axis to margin fining) to define a range of development predictions for uncertainty forecasting may simply pull estimates further from concepts of geological reality.

The simple models of internal channel belt heterogeneity developed for this study focused on the impact of facies variations within storeys and how these scales of heterogeneity link up to impact flow paths through channel belts and connected channel belts. It is not claimed that this collection of models provides a full assessment of heterogeneity patterns within all types of channel belts but rather that they demonstrate that this scale of heterogeneity can be a significant uncertainty in the assessment of some types of fluvial channel belt reservoirs. The models suggest that the geometry, shape and stacking of storeys, spatial variations in channel belt thickness and vertical grain-size trends and permeability contrasts between the different substoreys within a belt may all have significant impact on subsurface flow prediction. Vertical aggradation of a channel as it migrates to deposit a storey also clearly had very important influence on subsurface flow but this is difficult to constrain in the subsurface.

Flow behaviour through other types of fluvial channel belts may be impacted more by different types of heterogeneities than those that were the focus of this study. Preliminary modelling suggests that for a braided river with lower sinuosity channels, downstream-accretion-dominated bars and deeper junction scours, a key variable is the

extent to which coarser-grained thalweg lags extend downstream into adjacent bar head areas. In channel belts with strong lithic contrasts across individual depositional beds, modelling the impact of finer-scale mud drapes may be a key type of heterogeneity (Willis & Tang, 2010, Fustic *et al.*, 2013, Hassanpour *et al.*, 2013; Alpak & Barton, 2014).

A problem with smaller-scale heterogeneity models is that information on how these features vary in 3D across the larger deposits is still lacking. Commonly, it is assumed that smaller-scale features have stationary distribution throughout the volume being modelled but this is typically not the case in fluvial deposits. Dune-scale grain sorting changes upwards as cross-strata get thinner and average grain-size fines. Although an isolated shale drape with local holes many have little influence of subsurface flow, where many such drapes are spaced closely on inclined beds within a substorey, they may impose a major baffle. A continuing problem is that there has been observation of many scales of heterogeneity exposed in 2D outcrops but little information on how they should be specifically distributed in a 3D model for subsurface flow assessment. Without subsurface flow simulation, researchers have no idea as to which observed heterogeneities are important in reservoir characterisation but without better constraints on the actual 3D heterogeneity patterns within fluvial channel belts it is impossible to determine which subsurface flow models are reasonable.

CONCLUSIONS

1. This study supports previous conclusions that thresholds in channel belt connectivity with increasing net to gross are reasonably predicted by Percolation Theory and are not particularly sensitive to changes in channel belt geometry. This study does not support the contention that reservoir architecture should be considered an "insignificant uncertainty in many appraisal and development studies" as concluded by Larue & Hovadik (2008). The key geologic uncertainties may not be the geometry and static connectivity of reservoir bodies, but rather the architecture of internal facies patterns that produce spatial variations in the continuity of permeability contrasts and associated subsurface flow rates.

2. Recovery from different channel belts deposited by the same river (same discharge, same channel geometry, same mean grain-size and same final sinuosity at channel segment cutoff and abandonment) was predicted to vary from 25 to 90%. Recovery was better when storeys grew by only expansion (sinuosity increase) because channel downstream translation amalgamates basal coarse grained lags formed in adjacent channel bends. Downstream continuous coarse-grained lags tended to make reservoir sweep fronts more unstable and thus preferentially bypass lower quality facies before injector fluid breakthrough to producer wells. In general, more heterolithic belts, characterised by dynamic Lorenz coefficients defined from streamline subsurface flow simulation, tended to have lower recovery and average production flux. The most heterolithic belts were those with storeys formed by channels with moderate to high downstream translation rates, few storey rows, low storey swing and muddy channel abandonment fills. The most homogeneous belts are those with high storey swing, low continuity basal scour lags and sandy abandonment fills.

3. Vertical aggradation of channels while they migrate significantly increases recovery and flux by moving coarse grained lags higher within belts and opening more pathways around finer-grained abandonment fill deposits. It is difficult to define net channel belt aggradation in ancient deposits given much larger-scale cut and fill patterns associated with the migration of individual storeys. Combining downstream channel translation with vertical aggradation can increase flux by an order of magnitude, potentially allowing injected water breakthrough before any significant recovery is realised.

4. Vertical connection of channel belts tends to increase recovery but has significantly less influence on production than vertical storey aggradation within a channel belt. The greatest contrasts between connected and unconnected belts occur when their bases are offset by one-half channel-belt thickness. In that position the coarse grained basal lags of the upper belt cross cuts finer-grained facies in the lower in a way the tends to equalise the flood front advance though the reservoir. Where basal lags of the two superimposed belts align along the same horizon, there is increased flux along the base

of the reservoir and poorer recovery due to bypass of finer facies higher within the reservoir. Where belts with little vertical overlap are intersected by the same injector and producer wells they flood in a pattern more similar to that of separate channel belts.

5. Where different channel belts cross at higher angle, the vertical offset between the two belts had less influence on production behaviour compared with more parallel examples. When cross cutting channel belts are more completely offset, overall recovery can increase because the advancing flood from the injector wells is forced up thorough poorer quality facies of the lower belt to reach producer wells in the upper belt.

6. The channel belt models defined in this study incorporated a restricted number of heterogeneity types and scales and more studies are needed to broaden this assessment of intra-channel-belt heterogeneity effects on reservoir development.

ACKNOWLEDGEMENTS

Thanks to members of Chevron's Clastic Stratigraphy R&D team for discussions on fluvial reservoir characterisation: Bryan Bracken, Sean Connell, Kristy Milliken, Tobi Payenberg, Pete Sixsmith and Morgan Sullivan (team lead). Dennis Dull, Michael Pyrcz and Fabien Laugier provided advice on reservoir modelling. Sean Connell and Michael Pyrcz provided peer reviews of the original internal report. Suggestions by reviewers Stephen Hubbard and Allard Martinius and editor Massimiliano Ghinassi, helped strengthen the final presentation.

REFERENCES

Allen, J.R.L. (1979) Studies in fluviatile sedimentation: an elementary geometrical model for the connectedness of avulsion-related channel sand bodies. *Sed. Geol.*, **24**, 253–267.

Alpak, F.O. and **Barton, M.D.** (2015) Dynamic impact and flow-based upscaling of the estuarine point-bar stratigraphic architecture. *J. Petrol. Sci. Eng.*, **120**, 18–38.

Bridge, J.S. (1982) A revised mathematical model and FORTRAN IV program to predict flow in open-channel bends. *Comput. Geosci.*, **8**, 91–95.

Bridge, J.S. (1992) A revised model for water flow, sediment transport, bed topography and grain-size sorting in natural river bends: *Water Resour. Res.*, **28**, 999–1013.

Bridge, J.S. (1977) Flow, bed topography, grain-size and sedimentary structure in open channel bends: A three-dimensional model. *Earth Surf. Proc. Land.*, **2**, 401–416.

Bridge, J.S. (2003) *Rivers and Floodplains; Forms, Processes and Sedimentary Record.* Blackwell Publishing, Oxford. 491 pp.

Bridge, J.S. and **Diemer, J.A.** (1983) Quantitative interpretation of an evolving ancient river system. *Sedimentology*, **30**, 599–623.

Bridge, J.S. and **Leeder, M.R.** (1979) A simulation model of alluvial stratigraphy. *Sedimentology*, **26**, 617–644.

Bridge, J.S. and **Mackey, S.D.** (1993a) A revised alluvial stratigraphy model. In: *Alluvial Sedimentation* (Eds M. Marzo and C. Puigdefabregas), *Int. Assoc. Sedimentol. Spec. Publ.*, **17**, 319–336.

Bridge, J.S. and **Mackey, S.D.** (1993b) A theoretical study of fluvial sandstone body dimensions. In: *The Geological Modelling of Hydrocarbon Reservoirs* (Eds S. Flint and I.D. Bryant), *Int. Assoc. Sedimentol. Spec. Publ.*, **15**, 213–236.

Bridge, J.S. and **Tye, R.S.** (2000) Interpreting the dimensions of ancient fluvial channel bars, channels and channel belts from wire-line logs and cores. *AAPG Bull.*, **84**, 1205–1228.

Clark, J.D. and **Pickering, K.T.** (1996) Architectural elements and growth patterns of submarine channels: Application to hydrocarbon exploration. *Am. Assoc. Pet. Geol. Bull.* **80**, 194–221.

Dalman, R., **Jan Weltjea, G.** and **Karamitopoulosa, P.** (2015) High-resolution sequence stratigraphy of fluvio–deltaic systems: Prospects of system-wide chronostratigraphic correlation. *Earth Planet. Sci. Lett.*, **412**, 10–17.

Donselaar, M.E. and **Overeem, I.** (2008) Connectivity of fluvial point-bar deposits: An example from the Miocene Huesca fluvial fan, Ebro Basin, Spain. *AAPG Bull.*, **92**, 1109–1129.

Dreyer, T., **Falt, L.-M.**, **Hoy, T.**, **Knarud, R.**, **Steel, R.** and **Cuevas J.-L.** (1993) Sedimentary architecture of field analogues for reservoir information (SAFARI): a case study of the fluvial Escanilla formation, Spanish pyrenees. In: *The Geological Modelling of Hydrocarbon Reservoirs and Outcrop Analogues* (Eds S.S. Flint, I.D. Bryant), Blackwell Scientific Publications. 57–80.

Flood, Y.S. and **Hampson, G.J.** (2015) Quantitative analysis of the dimensions and distribution of channelised fluvial sandbodies within a large-scale outcrop dataset: Upper Cretaceous Blackhawk Formation, Wasatch Plateau, Central Utah, *USA. J. Sed. Res.*, **85**, 315–336.

Fustic, M., **Thurston, T.**, **Al-Dilwe, A.**, **Leckie, D.A.** and **Cadiou, D.** (2013) Modelling by constraining stochastic simulation to deterministically interpreted three-dimensional geobodies: case study from Lower Cretaceous McMurray Formation, Long Lake Steam-Assisted Gravity Drainage project, northeast Alberta. In: *Heavy-oil and Oil-sand Petroleum Systems in Alberta and Beyond* (Eds F.J. Hein, D. Leckie, S. Larter and J.R. Suter), *AAPG Stud. Geol.*, **64**, 565–604.

Hajek, E.A. and **Heller, P.L.** (2012) Low-depth scaling in alluvial architecture and nonmarine sequence stratigraphy: Example from the Castlegate Sandstone, central Utah, U.S.A. *J. Sed. Res.*, **82**, 121–130.

Hajek, E.A., **Heller, P.L.** and **Sheets, B.A.** (2010) Significance of channel-belt clustering in alluvial basins. *Geology*, **38**, 535–538.

Hassanpour, M., **Pyrcz, M.J.** and **Deutsch, C.V.** (2013) Improved Geostatistical Models of Inclined Heterolithic Strata for McMurray Formation, Alberta, Canada. *AAPG Bull.*, **97**, 1209–1224.

Heller, P.L. and **Paola, C.** (1996) Downstream changes in alluvial architecture: an exploration of controls on channel-stacking patterns. *J. Sed. Res.*, **66**, 297–306.

Hovadik, J.M. and **Larue, D.K.** (2007) Static characterisations of reservoirs: refining the concepts of connectivity and continuity. *Petrol. Geosci.*, **13**, 195–211.

Issautier, B.S., **Viseur S.**, **Audigane, P.** and **Nindre, Y.-M.** (2014) Impacts of fluvial reservoir heterogeneity on connectivity: Implications in estimating geological storage capacity for CO2. *International Journal of Greenhouse Gas Control*, **20**, 333–349.

Jones, A., **Doyle, J.**, **Jacobsen, T.** and **Kjønsvik, D.** (1995) Which sub-seismic heterogeneities influence water-flood performance? A case study of a low net-to-gross fluvial reservoir. In: *New Developments in Improved Oil Recovery* (Ed. H.J. Haan), *Geol. Soc. Spec. Publ.*, **84**, 5–18.

Jordon, D.W. and **Pryor, W.A.** (1992) Hierarchical levels of heterogeneity in the Mississippi River Meander belt and application to reservoir systems. *AAPG Bull.*, **76**, 1601–1624.

Karssenberg, D. and **Bridge, J.S.** (2008) A three-dimensional numerical model of sediment transport, erosion and deposition within a network of channel belts, floodplain and hill slope: extrinsic and intrinsic controls on floodplain dynamics and alluvial architecture. *Sedimentology*, **55**, 1717–1745.

King, P.R. (1990) The connectivity and conductivity of overlapping sand bodies. In: *North Sea Oil and Gas Reservoirs 11* (Eds A.T. Buller, E. Berg, O. Hjelmeland, J. Kleppe, O. Torsaeter and J.O. Aasen), Graham & Trotman, The Norwegian Institute of Technology, 353–362.

King, P.R., **Buldyrev, S.V.**, **Dokholyan, N.V.**, **Havlin, S.**, **Lee, Y.**, **Paul, G.** and **Stanley, H.E.** (1999) Applications of statistical physics to the oil industry: Predicting oil recovery using percolation theory. *Phys. Rev. A*, **274**, 60–66.

Jordon, D.W. and **Pryor, W.A.** (1992) Hierarchical levels of heterogeneity in the Mississippi River Meander belt and application to reservoir systems. *AAPG Bull.*, **76**, 1601–1624.

Keogh, K.J., **Martinius, A.W.** and **Oslanda, R.** (2007) The development of fluvial stochastic modelling in the Norwegian oil industry: A historical review, subsurface implementation and future directions. *Sed. Geol.*, **202**, 249–268.

Langbein, W.B. and **Leopold, L.B.** (1966) River Meanders — Theory of Minimum Variance. *US Geol. Surv. Prof. Pap.*, **422-H**, 15 pp.

Larue, D.K. and **Friedmann, F.** (2005) The controversy concerning stratigraphic architecture of channelised reservoirs and recovery by waterflooding. *Petrol. Geosci.*, **11**, 131–146.

Larue, D.K. and **Hovadik, J.** (2006) Connectivity of channelised reservoirs: A modelling approach: *Petrol. Geosci.*, **12**, 291–308.

Larue, D.K. and Hovadik, J. (2008) Why is reservoir architecture an insignificant uncertainty in many appraisal and development studies of clastic channelised reservoirs? *J. Petrol. Geol.*, **31**, 337–366.

Leeder, M.R. (1978) A quantitative stratigraphic model for alluvium with special reference to channel deposit density and interconnectedness. In: *Fluvial Sedimentology* (Ed. A.D. Miall), *Can. Soc. Petrol. Geol. Mem.*, **5**, 87–596.

Lopez, S., Cojan, I., Rivoirard, J. and Galli, A. (2009) Process-based stochastic modelling: Meandering channelised reservoirs, *Int. Assoc. Sedimentol. Spec. Publ.*, **40**, 139–144.

Mackey, S.D. and Bridge, J.S. (1995) Three-dimensional model of alluvial stratigraphy: theory and application. *J. Sed. Res.*, **65**, 7–31.

McHargue, T. Pyrcz, M.J. Sullivan, M.D., Clark, J.D., Fildani, A., Romans, B.W., Covault, J.A., Levy, M., Posamentier, H.W. and Drinkwater, N.J. (2011) Architecture of turbidite channel systems on the continental slope: Patterns and predictions. *Mar. Petrol. Geol.* **28**, 728–743.

Meirovitz, C.D., Stright, L., Hubbard, S.M. and Romans, B. (2016) The Influence of Intra- and Inter-Channel Architecture in Selecting Optimal Gridding for Field-Scale Reservoir Simulation. *AAPG Datapages, Search and Discovery Article* #41929.

Miall, A.D. (2013) *Fluvial Depositional Systems*. Springer-Verlag, Inc., Heidelberg, 316 pp.

Miall, A.D. (1996) *The Geology of Fluvial Deposits: Sedimentary Facies, Basin Analysis and Petroleum Geology*. Springer-Verlag, Inc., Heidelberg, 582 pp.

Mohrig, D., Heller, P.L., Paola, C. and Lyons, W.J. (2000) Interpreting avulsion process from ancient alluvial sequences: Guadalupe–Matarranya system (northern Spain) and Wasatch Formation (western Colorado). *Geol. Soc. Am. Bull.*, **112**, 1787–1803.

Møyner O., Krogstad, S. and Lie, K.A. (2014) The application of flow diagnostics for reservoir management. *Soc. Petrol. Eng.*, DOI: 10.2118/171557-PA; Document ID: SPE-171557-PA, 306–323

Natvig, J.R. and Lie, K.A. (2008) Fast computation of multiphase flow in porous media by implicit discontinuous Galerkin schemes with optimal ordering of elements., *J. Comput. Phys.*, **227**, 10108–10124.

Natvig, J.R., Lie, K.A., Eikemo, B. and Berre, I. (2007) An efficient discontinuous Galerkin method for advective transport in porous media. *Adv. Water. Resour.*, **30**, 2424–2438.

Nicholas, A.P., Ashworth, P.J., Sambrook Smith, G.H. and Sandbach, S.D. (2013) Numerical simulation of bar and island morphodynamics in anabranching megarivers, *J. Geophys. Res.: Earth Surface*, **18**, 2019–2044.

Parker, G. and Andrews, E.D. (1986) On the time development of meander bends. *J. Fluid Mech.*, **162**, 139–156.

Pyrcz, M., Boisvert, J. and Deutsch, C. (2009) ALLUVSIM: A program for event-based stochastic modelling of fluvial depositional systems. *Comput. Geosci.*, **35**, 1671–1685.

Pyrcz, M.J., Sech, R.P., Covault, J.A., Willis, B.J., Sylvester, Z. and Sun, T. (2015) Stratigraphic rule-based reservoir modelling. *Bull. Can. Petrol. Geol.*, **63**, 287–303.

Sadeghnejad, S., Masihi, M., King, P.R., Shojaei, A. and Pishvaie M. (2011a) A Reservoir Conductivity Evaluation Using Percolation Theory. *Petroleum Science and Technology*, **29**, 1041–1053.

Sadeghnejada, S., Masihi, M., Pishvaie, M., Shojaei A. and Kin, P.R. (2011b) Utilization of percolation approach to evaluate reservoir connectivity and effective permeability: A case study on North Pars gas field. *Scientia Iranica*, **18**, 1391–139.

Schuurman, F., Kleinhans, M.G. and Marra, W.A. (2013) Physics-based modelling of large braided sand-bed rivers: bar pattern formation, dynamics and sensitivity. *J. Geophys. Res.*, **118**, 2509–2527.

Shahvali, M., Mallison, B., Wei, K. and Gross, H. (2012) An alternative to streamlines for flow diagnostics on structured and unstructured grids.: *Soc. Pet. Eng. J.*, **17**, 768–778.

Shook, M. (2009) A Robust Measure of Heterogeneity for Ranking Earth Models: The F PHI Curve and Dynamic Lorenz Coefficient. Society of Petroleum Engineers Annual Technical Conference and Exhibition, 4–7 October, New Orleans, Louisiana, *SPE* 124625, 1–13.

Straub, K.M., Paola, C., Mohrig, D., Wolinsky, M.A. and George, T. (2009) Compensational stacking of channelised sedimentary deposits. *J. Sed. Res.*, **79**, 673–688.

Strong, N., Sheets, B.A., Hickson, T.A. and Paola, C. (2005) A mass-balance framework for quantifying downstream changes in fluvial architecture. In: *Fluvial Sedimentology VII* (Eds M. Blum, S. Marriott and S. Leclair), *Int. Assoc. Sedimentol. Spec. Publ.*, **35**, 243–253.

Sun, T., Jøssang, T., Meakin, P. and Schwarz, K. (1996) A simulation model for meandering rivers. *Water Resour. Res.*, **32**, 2937–2954.

Van de Lageweg, W.I. Schuurman, F. Cohen, K.M., Van Dijk, W.M. Shimizu Y. and Kleinhans, M.G. (2016) Preservation of meandering river channels in uniformly aggrading channel belts. *Sedimentology*, **62**, 1–23.

Villamizar, C.A., Hampson, G.J., Flood, Y.S. and Fitch, P.J.R. (2015) Object-based modelling of avulsion-generated sandbody distributions and connectivity in a fluvial reservoir analogue of low to moderate net-to-gross ratio. *Petrol. Geosci.*, **21**, 249–270.

Wang, Y., Straub, K.M. and Hajek, E.A. (2011) Scale-dependent compensational stacking: An estimate of autogenic time scales in channelised sedimentary deposits. *Geology*, **39**, 811–814.

Willems, C.J.L., Nick, H.M., Donselaar, M.E., Jan Weltje, G. and Bruhn, D.F. (2017) On the connectivity anisotropy in fluvial Hot Sedimentary Aquifers and its influence on geothermal doublet performance. *Geothermics*, **65**, 222–233.

Willis, B.J. (1993b) Ancient river systems in the Himalayan foredeep, Chinji Village area, northern Pakistan, *Sed. Geol.*, **88**, 1–76.

Willis, B.J. (1993a) Bedding geometry of ancient point bar deposits. In: *Alluvial Sedimentation* (Eds M. Marzo and C. Puigdefabregas), *Int. Assoc. Sedimentol. Spec. Publ.*, **17**, 101–114.

Willis, B.J. (1989) Paleochannel reconstructions from point bar deposits: a three-dimensional perspective.: *Sedimentology*, **36**, 757–766.

Willis B.J. and Tang, H. (2010) Three-dimensional connectivity of point-bar deposits. *J. Sed. Res.*, **80**, 440–454.

Index

Fluvial Meanders and Their Sedimentary Products in the Rock Record, First Edition.
Edited by Massimiliano Ghinassi, Luca Colombera, Nigel P. Mountney and Arnold Jan H. Reesink.
© 2019 International Association of Sedimentologists. Published 2019 by John Wiley & Sons Ltd.